T0135319

Lecture Notes in Electrical Engineering

Volume 460

"Lecture Notes in Electrical Engineering (LNEE)" is a book series which reports the latest research and developments in Electrical Engineering, namely:

- Communication, Networks, and Information Theory
- Computer Engineering
- Signal, Image, Speech and Information Processing
- Circuits and Systems
- Bioengineering

LNEE publishes authored monographs and contributed volumes which present cutting edge research information as well as new perspectives on classical fields, while maintaining Springer's high standards of academic excellence. Also considered for publication are lecture materials, proceedings, and other related materials of exceptionally high quality and interest. The subject matter should be original and timely, reporting the latest research and developments in all areas of electrical engineering.

The audience for the books in LNEE consists of advanced level students, researchers, and industry professionals working at the forefront of their fields. Much like Springer's other Lecture Notes series, LNEE will be distributed through Springer's print and electronic publishing channels.

More information about this series at http://www.springer.com/series/7818

Yingmin Jia · Junping Du
Weicun Zhang
Editors

Proceedings of 2017 Chinese Intelligent Systems Conference

Volume II

 Springer

Editors
Yingmin Jia
Beihang University
Beijing
China

Junping Du
Beijing University of Posts
  and Telecommunications
Beijing
China

Weicun Zhang
University of Science
  and Technology Beijing
Beijing
China

ISSN 1876-1100   ISSN 1876-1119 (electronic)
Lecture Notes in Electrical Engineering
ISBN 978-981-13-4892-1   ISBN 978-981-10-6499-9 (eBook)
https://doi.org/10.1007/978-981-10-6499-9

Printed on acid-free paper

This Springer imprint is published by Springer Nature
The registered company is Springer Nature Singapore Pte Ltd.
The registered company address is: 152 Beach Road, #21-01/04 Gateway East, Singapore 189721, Singapore

Contents

Adaptive Tracking Control of a Single-Link Flexible Robotic Manipulator System with Unmodeled Dynamics and Motion Constraint

Ningning Wang and Tianping Zhang

Abstract This paper studies the tracking control problem of a single-link flexible robotic manipulator system with unmodeled dynamics and motion constraint. A dynamic surface control scheme is proposed to design the adaptive controller ensuring both desired tracking performance and constraint satisfaction. A virtual state observer is constructed to estimate the unmeasured state signals. The RBF NNs are used to approximate unknown functions. Dynamic signal and nonlinear mapping are introduced to deal with the dynamic uncertainties and solve motion constraint problem, respectively. A simulation study is presented to verify the effectiveness of the proposed control approach.

Keywords Flexible robotic manipulator · Unmodeled dynamics · Motion constraint · Nonlinear mapping

1 Introduction

In the past few decades, the tracking control research of robotic system has drawn greater attention. Many control methods have been developed for the tracking control problem of robotic manipulator system, such as the feedback linearization method [1, 2], the adaptive sliding mode technique [3], the adaptive backstepping technique [4], and the PD control method [5].

A limitation of above mentioned literatures is that they require that the velocity on the link side or motor side is measurable. Whereas, due to the restrictions on economy and technology, velocity sensors are often omitted in practical robotic systems. To solve the problem of the velocity is unmeasurable, some appropriate observers have been adopted to design robotic systems, such as full-order observer [6], extended Kalman filter observer [7], and K-filters observer [8–10]. Moreover, it

N. Wang · T. Zhang (✉)
Department of Automation, College of Information Engineering,
Yangzhou University, Yangzhou 225127, China
e-mail: tpzhang@yzu.edu.cn

© Springer Nature Singapore Pte Ltd. 2018
Y. Jia et al. (eds.), *Proceedings of 2017 Chinese Intelligent Systems Conference*,
Lecture Notes in Electrical Engineering 460,
https://doi.org/10.1007/978-981-10-6499-9_1

is assumed that the torques can be directly applied to the robot links, and the actuator dynamics is ignored. But the actuator dynamics is an integral part of whole electromechanical system, especially under the circumstance of high-speed movement and heavy load variations. To ensure the high performance of the system, the actuator dynamics should be taken into account the design process. In [11], a robust tracking controller was developed for a class of electrically driven flexible-joint robots with time-varying parametric uncertainties and external disturbances. In [12], an optimal adaptive control was presented for a class of MIMO robot manipulators with uncertain dynamics. In [13], two different control methods were proposed for a flexible joint robot system with unmodeled dynamics. In addition to velocity measurement and actuator dynamics, another problem that can't be ignored is how to avoid the collisions between the manipulated object and the nearby surroundings. This demands the operating region is predefined, rather than infinitely large. By introducing barrier Lyapunov function (BLF), a coordinated fuzzy control was studied for a class of robotic system with actuator hysteresis and motion constraint in [14]. Besides BLF approach, nonlinear mapping (NM) is another effective approach to solve output and state constrained problems [15, 16]. However, the motion constraint has been completely ignored in most robotic systems, only a few of robotic systems have been taken it into account.

Motivated by above analysis, this paper studies a single-link flexible robotic manipulator system with unmodeled dynamics and motion constraint. Only the link position is measurable and a virtual state observer is constructed to estimate the unmeasured state signals. By introducing dynamic signal and nonlinear mapping, an adaptive dynamic surface tracking control scheme is proposed for a single-link flexible robotic manipulator system with unmodeled dynamics and motion constraint.

2   Problem Formulation and Preliminaries

We consider a single-link robotic manipulator system [8, 10], the dynamic equation of the system is described as follows:

$$
\begin{cases}
J_1 \ddot{q}_1 + F_1 \dot{q}_1 + K(q_1 - q_2/N) + mgd \cos q_1 = 0 \\
J_2 \ddot{q}_2 + F_2 \dot{q}_2 - K(q_1 - q_2/N)/N = K_t i \\
LDi + Ri + K_b \dot{q}_2 = u
\end{cases}
\tag{1}
$$

where q_1 and q_2 are the link angular position and motor shaft angular position, respectively. i and u are the armature current and voltage, respectively. J_1 and J_2 are the inertias. K, K_t and K_b are the constants of the spring, torque, and back-emf, respectively. F_1 and F_2 are the viscous friction constants. R and L are the armature resistance and inductance. M is the link mass, d is the position of the link's barycenter, N is the gear ratio and g is the gravity acceleration.

We assume only the link position q_1 is measurable. Taking state unmodeled dynamics into account, inspired by [8], let $q_1 = y$, the original single-link robotic manipulator system can be described as the following output feedback form:

$$\begin{cases} \dot{z} = q(z, y) \\ \dot{x}_i = x_{i+1} + f_i(y) + \Delta_i(z, y, t), \ i = 1, 2, 3, 4 \\ \dot{x}_5 = f_5(y) + b_0 u + \Delta_5(z, y, t) \\ y = x_1 \end{cases} \tag{2}$$

where $x_1, \ldots, x_5$ are the unmeasured states, $u \in R$ is the control input. $z \in R^{n_0}$ is the state unmodeled dynamics, $q(z, y)$ is an unknown Lipschitz function, Unknown smooth nonlinear $\Delta_1(z, y, t), \ldots, \Delta_5(z, y, t)$ are the dynamic disturbances. Unknown nonlinear functions $f_1(y) = -\left(\frac{R}{L} + \frac{F_1}{J_1} + \frac{F_2}{J_2}\right) y$, $f_2(y) = -\left[\frac{R}{L}\left(\frac{F_1}{J_1} + \frac{F_2}{J_2}\right) + \frac{K_b K_t}{L J_2} + \left(\frac{K}{J_1} + \frac{K}{J_2 N^2} + \frac{F_1 F_2}{J_1 J_2}\right)\right] y - \frac{mgd}{J_1} \cos y$, $f_3(y) = -\left[\frac{R}{L}\left(\frac{K}{J_1} + \frac{K}{J_2 N^2} + \frac{F_1 F_2}{J_1 J_2}\right) + \frac{F_1 K}{J_1 J_2 N^2} + \frac{F_2 K}{J_1 J_2} + \frac{K_b F_1 K_t}{L J_1 J_2}\right] y - \left(\frac{R}{L} + \frac{F_2}{J_2}\right) \frac{mgd}{J_1} \cos y$, $f_4(y) = -\left[\frac{R}{L}\left(\frac{F_1 K}{J_1 J_2 N^2} + \frac{F_2 K}{J_1 J_2}\right) + \frac{K_b K_t K}{L J_1 J_2}\right] y - \left(\frac{K}{N^2} + \frac{R F_2}{L} + \frac{K_b K_t}{L}\right) \frac{mgd}{J_1 J_2} \cos y$, $f_5(y) = -\frac{R}{L} \frac{mgdK}{J_1 J_2 N^2} y$, $b_0 = K_t K / (J_1 J_2 N L) \neq 0$ is a known constant. The output y is required to remain in the set $\Omega_1 = \{y: -k_{c1} < y < k_{c2}\}, \forall t \geq 0$, where k_{c1} and k_{c2} are two known positive design constants.

The control goal is to design controller u for system (2), such that the output y tracks the desired trajectory y_d, while the constraint $y \in \Omega_1$ is not violated.

Assumption 1 For $\forall k_{c2} > k_{c1} > 0$, there exist positive constants Y_0 and Y_1, such that $-k_{c1} < Y_1 \leq y_d \leq Y_2 < k_{c2}, \forall t \geq 0$.

Assumption 2 There exist unknown nonnegative continuous function $\phi_{i1}(\cdot)$ and unknown nonnegative monotonic increasing continuous function $\phi_{i2}(\cdot)$, such that inequality $|\Delta_i(z, y, t)| \leq \phi_{i1}(|y|) + \phi_{i2}(\|z\|)$ holds.

Definition 1 ([17]). For system $\dot{z} = q(z, y)$, if there exist Lyapunov function $V(z)$, class K_∞ functions $\underline{\alpha}(\cdot)$ and $\bar{\alpha}(\cdot)$, known positive constants c and d, and known class K_∞ function $\gamma(\cdot)$ satisfying

$$\underline{\alpha}(\|z\|) \leq V(z) \leq \bar{\alpha}(\|z\|) \tag{3}$$

$$\frac{\partial V(z)}{\partial z} q(z, y) \leq -c V(z) + \gamma(|y|) + d \tag{4}$$

then system $\dot{z} = q(z, y)$ is called to be exponentially input-state-practically stable (exp-ISpS), $V(z)$ is a exp-ISpS Lyapunov function of system $\dot{z} = q(z, y)$.

Assumption 3 System $\dot{z} = q(z, y)$ is exp-ISpS, i.e. inequalities (3) and (4) hold.

Lemma 1 ([17]). *If V is an exe-ISpS Lyapunov function of the subsystem $\dot{z} = q(z, y)$, then for any constant $\bar{c} \in (0, c)$, any initial condition $t_0 > 0$, any function $z_0 = z(t_0)$, $r_0 > 0$ and $\bar{\gamma}(|y|) \geq \gamma(|y|)$, there exists a finite $T_0 = \max\{0,$*

$\ln[V(z_0)/r_0]/(c-\bar{c})\} \geq 0$, *a nonnegative function $D(t_0,t)$ and a signal described by* $\dot{r} = -\bar{c}r + \bar{\gamma}(|y|) + \bar{d},\ r(t_0) = r_0,$ *such that $D(t_0,t) = 0$ for all $t \geq t_0 + T_0$ and* $V(z) \leq r(t) + D(t_0,t)$. *Without loss of generality, we assume that $\bar{\gamma}(|y| = \gamma(|y|)$.*

Lemma 2 *For real continuous function $f(x,y)$, inequality $|f(x,y)| \leq \varphi(x) + \vartheta(y)$ holds, where $x \in R^m,\ y \in R^n,\ \varphi(x)$ and $\vartheta(y)$ are positive smooth scalar functions.*

3   K-Filters and Observer Design Based on RBF NNs

Define the compact set $\Omega_y = \{y|\,|y| \leq M_y\} \subset R$, the unknown function $f_i(y)$ $(i = 1,\ldots,5)$ will be approximate by radial basis function neural networks (RBF NNs) on the compact set Ω_y, i.e. $f_i(y) = \theta_i^{*T}\phi_i(y) + \delta_i(y)$, where $M_y > 0$ is an unknown constant, $\delta(y)$ is the approximation error, $M_i > 1$ is the neural network node number, $\phi_i(y) = [\phi_{i1}(y), \ldots, \phi_{iM_i}(y)]^T \in R^{M_i}$ is a known smooth vector function, the basis function $\phi_{ij}(y) = \exp[-(y-\mu_{ij})^2/b_{ij}^2]$, where $1 \leq i \leq 5,\ 1 \leq j \leq M_i,\ \mu_{ij}$ and b_{ij} are the center of the receptive field and the width of the Gaussian function, respectively. Define the optimal weight vector $\theta_i^* = \arg\min\limits_{\theta_i \in R^{M_i}}[\sup\limits_{y \in \Omega_y}|\theta_i^T\phi_i(y) - f_i(y)|]$.

To facilitate the K-filters design, system (2) can be rewritten as follows:

$$\begin{cases} \dot{z} = q(z,y,t) \\ \dot{x} = Ax + \Phi^T(y)\theta + \delta(y) + e_5 b_0 u + \Delta(z,y,t) \\ y = x_1 \end{cases} \tag{5}$$

where $\quad x = \begin{bmatrix} x_1 \\ \vdots \\ x_5 \end{bmatrix},\quad A = \begin{bmatrix} 0 & I_4 \\ 0 & 0 \end{bmatrix},\quad \Phi^T(y) = \begin{bmatrix} \phi_1^T(y) & 0 & 0 \\ 0 & \ddots & 0 \\ 0 & 0 & \phi_5^T(y) \end{bmatrix} \in R^{5 \times M},$

$M = \sum\limits_{i=1}^{5} M_i,\ \theta = [\theta_1^*, \ldots, \theta_5^*]^T,\ \delta(y) = [\delta_1(y), \ldots, \delta_5(y)],\ \Delta(z,y,t) = [\Delta_1(z,y,t), \ldots, \Delta_5(z,y,t)]^T,\ e_5 = [0,0,0,0,1]^T$.

Based on [8], we design K-filters as $\dot{\zeta} = A_0\zeta + Ly,\ \dot{\Xi}^T = A_0\Xi^T + \Phi^T(y),$ $\dot{\lambda} = A_0\lambda + e_5 u$, where $L = [l_1, \ldots, l_5]^T,\ e_1 = [1,0,0,0,0]^T,\ A_0 = A - Le_1^T$ is a Hurwitz matrix, i.e. $PA_0 + A_0^T P = -hI,\ P = P^T > 0$, where h is a positive design constant.

Construct the virtual state estimation as $\hat{x} = \zeta + \Xi^T\theta + b_0\lambda$, and define the virtual state estimate error as $\varepsilon = x - \hat{x}$, then

$$x = \zeta + \Xi^T \theta + b_0 \lambda + \varepsilon \tag{6}$$

$$\dot{\varepsilon} = A_0 \varepsilon + \delta(y) + \Delta(z, y, t) \tag{7}$$

According to (6), we get

$$x_2 = b_0 \lambda_2 + \zeta_2 + \Xi_{(2)}^T \theta + \varepsilon_2 \tag{8}$$

where $\Xi_{(2)}^T$ represents the second row of the matrix Ξ^T, λ_2, ζ_2 and ε_2 denote the second element of λ, ζ and ε, respectively.

Substituting (8) into (2) yields

$$\dot{y} = b_0 \lambda_2 + \zeta_2 + \bar{\omega}^T \theta + \varepsilon_2 + \delta_1(y) + \Delta_1(z, y, t) \tag{9}$$

where $\bar{\omega}^T = \Xi_{(2)} + \Phi_{(1)}^T$, $\Phi_{(1)}^T$ represents the first row of the matrix Φ^T.

Through the above analysis, we have the following 5-order system:

$$\begin{cases} \dot{y} = b_0 \lambda_2 + \zeta_2 + \bar{\omega}^T \theta + \varepsilon_2 + \delta_1(y) + \Delta_1(z, y, t) \\ \dot{\lambda}_i = -l_i \lambda_1 + \lambda_{i+1}, \ i = 2, 3, 4 \\ \dot{\lambda}_5 = -l_5 \lambda_1 + u \end{cases} \tag{10}$$

4 Nonlinear Mapping

To handle output constrained problem, we introduce the following one to one nonlinear mapping:

$$s_1 = \ln (k_{c1} + y)/(k_{c2} - y) \tag{11}$$

Evidently though, when $s_1 \in R$, we can obtain $y \in \Omega_y = \{y: -k_{c1} < y < k_{c2}\}$.

It is easy to obtain that the inverse mapping of (11) is

$$y = (k_{c2} e^{s_1} - k_{c1})/(e^{s_1} + 1) = k_{c2} - (k_{c1} + k_{c2})/(e^{s_1} + 1) \tag{12}$$

Differentiating s_1 with respect to time t, and substituting (10) into it, we obtain

$$\dot{s}_1 = k(s_1)[b_0 \lambda_2 + \zeta_2 + \bar{\omega}^T \theta + \varepsilon_2 + \delta_1(y) + \Delta_1(z, y, t)] \tag{13}$$

where $k(s_1) = (e^{s_1} + e^{-s_1} + 2)/(k_{c1} + k_{c2}) > 0$.

5 Adaptive Tracking Control Design

To facilitate the design below, some notations are presented below: $\bar{z}_i = [z_1, \ldots, z_i]^T$, $\bar{y}_j = [y_2, \ldots, y_j]^T$, $1 \leq i \leq 5$, $2 \leq j \leq 5$, where $y_j = \omega_j - \alpha_{j-1}$, z_i, α_{j-1} and ω_j will be given later.

Define some Lyapunov functions $V_\varepsilon = \varepsilon^T P \varepsilon$, $V_{z_i} = z_i^2/2$, $V_1 = V_\varepsilon + V_{z_1} + r/\lambda_0$, where $i = 1, \ldots, 5$, $\lambda_0 > 0$ is a design constant.

Differentiating V_ε with respect to time t, substituting (7) into it yields

$$\dot{V}_\varepsilon \leq -(h-2)\varepsilon^T \varepsilon + \sum_{j=1}^{5} \|P\|^2 \delta_j^2(y) + \sum_{j=1}^{5} 2\|P\|^2 \left(\phi_{j1}^2(|y|) + \phi_{j2}^2(\|z\|) \right) \tag{14}$$

Based on Assumption 3 and Lemma 1, it yields

$$\|z\| \leq \alpha_1^{-1}(r + D(t_0, t)) \tag{15}$$

$$\phi_{j2}^2(\|z\|) \leq \phi_{j2}^2 \circ \alpha_1^{-1}(r + D(t_0, t)) \tag{16}$$

where $\phi_{j2}^2 \circ \alpha_1^{-1}(\cdot) = \phi_{j2}^2(\alpha_1^{-1}(\cdot))$, $0 \leq D(t_0, t) \leq D_0$. Since D_0 is a constant, $\phi_{j2}^2 \circ \alpha_1^{-1}(\cdot)$ is a nonnegative continuous, and based on Lemma 2, we get

$$\sum_{j=1}^{5} 2\|P\|^2 \phi_{j2}^2(\|z\|) \leq \|P\|^2 \varphi_0(r) + \|P\|^2 \vartheta_0(D(t_0, t)) \leq \|P\|^2 \varphi_0(r) + \|P\|^2 \vartheta_0^* \tag{17}$$

where $\varphi_0(\cdot)$, $\vartheta_0(\cdot)$ are unknown continuous functions, ϑ_0^* is an unknown constant. Substituting (17) into (14) yields

$$\dot{V}_\varepsilon \leq -(h-2)\varepsilon^T \varepsilon + \sum_{j=1}^{5} \|P\|^2 \delta_j^2(y) + \sum_{j=1}^{5} 2\|P\|^2 \phi_{j1}^2(|y|) + \|P\|^2 \varphi_0(r) + \|P\|^2 \vartheta_0^*$$

$$\tag{18}$$

Step 1 Let $\omega_1 = \hat{y}_d = \ln(k_{c1} + y_d)/(k_{c2} - y_d)$, define $z_1 = s_1 - \omega_1$.
Differentiating z_1 with respect to time t, in view of (13) yields

$$\dot{z}_1 = k(s_1)\left[b_0 \lambda_2 + \zeta_2 + \bar{\omega}^T \theta + \varepsilon_2 + \delta_1(y) + \Delta_1(z, y, t) \right] - \dot{\hat{y}}_d \tag{19}$$

Design the virtual control law α_1 as follows:

$$\alpha_1 = \frac{1}{b_0}\left(-\frac{k_1 z_1}{k(s_1)} - \zeta_2 - \bar{\omega}^T \theta - \frac{z_1 \theta_0}{2a_0^2 k(s_1)} \|\psi(X)\|^2 \right) \tag{20}$$

where k_1 and a_0 are positive design constants, θ_0 and θ respectively are the estimation value of θ_0 and θ at time t, and define $\tilde{\theta}_0 = \theta_0 - \hat{\theta}_0$, $\tilde{\theta} = \theta - \hat{\theta}$, $\theta_0 = \|W\|^2$. X and $\psi(X)$ will be defined later.

Introduce a first order filter as follows:

$$\tau_2 \dot{\omega}_2 + \omega_2 = \alpha_1, \quad \omega_2(0) = \alpha_1(0) \tag{21}$$

Let $y_2 = \omega_2 - \alpha_1$, then $\dot{\omega}_2 = y_2/\tau_2$, $\lambda_2 = z_2 + y_2 + \alpha_1$. Differentiating V_{z_1} with respect to time t, in view of (19) and (20), we obtain

$$
\begin{aligned}
\dot{V}_{z_1} = {}& -k_1 z_1^2 + b_0 k(s_1) z_1 z_2 + b_0 k(s_1) z_1 y_2 + k(s_1) z_1 \bar{\omega}^T \theta - z_1^2 \theta_0 \|\psi(X)\|^2/2a_0^2 \\
& + k(s_1) z_1 \varepsilon_2 + k(s_1) z_1 \delta_1(y) + k(s_1) z_1 \Delta_1(z, y, t) - z_1 \dot{y}_d
\end{aligned}
\tag{22}
$$

Based on Assumption 2 and Young's inequality, we obtain

$$
k(s_1) z_1 \Delta_1 \le \frac{k^2(s_1) z_1^2 \phi_{11}^2(|y|)}{a_{11}^2} + \frac{k^2(s_1) z_1^2}{a_1^2} \varphi_1^2(r) + k^2(s_1) z_1^2 + \frac{a_{11}^2}{4} + \frac{a_1^2}{4} + \frac{\vartheta_1^*}{4} \tag{23}
$$

Differentiating V_1 with respect to time t, substituting (18), (22) and (23) into it, and using Young's inequality, we have

$$
\begin{aligned}
\dot{V}_1 \le {}& -(h-9/4)\varepsilon^T \varepsilon - k_1 z_1^2 + b_0^2 z_2^2/4 + b_0^2 y_2^2/4 + k(s_1) z_1 \bar{\omega}^T \theta - z_1^2 \theta_0 \|\psi(X)\|^2/2a_0^2 \\
& + z_1 H(X) + Q(y, r) - \bar{c}r/\lambda_0 + \bar{d}/\lambda_0 + a_{11}^2/4 + a_1^2/4 + \vartheta_1^*/4 + \|P^2\| \vartheta_0^*
\end{aligned}
\tag{24}
$$

where $H(X) = 5k^2(s_1)z_1 + k^2(s_1)z_1 \phi_{11}^2(|y|)/a_{11}^2 + k^2(s_1)z_1 \varphi_1^2(r)/a_1^2 - \dot{y}_d$, $X = [y, r, y_d, \dot{y}_d]^T$, $Q(y, r) = \sum_{j=1}^{5} \|P\|^2 \delta_j^2(y) + \sum_{j=1}^{5} 2\|P\|^2 \phi_{j1}^2(|y|) + \|P\|^2 \varphi_0(r) + \frac{\delta_1^2(y)}{4} + \frac{\bar{r}(|y|)}{\lambda_0}$.

We will utilize RBF NNs to estimate the above unknown function $H(X)$ on the compact set $\Omega_X = \{X| \|X\| \le M_X\} \subset R^4$, i.e. $H(X) = W^{*T} \psi(X) + B(X)$, where M_X is an unknown positive constant, W is an adjustable parameter weight vector, $\psi(X)$ is a basis function vector, and $B(X)$ is the approximation error. Using $\theta_0 = \|W\|^2$ and Young's inequality, yields

$$z_1 W^T \psi(X) \le z_1^2 \theta_0 \|\psi(X)\|^2/(2a_0^2) + a_0^2/2 \tag{25}$$

$$z_1 B(X) \le z_1^2 + B^2(X)/4 \tag{26}$$

where a_0 is a positive design constant.

There exists a continuous function $\kappa(y, r, y_d, \dot{y}_d) > 0$ satisfying $|B_1(X)| \le \kappa$.

Applying (25)–(26) into (24), we obtain

$$\dot{V}_1 \leq -(h-9/4)e^T\varepsilon - (k_1-1)z_1^2 + b_0^2 z_2^2/4 + b_0^2 y_2^2/4 + k(s_1)z_1\bar{\omega}^T\theta$$
$$- \bar{c}r/\lambda_0 + z_1^2\theta_0\|\psi(X)\|^2/2a_0^2 + \kappa^2/4 + Q(y,r) + c_1 \tag{27}$$

where $c_1 = \bar{d}/\lambda_0 + a_0^2/2 + a_{11}^2/4 + a_1^2/4 + \vartheta_1^*/4 + \|P^2\|\vartheta_0^*$.

From $\dot{y}_2 = -y_2/\tau_2 - \dot{\alpha}_1$, we obtain that there exists a nonnegative continuous function $\eta_2(\bar{z}_2, y_2, \theta_0, \theta, r, \zeta, \lambda_2, y_d, \dot{y}_d, \ddot{y}_d)$ satisfying $|\dot{y}_2 + y_2/\tau_2| \leq \eta_2$, then we obtain $y_2(\dot{y}_2 + y_2/\tau_2) \leq |y_2||\dot{y}_2 + y_2/\tau_2| \leq |y_2|\eta_2$, Furthermore,

$$y_2\dot{y}_2 \leq -y_2^2/\tau_2 + y_2^2 + \eta_2^2/4. \tag{28}$$

Step i $(2 \leq i \leq 4)$ Define $z_i = \lambda_i - \omega_i$, then

$$\dot{z}_i = \lambda_{i+1} - l_i\lambda_1 - \dot{\omega}_i \tag{29}$$

Design the virtual control law α_i as follows:

$$\alpha_i = -k_iz_i + l_i\lambda_1 + \dot{\omega}_i \tag{30}$$

where k_i is a positive design constant.

Introduce a first-order filter as follows:

$$\tau_{i+1}\dot{\omega}_{i+1} + \omega_{i+1} = \alpha_i, \omega_{i+1}(0) = \alpha_i(0) \tag{31}$$

Let $y_{i+1} = \omega_{i+1} - \alpha_i$ then $\dot{\omega}_{i+1} = -y_{i+1}/\tau_{i+1}$, $\lambda_{i+1} = z_{i+1} + y_{i+1} + \alpha_i$. Differentiating V_{z_i} with respect to time t, in view of (29)–(30) and using Young's inequality, it yields

$$\dot{V}_{z_i} \leq -(k_i-2)z_i^2 + z_{i+1}^2/4 + y_{i+1}^2/4 \tag{32}$$

From $\dot{y}_{i+1} = -y_{i+1}/\tau_{i+1} - \dot{\alpha}_i$, we know that there exists a nonnegative continuous function $\eta_{i+1}(\bar{z}_{i+1}, \bar{y}_{i+1}, \theta_0, \theta, r, \zeta, \lambda_2, \varepsilon_2, y_d, \dot{y}_d, \ddot{y}_d)$ satisfying $|\dot{y}_{i+1} + y_{i+1}/\tau_{i+1}| \leq \eta_{i+1}$, then we obtain $y_{i+1}(\dot{y}_{i+1} + y_{i+1}/\tau_{i+1}) \leq |y_{i+1}||\dot{y}_{i+1} + y_{i+1}/\tau_{i+1}| \leq |y_{i+1}|\eta_{i+1}$,

$$y_{i+1}\dot{y}_{i+1} \leq -y_{i+1}^2/\tau_{i+1} + y_{i+1}^2 + \eta_{i+1}^2/4 \tag{33}$$

Step 5 Define $z_5 = \lambda_5 - \omega_5$, then

$$\dot{z}_5 = u - l_5\lambda_1 - \dot{\omega}_5 \tag{34}$$

Design the control law u as follows:

$$u = -k_5 z_5 + l_5 \lambda_1 + \dot{\omega}_5 \tag{35}$$

where k_5 is a positive design constant.

Differentiating V_{z_5} with respect to time t, substituting (34)–(35) into it and using Young's inequality, yields

$$\dot{V}_{z_5} = -k_5 z_5^2 \tag{36}$$

Design the adaptation laws of θ and θ_0 as follows:

$$\dot{\theta} = \gamma_1 [k(s_1)\bar{\omega} z_1 - \sigma_1 \theta] \tag{37}$$

$$\dot{\theta}_0 = \gamma_2 [z_1^2 \|\psi(X)\|^2 / (2a_0^2) - \sigma_2 \theta_0] \tag{38}$$

where $\gamma_1,$, γ_2, σ_1 and σ_2 are positive design constants.

6 Stability Analysis

Define the compact set $\Omega = \{(\bar{z}_5, \bar{y}_5, \theta_0, \theta, r, \varepsilon): V \leq p\} \subset R^{N_0}$, $\Omega_d = \{(y_d, \dot{y}_d, \ddot{y}_d): y_d^2 + \dot{y}_d^2 + \ddot{y}_d^2 \leq B_0\}$, where p and B_0 are positive design constants, $N_0 = 18$. When ζ and λ_2 are bounded, the continuous functions κ, $Q(y, r)$ and η_{i+1} have maximum value M_0, N_1 and N_{i+1} compact set $\Omega \times \Omega_d$, respectively.

Define Lyapunov function $V = V_1 + \sum_{i=2}^5 V_{z_i} + \sum_{i=2}^5 y_i^2 / 2 + \theta^T \theta / (2\gamma_1) + \theta_0^2 / (2\gamma_2)$.

Theorem 1 *Consider the closed-loop system consisting of the nonlinear plant (2) under Assumptions 1–3, the controller (35), adaptive laws (37)–(38). For any bounded initial conditions satisfying $V(0) \leq p$ and $y(0) \in \Omega_1$, there exist constants $\alpha_0, k_1, k_i, \tau_i, h_j$, such that all the signals in the closed-loop system remain bounded while the output constraint $y \in \Omega_1$ is not violated, and $\alpha_0, k_1, k_i, \tau_i, h_j$ satisfying $\alpha_0 = \min\{\bar{c}, \gamma_1 \sigma_1, \gamma_2 \sigma_2\}$, $k_1 \geq 1 + \alpha_0/2$, $k_i \geq b_0^2/4 + 9/4 + \alpha_0/2$, $\tau_i^{-1} \geq b_0^2/4 + 5/.4 + \alpha_0/2, 2 \leq i \leq 5$ $h_j \geq 13/4 + \alpha_0 \lambda_{\max}(P)$, $1 \leq j \leq 5$.*

Proof Similar to the discussion in [18], Theorem 1 can be established.

Fig. 1 Output y (dashed line) and desired trajectory y_d (solid line)

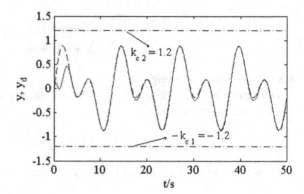

7   Simulation Results

In this section, we present a simulation study to verify the proposed control scheme in Sect. 5. In the simulation, the system parameters are chosen as $J_1 = 1.625$ kg m^2, $J_2 = 1.625$ kg m^2, $R = 0.5\,\Omega$, $K_t = 0.9$ N m/A, $K = 0.5868$, $K_b = 0.9$ N m/A, $m = 4.34$ kg, $L = 25 \times 10^{-3}$ H, $g = 9.8$ N/kg, $F_1 = 1.625 \times 10^{-2}$ Nm s/rad, $F_2 = 1.625 \times 10^{-2}$ Nm s/rad, $N = 2$, $d = 0.5$ m. The state unmodeled dynamics and disturbances are chosen as $q(z, y) = -z + y^2 + 0.5$, $\Delta_1 = 2z$, $\Delta_2 = 0.5z$, $\Delta_3 = yz$, $\Delta_4 = 0.1 \cos t + z$, $\Delta_5 = 0.1 \sin t + yz$.

The desired trajectory is set as $y_d = 0.5 \sin(t) + 0.5 \sin(0.5t)$, the output y is required to satisfy the output constraint $-k_{c1} < y < k_{c2}$, where $k_{c1} = 1.2, k_{c2} = 1.2$.

Simulation results are shown in Figs. 1, 2 and 3. It is observed that a satisfactory tracking performance is achieved, and the output constraint is not violated.

Fig. 2 Tracking error $y - y_d$

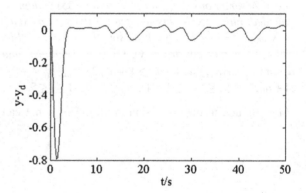

Fig. 3 Control signal u

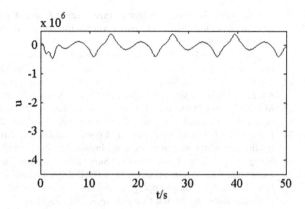

8 Conclusion

In this paper, a dynamic surface control scheme is proposed for a single-link flexible robotic manipulator system with unmodeled dynamics and motion constraint. Simulation results show the proposed control scheme is efficient.

Acknowledgements This work was partially supported by the National Natural Science Foundation of China (61573307, 61473250 and 61473249) and Yangzhou University Top-level Talents Support Program.

References

1. Bedrossian NS, Spong MW. Feedback linearization of robot manipulators and riemannian curvature. J Robot Syst. 1995;12(8):541–52.
2. Rodriguez H, Astolfi A, Ortega R. On the construction of static stabilizers and static output trackers for dynamically linearizable systems, related results and applications. Int J Control. 2006;79:1523–37.
3. Pukdeboon C. Lyapunov optimizing sliding mode control for robot manipulators. Appl Math Sci. 2013;7(63):3123–39.
4. Bang JS, Shim H, Park SK, Seo JH. Robust Tracking and vibration suppression for a two-inertia system by combining backstepping approach with disturbance observer. IEEE Trans Ind Electron. 2010;57:3197–206.
5. Luca AD, Siciliano B, Zollo L. PD control with on-line gravity compensation for robots with elastic joints: theory and experiments. Automatica. 2005;41:1809–19.
6. Chang YC, Chen BS, Lee TC. Tracking control of flexible joint manipulators using only position measurements. Int J Control. 1996;64:567–93.
7. Lightcap CA, Banks SA. An extended Kalman filter for real-time estimation and control of a rigid-link flexible-joint manipulator. IEEE Trans Control Syst Technol. 2010;18:91–103.
8. Kristic M, Kanellakopoulos I, Kokotovic PV. Nonlinear and adaptive control design. New York: Wiley; 1995.
9. Kim MS, Lee JS. Adaptive tracking control of flexible-joint manipulators without over parametrization. J Robot Sys. 2004;21:369–79.

10. Li YM, Tong SC, Li TS. Adaptive fuzzy output feedback control for a single-link flexible robot manipulator driven DC motor via backstepping. Nonlinear Anal: Real World Appl. 2013;14:483–94.
11. Chang YC, Yen HM. Robust tracking control for a class of electrically driven flexible-joint robots without velocity measurements. Inter J Control. 2012;85(2):194–212.
12. Yang RX, Yang CG, Mou Chen M, et al. Discrete-time optimal adaptive RBFNN control for robot manipulators with uncertain dynamics. Neurocomputing. 2017;234:107–15.
13. Mario RN, Gilberto OO, et al. On the robust trajectory tracking task for flexible-joint robotic arm with unmodeled dynamics. IEEE Access. 2016;4:7816–27.
14. Liu Z, Chen C, Zhang Y, Chen CLP. Coordinated fuzzy control of robotic arms with actuator nonlinearities and motion constraints. Inform Sci. 2015;296:1–13.
15. Meng WC, Yang QM, Si SN, Sun YX. Adaptive neural control of a class of output-constrained non-affine systems. IEEE Trans Cybern. 2016;46(1):85–95.
16. Guo T, Wu WC. Backstepping control for output-constrained nonlinear systems based on nonlinear mapping. Neural Comput Appl. 2014;25:166–1674.
17. Jiang ZP, Praly L. Design of robust adaptive controllers for nonlinear systems with dynamic uncertainties. Automatica. 1998;34(7):825–40.
18. Xia XN, Zhang TP. Adaptive output feedback dynamic surface control of nonlinear systems with unmodeled dynamics and unknown high-frequency gain sign. Neurocomputing. 2014;143:312–21.

The Intelligent Flight Control of Quadrotor in Tunnel Based on Simple Sensors

Rui Li and Yingjing Shi

Abstract This paper studies the automation flight control problem of a quadrotor in the tunnel. The LED lamp belt is installed in the tunnel and two moving points are the tracking target of the quadrotor. The velocity-control-mode is first well done before the tracking control algorithm is considered. A control law is then designed to achieve tracking task in the straight tunnel. When the parameter is properly designed, the quadrotor still can perform tracking task for the moving points even in the bent tunnel. Moreover, we prove that the collision between the quadrotor and the tunnel can be avoided by using the proposed control law. The validity of the proposed control algorithm is also demonstrated through numerical simulations.

Keywords Quadrotor · Automation flight · Tracking · Collision avoidance

1 Introduction

With the potential, both in military and civilian applications, for reconnaissance, surveillance, disaster management and emergency aid, unmanned aerial vehicles (UAVs) have gained great interesting among the research community. An important subset of UAVs is quadrotors, which have become popular recently due to their small size and maneuverability.

The development of quadrotor navigation and control technologies in indoor or enclosed environments has been an active area of research. Thomas et al. [1] developed a nonlinear vision-based controller for trajectory tracking in the image space, which made a step towards enabling grasping maneuvers using vision. In [2], the authors proposed a stochastic differential equation-based exploration algorithm

R. Li · Y. Shi (✉)
School of Automation Engineering, University of Electronic
Science and Technology of China, Chengdu 611731, China
e-mail: shiyingjing@163.com

R. Li
e-mail: Lirui@uestc.edu.cn

© Springer Nature Singapore Pte Ltd. 2018
Y. Jia et al. (eds.), *Proceedings of 2017 Chinese Intelligent Systems Conference*,
Lecture Notes in Electrical Engineering 460,
https://doi.org/10.1007/978-981-10-6499-9_2

to enable exploration in three-dimensional indoor environments with a payload constrained micro-aerial vehicle. Budiyono et al. [3] presented control system and collision detection system design for a quadrotor using a Kinect sensor. There is other work on control or navigation system design based on the indoor localization system. Examples include: balancing a pendulum [4]; aggressive maneuvers, such as flight through windows [5] or flips [6]; ball juggling with a racket [7] and so on. However, the indoor localization system is usually expensive and does not suit for narrow and dark environment.

Recently, the automation flight control of the quadrotor in tunnel has attracted much attention due to the important applications including exploration, surveillance, rescue, etc. However, the automation flight control of the quadrotor in the tunnel is difficult due to the limited space, dark environment, weak signal, etc. So far, the study on the automation flight control of the quadrotor in the tunnel environment is still in the early stage.

In this paper, we take a step to investigate the navigation and control problem of the quadrotor in the tunnel. Since the automation flight with known environment is a critical step for fully automation, we carry out the investigation on how to perform the automation flight in the known environment, where LED lamp belt is installed in the tunnel to simulate a known environment. The control system is designed to track two moving points on LED lamp belt in the tunnel, which can be viewed as a coordination problem of the multi-agent system and is solved by using the coordination control method.

The study on coordination control of multi-agent systems, based on a distributed control scheme, has received a great deal of attention and has interested many researchers due to their widespread applications in various real-world multi-agent systems. In the area of cooperative control of multi-agent systems, tracking is an important and fundamental problem, which has been extensively investigated. The authors in [8] and [9] proposed and analyzed a consensus tracking algorithm under a variable undirected network topology. In [10] and [11], the authors proposed a proportional-and-derivative-like consensus tracking algorithm under a directed network topology in both continuous-time and discrete-time settings. In [12], the authors studied a leader-follower consensus tracking problem with time-varying delays. In [13], the author studied a flocking algorithm under the assumption that the leader's velocity is constant and is available to all followers. In [14], the authors studied the problem of flight formation and trajectory tracking control for a group of mini rotorcraft. Schoellig et al. [15] presented an optimization-based iterative learning approach for trajectory tracking and the approach was successfully applied to quadrotor vehicles.

In this paper, we first describe the tracking problem of the quadrotor. Then we show the velocity controller design for the quadrotor. Base on the velocity controller, we establish the control algorithm, achieving the tracking task for the LED lamp either in the straight part of the tunnel or the curve part of the tunnel.

2 Problem Formulation

IMU has been the standard configuration for the intelligent aircraft and the main tasks of the intelligent aircraft are early-warning and reconnaissance, which are usually achieved with the help of the camera. Thus, this paper considers the problem of control system design for the quadrotor equipping with IMU and camera to achieve intelligent flight in the tunnel.

Suppose that there is no curve in the vertical section of the tunnel, and the LED lamp belt is installed in the tunnel, where r_1, r_2 are two moving points on the LED lamp belt and the velocity of r_1, r_2 is constant and known. We aim to design the control system by which the quadrotor flights keeping the same height and certain distant with the lamp belt. Suppose that the quadrotor looks at the two points r_1, r_2 simultaneously, $r = r_2 - r_1$, $\bar{r} = (r_1 + r_2)/2$, and $\hat{r}$ is the adjoint point of $\bar{r}$, satisfying

$$
\hat{r} = \begin{cases} \bar{r} + d\frac{r \times z}{\|r \times z\|}, & \|r \times z\| \neq 0, \\ \bar{r} + d\frac{(r \times z)^-}{\|(r \times z)^-\|}, & \|r \times z\| = 0. \end{cases} \tag{1}
$$

Here, $z = \begin{bmatrix} 0 & 0 & 1 \end{bmatrix}^T$; $(r \times z)^-$ is the left limit of $r \times z$ when the quadrotor enters the vertical tunnel.

Remark 1 Since there is no curve in the vertical segment of the tunnel, that is, $(r \times z)/\|r \times z\|$ is same when the quadrotor just enters the vertical segment and just departs from the vertical segment.

3 Velocity Controller Design

In order to perform tracking task for moving points in the tunnel, we first need to achieve the velocity control of the quadrotor. In this paper, the PID algorithm is applied to design the velocity controller of the quadrotor.

Based on the well-done attitude control, we solve the expected flight velocity via PID algorithm to obtain the anticipant axial acceleration. Then, we transform the acceleration to body axis system by using rotation matrix conversion, which is the control input. The transform relationship is given as follows.

$$
\begin{cases} \theta_{des} = \arctan\left(\frac{a_x}{a_z}\right) \\ \phi_{des} = -\arctan\left(\frac{a_y \cos(\theta)}{a_z}\right) \end{cases}
$$

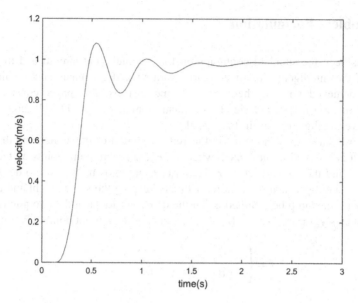

Fig. 1 Simulation result of the velocity controller

Here, $(\theta_{des}, \phi_{des})^T$ stands for the expected value of the pitching angle and rolling angle, $(a_x, a_y, a_z)^T$ is the value of the acceleration when the earth coordinate system is converted to the temporary coordinate system (the temporary coordinate system has the same heading angle with the body axes coordinate system, but pitching angle and rolling angle under the temporary coordinate system are zero). The simulation result of the velocity controller is given below.

From Fig. 1 we can find that the quadrotor can tack the expected velocity fast and stably. Thus it can be used for the design of the control algorithm to achieve tracking of the adjoint point.

4 Tracking Control

When achieving velocity control for quadrotor, we introduce the following control law to perform tracking task for the moving points:

$$\dot{x} = v\frac{r}{\|r\|} + K_v(\hat{r} - x) \tag{2}$$

Here, x is the position of the quadrotor; $v, r, \hat{r}$ are given as mentioned earlier; K_v is larger than zero, which is an unknown value to be determined. In order to avoid collision with the tunnel, K_v is usually selected according to the velocity v of LED lamp, the distance d between adjoint point and lamp belt, and the curvature of the tunnel.

4.1 The Theoretical Analysis of the Tracking Problem

Theorem 1 *For the straight part of the tunnel, when the control method given by Eq. (2) is applied, the quadrotor can achieve accurate tracking for the adjoint point.*

Proof For the straight tunnel, considering Eq. (1), we have

$$\hat{r} = \frac{\dot{r}_1 + \dot{r}_2}{2} = v \frac{r}{\|r\|}.$$

Thus, it can be seen from Eq. (2) that

$$\dot{x} - \hat{r} = -K_v(x - \hat{r}).$$

Let $e = x - \hat{r}$. We have $\dot{e} = -K_v e$, $K_v > 0$. Obviously, the error between the quadrotor and the adjoint is exponentially convergent.

Theorem 2 *Suppose that the curvature radius of the tunnel is ρ, the moving velocity of the LED lamp belt is v, and the distance between the adjoint point and the LED lamp belt is d. Then, the quadrotor is able to keep the flight distance with the adjoint point not more than d when we choose $K_v \geq vd/\rho l$.*

Proof Since the maximal deviation between the velocity of adjoint point and the velocity of LED lamp occurs in the horizontal tunnel, thus we have

$$\hat{r} = v \frac{\rho \pm d}{\rho} \frac{r}{\|r\|}, \tag{3}$$

where $\pm$ denotes that $\pm(\bar{r} - \hat{r})$ points the center of the turning circle. From Eq. (2), we have

$$\dot{x} = v \left(1 \pm \frac{d}{\rho} \mp \frac{d}{\rho} \right) \frac{r}{\|r\|} + K_v(\hat{r} - x).$$

Substituting Eq. (3) into the above equation, we obtain that

$$\dot{x} - \hat{r} = \mp \frac{vd}{\rho} \frac{r}{\|r\|} + K_v(\hat{r} - x).$$

Let $e = x - \hat{r}$, we get

$$\dot{e} = \mp \frac{vd}{\rho} \frac{r}{\|r\|} - K_v e.$$

Therefore, the quadrotor can keep the flight distance with adjoint point not more than l, only if

$$\|K_v e\| \geq \left\| \frac{vd}{\rho} \frac{r}{\|r\|} \right\|,$$

when $\|e\| \geq l$, that is, $K_v \geq vd/\rho l$.

Theorem 3 *Suppose that the distance between the initial position of the quadrotor and the adjoint point is less than d. If the control method given by Eq. (2) is applied, the collision between the quadrotor and the tunnel can be avoided.*

Proof We only need to illustrate that under the assumption of the theorem, with the control method given by Eq. (2), the quadrotor can avoid collision with the tunnel either for the straight tunnel or curved tunnel.

In the straight tunnel, the distance between the quadrotor and the adjoint point is exponentially convergent, thus the theorem is clearly established. For the curved tunnel, if let $l = d$ in Theorem 2, it can be easily obtained that the quadrotor can keep the flight distance not more than d from the adjoint point, that is, $K_v \geq v/\rho$. Thus, the quadrotor can avoid collision with the tunnel.

4.2 Simulation

The simulation is carried out in Matlab environment, where fixed-step simulation is applied and ODE solver applies fourth-order Runge-Kutta method.

The first simulation considers the situation in the spiraling tunnel. The curvature radius of the tunnel is 3 m, the velocity of the LED lamp is 1 m/s, the distance between the adjoint point and the LED lamp belt is 0.5 m, and the distance between two LED lamps is 0.5 m. The initial position of the LED lamp is set at origin, the initial position coordinate of the quadrotor is $[-1 \quad -1 \quad 0.5]^T$, and K_v is set to be 2.

The simulation result is shown in Fig. 2, where the blue line stands for the trajectory of the LED lamp, the green line stands for the trajectory of the adjoint point, the red line stands for the trajectory of the quadrotor, the hollow circle stands for the starting position of the simulation, and solid circle stands for the stopping position of the simulation. It can be seen from the simulation that, the quadrotor can track the adjoint point perfectly, which indicates that the quadrotor can safely flight in the tunnel.

The second simulation considers the situation in the vertical tunnel. The first part of the tunnel is an arc-shaped climbing course with the curvature radius as 3 m and

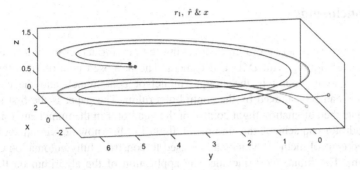

Fig. 2 The simulation curve of the spiraling tunnel

the arc length as 3 m, and the subsequent part is the straight tunnel being tangent to the circular arc. The data of the LED and the quadrotor are as same as what given in the above simulation, and $K_v = 1$ (Fig. 3).

From the simulation result, we can find that the quadrotor can track the adjoint point with the complicated environment. Since the tracking error between the quadrotor and adjoint point is exponentially convergent, then the quadrotor can well track the adjoint point in the straight part. It can be seen from the simulation that, the quadrotor can safely flight in the tunnel.

Fig. 3 The simulation of the vertical tunnel

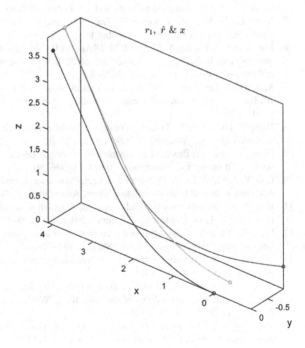

5 Conclusions

We have proposed an algorithm for tracking the moving target in the tunnel by a quadrotor equipping with IMU and camera. The proposed control algorithm can track the moving point in the straight tunnel or bent tunnel, and the collision avoidance can be achieved by the control algorithm. The paper is the first step to investigate the automation flight control of the quadrotor in the tunnel environment. The quadrotor can achieve the automation flight for the known environment by the proposed control method, which can be used to construct fully automation control algorithm. The future work includes the application of the algorithm on the real quadrotor plat and the fully automation flight control algorithm design in tunnel.

References

1. Thomas J, Loianno G, Polin J, Sreenath K, Kumar V. Toward autonomous avian-inspired grasping for micro aerial vehicles. Bioinspir Biomimetics. 2014;9:1–15.
2. Shen S, Michael N, Kumar V. Autonomous indoor 3D exploration with a micro-aerial vehicle. In: 2012 IEEE international conference on robotics and automation, 2012, Saint Paul, Minnesota, USA, 14–18 May 2012. p. 9–15.
3. Budiyono A, Lee G, Kim GB, Park J, Kang T, Yoon KJ. Control system design of a quad-rotor with collision detection. Aircraft Eng Aerosp Technol Int J. 2015;87(1):59–66.
4. Hehn M, D'Andrea R. A flying inverted pendulum. In: Proceedings of 2011 IEEE international conference on robotics and automation (ICRA), 2011. p. 763–70.
5. Mellinger D, Michael N, Kumar V. Trajectory generation and control for precise aggressive maneuvers with quadrotors. Int J Robot Res. 31(5):664–74.
6. Lupashin S, Schuollig A, Sherback M, D'Andrea R. A simple learning strategy for high-speed quadrocopter multi-flips. In: Proceedings of 2010 IEEE international conference on robotics and automation (ICRA), 2010. p. 1642–8.
7. Muuller M, Lupashin S, D'Andrea R. Quadrocopter ball juggling. In: Proceedings of 2011 IEEE/RSJ international conference on intelligent robots and systems (IROS), Sept 2011. p. 5113–20.
8. Hong Y, Hu J, Gao L. Tracking control for multi-agent consensus with an active leader and variable topology. Automatica. 2006;42(7):1177–82.
9. Hong Y, Chen G, Bushnell L. Distributed observers design for leader-following control of multi-agent networks. Automatica. 2008;44(3):846–50.
10. Cao Y, Ren W, Li Y. Distributed discrete-time coordinated tracking with a time-varying reference state and limited communication. Automatica. 2009;45(5):1299–305.
11. Ren W. Consensus tracking under directed interaction topologies: algorithms and experiments. IEEE Trans Control Syst Technol. 2010;18(1):230–7.
12. Peng K, Yang Y. Leader-following consensus problem with a varying-velocity leader and time-varying delays. Phys A. 2009;388(2–3):193–208.
13. Olfati-Saber R. Flocking for multi-agent dynamic systems: algorithms and theory. IEEE Trans Autom Control. 2006;51(3):401–20.
14. Guerrero JA, Castillo P, Challal Y. Trajectory tracking for a group of mini rotorcraft flying in formation. In: The proceedings of the 18th IFAC World Congress, 2011, Milano, Italy, 28 Aug–2 Sept 2011. p. 6331–6336.
15. Schoellig AP, Mueller FL, D'Andrea R. Optimization-based iterative learning for precise quadrocopter trajectory tracking. Auton Robot. 2012;33:103–27.

A 3D Printing Task Packing Algorithm Based on Rectangle Packing in Cloud Manufacturing

Zhen Zhao, Lin Zhang and Jin Cui

Abstract In Cloud Manufacturing environment, massive 3D Printing services in various types provide users with the ability of mass customization. The large number of 3D printing tasks bring more challenges on printing task scheduling to improve 3D printers' utilization and thus saving time and materials. This paper establishes the model of 3D printing process in different types and figures out the existence of auxiliary processes. Aiming at decreasing the ratio of auxiliary process time consumption, this paper develops an algorithm derived from the rectangle packing problem to pack printing tasks whose model size are relatively small into one task and print them all in one 3D printing process. Experiments show that the ratio of auxiliary process time consumption significantly reduced by this algorithm.

Keywords Cloud manufacturing · 3D printing · Rectangle packing

1 Introduction

3D printing is a highly digitized technology in additive manufacturing (AM) [1]. At present, 3D printing technology has been widely used in prototyping, medical, education and other fields [2, 3]. The combination of 3D printing and cloud manufacturing also makes the era of mass customization closer to us. Mapping 3D printing tasks to 3D printing services appropriately and optimizing 3D printing services for cloud manufacturing environments become the focus of research [4, 5]. Brett P. Conner evaluated the product from three dimensions: complexity, customization and production volume to select the processing [6]. Yicha Zhang

Z. Zhao · L. Zhang (✉) · J. Cui
School of Automation Science and Electrical Engineering,
Beihang University, Beijing 100191, China
e-mail: zhanglin@buaa.edu.cn

Z. Zhao · L. Zhang · J. Cui
Engineering Research Center of Complex Product Advanced Manufacturing Systems,
Ministry of Education, Beijing 100191, China

© Springer Nature Singapore Pte Ltd. 2018
Y. Jia et al. (eds.), *Proceedings of 2017 Chinese Intelligent Systems Conference*,
Lecture Notes in Electrical Engineering 460,
https://doi.org/10.1007/978-981-10-6499-9_3

proposed a method based on measuring knowledge value extracted from RP processes [7]. Ce's research uses triangular intuitionistic fuzzy numbers to quantify the description of fuzzy content, and then uses TOPSIS method to calculate the process selection results [8].

On the other hand, Vanek's PackMeger framework divides 3D model shell into segments to save material and time consumption [9]. Kai-Yin Fok's research improves the printing process by optimizing motion paths of the printing nozzle [10]. Due to process characteristics of 3D printing, which will be described in Sect. 2, its efficiency goes low when printing small size products. This paper proposes a 3D printing task packing algorithm based on rectangle packing and 3D printing constraints to improve the overall efficiency of 3D printers.

Subsequent sections are organized as follows: Sect. 2 gives process models of different types of 3D printing. Section 3 introduces the model of 3D printer and 3D printing task for our task packing algorithm. Section 4 introduces rectangle packing problem and its solutions. This leads to the theme of this section: 3D printing task packing algorithm. Section 5 gives a simulation experiment and contrasts algorithm with and without packing process. Section 6 shows the conclusion and future work of this research.

2 Working Process of 3D Printing

3D printing technology mainly has the following categories: Fused Deposition Modeling (FDM), Selected Laser Sintering (SLS), Stereo Lithography Appearance (SLA), Digital Light Processing (DLP) and Laminated Object Manufacturing (LOM) [11]. Three of them are most commonly used: FDM, SLA and DLP.

2.1 FDM

FDM contains four processes shown in Fig. 1. Heating, cooling and pick up and cleaning processes are what we called "auxiliary process" for these are just pre and post treatment that do not contribute to the construction of 3D models. When printing a small size model, the sum of auxiliary processes could be larger than printing time. This may cause great time waste under cloud manufacturing environment.

Fig. 1 Processes of FDM

Fig. 2 Processes of SLA

Fig. 3 Processes of DLP

2.2 SLA

SLA mainly contains three processes shown in Fig. 2. Similar to FDM, printing is the only working process and its time consumption is determined by dedicated execution file generated by slice software. The other processes can be considered as constant time consumption. Its total time consumption is a linear function of the number of layers after model slicing.

2.3 DLP

As shown in Fig. 3, since DLP and SLA are similar in terms of technical principles, they have same processes. However, the printing process in DLP consumes constant time for it uses a light surface created by DLP projector instead of a laser beam to cure photosensitive resin. This makes the printing time of one layer a constant value.

In general, there are two ways to increase the ratio of working process: Increasing working time and decreasing auxiliary time. But it is extremely difficult to reduce time consumption of auxiliary processes with current 3D printing technology. Therefore, the key is increasing working time. An effective method is to pack 3D printing tasks into a package and build them all in one 3D printing cycle. Tasks in the same package share auxiliary processes and other parameters. Working time increases while auxiliary time stays unchanged in one 3D printing cycle, thus increasing the ratio of working process.

3 Model of 3D Printer and 3D Printing Task

According to Yicha's research [7], there are six attributes in describing a 3D printing process: part accuracy, part surface roughness, tensile strength, elongation, part cost and build time. Ce's research [8] selected the same five attributes as Yicha's research and removed the last attribute, build time. Considering features of

Table 1 Six dimensions to describe 3D printers and tasks

Dimension	3D printer	3D printing task
Size (width, depth)	Size of molding platform	Size of 3D model
Height	Height of molding space	Height of 3D model
Color	Material color, an element of the color set	Acceptable material colors, an subset of the color set
Accuracy	Extruder diameter (FDM) Laser diameter (SLA) Projector pixel size (DLP)	Minimum acceptable accuracy of the 3D printer
Layer height	Minimum layer height that a 3D printer can print	Maximum layer height of the production
Material	Material type selected from $\{ABS, PLA, TPU, UVCR, FUVCR\}$	Material requirement from $\{flexible, inflexible\}$

3D printing in cloud manufacturing and all the methods above, 3D printers and tasks are described in six dimensions as Table 1.

The parameter vectors of 3D printer (P_i) and 3D printing task (M_k) defined as follow:

$$P_i = \left\{ p_{i-width}, p_{i-depth}, p_{i-height}, p_{i-color}, p_{i-accuracy}, p_{i-layerheight}, p_{i-material} \right\}$$
$$M_k = \left\{ m_{k-width}, m_{k-depth}, m_{k-height}, m_{k-color}, m_{k-accuracy}, m_{k-layerheight}, m_{k-material} \right\}$$

4 The 3D Printing Task Packing Algorithm (3DTPA)

4.1 Background

In this paper, we take advantage of ideas of rectangle packing algorithms to deal with the task packing problem of 3D printing tasks in a cloud manufacturing environment. A variant of rectangle packing problem described as follow: Given a set of rectangles and a bounding box, package the rectangles into the bounding box while minimizing the number of bounding boxes [12]. The 3D printing task packing problem in a cloud environment conforms to this problem. Under current circumstance, we also need to consider other constraints like model height and color.

The optimization of the rectangle packing problem is a NP-hard problem [13]. Eric Huang's absolute placement approach performs well on a variety of benchmarks [14]. But this method does not apply to our issue for the bottom size of model is not as regular as the benchmark. A novel heuristic algorithm presented by Lei Wu performs well on solving the problem of drilling platform layout [15]. This algorithm starts placing rectangles from the center of bounding box, which is important for placing 3D models on the molding platform. The 3D printing task packing algorithm presented in this paper draws on the idea of this heuristic algorithm.

4.2 Definitions in 3DTPA

Definition 1 (*Task package*). A typical task package is a combination of multiple tasks. But a single task can also constitute a task package. A 3D printer performs a printing cycle on a task package. Tasks in the same package share the following attributes: color, accuracy, layer height and material type. The relationship between task, task package and task queue is shown in Fig. 4. We define Q_i as the task queue of P_i, TP_{ij} as task package j in Q_i, n_{ij} as the number of tasks in TP_{ij}.

Definition 2 (*Layout*). In a task package, a series of task models are placed on the molding platform. The placement of the models is a layout. As shown in Fig. 5, the position of model k is represented by two coordinates $(x_{k1}, y_{k1}), (x_{k2}, y_{k2})$.

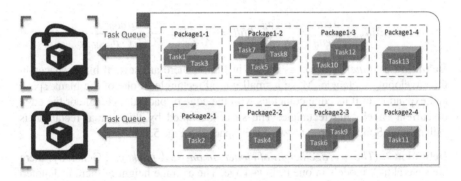

Fig. 4 Relationship between task, task package and task queue

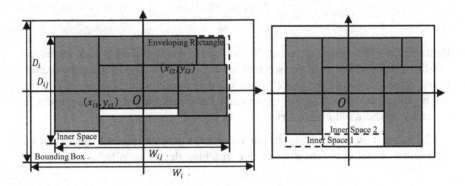

Fig. 5 Layout, enveloping rectangle and bounding box

Definition 3 (*Enveloping Rectangle and Bounding Box*). As shown in Fig. 5, the rectangle with minimum area that contain current layout is an enveloping rectangle denoted by ER_{ij}. The depth and width of ER_{ij} are denoted by D_{ij} and W_{ij}. The area of ER_{ij} is denoted by $Area_{ij}$. The rectangle of molding platform of P_i is a bounding box denoted by Box_i. D_i, W_i and $Area_i$ denote the depth, width and area of Box_i.

Definition 4 (*Aspect Ratio*). The aspect ratio of ER_{ij} is denoted by Nar_{ij} and the aspect ratio of Box_i is denoted by Nar_i. Their values are as follows:

$$Nar_{ij} = \frac{D_{ij}}{W_{ij}}, Nar_i = \frac{D_i}{W_i} = \frac{p_{i-depth}}{p_{i-width}}$$

Definition 5 (*Filling Rate*). For one task package TP_{ij}, the sum of areas of all packed rectangles is denoted by $area_{ij}$. The ratio between $area_{ij}$ and $Area_{ij}$ is called filling rate of enveloping rectangle denoted by $Efill_{ij}$. The ratio between $area_{ij}$ and $Area_i$ is called filling rate of molding platform denoted by $Pfill_{ij}$. These are calculated as follows:

$$area_{ij} = \sum_{M_k \in TP_{ij}} m_{k-width} \cdot m_{k-depth}$$

$$Efill_{ij} = \frac{area_{ij}}{Area_{ij}} = \frac{area_{ij}}{D_{ij} \cdot W_{ij}} Pfill_{ij} = \frac{area_{ij}}{Area_i} = \frac{area_{ij}}{p_{i-width} \cdot p_{i-depth}}$$

Definition 6 (*Inner Space*). As shown in Fig. 5, there are several blank spaces in the enveloping rectangle. Make a small virtual rectangle in one of the blank space and expand it until its edge coincides with the edge of packed rectangle or the edge of current enveloping rectangle. The space occupied by this virtual rectangle is called Inner Space. There are two inner spaces in Fig. 5.

Definition 7 (*The average height of models in one task package*). Generally, there are several task models in one task package. The average height of them is denoted by $\bar{h}_{ij}$ and it is calculated as follow:

$$\bar{h}_{ij} = \frac{\sum_{M_k \in TP_{ij}} m_{k-height}}{n_{ij}}$$

Definition 8 (*Matching Degree of Bottom Size*). When placing a task model on molding platform, first consider whether inner space is enough to place the model. If not, consider placing the model to the edge of enveloping rectangle. If there is still no space for the task model, create a new task package and place the model in the center of the molding platform.

The calculation of matching degree varies with different placing modes. When placing the model in the inner space, the matching degree is as follow:

$$g_i\left(m_{k-width}, m_{k-depth}\right) = 1$$

When placing the model to the edge of enveloping rectangle, the matching degree is as follow:

$$g_i\left(m_{k-width}, m_{k-depth}\right)$$

$$= max\left\{\frac{min\left(W_{ij}, m_{k-width}\right)}{max\left(W_{ij}, m_{k-width}\right)}, \frac{min\left(D_{ij}, m_{k-width}\right)}{max\left(D_{ij}, m_{k-width}\right)}, \frac{min\left(W_{ij}, m_{k-depth}\right)}{max\left(W_{ij}, m_{k-depth}\right)}, \frac{min\left(D_{ij}, m_{k-depth}\right)}{max\left(D_{ij}, m_{k-depth}\right)}\right\}$$

When placing the model in the center of the molding platform, the matching degree is as follow:

$$g_i\left(m_{k-width}, m_{k-depth}\right) = \frac{min\{Nar_k, Nar_k\} \cdot area_k}{max\{Nar_k, Nar_i\} \cdot Area_i}$$

Definition 9 (*Matching Degree of Height*). Assuming that there are two task models with huge height difference in one task package. The shorter one cannot be picked up until the taller one finishes printing. So we tend to put models whose height are close to each other into the same task package for 3D printing and the matching degree of height is calculated as follow:

$$g_i\left(m_{k-height}\right) = 1 - \frac{\left|m_{k-height} - \bar{h}_{ij}\right|}{max\left\{m_{k-height}, \bar{h}_{ij}\right\}}$$

Definition 10 (*Matching Degree of Color*). This paper chooses five colors that are widely used in 3D printing as the elements of the color set. This set is as follow:

$$U_{color} = \{Black, White, Red, Yellow, Blue\}$$

Typically, the value of $p_{i-color}$ and $m_{k-color}$ are selected from the color set. But $m_{k-color}$ has an extra option: Random. This option indicates that any color is acceptable for the task model. So the matching degree of color is calculated as follow:

$$g_{i-color} = \begin{cases} 1 & p_{i-color} \in m_{k-color} \; or \; m_{k-color} = Random \\ 0 & others \end{cases}$$

Definition 11 (*Matching Degree of Accuracy, Layer and Material*). These three attributes are constraints of 3D printing task like color. The matching degrees of them are calculated as follow:

$$g_{i-accuracy} = \begin{cases} 1 & m_{k-accuracy} \geq p_{i-accuracy} \\ 0 & m_{k-accuracy} < p_{i-accuracy} \end{cases}$$

$$g_{ij-layerheight} = \begin{cases} 1 & m_{k-layerheight} \geq p_{i-layerheight} \\ 0 & m_{k-layerheight} < p_{i-layerheight} \end{cases}$$

$$g_i(m_{k-material}) = \begin{cases} 1 & \begin{array}{c} m_{k-material} = flexible \ and \ p_{i-material} \in \{TPU, FUVCR\} \\ or \\ m_{k-material} = inflexible \ and \ p_{i-material} \in \{ABS, PLA, UVCR\} \end{array} \\ 0 & others \end{cases}$$

Definition 12 (*Overall Matching Degree*). The overall matching degree is the product of all the attribute matching degrees. The calculation is as follow:

$$G_{ij} = g_{ij-size} \cdot g_{ij-height} \cdot g_{i-color} \cdot g_{i-accuracy} \cdot g_{i-layerheight} \cdot g_{i-material}$$

4.3   Steps of the Task Packing Algorithm

Step1: Calculate matching degree of each task for each 3D printer without packing the them, the number of matching degrees for each task is the same as the number of 3D printers. Sort the tasks in descending order of model bottom area.

Step2: For a task, traverse all existing task packages and calculate its matching degree for all available task packages, then find the appropriate place in each available task package for the task. There are 3 schemes for placing task model:

(1) If there exist available inner space for model, place it in the inner space.

(2) If there is no available inner space, find out whether there exist enough space for placing model beside the enveloping rectangle and try to place the model to the edge of enveloping rectangle. Make sure the matching degree is the highest one. Recalculate the enveloping rectangle after packing.

(3) If there is no available space on the molding platform in all task packages of a 3D printer, create a new task package and place it in the center of the molding platform.

Step3: Select the task package that get the highest fit degree and put the model in the appropriate place on the molding platform.

Table 2 Parameters of virtual printers (mm)

Name	P_{width}	P_{depth}	P_{height}	P_{color}	$P_{accuracy}$	$P_{layerheight}$	$P_{material}$
Ultimaker (FDM)	220	210	230	Red	0.4	0.1	PLA
PrintRite (FDM)	280	180	180	Yellow	0.3	0.2	TPU
RepRap (FDM)	180	100	100	Black	0.5	0.2	ABS
Xiaofang (SLA)	130	130	180	Blue	0.1	0.05	FUVCR
ProjectD (DLP)	100	76	150	White	0.05	0.05	UVCR

5 Simulation Experiments

The host software of different types of 3D printers developed by their manufacturers and other developers has the feature of virtual printing. These sorts of software such as OctoPrint (FDM), PreForm (SLA) and NanoDLP (DLP) are used as simulation environment of our experiments to evaluate the time consumption of each 3D printer. And according to the parameters listed in the official websites of the major 3D printer manufacturers like Ultimaker, we generated five virtual 3D printers with the following parameters in Table 2.

We collected 30 models from the Internet and generated 30 tasks according to these models. Assigned these tasks to the five virtual 3D printers and generated task packages. One of the task packages is shown as Fig. 6.

Assigning the same 30 printing tasks to the 3D printers without packing them and the difference between packing and unpacking is shown in Fig. 7.

In the case of parallel printing of five 3D printers, the total time consumption of packing approach is 8.5% less than unpacking approach. For some 3D printers, you can even save more than 50% of the time.

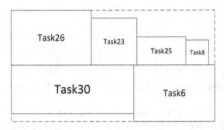

Fig. 6 One task package in results

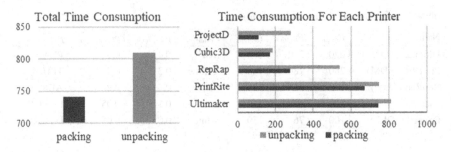

Fig. 7 Time comparison between packing and unpacking

6   Conclusions and Future Work

The 3D printing task packing algorithm presented in this paper provides a new idea on solving 3D printing process selection and optimized 3D printing efficiency. This approach packs tasks into one package and build them in the same printing cycle to increase the ratio of working process. Six dimensions determine the matching degree of printing task and 3D printer. The approach of packing models is based on rectangle packing solutions and 3D printing constraints. Simulation experiment shows that the packing algorithm can save over 50% of the time in some cases compared to unpacked situations.

In the future, the task packing algorithm needs to be further tested in the cloud platform. More dimensions and details of 3D printer and 3D printing service should be considered on calculating matching degree. The approach of packing task models also need to be further optimized.

Acknowledgements   The research is supported by National Natural Science Foundation of China (No. 61374199) and National High-Tech Research and Development Plan of China under Grant No. 2015AA042101.

References

1. Gibson I, Rosen DW, Stucker B. Additive manufacturing technologies. New York: Springer; 2010.
2. Berman B. 3-D printing: the new industrial revolution. Bus Horiz. 2012;55(2):155–62.
3. Canessa E, Fonda C, Zennaro M, et al. Low-cost 3D printing for science, education and sustainable development. Low-cost 3D Print. 2013;11.
4. Zhang L, Mai J, Li BH, et al. Future manufacturing industry with cloud manufacturing. In: Cloud-based design and manufacturing (CBDM). Springer International Publishing; 2014. p. 127–52.
5. Mai J, Zhang L, Tao F, et al. Customized production based on distributed 3D printing services in cloud manufacturing. Int J Adv Manuf Technol. 2016;84(1–4):71–83.
6. Conner BP, Manogharan GP, Martof AN, et al. Making sense of 3-D printing: creating a map of additive manufacturing products and services. Addit Manuf. 2014;1:64–76.

7. Zhang Y, Xu Y, Bernard A. A new decision support method for the selection of RP process: knowledge value measuring. Int J Comput Integr Manuf. 2014;27(8):747–58.
8. Shi C, Zhang L, Mai J, et al. 3D printing process selection model based on triangular intuitionistic fuzzy numbers in cloud manufacturing. Int J Model Simul Sci Comput. 2016; 1750028.
9. Vanek J, Galicia JA, Benes B, et al. Packmerger: a 3D print volume optimizer. Comput Graph Forum. 2014;33(6):322–32.
10. Fok KY, Ganganath N, Cheng CT, et al. A 3D printing path optimizer based on Christofides algorithm. In: 2016 IEEE international conference on consumer electronics-Taiwan (ICCE-TW). IEEE; 2016. p. 1–2.
11. Horvath J. Mastering 3D printing. Apress; 2014.
12. Korf RE, Moffitt MD, Pollack ME. Optimal rectangle packing. Ann Oper Res. 2010;179 (1):261–95.
13. He K, Ji P, Li C. Dynamic reduction heuristics for the rectangle packing area minimization problem. Eur J Oper Res. 2015;241(3):674–85.
14. Huang E, Korf RE. Optimal rectangle packing: an absolute placement approach. J Artif Intell Res. 2013;46(1):47–87.
15. Wu L, Zhang L, Xiao WS, et al. A novel heuristic algorithm for two-dimensional rectangle packing area minimization problem with central rectangle. Comput Ind Eng. 2016;102:208–18.

Exponential Stability of Switched Stochastic Systems with Multiple ADT Switching

Hongji Ma and Yingmin Jia

Abstract This paper addresses the mean square exponential stability of switched stochastic Itô linear systems. By use of $\mathcal{H}$-representation method and mode-dependent average dwell-time (MDADT) switching, a new stability criterion is presented for the considered models. It is shown that even if all its subsystems are essentially unstable, a set of appropriately designed MDADT switchings can guarantee the switched stochastic system to be exponentially stable in mean square. An illustrative example is supplied to show the effectiveness of the obtained results.

Keywords Stability · Switched stochastic systems · Mode-dependent average dwell time · Generalized Lyapunov operator

1 Introduction

Switching phenomena are wide spread in the real world and switched systems can serve as efficient mathematical models for a large variety of engineering faculties subject to structural switching. Therefore, considerable attention has been paid to the analysis and synthesis of switched systems in recent decades; e.g., see [2, 6, 7, 10–12, 17–19, 22] for the latest progress. In the current literature, many kinds of switching rules have been considered, such as random switching with certain transition probability [13], state-dependent switching [1, 16], and time-controlled switching [6, 7, 24–26].

In recent years, the control theory of switched diffusion processes has become a popular topic; see [5, 9, 14, 20] and the references therein. In [3], by use of multiple Lyapunov function method, global uniform asymptotic stability in probability was

H. Ma (✉)
College of Mathematics and Systems Science, Shandong University
of Science and Technology, Qingdao 266590, China
e-mail: ma_math@163.com

Y. Jia
The Seventh Research Division, Beijing University of Aeronautics
and Astronautics, Beijing 100191, China

© Springer Nature Singapore Pte Ltd. 2018
Y. Jia et al. (eds.), *Proceedings of 2017 Chinese Intelligent Systems Conference*,
Lecture Notes in Electrical Engineering 460,
https://doi.org/10.1007/978-981-10-6499-9_4

investigated for switched stochastic systems with multiplicative white noise. Further, by exploiting a stochastic comparison principle, [23] derived a sufficient condition for the stochastic input-to-output (I-O) stability of switched stochastic nonlinear systems. Note that the stability analysis of [4, 9, 23] was performed under the DT or ADT switching, and a common feature of these works is that all stochastic subsystems are required to be stochastically stable (or I-O stable). However, this assumption is sometimes too ideal and strict to be satisfied in practice. This study is devoted to deal with switched stochastic systems whose subsystems may be all essentially unstable [8].

In this paper, we will investigate the exponential stability of switched stochastic Itô systems. Different from the available literature, all subsystems involved in the underlying switched stochastic model are allowed to be essentially unstable. In order to develop a suitable switching law to stabilize the switched stochastic dynamics, the so-called $\mathcal{H}$-representation method [21] is utilized to set up a bridge between the stochastic Itô system and deterministic system. The main result will propose a sufficient condition for the mean square exponential stability of switched stochastic linear Itô systems whose subsystems may be all essentially unstable, which improves [9] in the sense that the existence of common dwell time is no more necessary. Besides, a numerical example with simulations is also presented to illustrate the validity of the proposed theoretical result.

Notations. $R(R_+)$: the set of all (nonnegative) real numbers; R^n: the set of n-dimensional real vectors; $R^{m \times n}$: the set of all $m \times n$ real matrices; A': the transpose of a matrix (vector) A; $\| \cdot \|$: Euclidean norm of a vector or spectral norm of a matrix; I_n: the $n \times n$ identity matrix; $\mathbb{Z}_+ := \{0, 1, 2, \ldots\}$; C: the set of all complex numbers; C^n: the n-dimensional complex space; $C^-(C^+)$: the open left(right)-half complex plane; $C^{-,0}$: the closed left-half complex plane; $\oplus$: direct sum of two subspaces.

2 Preliminaries

Let $(\Omega, \mathcal{F}, \mathcal{P}; \mathcal{F}_t)$ be a filtered probability space with independent standard one-dimensional Brownian motions $w_k(t)$ $(k = 1, \ldots, r)$ and the natural filter $\mathcal{F}_t$ generated by $\{w_k(s), 0 \leq s \leq t, 1 \leq k \leq r\}$. Consider the following switched stochastic Itô linear systems with state-dependent noises:

$$\begin{cases} dx(t) = A_{\sigma(t)} x(t) dt + \sum_{k=1}^{r} C_{\sigma(t)}^k x(t) dw_k(t), \\ x(0) = x_0 \in R^n, \ t \in R_+, \end{cases} \tag{1}$$

where $x(t) \in R^n$ represents the system state, and the switching signal $\sigma(t) : R_+ \to D = \{1, 2, \ldots, d\}$ is a piecewise constant function satisfying $\sigma(t) = \lim_{s \to t_+} \sigma(s)$. For any given $i \in D$, A_i and $\{C_i^k\}_{k=1}^r$ are assumed to be $n \times n$ real matrices. By [15],

$$\begin{cases} dx(t) = A_i x(t)dt + \sum_{k=1}^{r} C_i^k x(t)dw_k(t), \\ x(0) = x_0 \in R^n, \; t \in R_+, \end{cases} \tag{2}$$

has a unique strong solution on any finite horizon and thus, the switched stochastic system (1) admits a solution in the Caratheodory sense.

Definition 2.1 [24]: Given $i \in D$ and $t > s \geq 0$, let $\mathcal{N}_i(s,t)$ be the activating numbers of the i-th subsystem during the time interval $[s,t]$ and $\mathcal{T}_i(s,t)$ indicate the running time of the i-th subsystem within the interval $[s,t]$. The switching signal $\sigma(t)$ is called a mode-dependent average dwell-time switching if there admit two set of positive numbers $\{\mathcal{N}_i^0\}_{i=1}^d$ and $\{\tau_i\}_{i=1}^d$ satisfying

$$\mathcal{N}_i(s,t) \leq \mathcal{N}_i^0 + \mathcal{T}_i(s,t)/\tau_i, \tag{3}$$

where τ_i and $\mathcal{N}_i^0$ are called the mode-dependent average dwell time and the mode-dependent chatter bound of the mode i, respectively. $\qquad\square$

In the sequel, we denote by $S[\tau_i, \mathcal{N}_i^0 | i \in D]$ the set of all mode-dependent average dwell-time switching signals $\sigma(t)$ defined in Definition 2.1.

Definition 2.2 The switched stochastic linear system (1) is called exponentially stable in mean square, if there exist $\alpha > 0$ and $\beta > 0$ such that for any $t \geq t_0 \geq 0$, there holds

$$E\|x(t)\|^2 \leq \alpha e^{-\beta(t-t_0)} \|x_0\|^2, \tag{4}$$

where $x(t)$ is the state trajectory of (1) originating from $x(t_0) = x_0 \in R^n$. $\qquad\square$

In this paper, we aim to design a switching law $\sigma(t) \in S[\tau_i, \mathcal{N}_i^0 | i \in D]$ such that (1) can be stabilized even when all its subsystems are unstable (in stochastic sense).

Next, we will recall some fundamental results about the stability analysis of linear stochastic Itô systems. Let S_n be the set of all $n \times n$ symmetric matrices whose entries may be complex, and $\{\mathcal{L}_i\}_{i \in D}$ be a sequence of generalized Lyapunov operators defined as follows:

$$\mathcal{L}_i : \; X \in S_n \mapsto A_i X + X A_i' + \sum_{k=1}^{r} C_i X C_i' \in S_n. \tag{5}$$

Based on the spectral location of $\mathcal{L}_i$, there is the following stability criterion of stochastic systems, which has been proven in [8].

Lemma 2.1 Set $\Lambda(\mathcal{L}_i) = \{\lambda \in C | \exists 0 \neq X \in S_n, \mathcal{L}_i(X) = \lambda X\}$. Then, System (2) is asymptotically mean square stable iff $\Lambda(\mathcal{L}_i) \in C^-$; System (2) is critically stable iff $\Lambda(\mathcal{L}_i) \in C^{-,0}$; System (2) is essentially unstable iff $\Lambda(\mathcal{L}_i) \cap C^+ \neq \phi$. $\qquad\square$

Let $x(t)$ be the state of (2) and $X_i(t) = E[x(t)x(t)']$. Then, by use of Itô's formula, we have the following ordinary differential equation:

$$\begin{cases} \dot{X}(t) = A_i X(t) + X(t)A_i' + \sum_{k=1}^{r} C_i^k X(t) C_i^{k'}, \\ X(t_0) = x_0 x_0' \in S^n, \ t \in R_+, \end{cases} \tag{6}$$

By comparing (6) with (5), it is easy to derive that $\dot{X}(t) = \mathcal{L}_i(X(t))$ and $X(t) = e^{\mathcal{L}_i(t-t_0)}X(0)$. Here, the operator $e^{\mathcal{L}_i t}$ is expressed by

$$e^{\mathcal{L}_i t} = \sum_{k=0}^{\infty} \frac{\mathcal{L}_i^k t^k}{k!}. \tag{7}$$

In view of $\|X(t)\| \le E\|x(t)\|^2 \le n\|X(t)\|$, (2) is exponentially mean square stable iff (6) is exponentially stable.

Below, we will decompose the linear space S_n into a finite number of subspaces with respect to the eigenvalues of the operator $\mathcal{L}_i$. It can be shown by some standard knowledge of linear algebra and hence is omitted here.

Proposition 2.1 Assume that $\Lambda(\mathcal{L}_i) = \{\lambda_1^{(i)}, \dots, \lambda_l^{(i)}\}$ ($l \le \frac{n(n+1)}{2}$) and the algebraic multiplicity of $\lambda_j^{(i)}$ is $v_j^{(i)}(\ge 1)$, then the linear space S_n can be represented as the following direct sum:

$$S_n = \mathbb{S}_1^{(i)} \oplus \dots \oplus \mathbb{S}_l^{(i)}, \tag{8}$$

where the $\mathbb{S}_j^{(i)}$ is a $v_j^{(i)}$-dimensional subspace of S_n spanned by the generalized eigenvectors corresponding to $\lambda_j^{(i)}$. Moreover, $\mathcal{L}_i \mathbb{S}_j^{(i)} \subseteq \mathbb{S}_j^{(i)}$ and $e^{\mathcal{L}_i t} \mathbb{S}_j^{(i)} \subseteq \mathbb{S}_j^{(i)}$.

Based on the preceding discussion, S_n can be split into two matrix subspaces associated with the spectrum of $\mathcal{L}_i$:

$$\mathbb{S}_+^{(i)} = \sum_{\substack{Re(\lambda_j^{(i)}) \ge 0, \\ \lambda_j^{(i)} \in \Lambda(\mathcal{L}_i)}} \mathbb{S}_j^{(i)}, \quad \mathbb{S}_-^{(i)} = \sum_{\substack{Re(\lambda_j^{(i)}) < 0, \\ \lambda_j^{(i)} \in \Lambda(\mathcal{L}_i)}} \mathbb{S}_j^{(i)}, \tag{9}$$

where $\mathbb{S}_j^{(i)}$ is the same as in (8). From Proposition 2.1, it follows that $\mathbb{S}_+^{(i)}$ and $\mathbb{S}_-^{(i)}$ are both $e^{\mathcal{L}_i t}$-invariant subspaces and $S_n = \mathbb{S}_+^{(i)} \oplus \mathbb{S}_-^{(i)}$. In the sequel, we denote

$$\lambda_+^{(i)} = \sup_{\substack{Re(\lambda_j^{(i)}) \ge 0, \\ \lambda_j^{(i)} \in \Lambda(\mathcal{L}_i)}} Re(\lambda_j^{(i)}), \quad \lambda_-^{(i)} = - \sup_{\substack{Re(\lambda_j^{(i)}) < 0, \\ \lambda_j^{(i)} \in \Lambda(\mathcal{L}_i)}} Re(\lambda_j^{(i)}). \tag{10}$$

Based on Proposition 2.1, we can get the following result.

Proposition 2.2 For system (6), if $\lambda_-^{(i)} > \alpha^{(i)} > 0$ and $\beta^{(i)} > \lambda_+^{(i)} \ge 0$, then there exists $\varepsilon^{(i)} > 0$ such that

$$\|e^{\mathcal{L}_i t}\|_{S_n} \leq e^{\varepsilon^{(i)} + \beta^{(i)} t}, \quad \|e^{\mathcal{L}_i t}\|_{\mathbb{S}_-^{(i)}} \leq e^{\varepsilon^{(i)} - \alpha^{(i)} t}, \tag{11}$$

where $\|e^{\mathcal{L}_i t}\|_S := \sup\limits_{0 \neq X \in S} \dfrac{\|e^{\mathcal{L}_i t}(X)\|}{\|X\|}$ *for any subspace* $S \subseteq S_n$.

3 Stability Analysis

In this section, we will focus on the stability of the switched stochastic system (1). Different from the existing works on the stability analysis of switched stochastic systems, we will demonstrate that even if all the subsystems of (1) are essentially unstable, the switched dynamics may be still exponentially stable in mean square under some suitable MDADT switchings.

Theorem 3.1 *Let* $\lambda_-^{(i)} > \alpha^{(i)} > 0$, $\beta^{(i)} > \lambda_+^{(i)} \geq 0$ *(see (10) for* $\lambda_+^{(i)}, \lambda_-^{(i)}$*),* $\delta^{(i)} > 0$ *and* $\varepsilon^{(i)} > 0$ $(i \in D)$ *be given. If the state space of* $\sigma(t)$ *can be decomposed as* $D = D_1 \cup D_2$ *such that*

$$\mathbb{S}_1 = \bigcap_{i \in D_1} \mathbb{S}_-^{(i)}, \quad \mathbb{S}_2 = \bigcap_{i \in D_2} \mathbb{S}_-^{(i)} \tag{12}$$

are both $\mathcal{L}_i$*-invariant for* $i \in D$ *and* $\mathbb{S}_3 = \sum\limits_{i \in D_1} \mathbb{S}_+^{(i)}$ *satisfies* $\mathbb{S}_3 \subseteq \mathbb{S}_2$, *then the MDADT switching signal* $\sigma(t) \in S[\tau_i, \mathcal{N}_i^0 | i \in D]$ *that fulfills the following conditions:*

$$\sum_{i \in D_1} \left(\beta^{(i)} + \frac{\varepsilon^{(i)}}{\tau_i} + \delta^{(i)} \right) T_i(0, T) \leq \sum_{i \in D_2} \left(\alpha^{(i)} - \frac{\varepsilon^{(i)}}{\tau_i} - \delta^{(i)} \right) T_i(0, T), \ \forall T > 0, \tag{13}$$

$$\sum_{i \in D_2} \left(\beta^{(i)} + \frac{\varepsilon^{(i)}}{\tau_i} + \delta^{(i)} \right) T_i(0, T) \leq \sum_{i \in D_1} \left(\alpha^{(i)} - \frac{\varepsilon^{(i)}}{\tau_i} - \delta^{(i)} \right) T_i(0, T), \ \forall T > 0, \tag{14}$$

can guarantee the switched stochastic system (1) to be exponentially mean square stable.

Proof It is clear that all of $\mathbb{S}_j$ $(j = 1, 2, 3)$ are subspaces of S_n. Moreover, the complementary space of $\mathbb{S}_1$ can be computed as follows:

$$\mathscr{C}(\mathbb{S}_1) = \mathscr{C}\left(\bigcap_{i \in D_1} \mathbb{S}_-^{(i)} \right) = \sum_{i \in D_1} \mathscr{C}(\mathbb{S}_-^{(i)}) = \sum_{i \in D_1} \mathbb{S}_+^{(i)} = \mathbb{S}_3. \tag{15}$$

Bearing in mind $\mathbb{S}_1 \cap \mathbb{S}_3 = \{0\}$, (15) implies that

$$S_n = \mathbb{S}_1 \oplus \mathbb{S}_3. \tag{16}$$

Given a $T > 0$ that is sufficiently large, let $0 = t_0 < t_1 < \ldots < t_j < t_{j+1} < \ldots < t_{\mathcal{N}_\sigma(0,T)} < T$ be the switching instants during the time interval $[0, T]$, where $\mathcal{N}_\sigma(0, T) := \sum_{i=1}^d \mathcal{N}_i(0, T)$ denotes the total switching times of the switching signal $\sigma(t)$. Below, we will show that System (1) is exponentially stable in mean square under the switching rule constrained by (13), (14). By (16), for any initial state $x(0) \in R^n$, the symmetric matrix $X(0) = x(0)x(0)'$ can be uniquely represented as follows:

$$X(0) = \hat{X}(0) + \bar{X}(0), \quad \hat{X}(0) \in \mathbb{S}_1, \quad \bar{X}(0) \in \mathbb{S}_3. \tag{17}$$

So, the solution of (6) can be decomposed into two parts:

$$X(T) = e^{\mathcal{L}_{\sigma(t_{\mathcal{N}_\sigma(0,T)})}(T - t_{\mathcal{N}_\sigma(0,T)})} \ldots e^{\mathcal{L}_{\sigma(t_j)}(t_{j+1} - t_j)} \ldots$$
$$\times e^{\mathcal{L}_{\sigma(t_0)}(t_1 - t_0)}(\hat{X}(0) + \bar{X}(0)) = \hat{X}(T) + \bar{X}(T). \tag{18}$$

For the sake of subsequent discussion, let $\mathcal{I}_1$ and $\mathcal{I}_2$ indicate the sets of $\{j | \sigma(t_j) \in D_1, j \in \{1, 2, \ldots, \mathcal{N}_\sigma(0, T)\}\}$ and $\{j | \sigma(t_j) \in D_2, j \in \{1, 2, \ldots, \mathcal{N}_\sigma(0, T)\}\}$, respectively. Thus, by the $\mathcal{L}_i$-invariance of $\mathbb{S}_1$ and (18), we can give the following estimation for the norm of $\hat{X}(T)$:

$$\|\hat{X}(T)\| = \|e^{\mathcal{L}_{\sigma(t_{\mathcal{N}_\sigma(0,T)})}(T - t_{\mathcal{N}_\sigma(0,T)})} \ldots e^{\mathcal{L}_{\sigma(t_0)}(t_1 - t_0)}\hat{X}(0)\|$$
$$\leq \prod_{j \in \mathcal{I}_2} \|e^{\mathcal{L}_{\sigma(t_j)}(t_{j+1} - t_j)}\|_{\mathbb{S}_n} \prod_{j \in \mathcal{I}_1} \|e^{\mathcal{L}_{\sigma(t_j)}(t_{j+1} - t_j)}\|_{\mathbb{S}_1} \|\hat{X}(0)\|.$$

Note that $\mathbb{S}_1 \subseteq \mathbb{S}_-^{(i)}$ for $i \in D_1$ and $\mathcal{I}_j = D_j$ $(j = 1, 2)$ for sufficiently large $T > 0$. By Proposition 2.2, it is yielded that

$$\|\hat{X}(T)\| \leq \prod_{\sigma(t_j) \in D_2} \|e^{\mathcal{L}_{\sigma(t_j)}(t_{j+1} - t_j)}\|_{\mathbb{S}_n} \cdot \prod_{\sigma(t_j) \in D_1} \|e^{\mathcal{L}_{\sigma(t_j)}(t_{j+1} - t_j)}\|_{\mathbb{S}_-^{(\sigma(t_j))}} \|\hat{X}(0)\|$$
$$\leq \prod_{i \in D_2} e^{\left(\varepsilon^{(i)} \mathcal{N}_i(0,T) + \beta^{(i)} T_i(0,T)\right)} \cdot \prod_{i \in D_1} e^{\left(\varepsilon^{(i)} \mathcal{N}_i(0,T) - \alpha^{(i)} T_i(0,T)\right)} \|\hat{X}(0)\|.$$

Applying the switching laws (3) and (13) to the above, it follows that

$$\|\hat{X}(T)\| \leq e^{\left(\sum_{i \in D_2} \beta^{(i)} T_i(0,T) - \sum_{i \in D_1} \alpha^{(i)} T_i(0,T) + \sum_{i \in D} \frac{\varepsilon^{(i)}}{\tau_i} T_i(0,T)\right)}$$
$$\times e^{\sum_{i \in D} \varepsilon^{(i)} \mathcal{N}_i^0} \|\hat{X}(0)\|$$
$$\leq e^{\sum_{i \in D} \varepsilon^{(i)} \mathcal{N}_i^0} \cdot e^{\left(-\sum_{i \in D} \delta^{(i)} T_i(0,T)\right)} \|\hat{X}(0)\|$$
$$\leq e^{\sum_{i \in D} \varepsilon^{(i)} \mathcal{N}_i^0} \cdot e^{-\delta T} \|\hat{X}(0)\|, \quad \delta = \min_{i \in D} \delta^{(i)} > 0. \tag{19}$$

On the other hand, recalling the assumption of $\mathbb{S}_3 \subseteq \mathbb{S}_2$, there holds $\bar{X}(0) \in \mathbb{S}_2$. Thus, for the norm of $\bar{X}(T)$, it is got from the $\mathcal{L}_i$-invariance of $\mathbb{S}_2$ that

$$\|\bar{X}(T)\| = \|e^{\mathcal{L}_{\sigma(t_{N_\sigma(0,T)})}(T - t_{N_\sigma(0,T)})} \dots e^{\mathcal{L}_{\sigma(t_0)}(t_1 - t_0)} \bar{X}(0)\|$$

$$\leq \prod_{j \in \mathcal{I}_1} \|e^{\mathcal{L}_{\sigma(t_j)}(t_{j+1} - t_j)}\|_{S_n} \prod_{j \in \mathcal{I}_2} \|e^{\mathcal{L}_{\sigma(t_j)}(t_{j+1} - t_j)}\|_{S_2} \|\hat{X}(0)\|.$$

As before, $\mathcal{I}_i = \mathcal{D}_i$ ($i = 1, 2$) when $T > 0$ is sufficiently large. Since $\mathbb{S}_2 \subseteq \mathbb{S}_-^{(i)}$ for $i \in \mathcal{D}_2$, we can deduce from Proposition 2.2 that

$$\|\bar{X}(T)\| \leq \prod_{\sigma(t_j) \in \mathcal{D}_1} \|e^{\mathcal{L}_{\sigma(t_j)}(t_{j+1} - t_j)}\|_{S_n} \cdot \prod_{\sigma(t_j) \in \mathcal{D}_2} \|e^{\mathcal{L}_{\sigma(t_j)}(t_{j+1} - t_j)}\|_{\mathbb{S}_-^{(\sigma(t_j))}} \|\bar{X}(0)\|$$

$$\leq \prod_{i \in \mathcal{D}_1} e^{\left(\varepsilon^{(i)} \mathcal{N}_i(0,T) + \beta^{(i)} T_i(0,T)\right)} \cdot \prod_{i \in \mathcal{D}_2} e^{\left(\varepsilon^{(i)} \mathcal{N}_i(0,T) - \alpha^{(i)} T_i(0,T)\right)} \|\bar{X}(0)\|. \qquad (20)$$

Implementing the switching rules (3) and (14) on (20), it turns out that

$$\|\bar{X}(T)\| \leq e^{\left(\sum_{i \in \mathcal{D}_1} \beta^{(i)} T_i(0,T) - \sum_{i \in \mathcal{D}_2} \alpha^{(i)} T_i(0,T) + \sum_{i \in \mathcal{D}} \frac{\varepsilon^{(i)}}{\tau_i} T_i(0,T)\right)} \cdot e^{\sum_{i \in \mathcal{D}} \varepsilon^{(i)} \mathcal{N}_i^0} \|\bar{X}(0)\|$$

$$\leq e^{\sum_{i \in \mathcal{D}} \varepsilon^{(i)} \mathcal{N}_i^0} \cdot e^{\left(-\sum_{i \in \mathcal{D}} \delta^{(i)} T_i(0,T)\right)} \|\bar{X}(0)\| \leq e^{\sum_{i \in \mathcal{D}} \varepsilon^{(i)} \mathcal{N}_i^0} \cdot e^{-\delta T} \|\bar{X}(0)\|, \qquad (21)$$

where $\delta > 0$ is the same as in (19). Now, summarizing (19) and (21), we arrive at that for any $X(0) \in S_n$, the switching signal $\sigma(t) \in S[\tau_i, \mathcal{N}_i^0 | i \in \mathcal{D}]$ specified by (13), (14) can guarantee the solution of (6) to satisfy

$$\|X(T)\| \leq \tilde{C} e^{-\delta T} \|X(0)\|, \quad X(0) \neq 0, \qquad (22)$$

where $\tilde{C} = \frac{\|\hat{X}(0)\| + \|\bar{X}(0)\|}{\|X(0)\|} \cdot e^{\sum_{i \in \mathcal{D}} \varepsilon^{(i)} \mathcal{N}_i^0} > 0$. Combined with $\|X(t)\| \leq E\|x(t)\|^2 \leq n\|X(t)\|$, (22) shows that for sufficiently large $T > 0$,

$$E\|x(T)\|^2 \leq n\tilde{C} e^{-\delta T} E\|x(0)\|^2, \quad x(0) \in R^n, \qquad (23)$$

which means that System (1) is exponentially stable in mean square. This completes the proof. □

Remark 3.1 In previous literature [5, 9, 14], in order to guarantee the stability of switched diffusion processes, it generally requires all (or at least part) subsystems to be asymptotically mean square stable, while Theorem 3.1 releases this constraint by allowing that all subsystems of (1) may be essentially unstable. Besides, the $\mathcal{L}_i$-invariance of $\mathbb{S}_1(\mathbb{S}_2)$ can be naturally achieved when this intersection coincides with a subspace spanned by some common generalized eigenvectors corresponding to distinct eigenvalues of $\mathcal{L}_i$. □

Example 3.1 Consider a switched RLC electric circuit represented by the following switched stochastic system:

$$dx(t) = \begin{bmatrix} \frac{R_i}{\Gamma_i} & -\frac{1}{\Gamma_i} \\ \frac{1}{\Phi_i} & 0 \end{bmatrix} x(t)dt + \begin{bmatrix} \gamma_i & 0 \\ 0 & 0 \end{bmatrix} x(t)dw(t). \tag{24}$$

To illustrate the proposed theoretical result, we take the parameters of (24) as: $\Gamma_1 = 4.17, \Phi_2 = 3.02, R_1 = 0.14, \gamma_1 = 0.707, \Gamma_2 = 1.09, \Phi_2 = 2.16, R_1 = 0.23$ and $\gamma_2 = 1.22$. By Proposition 2.1, it can be computed that $\Lambda(\mathcal{L}_1) = \{0.2312, 0.1517, -0.4516\}$ and $\Lambda(\mathcal{L}_2) = \{-1.0911, -1.2137, 0.1502\}$. Thus, two subsystems are both essentially unstable.

Below, if we split D into $D = D_1 \cup D_2$ with $D_1 = \{1\}$ and $D_2 = \{2\}$, then there is $\mathbb{S}_1 = span\{U_3\}$ and $\mathbb{S}_2 = span\{U_1, U_2\}$ with $\varphi(U_1) = (0.5632\ 0.2357\ 0.4704)'$, $\varphi(U_2) = (0.9701\ -0.4252\ 0.9428)'$ and $\varphi(U_3) = (-0.6748\ 0.5904\ 0.4419)'$. Further, it can be directly testified that $\mathbb{S}_1$ and $\mathbb{S}_2$ are noth $\mathcal{L}_i$-invariant ($i = 1, 2$) and $\mathbb{S}_3 = \mathbb{S}_2$.

According to Theorem 3.1, for the given $\varepsilon^{(i)} = \delta^{(i)} = 0.01$ ($i = 1, 2$), we can take $\alpha^{(1)} = 0.4$, $\alpha^{(2)} = 1$ and $\beta^{(i)} = 0.25$ ($i = 1, 2$). Moreover, by (13)–(14), the parameters of MDADT switching signal may be designed as $\mathcal{N}_i^0 = 0$ ($i = 1, 2$), $\tau_1 = 2$ and $\tau_2 = 1$ with $0.667 < \frac{T_1(0,T)}{T_2(0,T)} < 3.8$ for sufficiently large $T > 0$. In the simulations, the following switching law is adopted:

$$\sigma(t) = \begin{cases} 1, & 6l \le t < 6l + 0.6,\ 6l + 2.4 \le t < 6l + 3.8,\ l \in \mathbb{Z}_+, \\ 2, & \text{otherwise.} \end{cases} \tag{25}$$

Fig. 1a demonstrates the switched signal of (25), and Fig. 1b displays 50 random trajectories of the switched stochastic system with the initial state $(x_1(0), x_2(0)) = (40, -80)$. Obviously, the mean values denoted by the blue lines indicate that (24) is exponentially stable in mean square under the switching (25).

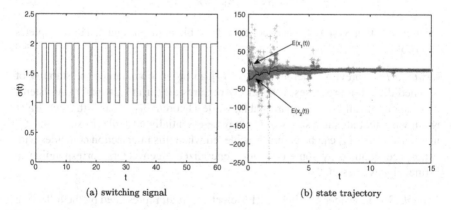

(a) switching signal   (b) state trajectory

Fig. 1 State responses of switched stochastic subsystems

4 Conclusions

By means of the $\mathcal{H}$-representation and MDADT switching, a stability criterion has been developed for the switched stochastic system with possible all unstable modes to be exponentially stable. In comparison with existing studies, this result not only removes the stability prerequisite on the subsystems, but also improves the flexibility of the switching law. Moreover, it is expected that this stability analysis will have potential applications in the related control issues, such as L_2-gain analysis and H_∞ control of switched stochastic systems [14], which no doubt deserve a further research.

Acknowledgements This work was supported by the Natural Science Foundation of Shandong Province (ZR2016FM16), the National Basic Research Program of China (973 Program: 2012CB821201) and the NSFC (61327807), the Scientific Research Foundation of Shandong University of Science and Technology for Recruited Talents (No. 2016RCJJ031) and SDUST Research Fund (No.2015TDJH105).

References

1. Allerhand L, Shaked U. Robust state dependent switching of linear systems with dwell time. IEEE Trans Autom Control. 2013;58(4):994–1001.
2. Briat C. Convex conditions for robust stability analysis and stabilization of linear aperiodic impulsive and sampled-data systems under dwell-time constraints. Automatica. 2013;49(1):3449–557.
3. Chatterjee D, Liberzon D. Stability analysis of deterministic and stochastic switched systems via a comparison principle and multiple Lyapunov functions. SIAM J Control Optim. 2006;45(1):174–206.
4. Feng W, Tian J, Zhao P. Stability analysis of switched stochastic systems. Automatica. 2011;47(1):148–57.
5. Filipovoc V. Exponential stability of switched stochastic systems. Tranctions Inst Meas Control. 2009;31(2):205–12.
6. Geromel JC, Colaneri P. H_∞ and dwell time specifications of continuous-time switched linear systems. IEEE Trans Autom Control. 2010;55(1):207–12.
7. Haspanha JP. Uniform stability of switched linear systems: extensions of Lasalle's invariance principle. IEEE Trans Autom Control. 2004;49(4):470–82.
8. Hou T, Zhang W, Ma H. Essential instability and essential destabilization of linear stochastic systems. IET Control Theory Appl. 2011;5(3):334–40.
9. Huang R, Lin Y, Ge SS, Lin Z. H_∞ stabilization of switched linear stochastic systems under dwell time constraints. Int J Control. 2012;85(9):1209–17.
10. Ji Z, Wang L, Guo X. On controllability of switched linear systems. IEEE Trans Autom Control. 2008;53(3):796–801.
11. Lin H, Antsaklis PJ. Switching stabilizability of continuous-time uncertain switched linear systems. IEEE Trans Autom Control. 2007;54(4):633–46.
12. Lin H, Antsaklis PJ. Stability and stabilization of switched linear systems: a survey of recent results. IEEE Trans Autom Control. 2009;54(2):308–22.
13. Ma H, Jia Y. H_2 control of discrete-time periodic systems with Markovian jumps and multiplicaitve noise. Int Control. 2013;6(10):1837–49.

14. Ma H, Jia Y.: Stochastic bounded real lemma for switched stochastic systems with average dwell time, In: Proceedings of the 53rd IEEE Conference on Decision and Control, Los Angeles, USA, 2014. pp. 216–4221.
15. Mao X. Stochastic Differential Equations and Applications. London: Horwood; 1997.
16. Persis CD, Santis RD, Morse AS. Switched nonlinear systems with state-dependent dwell-time. Syst Control Lett. 2003;50(4):291–302.
17. Sun ZD, Ge SS. Switched Linear Systems - Control and Design. London: Springer; 2004.
18. Wang Y, Gupta V, Antsaklis PJ. On passivity of a class of discrete-time switched nonlinear systems. IEEE Trans Autom Control. 2014;59(3):692–702.
19. Zhang L, Gao H. Asynchronously switched control of switched linear systems with average dwell time. Automatica. 2010;46(5):953–8.
20. Zhang W, Hu J, Lian J. Quadratic optimal control of switched linear stochastic systems. Syst Control Lett. 2010;59(11):736–44.
21. Zhang W, Chen BS. $\mathcal{H}$-representation and applications to generalized Lyapunov equations and linear stochastic systems. IEEE Trans Autom Control. 2012;57(12):3009–22.
22. Zhao J, Hill DJ. On stability, $\mathcal{L}_2$-gain and H_∞ control for switched systems. Automatica. 2008;44(5):1220–32.
23. Zhao P, Feng W, Kang Y. Stochastic input-to-state stability of switched stochastic nonlinear systems. Automatica. 2012;48(10):2569–76.
24. Zhao X, Zhang L, Shi P, Liu M. Stability and stabilization of switched linear systems with mode-dependent average dwell time. IEEE Trans Autom Control. 2012;57(7):1809–15.
25. Zhao X, Zhang L, Shi P, Liu M. Stability of switched positive linear systems with average dwell time switching. Automatica. 2012;48(6):1132–7.
26. Zhao X, Zheng X, Niu B, Liu L. Adaptive tracking control for a class of uncertain switched nonlinear systems. Automatica. 2015;52(1):185–91.

Dual Color Image Blind Watermarking Algorithm Based on Compressive Sensing

Fan Liu, Hongyong Yang and Gaohuan Lv

Abstract As a new information processing theory, Compressive Sensing (CS) should be further researched on the applications of digital watermarking. This paper presents a novel blind digital watermarking scheme embedding a color watermark into a color host image. By using Compressive Sensing theory, Gaussian random matrix is carried out to observe the sparse image and watermark is embedded in compressed observation domain. Finally a smooth norm algorithm is used to reconstruct watermarked image. The watermark extraction process is the blind extraction. The experimental results indicate that the algorithm meets the requirements of the watermark invisibility, robustness and security.

Keywords Compressive sensing · Information hiding · Digital watermarking · Blind extraction

1 Introduction

With the rapid development of information technology, an amount of transferred data and high sampling frequency make the transfer, storage and processing of mass data into one of the main bottlenecks in the field of information. The theory of Compressive Sensing (CS) proposed by Candes et al. [1] and Donoho [2] is a kind of new signal acquisition and reconstruction theory which makes full use of signal sparsity or compressibility. CS has been widely concerned because of the characteristics of low sampling rate, low speed and high reconstruction quality. Related

F. Liu · H. Yang (✉) · G. Lv
School of Information and Electrical Engineering, Ludong University,
Yantai 264025, Shandong, China
e-mail: yhy9919@163.com

F. Liu
e-mail: jsgyliufan@163.com

G. Lv
e-mail: lvgh@ldu.edu.cn

© Springer Nature Singapore Pte Ltd. 2018
Y. Jia et al. (eds.), *Proceedings of 2017 Chinese Intelligent Systems Conference*,
Lecture Notes in Electrical Engineering 460,
https://doi.org/10.1007/978-981-10-6499-9_5

works involve many fields, such as signal processing, pattern recognition, wireless communications, optical imaging, radar remote sensing and so on.

Digital watermarking, as a kind of information hiding technology, effectively curbs the pirate and tort. At present most of the watermarking algorithms focus on the spatial domain and transform domain. Compressive Sensing domain is a new kind of watermarking "transform domain". Chen et al. [3] put CS transform on the gray host image to embed watermark by modifying the coefficients of perception domain, and then recovered the watermarked image by reconstruction algorithm; Sui et al. [4] and Liu et al. [5] compressed the watermark image by CS, which greatly reduces the watermark information, and then a small amount of embedded information could reconstruct the original image watermarking very well. However, these algorithms take the binary or gray image as watermark image. It is because the amount of information of color digital image is 3 times than the gray image with same size and binary image is 24 times, which greatly increases the difficulty of watermark embedding [6–10]. In addition, most of watermarking algorithms are not blind extraction, which requires the participation of the original host image or the original watermark image during extracting.

Based on Compressive Sensing theory and combined with Arnold scrambling transformation and blocking theory, this paper proposes a novel double color image blind watermarking scheme. According to CS, firstly, the original host image is executed sparse transform and Gaussian random matrix is selected to measure the sparse image, and then coefficients of CS domain are modified to embed watermark. Finally, smooth l_0 norm (SL0) algorithm is used to reconstruct watermarked image. The original host image and the original watermark image are not required in the extraction process. Plenty of experiments prove that the algorithm has good watermark invisibility and strong robustness.

2 Compressive Sensing

Traditional methods of signal sampling follows the Nyquist-Shannon Theorem, which points out that the sampling frequency should be higher twice than the bandwidth of signal in order to make signal get accurate reconstruction. CS, combining compression with sampling, is a new research direction in the field of signal processing. CS mainly includes the sparse representation of signal, measurement sampling and reconstruction algorithm.

For an orthonormal basis matrix ψ, $\psi = \{\varphi_n\}_{n=1}^{N}$, if original signal $X \in R^N$ can be expressed as a linear combination of K basis vectors, which is expressed as

$$X = \psi x = \sum_{m=1}^{N} x_m \varphi_m = \sum_{m=1}^{K} x_{m_i} \varphi_{m_i} \tag{1}$$

where $x_m = \langle X, \varphi_m \rangle = \varphi_m^T X$ is the transform coefficient vector. If only $K (K \ll N)$ elements of the coefficient vector are nonzero, the signal is called as K-sparse.

The compressed value denoted as a vector $Y \in R^M$ can be got by transforming sparse signal to low-dimensional space with the help of random matrix $\Phi = [\varphi_1, \varphi_2, \ldots, \varphi_m]^T$, which is expressed as

$$Y = \Phi X = \Phi \psi x = Ax \tag{2}$$

where Φ is a measurement matrix of size $M \times N$, $A = \Phi \psi$ is sensor matrix.

Signal reconstruction process is the signal recovery process from undersampling. It can be got infinitely many solutions obviously because the dimension of Y is far below dimension of x. But original signal x is K-sparse. The original signal can be recovered accurately from measurements Y by solving the optimization l_0 norm problem, which is expressed as

$$\hat{x} = \arg \min \|x_0\|, \text{ subject to } Y = AX \tag{3}$$

Candes et al. [1] and Donoho [2] indicated that in order to accurately reconstruct K-sparse signal x, measurement time M should satisfy $M = O(K1g(N))$, and measurement matrix should meet the restricted isometry property (RIP), which is expressed as

$$1 - \delta \leq \frac{\|\Phi x\|_2^2}{\|x\|_2^2} \leq 1 + \delta \tag{4}$$

where δ is isometry constant.

Donoho [2] indicated that the problem of optimization process is an NP-hard problem. According to this, many effective reconstruction algorithms have been proposed such as matching pursuit [3], minimum norm method [4], and threshold iteration method. Due to good performance such as high convergence speed, smooth l_0 norm algorithm [11] is adopted in the proposed watermarking scheme.

3 Algorithm Description

3.1 Watermark Preprocessing

The watermark preprocessing is depicted as shown Fig. 1. The detailed process is formulated as follows:

(1) The 3-D color watermark image W is decomposed into three independent components, which are Red, Green and Blue component.
(2) Each component is scrambled by Arnold transform with the secret key *Ka*.

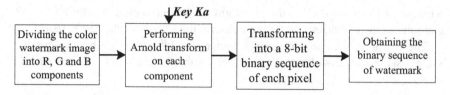

Fig. 1 Watermark preprocessing

(3) The pixel of each scrambled component is converted to 8-bit binary sequence
    one by one, and then the obtained binary sequence is combined to one
    dimensional watermark information.

3.2 Watermark Embedding Procedure

The watermark embedding procedure is depicted as shown Fig. 2. The main steps
are formulated as follows:

(1) The color original host image H is divided into Red, Green and Blue compo-
    nent image.
(2) The three component images are individually measured by CS with the help of
    measurement matrix Φ. The compressed result is called sensor coefficient matrix
    Y_1. Then the coefficient matrix is partitioned into 4×4 non-overlapping blocks.
(3) Watermark information is embedded by modifying the biggest energy element
    of each block according to Eq. (5).

$$
y_{max}^* = \begin{cases} y_{max} - \mod(y_{max}, K) + 0.75 \times K, w = {}'1' \\ y_{max} - \mod(y_{max}, K) + 0.25 \times K, w = {}'0' \end{cases} \tag{5}
$$

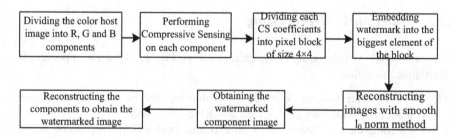

Fig. 2 Flowchart of watermark embedding procedure

where w is the embedded watermark information, y_{max} is the most energy element, y_{max}^* is the result of embedded watermark, K is embedding intensity, and mod() is the modulus function.

(4) All of the watermark information are embedded into the coefficient matrix by repeating step (3) of the embedding process, which can obtain the watermarked coefficient matrix Y_2.
(5) The watermarked component image can be recovered from the coefficient matrix Y_2 by using smooth l_0 norm algorithm.
(6) The watermarked host image H^* is obtained by recombining the watermarked Red, Green, Blue component image.

3.3 Watermark Extraction Procedure

The proposed extraction method belongs to the blind extraction method, which can extract the watermark from the watermarked image without the help of the original host image or the original watermark image. The watermark extraction procedure is shown in Fig. 3. The main steps are formulated as follows:

(1) The watermarked image H^* is decomposed into three independent components, which are Red, Green and Blue component image.
(2) The three component images are individually measured by CS with the help of measurement matrix Φ. The compressed result is called sensor coefficient matrix Y_1^*. Then the coefficient matrix is partitioned into 4×4 non-overlapping blocks.
(3) Watermark information w^* is extracted from the biggest energy element y_{max}^* of each block according to Eq. (6).

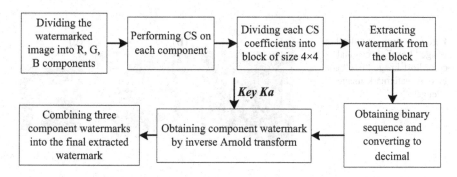

Fig. 3 Flowchart of watermark extraction procedure

$$w^* = \begin{cases} 0, & \mod (y^*_{max}, K) < K/2 \\ 1, & \mod (y^*_{max}, K) \geq K/2 \end{cases} \tag{6}$$

(4) All of the watermark information are extracted from the coefficient matrix by repeating step (3) of the extraction process.

(5) The extracted watermark sequence is converted to decimal pixels and reshaped to component watermarks.

(6) Each component watermark is performed Arnold inverse transform with the secret key Ka, and the extracted watermark image W^* is obtained by recombining the reshaped Red, Green, Blue component watermark.

4 Experimental Simulations and Analysis

Plenty of simulation experiments are made to validate the above algorithm. In the experiments, two color images "Lena" and "F16" with 512×512 pixels are served as the original host images which are selected from USC-SIPI image databases while a color image "Peugeot" with size 32×32 is taken as the watermark image. Three images are shown in Fig. 4. Wavelets basis and Gaussian random matrix are selected as signal sparse basis matrix and the measurement matrix respectively. The compression ratio is set to 0.8, the watermark embedding intensity K is set to 30, and SL0 is served as reconstruction algorithm.

Imperceptibility and robustness are generally used to evaluate the performance of the proposed watermarking algorithm. The imperceptibility of watermark is the ability of the host image to hide information. This paper adopts the traditional Peak Signal-to-Noise Ratio (PSNR) to calculate the level of similarity between the original host image (H) and the watermarked image (H*). In addition, Normalized Cross-correlation (NC) is served as evaluation criterion of robustness [8–10].

Fig. 4 **a** Original color host image "Lena", **b** original color host image "F16" and **c** color watermark image "Peugeot"

(a)   **(b)**   **(c)**

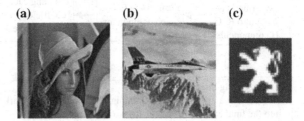

4.1 Imperceptibility Test

More generally, the deformation of the host image may have good performance of imperceptibility when the PSNR value of the watermarked image is more than 30.00 dB. The watermarked image of "Lena" and "F16" whose PSNR value both reach 32 dB are depicted in Fig. 5, where there is no perceptual degradation observed from the point view of human visual characteristics. The reconstructed color watermark image "Peugeot" whose NC value is 1.0 is illustrated in Fig. 5, from which the difference between the original watermark and the corresponding reconstruction cannot be discerned visually.

4.2 Robustness Test

4.2.1 Common Attack Test

JPEG compression, Gaussian noise, salt and pepper noise and Gaussian low-pass filtering are adopted to verify the robustness of the proposed algorithm. According to the results of robustness test under different attacks shown in Table 1, the NC value of extracted watermark after attacks is still above 0.8, which indicates that the proposed algorithm has strong robustness.

Besides using NC as a measurement, parts of extracted watermark images of visual perception under different attacks are given in Table 2.

4.2.2 Comparison with Similar Algorithm

The major difference between the algorithm of this paper and other CS watermarking algorithm is to embed color image as watermark into color host image. The process of color image is more complex than grayscale image because the

	Lena	F16
Watermarked image		
PSNR（dB）/SSIM	32.1327/0.9204	32.2962/0.9295
Extracted watermark		
NC	1.0000	1.0000

Fig. 5 Unattacked watermarked image and extracted watermark

Table 1 Robustness test under different attacks with different parameters

Attack method		The NC value of different images under different attacks				
JPEG compression	Compression factor	100	95	90	85	80
	Lena	1.0000	0.9996	0.9573	0.8615	0.8294
	F16	1.0000	0.9997	0.9524	0.8719	0.8229
Salt and pepper noise	Noise density	0.001	0.002	0.003	0.004	0.005
	Lena	0.9301	0.8373	0.7936	0.7695	0.7325
	F16	0.9228	0.8440	0.7949	0.7640	0.7322
Gaussian noise	Standard deviation	0.0001	0.0002	0.0003	0.0004	0.0005
	Lena	0.9980	0.9677	0.9239	0.8980	0.8683
	F16	0.9956	0.9697	0.9412	0.8802	0.8600
Gaussian low-pass filtering	Sigma (3×3)	0.1	0.2	0.3	0.4	0.5
	Lena	1.0000	1.0000	1.0000	0.9990	0.8158
	F16	1.0000	1.0000	1.0000	0.9998	0.7913

Table 2 Extracted watermark images under different attacks

Attacks method		Extracted watermark from Lena	NC	Extracted watermark from F16	NC
JPEG compression	Compression factor 90%		0.9573		0.9524
Salt and pepper noise	Noise density 0.001		0.9302		0.9228
Gaussian noise	Standard deviation 0.0002		0.9679		0.9697
Gaussian low-pass filtering	Sigma 0.3 (3×3)		1.0000		1.0000

information volume can reach three times than gray watermark. Comparison with scheme in [3] which is also based on CS, the robustness of the proposed scheme is stronger. The experimental results under the same attack condition are shown in Table 3.

Table 3 Performance comparison with scheme in [3] under different attacks

Attack method	Parameters	Scheme in [3]	Proposed scheme
JPEG compression	95	0.6911	0.9996
	85	<0.5	0.8615
Gaussian noise	0.0001	0.6940	0.9985
	0.0003	<0.5	0.9239
Gaussian low-pass filtering	0.1	0.9998	1
	0.5	0.7487	0.8158

5 Conclusions

In this paper, Compressive Sensing is introduced into digital watermarking and a novel dual color image blind watermarking scheme based on CS is presented. After sparse transform, the original host image is measured by Gaussian random matrix, and then coefficients of CS domain are modified to embed watermark. Finally, smooth l_0 norm algorithm is used to recover watermarked images. Without resorting to the original host image or original watermark, the embedded watermark can be extracted from the different attacked images. Experimental results have shown that this algorithm not only provides good performance in invisibility but also has strong robustness in the operation of common image processing. In addition, Gaussian random matrix, measurement matrix, takes a key role to ensure the security of the watermark. Blind extraction reduces the limitation of the watermark certification and broadens the watermarking's background in application. In conclusion, though this algorithm has good performance in the process of watermark, it could be further researched in depth in terms of sparsity and the compression ratio, which will be the emphasis of watermarking in the future.

Acknowledgements The authors would like to thank support from the National Natural Science Foundation of China (under grant 61673200, 61603172, 61771231, 61471185, and 61472172), Natural Science Foundation of Shandong Province (under grant ZR2017MF 010), Department of Science and Technology of Shandong Province (under grant 2015GSF116001), Key Science and Technology Plan Projects of Yantai City (under grant 2016ZH057).

References

1. Candes EJ, Romberg J, Tao T. Robust uncertainty principles: exact signal reconstruction from highly incomplete frequency information [J]. IEEE Trans Inf Theory. 2006;52(2):489–509.
2. Donoho DL. Compressive sensing [J]. IEEE Trans Inf Theory. 2006;52(4):1289–306.
3. Chen GF, Guo SX, Yang LI, et al. Digital image watermark algorithm based on compressive sensing [J]. Modern Electron Tech. 2012.
4. Sui L, Zhou B, Wang Z, et al. An optical color image watermarking scheme by using compressive sensing with human visual characteristics in gyrator domain[J]. Opt Lasers Eng. 2017;92:85–93.

5. Liu H, Xiao D, Zhang R, et al. Robust and hierarchical watermarking of encrypted images based on Compressive Sensing [J]. Signal Process Image Commun. 2016;45(C):41–51.
6. Su Q, Niu Y, Zou H, et al. A blind dual color images watermarking based on singular value decomposition[J]. Appl Math Comput. 2013;219(16):8455–66.
7. Su Q, Niu Y, Wang G, et al. Color image blind watermarking scheme based on QR decomposition[J]. Sig Process. 2014;94(1):219–35.
8. Su Q, Niu Y, Liu X, et al. A blind dual color images watermarking based on IWT and state coding[J]. Opt Commun. 2012;285(7):1717–24.
9. Su Q. Novel blind colour image watermarking technique using Hessenberg decomposition. IET image processing, 11/2016, 10(11), 817– 829.
10. Su Q, Wang G, Lv G, et al. A novel blind color image watermarking based on Contourlet transform and Hessenberg decomposition [J]. Multimedia Tools Appl. 2016;1–21.
11. Mohimani H, Babaie-Zadeh M, Jutten C. A fast approach for overcomplete sparse decomposition based on smoothed l_0 norm [J]. IEEE Trans Signal Process. 2008;57 (1):289–301.

Ground Moving Target Indication Based on Doppler Spectrum in Synthetic Aperture Radar Images

Yuexiang Wang, Hongyong Yang and Gaohuan Lv

Abstract Two moving target detection methods in a given synthetic aperture radar (SAR) imagery are proposed based on the theory of SAR imaging. The methods are performed by using the phase information of the image patch by patch. First, an average Doppler spectrum is achieved from the stationary parts in the image and used as a standard one in the successive computations. Then an average Doppler spectrum of a selected image is computed and used to detect moving targets with random directions by correlation and normalized difference computation. Finally, the effectiveness and practicability are validated by theoretical and field data.

Keywords Synthetic aperture radar · Moving target detection · Doppler centroid estimation

1 Introduction

Synthetic aperture radar (SAR) has been widely used both in military and civilian applications due to its all-weather and high resolution imaging for the ground surface [1–5], and moving target detection in SAR images has been an active research topic in the fields of SAR-GMTI [2–10]. Due to the difficulty of detecting moving targets in a single SAR amplitude imagery, typical research methods are based on its phase information, such as auto-focusing and Doppler center detection.

Fienup broke a SAR image into several small patches, and then refocusing operations are performed on each patch by using the slicing average method. Sharpness ratio between the original and refocused patches is used as a criterion to determine

Y. Wang (✉) · H. Yang · G. Lv
School of Information and Electrical Engineering, Ludong University,
Yantai 264025, Shandong, China
e-mail: qian31320@163.com

H. Yang
e-mail: yhy9919@163.com

G. Lv
e-mail: lvgh@ldu.edu.cn

© Springer Nature Singapore Pte Ltd. 2018
Y. Jia et al. (eds.), *Proceedings of 2017 Chinese Intelligent Systems Conference*,
Lecture Notes in Electrical Engineering 460,
https://doi.org/10.1007/978-981-10-6499-9_6

whether the moving targets exist or not [3]. However, the autofocusing is a process of parameter estimation, it is sensitive to noise and clutter. Weihing built focusing filters with different Doppler slopes to construct a image stack [4]. For a given distance and azimuth, the peak across the image stack location can be used to determine the position where the moving target appears. The algorithm is computationally complex in that it takes a long time to establish the image stack. Baumgartner and Krieger showed a SAR-GMTI and parameter estimation method based on a single channel and multi-channel SAR. It is suitable for real-time traffic monitoring due to its less computation complexity [5]. However, it needs the prior knowledge of ground path. Dias and Marques using Doppler spectral width method to detect moving target with azimuth velocity component, but this method is not robust against the ground background clutter [11, 12]. We developed two methods, symmetric defocusing and symmetric Doppler views, to detect moving target with azimuth and range velocity component, respectively. Both the methods have their own shortcomings, they can only detect moving target with azimuth and range velocity.

In this paper, we proposed two methods named average correlation coefficient method and normalized Doppler spectrum difference method to detect moving target in SAR images. The methods based on the fact that the Doppler spectrum of a moving target shows different characteristics compared to that of stationary background. The differences are Doppler width and Doppler centroid shift. The Doppler width varies with azimuth velocity, while the Doppler centroid shift is caused by radial velocity. The Doppler spectrum differences can be used to detect targets moving along any direction. Experiments showed the feasibility and effectiveness of the two proposed methods.

2   Fundamentals

SAR acquires target images using relative motion between the radar and target. SAR can be space borne or airborne [11]. In different applications, there are multiple imaging mode of SAR, such as stripmap, spotlight and scanning. Stripmap mode SAR includes side-looking SAR and squint SAR [13]. This paper will address the side-looking SAR. However, the proposed methods can also be applied to other mode.

Figure 1 depicts a typical geometric relationship between a moving target and a moving radar. When $t = 0$, the radar is at $(0, 0)$ and moves along the x axis with a constant velocity V_p. A point scatterer is located at (x_0, R_0), moving with a radial velocity component V_r and azimuth velocity component V_a.

Neglecting the antenna pattern, the azimuth signal from the moving scatterer can be written as

$$s(t) = \sigma \exp\left[-j\frac{4\pi r(t)}{\lambda}\right] \tag{1}$$

where t represents the slow time, σ is reflectivity coefficient of the scatterer, $r(t)$ represents the distance between the scatterer and the radar, λ is the wavelength of

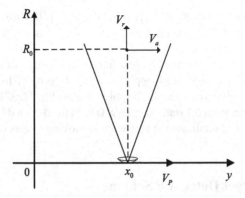

Fig. 1 Scheme for GMTI in SAR images

carrier, and t_0 is the moment when the scatterer is in radar's broadside,

$$t_0 = \frac{x_0}{V_p - V_a} \tag{2}$$

The distance $r(t)$ can be written as

$$r(t) = \sqrt{(R_0 + V_r t)^2 + [x_0 - (V_p - V_a)t]^2} \tag{3}$$

Equation (3) can be approximated by a second-order Taylor series, i.e.,

$$r(t) = R_0 + \frac{(V_p - V_a)^2}{2R_0}(t - t_0)^2 + V_r t \tag{4}$$

Substituting Eq. (4) into Eq. (1), and neglecting constant phase item, the chirp signal takes the form

$$s(t) = \sigma \exp\left[-j\pi\frac{2(V_p - V_a)^2}{\lambda R_0}(t - t_0)^2\right] \exp\left(-j\frac{4\pi V_r t}{\lambda}\right), |t - t_0| \leq \frac{T}{2} \tag{5}$$

According to the stationary phase principle and neglecting the constant amplitude and constant phase terms, the Fourier transform of $s(t)$ can be written as

$$S(f) = \sigma \exp\left[j\pi\frac{\lambda R_0}{2(V_p - V_a)^2}(f - f_D)^2\right] \exp\left[-j2\pi(f - f_D)(t_0 - \Delta t)\right], |f| \leq \frac{B}{2} \tag{6}$$

where B is Doppler bandwidth, that is $B = 2(V_p - V_a)/D$ with D being antenna length, f_D is Doppler frequency shift with $f_D = -2V_r/\lambda \ \Delta t = -V_r R_0/(V_p V_a)$. $S(f)$ is called the Doppler spectrum.

From Eq. (6), it can be seen that Doppler frequency centroid of a stationary target is zero, while for moving target, it's Doppler centroid shifts by f_D. In addition, moving target's azimuth velocity will affect its Doppler bandwidth. Thus, by comparing the standard Doppler spectrum (of stationary target) with the detected Doppler spectrum (of moving target), it can indicate the existence of moving targets in a SAR imagery.

3 Moving Target Detection Scheme

The detection scheme is showed in Fig. 2. Firstly, the given SAR image is divided into small patches with the same size, then the Doppler spectrum is computed patch by patch, and the Doppler spectrum is averaged and normalized. The normalized Doppler spectrum is compared with a standard Doppler spectrum by using two methods, one is the average correlation coefficient method and the other is the normalized difference method. The threshold values are achieved by using the constant false alarm detection method. As a result, the moving targets in the images are detected.

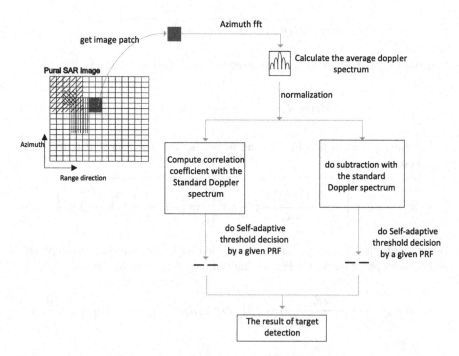

Fig. 2 The scheme flow chart of the two moving target detection methods

3.1 Average Correlation Coefficient Method

Assuming the normalized standard Doppler spectrum is expressed by $S_0(n)$, $n = 1,2,\ldots,N$, and a image patch I(m,n) is composed of M range cells times N azimuth cells, by taking the Fourier transform of $I(n, m)$ along the side of azimuth direction, the Doppler spectrum of the patch can be expressed by

$$S(k, m) = F_n[I(n, m)], \ k = 1, 2, \ldots, N \tag{7}$$

Taking the average of the amplitude of S(k,m), i.e.,

$$\bar{S}(k) = \frac{1}{M} \sum_{m=1}^{M} |S(k, m)| \tag{8}$$

and normalizing $\bar{S}(k)$, the normalized Doppler spectrum of a detected patch is gotten as

$$\tilde{S}(k) = \frac{N\bar{S}(k)}{\sqrt{\sum_{k=1}^{N} |\bar{S}(k)|^2}} \tag{9}$$

The correlation coefficient between the standard Doppler spectrum and the Doppler spectrum of every image patch is defined by

$$\rho(S_0, \tilde{S}) = \frac{\sum_{k=1}^{N} S_0(k)\tilde{S}(k)}{\sqrt{\sum_{k=1}^{N} S^2(k) \sum_{k=1}^{N} \tilde{S}^2(k)}} \tag{10}$$

When the calculated correlation coefficient is greater than a threshold value, the detected patch will be determined as background, otherwise it will be determined as a moving one.

3.2 Normalized Doppler Spectrum Difference Method

The difference between the standard Doppler spectrum and the detected Doppler spectrum is defined by

$$D(S_0, \tilde{S}) = \sum_{k=1}^{N} |S_0(k) - \tilde{S}(k)| \tag{11}$$

The smaller $D(S_0, \tilde{S})$ is, the smaller the difference between Doppler spectrum of image patches and standard Doppler spectrum is, and thus the detected patch looks more like background. When the difference is greater than a threshold value, a moving target is determined.

3.3 Adaptive Threshold Selection

After the correlation coefficient ρ or Doppler spectrum difference D is computed, the first step is to determine the threshold ρ_{TH} or D_{TH}. We adopt the constant false alarm method to determine the threshold value adaptively. For correlation coefficient, assuming the probability density function of ρ is $p_1(x)$, and the constant false alarm rate is P_{FA}, then ρ_{TH} is computed by

$$P_{FA} = \int_{\rho_{TH}}^{+\infty} p_1(x)dx \qquad (12)$$

In the same way, if the probability density function of D is $p_2(x)$, the detection threshold D_{TH} can be computed by

$$P_{FA} = \int_{D_{TH}}^{+\infty} p_2(x)dx \qquad (13)$$

In actual processing, the computed correlation coefficient and spectrum difference set can be rearranged to a form of histogram. For example, correlation coefficient could be divided into K intervals, and for each interval, the corresponding value is

$$\hat{\rho}(k) = (1 + 0.5)\Delta\rho, k = 1, 2, \ldots, K \qquad (14)$$

where $\Delta\rho = (\max(\rho) - \min \rho)/K$. Assuming the kth interval has $N_c(k)$ values, and the total number of ρ is S_c, so the threshold selection algorithm is described as follows,

```
num=0;
for k=K:-1:1
    num=num+N_c(k);
    if num/S_c > P_FA,
        break;
    end
end
```

The corresponding threshold value is $\rho_{TH} = \hat{\rho}(k)$. Doppler spectrum difference can be determined in the same way as above, thus, moving targets can be detected adaptively.

4 Experimental Results

We used a patch of field data to validate the proposed two methods. The field data is shown in Fig. 3a, and its corresponding range Doppler map is shown in Fig. 3b.

It can be seen that there are two moving targets in the given SAR image. The size of selected image patch is 64×32 in range and azimuth direction, respectively. After computation, the standard Doppler spectrum is shown in Fig. 4.

When false alarm rate $P_{FA} = 0.01$, the detection result based on correlation coefficient and Doppler spectrum difference are shown in Figs. 5a, b, respectively.

Figure 6 shows the detection results of correlation coefficient method and Doppler spectrum difference method for $P_{FA} = 0.02$, respectively.

It can be seen that all these two methods can correctly detect moving targets. When P_{FA} is low, the detected target area is small. It has the two reasons. One lies in that parts of moving target is weak in the whole image, so the correlation coefficient or spectrum difference can't reach the decision threshold. The other one is due to

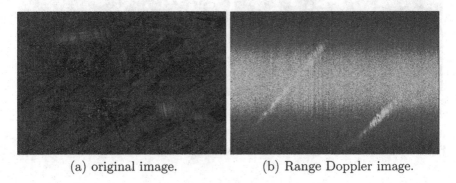

(a) original image.   (b) Range Doppler image.

Fig. 3 An original image and its range Doppler image

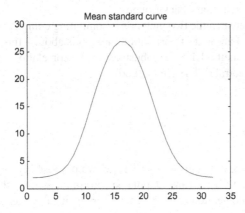

Fig. 4 Standard Doppler spectrum

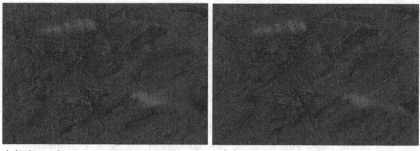

(a) Correlation coefficient method.   (b) Dopplers pectrum difference
                                       method.

Fig. 5 When $P_{FA} = 0.01$, the result of moving target detection

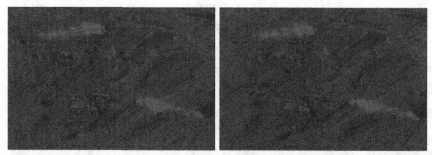

(a) Correlation coefficient method.   (b) Doppler spectrum difference
                                       method.

Fig. 6 Effects of moving target detection effect when $P_{FA} = 0.02$

stronger stationary targets embedded in the image of a moving target, and it leads to
the appearance of a stationary background.

The difference between the two methods is computing complexity. In our computation, the calculated amount of correlation method is about 4 times of the Doppler
spectrum difference method, but the robustness of the correlation coefficient method
is better than the spectral difference method.

5 Conclusions

To detect moving target in a given SAR image, we proposed correlation two methods named coefficient method and Doppler spectrum method. Both the methods are
based on the idea of detection patch by patch. Firstly, it calculates standard Doppler
spectrum, and then calculates average Doppler spectrum patch by patch. By com-

paring the standard Doppler spectrum and the detected one, moving targets are indicated.

However, the threshold is global, so when the false alarm rate is small and target signal is weak, leak detection may occur. In addition when the velocity of target is quite slow, its Doppler spectrum is close to the standard one, and thus the correlation coefficient or spectrum Doppler difference is also small, leak detection can also occur. This problem can be solved by the local threshold selection method.

Acknowledgements The authors would like to thank support from the Natural Science Foundation of China (under grant 61673200, 61603172, 61471185, and 61472172).

References

1. Lv G, Wang J, Liu X. Ground moving target indication in SAR images by symmetric defocusing. IEEE Geosci Remote Sens Lett. 2013;10(2):241–5.
2. Lv G, Li Y, Zhang Y, Wang G. Ground moving target indication in SAR images with symmetric Doppler views. IEEE Trans Geosci Rmote Sens. 2016;54(1):533–43.
3. Fienup JR. Detecting moving targets in SAR imagery by focusing. IEEE Trans Aerosp Electron Syst. 2001;37(3):794–809.
4. Weihing D, Hinz S, Meyer F, Suchandt S, Bamler R. Detecting moving targets in dual-channel high resolution spaceborne SAR images with a compound detection scheme. International: Geoscience and Remote Sensing Symposium. IGARSS: IEEE; 2007. p. 4818–21.
5. Baumgartner SV, Krieger G. Real-time road traffic monitoring using a fast a priori knowledge based SAR-GMTI algorithm, Geoscience and Remote Sensing Symposium. IEEE Xplore; 2010. pp. 1843–1846.
6. Wei B, Zhu D, Di WU. A SAR-GMTI approach based on moving target focusing. J Electron Inf Technol. 2006.
7. Guo HW, Zhang YL, Zhang XR. Research on airborne synthetic aperture radar imaging algorithm of moving targets. Syst Eng Electron. 2006;28(8):1164–8.
8. Zhu YP. A new method for estimating the rotation angle of ISAR image based on phase matching. Signal Process. 2009;25(4):679–84.
9. Cumming IG, Wong HC. Digital processing of synthetic aperture radar data : algorithms and implementation, International Radar Conference; 1980. pp. 168–175.
10. Raney RK. Synthetic aperture imaging radar and moving targets. IEEE Trans Aerosp Electron Syst. 1971;7(3):499–505.
11. Fienup JR, Miller JJ. Aberration correction by maximizing generalized sharpness metrics. J Opt Soc Am A Opt Image Sci Vis. 2003;20(4):609–20.
12. Gierull CH. Ground moving target parameter estimation for two-channel SAR. 2006;153(3):224–233.
13. Dias JMB, Marques PAC. Multiple moving target detection and trajectory estimation using a single SAR sensor. IEEE Trans Aerosp Electron Syst. 2003;39(2):604–24.
14. Zhu S, Liao G, Qu Y, Zhou Z. Ground moving targets imaging algorithm for synthetic aperture radar. IEEE Trans Geosci Remote Sens. 2011;49(1):462–77.

Handwritten Digit Recognition Based on Improved BP Neural Network

Yawei Hou and Huailin Zhao

Abstract Due to the different writing habits, it is difficult to achieve the recognition of handwritten numbers. The artificial neural network has been widely used in character recognition because of its strong self-learning ability, adaptive ability, classification ability, fault tolerance and fast recognition. BP neural network is used to identify handwritten numerals in this paper. In order to obtain a higher correct rate, this paper improves the traditional BP neural network and experiments with the MNIST data set on the MATLAB simulation platform. The experimental results show that the improved network converge faster and the classification is more accurate.

Keywords BP neural network · Handwritten digit recognition · MATLAB · Principal component analysis

1 Introduction

The optical character recognition technology includes handwritten character recognition and printed character recognition. Handwritten digital recognition as part of handwritten character recognition is also a very important research direction. Handwritten digital recognition mainly identifies 0–9 of 10 characters, and the category of classification is much less than the optical character recognition [1]. In recent years, along with the development of computer technology and pattern recognition technology, the handwritten digital recognition has been widely used in postal code, financial value identification, tax form recognition, e-commerce digital processing, and even student achievement recognition [2]. Although the classifier

Y. Hou (✉) · H. Zhao
School of Electrical and Electronic Engineering, Shanghai Institute of Technology,
Shanghai 201418, China
e-mail: houyawei2017@163.com

H. Zhao
e-mail: zhao_huailin@yahoo.com

© Springer Nature Singapore Pte Ltd. 2018   63
Y. Jia et al. (eds.), *Proceedings of 2017 Chinese Intelligent Systems Conference*,
Lecture Notes in Electrical Engineering 460,
https://doi.org/10.1007/978-981-10-6499-9_7

has been further enriched, but the researchers still didn't find a way to achieve the perfect effect of the algorithm.

The artificial neural network has been widely used in character recognition because of its strong self-learning ability, adaptive ability, classification ability, fault tolerance and fast recognition [3]. BP neural network is used to identify handwritten numerals in this paper. The extracting eigenvectors method depending on principal component analysis. We use mini-batch gradient descent algorithm to accelerate learning. At the same time, the number of hidden nodes and the learning rate is improved. Identification by using MNIST database, confirmed the network structure in this paper can obtain better classification results.

Although the use of neural network is more and more widely, but its convergence has been one of the bottleneck of its development. The problem of convergence includes two categories. The first problem is that neural networks are difficult to converge quickly. The second problem is that the neural network is difficult to converge to the global optimal solution. In order to make the neural network fast convergence, we can reduce the dimension of the network. Therefore, this paper uses the principal component analysis to reduce the input dimension.

2 Principal Component Analysis

The principal component analysis [4] (PCA) is a data dimension reduction algorithm that can greatly improve the learning speed of unsupervised features. The character recognition is the recognition of the image. There is a correlation between the pixels in the image, which makes the data with a certain degree of redundancy. Because of the correlation between pixels in the image, the PCA algorithm can be used to compress the input data. The PCA algorithm can retain the large variation of the elements and reduce the computational complexity of the network. The error generated by the PCA algorithm is very small.

In PCA, the number of principal components has a large impact on the final result. If the number of principal components is too large, the dimension of the data is still very high. So it can not achieve the purpose of reducing the dimension of the data. On the contrary, if the data is too small the approximation error is too large.

Generally, let λ_1, λ_2, ..., λ_n be the eigenvalue of the covariance matrix, and the eigenvalues corresponding to the eigenvectors u_j is λ_j. Equation 1 represents the percentage of the retained eigenvalues that account for the total eigenvalue.

$$\frac{\sum_{j=1}^{k} \lambda_j}{\sum_{j=1}^{n} \lambda_j} \tag{1}$$

The former k elements are preserved, that is, the former k eigenvalues are preserved. For image processing, according to experience, the number of principal components is chosen to satisfy the minimum value k in Eq. 2. A customary rule of thumb is to choose k to preserve the variance of 99%, and select the minimum k value that satisfies the following formula:

$$\frac{\sum_{j=1}^{k} \lambda_j}{\sum_{j=1}^{n} \lambda_j} \geq 0.99 \tag{2}$$

0.99 in Eq. 2 is to select k components to retain 99% variance. In practical applications, the choice of k can also be retained 90–99% of the variance range.

3   The Construction of BP Network

BP neural network is a multi-layer feedforward neural network. The main feature of the network is the signal forward transmission and the error back propagation. In the forward transfer, the input signal is processed from the input layer through the hidden layer until the output layer. Neuronal states of each layer only affect the underlying neuronal state. If the output layer is not the desired output, it is transferred to the reverse propagation. Adjust the network weights and thresholds according to the prediction error, so that the BP neural network prediction output is constantly approaching the desired output. The topology of BP neural network is shown in Fig. 1.

BP neural network using the back-propagation algorithm and the algorithm can be described as follows:

(1) Input x: set the corresponding activation value for the input layer a^1.
(2) Forward propagation: calculate the corresponding weighted input z^l and the activation value a^l for each $l = 2, 3, \ldots, L$.

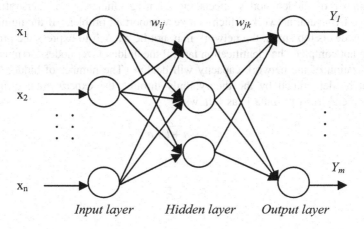

Fig. 1 The topology of BP neural network

$$z^l = w^l a^{l-1} + b^l \tag{3}$$

$$a^l = \sigma(z^l) \tag{4}$$

In the Eq. 4, $\sigma(.)$ represents the activation function.

(3) Calculation of output layer error δ^L

$$\delta^L = \nabla_a C \bullet \sigma'(z^L) \tag{5}$$

where $\bullet$ represents the Hadamard product.

(4) Reverse error propagation: calculate δ^l for each $l = L - 1, L - 2, ..., 2$.

$$\delta^l = ((w^{l+1})^T \delta^{l+1}) \bullet \sigma'(z^L) \tag{6}$$

(5) Output: the gradient of the cost function is given by Eqs. 7 and 8.

$$\frac{\partial C}{\partial w_{jk}^l} = a_k^{l-1} \delta_j^l \tag{7}$$

$$\frac{\partial C}{\partial b_j^l} = \delta_j^l \tag{8}$$

In order to improve the performance of the network, the network is improved on the basis of the standard BP algorithm in this paper.

3.1 Determination of the Number of Hidden Nodes

The number of hidden nodes affects the learning efficiency and generalization ability of BP neural network, which is a very important problem. If the number of hidden nodes is too small, the network may not be enough to express the problem and can not complete the identification task; if the hidden layer nodes too much, the generalization of the network capacity will decline. The number of hidden nodes can not be determined by the theory, only through the experimental estimates. Common empirical formula is as follows:

$$m = \sqrt{n+l} + a \tag{9}$$

$$m = \log 2^n \tag{10}$$

$$m = \sqrt{nl} \tag{11}$$

where n is the number of input nodes, l is the number of output nodes and m is the number of hidden nodes. a is a constant between 1 and 10. In general, the number of hidden nodes is determined according to Eqs. 9–11. However, in the experiment, it is found that the number of hidden nodes obtained by empirical formula is not optimal compared with other hidden nodes. Therefore, this paper based on a large number of experiments to determine the number of hidden nodes.

3.2 Momentum Factor

When the standard BP algorithm adjusts the connection weights, the gradient direction of the previous adjustment isn't considered. So the network often shocks in the training process and the rate of convergence is slow [5]. In order to reduce the oscillation and improve the training speed of the network, it can add the momentum item when the connection weights are adjusted. Equation 12 gives the network weights adjustment formula after adding the momentum term.

$$\Delta W(t) = \eta \delta X + \alpha \Delta W(t-1) \tag{12}$$

In the Eq. 12, W represents a certain layer matrix, X represents a certain layer input vector, α is called the momentum coefficient, $0 < \alpha < 1$.

3.3 Learning Rate Adaptive Adjustment Algorithm

Different learning rates can have a great impact on BP neural networks. If the learning rate is too large, the network training is not convergence; in contrast, the network training time is too long. Therefore, you need to choose a best learning rate for the BP algorithm. In order to achieve the best learning rate, the learning rate must be adjusted during the training process.

There are many ways to modify the learning rate. The method used in this paper is to automatically adjust the learning rate by checking whether the correction value of the network weights reduces the error function [6]. So that the network is always trained with the maximum acceptable learning rate.

$$\eta(t+1) = \begin{cases} 1.05\eta(t) & E(t) < E(t-1) \\ 0.7\eta(t) & E(t) > 1.04E(t-1) \\ \eta(t) & other \end{cases} \tag{13}$$

Using the learning rate adaptive adjustment algorithm which given in Eq. 13, it is possible to shorten the training time of the network and make the learning algorithm more reliable.

4 Experimental Results and Analysis

In this paper, a three-layer BP neural network is used and select the Xavier initialization method. Feature extraction used the PCA method and added momentum factors to the network. The learning rate adaptive adjustment algorithm and the mini-batch gradient descent method are used to speed up the learning of the network. Experimental computer cpu model is intel i7 7700 K which frequency is 4.2 GHz.

The data set used in this paper is a recognized MNIST data set which included 60,000 training samples and 10,000 test samples. In order to improve the performance of the network, some of the images in the MNIST data set were rotated in this paper. Partial training data image as shown in Fig. 2.

According to the above description, this paper first gives the recognition rate in the number of different principal components, the number of different hidden layers of neurons. At the same time, this paper considers the influence of mini-batch value on the network in the mini-batch gradient descent algorithm. Therefore, the network recognition rate in different mini-batch is compared in Tables 1 and 2.

The data in Tables 1 and 2 are analyzed synthetically. The final selection of 100 hidden layer nodes, the number of principal components in the choice of 95% variance, mini-batch selected as 100 in this paper.

According to the values of the selected network parameters, the neural network is established. In order to prove that the established network is superior to the traditional neural network, Table 3 shows the contrast between the standard neural network and the network established in this paper. Table 4 gives the experimental results for feature extraction and no feature extraction.

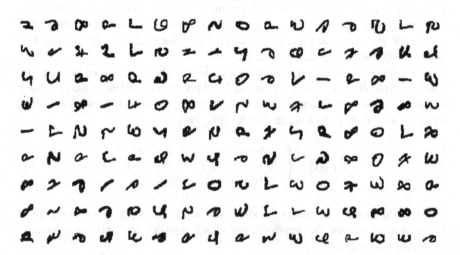

Fig. 2 Partial training data image

Table 1 The recognition rate of the hidden neurons is 30

Variance mini-batch	0.9	0.95	0.99
10	94.99%	95.59%	95.41%
100	95.52%	95.59%	95.60%
1000	96.04%	96.05%	96.06%

Table 2 The recognition rate of the hidden neurons is 100

Variance mini-batch	0.9	0.95	0.99
10	97.48%	97.56%	97.75%
100	97.40%	97.77%	96.94%
1000	97.49%	97.67%	95.92%

Table 3 Comparison of standard neural networks with the networks established in this paper

	Standard network	The method of this article
The time for an iteration (s)	0.3	0.4
Network recognition rate (%)	88.14	97.77

Table 4 Comparison of PCA feature extraction and no PCA feature extraction

	No PCA feature extraction	PCA feature extraction
The time for an iteration (s)	1.2	0.4
Network recognition rate (%)	97.73	97.77

According to a large number of experiments, the network iteration period is selected as 100 times in the improved network in this paper. But for the standard BP neural network, the network iteration period is selected as 1000 times.

By analyzing Table 3, it can be found that the recognition rate is 9.63% higher than the standard BP network. The convergence time of the network is 1/7 of the standard network.

By analyzing Table 4, it can be found that the recognition rate of the network is basically equal.

However, after using the PCA feature extraction, the network training time is 40 s, which is 1/3 of the training time of no PCA feature extraction. it greatly improves the training time of the network.

5 Conclusion

The effectiveness of the proposed algorithm is demonstrated through experiments using a MNIST dataset. Not only the network training fast, and the network recognition rate than the standard BP neural network is also high. As the network designed in this paper is fast, it can be applied to the real-time processing.

References

1. Shuying Y. Image recognition and project practice. Beijing: Publishing House of Electronics Industry; 2014.
2. Basu S, Das N, Sarkar R, et al. Recognition of Numeric postal codes from multi-script postal address block. In: International conference on pattern recognition and machine intelligence. Springer; 2009. p. 381–6.
3. Impedovo S, Pirlo G, Modugno R, et al. Zoning methods for hand-written character recognition: an overview. In: International Conference on Frontiers in Handwriting Recognition. IEEE Computer Society;2010. p. 329–34.
4. Chenglong Y. Features selection algorithm based on PCA. Comput Technol Dev. 2011;4:123–5.
5. Feiyan Z. A research on the improvement of artificial fish algorithm and its optimization on BP neural network. Hunan: Hunan University of Science and Technology; 2015.
6. Meixian W, Xueliang Z, Shuhua W. Guo qin. Double self-adaptive learning rate algorithm for BPNN. Modern Manuf Eng. 2005;10:29–32.

Research on Cable Fault Location Algorithm Based on Improved HHT

Shu Tian, Shasha Li and Zhan Zhang

Abstract The accurate calibration of the fault traveling wave head is the key to realize the cable fault traveling wave ranging. Modal aliasing will appear when the traveling wave ranging method based on HHT carries on empirical mode decomposition, so that the ranging accuracy is low. In order to solve this problem, an improved HHT algorithm based on ensemble empirical mode decomposition (EEMD) is put forward. At first, the EEMD algorithm is used to extract the inherent modal function of the fault traveling wave head. Secondly, the instantaneous frequency is calculated by the Hilbert transform. Thirdly, the traveling wave head is accurately calibrated by the abrupt change point of the instantaneous frequency. Then use the double-terminal fault ranging algorithm to achieve more accurate fault location. Finally, the 10 kV distribution network model based on the cable line is built in PSCAD/EMTDC software. A large number of simulation results show that the proposed method is feasible, and it is more precise than traditional methods.

Keywords Band aliasing · Ensemble empirical mode decomposition (EEMD) · Traveling-wave fault location · PSCAD software simulation

1 Introduction

With the power cable widely applied in the distribution network, the problems of cable fault location need to be resolved urgently. The traditional cable fault location is dominated by the offline method, and the offline method is difficult to meet the

S. Tian · S. Li (✉) · Z. Zhang
School of Electrical Engineering and Automation,
Henan Polytechnic University, Jiaozuo, Henan, China
e-mail: 359036331@qq.com

S. Tian
e-mail: tianshu@hpu.edu.cn

Z. Zhang
e-mail: zhangzhan@hpu.edu.cn

© Springer Nature Singapore Pte Ltd. 2018
Y. Jia et al. (eds.), *Proceedings of 2017 Chinese Intelligent Systems Conference*,
Lecture Notes in Electrical Engineering 460,
https://doi.org/10.1007/978-981-10-6499-9_8

needs of the smart grid development because of the complexity of the external equipments, the limited scope of application, the low ranging accuracy and other reasons [1]. Online fault traveling wave location has become the research hotspot because of its simple principle and strong practicability. And the key of traveling wave location is the accurate calibration of the wave head. Many scholars use wavelet analysis to calibrate the traveling wave head, but the wavelet is not adaptive, and the result is affected by the wavelet types and the decomposition scales [2, 3]. Hilbert–Huang Transform (HHT) is a new method developed in recent years for analyzing non-stationary signals. It is composed of two parts: Empirical Mode Decomposition (EMD) and Hilbert Transform. Complex adaptive signals are decomposed into a finite number of meaningful instantaneous frequency, and amplitude or frequency modulated high-frequency and low-frequency Inherent Mode Function (IMF) by EMD. The frequency components contained in each IMF component vary with both the sampling frequency and the signal itself. So HHT is suitable for the analysis of fault traveling wave signals [4–7]. But the modal aliasing occurs when HHT algorithm performs mode decomposition, resulting in low ranging accuracy. Huang came up with (Ensemble Empirical Mode Decomposition) EEMD to suppress the modal aliasing. It is a noise-assisted data analysis method, which can recover the essence of the signals well [8–10].

Based on this, an improved HHT cable fault location method based on EEMD and Hilbert transform is put forward. It is verified by PSCAD model simulation that the algorithm is feasible and the ranging accuracy is high, which meets the demands of engineering practice.

1.1 Empirical Mode Decomposition (EMD)

Empirical Mode Decomposition is to decompose a signal $x(t)$ into the sum of several inherent modal components and the margin:

$$x(t) = \sum_{i=1}^{n} c_i + r_n \tag{1}$$

Modal aliasing phenomenon means that the same IMF component contains different scale components, which directly leads to the mixing IMF lacking sufficient physical meaning, so that the subsequent time-frequency distribution will be confused [11]. A simulation signal is used to verify the presence of modal aliasing phenomenon in EMD. As shown in Fig. 1, the test signal x is superimposed by a sinusoidal signal a and an intermittent signal b. The signal is decomposed by EMD, and the results are shown in Fig. 2.

It can be seen from Fig. 2 that there exists obvious modal aliasing phenomenon in the IMF1 after EMD decomposition.

Fig. 1 Test signal and its composition

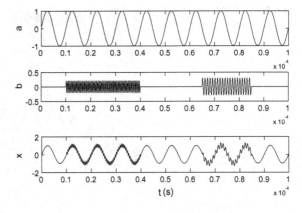

Fig. 2 EMD results of the test signal

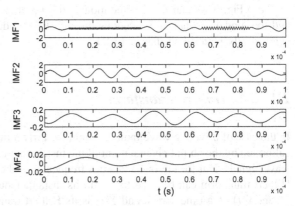

1.2 Ensemble Empirical Mode Decomposition (EEMD)

The essence of the EEMD algorithm is to superimpose the Gaussian white noise on the original signal, perform multiple EMD decomposition, and take the mean of the IMF component as the final result. The algorithm makes use of the statistical property of Gaussian white noise, so that the noisy signal has continuity at different frequency scales and the problem of modal aliasing is solved effectively [8].

The specific algorithms are as follows:

(1) Add a white noise to the analyzed signal $x(t)$;
(2) Decompose the noisy signal; then get each IMF;
(3) Repeat steps (1) and (2), but adds different white noise each time;
(4) Take the mean value of each IMF component obtained by multiple decomposition as the final result.

The test signal is decomposed by EEMD, and the results are shown in Fig. 3.

Fig. 3 The EEMD results of the test signal

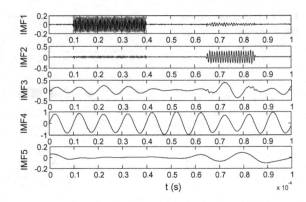

From Fig. 3 we can see that the modal aliasing phenomenon is well suppressed, and the modal distinction is obvious through EEMD decomposition. The effect is significantly better than that of EMD.

1.3 The Hilbert Transform

In the case of traveling wave fault location, the arrival time of the wave head can be calibrated by the abrupt change point of the instantaneous frequency of the traveling wave signals. But for the instantaneous frequency of the non-stationary signals, the Hilbert transform can be used to obtain its instantaneous frequency.

Set $X(t)$ as a time series, and $Y(t)$ is its Hilbert transform, that is:

$$Y(t) = \frac{1}{\pi} \int_{-\infty}^{+\infty} \frac{X(\tau)}{t - \tau} d\tau \tag{2}$$

The inverse transformation is:

$$X(t) = \frac{1}{\pi} \int_{-\infty}^{+\infty} \frac{Y(\tau)}{\tau - t} d\tau \tag{3}$$

Get the analytical signal:

$$Z(t) = X(t) + iY(t) = A(t)e^{i\theta(t)} \tag{4}$$

where: $A(t)$ is the instantaneous amplitude, $A(t) = \sqrt{X^2(t) + Y^2(t)}$

$\theta(t)$ is the phase, $\theta(t) = \arctan\dfrac{Y(t)}{X(t)}$.

The instantaneous frequency can be defined as

$$f(t) = \frac{1}{2\pi} \frac{d\theta(t)}{dt} \tag{5}$$

After the EEMD of the acquired fault traveling wave signal, a series of inherent modal components that just containing a kind of vibration mode are obtained. Then the instantaneous frequency can be obtained by the Hilbert transform. Thus the fault traveling wave head can be accurately calibrated by the abrupt change point of the instantaneous frequency.

1.4 Traveling Wave Fault Location Algorithm Based on the Improved HHT

The transient signals generated by the cable short- circuit fault have a wide-area frequency band. Due to the influence of the frequency dispersion effect, the propagation velocity of the traveling wave signal varies with different modulus and different frequency, which causes the distortion of the traveling wave head in the propagation process. So that there will have large error in the ranging results. In this paper, we adopt the traveling wave aerial mode components of smaller dispersion to calibrate the wave head [12–14].

1.5 The Extraction of Traveling Wave Aerial Mode Components

The complex electromagnetic coupling relationship lies in the three-phase of cables, which causes distortion in the spread of the fault traveling wave to reduce the ranging accuracy. Therefore, it is necessary to decouple the extracted fault traveling wave signals. At present, the phase-mode conversion technology is mainly used for decoupling. Commonly used methods include Symmetrical components transformation, Clarke transformation, Karenbauer transformation, and Wedpohl transformation and so on. However, the modes calculated by the phase-mode transformation matrix under time-domains have the problem of single modes can not reflect all fault types [15]. In order to overcome the above shortcomings, this paper uses a new phase-mode transformation matrix to complete the decoupling [16].

The decoupling process is as follows:

$$\begin{bmatrix} I_0(n) \\ I_\alpha(n) \\ I_\beta(n) \end{bmatrix} = [S]^{-1} \begin{bmatrix} I_a(n) \\ I_b(n) \\ I_c(n) \end{bmatrix} \tag{6}$$

$$[S]^{-1} = \begin{bmatrix} 1 & 1 & 1 \\ 1 & 2 & -3 \\ 1 & -3 & 2 \end{bmatrix} \tag{7}$$

$I_a(n)$, $I_b(n)$, $I_c(n)$ are the three-phase current of the lines respectively. $I_0(n)$, $I_\alpha(n)$, $I_\beta(n)$ are the 0 mode, α mode and β mode current component after decoupling. Where the α mode and β mode are called the aerial mode components. Further the wave velocity equation of traveling wave under different modulus can be obtained:

$$\begin{cases} v_0 = \frac{1}{\sqrt{L_0 C_0}} \\ v_1 = \frac{1}{\sqrt{L_1 C_1}} \end{cases} \tag{8}$$

Among them, L_0, C_0 and L_1, C_1 stand for the zero mode and aerial mode parameters of the cable lines.

1.6 The Concrete Realization of the Algorithm of Fault Location Based on Traveling Wave

The traveling wave voltage and current will change sharply when the traveling wave generated by the fault point arrives at the measuring point. The wave head of the traveling wave will manifest as high frequency mutation in the time-frequency diagram, and the mutation point is the arrival time of the wave head. The EEMD decomposition of the aerial mode of the double-terminal traveling wave is carried out, and the instantaneous frequency is obtained by extracting the respective IMF1 components by Hilbert transform. Then the first frequency mutation point on the time-frequency diagram corresponds to the arrival time of each fault traveling wave head. At last the fault location is carried out through the double-terminal fault ranging algorithm, details are as follows:

$$x = \frac{(t_1 - t_2)v + l}{2} \tag{9}$$

Among them, t_1 and t_2 are the arrival time to the two measuring point of the fault traveling wave respectively, v is the aerial mode wave speed of the traveling wave, l is the total length of the cable, and x is the cable fault distance calculated.

1.7 The Construction of the Simulation Model

In order to achieve accurate simulation of cable lines and its running environment, on the basis of numerous simulation experiments, this paper chooses the more accurate PSCAD/EMTDC software to build the 10 kV distribution network model based on cable lines

1.8 The Simulation Example

Set the cable length of 12 km,occur single-phase ground fault from the head-terminal of 1 km, 3 km, 6 km, 7 km, 9 km, 10 km, 11 km respectively,and set different transition resistances for simulation calculations. The simulation time is 0.05 s; fault occurs at 0.02 s; the sampling frequency is 1 MHz. According to the parameters mentioned above and reference [17], the speed of traveling wave aerial mode in this model can be calculated to be 198.26 m/μs. Next make specific analysis of the fault point from the head-terminal of 3 km, the transition resistance of 0.1 Ω:

(1) The two terminals of the cable are labeled as A, B. Use the new phase-mode transformation matrix to decouple the three–phase current traveling wave signal of the double-terminal, and get its aerial mode components. Finally extract the current traveling wave aerial mode components of t = 0.0198–0.0204 s to take analysis. The aerial mode components of the double-terminal traveling wave are shown in Fig. 4.

Fig. 4 Doubled-ended traveling wave aerial mode components

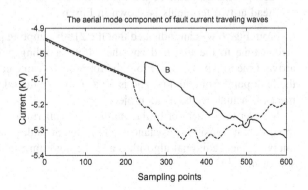

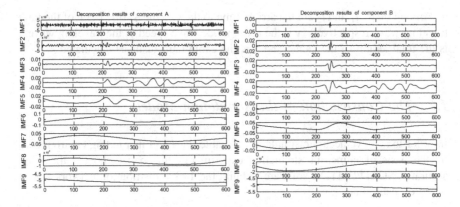

Fig. 5 EEMD results of the terminal traveling wave mode component

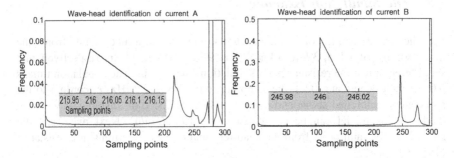

Fig. 6 The time calibration of the wave head arrives at both ends

(2) The EEMD decomposition of the extracted double-terminal aerial mode components was carried out, and the decomposition results are shown in Fig. 5.

(3) The time-frequency graph is obtained by Hilbert transform of the double- ended IMF1 component, and the arrival time of the wave head is calibrated by the abrupt change point of the instantaneous frequency. Calibration of the wave head at two terminals is shown in Fig. 6.

From Fig. 6 we can indicated that the 216th sampling point can be marked as the arrival time to the first end and the 246th sampling point can be marked as the arrival time to the last end. From the formula (9) can calculated that the distance of the fault point from the first end is 3026.1 m, and the relative error is 0.87%. It can meet the actual needs of the project.

In order to further verify the validity and superiority of the proposed algorithm, the different transition resistances and different fault positions are simulated separately by the traditional algorithm and the algorithm in this paper. The results are shown in Table 1.

Table 1 Results of fault location

Fault distance (km)	Transition resistance (Ω)	Location result (km)			
		Improved HHT	Relative error	Traditional HHT	Relative error (%)
1	0.1	1.0435	4.35%	1.0448	4.48
	10	1.0435	4.35%	1.0463	4.63
	100	0.9444	5.56%	1.0445	4.45
3	0.1	3.0261	0.87%	3.1252	4.174
	10	3.1252	4.17%	3.1252	4.174
	100	3.0261	0.87%	2.9765	0.782
6	0.1	6.0991	1.65%	6.1487	2.478
	10	6.0991	1.65%	6.1487	2.478
	100	6	0	6.0991	1.65
7	0.1	7.0904	1.291%	7.1399	1.999
	10	7.0904	1.291%	7.1399	1.999
	100	7.0904	1.291%	7.1499	2.141
9	0.1	9.0730	0.811%	9.1226	1.362
	10	9.0730	0.811%	9.1226	1.362
	100	9.0235	0.261%	9.1325	1.472
10	0.1	9.9652	0.348%	10.113	1.139
	10	9.9652	0.348%	10.113	1.139
	100	9.7669	2.331%	10.124	1.238
11	0.1	11.254	2.309%	11.303	2.755
	10	10.957	0.391%	11.155	1.409
	100	11.155	1.409%	11.155	1.409

As shown in Table 1, the proposed fault location algorithm is basically not affected by the fault resistances, and the fault location is effective under at fault distances. When the fault point approaches the end point, the ranging accuracy is slightly decreased, but the maximum relative error is no more than 5%. Compared with the traditional algorithm based on HHT, the algorithm in this paper can realize the calibration of the fault traveling wave head more accurately, and the ranging accuracy is higher, which satisfies the actual requirements of the project.

2 Conclusions

In order to solve the problem that the ranging algorithm based on HHT is prone to generate modal aliasing when the mode decomposition is performed, the ranging accuracy is not high due to the inaccurate calibration of the traveling wave head. An improved HHT algorithm based on ensemble empirical mode decomposition (EEMD) is proposed in this paper. The traveling wave is decomposed by EEMD, extraction of the first IMF component obtain the instantaneous frequency by Hilbert transform for accurate calibration the arrival time of the fault traveling wave head through the abrupt change point of the instantaneous frequency. At last, the accurate location of the cable fault point is realized by the double-terminal ranging algorithm. Considering the influence of frequency-dependent characteristic of the cable lines parameters on the traveling wave propagation properties, the 10 kV distribution system model based on the frequency-dependent cable models is established by using PSCAD/EMTDC software. By using the traditional HHT ranging algorithm and the improved HHT ranging algorithm in this paper, the fault conditions of different fault locations and different transition resistors are fully simulated and calculated. The simulation results show that compared with the traditional algorithm, the improved HHT ranging algorithm in this paper can inhibit the modal aliasing better and has more accurate ranging accuracy, and comply with the requirements of engineering practice.

References

1. Ji T. Study on fault location of distribution feeders based on transient traveling waves [D]. Shandong University.
2. Yi F, Jun DX, Zeming JD. The application of wavelet transform in cable fault location [J]. High Volt Eng. 2000;26(4):9–10.
3. Dong A, Su Y. Research of fault location of underground power cable based on multi-wavelet [J]. Ind Mine Autom 2014;05:37–42.
4. Wenfan L, Zhigang L, Wanlu S. Development of power quality detection system based on HHT[J]. Power Syst Prot Control. 2011;23:123–7.
5. Sima W, Wang J, Yang Q, Xie B. Application of Hilbert-Huang transform to power system over-voltage recognition [J]. High Volt Eng. 2010;06:1480–1486.
6. Song Z, Baohui Z. Application of Hilbert-Huang transformation in distance protection [J]. Autom Electr Power Syst. 2010;19:69–74.
7. Baoping T, Youming Z, Fabin C. Research on non-stationary signal analyzer based on HHT [J]. Chin J Sci Instru. 2007;01:29–33.
8. Li C, Kong F, Huang W, Chen H, Whang C, Yuan Z. Rolling bearing fault diagnosis based on EEMD and Laplace wavelet [J]. J Vibr Shock. 2014;03: 63–69+88.
9. Chen P, Lu Y. Research on fault diagnosis method of rolling bearing based on EEMD and Hilbert [J]. Electr Power Sci Eng. 2013;09:70–3.
10. Yongxiang Z, Jieping Z, Shuai Z. Rolling element bearing feature extraction based on EEMD [J]. J Naval Univ Eng. 2014;06:90–4.
11. Zhao L, Liu X, Qin S, Ju P, Zhao F. Use of masking signal to improve empirical mode decomposition [J]. J Vibr Shock. 2010;09:13–17+239.
12. Wang J, Dong X, Shi S. Traveling wave transmission research for overhead lines of radial distribution power systems considering frequency characteristics [J]. Proc CSEE. 2013;22:96–102+16.
13. Shu H, Tian X, Dong J, Yang Y. A single terminal cable fault location method based on fault characteristic frequency band and TT transform [J]. Proc CSEE. 2013;22:103–112+17.
14. Qin J, Chen X, Zheng J. Study on dispersion of travelling wave in transmission line [J]. Proc CSEE. 1999;09:28–31+36.

15. Jinghan H, Biao Z, Fan Y, Zaojun O. Study on the application of the decoupling transformation in power system transient protection [J]. J Beijing Jiao Tong Univ. 2006;05:101–4.
16. Guobing S, Sen L, Xiaoning K, Desheng Z, Zhongli Y, Jiale S. A novel phase-mode transformation matrix [J]. Autom Electr Power Syst. 2007;14:57–60.
17. Zhong T, Wei G. Calculation and research of single-core XLPE power cable distribution parameters [J]. East China Electr Power. 2014;03:534–40.

Spherical Panorama Stitching Based on Feature Matching and Optical Flow

Benzhang Wang, Fan Tang and Hengzhu Liu

Abstract In recent yeas, 360° spherical panorama images and videos have seen huge adoption in virtual reality research. Image mosaic is the main technology to stitch the little scale images to a large scale panoramic image. Because of the gross distortion in the edge of fisheye lens, the misregistration on the corrected images leads to obvious ghosting after stitched. In this paper, we use pyramid LK optical flow algorithm to reduce the misregistration areas by remapping with interpolation algorithm. Additionally, we use sample point algorithm to adjust the brightness of images for minimize visual seams. Experiments show convincing evidence that the effect of our spherical panorama stitching improves significantly.

Keywords Spherical panorama · Image mosaic · Feature matching · Optical flow

1 Introduction

Recently, the development and application of 360° spherical panorama images and videos become a new hotspot in virtual reality and computer vision. Panorama image in the spherical projection is meant to be viewed as the image is wrapped into a sphere and viewed from within. It can provides 360 horizontal by 180 vertical filed of views generally by stitching at least two fisheye images with overlapping fields of views.

Panoramic image stitching is to transform the many overlap images to the same coordinate system based projection model and align them, then form a panorama image using composition algorithm [1]. It usually can be divided into two main components: image registration and image composition [2]. Image registration or

B. Wang (✉) · F. Tang · H. Liu
College of Computer, National University of Defense Technology,
Changsha 410073, China
e-mail: benzhang_wang@163.com

© Springer Nature Singapore Pte Ltd. 2018
Y. Jia et al. (eds.), *Proceedings of 2017 Chinese Intelligent Systems Conference*,
Lecture Notes in Electrical Engineering 460,
https://doi.org/10.1007/978-981-10-6499-9_9

image alignment algorithms are well studied and mainly fall into two categories: intensity-based and feature-based [3]. Intensity-based methods use some statistical information of the images themselves to measure the similarity, while the more robust feature-based methods firstly find correspondence matched features between images that consequently decides the remap geometrical transformation parameters. SIFT-based image registration methods are widely applied for its scale-invariant to detect and describe local features [4]. Generally, methods based on feature stitching can stitch images well in majority regions, but there may still be local image misregistration, due to deviations from the idealized parallax-free camera model included camera coordinate translation, lens distortion and the malposition of the optical center.

To improve the misregistration problem and minimize the blending visible seams, in this work we will use optical flow algorithm to remap the initial aligned images based on the SIFT feature image registration algorithm and brightness adjustment with sample point algorithm. Then we use fade in and fade out composition algorithm to form the seamless panorama image.

2   Related Theory

Spherical projection model. Spherical panorama is the most close to the human eye model and can provides 360 horizontal by 180 vertical filed of views. It projects the panorama image to the spherical surface centered observer, according to the views of the observer to show the local scene, and all the scenes are continuous. We know the coordinate of sphere panorama is three-dimensional, and the coordinate in image is two-dimensional. So we need to solve how to map a point on the sphere to a plane coordinate system. the most suitable mapping method is the longitude and latitude method. For a point p on the spherical panorama, we will map its longitude to the horizontal coordinate and latitude to the vertical coordinate of the plane rectangle, and finally form a 2:1 plane panorama image (Fig. 1).

Optical Flow algorithm. Optical flow is the description of motion information of brightness patterns in an image, which can assess motion between two frame images using the temporal and spatial information. In 1981, Lucas and Kanade proposed an algorithm [5, 6] to compute dense optical flow. The method is based on three assumptions as below:

Brightness constancy. A pixel from the image of an object in the scene does not change in appearance as it (possibly) moves from frame to frame.

Temporal persistence. The image motion changes slowly in time. In practice, this means the brightness of a particular point in the image plane is constant to the motion of the brightness pattern.

Spatial coherence. Neighboring points in a scene belong to the same surface, have similar motion, and project to nearby points on the image plane.

Fig. 1 The description of spherical panorama image

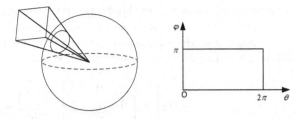

Lucas-Kanade optical flow has become one of the most widely used technique in computer vision, but in fact it works very unsatisfactory for this reason: when we want to catch large motions in scene, we need a large window in image that breaks the coherent motion assumption unconsciously. To circumvent this problem, an intuitive way is to reduce the image size. So the optical flow can be refined down the levels of the image pyramid [7].

3　Image Initial Registration

Before blending images, it is necessary to align the images by confirming the position between images. We need to establish the mathematical relationships that map pixel coordinate from one image to another. For standard lens images, the mathematical model is usually established by affine transformation, and the least squares method is used to optimize the parameters. Fish-eye images alignment is relatively complex, because it is not two-dimensional image essentially. Before aligning images, fish-eye image should be corrected according to its projection model [8]. After correction, we can establish the geometric transformation relation based feature method to remap another image to the reference image coordinate. In this work, we use SIFT-based method to find and describe match points. The k-d tree search strategy is introduced to improve the efficiency and the speed of matching. RANSAC [9] is used to erase the error matched keypoints. Then we project the matched pair points to the spherical surface according the fisheye lens projection model.

4　Registration Refine by Using Optical Flow

In fact, there are still small differences between two images after aligned, which derived from the distortion especially on the fisheye lens, and other external scene motion factors such as tree branch waves with the wind or people moves in and out pictures. These differences lead to a problem that the "ghosting" will appears on the merged image. To deal with this situation, we should try to reduce the differences between two images. Using the optical flow information to remap two images is a feasible way.

Affected by ambient light, there still exists the differences in brightness and color between the images after auto exposure and white balance process. These differences are contrary to the brightness constancy of optical flow. So we should first adjust the brightness of the images. Sample point [10] is one of ways to adjust image brightness. The three image channels RGB are handled separately, and the differences in brightness and color are approximated by a linear equation:

$$\begin{bmatrix} R_2 \\ G_2 \\ B_2 \end{bmatrix} = \begin{bmatrix} a_r & 0 & 0 \\ 0 & a_g & 0 \\ 0 & 0 & a_b \end{bmatrix} \cdot \begin{bmatrix} R_1 \\ G_1 \\ B_1 \end{bmatrix} + \begin{bmatrix} b_r \\ b_g \\ b_b \end{bmatrix}$$

where a and b are linear transformation parameters. The sample points are chosen from the SIFT match feature pairs. MSAC [11] is used to refine the sample points. Regression analysis is used to compute the linear transform parameters of the three channels. The details of the algorithm can be find in [11].

After adjusted the brightness and computed the optical flow of images, we can refine image registration. The left image in the overlap region is denoted as L and the right one is denoted by R. The optical flow $D_1(d_x^1, d_y^1)$ of the left image L to the right image R is obtained by the pyramid LK algorithm. We also compute the optical flow D_2 (d_x^2, d_y^2) of the right image R to the left image L. An image point $\mathbf{u} = (u_x, u_y)$ on the image L and an image point $\mathbf{v} = (v_x, v_y)$ on the image R satisfy the equations $\mathbf{v} = \mathbf{u} + \mathbf{d} = (u_x + d_x^1, u_y + d_y^1)$ and $\mathbf{u} = \mathbf{v} + \mathbf{d} = (v_x + d_x^2, v_y + d_y^2)$. Then we use the remap(p, mode) function, which means the value at the point p is re-sampling by the interpolation method, to obtain a smoother image. Here we use bicubic interpolation mode to keep a smoother image edge. The remap of two images can be defined as:

$$R_{remap}(x, y) = remap(R(x + \omega d_x^2, y + \omega d_y^2), INTER)$$
$$R_{remap}(x, y) = remap(L(x + (1 - \omega)d_x^1, y + (1 - \omega)d_y^1), INTER)$$

where INTER represents the interpolation mode and $\omega = x/w$ is the weight coefficient.

5   Experiment and Results

We use panorama camera with four wide angel fisheye lens surrounded horizontally to take four pictures at the same frame. After distortion correction of the all four fisheye images, we get four corrected images. Because of the sunlight, we can see that visible exposure differences are appeared between the four images. We use the sample point method to adjust the brightness of the four images. The result shows in Fig. 2.

We use the optical flow to remap the images, and composite them to create a panorama image. Compared without refining by optical flow, it can reduce ghosting obviously. The results of the image mosaic composited by fade in and fade out method has been shown in Fig. 3.

Fig. 2 Brightness adjustment of the corrected images

Without refined   Refined

Fig. 3 The result shows that ghosting has been reduced efficiently

6 Conclusions

In this paper, we have introduced a method based on feature matching and optical flow to get spherical panorama taken from fish-eye lens. This method has a promising effect to deal with exposure difference and ghosting reduction. But it is out of work to deal with large motion in the overlap regions like as car driving. In a word, it provides the comprehensive significance and values in image mosaic research and application.

References

1. Shum HY. Panoramic image mosaics. Int J Comput Vis. 2000;36(2):101–30.
2. Szeliski R. Image alignment and stitching: a tutorial. Found Trends Comput Graph Vis. 2007;2(1):1–104.
3. Shum HY, Szeliski R. Systems and experiment paper: construction of panoramic image mosaics with global and local alignment. Int J Comput Vis. 2000;36(2):101–30.
4. Lowe DG. Distinctive image features from scale-invariant keypoints. Int J Comput Vis. 2004;60(2):91–110.
5. Black MJ, Jepson AD. Eigentracking: robust matching and tracking of articulated objects using a view-based representation. Int J Comput Vis. 1998;26(1):63–84.
6. Lucas BD, Kanade T. An iterative image registration technique with an application to stereo vision. International joint conference on artificial intelligence. Morgan Kaufmann Publishers Inc;1981. p. 674–79.
7. Bouguet JY. Pyramidal implementation of the Lucas Kanade feature tracker description of the algorithm. Opencv Doc. 1999;22(2):363–81.
8. Kannala J, Brandt SS. A generic camera model and calibration method for conventional, wide-angle, and fish-eye lenses. IEEE Trans Pattern Anal Mach Intell. 2006;28(8):1335–40.
9. Mach CAC. Random sample consensus: a paradigm for model fitting with application to image analysis and automated cartography;1981.
10. Zhao H, Chen H, Yu H. Image mosaic based on automatic brightness and white blance ajustment. Science paper online;2005. p. 12 [in Chinese].
11. Capel DD. Image mosaicing and super-resolution. 2001;152(4):653–58 Vol.2.

Longitudinal Control of Unmanned Powered Parafoil with Precise Control Gain

Sai Chen, Qinglin Sun, Shuzhen Luo and Zengqiang Chen

Abstract Unmanned powered parafoil is a complex nonlinear system. In this paper, a novel approach based on active disturbance rejection control (ADRC) with precise control gain is constructed for unmanned powered parafoil to reach the precise reference altitude. We first outline the dynamic model of unmanned powered parafoil. Moreover, the longitudinal altitude controller is introduced, where the extended state observer (ESO) estimates the total disturbances involving model uncertainties, internal coupling and external wind disturbance. Furthermore, the highlight of paper, is that the control gain is directly obtained from the system model rather than a trial value, which can optimize the state error feedback (SEF) and enhance the stability and disturbance-rejection of the controller. After that, the introduction of semi-physical platform is presented and the experimental results are analyzed. The experiment results verify the efficiency of this control approach.

Keywords Active disturbance rejection control · Unmanned powered parafoil · Longitudinal control · Precise control gain · Wind disturbance

1 Introduction

Unmanned powered parafoil (UPP) is a kind of flexible wing vehicle, consisting of traditional parafoil system and power plant. Because of its high ratio of lift and drag, perfect stability and operability, UPP has been applied in both military and civil areas, such as precision aerial delivery systems and aerial photography [1, 2]. Based on such advantages, UPP will expand applications in many fields [3–5]. In all this projects, fulfilling the task of tracking the target trajectory is the most important work [6, 7].

Some remarkable results have been investigated about UPP studies. Slegers created a simplified six-degree-of-freedom (dof) model and designed a trajectory

S. Chen · Q. Sun(✉) · S. Luo · Z. Chen
College of Computer and Control Engineering, Nankai University,
Tianjin 300350, China
e-mail: sunql@nankai.edu.cn

© Springer Nature Singapore Pte Ltd. 2018
Y. Jia et al. (eds.), *Proceedings of 2017 Chinese Intelligent Systems Conference*,
Lecture Notes in Electrical Engineering 460,
https://doi.org/10.1007/978-981-10-6499-9_10

tracking controller based on the model predictive strategy [8]. Aoustin designed a nonlinear control law based on the partial feedback linearization for controlling the longitudinal motion [9]. Ochi identified parameters and reduced a given plant model to a second-order system for the designed eight-dof powered parafoil model, and created a three-term PID controller [10]. Tao proposed a homing trajectory planning scheme based on the linear active disturbance rejection control of unmanned powered parafoil [7]. The total disturbance was estimated via ESO using input and output information, and compensated by real-time dynamic feedback in ADRC. However, in the existing studies, the control gain is mostly obtained by trial and error method, and the robustness of the control strategy under the wind disturbance and model uncertainties is not considered.

Therefore, in this paper, a precise longitudinal control approach of UPP based on ADRC and accurate control gain is proposed. In order to enhance the anti-disturbance ability and tracking precision simultaneously, the linear ESO and state error feedback with precise control gain are constructed. The controller designed in this paper can work well under various atmospheric wind environments. The validity of controller is verified by the semi-physical experiments.

2 Mathematical Model

The existing studies of the dynamic model for parafoil system mainly focused on three, four, and six dof models [11]. In the UPP, the relative pitch angle and yaw angle between the parafoil and payload should be taken into account, since they affect the payload attitude, which in turn determines the thrust direction. In the case where apparent mass and the relative motions are involved, the eight-dof model of UPP based on Kirchhoff motion equation is built and described.

In order to facilitate analysis, three main coordinate systems to be used are established: geodetic coordinate $O_d x_d y_d z_d$, parafoil coordinate $O_s x_s y_s z_s$ and payload coordinate $O_w x_w y_w z_w$, shown as Fig. 1.

2.1 Motion Equations of Parafoil and Payload

The forces acting on the parafoil and payload include aerodynamic force, gravity, tension of suspension lines, and thrust provided by the power plant. Since the gravity and thrust can be reasonably assumed to act upon the mass center of payload, the angular momentums due to gravity and thrust are negligible. Considering the momentum theorem and angular momentum theorem, we analyze the forces on the parafoil and payload respectively:

$$\frac{\partial P_w}{\partial t} + W_w \times P_w = F_w^{aero} + F_w^G + F_w^t + F_w^{th} \tag{1}$$

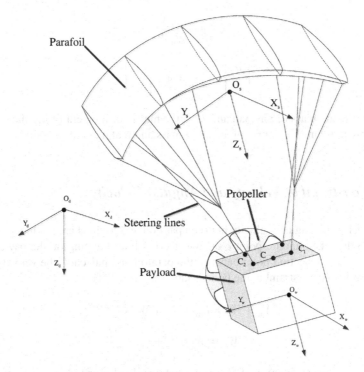

Fig. 1 Structure of unmanned powered parafoil

$$\frac{\partial H_w}{\partial t} + W_w \times H_w = M_w^{aero} + M_w^f + M_w^t \tag{2}$$

$$\frac{\partial P_s}{\partial t} + W_s \times P_s = F_s^{aero} + F_s^G + F_s^t \tag{3}$$

$$\frac{\partial H_s}{\partial t} + W_s \times H_s + V_s \times P_s = M_s^{aero} + M_s^G + M_s^t + M_s^f \tag{4}$$

where P and H are the momentum and angular momentum respectively; F and M denote the force and moment acting on the parafoil and payload respectively; $V = \begin{bmatrix} u & v & w \end{bmatrix}^T, W = \begin{bmatrix} p & q & r \end{bmatrix}^T$ denote the velocity and angular velocity respectively; subscript w, s denote the payload coordinate and parafoil coordinate respectively; *areo* is the aerodynamic force; t is the tension of suspension lines; G is the gravity; f is the friction; *th* is the thrust, $F_w^{th} = \begin{bmatrix} T_x & 0 & 0 \end{bmatrix}^T$ acting on the payload and the direction is along the positive x-axis. $\times$ means the cross-product of two vectors.

Considering the apparent mass and the caused moment of inertia, the momentum and angular momentum of two bodies are summarized as follows:

$$\begin{cases} P_w = m_w V_w \\ H_w = J_w W_w \end{cases} \tag{5}$$

$$\begin{bmatrix} P_s \\ H_s \end{bmatrix} = \begin{bmatrix} A_a + A_r \end{bmatrix} \begin{bmatrix} V_s \\ W_s \end{bmatrix} \tag{6}$$

where m_w is the mass of payload and J_w is the matrix of moment of inertia; A_a and A_r represent the inertia matrix of apparent mass and real mass of parafoil.

2.2   Constraint of Velocity and Angular Velocity

The velocity and angular velocity between parafoil and payload are not independent of each other. The middle point c of two steering lines hanging on the payload is treated as the connection point between the parafoil and payload. The velocity and angular velocity constraint at the point c satisfies:

$$V_w + W_w \times L_{wc} = V_s + W_s \times L_{sc} \tag{7}$$

$$W_w = W_s + \tau_s + \kappa_w \tag{8}$$

where the distances from parafoil centroid and payload centroid to the point c are expressed as L_{sc} and L_{wc} respectively; $\tau_s = \begin{bmatrix} 0 & 0 & \psi_r \end{bmatrix}^T$, $\kappa_w = \begin{bmatrix} 0 & \theta_r & 0 \end{bmatrix}^T$, ψ_r and θ_r denote relative yaw angle and pitch angle respectively.

Based on above formulas, the eight-dof mathematical model of UPP can be well built. For detailed modeling process, please refer to the literature [5].

3   Longitudinal Altitude Controller Using ADRC with Precise Control Gain

3.1   Precise Control Gain

For the complex nonlinear characteristic of the UPP system, the control gain is mostly obtained from trial and error method, probably resulting in the loss of tracking precision. Then, in this section, this parameter is obtained precisely by the longitudinal dynamic of the UPP rather than a trial value.

According to the transformation of position and velocity between geodetic and parafoil coordinate, the vertical velocity and vertical acceleration of the geodetic coordinate can be obtained:

Note: for arbitrary angle α, $\sin \alpha \equiv s_\alpha$, $\cos \alpha \equiv c_\alpha$.

$$\dot{H} = T_{d-s}^T V_s = u_s s_\theta - v_s s_\phi c_\theta - w_s c_\phi c_\theta \tag{9}$$

$$\ddot{H} = \dot{u}_s s_\theta + (u_s c_\theta + v_s s_\phi s_\theta + w_s c_\phi s_\theta)\dot{\theta} - \dot{v}_s s_\phi c_\theta$$
$$- \dot{w}_s c_\phi c_\theta + (w_s c_\theta s_\phi - v_s c_\theta c_\phi)\dot{\phi} \tag{10}$$

where T_{d-s} is the transformation from the geodetic to parafoil coordinateis, which is represented by Euler angles (ϕ, θ, ψ). Since the principle of force interaction, the tension of suspension lines satisfies,

$$F_s^t = -T_{w-s}F_w^t \tag{11}$$

where T_{w-s} is the transformation from the payload to parafoil coordinate which is represented by ψ_r and θ_r. From Eq. 10, the relation between velocity components and thrust is required. Thus, substituting Eq. 11 into Eqs. 1 and 3 yields:

$$A_1\dot{V}_s + A_2\dot{W}_s + T_{w-s}m_w\dot{V}_w = T_{w-s}(F_w^{aero} + F_w^G + F_w^{th}) - T_{w-s}W_w \times (m_w V_w)$$
$$+ F_s^G + F_s^{aero} - W_s \times (A_1 V_s + A_2 W_s) \tag{12}$$

where $A_i(i = 1, \cdots, 4)$ represents the third-order submatrix of $(A_a + A_r)$.

In order to obtain the direct relationship between the control input and the altitude with the variable f_1, f_2, f_3 being irrelevant portions from Eq. 12, Eq. 12 can be rewritten as:

$$\begin{cases} \dot{u}_s = \dfrac{\cos\theta_r \cos\psi_r}{m_s + m_{a,11}} T_x + f_1 \\[2mm] \dot{v}_s = \dfrac{\cos\theta_r \sin\psi_r}{m_s + m_{a,22}} T_x + f_2 \\[2mm] \dot{w}_s = \dfrac{-\sin\theta_r}{m_s + m_{a,33}} T_x + f_3 \end{cases} \tag{13}$$

where m_s denotes parafoil mass; $m_{a,11}, m_{a,22}, m_{a,33}$ denote components in the apparent mass matrix.

The relationship between control input and output can be required by substituting Eq. 13 into 10.

$$\ddot{H} = f + bT_x \tag{14}$$

$$b = \frac{\sin\theta \cos\theta_r \cos\psi_r}{m_s + m_{a,11}} - \frac{\sin\phi \cos\theta \cos\theta_r \sin\psi_r}{m_s + m_{a,22}} + \frac{\cos\phi \cos\theta \sin\theta_r}{m_s + m_{a,33}} \tag{15}$$

From Eq. 15, the accurate control gain is obtained. Now the control gain b is applied to ESO and state error feedback for improving the control performance.

3.2 Principle of ADRC

ESO and SEF control law are utilized in ADRC, and the unknown system can be reduced to an integral series system. For the traditional second-order controlled object, the mathematical expression can be written as follow.

$$\ddot{y} = f(\dot{y}, y, w) + bu \tag{16}$$

where y is the output, u is the input, b is the control gain of input, which is set as a trial value mostly in the UPP, w is the external disturbance, f is the unknown function of the system.

Based on the disturbance concept in the ADRC, unknown disturbance f can be estimated and then cancelled. Thus, we choose to estimate f in real time via an observer instead of relying on its mathematical representation. So we define,

$$h = \dot{f}(\dot{y}, y, w) \tag{17}$$

And we convert Eq. 16 into the extended state space form:

$$\begin{cases} \dot{x}_1 = x_2 \\ \dot{x}_2 = x_3 + bu \\ \dot{x}_3 = h \\ y = x_1 \end{cases} \tag{18}$$

Because h is unknown and not available, it can be assumed to be zero. The total disturbance of the system can be observed by linear ESO as shown in Eq. 19.

$$\begin{cases} e_1 = y - z_1 \\ \dot{z}_1 = z_2 - \beta_1 e_1 \\ \dot{z}_2 = z_3 + bu - \beta_2 e_1 \\ \dot{z}_3 = -\beta_3 e_1 \end{cases} \tag{19}$$

where z_i is the observed states of x_i; $L = \begin{bmatrix} \beta_1 & \beta_2 & \beta_3 \end{bmatrix}$ is the ESO error feedback gain matrix. Gao simplifies the ESO parameterization, by assigning the characteristic roots of observer at $-\omega_0$, and defines $L = \begin{bmatrix} 3\omega_0 & 3\omega_0^2 & \omega_0^3 \end{bmatrix}$ [12].

With the parameter adjustment, x_i can be accurately estimated by ESO. Ignoring the estimation error in z_3, and the control quantity is designed as:

$$u = u_0 - \frac{z_3}{b} \tag{20}$$

Substituting Eq. 20 into 16, the system can be expressed as:

$$\ddot{y} = bu_0 \tag{21}$$

Utilizing linear PD form of the SEF control law, we have:

$$u_0 = k_p(r - z_1) + k_d(\dot{r} - z_2) \tag{22}$$

where k_p and k_d denote the control parameters respectively.

Referring to Eq. 21, the nonlinear UPP system can be transformed into a linear system with the proper parameters of ESO and SEF control law. With this method the longitudinal altitude controller is constructed.

4 Semi-physical Experiment

In this section, the semi-physical experiments are carried out. The parameters of UPP are shown in Table. 1. The following is the introduction of semi-physical experimental setup and the analysis of experimental results.

The semi-physical experiment platform is composed of the host computer and the lower computer. The host computer consists of the dynamic model of UPP and various wind disturbance models established in the MATLAB. The lower computer is the embedded ARM microprocessor executing the ADRC algorithm mainly. The system structure is shown as Fig. 2. From Figs. 2, 4 is the thrust motor; 5 is the propeller; 7 is the motor driver; 8 is the ARM microprocessor; 9 and 10 is serial line, as the communication protocol between the controller and model. The other parts are horizontal tracking device.

The working conditions are set as follows: the initial position is $(0, -300, 2000)$m, and the initial velocity $V_s = (14.9, 0, 2.1)$ m/s; the reference altitude is 1950 m. The transverse average wind with a speed of 3 m/s and direction along the y-axis is added; NASA's classic gust model [13] is also added into the environment. The simulation time is 250 s and simulation step is 0.025s. The ADRC controller parameters are tuned to be: $\omega_o = 30$, $k_p = 0.0134$, $k_d = 0.3$.

Through the flight experiment data the time-varying b is acquired. Figure 3 indicates curve of the control gain b, the gain b becomes stable gradually after an initial fluctuation. The value is about 0.013 finally. Therefore, in the semi-physical experiments, b is selected as the stable value.

Table 1 Parameters of unmanned powered parafoil

Span	10.5/m	Chord	3.1/m
Aspect ratio	3.4	Area of canopy	$33.0/m^2$
Length of lines	6.8/m	Rigging angle	$10/^\circ$
Mass of canopy	10/kg	Mass of payload	80/kg
Characteristic area of payload	$0.6/m^2$	Thrust	0–400/N

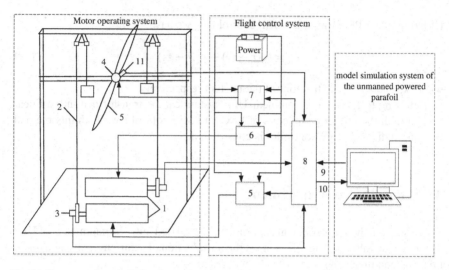

Fig. 2 Structure of semi-physical experiment platform

Fig. 3 Precise control gain b

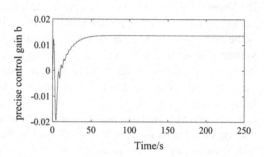

To simulate the disturbance under the actual flight environment, the wind disturbance of Fig. 4a, b is set as follows: the constant wind with a speed of 3 m/s is added at 50 s, and gust wind is added at 100 s along the y-axis, which the velocity is set to 5 m/s, the duration is 15 s. The gust wind affects the parafoil altitude by less than 0.7 m. The control quantity is quickly adjusted to reduce the error and suppress the disturbance of gust, and restore stability rapidly. The controller can track the reference altitude well, and the control quantity goes stable gradually. The average error finally converges to zero.

The wind environment of Fig. 4c, d is set as: the mixture of constant and turbulent wind are added at 50 s persistently. Even under the influence of various wind, the UPP can achieve reference altitude without overshoot. The adjustment time is shorter, the thrust is regulated to keep the flight path. There is no obvious oscillation on the whole control process.

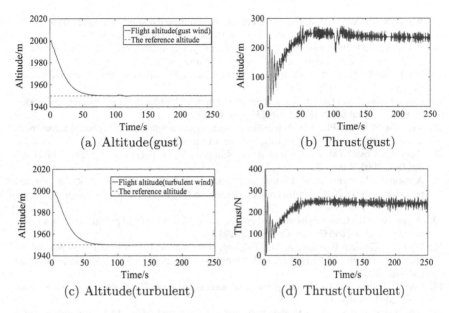

Fig. 4 Altitude and thrust under different wind disturbance

5 Conclusion

Aiming at the nonlinear characteristic of unmanned powered parafoil, this paper proposes an accurate longitudinal control method based on ADRC with precise control gain. Rather than the trial and error method or experience value, the control gain is accurately obtained by the longitudinal dynamics of the unmanned powered parafoil system. Through semi-physical experiments with various wind disturbance, this control approach realizes the altitude control precisely with zero margin for error. This research provides an effective method for the trajectory tracking study of unmanned powered parafoil.

Acknowledgements This work is supported by National Natural Science Foundation of China under Grant (61273138, 61573197), National Key Technology R and D Program under Grant (2015BAK06B04), and the key Technologies R and D Program of Tianjin under Grant (14JCZDJC39300).

References

1. Zhang M, Nie H, Zhu R. Stochastic optimal control of flexible aircraft taxiing at constant or variable velocity. Nonlinear Dyn. 2010;62(1):485–97.
2. Liu H, Guo L, Zhang Y. An anti-disturbance pd control scheme for attitude control and stabilization of flexible spacecrafts. Nonlinear Dyn. 2012;67(3):2081–8.

3. Watanabe M, Ochi Y. Modeling and simulation of nonlinear dynamics of a powered paraglider. AIAA Guidance,Navigation and Control Conference and Exhibit; 2008. p. 18–21.

4. Yakimenko OA. Precision aerial delivery systems: modeling, dynamics, and control. Reston, Virginia: American Institute of Aeronautics and Astronautics; 2015.

5. Zhu E, Sun Q, Tan P, et al. Modeling of powered parafoil based on kirchhoff motion equation. Nonlinear Dyn. 2015;79(1):617–29.

6. Gao HT, Tao J, Sun QL. Design and optimization in multiphase homing trajectory of parafoil system. J Central South Univ. 2016;23(6):1416–26.

7. Tao J, Sun QL, Tan PL, et al. Active disturbance rejection control(ADRC)-based autonomous homing control of powered parafoils. Nonlinear Dyn. 2016;86(3):1461–76.

8. Slegers N, Costello M. Model predictive control of a parafoil and payload system. J Guid Control Dyn. 2005;28:816–21.

9. Aoustin Y, Martynenko Y. Control algorithms of the longitude motion of the powered paraglider. ASME 2012 11th Biennial Conference on Engineering Systems Design and Analysis; 2012 pp. 775–784.

10. Watanabe YOK. Linear dynamics and pid flight control of a powered paraglider. AIAA Guidance, Navigation, and Control Conference; 2013.

11. Pollini L, Giulietti F, Innocenti M. Modeling, simulation and control of a wing parafoil for atmosphere to ground flight. AIAA Modeling and Simulation Technologies Conference and Exhibit; 2006.

12. Gao Z. Scaling and bandwidth-parameterization based controller tuning. Proc Am Control Conf. 2003;6:4989–96.

13. Adelfang S, Smith O. Gust Models for launch vehicle ascent. 36th AIAA Aerospace Science Meeting and Exhibit; 1998.

A Location Estimation Method on Man and Vehicle in Coke Oven

Tianyang Yu, Xiaobin Li and Haiyan Sun

Abstract In this paper, a moving target localization method in ultra wide band (UWB) technology from time difference of arrival (TDOA) and angle of arrival (AOA) measurements is proposed. This method provides much higher precise estimate of target position than the single measure method. The UWB radar system which can be used with advantage of tracking of persons or vehicle moving behind obstacles. Simulation studies are given to confirm the approximate efficiency. A Field measurement is conducted in order to verify the accurately approximating impact on a person walking outdoors, which provides a great experiment result.

Keywords Four main coke oven vehicles · Moving targets · Localization method

1 Introduction

With a strong push forward environmental pollution control, state of the rapid development of steel and coking industry made a series of claims about the presence of pollution and security issues. To reduce the risk of contamination and safety hazards of production workers, the unmanned technology of coke oven locomotive is the key to accomplish it. And the key that unmanned technology of coke oven locomotive could be implemented is the anti-collision of man and machine.

Until now, among the researches on the automatic operation of coke oven locomotive, the literature [1] discussed the methods of the locomotive address detection and process control which based on induction radio technology are applying to safety interlock control and location control. And a new method which is suitable for improving the position detection resolution at the literature [2]. The paper [3]

T. Yu · X. Li(✉)
School of Electrical and Electronic Engineering, Shanghai Institute
of Technology, Shanghai 201418, China
e-mail: lixiaobinauto@163.com

H. Sun
School of Ecological Technology and Engineering, Shanghai Institute
of Technology, Shanghai 201418, China

© Springer Nature Singapore Pte Ltd. 2018
Y. Jia et al. (eds.), *Proceedings of 2017 Chinese Intelligent Systems Conference*,
Lecture Notes in Electrical Engineering 460,
https://doi.org/10.1007/978-981-10-6499-9_11

introduces about the interlocking control and automatically working control conditions and application of systems of coke oven locomotive.

The researches on the wireless orientation also have some latest development. The literature [4] proposed one kind of attitude identification method of road header based on ultra wide band ranging technique and conducted experiments to verify the accuracy ranging impact on the system including signal delay. A through-wall measurement was realized and corresponding signals acquired by two different UWB radar systems have been processed at the literature [5]. At the literature [6] proposed the method of calculating the target position by mixing with TDOA and AOA location.

In the former study, the problems of the locomotive address detection and process control have been resolved by the method based on induction radio technology. And except the presence of obstacles, the UWB ranging technique had higher precise at the narrow roadway. For coke oven locomotive which has a serious problem in the receipt of the signal, it still needs investigating.

Against the propagation obstacles in environment of coke oven locomotive which has unmanned and anti-collision features, when using UWB technology positioning, the obstruction of the signal seriously affected the positioning and ranging of man and vehicle, a kind of intelligent measuring and location method based on jointing TDOA and AOA is proposed in this article. In the coke oven production site has been used to verify the reliability of the method, accuracy and stability.

2 Positioning and Measuring Method

When the moving target is positioned, the more the measurement information using, hybrid positioning method can be used to improve the accuracy. In TDOA positioning, at least three base stations to achieve the targets, the accuracy of positioning a large impact received NLOS. The AOA positioning method using the angle information can be measured at least two target base stations. The use of AOA positioning in a wide range of areas requires high measurement accuracy for angles.

According to locate their strengths TDOA and AOA, the method of hybrid positioning of TDOA/AOA is used to locate man and vehicles. Through the distance measurement, angle measurement, the establishment of positioning model and the establishment of positioning algorithm to achieve the process.

2.1 Hybrid Positioning Model

Assume that the positions of the M base stations are known to be (x_1, y_1), (x_2, y_2) to (x_M, y_M) and the coordinate of the moving target is (x, y). Assume that each base station can receive a signal transmitted moving target with TOA value while the signal can be measured in a straight path to the base station and the angle of AOA values. Where the time of arrival TOA at which the signal arrives at the base station can be described by the distance d_i, the angle AOA value can be described by a horizontal angle α.

$$d_i^2 = \left(x - x_i\right)^2 + \left(y - y_i\right)^2 \tag{1}$$

$$\alpha_i = \tan^{-1}\left(\frac{y - y_i}{x - x_i}\right) \tag{2}$$

where $i = 1, 2, \cdots, M$.

The hybrid positioning model is shown below,

$$Y = f(X) + V \tag{3}$$

And $X = [x, y]^T$, $Y = [\alpha_1, \alpha_2 \ldots \alpha_M, d_1, d_2 \ldots d_M]^T$, $V = [v_{\alpha_1} \ldots v_{\alpha_M}, v_{d_1} \ldots v_{d_M}]^T$. Solve the vector equation can be derived estimates a moving target.

2.2 Hybrid TDOA/AOA Location Algorithm

Suppose the serving base station is defined as the value calculated TDOA measurement reference base station and set the service base station to BS1, TDOA measured value corresponding to a distance difference is $d_{i,1} = d_i - d_1$. The Eq. (1) can be rewritten as follow,

$$d_{i,1}^2 + 2d_{i,1}d_1 = -2x_{i,1}x - 2y_{i,1}y + K_i + K_1 \tag{4}$$

with $i = 2, 3, \cdots, M$ and $x_{i,1} = x_i - x_1$, $y_{i,1} = y_i - y_1$, $K_i = x_i^2 + y_i^2$, $K_1 = x_1^2 + y_1^2$. While assuming that the serving base station BS1 can always provide the AOA measured value of the moving target, the equation as below can be listed.

$$\tan \alpha = \frac{y - y_1}{x - x_1} \tag{5}$$

Unknown variable is $z_a = \left[x^T, d_1\right]^T$. Using Eqs. (4) and (5), we can establish a linear system of equations as variables,

$$h = Gz_a \tag{6}$$

And the covariance matrix of the error vector ψ is

$$\psi = h - Gz_a^0 \tag{7}$$

with

$$h = \begin{bmatrix} d_{2,1}^2 - K_2 + K_1 \\ d_{3,1}^2 - K_3 + K_1 \\ \vdots \\ d_{M,1}^2 - K_M + K_1 \\ x_1 \tan \alpha - y_1 \end{bmatrix}, G = - \begin{bmatrix} 2x_{2,1} & 2y_{2,1} & d_{2,1} \\ 2x_{3,1} & 2y_{3,1} & d_{3,1} \\ \vdots & \vdots & \vdots \\ 2x_{M,1} & 2y_{M,1} & d_{M,1} \\ -\tan \alpha & 1 & 0 \end{bmatrix}$$

z_z^0 is the vector value of the actual position of the moving target. The weighted least squares algorithm WLS is used to solve. The covariance matrix Q of the measured values of TDOA and AOA is used instead of the covariance matrix of the error vector ψ.

$$z_a = \left(G^T Q^{-1} G\right)^{-1} Q^{-1} h \tag{8}$$

The sequence (x, y) in z_a is the approximate estimated position of the moving target. I.e. the first WLS estimate. Supposing each TDOA and AOA measurements independent of one another, the matrix Q in the above equation is a diagonal matrix.

$$Q = diag \left\{ \sigma_{2,1}^2, \sigma_{3,1}^2, \ldots, \sigma_{M,1}^2, \sigma_\alpha^2 \right\} \tag{9}$$

$\sigma_{i,1}, \sigma_\alpha$ are the standard deviations of the TDOA measurements and the AOA measurements, respectively. To ensure the consistency of the dimensions in Q and assume that the accuracy of AOA measurements is high while the standard deviation of the error measured by AOA is very small, AOA corresponds to the standard deviation of the distance error is

$$\sigma_\alpha = \bar{d_1} \times std\alpha \tag{10}$$

where $\bar{d_1}$ is the distance between the moving target and the serving base station BS1.Since the unknown variable d_1 is related to (x, y) in vector z_a. The covariance matrix with the matrix Q instead of the measurement error vector will bring some deviations. In order to get a more accurate estimate of the location, We use a similar process to the Chan algorithm [7]. For the measurement error vector ψ, the error vector corresponding to the M-1 TDOA measurements can be expressed as

$$\begin{cases} \psi_{tdoa} = B_{tdoa} n \\ B_{tdoa} = diag\{d_2^0, d_3^0 \cdots, d_M^0\} \end{cases} \tag{11}$$

when the TDOA measurement error is relatively small. The amount is the actual distance between the moving target and the *ith* reference base station and n is the measurement of noise in units of distance. When the AOA error is relatively small, the error corresponding to the AOA measured value observed by the serving base station BS1 is

$$\psi_\alpha = \delta_\alpha \tag{12}$$

The covariance matrix of the total measurement error, which is composed of TDOA and AOA measurements, is thus obtained

$$\begin{cases} \Psi = \mathrm{E}\left(\psi\psi^T\right) = BQB \\ B = diag\{d_2^0, d_3^0, \ldots, d_M^0, 1\} \end{cases} \tag{13}$$

The new matrix B can be calculated by using the value calculated by Eq. (8) and then use the weighted least squares algorithm to obtain an improved estimated position in the Eq. (14) which is the results of the second WLS estimate.

$$z_a = \left(G^T\Psi^{-1}G\right)^{-1}\Psi^{-1}h \tag{14}$$

In order to get much more accurate position estimate, the covariance matrix of the estimated position z_a is first calculated,

$$\mathrm{cov}(z_a) = \left(G^T\Psi^{-1}G\right)^{-1} \tag{15}$$

According to the relationship, Eq. (3), between distance and coordinates, we can construct a new error vector ψ'.

$$\psi' = h' - G'z_a' \tag{16}$$

with

$$h' = \begin{bmatrix} (z_{a,1} - x_1)^2 \\ (z_{a,2} - y_1)^2 \\ z_{a,3}^2 \end{bmatrix}, G' = \begin{bmatrix} 1 & 0 \\ 0 & 1 \\ 1 & 1 \end{bmatrix}, z_a' = \begin{bmatrix} (x - x_1)^2 \\ (y - y_1)^2 \end{bmatrix}_a^0.$$

The final third WLS estimate is

$$z_a' = \left(G'^T\Psi'^{-1}G'\right)^{-1}\Psi'^{-1}h' \tag{17}$$

with

$$\begin{cases} \Psi' = \mathrm{E}\left(\psi'\psi'^T\right) = 4B'\mathrm{cov}(z_a)B' \\ B' = diag\{x^0 - x_1, y^0 - y_1, d_1^0\} \end{cases} \tag{18}$$

can be calculated by the formula (16) approximation, the final moving target positioning calculation results is

$$\hat{X} = \sqrt{z_a'} + \begin{bmatrix} x_1 \\ y_1 \end{bmatrix} \text{ or } \hat{X} = -\sqrt{z_a'} + \begin{bmatrix} x_1 \\ y_1 \end{bmatrix} \tag{19}$$

The ambiguity in (19) can be eliminated by the a priori information of the locating area. For example, if the moving target reaches a certain base station, the distance is greater than the maximum distance of the base station in the positioning area. The maximum likelihood estimation of the moving target position is completed under the premise that the TDOA and AOA measurement errors are subject to the Gaussian distribution of zero mean.

3 Simulation and Results Analysis

In order to test and compare the positioning performance of the coke oven in the wireless network channel environment, a simulation test was conducted. There are four base stations participating in TDOA measurements, which contain a serving base station. Let the TDOA measurement error caused by the measurement system obey the Gaussian distribution with the mean zero and the variance of 1 m and the AOA measured error provided by the serving base station obeys the Gaussian distribution with the mean 0 and the standard deviation of 0.01 rad. In Matlab, the following experiment are

(1) The number of base stations participating in positioning is 4;

(2) The 20 m * 30 m plane area was selected as the locating area. The coordinates of the four base stations are A (0,0), B (0,30), C (20,0), D (20,30);

(3) Assuming that the measured object is moved from x = 1 m to x = 30 m at a speed of 1 m/s along the x-positive direction, the positioning frequency is 1 Hz;

(4) The error variance of TDOA is 1 m, and the standard deviation of AOA is 0.01 rad.

The experimental results are shown in Fig. 1.

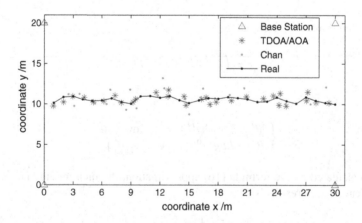

Fig. 1 Chan algorithm and TDOA/AOA algorithm positioning contrast

As can be seen from Fig. 1, TDOA/AOA hybrid positioning algorithm is better than Chan algorithm in the case of the same TDOA noise interference. It is proved that the hybrid positioning technology with different feature measurements to locate the moving targets has higher positioning accuracy than the single positioning technology. It can be seen from Fig. 2, Chan algorithm in the x direction of the maximum positioning error of 0.91 m and TDOA/AOA hybrid algorithm in the X direction of the maximum positioning error of 0.60 m; It can be seen from Fig. 3, Chan algorithm in the Y direction of the maximum positioning error of 1.27 m, TDOA/AOA algorithm in the Y direction of the maximum positioning error of 0.95 m. It can be

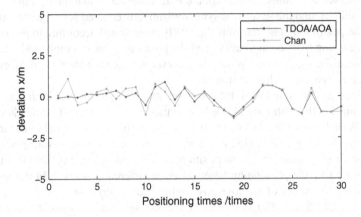

Fig. 2 Chan algorithm and TDOA/AOA algorithm X positioning bias comparison

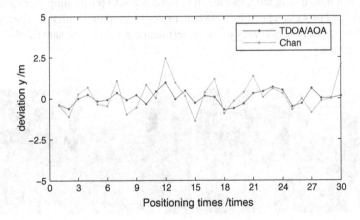

Fig. 3 Chan algorithm and TDOA/AOA algorithm Y positioning bias comparison

concluded that the accuracy of AOA measurement is better than that of Chan algorithm. The TDOA/AOA hybrid positioning algorithm has better performance than Chan algorithm.

4 Practical Application Results

The whole positioning system consists of four positioning base stations and positioning tags, POE Ethernet switches, positioning servers. The base station and the location server are connected through the POE Ethernet switch through the TCP/IP protocol cluster, and the same time synchronization is ensured between the base stations. The positioning tag transmits the UWB pulse signal according to its configuration according to its configuration, and the positioning base station sends the time of receiving the signal to the positioning server to calculate the time difference of the signal arriving at each base station.

Field test as shown in Fig. 4, the four base stations used for positioning are fixed on the same height of the tripod, the base station from the ground perpendicular to the height of 1.70 m, the label worn on the chest of the test personnel to ensure that with the positioning base station in a horizontal plane. The test personnel wear the positioning tag to move within the positioning area and measure the coordinate values measured by the positioning base station at different positions. In the test, the calibration coordinates of the four positioning base stations are: (0,0), (22.145,0), (0,34.237), (21.853,35.097), in meters. In the test, select a location (11.265,10.448) to locate, to verify the TDOA/AOA hybrid algorithm positioning accuracy. In the 20 m * 30 m positioning area, located in (11.265,10.448) positioning label positioning results and processing results shown in Table 1. Compared with the experimental results, the positioning accuracy of the experiment is basically the same as the experimental accuracy.

Fig. 4 Live test photos

Table 1 Positioning of the measurement results and processing results

Calibration coordinate/m	X	ΔX	Y	ΔY
(11.265,10.448)	10.979	−0.289	9.560	−0.888
(11.265,10.448)	10.946	−0.322	9.673	−0.775
(11.265,10.448)	10.960	−0.308	9.817	−0.631
(11.265,10.448)	11.008	−0.260	9.935	−0.513
(11.265,10.448)	11.056	−0.212	10.055	−0.393
(11.265,10.448)	11.158	−0.110	10.285	−0.163
(11.265,10.448)	11.218	−0.050	10.386	−0.062
(11.265,10.448)	11.307	0.039	10.535	0.087
(11.265,10.448)	11.359	0.091	10.655	0.207
(11.265,10.448)	11.400	0.132	10.757	0.309
(11.265,10.448)	11.441	0.173	10.859	0.411
(11.265,10.448)	11.464	0.196	10.975	0.527
(11.265,10.448)	11.432	0.164	10.856	0.408
(11.265,10.448)	11.241	−0.027	10.340	−0.108

5 Conclusion

This paper presents a method of using UWB technology to locate mobile targets based on TDOA/AOA. The combination of TDOA/AOA hybrid positioning method, which can provide better performance than traditional wireless communication technology and single positioning method, can be improved by using ultra wide band technology in multi-path environment. The results of simulation experiment and field test show that the proposed method of TDOA/AOA hybrid UWB technique can meet the requirement of positioning accuracy in coke oven [8].

References

1. JIMing P, LONG W, DING H. Application of orientation system for coke-oven's four movable cars. Metall Ind Autom. 2010;34(2):24–8.
2. Chen J, Yang X. Method for improving position detection resolution by induction radio. J Electron Meas Instrum. 2014;28(9):943–50.
3. Yang H. The study of 4-coke oven control systems and wireless communication. University of Science and Technology Liaoning 2014.
4. Shi-chen FU, Yi-ming LI. Accuracy testing experiment in narrow roadway based on TW-TOF ranging technique of UWB signals. Coal Technol. 2017;36(3):246–8.
5. Jana R, Dušan K. Experimental comparison of two UWB radar systems for through-wall tracking application. Acta Electrotechnica et Informatica. 2012;12(2):59–66.
6. Congfeng L, Jie Y, Fengshuai W. Joint TDOA and AOA location algorithm. J Syst Eng Electron. 2013;24(2):183–8.

7. Chan YT, Ho KC. A simple and efficient estimator for hyperbolic location. IEEE Trans Signal Process. 1995;42(8):1905–15.
8. TU XD. Research and performance analysis of UWB positioning algorithm on single-base station and multi-base station. Ocean University of China 2012.

An Interpolation Method of Soil Erosion Based on Flexible Factor

Qingfei Xia, Jiapeng Xiu, Zhengqiu Yang and Chen Liu

Abstract Interpolation is a model to estimate value of an unknown point with a set of known points. In order to get the value of the soil erosion with high precision, it is necessary to develop a high precision soil erosion interpolation method. The paper introduces an interpolation method based on Flexible Factor, called Flexible Factor Based Inverse Distance Weighting (FFBIDW) and two popular classic interpolation methods called Ordinary Kriging (OK) and Inverse Distance Weighting (IDW). Considering the precipitation's effect to the soil erosion, FFBIDW has more accurate result. FFBIDW will be checked by a series of experiments with Yunnan's soil erosion data and have a compare to OK and IDW using RMSE. The test results demonstrate FFBIDW's correctness and feasibility.

Keywords Interpolation · Flexible factor · Soil erosion · Inverse distance weighting

1 Introduction

According to the third soil erosion survey data released by China, the main types of soil erosion in China are water erosion, wind erosion and freeze-thaw erosion. The total area of soil erosion is over 3.56 million km^2, covering 37% of China [1]. Water erosion is the most serious type, with a total area of 1.61 million km^2 and is

Q. Xia (✉) · J. Xiu · Z. Yang · C. Liu
School of Software Engineering, Beijing University of Posts and Telecommunications, Beijing 100876, China
e-mail: xiaqingfei2011@hotmail.com

J. Xiu
e-mail: xiujiapeng@bupt.edu.cn

Z. Yang
e-mail: zqyang@bupt.edu.cn

C. Liu
e-mail: lchen@bupt.edu.cn

© Springer Nature Singapore Pte Ltd. 2018
Y. Jia et al. (eds.), *Proceedings of 2017 Chinese Intelligent Systems Conference*,
Lecture Notes in Electrical Engineering 460,
https://doi.org/10.1007/978-981-10-6499-9_12

most widely distributed in China. Continuous data is necessary to do research on soil erosion. One of the possible way is to found high density and even distributed sampling point. However, it is difficult to do so because of the limit of topography and economic condition. Also it is unpractical for the countries that have Large area like China. To obtain the data from the unsampled places, interpolation is the More commonly used and feasible way to achieve the goal. Investigating the degree of soil erosion across the country is important for the ecological environment. Interpolation is a thought and different methods can lead to different results.

Interpolation is based on samples that can be used to represent the features of the region, using known data to predict the values of the unknown place [2]. Inverse Distance Weighting (IDW) and Kriging are the most popular Interpolation methods in many scientific and application fields. Many Scholars analyzed the difference between IDW and Kriging in the different scopes. Wang Cuiying compared the difference between Kriging and IDW in the regional precipitation and find that Kriging is more accurate [3]. Fan Yujie compared the interpolation precision of each month using 50 years' precipitation data and find that Kriging is more accurate in wet season but IDW is better in low water period [4]. Shahab Shahbeik compared IDW and Kriging methods based on error estimation in the Dardevey iron ore deposit, NE Iran and find that Kriging has less error estimation and its result is more reliable [5].

However, the classic interpolation methods can not fit all of the fields very well. For soil erosion, the classic methods don't consider any factors that can affect the result and make more deviation. Scholars have done a lot of research on spatial interpolation methods these years, developing a variety of spatial interpolation methods and optimizing the classical spatial interpolation method for various fields. Ian A. Nalder estimated 30-year averages of monthly temperature and precipitation at specific sites in western Canada and propose a new method called gradient-plus-inverse distance squared [6]. Narushige Shiode proposes a network-based IDW and OK and improves accuracy and performance [7]. Babak O. presents an approach to integrate statistical controls such as minimum error variance into IDW [8]. Li Mian presents an IDW interpolation method with choosing convex hull points adaptively and provides a more accuracy result [9]. Each has its own suitable fields. The paper will explore the problem of optimizing IDW using a Flexible Factor for soil erosion and propose a more precise result.

2 Related Concepts

2.1 Inverse Distance Weighting

IDW is a commonly used interpolation method with a set of known points. The values of the unknown points are depending on the value of the known points with their weights. The weight depends on the distance between the interpolated point and the interpolating point. The weight will be larger if the interpolating point is closer to the interpolated point, because their similarity is higher, otherwise it will be smaller.

Suppose that z(x, y) represents the value of the unknown point and (x, y) is the coordinate of the point. (x_i, y_i) is the attribute value of the i-th sample point whose coordinate is (x_i, y_i). d_i represents the distance between point (x, y) and point (x_i, y_i). λ_i is the weight that point (x_i, y_i) affects point (x, y). n is the number of the known points used in interpolation. Then the equation of the IDW can be represented as following:

$$z(x, y) = \begin{cases} \frac{\sum_{i=1}^{n} \lambda_i z(x_i, y_i)}{\sum_{i=1}^{n} \lambda_i}, & d_i \neq 0 \\ z(x_i, y_i), & d_i = 0 \end{cases} \tag{1}$$

where

$$d_i = \sqrt{(x - x_i)^2 + (y - y_i)^2} \tag{2}$$

and

$$\lambda_i = d_i^{-\rho} \tag{3}$$

Note that ρ is the power number of the IDW and is always a positive number. The power number can control the effect of the weight of the known points by distance. If the power number is higher, the closer known points' effects are larger and the farther ones are smaller.

2.2 Kriging

Kriging is also called as the best linear unbiased prediction. Kriging is developed by Georges Matheron and is named after South Africa's scientist Danie G. Krige. Statistically speaking, Kriging is a method to get unbiased prediction from the variable correlation and variability. In terms of interpolation, Kriging is a method to get unbiased prediction from data distributed in the space [10]. The kernel of Kriging is the semi-variance model.

Ordinary Kriging (OK) is the most popular Kriging method. Suppose that x_0 is the unknown point and $z(x_i)$ is the value of the i-th known point. n is the number of the known points. Then the equation of OK is as following:

$$z(x_0) = Y' W^{-1} B \tag{4}$$

where

$$W = \begin{bmatrix} cov(x_1, x_1) & \cdots & cov(x_1, x_n) & 1 \\ cov(x_2, x_1) & \cdots & cov(x_2, x_n) & 1 \\ \cdots & \cdots & \cdots & \cdots \\ cov(x_n, x_1) & \cdots & cov(x_n, x_n) & 1 \\ 1 & 1 & 1 & 0 \end{bmatrix} \tag{5}$$

and

$$B = \begin{bmatrix} cov(x_0, x_1) \\ cov(x_0, x_2) \\ \cdots \\ cov(x_0, x_n) \\ 1 \end{bmatrix} \tag{6}$$

and

$$Y = \begin{bmatrix} z(x_1) \\ z(x_2) \\ \cdots \\ z(x_n) \\ 0 \end{bmatrix} \tag{7}$$

where $cov(x, y)$ is the covariance function, the equation is as following:

$$cov(x, y) = E(xy) - E(x)E(y) \tag{8}$$

2.3 RMSE

Root Mean Square Error (RMSE) is usually used to evaluate the error between the estimated value and the actual value. RMSE can gather error of all the predicted values. RMSE is an important norm to represent the deviation of the data and it's widely used. If the value of RMSE is higher, the precision is lower. However, the range of RMSE is not steady and it depends on the span of the data. Thus it is meaningless to use RMSE to compare the accuracy of different data types.

Suppose that $\hat{z}_i$ is the i-th interpolation value, z_i is the i-th actual value, n is the number of points. The equation of RMSE is as following:

$$RMSE = \sqrt{\frac{\sum_{i=1}^{n}(\hat{z}_i - z_i)^2}{n}} \tag{9}$$

3 FFBIDW

Flexible Factor based Inverse Distance Weighting (FFBIDW) is a method based on the classic IDW with a flexible factor involved to improve the interpolation precision of soil erosion. The quantity of soil erosion range from tens to thousands, making the error of interpolation very high. The factor must have some effects to soil erosion, for example, the precipitation.

The level of soil erosion depends on many factors, for example, precipitation, biologic and terrain. Precipitation is a very important factor. Precipitation has some potential contacts with soil erosion, as shown in Fig. 1. The peak of the soil erosion is usually corresponding to the peak of the precipitation. However, the quantity of soil erosion don't have direct relation to the quantity of precipitation. Higher precipitation doesn't mean Higher soil erosion. Therefore, precipitation reach the requirement to be a flexible factor.

Scholars have done a lot to improve the precision of IDW. Duan Ping uses the distance search strategy to select referent points, this method reduces the 'buphthalmos' phenomenon and has higher accuracy [11]. Fan Zide develops a new method called ACGIDW considering the effect of complex topography factors to offer a more accurate result [12]. The equation of ACGIDW is as following:

$$Z = \frac{\sum_{i=1}^{m} \left[Z_i + (X - X_i)C_x + (Y - Y_i)C_y \right] d_i^{-\alpha_i}}{\sum_{i=1}^{m} d_i^{-\alpha_i}} \tag{10}$$

C_x and C_y is the partial regression coefficient, d_i is the distance between (X, Y) and (X_i, Y_i), α_i is the distance-decay parameters. α_i is determined by the standardized partial nearest index μ_R. The equation is

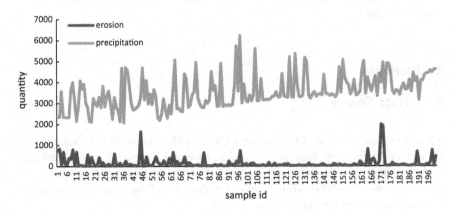

Fig. 1 The quantity of soil erosion and precipitation in some sample places

$$\mu_R = \begin{cases} 0, & R(S_0) < R_{min} \\ 0.5 + 0.5 sin\left[\dfrac{\pi\left(R(S_0) - \frac{R_{max}+R_{min}}{2}\right)}{R_{max} - R_{min}}\right], & R_{min} \leq R(S_0) \leq R_{max} \\ 1 & R(S_0) > R_{max} \end{cases} \quad (11)$$

$R(S_0)$ is the partial nearest index, R_{min} is the minimum value of $R(S_0)$, R_{max} is the maximum value of $R(S_0)$.

However, these available methods has some problems. Some methods just consider to adjust the inner influences and ignore the outside effects. Some methods consider too much outside effects and make the calculation too complex. FFBIDW is designed to provide a simple and effective method to improve the accuracy.

Consider that if the unknown point has more precipitation than a known point, it's very probable that the soil erosion of the unknown point is higher than the known one, and the weight of the known point should be increased, and vice versa. Suppose that s_i is the flexible factor of the i-th known point, σ is the power number of the flexible factor, the equation of FFBIDW is as following:

$$z(x, y) = \frac{\sum_{i=1}^{n} z_i d_i^{-\rho} s_i^{-\sigma}}{\sum_{i=1}^{n} d_i^{-\rho} s_i^{-\sigma}} \quad (12)$$

where

$$s_i = \frac{r_i}{r} \quad (13)$$

r_i is the precipitation of the i-th known point, r is the precipitation of the unknown point.

4 Example

4.1 Data Source

Yunnan province is located in southwest China, with Area of 390 thousand square kilometers. Yunnan has large elevation difference and complex Terrain. Most area of the province has rich precipitation. Tens of precipitation stations and thousands of samples are distributed all over the province. All of the data is based on the measured data from the stations and samples.

4.2 Procedures

The experiment is selected to check the correctness of the FFBIDW and indicate the difference of OK, IDW and FFBIDW. First, 200 samples are chosen randomly from the samples in Yunnan, as shown in Fig. 2, the densities among different parts have sensible disparity. The north and south parts' densities are sparse and the center place has more conferted samples. Then, 15%, 30%, 45%, 60%, 75% of the sample rate are chosen to represent different sample rates and will be used in each test. What's more, the precipitation of each sample place is Interpolated by Kriging with all of the precipitation stations in Yunnan. Last, compute the result and compare the difference of each test case.

4.3 Results Analysis

200 samples in total, choose 15%, 30%, 45%, 60%, 75% of samples randomly and compare the difference of results with IDW and FFBIDW. The power of IDW and FFBIDW ρ is 2, the power of flexible factor σ is 2. Calculate the RMSE of the test cases. The result is shown in Table 1.

Fig. 2 The distribution of the 200 samples in Yunnan

Table 1 The comparison between IDW and FFBIDW

Sample rate (%)	RMSE	
	IDW ($\rho = 2$)	FFBIDW ($\rho = 2, \sigma = 2$)
15	269.0079	265.7357
30	227.5702	221.9634
45	219.1548	205.8726
60	208.7513	200.0178
75	206.3657	206.6644

The line graph of the data from Table 1 is shown in Fig. 3. When the sample rate is too low, FFBIDW has only a little difference with IDW and the precision between FFBIDW and IDW is very close. However, when the sample rate is too high, FFBIDW gets a worse precision than IDW. When the sampling rate is not too high or too low, FFBIDW has higher precision. But the improvements is not very obviously because of the low power of flexible factor σ, representing that the effect of flexible factor is low.

Figure 3 shows that the flexible factor really has effects on the precision. The degree of the effect depends on the power of flexible factor σ. Table 2 shows the RMSE of the different sample rates when σ changes. Each data of the same sample rate is same, the only difference is σ.

As is shown in Fig. 4. The value of the power number of flexible factor has very important effect on the precision. When the sample rate is too high or too low, the effect is not very obvious, but can also improve some accuracy. When the sample rate is in middle, especially around 50%, FFBIDW makes a huge improvement. Also, the result shows that the power number of flexible factor has a best value. Too large value or too small value will reduce the effect of FFBIDW.

In this case, the best power number of flexible factor is 10, as is shown in Fig. 4. When the power number is 0, FFBIDW will regress to the classic IDW. When the sample rate is 45% and σ is 10, FFBIDW makes a big improvement. In order to prove that the improvement wasn't caused by occasionally special data, 5 sets of tests were made, as shown in Table 3, every set of data was selected randomly. Then compute the improvements of each sample rate between IDW and FFBIDW

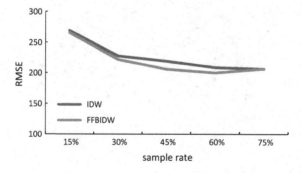

Fig. 3 Line graph of Table 1

Table 2 The difference when using different power number in FFBIDW

σ	RMSE				
	15%	30%	45%	60%	75%
0	269.007	227.570	219.154	208.75	206.365
2	265.735	221.963	205.872	200.017	206.664
4	263.461	217.416	186.019	187.821	206.538
6	261.451	213.364	161.084	174.249	205.596
8	259.391	210.481	139.812	165.122	203.563
10	258.997	210.391	129.857	162.600	200.937
12	262.151	213.424	128.730	164.211	198.790

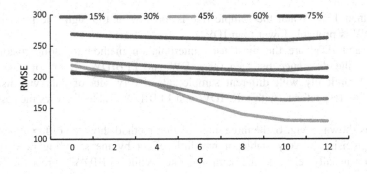

Fig. 4 Line graph of Table 2

Table 3 The precision improvements in different sample rates

No.	Precision improvements				
	15%	30%	45%	60%	75%
1	−0.025	0.011	0.104	0.043	0.083
2	0.116	0.014	0.064	0.082	0.003
3	0.182	0.070	0.024	0.175	−0.007
4	0.119	0.117	0.280	0.059	0.007
5	0.022	0.227	0.184	0.029	0.133

and compare the average improvements of each sample rate. Figure 5 shows the result.

As is shown in Fig. 5, when sample rate is lower than 30%, the average precision improves slowly and the difference is very small. When sample rate is higher than 60%, the average precision decreases quickly. 45% is the best sample rate and the average precision is obviously higher than others. However, as is shown in Table 3, the precision improvement is not strictly related to the sample rate for a specific data, for example, in the row of No. 2, the precision improvement of 45% is

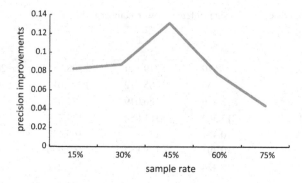

Fig. 5 The average precision improvements of different sample rates

lower than 15%. When the sample rate is too low or too high, the precision of FFBIDW is probable lower than IDW.

OK and IDW are the most used interpolation method for soil erosion. To compare the difference between OK, IDW and FFBIDW, 5 sets of data were selected randomly with different sample rate. Every sets of data was used to interpolate respectively with OK, IDW and FFBIDW. Table 4 shows the result of the test.

As is shown in Fig. 6, the three interpolation method show the different features. OK's precision is very stable, it has little effect by the sampling rate. IDW's precision usually relies on the sampling rate. While FFBIDW's precision highly

Table 4 Comparison between OK, IDW and FFBIDW

Sample rate (%)	RMSE		
	OK	IDW ($\rho = 2$)	FFBIDW ($\rho = 2, \sigma = 10$)
15	268.108	269.007	258.997
30	263.107	227.570	210.391
45	260.096	219.154	129.857
60	235.034	208.751	162.600
75	249.074	206.365	200.937

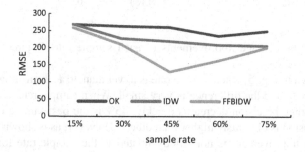

Fig. 6 Line graph of Table 4

depends on the sampling rate. Only if the sampling rate is neither too high nor too low, FFBIDW can work fine and it's precision is higher than OK and IDW.

5   Conclusion

Based on the Interpolation of the Yunnan's soil erosion, introducing flexible factor can strikingly improve the precision of soil erosion. Considering the limitations of the present interpolation methods, the paper introduces FFBIDW. FFBIDW can adjust the weight of IDW dynamically under the influence of flexible factor and improves the precision and adaptability of IDW. With the verification of several sets of data and the Comparison between OK and IDW, FFBIDW is practicable and advantageous. However, FFBIDW just considers one flexible factor in the paper and it limit the scope of application. The future work is how to introduce multiple flexible factors.

References

1. Li Z, Cao W, Liu B, Luo Z. Current status and developing trend of soil erosion in China. Sci Soil Water Conserv. 2008;01:57–62.
2. Dong X, Bo H, Deng X, et al. Rainfall spatial interpolation methods and their applications to Qingjiang River basin. J China Three Gorges Univ. (Natural Sciences). 2009;31(6):6–10.
3. Wang C, Yang Q, Guo W, Yao Z, Li W. Selection of the spatial interpolation methods for precipitation in the regional soil Erosion. Res Soil Water Conserv. 2008;02:88–91.
4. Fan Y, Yu X, Zhang H, Song M. Comparison between Kirging interpolation method and inverse distance weighting interpolation for precipitation data analysis: taking Lijiang River Basin as a study case. J China Hydrol. 2014;06:61–66.
5. Shahbeik S, Afzal P, Moarefvand P, Qumarsy M. Comparison between Ordinary Kriging (OK) and Inverse Distance Weighted (IDW) based on estimation error. Case study: Dardevey iron ore deposit, NE Iran. Arab J Geosci. 2014;7(9):3693–704.
6. Nalder IA, Wein RW. Spatial interpolation of climatic normals: test of a new method in the Canadian boreal forest. Agric For Meteorol. 1998;92(4):211–25.
7. Shiode N, Shiode S. Street-level spatial interpolation using network-based IDW and Ordinary Kriging. Trans GIS. 2011;15(4):457–77.
8. Babak O, Deutsch C. Statistical approach to inverse distance interpolation. Stoch Environ Res Risk Assess. 2009;23(5):543–55.
9. Li M, Zhao H, Bao C. Application of an improving IDW interpolation in seawater quality assessment. Marine Environ Sci. 2014;02:258–61.
10. Su S, Lin AW, Liu QH. Ordinary Kriging algorithm applied for interpolation. J South Yangtze Univ. 2004;01:18–21.
11. Duan P, Sheng Y, Li J, Lv H, Zhang S. Adaptive IDW interpolation method and its application in the temperature field. Geogr Res. 2014;08:1417–6
12. Fan Z, Li J, Deng M. An adaptive inverse-distance weighting spatial interpolation method with the consideration of multiple factors. Geomat Inf Sci Wuhan Univ. 2016;06:842–7.

Adaptive Maximum a Posteriori Filtering for Relative Attitude and Position Estimation

Kai Xiong and HaoYu Zhang

Abstract In the presented algorithm, the Gaussian maximum a posteriori (MAP) filter and the traditional extended Kalman filter (EKF) are implemented in parallel to obtain the adaptive ability. One of the elemental filters, the EKF yields high precision in the scenario with low process noise, whereas the other elemental filter, the Gaussian MAP filter is adopted to utilize the measurement maximally in the presence of high process noise. The state estimates of the parallel filters are combined automatically based on the confidence for the underlying situation, such that the presented algorithm can adapt to different operation scenarios. The presented algorithm can provide precise relative attitude and position knowledge between two spacecrafts. It is applicable for many space missions, such as spacecraft formation flying, autonomous rendezvous docking and failed satellite removal. This is the first paper that presents the adaptive MAP estimator based on parallel multiple filters for spacecraft relative navigation.

Keywords Adaptive filter · MAP estimation · EKF · Spacecraft navigation

1 Introduction

The goal of filtering is to recursively estimate a vector-valued Markov process from noisy measurements. The measurement values are often inaccurate due to the instrumental errors or circumstance disturbances. It is commonly assumed that the

This study was supported in part by China Natural Science Foundation (61573059) and Beijing Natural Science Foundation (4162070).

K. Xiong · H. Zhang (✉)
Science and Technology on Space Intelligent Control Laboratory,
Beijing Institute of Control Engineering, Beijing 100190, China
e-mail: 2573767066@qq.com

K. Xiong
e-mail: tobelove@yeah.net

© Springer Nature Singapore Pte Ltd. 2018
Y. Jia et al. (eds.), *Proceedings of 2017 Chinese Intelligent Systems Conference*,
Lecture Notes in Electrical Engineering 460,
https://doi.org/10.1007/978-981-10-6499-9_13

121

measurement error sequence is subject to Gaussian distribution whose statistical property is described by its mean and covariance. The Kalman filter (KF) was one of the first filtering methods developed and it has been the dominative filter type for the past 40 years [1–3].

The KF provides the minimum mean square error (MMSE) for linear systems with Gaussian noise and can be executed within a computer iteratively without any extensive modification. The high computational efficiency has led to the widespread popularity of the KF and its variants. In order to implement the KF recursion for nonlinear systems, some approximations to the KF have been developed, such as the extended Kalman filter (EKF) [4], unscented Kalman filter (UKF) [5] and cubature Kalman filter (CKF) [6].

A problem of the KF's approximations, e.g. EKF, UKF or CKF, is that the update step for nonlinear system is based on linear equations. This linear update may lead to unsatisfactory estimates when the measurement equation is nonlinear, as the information in measurement is not utilized maximally. To deal with this problem, the maximum a posteriori (MAP) estimator is developed to improve the performance of the update step. A well-known Gaussian MAP filter is the iterated EKF (IEKF), which is based on the Gaussian-Newton optimization method [7]. An iterative version of the ensemble Kalman (IEnKF), which is a Monte Carlo method for state estimation of nonlinear models, is developed by Lorentzen and Nevdal [8]. Two Gaussian MAP filtering algorithms based on Kalman optimization (KO) are presented in [9].

In this paper, we focus on the design of an adaptive MAP filtering algorithm for nonlinear stochastic systems. The basic idea is to use two elemental filters in parallel, such that the filtering algorithm can adapt to different situations. One of the elemental filters, an EKF, is well suited for the scenario with low process noise, whereas the second elemental filter, the Gaussian MAP filter, adapts to the scenario with high process noise. For the implementation of the adaptive MAP filter, the two elemental filters are run individually to provide multiple estimates. The overall state estimate is provided through a weighted sum of the two elemental filters' estimates. The weight, which is calculated based on the measurement innovation, gives an indicator whether the corresponding elemental filter is the appropriate one for the certain scenario. The state estimate from the elemental filter with the higher likelihood is over-weighted. The weight update approach is similar to the multiple-model adaptive estimation (MMAE) [10]. The adaptive MAP filter is a suitable methodology to deal with the relative attitude and position estimation problem for a space target whose attitude stability is uncertain.

2   System Model

Consider a nonlinear stochastic system with additive Gaussian noise,

$$\mathbf{x}(k) = \mathbf{f}(\mathbf{x}(k-1)) + \mathbf{w}(k) \tag{1}$$

$$\mathbf{y}(k) = \mathbf{h}(\mathbf{x}(k)) + \mathbf{v}(k) \tag{2}$$

where $\mathbf{x}(k)$ is the state vector at time index k, $\mathbf{y}(k)$ is the measurement vector, $\mathbf{f}(\cdot)$ stands for the state transition function, $\mathbf{h}(\cdot)$ represents the measurement function, $\mathbf{w}(k) \sim N(0, \mathbf{Q}(k))$ is the process noise, and $\mathbf{v}(k) \sim N(0, \mathbf{R}(k))$ is the measurement noise, where $N(\cdot, \cdot)$ indicates a Gaussian distribution. $\mathbf{Q}(k)$ and $\mathbf{R}(k)$ are process noise covariance matrix and measurement noise covariance matrix respectively. It is assumed that the process noise, the measurement noise and the initial state value are independent of each other.

3　Adaptive MAP Filter

The Gaussian MAP filter is superior to the EKF for the system with high process noise, whereas it is worse than the EKF for the system with low process noise. Although the unfavorable effect of the noise can be partly eliminated by fine tuning the filtering parameters (such as the stopping criteria of the iterations), it is difficult to achieve superior performance in both cases by using the same filter. To cope with this difficulty, we present an adaptive MAP filtering algorithm which inherently combines the state estimates of the EKF and the Gaussian MAP filter, such that the estimator can adapt to both high and low process noise.

The adaptive MAP filter for the system shown in (1) and (2) is developed, where two elemental filters, the EKF and the Gaussian MAP filter, are run together to provide the individual state estimate. Then the two elemental filters' outputs are combined in a weighted fashion to produce the overall state estimate. The entire algorithm for time step $k = 0, 1, 2, \ldots$ works as follows.

Step 1: Initialization

The state and covariance estimates of the two elemental filters are initialized as $\widehat{\mathbf{x}}^{(1)}(0) = \widehat{\mathbf{x}}^{(2)}(0)$ and $\mathbf{P}^{(1)}(0) = \mathbf{P}^{(2)}(0)$. The superscript is introduced to distinguish the different elemental filters. The corresponding weights are set as $\omega^{(1)}(0) = \omega^{(2)}(0) = 0.5$.

Step 2: Parallel filtering

The EKF and the Gaussian MAP filter are run in parallel to predict and update the state estimate. The first elemental filter is the EKF with the equations

$$\hat{\mathbf{x}}^{(1)}(k|k-1) = \mathbf{f}\left(\hat{\mathbf{x}}^{(1)}(k-1)\right) \tag{3}$$

$$\mathbf{P}^{(1)}(k|k-1) = \mathbf{F}^{(1)}(k)\mathbf{P}^{(1)}(k-1)\mathbf{F}^{(1)}(k)^{\mathrm{T}} + \mathbf{Q}(k) \tag{4}$$

$$\widehat{\mathbf{x}}^{(1)}(k) = \widehat{\mathbf{x}}^{(1)}(k|k-1) + \mathbf{K}^{(1)}(k)\left[\mathbf{y}(k) - h\left(\widehat{\mathbf{x}}^{(1)}(k|k-1)\right)\right] \tag{5}$$

$$\mathbf{K}^{(1)}(k) = \mathbf{P}^{(1)}(k|k-1)\mathbf{H}^{(1)}(k)^{\mathrm{T}}\left[\mathbf{H}^{(1)}(k)\mathbf{P}^{(1)}(k|k-1)\mathbf{H}^{(1)}(k)^{\mathrm{T}} + \mathbf{R}(k)\right]^{-1} \tag{6}$$

$$\mathbf{P}^{(1)}(k) = \left[\mathbf{I} - \mathbf{K}^{(1)}(k)\mathbf{H}^{(1)}(k)\right]\mathbf{P}^{(1)}(k|k-1)\left[\mathbf{I} - \mathbf{K}^{(1)}(k)\mathbf{H}^{(1)}(k)\right]^{\mathrm{T}} + \mathbf{K}^{(1)}(k)\mathbf{R}(k)\mathbf{K}^{(1)}(k)^{\mathrm{T}} \tag{7}$$

where the Jacobian matrices are calculated as $\mathbf{F}^{(1)}(k) = \frac{\partial \mathbf{f}(\mathbf{x})}{\partial \mathbf{x}}\Big|_{\mathbf{x} = \widehat{\mathbf{x}}^{(1)}(k-1)}$ and $\mathbf{H}^{(1)}(k) = \frac{\partial h(\mathbf{x})}{\partial \mathbf{x}}\Big|_{\mathbf{x} = \widehat{\mathbf{x}}^{(1)}(k|k-1)}$.

The second elemental filter is the Gaussian MAP filter with the prediction step that is similar to the EKF

$$\widehat{\mathbf{x}}^{(2)}(k|k-1) = \mathbf{f}\left(\widehat{\mathbf{x}}^{(2)}(k-1)\right) \tag{8}$$

$$\mathbf{P}^{(2)}(k|k-1) = \mathbf{F}^{(2)}(k)\mathbf{P}^{(2)}(k-1)\mathbf{F}^{(2)}(k)^{\mathrm{T}} + \mathbf{Q}(k) \tag{9}$$

with $\mathbf{F}^{(2)}(k) = \frac{\partial \mathbf{f}(\mathbf{x})}{\partial \mathbf{x}}\Big|_{\mathbf{x} = \widehat{\mathbf{x}}^{(2)}(k-1)}$. To implement the update step, the KO recursion is performed with the prior estimates $\bar{\mathbf{x}}_0(k) = \widehat{\mathbf{x}}^{(2)}(k|k-1)$ and $\bar{\mathbf{P}}_0(k) = \mathbf{P}^{(2)}(k|k-1)$. The iteration procedure for $i = 1, 2, \ldots, N$ is given by

$$\bar{\mathbf{x}}_{i+1}(k) = \widehat{\mathbf{x}}^{(2)}(k|k-1) + \mathbf{L}_i(k)\left\{[\mathbf{y}(k) - h(\bar{\mathbf{x}}_i(k))] - \mathbf{H}_i(k)\left(\widehat{\mathbf{x}}^{(2)}(k|k-1) - \bar{\mathbf{x}}_i(k)\right)\right\} \\ - \alpha_i^2(\mathbf{I} - \mathbf{L}_i(k)\mathbf{H}_i(k))\widehat{\mathbf{P}}_i(k)\check{\mathbf{P}}_i(k)^{-1}\left(\widehat{\mathbf{x}}^{(2)}(k|k-1) - \bar{\mathbf{x}}_i(k)\right) \tag{10}$$

$$\bar{\mathbf{P}}_{i+1}(k) = \left(\bar{\mathbf{P}}_i(k)^{-1} + \alpha_i^{-2}\check{\mathbf{P}}_i(k)^{-1}\right)^{-1} \tag{11}$$

$$\mathbf{L}_i(k) = \widehat{\mathbf{P}}_i(k)\mathbf{H}_i(k)^{\mathrm{T}}\left(\mathbf{H}_i(k)\widehat{\mathbf{P}}_i(k)\mathbf{H}_i(k)^{\mathrm{T}} + \mathbf{R}(k)\right)^{-1} \tag{12}$$

$$\widehat{\mathbf{P}}_i(k) = \left(\mathbf{P}^{(2)}(k|k-1)^{-1} + \alpha_i^2\bar{\mathbf{P}}_i(k)^{-1}\right)^{-1} \tag{13}$$

$$\check{\mathbf{P}}_i(k) = \left(\mathbf{P}^{(2)}(k|k-1)^{-1} + \mathbf{H}_i(k)^{\mathrm{T}}\mathbf{R}(k)^{-1}\mathbf{H}_i(k)\right)^{-1} \tag{14}$$

with $\mathbf{H}_i(k) = \frac{\partial \mathbf{h}(\mathbf{x})}{\partial \mathbf{x}}\big|_{\mathbf{x} = \bar{\mathbf{x}}_i(k)}$. When the KO recursion stops, we obtain the posterior state and covariance estimates $\hat{\mathbf{x}}^{(2)}(k) = \bar{\mathbf{x}}_N(k)$ and $\mathbf{P}^{(2)}(k) = \check{\mathbf{P}}_N(k)$ for the second elemental filter.

Step 3: Weight update

The weights $\omega^{(\tau)}(k)$ $(\tau = 1, 2)$ for the two elemental filters are calculated using the innovations $\tilde{\mathbf{y}}^{(\tau)}(k)$ and normalized such that they sum up to one. This process is formulated as

$$\omega^{(1)}(k) = \frac{\omega^{(1)}(k-1)\Lambda^{(1)}(k)}{\omega^{(1)}(k-1)\Lambda^{(1)}(k) + \omega^{(2)}(k-1)\Lambda^{(2)}(k)},$$

$$\omega^{(2)}(k) = \frac{\omega^{(2)}(k-1)\Lambda^{(2)}(k)}{\omega^{(1)}(k-1)\Lambda^{(1)}(k) + \omega^{(2)}(k-1)\Lambda^{(2)}(k)}$$

(15)

with the likelihood of each elemental filter calculated as

$$\Lambda^{(\tau)}(k) = \frac{1}{\sqrt{|2\pi\mathbf{C}(k)|}}\exp\left[-\frac{1}{2}\left(\tilde{\mathbf{y}}^{(\tau)}(k)\right)^{\mathrm{T}}\mathbf{C}(k)^{-1}\tilde{\mathbf{y}}^{(\tau)}(k)\right] \quad (16)$$

where $\tilde{\mathbf{y}}^{(\tau)}(k) = \mathbf{y}(k) - \mathbf{h}\left(\hat{\mathbf{x}}^{(\tau)}(k|k-1)\right)$ is the measurement innovation, and $\mathbf{C}(k)$ is a positive definite tuning matrix. Similar likelihood has been used in MMAE algorithms by Xiong [10]. It can be seen from (15) that the elemental filter with lower innovation receives larger weight. It can be proved that the weight that corresponds to the filter with lower estimation error converges to 1, while the other converges to 0, such that the appropriate elemental filter will be chosen adaptively to dominate the estimator.

Step 4: Combination

The state estimates $\hat{\mathbf{x}}^{(\tau)}(k)$ of the two elemental filters are linearly combined with their corresponding weight $\omega^{(\tau)}(k)$, yielding the final state estimate $\hat{\mathbf{x}}(k)$ of the adaptive MAP filter, i.e.,

$$\hat{\mathbf{x}}(k) = \omega^{(1)}(k)\hat{\mathbf{x}}^{(1)}(k) + \omega^{(2)}(k)\hat{\mathbf{x}}^{(2)}(k). \quad (17)$$

Steps 2–4 are run recursively to update the state estimate and the weight for each elemental filter. The overall state estimation error covariance is not used in the recursive procedure.

4 Convergence Analysis

To demonstrate the feasibility of the presented algorithm, the weight convergence property of the adaptive MAP filter is studied for the considered nonlinear stochastic system. We expect that for the filter to converge, one of the weights is nearly equal to 1, and the other is close to 0. The covariance matrix of the state estimation error is defined as

$$\Sigma^{(\tau)}(k) = E\left[\tilde{\mathbf{x}}^{(\tau)}(k)\left(\tilde{\mathbf{x}}^{(\tau)}(k)\right)^T\right] \tag{18}$$

where $\tilde{\mathbf{x}}^{(\tau)}(k) = \mathbf{x}(k) - \tilde{\mathbf{x}}^{(\tau)}(k)$ is the estimation error. Accordingly, the prediction error is defined as $\tilde{\mathbf{x}}^{(\tau)}(k|k-1) = \mathbf{x}(k) - \tilde{\mathbf{x}}^{(\tau)}(k|k-1)$.

We evaluate the weights of the elemental filters in the case that the estimate of the first elemental filter (which is the EKF) is more accurate than that of the second elemental filter (which is the Gaussian MAP filter), i.e., $\Sigma^{(1)}(k) < \Sigma^{(2)}(k)$ for $k = 0, 1, 2, \ldots$.

From (15), the ratio of the weights is given by

$$\frac{\omega^{(2)}(n)}{\omega^{(1)}(n)} = \frac{\omega^{(2)}(n-1)\Lambda^{(2)}(n)}{\omega^{(1)}(n-1)\Lambda^{(1)}(n)}. \tag{19}$$

As the weights for the parallel filters are initialized as $\omega^{(1)}(0) = \omega^{(2)}(0) = 0.5$, the equation is written as

$$\frac{\omega^{(2)}(n)}{\omega^{(1)}(n)} = \exp\left\{-\frac{1}{2}\sum_{k=1}^{n}\left[\left(\tilde{\mathbf{y}}^{(2)}(k)\right)^T\mathbf{C}(k)^{-1}\tilde{\mathbf{y}}^{(2)}(k) - \left(\tilde{\mathbf{y}}^{(1)}(k)\right)^T\mathbf{C}(k)^{-1}\tilde{\mathbf{y}}^{(1)}(k)\right]\right\} \tag{20}$$

Let

$$\alpha^{(2)}(n) = \sum_{k=1}^{n}\left[\left(\tilde{\mathbf{y}}^{(2)}(k)\right)^T\mathbf{C}(k)^{-1}\tilde{\mathbf{y}}^{(2)}(k) - \left(\tilde{\mathbf{y}}^{(1)}(k)\right)^T\mathbf{C}(k)^{-1}\tilde{\mathbf{y}}^{(1)}(k)\right] \tag{21}$$

For simplicity, the tuning matrix $\mathbf{C}(k)$ is set as a diagonal matrix $\mathbf{C}(k) = \sigma_v^2\mathbf{I}$. The Eq. (22) becomes

$$\alpha^{(2)}(n) = \frac{1}{\sigma_v^2}\sum_{k=1}^{n}\left\{\text{tr}\left[\tilde{\mathbf{y}}^{(2)}(k)\left(\tilde{\mathbf{y}}^{(2)}(k)\right)^T - \tilde{\mathbf{y}}^{(1)}(k)\left(\tilde{\mathbf{y}}^{(1)}(k)\right)^T\right]\right\}. \tag{22}$$

The expectation of $\alpha^{(2)}(n)$ is formulated as

$$E\left(\alpha^{(2)}(n)\right) = \frac{1}{\sigma_v^2} \sum_{k=1}^{n} \left[\text{tr}\left(\boldsymbol{\Omega}^{(2)}(k) - \boldsymbol{\Omega}^{(1)}(k)\right)\right] \qquad (23)$$

where

$$\boldsymbol{\Omega}^{(\tau)}(k) = E\left[\widetilde{\mathbf{y}}^{(\tau)}(k)\left(\widetilde{\mathbf{y}}^{(\tau)}(k)\right)^{\mathrm{T}}\right] \qquad (24)$$

In terms of the linearized model [11],

$$\widetilde{\mathbf{y}}^{(\tau)}(k) = \mathbf{H}(k)\mathbf{F}(k)\widetilde{\mathbf{x}}^{(\tau)}(k-1) + \mathbf{H}(k)\mathbf{w}(k) + \mathbf{v}(k). \qquad (25)$$

Then we can know that

$$\boldsymbol{\Omega}^{(\tau)}(k) = \mathbf{H}(k)\mathbf{F}(k)\boldsymbol{\Sigma}^{(\tau)}(k-1)\mathbf{F}(k)^{\mathrm{T}}\mathbf{H}(k)^{\mathrm{T}} + \mathbf{H}(k)\mathbf{Q}(k)\mathbf{H}(k)^{\mathrm{T}} + \mathbf{R}(k) \qquad (26)$$

It is easy to see from (26) that the value of the innovation covariance $\boldsymbol{\Omega}^{(\tau)}(k)$ depends on the estimation error covariance $\boldsymbol{\Sigma}^{(\tau)}(k-1)$. As it is assumed that $\boldsymbol{\Sigma}^{(1)}(k) < \boldsymbol{\Sigma}^{(2)}(k)$, we obtain

$$\boldsymbol{\Omega}^{(1)}(k) < \boldsymbol{\Omega}^{(2)}(k) \qquad (27)$$

for $k = 1, 2, \cdots$. Let

$$\Delta_{\min}^{(2)} = \min_k\left[\text{tr}\left(\boldsymbol{\Omega}^{(2)}(k) - \boldsymbol{\Omega}^{(1)}(k)\right)\right]. \qquad (28)$$

It can be seen from (27) and (28) that $\Delta_{\min}^{(2)} > 0$. From (23) and (28),

$$E\left(\alpha^{(2)}(n)\right) \geq \frac{1}{\sigma_v^2}\Delta_{\min}^{(2)}n. \qquad (29)$$

As $\Delta_{\min}^{(2)}$ is a positive constant, it indicates that $\alpha^{(2)}(n)$ will increase as time goes on. From the recursive relation in (20), we expect that as $n \to \infty$, $\omega^{(2)}(n) \to 0$. Thus, we infer that as $n \to \infty$, $\omega^{(1)}(n) \to 1$.

From the prior analysis, we found that the adaptive MAP filter will converge to the first elemental filter (which is the EKF) adaptively if the state estimate of the first elemental filter is more accurate than that of the second one. In other words, the appropriate elemental filter can be chosen based on the presented adaptive law to suit for different situations.

5 Relative Attitude and Position Estimation

Many space missions, such as spacecraft formation flying, autonomous rendezvous docking and failed satellite removal, rely on precise relative attitude and position knowledge between two spacecrafts [12, 13]. The binocular vision based navigation sensor can provide reliable measurement information for the relative spacecraft attitude and position estimation system. The binocular vision based navigation sensor is mounted on a chase spacecraft, and it is used to observe a target spacecraft. The measurements of the navigation sensor are relative position vectors of the feature points on the target spacecraft. Suppose that the locations of the feature points in the target spacecraft's body frame are known a priori. The chase spacecraft acquires its own absolute position and attitude information by using the global positioning system (GPS), star sensor and gyroscopes onboard the spacecraft. As the target spacecraft may be a failed satellite, and the gyroscopes measurement (or measurement from other attitude sensor) on a failed satellite is not achievable, the scheme presented in this paper does not rely on the gyroscope measurement from the target spacecraft. The dynamic model is adopted to predict the attitude of the target spacecraft. The magnitude of the process noise depends on the attitude stability of the target spacecraft. The attitude stability is dissimilar for different target spacecraft, and it is often unknown for the practitioners before the implement of the space mission.

The filtering algorithm is designed based on the system model to determine the relative position and attitude between the target spacecraft and the chase spacecraft. A dynamic system has the following components: the state, the measurement, the state propagation equation and the measurement equation. For the considered system, relative position $\boldsymbol{\rho}$, relative velocity $\dot{\boldsymbol{\rho}}$, relative attitude $\mathbf{q}_{ct}$ and target spacecraft absolute angular velocity $\boldsymbol{\omega}_{tb}$ are defined as the state vector. The state propagation equation and the measurement equation are nonlinear. To provide a set of linear equations for the covariance propagation, we employ the linearization approach. The error-state dynamics equation is given by

$$\Delta \dot{\mathbf{x}} = \mathbf{F} \Delta \mathbf{x} + \mathbf{w} \tag{30}$$

with

$$\Delta \mathbf{x} = \begin{bmatrix} \Delta \boldsymbol{\rho}^{\mathrm{T}} & \Delta \dot{\boldsymbol{\rho}}^{\mathrm{T}} & \Delta \mathbf{q}_{ct}^{\mathrm{T}} & \Delta \boldsymbol{\omega}_{tb}^{\mathrm{T}} \end{bmatrix}^{\mathrm{T}} \tag{31}$$

$$\mathbf{F} = \begin{bmatrix} \mathbf{F}_{\rho 6 \times 6} & \mathbf{0}_{6 \times 6} \\ \mathbf{0}_{6 \times 6} & \mathbf{F}_{q 6 \times 6} \end{bmatrix} \tag{32}$$

$$\mathbf{F}_{\rho 6 \times 6} = \begin{bmatrix} \mathbf{0}_{3 \times 3} & \mathbf{I}_{3 \times 3} \\ \boldsymbol{\Phi}_{21} & \boldsymbol{\Phi}_{22} \end{bmatrix} \tag{33}$$

$$\mathbf{\Phi}_{21} = \begin{bmatrix} 0 & 0 & 0 \\ 0 & -\omega_0^2 & 0 \\ 0 & 0 & 3\omega_0^2 \end{bmatrix}, \mathbf{\Phi}_{22} = \begin{bmatrix} 0 & 0 & 2\omega_0 \\ 0 & 0 & 0 \\ -2\omega_0 & 0 & 0 \end{bmatrix} \tag{34}$$

$$\mathbf{F}_{q6 \times 6} = \begin{bmatrix} -[\mathbf{\omega}_{cb} \times] & -0.5\mathbf{A}(\mathbf{q}_{ct}) \\ \mathbf{0}_{3 \times 3} & \mathbf{I}_t^{-1}\{[(\mathbf{I}_t \mathbf{\omega}_{tb}) \times] - [\mathbf{\omega}_{tb} \times]\mathbf{I}_t\} \end{bmatrix} \tag{35}$$

where $\Delta \mathbf{x}$ is the error-state vector, $\mathbf{w}$ is the process noise, $\Delta \mathbf{\rho}$ is the relative position error vector, $\Delta \mathbf{q}_{ct}$ is the vector part of the relative attitude quaternion error that represents the rotation from the estimated relative quaternion to the true relative quaternion, $\Delta \mathbf{\omega}_{tb}$ is the angular velocity error of the target spacecraft, ω_0 is the orbital angular velocity of the target spacecraft, which can be calculated by using the known orbit information of the chase spacecraft and the relative position esti-mate, $\mathbf{\omega}_{cb}$ is the known angular velocity of the chase spacecraft, $\mathbf{q}_{ct}$ is the relative attitude quaternion that maps the target spacecraft body frame to the chase space-craft body frame, $\mathbf{I}_t$ is the known inertia matrix of the target spacecraft. The attitude matrix $\mathbf{A}(\mathbf{q}_{ct})$ is related to the relative attitude quaternion by

$$\mathbf{A}(\mathbf{q}_{ct}) = \mathbf{\Xi}(\mathbf{q}_{ct})^T \mathbf{\Psi}(\mathbf{q}_{ct}) \tag{36}$$

with

$$\mathbf{\Xi}(\mathbf{q}_{ct}) = \begin{bmatrix} q_{ct4}\mathbf{I}_{3 \times 3} + [\mathbf{\varrho}_{ct} \times] \\ -\mathbf{\varrho}_{ct}^T \end{bmatrix} \tag{37}$$

$$\mathbf{\Psi}(\mathbf{q}_{ct}) = \begin{bmatrix} q_{ct4}\mathbf{I}_{3 \times 3} - [\mathbf{\varrho}_{ct} \times] \\ -\mathbf{\varrho}_{ct}^T \end{bmatrix} \tag{38}$$

for the relative attitude quaternion defined by $\mathbf{q}_{ct} = \begin{bmatrix} \mathbf{\varrho}_{ct}^T & q_{ct4} \end{bmatrix}^T$ with $\mathbf{\varrho}_{ct} = \begin{bmatrix} q_{ct1} & q_{ct2} & q_{ct3} \end{bmatrix}^T$. The linearized state-space model shown in (49) is derived by using the nonlinear relative attitude kinematics equation and the well-known CW (Clohessy and Wiltshire) equation. As similar expressions can be found in [12], the nonlinear differential equations are not repeated here for brevity.

The binocular vision based navigation sensor consists of two optical cameras mounted with a fixed distance and known location, which is used to measure the relative position vectors of the feature points on the target spacecraft. Without loss in generality, it is assumed that the sensor frame coincides with the chase spacecraft body frame. The measurement model is written as

$$y = \begin{bmatrix} \mathbf{A}(\mathbf{q}_{ct})\boldsymbol{\rho}_{in1} - \mathbf{C}_o^b\boldsymbol{\rho} \\ \mathbf{A}(\mathbf{q}_{ct})\boldsymbol{\rho}_{in2} - \mathbf{C}_o^b\boldsymbol{\rho} \\ \vdots \\ \mathbf{A}(\mathbf{q}_{ct})\boldsymbol{\rho}_{inM} - \mathbf{C}_o^b\boldsymbol{\rho} \end{bmatrix} + \mathbf{v} \tag{39}$$

where $\mathbf{y}$ is the measurement of the relative position vector of the feature points on the target spacecraft, $\boldsymbol{\rho}_{in1}$, $\boldsymbol{\rho}_{in2}$, ..., $\boldsymbol{\rho}_{inM}$ are the known locations of the feature points within the target spacecraft's body frame, M is the number of the feature points, $\mathbf{C}_o^b$ is the attitude matrix from the target spacecraft orbital frame to the chase spacecraft body frame, which can be calculated by using the known orbit and attitude information of the chase spacecraft and the relative position estimate of the target spacecraft, $\boldsymbol{\rho}$ is the relative position vector between the target spacecraft and the chase spacecraft, $\mathbf{v}$ is the measurement noise. With the error-state vector defined in (50), the linearized measurement equation is derived as

$$\Delta y = \mathbf{H}\Delta \mathbf{x} + \mathbf{v} \tag{40}$$

where Δy is the measurement innovation. The measurement sensitivity matrix $\mathbf{H}$ is derived as:

$$\mathbf{H} = \begin{bmatrix} -\mathbf{C}_o^b & 0_{3\times3} & 2[\mathbf{A}(\mathbf{q}_{ct})\boldsymbol{\rho}_{in1} \times] & 0_{3\times3} \\ -\mathbf{C}_o^b & 0_{3\times3} & 2[\mathbf{A}(\mathbf{q}_{ct})\boldsymbol{\rho}_{in2} \times] & 0_{3\times3} \\ \vdots & \vdots & \vdots & \vdots \\ -\mathbf{C}_o^b & 0_{3\times3} & 2[\mathbf{A}(\mathbf{q}_{ct})\boldsymbol{\rho}_{inM} \times] & 0_{3\times3} \end{bmatrix}. \tag{41}$$

The EKF, the Gaussian MAP filter and the adaptive MAP filter algorithms can now be implemented using the linearized model for the covariance propagation as well as the original nonlinear model for the state propagation.

6 Simulation Results

In the simulation, the two spacecraft are assumed to fly at a 7100 km near-circular earth orbit with an inclination of 48°. The distance between the target spacecraft and the chase spacecraft is about 100 m. The absolute angular velocity of the target spacecraft is 2°/s. The true orbit data of the target spacecraft and the chase spacecrafts are produced using a numerical orbit propagator. With the simulated orbit and attitude data, the time history of the measurements is generated according to the measurement model shown in (39). The filtering algorithms operate on the measurements and produce state estimates. The estimation error curves are obtained by comparing the estimates with the true states.

The simulation time for the relative attitude and position estimation is about 3 orbital periods of the target spacecraft. The update rate for the binocular vision based navigation sensor is 1 s, and the standard deviation of the measurement noise is 1 cm. Assume that three feature points are identified by the navigation sensor. The locations of the feature points within the target spacecraft's body frame is given by

$$\boldsymbol{\rho}_{\text{in1}} = [\, -1 \quad 0 \quad 0\,]^{\text{T}}\text{m}, \ \boldsymbol{\rho}_{\text{in2}} = [\,1 \quad 0 \quad 0\,]^{\text{T}}\text{m}, \ \boldsymbol{\rho}_{\text{in3}} = [\,0 \quad 1 \quad 0\,]^{\text{T}}\text{m}$$

The initial state errors are sampled from zero-mean distribution with the covariance matrix given by

$$\mathbf{P}(0) = \begin{bmatrix} \sigma_\rho^2 \mathbf{I}_{3\times3} & & & \\ & \sigma_{\dot{\rho}}^2 \mathbf{I}_{3\times3} & & \\ & & \sigma_q^2 \mathbf{I}_{3\times3} & \\ & & & \sigma_\omega^2 \mathbf{I}_{3\times3} \end{bmatrix}$$

where $\sigma_\rho = 10$ m, $\sigma_{\dot{\rho}} = 0.01$ m/s, $\sigma_q = 0.5°$, $\sigma_\omega = 0.1°/\text{s}$. In this simulation, we analyze two scenarios that differ in the process noise covariance. For scenario 1, the standard deviation of the process noise for the absolute angular velocity of the target spacecraft is 0.001°/s. For scenario 2, the corresponding process noise standard deviation is 0.1°/s.

In the case that the process noise is rather low, the EKF and the adaptive MAP filter have roughly the same performance, and outperform the Gaussian MAP filter. As the error curve of the EKF coincides with that of the adaptive MAP filter, it is not included in Fig. 1. For the Gaussian MAP filter, the iteration performs several

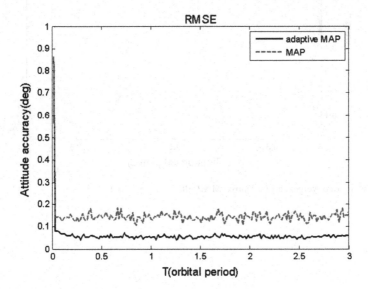

Fig. 1 RMS relative attitude error curves against time for scenario 1

corrections with the same measurement. Thus, it is sensitive to the measurement noise, and this leads to a higher estimation error of the filter.

The weights of the two elemental filters in the adaptive MAP filter for scenario 1 are plotted versus time in Fig. 2. As can be seen in the figure, the weight for the EKF converges toward 1, while the weight for the Gaussian MAP filter converges to 0. It illustrates that the better elemental filter dominants the estimator.

Next, we compare the performance of the filtering algorithms in scenario 2. The estimation results of the EKF and the adaptive MAP filter are shown in Fig. 3. In the presence of relatively high process noise, the adaptive MAP filter is still the better algorithm, while the EKF has a higher estimation error. This result is understood in that the measurement information is used adequately in the adaptive MAP filter. The error curve of the Gaussian MAP filter is similar to that of the adaptive MAP filter and it is not presented here.

Figure 4 shows the convergence of the weights for the adaptive MAP filter in scenario 2. The adaptive MAP filter converges to the second elemental filter (which is the Gaussian MAP filter) consequently. As the Gaussian MAP filter is effective to utilize the measurement information maximally in this scenario, the simulation result shows that the appropriate elemental filter can be chosen adaptively.

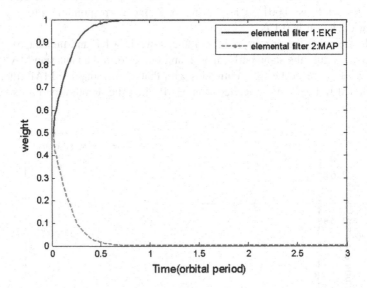

Fig. 2 Weight convergence of adaptive MAP filter for scenario 1

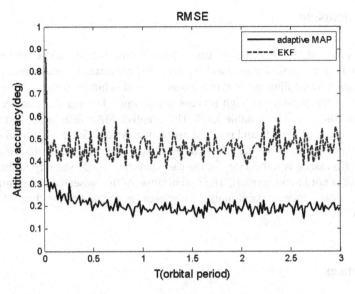

Fig. 3 RMS relative attitude error curves against time for scenario 2

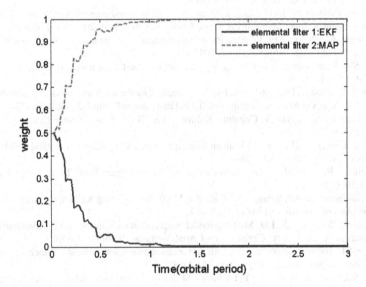

Fig. 4 Weight convergence of adaptive MAP filter for scenario 2

7 Conclusion

The adaptive MAP filter, which can adapt to different operation scenarios, is developed for the nonlinear stochastic system. The presented algorithm outperforms the Gaussian MAP filter for low process noise, and achieves better accuracy than the EKF in the presence of high process noise, while keeping the number of the elemental filters on a reasonable level. The adaptive MAP filter is applied for the relative spacecraft attitude and position estimation system. Simulation results show that the appropriate elemental filter can be chosen adaptively, and the overall filtering performance is satisfactory in the case that the attitude stability of the target spacecraft is not known a priori. The optimization of the presented algorithm in the case of variant measurement noise covariance is an interesting line of future research.

References

1. Wang R, Liu J, Xiong Z, Zeng Q. Double-layer fusion algorithm for EGI-based system. Aircraft Eng Aerosp Technol. 2013;85(4):258–66.
2. Zhou G, Pelletier M, Kirubarajan T, Quan T. Statically fused converted position and doppler measurement Kalman filters. IEEE Trans Aerosp Electron Syst. 2014;50(1):300–18.
3. Panferov AI, Ponomarev V. Algorithms for estimation of parameters of micromechanical gyros. Aircraft Eng Aerosp Technol. 2015;87(6):557–62.
4. Kay SM. Fundamentals of statistical signal processing: estimation theory. Englewood Cliffs, NJ: Pientice-Hall; 1993.
5. Julier S, Uhlmann J, Durrant-Whyte HF. A new method for the nonlinear transformation of means and covariances in filters and estimators. IEEE Trans Autom Control. 2000;45(3):477–82.
6. Arasaratnam I, Haykin S. Cubature Kalman filter. IEEE Trans Autom Control. 2009;54 (6):1254–69.
7. Bell B, Cathey F. The iterated Kalman filter update as a Gauss-Newton method. IEEE Trans Autom Control. 1993;38(2):294–7.
8. Lorentzen RJ, Nevdal G. An iterative ensemble Kalman filter. IEEE Trans Autom Control. 2011;56(8):1990–5.
9. Garcia-Fernandez AF, Svensson L. Gaussian MAP filtering using Kalman optimization. IEEE Trans Autom Control. 2015;60(5):1336–49.
10. Xiong K, Wei CL, Liu LD. Multiple-model adaptive estimation for space surveillance with measurement uncertainty. Optim Control Appl Methods. 2016;37:404–23.
11. Kwon SJ. Robust Kalman filtering with perturbation estimation process for uncertain system. IEE Proc Control Theory Appl. 2006;153(5):600–6.
12. Kim SG, Crassidis JL, Cheng Y, Fosbury AM. Kalman filtering for relative spacecraft attitude and position estimation. J Guid Control Dyn. 2007;30(1):133–43.
13. Zhang L, Yang H, Zhang S, Cai H. Kalman filtering for relative spacecraft attitude and position estimation: a revisit. J Guid Control Dyn. 2014;37(5):1706–11.

A Fourth-Order Current Adaptive Model for Online Denoising by Kalman Filter

Sheng-lun Yi, Xue-bo Jin, Ting-li Su and Qiang Cai

Abstract Dealing with noisy time series is a significant task in many data-driven real-time applications. In order to improve the performance of time series data, an important pre-processing step is the online denoising of data before performing any action. In this paper, a novel method was proposed to dispose the noisy time series data based on a fourth-order current adaptive model (FCAM), which can capture the feature of high unstable time series data. The proposed model consists of two parts. The first one is to estimate the system state within FCAM. The second one is to update the adaptive parameter in the FCAM based on the Yule-Walker algorithm. Finally, the favorable denoising effectiveness of the method was verified by the simulation experiment.

Keywords Fourth-order current adaptive model (FCAM) · Kalman filter · Yule-Walker algorithm · Time series data processing · Online denoising

1 Introduction

Real-time data-driven systems usually make use of the discrete valued time series data and their performance is highly dependent on the accuracy of such data. However, these systems may be subject to various kinds noise, which is the issues on account of the loss or delay during transmission, incorrect measurements or imperfect

S. Yi · X. Jin · T. Su · Q. Cai (✉)
School of Computer Information and Engineering, Beijing Technology and Business University, Beijing, China
e-mail: caiq@th.btbu.edu.cn

S. Yi
e-mail: yishenglun@st.btbu.edu.cn

X. Jin
e-mail: jinxuebo@btbu.edu.cn

T. Su
e-mail: sutingli@btbu.edu.cn

© Springer Nature Singapore Pte Ltd. 2018
Y. Jia et al. (eds.), *Proceedings of 2017 Chinese Intelligent Systems Conference*, Lecture Notes in Electrical Engineering 460, https://doi.org/10.1007/978-981-10-6499-9_14

data collection. Therefore, in order to improve the performance of these real-time data-driven systems, online denoising is crucial in practical systems such as disease case count prediction [1], financial forecasting [2] or tourism forecasting [3].

Focus on online denoising discrete denoising problem to deal with the issue of data correction in time-sensitive applications. For this purpose, Muneyasu et al. [4] proposed a novel type of edge-preserving smoothing filter to be applied to an image corrupted by impulsive and white Gaussian noise. Kim [5] provided a method of reducing noise in a medical image while maintaining structure characteristics within the medical image. Nevertheless, these methods can not obtain denoising effectiveness when the data possessed the high unstable fluctuation.

On the other hand, Kalman filter, as a popular estimation method, has been widely used in different kinds of online applications. However, the optimality of Kalman filters is highly depending on the assumptions that the linear dynamic system model is precisely obtained a prior, and the noise of process and measurement are zero-mean, jointly standard Gaussian with known covariance matrices. A good dynamic model of the system will certainly facilitate this information extraction to a great extent. Therefore, the stand or fall of the denoising effect is determined by the choice of dynamic model.

Various models have been developed over the past three decades. The simplest dynamic model is the so-called white-noise acceleration model [6]. The main attractive feature of this model is its simplicity. It is sometimes used when the time series is smooth or steady and the noise is white. Next, the slightly complicated model is the so-called Wiener-process acceleration model [6]. Afterwards, more and more models were proposed such as singer acceleration model (SAM) [7], mean-adaptive acceleration model (MACM) [8], asymmetrically distributed normal acceleration model (ADNAM) [9] and so on.

Based on the models, there are lots of modified models for different time series data, but all of them accessed a fixed parameter value. In addition, these models are in essence a priori model since these do not use online information about the time series data. Thus, they are inapplicable in the online denoising for high unstable time series data. Therefore, we proposed a fourth-order current adaptive model (FCAM). In this model, we can adjust the process matrix and process noise at same moments because its joint state-and-parameter estimation is to get accurate parameter (and state) estimation in complex situations using synchrophasor data [10, 11], where the original nonlinear parameter estimation problem is reformulated as two loosely-coupled linear subproblems: state estimation and parameter update, respectively.

Compared to previous work, the contribution of this notes is that we used a FCAM for online denoising, which can obtain the better denoising performance for the time series data. The comparison between our model and the third-order current adaptive model (TCAM) would be given in Sect. 4, and the results show that the developed FCAM can obtain the better denoising effect.

The structure of this paper is as follows. Section 2 came up with a fourth-order current adaptive model. The algorithm flow of the online denoising is proposed in

Sect. 3. The overview of the simulation experiment is provided in Sect. 4. Section 5 confirms the preponderance of the proposed method.

2   A Fourth-Order Current Adaptive Model

In this section, we introduced a fourth-order current adaptive model (FCAM), which utilized the fourth-order matrix to describe state vector. The FCAM differs from TCAM usually in smoothly transitional time series data, relying on whether the data variation is better described by a discrete time zero-mean Markov process. This model can preferably capture the change characters and the trends of the high unstable time series.

With regard to the model structure, state vector is always taken to be a fourth-order matrix as $x = [x, \dot{x}, \ddot{x}, \dddot{x}]$. Then we can get the discrete-time equivalent as the following:

$$x(k+1) = \Phi(k+1,k)x(k) + U(k)\overline{\ddot{x}}(k) + w(k) \tag{1}$$

where the $\overline{\ddot{x}}$ is the mean of $\dddot{x}$ in the current time period, whose value is set as the predicted value, i.e., $\dddot{x}(k|k+1)$, $w(k)$ is the Gaussian white noise with zero mean and variance of 'Q'. The process matrix such as $\Phi(k+1,k)$, $U(k)$ and the process noise $Q(k)$ are as the following:

$$\Phi(k+1,k) = \begin{bmatrix} 1 & T_0 & \frac{T_0^2}{2} & p_1 \\ 0 & 1 & T_0 & q_1 \\ 0 & 0 & 1 & r_1 \\ 0 & 0 & 0 & s_1 \end{bmatrix} \tag{2}$$

We have

$$
\begin{aligned}
p_1 &= \frac{2 - 2\alpha T_0 + \alpha^2 T_0^2 - 2e^{-\alpha T_0}}{2\alpha^3} \\
q_1 &= \frac{e^{-\alpha T_0} - 1 + \alpha T_0}{\alpha^2} \\
r_1 &= \frac{1 - e^{-\alpha T_0}}{\alpha} \\
s_1 &= e^{-\alpha T_0}
\end{aligned}
\tag{3}
$$

$$U(k) = \begin{bmatrix} \frac{6}{\alpha^3}(\alpha^3 T_0^3 - 3\alpha^2 - 6 + 6\alpha T_0 + 6e^{-\alpha T_0}) \\ \frac{1}{\alpha}\left(-T_0 + \frac{\alpha T_0^2}{2} + \frac{1-e^{-\alpha T_0}}{\alpha}\right) \\ T_0 - \frac{1-e^{-\alpha T_0}}{\alpha} \\ 1 - e^{-\alpha T_0} \end{bmatrix} \tag{4}$$

$$Q = 2\alpha \delta_w^2 \begin{bmatrix} q_{11} & q_{12} & q_{13} & q_{14} \\ q_{21} & q_{22} & q_{23} & q_{24} \\ q_{31} & q_{32} & q_{33} & q_{34} \\ q_{41} & q_{42} & q_{43} & q_{44} \end{bmatrix} \tag{5}$$

We have

$$q_{11} = \frac{1}{2\alpha^7}\left[\frac{\alpha^5 T_0^5}{10} - \frac{\alpha^4 T_0^4}{2} + \frac{4\alpha^3 T_0^3}{3} - 2\alpha^2 T_0^2 + 2\alpha T_0 - 3 + 4e^{-\alpha T_0} + 2\alpha^2 T_0^2 e^{-\alpha T_0} - e^{-2\alpha T_0}\right]$$

$$q_{12} = \frac{1}{2\alpha^6}\left[1 - 2\alpha T_0 + 2\alpha^2 T_0^2 - \alpha^3 T_0^3 + \frac{\alpha^4 T_0^4}{4} + e^{-2\alpha T_0} + 2\alpha T_0 e^{-\alpha T_0} - 2e^{-\alpha T_0} - \alpha^2 T_0^2 e^{-\alpha T_0}\right]$$

$$q_{13} = \frac{1}{2\alpha^5}\left[2\alpha T_0 - \alpha^2 T_0^2 - \frac{\alpha^3 T_0^3}{3} - 3 - 2e^{-2\alpha T_0} + 4e^{-\alpha T_0} + \alpha^2 T_0^2 e^{-\alpha T_0}\right]$$

$$q_{14} = \frac{1}{2\alpha^4}\left[1 + e^{-2\alpha T_0} - 2e^{-\alpha T_0} - 2\alpha^2 T_0^2 e^{-\alpha T_0}\right]$$

$$q_{22} = \frac{1}{2\alpha^5}\left[1 - e^{-2\alpha T_0} + \frac{2\alpha^3 T_0^3}{3} + 2\alpha T_0 - 2\alpha^2 T_0^2 - 4\alpha T_0 e^{-\alpha T_0}\right]$$

$$q_{23} = \frac{1}{2\alpha^4}\left[1 + \alpha^2 T_0^2 - 2\alpha T_0 + 2\alpha T_0 e^{-\alpha T_0} + e^{-2\alpha T_0} - 2e^{-\alpha T_0}\right]$$

$$q_{24} = \frac{1}{2\alpha^3}\left[1 - e^{-2\alpha T_0} - 2\alpha T_0 e^{-2\alpha T_0}\right]$$

$$q_{33} = \frac{1}{2\alpha^3}\left[4e^{-\alpha T_0} - e^{-2\alpha T_0} + 2\alpha T_0 - 3\right]$$

$$q_{34} = \frac{1}{2\alpha^2}\left[e^{-2\alpha T_0} + 1 - 2\alpha T_0\right]$$

$$q_{44} = \frac{1}{2\alpha}\left[1 - e^{-2\alpha T_0}\right]$$

$$\tag{6}$$

where α and δ_w^2 are the adaptive parameters to depict the variation size of the data waveform, and the adaptive parameters are updated by Yule-Walker algorithm [12] as the following

$$l_k(1) = l_{k-1}(1) + \frac{1}{k}\left[\ddot{x}(k|k)\ddot{x}(k-1|k-1) - l_{k-1}(1)\right] \tag{7}$$

$$l_k(0) = l_{k-1}(0) + \frac{1}{k}\left[\ddot{x}(k|k)\ddot{x}(k|k) - l_{k-1}(0)\right] \tag{8}$$

and

$$\delta_{\bar{x}w}^2(k) = l_k(0) - \alpha(k)l_k(1) \tag{9}$$

$$\alpha(k) = -\frac{1}{T}\ln\frac{l_k(1)}{l_k(0)} \tag{10}$$

$$\delta_w^2(k) = \frac{\delta_{\bar{x}w}^2(k)}{[(1 - (\frac{l_k(1)}{l_k(0)})^2]} \tag{11}$$

Then the parameter α and δ_w^2 are replaced by $\alpha(k)$ and $\delta_w^2(k)$ in Eqs. (9)–(11) and use Eqs. (2)–(6) to get the system parameter $\Phi(k+1, k)$, $U(k)$ and $Q(k)$.

3 Online Denoiser by Kalman Filter

Considering the discrete data-driven systems:

$$x(k+1) = \Phi(k+1, k)x(k) + U(k)\bar{x}(k) + w(k) \tag{12}$$

$$z(k+1) = H(k)x(k+1) + v(k+1) \tag{13}$$

where x is the state vector of the system to be estimated, whose initial value and covariance are known as x_0 and P_0. $\Phi(k+1, k)$ is the state-transition matrix. $U(k)$ is the system input matrix. $w(k)$ and $v(k)$ are the process noise and measurement noise respectively, and the variance of $v(k)$ is known (as R). Note that both $w(k)$ and $v(k)$ are white noise with zero mean and independent of the initial state x_0. $z(k)$ is the observation vector and $H(k)$ is the observation matrix.

The Kalman filtering considers the correlation between errors in the prediction and the measurements. The algorithm is in a predict-correct form, which is convenient for implementation:

(1) Initialization: $k = 0$

$$\hat{x}(0|0) = x_0 \, P(0|0) = P_0 \tag{14}$$

$$\alpha(0) = \alpha_0 \delta_w^2(0) = \delta_{w0}^2 \, l_0(0) = \ddot{x}_0 \cdot \ddot{x}_0 l_0(1) = \ddot{x}_0 \tag{15}$$

(2) Recursion: $k = k + 1$

   (a) Prediction:

$$\hat{x}(k+1|k) = \Phi(k+1,k)\hat{x}(k|k) + U(k)\bar{\bar{x}} \tag{16}$$

$$P(k+1|k) = \Phi(k+1,k)P(k|k)\Phi^T(k+1,k) + Q(k) \tag{17}$$

   (b) State update:

$$\hat{x}(k+1|k+1) = \hat{x}(k+1|k) + K(k+1)[y(k+1) - H(k+1)\hat{x}(k+1|k)] \tag{18}$$

$$K(k+1) = P(k+1|k)H^T(k+1)[H(k+1)P(k+1|k)H^T(k+1) + R]^{-1} \tag{19}$$

$$P(k+1|k+1) = [I - K(k+1)H(k+1)]P(k+1|k) \tag{20}$$

The algorithm works in a two-step process. In the prediction step, the Kalman filter produces estimates of the current state variables along with their uncertainties. Once the outcome of the next measurement (necessarily corrupted with some amount of error, including random noise) is observed, these estimates are updated using a weighted average, with more weight being given to estimates with higher certainty. Since the algorithm could run recursively, we can implement it step by step, that is, the denoised data can be obtained in real time.

Using the method described in this section, online denoising of time series data with the high unstable characteristic was then accomplished. The flow chart of the proposed method was given in Fig. 1. It can be seen that the method consists of two parts within a closed loop. The first one is to estimate the system state with the Kalman filter based on the FCAM, and the other is to update the adaptive parameter in the model by the Yule-Walker algorithm.

4   Simulation Experiment

In order to verify the effectiveness of the proposed algorithm, the simulation experiment would be come up with a set of simulation time series data, which possessed high unstable characteristic with colored noise. As we can be clearly seen from Fig. 2, the colored noise of the time series data is extraordinary huge.

In case 1, the online denoising algorithm based on the FCAM was used to deal with noisy time series. In case 2, we compared the proposed method with the method based on the TCAM.

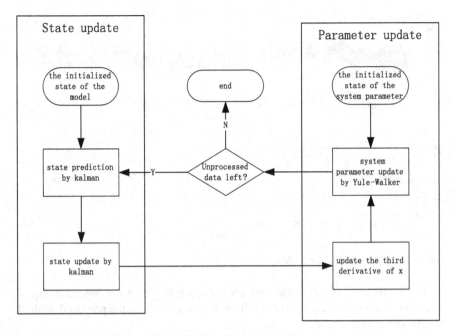

Fig. 1 The flow chart of the proposed online denoising method

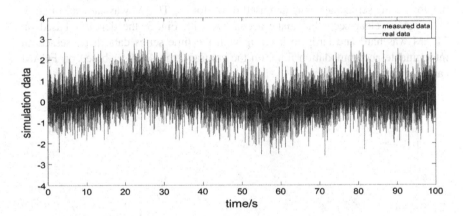

Fig. 2 Simulation data of measured value and real value

Case 1. Simulation experiment of Kalman filtering algorithm based on FCAM

In this paper, we utilized the proposed method to deal with the noisy time series data. The results in Fig. 3 demonstrated the commendable denoising effect of the proposed method in this paper. The useful information would be extracted by the recursive estimation.

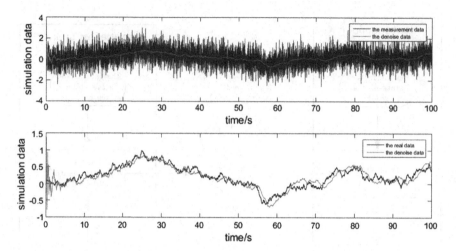

Fig. 3 Denoising effect of the FCAM

In order to better describe the error and compare the denoising precision, Fig. 4 gives the error of the FCAM. The results in Fig. 4 shows that the proposed method has the reasonable denoising effect.

Case 2. The denoising effect comparison of FCAM and TCAM

We compared FCAM with TCAM to deal with the online denoising problem for the noisy time series data. The denoised result for the TCAM was shown in Fig. 5.

It can be clear seen that make a contrast with Fig. 3, the denoiser based on TCAM was hard and tardy to keep up with the time series data. It possessed an interminable accommodation time, on the contrary, the denoiser based on FCAM can be quick adjustment.

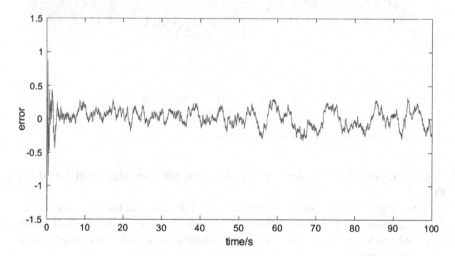

Fig. 4 Error for the FCAM

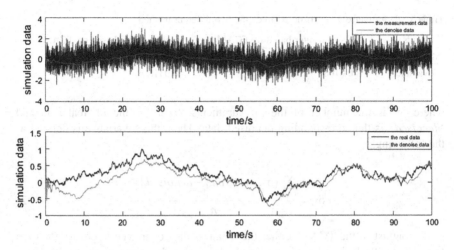

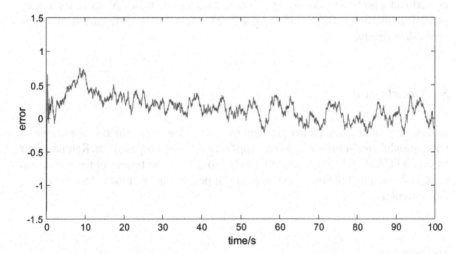

Fig. 5 Denoising effect of the TCAM

Fig. 6 Error for the TCAM

In addition, the error of the real data and the denoise data was shown in Fig. 6.

In order to evaluate the superiority of the proposed method in this paper, mean and covariance of the error were compared with Kalman filtering based on TCAM. Specifically, 'mean' here represents averaged absolute value of difference between the real data and the denoised data, i.e.,

$$\text{mean} = \frac{\sum_{i=1}^{n} |real_i - denoised_i|}{n} \tag{21}$$

Table 1 Performance comparison between the two different methods

Methods	Kalman filter based on FCAM	Kalman filter based on TCAM
Mean	0.1085	0.1197
Covariance	0.0168	0.0318

where 'n' is the number of the measurements, '$real_i$' is the ith real data and '$denoised_i$' is the corresponding denoised data. Then, the covariance is defined as the following:

$$\text{cov} = \frac{\sum_{i=1}^{n} (\text{mean} - |real_i - denoised_i|)^2}{n} \tag{22}$$

In contrast to the TCAM, either the mean or the covariance is greater than the former. The Table 1 is the summarize of the mean and covariance of two methods.

From the table above it can come to a conclusion that online denoising based on FCAM had a preferable denoising effect. It is because that TCAM had a regulatory process at the beginning, in contrast, FCAM can capture the feature of the time series data fleetly.

5 Conclusion

In this chapter, we studied the problem of online denoising which is applicable in high unstable real-time data driven applications. We proposed an Kalman filter based on FCAM and made use of this model to capture the feature of the time series data. It showed that the proposed method can perform the denoising task faster and more accurate.

References

1. Chakraborty P, Khadivi P, Lewis, B, et al. Forecasting a moving target: ensemble models for ILI case count predictions. In: Proceedings of the SIAM international conference on data mining; 2014, p. 262–70.
2. Sapankevych NI, Sankar R. Time series prediction using support vector machines: a survey. Comput Intell Mag IEEE. 2009;4(2):24–38.
3. Athanasopoulos G, Silva AD. Multivariate exponential smoothing for forecasting tourist arrivals to Australia and New Zealand. Monash Econ Bus Stat Work Papers; 2010:1737–39.
4. Muneyasu M, Wada Y, Hinamoto T. Realization of adaptive edge-preserving smoothing filters. Electron Commun Jpn. 2015;80(10):19–27.
5. Kim JH. Method for reducing noise in medical image: U.S. Patent 9,269,128; 2016:2–23.
6. Bar-Shalom Y, Li XR, Kirubarajan T. Estimation with applications to tracking and navigation: theory algorithms and software. Wiley; 2004.

7. Singer RA. Estimating optimal tracking filter performance for manned maneuvering targets. IEEE Trans Aerosp Electron Syst. 1970;4:473–83.
8. Kumar KSP, Zhou H. A 'current' statistical model and adaptive algorithm for estimating maneuvering targets. J Guid Control Dyn. 1984;7(5):596–602.
9. Kendrick JD, Maybeck PS, Reid JG. Estimation of aircraft target motion using orientation measurements. IEEE Trans Aerosp Electron Syst. 1981;2:254–60.
10. Bian X, Li XR, Chen H, et al. Joint estimation of state and parameter with synchrophasors—Part I: state tracking. IEEE Trans Power Syst. 2011;26(3):1196–208.
11. Bian X, Li XR, Chen H, et al. Joint estimation of state and parameter with synchrophasors—Part II: parameter tracking. IEEE Trans Power Syst. 2011;26(3):1209–20.
12. Jin XB, Du JJ, Bao J. Maneuvering target tracking by adaptive statistics model. J Chin Univ Posts Telecommun. 2013;20(1):108–14.

A Practical Approach for Test Equipment Perception and Virtualization in Cloud Manufacturing

Yuyan Xu, Lin Zhang, Xiao Luo and Han Zhang

Abstract As an advanced manufacturing paradigm, Cloud Manufacturing provides a new way to solve the problems in the testing process. In cloud manufacturing, resource perception and resource virtualization are the foundation of building an actual service platform. According to the characteristic of test equipment, a practical approach for test equipment perception and virtualization is proposed, which is based on cloud computing, Automatic Test Markup Language (ATML) description and data mapping. A test service platform prototype is given to validate the feasibility of this approach.

Keywords Cloud manufacturing · Test equipment · Resource perception · Resource virtualization

1 Introduction

Testing is an important part of manufacturing. Recently, with increasing demand for higher quality and shorter production time, testing has become the bottleneck of the development of the manufacturing industry, especially in electronic industry. Under current test mode, the test of product is performed by a dedicated ATS

Y. Xu (✉) · L. Zhang (✉) · X. Luo
School of Automation Science and Electrical Engineering, Beihang University,
Beijing 100191, China
e-mail: prophet_xu@buaa.edu.cn

L. Zhang
e-mail: zhanglin@buaa.edu.cn

Y. Xu · L. Zhang · X. Luo
Engineering Research Center of Complex Product Advanced Manufacturing Systems,
Ministry of Education, Beijing 100191, China

H. Zhang
Science and Technology on Special System Simulation Laboratory,
Beijing Simulation Center, Beijing 100854, China

© Springer Nature Singapore Pte Ltd. 2018
Y. Jia et al. (eds.), *Proceedings of 2017 Chinese Intelligent Systems Conference*,
Lecture Notes in Electrical Engineering 460,
https://doi.org/10.1007/978-981-10-6499-9_15

(Automatic Test System). Through the field research on electronic manufacturing enterprises, we found that the existing ATS exposes a lot of problems, such as low utilization rate of equipment, long line changing time and long development cycle. As an advanced emerging manufacturing paradigm, cloud manufacturing provides a new way of solving these problems.

Cloud Manufacturing [1, 2] is a service-oriented and networked manufacturing model. In this model, provider's manufacturing resources are accessed to Cloud Manufacturing platform, then with the help of virtualization technology, these resources form a giant virtual resource pool. After service encapsulation of these virtual resources, consumer can use these service to manufacture their own products. Apply this model in test industry will greatly improve the flexibility and agility of testing process. The premise of realizing all of these is to realize the perception and virtualization of test equipment [3]. In this paper, we propose an approach for test equipment perception and virtualization in Cloud Manufacturing environment. This approach is suitable for performing digital instruction test and analog test based on virtual instrument.

2   Architecture Overview

The architecture of Cloud Manufacturing has been widely proposed by researchers [1–3], with slightly adjustment, testing can be fit into Cloud Manufacturing seamlessly. Figure 1 shows the architecture of Cloud Manufacturing test service platform.

ATE (Automatic Test Equipment), TPS (Test Program Set), test instrument and among other test resources and capabilities are at the test resource layer. Test resource perception are the key for these resource to be accessed by the cloud platform. Virtual resource layer contains virtual resources of the platform. It is primarily responsible for the virtualization and encapsulation of the test resources. Our study focuses on these two layers.

Current research on resource perception and virtualization focus on production resource rather than test resource. On resource perception, previous researches usually use IoT (Internet of things) technology and emphases on monitoring or remote control of manufacturing resources [4, 5]. On resource virtualization, they use virtualization method in cloud computing for soft resource, and for the hard resource, these research focus on the description of resource [6]. However, test equipment is different from production equipment, it is a tight combination of hardware and software while it can be easily decoupled. The hardware of ATS (such as control computer, dedicated instrument, data acquisition equipment) are very common. For different test task, what need to change is the composition of these hardware and TPS. These characteristics of the test equipment, can make our study of perception and virtualization, not only concentrated in the state monitoring and remote control, but also through the perception and virtualization to decouple the test equipment into cloud manufacturing platform. So we can make use of the

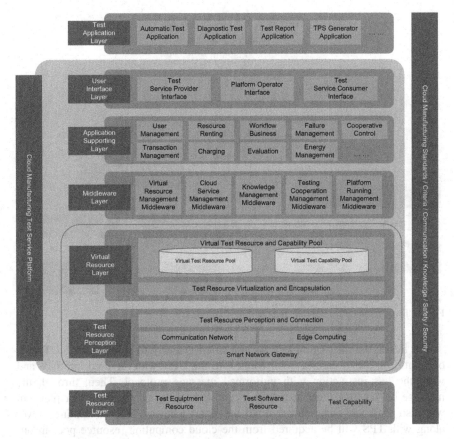

Fig. 1 The architecture of Cloud Manufacturing test service platform

service encapsulation and service composition technology in the cloud manufacturing platform to realize the more flexible and agile test flow.

3 Test Equipment Perception

In the actual production, ATS mainly has two kinds of tasks. One is for testing the UUT's (Unit under Test) digital output, the other is for testing the UUT's analog output. In the first case, ATS usually connects to UUT with USB, RS485, Ethernet or any other standard interface, the control computer in the ATS sends and receives command through these interfaces to perform the test. For the second case, ATS usually using dedicated instrument or data acquisition equipment to acquire the analog output signal from the UUT while sending instruction to the UUT through standard bus interface. These two kinds of task will be discussed next.

Our approach is shown in Fig. 2.

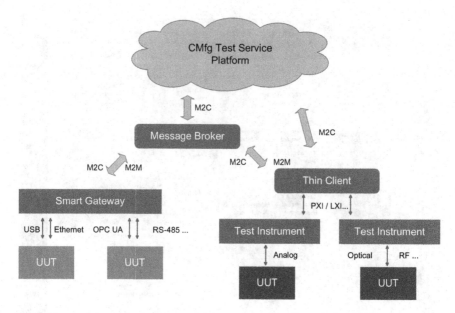

Fig. 2 Test equipment perception approach

For the digital instruction test, we use smart gateway to connect the UUT with our platform. Smart gateway will connect to the UUT with required interface and wrap the test instruction with uniformed message protocol. Then, through the message broker, these test instructions can be send to our platform. Our platform will also send test command through the smart gateway. The computing power along with TPS will be acquired from the cloud computing resource pool in our platform. In these case, smart gateway will only wrap and forward messages, thus no strong computing power is needed, so it can be a low cost common device which can be deployed easily.

For the analog signal test, we take advantage of the rapid development of test equipment. Nowadays, virtual instrument has been widely used because of its flexibility and similarity. It has simple hardware, and each kind of test function mainly realizes by software which running at control computer. For these instrument, we use standard bus interface (PXI, LXI, etc.) to connect them with a thin client computer. This computer has specific driver for those instruments, and connects to the cloud platform through a message broker. It will control the test instrument to perform the test via the instruction ordered by cloud platform, and generate the required data to the cloud. If necessary, such as sending large files, it can be directly connected with the platform. The analysis program and TPS will running in the cloud.

Through the approach mentioned above, the ATS actually has been decoupled into our platform. Apart from the data acquisition, most control and analysis task will be performed in the cloud. In these model, due to the presence of message broker, the communication between the devices can be facilitated, which facilitates the cooperation of the test task.

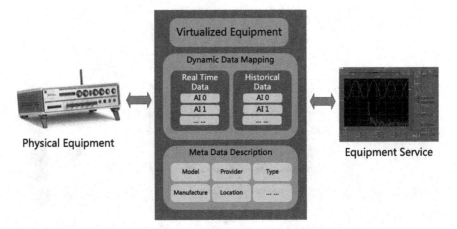

Fig. 3 Test equipment virtualization approach

4 Test Equipment Virtualization

In cloud manufacturing, the purpose of virtualization is to achieve the centralized management of distributed resources through the description of resources [6, 7]. In the field of test, there are also research on instrument virtualization [8–10]. These studies focus on the use of cloud computing virtualization technology to virtualize the test equipment. A typical study [9] is to install a test program on a virtual machine and connect the physical device remotely via technical means. According to the concept of cloud manufacturing, the method we proposed is to maintain a data mirror of physical equipment based on description (Fig. 3).

For the equipment description, we refer to the ATML standard [11], which is a standard for interfacing test system, and has been widely accepted in test industry. Through the description, we can map the properties of physical equipment to the data mirror. Data mirror is a logistic concept. User can acquire the real time data and historical data of physical equipment via this mirror. Equipment can also be controlled by this mirror. All the command given to the mirror will be sent to physical equipment simultaneously.

5 Platform Prototype

According to our approach, we have implemented a prototype of cloud manufacturing Test Service platform for some typical test equipment, as shown in Fig. 4. The emphasis here is on the test equipment perception and virtualization, so our platform focuses on these two parts, the other parts is relatively primitive.

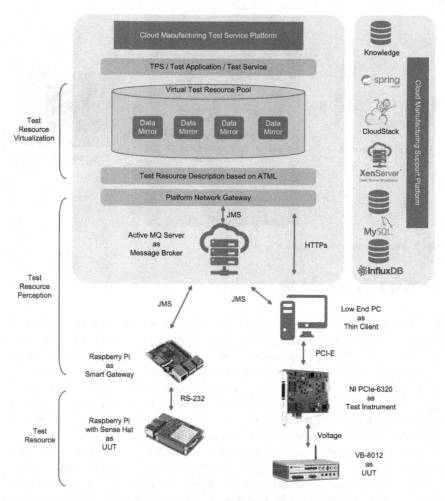

Fig. 4 The architecture of Cloud Manufacturing test service prototype platform

The platform prototype runs on a cloud manufacturing support platform. In addition to the various virtualized resources provided in the support platform, we introduce the time series database [12], InfluxDB, to store and retrieve a large number of timing data generated during the testing process.

For digital instruction test, we use Raspberry Pi equiped with Sense Hat as UUT, using Raspberry Pi as the smart gateway. The smart gateway communicates with the UUT via RS-232 interface, while using JMS to connect to the message broker.

For analog signal test, we use NI's VB-8012 as UUT. VB-8012 is a configurable experimental device, we use it to send the voltage signal with certain frequency waveform. Test equipment we have chosen is the NI's PCIe-6320 data acquisition card.

Fig. 5 Test equipment

```
1  <id:InstrumentDescription type="Module" uuid="{7c797140-f6d8-11cf-9fd6-00a024178a17}"
2      xmlns:id="urn:P-IEEE-1671.2:2008.02:InstrumentDescription"
3      xmlns:c="urn:IEEE-1671:2008.01:Common"
4      xmlns:hc="urn:IEEE-1671:2008.01:HardwareCommon"
5      xmlns:xsi="http://www.w3.org/2001/XMLSchema-instance"
6      xsi:schemaLocation="urn:P-IEEE-1671.2:2008.02:InstrumentDescription InstrumentDescription.xsd"
7      >
8      <c:Identification>
9          <c:ModelName>PCIe-6320</c:ModelName>
10     </c:Identification>
11     <id:Buses>
12         <id:Bus xsi:type="id:PCIe" vendorID="1093" deviceID="C4C4" numberOfLanes="3"/>
13     </id:Buses>
14     <hc:Interface>
15         <c:Ports>
16             <c:Port name="AI_0P" direction="Bi-Directional" type="Analog">
17                 <c:ConnectorPins>
18                     <c:ConnectorPin connectorID="AI_0" pinID="68"/>
19                 </c:ConnectorPins>
20             </c:Port>
21             <c:Port name="AI_0M" direction="Bi-Directional" type="Analog">
22                 <c:ConnectorPins>
23                     <c:ConnectorPin connectorID="AI_9" pinID="66"/>
24                 </c:ConnectorPins>
25             </c:Port>
26         <c:Ports>
```

Fig. 6 ATML description for PCIe-6320

It is connected to a low PC via standard PCI-E bus. This PC functions as a thin client, running client application and connects to message broker via JMS. It also connects to the platform gateway via HTTPs protocol to transfer the large amount of original signal data generated during the test.

Some of the test equipment involved in the experiment are shown in Fig. 5.

After the resource perception, we proceed the resource description based on ATML. We make descriptions for all the test instrument involved in the experiment. An example description for PCIe-6320 are given in Fig. 6.

Through this description, the platform can know all the information of the equipment. According to the information of the description file and perception layer, the platform maps the physical device to its data mirror, so as to realize the virtualization of the test equipment.

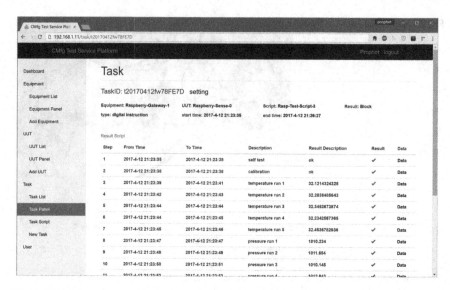

Fig. 7 Result page for one instruction test

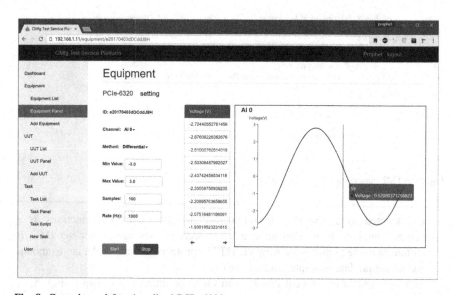

Fig. 8 Control panel for virtualized PCIe-6320

After virtualization and service encapsulation, consumer can use the test service on the platform. Figure 7 shows the history data of one instruction test. Figure 8 shows the equipment panel where user can take control of PCIe-6320 and acquire its real-time and historical data.

6 Conclusion

As an advanced manufacturing paradigm, cloud manufacturing provides a new way to solve the problems in the testing process. This paper presents a practical approach for test equipment perception and virtualization in cloud manufacturing, which lays the foundation for implementation of cloud manufacturing in test industry. At the end of this paper, a cloud manufacturing test service platform prototype is given, which supports some typical test equipment and can perform basic test.

Acknowledgements The research is supported by the National High-Tech Research and Development Plan of China under Grant No. 2015AA042101 and Fund of State Key Laboratory of Intelligent Manufacturing System Technology in China.

References

1. Zhang L, Luo Y, Tao F, Li BH, Ren L, Zhang X, et al. Cloud manufacturing: a new manufacturing paradigm. Enterp Inf Syst. 2012;8(2):1–21.
2. Bo-Hu LI, Zhang L, Wang SL, Tao F, Cao JW, Jiang XD, et al. Cloud manufacturing: a new service-oriented networked manufacturing model. Comput Integr Manuf Syst. 2010;16(1):1–7.
3. Adamson G, Wang L, Holm M, Moore P. Cloud manufacturing—a critical review of recent development and future trends. Int J Comput Integr Manuf. 2015; 1–34.
4. Tao F, Zuo Y, Xu LD, Zhang L. Iot-based intelligent perception and access of manufacturing resource toward cloud manufacturing. IEEE Trans Industr Inf. 2014;10(2):1547–57.
5. Qu T, Lei SP, Wang ZZ, Nie DX, Chen X, Huang GQ. Iot-based real-time production logistics synchronization system under smart cloud manufacturing. Int J Adv Manuf Technol. 2016;84(1):147–64.
6. Lei R, Lin Z, Zhang YB, Fei T, Luo YL. Resource virtualization in cloud manufacturing. Jisuanji Jicheng Zhizao Xitong/comput Integr Manuf Syst Cims. 2011;17(3):511–8.
7. Wang XV, Xu XW. Virtualize manufacturing Capabilities in the cloud: requirements and architecture. In: International Manufacturing Science and Engineering Conference Collocated with the, North American Manufacturing Research Conference. ASME;2013. pp. V002T02A002-V002T02A002.
8. Hu L, Xiao M. The future of automatic test system (ats) brought by cloud computing. 2009;412–4.
9. Heng-Jing HE, Zhao W, Huang SL. Research on the instrumentation virtualization in cloud computing environment. Electr Measur Instrum. 2014.
10. Ding C, Tang LW, Deng SJ. A structure of cloud test system for signal test. Key Eng Mater. 2016;693:1314–20.
11. 1671–2010—IEEE Standard for Automatic Test Markup Language (ATML) for Exchanging Automatic Test Equipment and Test Information via XML.
12. Lehmann M, Biørn-Hansen A, Ghinea G, Grønli TM, Younas M. Data analysis as a service: an infrastructure for storing and analyzing the internet of things. In: International Conference on Mobile Web and Information Systems 2015 August. Springer International Publishing;2015. p. 161–9.

Attitude Control of a Class of Quad-Rotor Based on LADRC

Jiawei Tang, Shuhua Yang and Qing Li

Abstract Based on the analysis of the working principle and dynamic characteristics of quad-rotor, its dynamic model is established. And the simulation platform of quad-rotor control system is developed in Matlab/Simulink environment. Considering the characteristics of nonlinear, strong coupling, under-actuated, and uncertain of the quad-rotor aircraft, the most widely used cascade PID controller is adopted in the simulation of the quad-rotor control system. To deal with the aircraft parameter uncertainties and external interferences, this paper designed a linear active disturbance rejection controller (LADRC), which realizes the real-time estimation and compensation of internal disturbances and external disturbances by using the extended state observer (ESO) to overcome the strong coupling, model uncertainties and external disturbances of the quad-rotor aircraft. Simulation results on attitude tracking and height control of the LADRC system are analyzed on the simulation platform and compared with the cascade PID control system. The simulation results show that the LADRC is superior to the cascade PID controller in terms of disturbance rejection.

Keywords Quad-rotor · Cascade PID · LADRC

1 Introduction

In recent years, the quad-rotor aircraft with its simple structure, good mobility, low cost and other advantages, is widely used in plant protection, aerial photography, detection and other civil and military fields [1]. Compared to fixed-wing aircraft, the movement of quad-rotor is more flexible. It can complete the vertical takeoff and landing, fixed hover, low-speed flight and indoor flight. The movement of the quad-rotor aircraft is carried out on the basis of the change of the attitude. For the

J. Tang (✉) · S. Yang · Q. Li
School of Automation and Electronic Engineering, University of Science & Technology
Beijing, Beijing 100083, China
e-mail: jw_tang0521@163.com

© Springer Nature Singapore Pte Ltd. 2018
Y. Jia et al. (eds.), *Proceedings of 2017 Chinese Intelligent Systems Conference*,
Lecture Notes in Electrical Engineering 460,
https://doi.org/10.1007/978-981-10-6499-9_16

quad-rotor aircraft attitude control, the current research has been carried out, including cascade PID control [2], back-stepping control [3], sliding mode control [4], robust control [5, 6], etc.

Among them, back-stepping control, sliding mode control and robust control need high model accuracy. However, it is difficult to obtain the accurate parameters of the aircraft due to the interference of environmental factors. Therefore, the cascade PID controller, which is independent of the model, is the most widely used in practical aircraft. However, there are still some defects in cascade PID control system, such as load change, strong wind and other disturbances will affect the control quality. In order to solve the above problems, ADRC technology has been widely concerned by the developers [7, 8].

The main idea of active disturbance rejection control is to count the internal disturbances and external disturbances of the system into the total perturbations, then estimate the total disturbance by the extended state observer and compensate it in real time. Professor Gao proposed a linear active disturbance rejection controller, which simplifies the conversion of the ADRC to only three parameters by adjusting the bandwidth, and the physical meaning of these parameters are clear [9]. So the LADRC is more suitable for actual system control applications.

Based on the analysis of the structure, flight principle and dynamic characteristics of the quad-rotor aircraft, linear active disturbance rejection controllers are designed for pitch, roll, yaw and height respectively. The simulation is carried out in the MATLAB/Simulink environment and the simulation results show that the attitude and height tracking and stabilization are achieved.

2 Dynamic Model of Quad-Rotor Aircraft

In order to describe the mathematical model of the quad-rotor aircraft, it is first necessary to establish two coordinate systems—the body coordinate system $B(X_b, Y_b, Z_b)$ and the earth coordinate system $E(X_e, Y_e, Z_e)$ as shown in the Figs. 1 and 2, respectively.

Quad-rotor aircraft is a complex mechanical system, which contains a lot of physical effects in the field of mechanics and aerodynamics.

The motion of the quad-rotor in space can be regarded as a rigid body motion with six degrees of freedom, and its motion equation can be divided into two parts: the horizontal motion in space and the rotation around the center of mass. According to the Newton-Euler equation, the motion of the quad-rotor can be expressed in the body coordinate system:

$$\begin{bmatrix} mI_{3*3} & 0 \\ 0 & J \end{bmatrix} \begin{bmatrix} \dot{v}_b \\ \dot{\omega} \end{bmatrix} + \begin{bmatrix} \omega \times mv_b \\ \omega \times J\omega \end{bmatrix} = \begin{bmatrix} F \\ T \end{bmatrix} \tag{1}$$

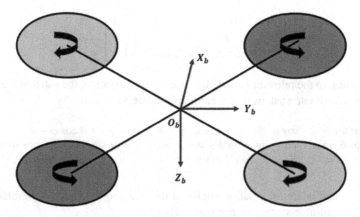

Fig. 1 Body coordinate system

Fig. 2 Earth coordinate system

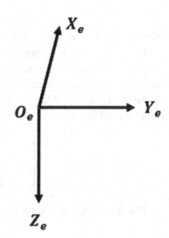

where m is the mass of the quad-rotor, $J = [J_x \quad J_y \quad J_z]^T$ is the rotary inertia of the three axes in the body coordinate system, $v_b = [u \quad v \quad w]^T$ is the speed of the quad-rotor in the body coordinate system along the three axis, $\omega = [p \quad q \quad r]^T$ is the angular velocity that around the three axes in the body coordinate system, $F = [F_x \quad F_y \quad F_z]^T$ is the composition of forces which the quad-rotor accepts along the three axis in the body coordinate system, $T = [\tau_x \quad \tau_y \quad \tau_z]^T$ is the torque of the quad-rotor around the three axes in the body coordinate system.

The angular motion equation of the quad-rotor is exactly the same as that of other types of air vehicles. The relationship between the Euler angular velocity and the three axis angular velocity is:

$$\begin{bmatrix} 1 & 0 & -\sin\theta \\ 0 & \cos\phi & \sin\phi\cos\theta \\ 0 & -\sin\phi & \cos\phi\cos\theta \end{bmatrix} \begin{bmatrix} \dot\phi \\ \dot\theta \\ \dot\psi \end{bmatrix} = \omega \tag{2}$$

According to the relevant knowledge of inertial navigation, the rotation matrix of the body coordinate system to the earth coordinate system is:

$$R = \begin{bmatrix} \cos\theta\cos\psi & \cos\psi\sin\theta\sin\phi - \sin\psi\cos\phi & \cos\psi\sin\theta\cos\phi + \sin\psi\sin\phi \\ \cos\theta\sin\psi & \sin\psi\sin\theta\sin\phi + \cos\psi\cos\phi & \sin\psi\sin\theta\cos\phi - \cos\psi\sin\phi \\ \sin\theta & \sin\phi\cos\theta & \cos\phi\cos\theta \end{bmatrix}$$

where φ, θ and ψ are the rotation angles of the body coordinate system relative to the earth coordinate system respectively. Then:

$$v_e = R v_b \tag{3}$$

where $v_e = [v_x \quad v_y \quad v_z]^T$ is the quad-rotor's speed in the earth coordinate system.

The quad-rotor is affected by the gravity–mg and the total pulling force of the propeller–F_T, where the gravity direction is the positive direction of the Z axis in the earth coordinate system and the pulling force direction is the negative direction of the Z axis in the body coordinate system. The gravity is decomposed along the three axis of the body coordinate system, and put it in the formula (1), then:

$$\begin{cases} \dot u = -g\sin\theta + rv - qw \\ \dot v = g\sin\varphi\cos\theta + pw - ru \\ \dot w = g\cos\varphi\cos\theta + qu - pv - F_T \end{cases} \tag{4}$$

The quad-rotor's external torque in the body coordinate system is:

$$\begin{cases} \tau_x = \frac{\sqrt 2}{2}l(T_1 - T_2 - T_3 + T_4) - J_r q(\Omega_1 - \Omega_2 + \Omega_3 - \Omega_4) \\ \tau_y = \frac{\sqrt 2}{2}l(T_1 - T_2 - T_3 + T_4) + J_r p(\Omega_1 - \Omega_2 + \Omega_3 - \Omega_4) \\ \tau_z = d(\Omega_1^2 - \Omega_2^2 + \Omega_3^2 - \Omega_4^2) \end{cases} \tag{5}$$

where T_i is the lift force of the ith propeller, ith is the speed of the ith propeller, J_r is the moment of inertia of the propeller and d is the inverse torque coefficient of the propeller.

There is a relationship between the lift and the speed of the propeller:

$$T_i = b\Omega_i^2 \tag{6}$$

where b is the lift coefficient of the propeller.

According to formula 1–6, the mathematical model of the quad-rotor aircraft can be expressed as:

$$
\begin{cases}
\dot{p} = \dfrac{(J_y - J_z)}{J_x}qr + \dfrac{J_rq(\Omega_1 - \Omega_2 + \Omega_3 - \Omega_4)}{J_x} + \dfrac{\frac{\sqrt{2}}{2}l(T_1 - T_2 - T_3 + T_4)}{J_x} \\[2mm]
\dot{q} = \dfrac{(J_z - J_x)}{J_y}pr - \dfrac{J_rp(\Omega_1 - \Omega_2 + \Omega_3 - \Omega_4)}{J_y} + \dfrac{\frac{\sqrt{2}}{2}l(T_1 - T_2 - T_3 + T_4)}{J_y} \\[2mm]
\dot{r} = \dfrac{(J_x - J_y)}{J_z}pq + d(\Omega_1^2 - \Omega_2^2 + \Omega_3^2 - \Omega_4^2) \\[2mm]
\dot{\phi} = p + (\sin\phi\,\tan\theta)q + (\cos\phi\,\tan\theta)r \\[1mm]
\dot{\theta} = q\cos\phi - r\sin\phi \\[1mm]
\dot{\psi} = (\sin\phi/\tan\theta)q + (\cos\phi/\tan\theta)r \\[1mm]
\dot{u} = -g\sin\theta + rv - qw \\[1mm]
\dot{v} = g\sin\varphi\,\cos\theta + pw - ru \\[1mm]
\dot{w} = g\cos\varphi\,\cos\theta + qu - pv - F_T \\[1mm]
\ddot{X} = R[u \quad v \quad w]^T
\end{cases}
\tag{7}
$$

where $X = [x \quad y \quad z]^T$ is the position of the quad-rotor aircraft under the earth system.

All the parameters in the model are shown in Table 1

Table 1 Simulation parameter of quad-rotor

Parameter symbol	Physical meaning	Value	Unit
m	Mass of the quad-rotor	1	kg
1	Distance from motor to centroid of quad-rotor	0.24	m
g	Gravity acceleration	9.81	m/s^2
b	Lift coefficient	5.42×10^{-5}	Ns^2
d	Resistance coefficient	1.1×10^{-6}	Ns^2
I_x	Rotary inertia of X axis	8.1×10^{-3}	kgm^2
I_y	Rotary inertia of Y axis	8.1×10^{-3}	kgm^2
I_z	Rotary inertia of Z axis	14.2×10^{-3}	kgm^2
J_r	Rotary inertia of propeller	1.04×10^{-4}	kgm^2

3   Attitude Control System Design

In this paper, using the incremental output in the attitude controller design. First, PWM wave with fixed duty cycle was transmitted to every motor to keep a base speed–R. Then the output of the controller is added or reduced on the basis of R. The speed that a pair of motors increases is equal to the other pair reduces. For this reason, it is possible to use a squared difference formula when calculating the torque. Thus, the non-linear relationship between the square of the speed and the lift is eliminated. The base speed R is set to the speed of four motors when the quad-rotor hovered steadily. That is $mg = 4bR^2$.

Put the parameters of model given in chapter two into above formula, it can be calculated that R = 212.6 rad/s.

When designing the controller one channel, assume that other channels are set to zero. Then the transfer function of the roll channel is:

$$G_r(s) = \frac{1.93}{(0.178s + 1)s^2}$$

Similarly, the transfer functions of the pitch channel and yaw channel are:

$$G_p(s) = \frac{1.93}{(0.178s + 1)s^2}, \; G_y(s) = \frac{0.132}{(0.178s + 1)s^2}.$$

As for the height channel, all the motors accelerate and decelerate simultaneously, so the square relationship can't be eliminated between the motor speed and lift. Thus the mathematical model of height channel is:

$$\ddot{z} = g - \frac{b(R + \sum \Delta\Omega_i)^2}{m}$$

3.1   Cascade PID Controller Design

As one of the earliest developed control strategies, PID control is widely used in production practice because of its simple algorithm, good robustness and high reliability. The cascade PID attitude controller connects the two PID controllers in series, where the output of the previous controller is used as the reference for the next controller. Inner loop named stabilization loop set the amount of change in attitude, responsible for coarse adjustment, and the outer loop named attitude loop set the attitude, responsible for fine tuning. The inner loop makes the response speed of the system be improved. Simultaneously it can suppress the disturbance in the inner loop, and enhance the anti-interference ability of the system. The structure of cascade PID control system is shown in Fig. 3.

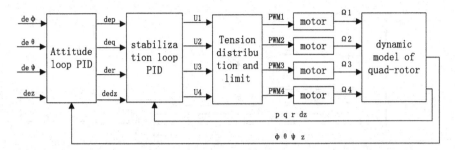

Fig. 3 The structure of cascade PID controller

In the cascade PID parameter tuning process, first disconnect the outer loop, set the inner loop. The inner loop is the follower system, its main role is to quickly track, resist interference, and the inner loop itself includes pure integral part. Therefore PD control is adopted.

In order to ensure the inner ring of the anti-interference ability of the system, we choose a larger value of $kp = 100$. The performance requirements of the inner loop system are: fast response and allow for a small amount of overshoot, so we select the value near the root locus separation point as the final closed-loop pole of the inner loop system, where $kd = 6$.

The inner loop PID parameters are determined and then connected to the outer loop, the outer loop also contains a pure integral link, so the outer loop can also be achieved without static error by PD controller. The same as the inner loop. First, increase the proportion coefficient to improve the response speed of the system. Then, increase the derivative time gradually to suppress overshoot. After repeated debugging, the final outer loop parameters are determined as $kp = 20$, $kd = 0.3$.

The parameter tuning process of the other three channels is similar to that of the roll channel, the details are omitted here due to space limit.

Finally, the parameters of the cascade PID controller are obtained, which are shown in Table 2

3.2 LADRC Controller Design

The core of linear active disturbance rejection controller is the linear extended state observer (ESO). The linear extended state observer makes use of the input and

Table 2 Parameters of cascade PID controller

	P_in	D_in	P_out	I_out	D_out
Roll	100	6		0	0.3
Pitch	100	6	20	0	0.3
Yaw	100	16	10	0	0.7
Z	100	0.8	2	1	0

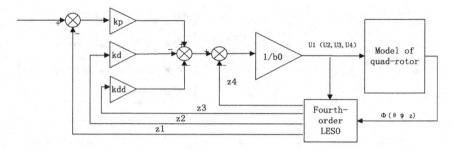

Fig. 4 Linear active disturbance rejection attitude controller

output information of the system to estimate the state of the system and the total disturbance. And then we design the linear state feedback to compensate this total disturbance. As the three attitude channels are third-order system, while the height channel can also be similar to the third-order system, so we select the fourth-order extended state observer. The linear active disturbance rejection attitude controller is shown in the Fig. 4:

Taking the roll channel as an example, the design process of the linear active disturbance rejection attitude controller is presented.

First, the state space expression of the roll channel is expressed and the linear expansion state observer is established as follows:

$$
\begin{cases}
\dot{x}_1 = x_2 \\
\dot{x}_2 = x_3 \\
\dot{x}_3 = x_4 + bu \\
\dot{x}_4 = w(t) \\
\phi = x_1
\end{cases}
\qquad
\begin{cases}
e_1 = z_1 - y \\
\dot{z}_1 = z_2 - \beta_1 e_1 \\
\dot{z}_2 = z_3 - \beta_2 e_1 \\
\dot{z}_3 = z_4 - \beta_3 e_1 + bu \\
\dot{z}_4 = -\beta_4 e_1
\end{cases}
$$

where x_3 is the extended new state variable.

According to the simple method that use the bandwidth to determine the ESO parameters proposed by Professor GAO in article [9], the characteristic equation of the observer is configured as $(s + \omega_0)^4$. Then, β_1, β_2, β_3, β_4 can be expressed as: $\beta_1 = 4\omega_0$, $\beta_2 = 6\omega_0^2$, $\beta_3 = 6\omega_0^3$, $\beta_4 = \omega_0^4$.

Where ω_0 is the parameter of linear ESO, it can be determined according to the requirement of system bandwidth or adjusted on-line.

In addition to ESO, the process of tuning state feedback parameters can also be simplified in using bandwidth. Similarly, state feedback parameter is expressed as: $kp = \omega_c^3$, $kd = -3\omega_c^2$, $kdd = -3\omega_c$.

Similar to ω_0, ω_c is another parameter can be adjusted online.

As for as compensation coefficient b_0, in general, the closer it near the open loop gain of the system, the better control effect we can get.

After repeated debugging, the LADRC parameters for each channel are determined, which are shown in Table 3

Table 3 Parameters of cascade LADRC

	ω_c	ω_0	b_0
Roll	20	500	10
Pitch	20	500	10
Yaw	10	200	0.7
Z	3	200	0.5

4 Simulation Experiment and Result Analysis

Set the initial values of the three attitude angles and height as 0, and set the input signal as step signal, where 0.4 rad in roll channel, 0.5 rad in pitch channel, 0.3 rad in yaw channel and −5 m in height channel (Since the positive direction of Z axis is downward, the −5 m represents the rise to 5 m). The system output curve as shown in Fig. 5 in the cascade PID controller and LADRC controller.

It can be seen from the figure that the two control methods can quickly track the attitude angle while lifting the height, and no overshoot, which cascade PID control system reach the target attitude faster, and LADRC system is smoother. Both of them can meet the needs of practical applications.

After the attitude is stable in previous experiment, add the moment disturbance at 3, 4 and 5 s in roll, pitch and yaw channel. The moment disturbance is square wave which amplitude is 1 N m, 0.5 N m and 0.1 N m respectively and width is 0.2 s. The result is shown in Fig. 6.

From the results shown in the figure, LAD controller's disturbance rejection is significantly better than the cascade PID controller. In addition, when dealing with the coupling of each channel of the quad-rotor, the LADRC controller of either channel can treat the control of the other channels as part of the total disturbance and compensate it. Therefore, the LADRC is better than the cascade PID controller in dealing with the coupling of each channel of the quad-rotor.

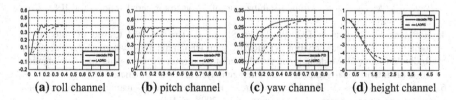

    (**a**) roll channel   (**b**) pitch channel   (**c**) yaw channel   (**d**) height channel

Fig. 5 The performance of cascade PID and LADRC in attitude tracking

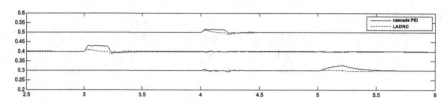

Fig. 6 The performance of cascade PID and LADRC in disturbance rejection

5 Conclusion

In this paper, two kinds of controllers, cascade PID and LADRC, are designed for the attitude control of the quad-rotor aircraft. And the influence of the quad-rotor's load change and the external environment on the control effect of the two controllers are discussed through theoretical modeling and simulation. The results show that the LADRC controller is superior to the cascade PID controller in dealing with the coupling and the disturbance rejection. In the future work, we will use the actual quad-rotor aircraft flight experiments to verify the theoretical and simulation results, so as to complete the transition from theory to practice.

Acknowledgements This work was supported by National Natural Science Foundation of China (No. 61520106010), National Key Technologies R&D Program (No. 2013BAB02B07) and National Natural Science Foundation of China (No. 61603362).

References

1. Paul GF, Thomas JG. Introduction to UAV system. Columbia, MD: UAV Systems; 1998.
2. Gao J. Design and implementation of control system for a quad rotor. Dalian University of Technology; 2015.
3. Tayebi A, McGilvray S. Attitude stabilization of a four-rotor aerial robot. Proc IEEE Conf Decis Control. 2004.
4. Hoffmann G, Rajnarayan DG, Waslander SL, et al. The Stanford testbed of autonomous rotorcraft for multi agent control (STARMAC). In: Digital avionics system conference; 2004.
5. Rodic A, Mester G. Modelling and simulation of quad-rotor dynamics and spatial navigation. In: IEEE international symposium on intelligent systems & informatics; 2011.
6. Amir MY, Abbass V. Modelling of quad rotor Helicopter dynamics. In: International conference on smart manufacturing application; 2008. p. 100–5.
7. Jingqing H. From PID technique to active disturbances rejection control technique. Control Eng China. 2002;9(3):13–8.
8. Yisha L, Shengxuan Y, Wei W. An active disturbance-rejection flight control method for quad-rotor unmanned aerial vehicles. Control Theor Appl. 2015;32(10):1351–60.
9. Gao Z. Scaling and bandwidth-parameterization based controller tuning. In: Proceedings of the American control conference, 2003. IEEE; 2003. p. 4989–96.

New Progress on Research of Weighted Multiple Model Adaptive Control

Yuzhen Zhang, Weicun Zhang and Jiqiang Wang

Abstract This paper is concerned with the stability of weighted multiple model adaptive control (WMMAC) system, which is a long-standing issue in the field of robust adaptive control. First, a new weighting algorithm is proposed with assured convergence under smooth conditions. Second, based on virtual equivalent system methodology, the stability results of WMMAC system for both linear time-invariant (LTI) and parameter jump linear plants are presented. The analysis method is independent of specific local control strategy and specific weighting algorithm.

Keywords WMMAC · Stability · Convergence · Virtual equivalent system

1 Introduction

"Robust" and "Adaptive" are two characteristics that all researchers in control theory field dream of, some literatures even compare it to the Holy Grail of the Christian world [1, 2]. Weighted multiple model adaptive control is an important method for robust adaptive control, which originally proposed by Magill in 1960s and applied to the state estimation of stochastic systems with uncertain parameters —Multiple Model Adaptive Estimation (MMAE) [3]. Soon afterwards, Lainiotis and Athans et al. proposed adaptive control using a multiple model method [4, 5]— Multiple Model Adaptive Control (MMAC), in which, each 'local' controller is designed according to LQG strategy. Subsequently, a lot of related research results were published [6–13] and the application range covers target tracking, medical engineering, and flight control, etc. In recent years, attention has been paid to the

Y. Zhang · W. Zhang (✉)
School of Automation & Electrical Engineering,
University of Science and Technology Beijing, 100083 Beijing, China
e-mail: weicunzhang@ustb.edu.cn

J. Wang
College of Energy and Power Engineering, Nanjing University of Aeronautics
and Astronautics, 210016 Nanjing, China

© Springer Nature Singapore Pte Ltd. 2018   167
Y. Jia et al. (eds.), *Proceedings of 2017 Chinese Intelligent Systems Conference*,
Lecture Notes in Electrical Engineering 460,
https://doi.org/10.1007/978-981-10-6499-9_17

study of weighted multiple model adaptive control, such as the robust multiple model adaptive control based on the combination of weighted multiple model and robust controller design [1, 14–18], multiple model adaptive fusion control that uses fuzzy rules [19, 20] and so on. While researches on MMAC of domestic scholars are later than abroad, especially before 1980s [21], some of the related researches are listed in Refs. [22–27].

WMMAC can be regarded as a MMAC based on soft switching and it has some similarities between the gain scheduling control and the T-S model-based fuzzy control. The MMAC based on soft switching is suitable for the stochastic control systems with uncertain parameters (limited numbers), which has been studied and applied in many fields, especially in aviation flight control, active suspension control and drug control in medicine etc. For example, control of arterial blood pressure and cardiac output can be achieved by the WMMAC, which can achieve better results than conventional PID control [7].

The stability problem of WMMAC systems is a challenging problem, which has not been well addressed for nearly half a century since the weighted multi-model method was proposed [1, 2, 19, 22–26, 28]. Recently, a new adaptive control system is designed by adopting a new weighting algorithm—virtual equivalent system method [29–32]. The stability of WMMAC is proved in Ref. [33] for the first time (linear time invariant discrete random controlled plant), and results are extended in Refs. [34–37], including improvement of the weighted algorithm, extending the scope of the controlled plant (parameter jump system, continuous system, colored noise system, the non-accurate modeling system, etc.).

The convergence of weighted algorithm is theoretical analyzed in Refs. [38–41], since the traditional weighting algorithm is based on the Kalman filter and the dynamic hypothesis test and the Bayesian posterior probability formula, so the analysis is complex and the requirements are harsh too, it can be summarized as follows: the output error signal generated by each model needs to be stable, ergodic, and it also can be distinguished [1].

It should be noted that the stability of the WMMAC refers to the situation where the input and output signals of the adaptive control system are bounded. The convergence of the WMMAC is that the input and output signals of the adaptive control system converge to that of the corresponding non-adaptive control system in the mean square sense with probability 1. In addition, the limit operations involved in this paper are in the sense of probability 1.

2 Classical Multiple Model Adaptive Control

Classical multiple model adaptive control (CMMAC) refers to the weighted multiple model adaptive control in early years. The state estimation and control is bound is a special character of it, which refers to decentralized control and global control are all dependent on state estimation, and the decentralized controller adopts LQG strategy. The principle of the system [1] is shown in Fig. 1a, KF represents

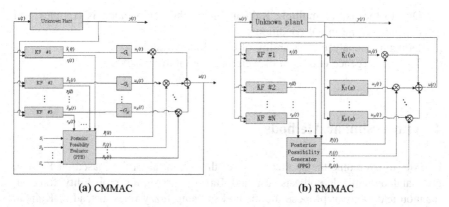

(a) CMMAC   (b) RMMAC

Fig. 1 Block diagram

the Kalman filter, the state estimation $\hat{x}_i(t)$ is given by decentralized Kalman Filter, and the overall state estimate is $\hat{x}(t/t) = \sum_{i=1}^{N} p_i(t)x_i(t/t)$. Each output of decentralized controller $u_i(t) = -G_i\hat{x}_i(t/t)$ is got according to the LQ gain, and $u(t) = \sum_{i=1}^{N} p_i(t)u_i(t)$ is the output of global control. The formula for calculating the posterior probability estimator (PPE) of the weighted $p_i(t)$ is complex and not given here [1, 3–5]. S_i is variance matrix of steady state residual $r_i(t)$ of the i-th Kalman filter.

3 Robust Multiple Model Adaptive Control

Robust multiple model adaptive control (RMMAC) which still belongs to WMMAC. More concretely, it is based on the CMMAC and combines the modern robust control system design method to design the decentralized controller. The structure of CMMAC is modified as follows: the decentralized controller of CMMAC uses the state feedback to generate the decentralized control variables, while the RMMAC decentralized controller uses the feedback of output. Besides, the Kalman filter in CMMAC is used for the calculation of the weights and the state feedback control of the decentralized controller. While in the RMMAC it is only used for the calculation of the weighted value, and the estimation subsystem is separated from the control part. The RMMAC system is shown in Fig. 1b.

Remark 1 The structure of RMMAC in Ref. [1] ignored the following points: the weighted values can only be calculated at discrete moments $p_i(k)$ and continuous weighted value $p_i(k)$ can be obtained through (zero order) retainer. As the classical weighted adaptive control, RMMAC has the following problems:

(1) Due to the use of multiple Kalman filters, the computation is complex and depends on the initial conditions;
(2) Difficult to analyze the stability of closed-loop system;
(3) The convergence condition of weighted algorithm is harsh.

4 Other Similar Methods

In order to overcome the shortcomings of the weight algorithm based on Kalman filter and dynamic hypothesis test and Bayesian posterior probability formula, Ioannou and Dumont propose the method of using fuzzy rules instead of Kalman filter and dynamic hypothesis test and the weights are calculated directly according to the membership function [19, 20], this approach is essentially a "Fuzzy Fusion" between multiple controllers, the calculation is relatively reduced and easy to implement. But, it is not easy to get the exact membership function of fuzzy rules.

In recent years, Han and Narendra have proposed the weighted fusion of multiple model parameters of the controlled plant, and then designed the adaptive controller on the model. Actually this weighted idea is introduced on the level of parameter estimation, and good theoretical results and simulation results are obtained [42]. On the basis of Ref. [43], N on-line identification models and a fixed model are used to simulate a class of nonlinear systems.

5 Development on Modeling and Weighting Algorithms

The main contents of the research on weighted multiple model adaptive control are three aspects: model selection, weighting algorithm and local controller strategy. Among them, the model selection belongs to the modelling problem, and the research of this problem is mainly related to the measurement of model difference [44, 45]. Decentralized control strategy has experienced from the early LQG control strategy to the current robust control strategy, including H_∞, H_2, and μ synthesis method, in fact, any robust control strategy can be adopted.

The early weighted algorithm [1, 3] and the later fuzzy fusion algorithm [19, 20] can be used in weighted algorithm aspect. New weighted algorithm is proposed in Ref. [33], the calculation of the weights is no longer dependent on the Kalman filter and the Bayesian posterior probability formula, but directly based on the online "performance" of the "local" model, some disadvantages of classical weighting algorithm are basically overcome through this algorithm. Its simple structure is shown in Fig. 2:

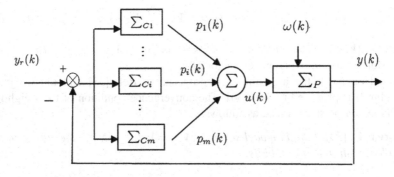

Fig. 2 Block diagram for WMMAC

6 MMAC Using New Weighted Algorithm

The following time-varying discrete linear stochastic system is considered for the realization of computer control.

$$A(q^{-1}, k)y(k) = q^{-d}B(q^{-1}, k)u(k) + w(k) \tag{1}$$

$$A(q^{-1}, k) = 1 + a_1(k)q^{-1} + \ldots + a_{na}(k)q^{-na} \tag{2}$$

$$B(q^{-1}, k) = b_0(k) + b_1(k)q^{-1} + \ldots + b_{nb}(k)q^{-nb} \tag{3}$$

where $na \geq 1; nb \geq 1$ and d is system delay, $y(k)$, $u(k)$, and $\omega(k)$ are the system output signal, input signal and interference noise signal respectively.

If the time-varying parameters of $A(q^{-1}, k)$, $B(q^{-1}, k)$ are piecewise constant function in sampling time k, the system can be regarded as parameters jump system. The MMAC method uses multiple models to cover the uncertainty of the system in order to overcome these shortcomings of traditional adaptive control. It can respond quickly to changes in system parameters when the system jumps.

First of all, the model set $M = \{M_i, i = 1, 2, \ldots, m\}$ is created for a controlled plant. And then the local controller C_i is designed respectively for each model in the model set M_i to get the controller set $C = \{C_i, i = 1, 2, \ldots, m\}$. The local controller can be designed according to any existing control strategy, which just to meet the stability requirements for M_i, such as pole configuration, predictive control, robust control (μ integrated approach, H^∞) and so on. In this paper, the pole configuration strategy is used in the simulation. The output of local controller is $u_i(k)$ and the output of global controller is $u_i(k) = \sum_{i=1}^{m} p_i(k)u_i(k)$.

The local controller is designed to stabilize the local model and track the bounded external reference input signal $y_r(k) > 0$.

The output error of the local model $e_i(k)$ is calculated at each sampling time in order to calculate the weighting value.

$$e_i(k) = y(k) - y_i(k) = y(k) - \phi^T(k-d)\theta_i \tag{4}$$

where $\quad y_i(k) = 0, e_i(k) = 0, \forall k < 0, \quad \phi^T(k-d) = [y(k-1), \ldots, u(k-d), \ldots]$ is regression vector, θ_i is parameter vector of M_i.

Based on $e_i(k)$, the algorithm I is proposed in Ref. [33] and the weighted algorithm II is proposed in Ref. [34]. The convergence condition of the weighting algorithm can be summarized as follows:

Theorem 1 [33]. *If M_j in model set $M = \{M_i, i = 1, 2, \ldots, N\}$ is the closest to the true plant with unity probability*:

$$\begin{cases} \sum_{p=1}^{k} e_j^2(p) < \sum_{p=1}^{k} e_i^2(p) \\ \frac{1}{k} \sum_{p=1}^{k} e_j^2(p) \to R_j \qquad k \ge d; i \ne j; R_j < R_i \\ \frac{1}{k} \sum_{p=1}^{k} e_i^2(p) \to R_i \end{cases} \tag{5}$$

where d is output delay of system, R_j is a constant and R_i is a constant or infinity; then the weighting algorithm I will converge according to probability 1, that is $\lim_{k \to \infty} p_j(k) = 1$, $\lim_{k \to \infty} p_i(k) = 0, i \ne j$.

Theorem 2 [34]: *If M_j in model set $M = \{M_i, i = 1, 2, \ldots, N\}$ is the closest to the true plant with unity probability, there is $\sum_{r=1}^{k} e_j^2(r) < \sum_{r=1}^{k} e_i^2(r), \forall k \ge d, i \ne j$, where d is output delay of system, then the weighted algorithm II will converge with probability 1, that is $\lim_{k \to \infty} p_j(k) = 1$, $\lim_{k \to \infty} p_i(k) = 0, i \ne j$.*

7   Stability and Convergence of WMMAC

In this section, the stability and convergence of a WMMAC system are discussed and the weighted algorithm II is used.

7.1   WMMAC for Time-Invariant Systems

The WMMAC of the invariant controlled plant described in the previous section has the following conclusion.

Theorem 3 [34]: *If WMMAC of time-invariant plants with weighted algorithm II has the following properties*

(1) *The model set contains the real plant model M_j;*
(2) *The control strategy makes each local controller and the corresponding model in the model set constitute a stable closed-loop system, and track the reference input signal;*

(3) *The model output error can be divided, that is*

$$\sum_{p=1}^{k} e_j^2(p) < \sum_{p=1}^{k} e_i^2(p), k \geq d, i \neq j$$

Then the WMMAC system is stable and convergent.

7.2 WMMAC of Variable Jump Controlled Plants

Theorem 4 [34]: *If WMMAC of parameter limited jump plant with weighted algorithm II has the following properties:*

(1) *The interval at which the controlled plant parameter changes is sufficiently long and eventually stops on a fixed model parameter M_j;*
(2) *The model set contains M_j;*
(3) *The control strategy makes each local controller and the corresponding model in the model set constitute a stable closed-loop system, and track the reference input signal;*
(4) *The model output error can be divided, that is $\sum_{p=1}^{k} e_j^2(p) < \sum_{p=1}^{k} e_i^2(p), k \geq d, i \neq j$*

Then the WMMAC system is stable and convergent.

Theorem 5 [34]: *If WMMAC of parameter infinity jump plant with weighted algorithm II has the following properties:*

(1) *The interval at which the controlled plant parameter changes is sufficiently long;*
(2) *The model set contains the real plant model for each stage, $M_j = \sum P$;*
(3) *The control strategy makes each local controller and the corresponding model in the model set constitute a stable closed-loop system, and track the reference input signal;*
(4) *The model output error can be divided, that is $\sum_{p=1}^{k} e_j^2(p) < \sum_{p=1}^{k} e_i^2(p), k \geq d, i \neq j$*

Then the WMMAC system is stable and convergent.

8 Further Research Directions and Conclusions

Although the stability has been proven, the WMMAC system still has a lot of issues to be addressed. The first is the model set: if the uncertain parameter vector in the system model is of high dimension and the requirement of system performance is high, there will be a large number of models. The number of models is determined based on the degree of improvement (percentage) of the performance indicators relative to the non-adaptive control system in Ref. [1] and the Vinnicombe measure [45] is used to determine the number of models in Ref. [44]. It is worthy of further study upon how to take the appropriate model set according to the appropriate criteria. Second, the weighted value needs to be reset if the system changes after the weighted value converged. It is a problem how to ensure that the weighted value of the reset is simple, fast and accurate. There is another way to set the weighted value of the "dead zone," that is, when the weighted value is less than a threshold, it no longer decreases. The third is anti-interference problem: although the two weighted algorithms described in this paper have better anti-noise ability, the algorithm based on the index function (4) is still inevitably introduced into the noise in essence. So it is still necessary to explore more practical index function and even weighted algorithms. Finally, the existing stability results need to be extended to time-varying non-linear and multivariable systems which even contain colored noises and so on.

The WMMAC and the fuzzy control based on T-S model using parallel compensation algorithm are local controller based on multiple linear models. The global controller is weighted by local controller. But the fuzzy control based on the T-S model has a membership function problem which finally plays a role in determining the weighted value. Compared with the fuzzy control based on T-S model, the WMMAC is more direct and easy to implement. On the other hand, since the weighted algorithm used in this paper is based on the "online" performance of each model to determine the weight of the corresponding controller, it is reasonable to believe that such weighting is more accurate than the fuzzy control based on the T-S model which uses the "offline" method to determine the weight calculated by the membership function. Of course, it should also be noted that fuzzy control of the weight calculation is not recursive. So its response is faster and there is no weight reset problem for parameter jumps.

In conclusion, the authors believe that WMMAC is very suitable for the control of processes with slow-changing parameters and random noise; and the T-S model based fuzzy control is preferable for control problems of fast-changing parameters with less noise interference.

Acknowledgements The author would like to thank the anonymous reviewers for their constructive and insightful comments for further improving the quality of this work. This work was supported by National Natural Science Foundation of China (No. 61520106010), National Key Technologies R&D Program (No. 2013BAB02B07) and National Natural Science Foundation of China (No. 61603362).

References

1. Fekri S, Athans M, Pascoal A. Issues, progress and new results in robust adaptive control. Int J Adapt Control Signal Process. 2006;20(10):519–79.

2. Ioannou P. CDC Semi-plenary: robust adaptive control: the search for the holy grail. In: Proceedings IEEE 47th Conference Decision and Control, Cancun, Mexico: Dec, 2008; p. 12–3.
3. Magill DT. Optimal adaptive estimation of sampled stochastic processes. IEEE Trans Automat Contr. 1965;10:434–9.
4. Lainiotis DG. Partitioning: a unifying framework for adaptive systems—i: estimation; ii: control. Proc IEEE. 1976;64:1126–43,1182–97.
5. Athans M, et al. The stochastic control of the F-8C aircraft using a multiple model adaptive control (MMAC) method—part I: equilibrium flight. IEEE Trans Automat Contr. 1977;22:768–80.
6. Lane DW, Maybeck PS. Multiple model adaptive estimation applied to the lambda urv for failure detection and identification. In: Proceedings IEEE 33rd Conference Decision and Control, Lake Buena Vista, FL; Dec. 1994. p. 678–83.
7. Yu C, Roy RJ, Kaufman H, Bequette BW. Multiple-model adaptive predictive control of mean arterial pressure and cardiac output. IEEE Trans Biomed Eng. 1992;39:765–78.
8. Moose RL, Van Landingham HF, McCabe DH. Modeling and estimation for tracking maneuvering targets. IEEE Trans Aerospace Elec Syst. 1979;15:448–56.
9. Li XR, Bar-Shalom Y. Design of an interacting multiple model algorithm for air traffic control tracking. IEEE Trans Contr Syst Tech. 1993;1:186–94.
10. Badr A, Binder Z, Rey D. Application of tracking multi-model control to a non-linear thermal process. Int J Syst Sci. 1990;21(9):1795–803.
11. Badr A, Binder Z, Rey D. Weighted multi-model control. Int J Syst Sci. 1992;23(1):145–9.
12. Nagib G, Gharieband W, Binder Z. Qualitative multi-model control using a learning approach. Int J Syst Sci. 1992;23(6):855–69.
13. Aly A, Badr A, Binder Z. Multi-model control of mimo systems: location and control algorithms. Int J Syst Sci. 1992;19(9):1687–98.
14. Fekri S, Athans M, Pascoal A. Robust multiple model adaptive control (rmmac): a case study. Int J Adapt Control Signal Process. 2007;21(1):1–30.
15. Aguiar A, Hassani V, Pascoal A, Athans M. Identification and convergence analysis of a class of continuous-time multiple-model adaptive estimators. In: Proceedings of the 17th IFAC World Congress, Seoul:Korea;2008.
16. Hassani V, Aguiar A, Pascoal A, Athans M. A performance based model-set design strategy for multiple model adaptive estimation. In: Proceedings of European Control Conference, Budapest:Hungary;2009.
17. Hassani V, Aguiar A, Athans M, Pascoal A. Multiple model adaptive estimation and model identification using a minimum energy criterion. In: Proceedings of American Control Conference, St. Louis:Missouri;2009.
18. Hassani V, Athans M, Pascoal A. An application of the rmmac methodology to an unstable plant. In: Proceedings of 17th Mediterranean Conference on Control and Automation, Thessaloniki:Greece;2009.
19. Kuipers M, Ioannou P. Multiple model adaptive control with mixing. IEEE Trans Autom Control. 2010;55(8):1822–36.
20. Sadati N, Dumont GA, Mahdavian HRF. Robust multiple model adaptive control using fuzzy fusion. In: 42nd South Eastern Symposium on System Theory (University of Texas at Tyler, Tyler). TX: USA; March 2010.
21. He WG. Multiple model adaptive control for a class of systems with uncertain parameters. Acta Automatica Sinica. 1988;3:191–8.
22. Li X, Wang W, Sun W. Multi-model adaptive control. Control Decis.2000.
23. Yuan X, Shi X. Research progress of adaptive control based on multiple models. J Shanghai Jiao Tong Univ. 1999;33(5):626–30.
24. Zheng Y, Wang X, Li S, et al. Multiple models direct adaptive decoupling controller for a stochastic system. Acta Automatica Sinica. 2010;36(9).
25. Wang SH, Shen J, Li YG. Multi-model control and its study progress. Ind Instrum Autom. 2008;1:13–7.

26. Hu G, Sun Y. Research progress and application of multi-model control method. Inf Control. 2004;33(1):72–6.
27. Dong Z, et al. Application of weighted multiple models adaptive controller in the plate cooling process. Acta Automatica Sinica. 2010;36(8):1144–50.
28. Narendra KS, Han Z. The changing face of adaptive control: the use of multiple models. Ann Rev Control. 2011;35(1):1–12.
29. Zhang W. On the stability and convergence of self-tuning control—virtual equivalent system approach. Int J Control. 2010;83(5):879–96.
30. Zhang W, Chu TG, Wang L. A new theoretical framework for self-tuning control. Int J Inf Technol. 2005;11(11):123–39.
31. Zhang W, Shanghai PR. The convergence of parameter estimates is not necessary for a general self-tuning control system—stochastic plant. In: Proceedings of 48th IEEE Conference on Decision and Control, China;2009.
32. Zhang W, Li XL, Choi JY. A unified analysis of switching multiple model adaptive control—virtual equivalent system approach. In: Proceedings of 17th IFAC Word Congress, Seoul: Korea;2008.
33. Zhang W. Stable weighted multiple model adaptive control of discrete-time stochastic plant. Int J Adapt Control Signal Process. 2013;27(7):562–81.
34. Zhang W. Further results on stable weighted multiple model adaptive control: discrete-time stochastic plant. Int J Adapt Control Signal Process,online published, May, 2015.
35. Zhang W. Stability of weighted multiple model adaptive control. Control Theory Appl. 2012;29(12):1657–60.
36. Zhang W. Weighted multiple model adaptive control of discrete-time stochastic system with uncertain parameters. Zidonghua Xuebao/Acta Automatica Sinica. 2015;41(3):541–50.
37. Zhang W, Liu JW, Hu GD. Stability analysis of robust multiple model adaptive control systems. Control Theory Appl. 2013;41(1):113–21.
38. Baram Y. Information, consistent estimation and dynamic system identification. Ph.D. Dissertation, MIT, Cambridge, MA, U.S.A;1976.
39. Baram Y, Sandell NR. An information theoretic approach to dynamical systems modeling and identification. IEEE Trans Autom Control. 1978;23(1):61–6.
40. Baram Y, Sandel NR. Consistent estimation on finite parameter sets with application to linear systems identification. IEEE Trans Autom Control. 1978;23(3):451–4.
41. Kehagias A. Convergence properties of the lainiotis partition algorithm. Control Comput. 1991;19(1):1–6.
42. Han Z, Narendra KS. New concepts in adaptive control using multiple models. IEEE Trans Autom Control. 2012;57(1):78–89.
43. Chen W, Sun J, Chen C, Chen J. Adaptive control of a class of nonlinear systems using multiple models with smooth controller. Int J Robust Nonlinear Control. 2013;25(6):865–77.
44. Mahdianfar H, Ozgoli S, Momeni HR. Robust multiple model adaptive control: modified using v-gap metric. Int J Robust Nonlinear Control. 2011;21:2027–63.
45. Vinnicombe G. Frequency domain uncertainty and the graph topology. IEEE Trans Autom Control. 1993;38(9):1371–83.

Research on Gait Recognition and Step Rate Detection Based on SVM and Auto-Correlation Analysis

Hao Wu and Xi-Sheng Li

Abstract In order to monitor the motion state of human body, wearable system is studied. The system can identify gait and detect step rate. By data acquisition system with acceleration sensor, acceleration data are collected under various moving states, and immediately transmitted via blue-tooth. After signals have been analyzed by pre-processing and relevant feature extraction, gait patterns are recognized through support vector machine (SVM) classifier. Compared with back propagation neural network (BPNN) classifier, results indicate that SVM classifier has advantages of simple design and high accuracy for the same data set. So SVM classifier is more suitable. According to crest-detection, auto-correlation analysis algorithm is proposed and tested. Research findings show that auto-correlation analysis can greatly improve the accuracy and adaptability of step rate measurement.

Keywords Acceleration sensor · Gait recognition · Step rate detection · Support vector machine (SVM) · Auto-correlation analysis algorithm · Pattern recognition

1 Introduction

With the continuous development of artificial intelligence technology, there has been significant research interest in the field of human motion pattern recognition. In a wide variety of sensors, accelerometer can excellently reflect our bodily movements in real time. So it is the best choice to judge the mode of motion [1, 2]. There are thousands of movement pattern recognition systems. However, it is hard to obtain a universal method for feature extraction and data processing of acceleration signals [3]. Generally speaking, in view of different objects and requirements, novel features should be sought to better match recognition model

H. Wu · X.-S. Li (✉)
School of Automation and Electrical Engineering, University of Science and Technology Beijing, Beijing 100083, China
e-mail: lxs@ustb.edu.cn

© Springer Nature Singapore Pte Ltd. 2018
Y. Jia et al. (eds.), *Proceedings of 2017 Chinese Intelligent Systems Conference*, Lecture Notes in Electrical Engineering 460, https://doi.org/10.1007/978-981-10-6499-9_18

177

and improve system's performance. Classification algorithm plays a decisive role in gait recognition [4]. Nowadays, many kinds of classifiers, such as Bayesian, Back Propagation Neural Network (BPNN) and support vector machine (SVM), have been applied to movement pattern recognition [5, 6].

Combining the current research on pattern recognition, acceleration sensors collect acceleration data of people's daily motion. Then signals are preprocessed by noise reduction and data segmentation. Next, we extract the appropriate features. After that, SVM classifier is designed and compared with BPNN classifier. Meanwhile, based on auto-correlation analysis algorithm, step-rate test is successfully implemented under different moving states and compared with crest-detection. Following the development concepts of smart healthcare, research on wearable body health monitoring system, which can realize synchronization between gait recognition and step-rate surveillance, bears a practical research value and also opens up the vast prospects in health monitoring, smart monitor, health care and other fields.

2 System Design

Through acceleration data acquisition module, storage and analysis about human daily movement has been successfully accomplished. System designed for gait recognition and step rate detection mainly consists of three main parts: motion data acquisition and transmission, mobile real-time monitoring and off-line data analysis. Figure 1 shows hardware block diagram. Among them, acquisition module with STM32F103C8T6 microcontroller collects motion data by MPU6050 accelerometer. Then, data are transmitted to mobile phone via Bluetooth and saved to SD card in text format. Eventually, the stored data make preparations for the next PC data analysis.

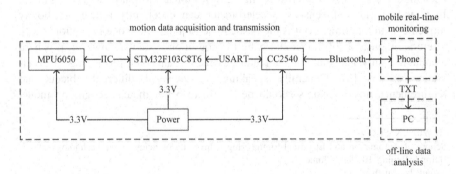

Fig. 1 Hardware system diagram

3 Experimental Methods and Processes

3.1 Data Acquisition

Data acquisition module with MPU6050 acceleration sensor is shown in Fig. 2. Figure 3 illustrates the positive direction of X, Y and Z axis in MPU6050 chip manual. When worn, the Y axis of the sensor is in direct alignment with the body's height. MPU6050 containing 16-bit ADC has two ranges: ±2 and ±16 g. In order to boost precision, ±2 g range is chosen.

Fig. 2 Acceleration sensor

Fig. 3 Orientation of MPU6050

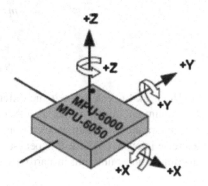

3.2 Data Pre-processing

On account of disturbances and sensor's peculiarity, signals are contaminated by multi-noises. First of all, we need to preprocess the data to reduce the influence of noise, and then lower the computational cost for the following feature judgment, gait recognition and step-rate analysis. Proper de-noising method is vital important to the fluency and redundancy of information processing for whole system.

In this study, to agree with the demand of real-time collection and large volume data, characteristics can't be extracted directly. Otherwise, a signal that contains multiple motions is most likely to be misjudged as a state. To avoid such an error, window processing method is adopted to segment the data, thus reducing the amount of computation at each time. Window size is focal point. Our body movement frequency is ordinarily between 0.5 and 10 Hz, so both direct current and high frequency signals above 10 Hz should be filtered before smoothing data. Complex filters may affect instantaneity, and then a first-order low-pass active filter is selected.

3.3 Feature Extraction

After preprocessing, we begin to extract and analyze special features. Because feature extraction is in the middle of the process, selection and extraction of features are directly related to classification ability. For acceleration data, this paper mainly introduces the mean, standard deviation, peak-to-peak, jump rate and Signal Magnitude Area (SMA) eigenvalues. The details are as follows.

Mean and standard deviation are the most common statistics. In formulas (1) and (2), m and σ represent the mean and standard deviation respectively, N is the number of sampling points of the calculation window, and a_i is the sampling value of corresponding acceleration axis at i moment.

$$m = \frac{1}{N} \sum_{i=1}^{n} a_i \tag{1}$$

$$\sigma = \frac{1}{N} \sum_{i=1}^{n} (a_i - m)^2 \tag{2}$$

The peak-to-peak value is the difference between the maximum wave crest and the minimum trough value. In other words, it indicates the difference between maximum and minimum acceleration. Peak-to-peak value is suitable for analyzing exercise intensity. Where P is peak-to-peak value. $F_{max}(N)$ and $F_{min}(N)$ denote the maximum and minimum in narrow window, respectively.

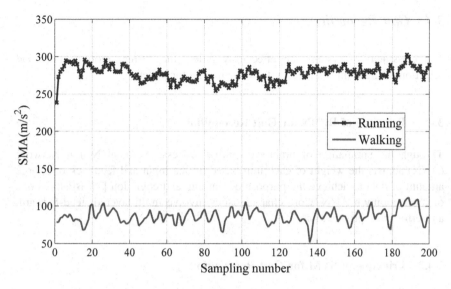

Fig. 4 Value of SMA in the condition of walking and running

$$P = F_{\max}(N) - F_{\min}(N) \tag{3}$$

The jump rate is the increase of signal in unit time, which reflects instantaneous change degree during exercise. In formula (4), $V(t)$ is the jump frequency and $F(t)$ is the acceleration at t moment. In addition, $Max(N)$ and $Min(N)$ represent the maxima and minima in the window, respectively.

$$V(t) = \frac{F(t) - Min(N)}{Max(N) - Min(N)} \tag{4}$$

SMA elucidates the degree of fluctuation of acceleration signals. The larger this value means the more intensive activities. Figure 4 shows the SMA values for walking and running. It can be seen that running is more volatile than walking. Where N is the points of calculating window. At i moment, $x(i)$, $y(i)$ and $z(i)$ respectively denote acceleration of X, Y and Z axis.

$$SMA = \sum_{t=1}^{N} (|x(i)| + |y(i)| + |z(i)|) \tag{5}$$

3.4   Gait Recognition

The classification system can effectively reflect the current motion state, and selection of algorithm is the key to recognize motor pattern.

3.4.1   Principle of BPNN for Gait Recognition

Through the simulation of brain's cognitive process, Artificial Neural Network (ANN) adjusts the weight of each unit based on the input and feedback of a large amount of data to achieve the purpose of learning and cognition [7]. BPNN is one of most common ANN. Core characteristics involve multi-layered, feed-forward, and error back propagation.

3.4.2   Principle of SVM for Gait Recognition

Support Vector Machine (SVM) is a kind of machine learning method regarding statistical learning theory as the core. It is based on the Vapnik Chervonenkis (VC) dimension theory. And the method takes the structural risk minimization as the criterion to improve the generalization ability of machine learning [8, 9]. SVM classifier has simple structure, focus on samples' self-information and provides better generalization [10]. Hence, it has an advantage of high accuracy in motion recognition for acceleration signals.

In SVM classifier, we use least squares method, select the Gaussian Radial Basis Function (RBF) as kernel function, and determine penalty factor C and variance g by cross validation [11].

3.5   Step Calculating Algorithm Based on Autocorrelation Analysis

Usually, the frequency of body movement is between 0.5 and 10 Hz. In order to get more details and make calculating algorithm more accurate, the sampling rate is selected as 100 Hz. When the sensor is worn, Y-axis should be parallel to our height. Among three-dimensional coordinate system, Y-axis data characteristics is most obvious through data analysis. Therefore, we mainly use Y axis data. Figure 5 shows the original data of Y-axis in walking state. It can be seen in the waveform that the acceleration goes up and down with the sampling points (or time). This phenomenon roughly reflects the movement of walking state. In order to improve the real-time and accuracy of the algorithm, acceleration signals need to be pre-processed. Results are shown in Fig. 6. It is observed that signal characteristics are more obvious after processing.

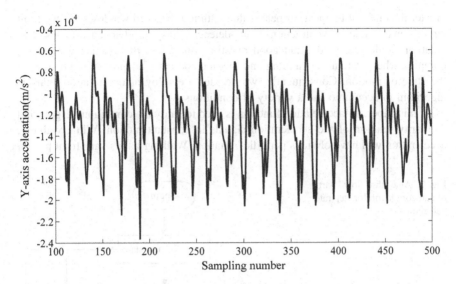

Fig. 5 Raw data of Y axis acceleration in the state of walking

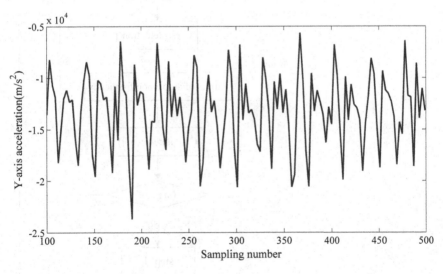

Fig. 6 Y axis acceleration data after processing in the state of walking

From the figure above, each waveform corresponds to one step. We can determine whether a new step is generated through peaks and troughs. The step rate testing algorithm is based on crest detection. Number of local extreme values is five both on left and right. In addition, some constraints should be added at the extremes to ensure completeness. The main limits here include magnitude and frequency. From Fig. 5, amplitude difference between the maximum and the local minimum is greater than half of the peak-to-peak value within designated windows. So we limit target's magnitude by subtraction. Crest-detection has the advantages of less calculation, facile realization, and good real-time operation. But Human behavior is complex when walking or running and this method can't adapt well to changes in the state of motion. Consequently, we decide to employ autocorrelation analysis algorithm to accommodate a variety of conditions.

In this paper, autocorrelation function is used to study the continuous degree and similarity relation of the acceleration signals in constant motion. While walking, acceleration will have obvious periodic variation. Walking habits vary from person

Fig. 7 Algorithm flow chart of autocorrelation for step rate detection

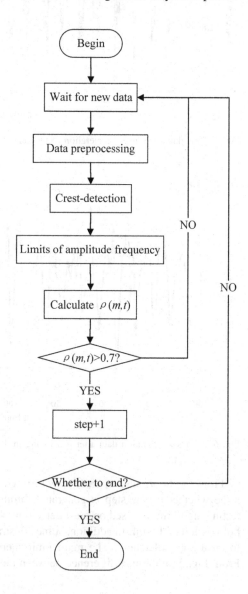

to person, which may cause different periodic states. But for one person, the stride frequency is stable in a certain range. By using these correlations, more accurate state judgments can be achieved.

According to above eigenvalues, such as mean and standard deviation, it is natural to get the autocorrelation function of acceleration sequence. Where $m(v, t)$ and $\sigma(v, t)$ represent the mean and standard deviation of the acceleration sequence a $(v + k)$, respectively. If sampling period is closer to a walk cycle, autocorrelation function $x(v, t)$ will be more similar to 1. For sampling period t, it is indispensable to make a concrete analysis of each person's specific situation. Autocorrelation coefficient $\rho(m, t)$ for a period of time is as follows.

$$x(v,t) = \frac{\sum_{k=0}^{t-1} [(a(v+k) - m(v,t)) \times (a(v+k+t) - m(v+t,t))]}{t \times \sigma(v,t) \times \sigma(v+t,t)} \tag{6}$$

$$\rho(m,t) = \max_{t=t_{\min}}^{t=t_{\max}} (x(v,t)) \tag{7}$$

By calculating $\rho(m, t)$, cycle time t that corresponds to max $\rho(m, t)$ is found from $t_{\min}$ to $t_{\max}$ and defined as sampling period. When the autocorrelation coefficient is greater than 0.7, possibility of walking is 90% and detection precision is higher [12]. So threshold of $\rho(m, t)$ is set to 0.7. Chart of the whole step rate detection algorithm is shown in Fig. 7.

4 Analysis of Experimental Results

4.1 Results and Comparative Analysis of Gait Recognition

In order to verify the accuracy of gait recognition above, which is based on SVM, the relevant data of daily movement are collected about 20 people including 10 males and 10 females. Experimenters are 24 years old on average and all in good health condition. They complete four states of movement: rest, walking, running, and jumping. Table 1 lists the confusion matrix for classifying motion data by SVM classifier.

Table 1 Confusion matrix for classifying motion data by SVM classifier

	Rest	Walk	Run	Jump
Rest	0.947	0.033	0.012	0.008
Walk	0.052	0.938	0.004	0.006
Run	0.024	0.021	0.924	0.031
Jump	0.013	0.024	0.018	0.945

Table 2 Confusion matrix for classifying motion data by BPNN classifier

	Static	Walk	Run	Jump
Static	0.883	0.062	0.025	0.030
Walk	0.040	0.878	0.056	0.026
Run	0.039	0.035	0.869	0.057
Jump	0.025	0.034	0.046	0.895

To further describe the reliability and superiority, SVM classifier is compared with BPNN classifier in performance. BPNN with 20 neurons have same training and testing samples with the SVM classifier. Then we test the results of training. Table 2 is the confusion matrix of BPNN classifier for classifying motion data.

Result of comparison suggests that BPNN classifier has an average classification accuracy of 88.1%, while SVM classifier is 93.8%. The latter is better than the former. So, SVM classifier can not only identify the different gait mode more accurately, but also reflect the current movement state more effectively.

4.2   Results and Comparative Analysis of Step Rate Test

The calculating effect of three different states of motion, slow, fast and the combination of these, are tested. Then, we compare autocorrelation with peak detection. The results are shown in Table 3. Average accuracy of peak detection is 87.5%, while autocorrelation analysis is 97.5%. The autocorrelation analysis increases accuracy by 10%. It can effectively detect the number of steps under different motion states and has great practicality and adaptability.

Table 3 Comparison of two algorithms for counting steps' number

Motion state	Actual number	Peak detection		Autocorrelation analysis	
		Step calculation	Accuracy %	Step calculation	Accuracy %
Slow	100	97	97	99	99
	100	102	98	99	99
Fast	100	80	80	96	96
	100	82	82	97	97
Combination	100	85	85	103	97
	100	83	83	97	97

5 Conclusions

The gait and step rate are important indexes that reflect characteristics of human life characteristics. They should be monitored at all times, especially for high-pressure groups. Following the development concepts of smart healthcare, wearable body health monitoring system, which can realize both gait recognition and step-rate surveillance, is under research. Firstly, the data acquisition system is designed with acceleration sensor, and data of daily motion states about 20 people are collected. SVM classifier is used to classify signals, and gait pattern recognition is successfully realized. At the meantime, performance of SVM and BPNN classifier is analyzed on the same data set. The results show that average classifying accuracy of BPNN classifier is 88.1%, while SVM classifier is 93.8%. Therefore, SVM classifier is more suitable for human gait recognition. Then, autocorrelation analysis algorithm is proposed based upon the crest detection and tested on experimental data. The results indicate that average accuracy of peak detection is 87.5%, and autocorrelation analysis algorithm is 97.5%. The autocorrelation analysis increases accuracy by 10% and improves the detection accuracy and adaptability. Study of this paper bears a practical research value and also opens up the vast prospects in health monitoring, smart monitor, health care and other fields.

Acknowledgements This work was supported by the National Natural Science Foundation of China Under Grants of 61273082.

References

1. Bonomi AG, Plasqui G, Goris AHC, et al. Improving assessment of daily energy expenditure by identifying types of physical activity with a single accelerometer. J Appl Physiol. 2009;107 (3):655–61.
2. Peng BZ, Lin DJ, Jin Cheng OU. Application of artificial intelligence in the field of sensor. J Transducer Technol. 2002;21(3):5–7.
3. Mantyjarvi J, Himberg J, Seppanen T. Recognizing human motion with multiple acceleration sensors IEEE International Conference on Systems, Man, and Cybernetics. IEEE Xplore,vol. 2, 2001. p. 747–2.
4. Kela J, Korpip Panu, et al. Accelerometer-based gesture control for a design environment. Pers Ubiquit Comput. 2006;10(5):285–99.
5. Maurer U, Rowe A, Smailagic A, et al. Location and activity recognition using ewatch: A wearable sensor platform. Ambient Intelligence in Everyday Life. Berlin Heidelberg: Springer; 2006. p. 86–102.
6. Preece SJ, John Yannis G, Kenney LPJ, et al. A comparison of feature extraction methods for the classification of dynamic activities from accelerometer data. IEEE Trans Biomed Eng. 2009;56(3):871–9.
7. Shi XL. Development and application of artificial neural network. J Chongqing Univ Sci Technol. 2006.
8. Brown M, Gunn SR, Lewis HG. Support vector machines for optimal classification and spectral unmixing. Ecol Model. 1999;120(2–3):167–79.

9. Hu G, Hu L, Li H, et al. Grid resources prediction with support vector regression and particle swarm optimization. Third International Joint Conference on Computational Science and Optimization. IEEE, 2010. p. 417–2.
10. Lu J, Zhang E. Gait recognition for human identification based on ICA and fuzzy SVM through multiple views fusion. Pattern Recogn Lett. 2007;28(16):2401–11.
11. Osuna EE. Support vector machines: training and applications. Massachusetts Institute of Technology, 1998.
12. Chen GL, Zhang YZ, Zhou Y. Realization of pedometer with auto-correlation analysis based on mobile phone sensor. J Chin Inert Technol. 2014;22(6):794–8.

Consensus of Mixed-Order Multi-agent Systems with Directed Topology

Hua Geng, Zengqiang Chen, Zhongxin Liu, Chunyan Zhang and Qing Zhang

Abstract In this paper, an in-depth study about the consensus problem of mixed-order multi-agent system with directed topology is performed. Specifically, this system is composed of two classes of agents respectively described by first-order and second-order dynamics. By the aid of state transformation, two different methods are proposed to solve the consensus. Still, two meaningful examples are provided to verify the effectiveness of the gained theoretical results. This paper is expected to establish a more realistic model and provide effective measures to solve the consensus problem.

Keywords Consensus · Multi-agent systems · Mixed-order · Directed topology
Variable transformation

1 Introduction

Distributed coordinate control of multi-agent systems has drawn greatly attention in past decades due to its potentially huge value. It has been an emerging research direction of control theory and came into widespread use, such as consensus [1], flocking [2], formation control [3], optimization-based operative control [4], etc.

H. Geng
School of Information Electrical Engineering, Hebei University
of Engineering, Handan 056038, China

Z. Chen(✉) · Z. Liu · C. Zhang
College of Computer and Control Engineering, Nankai University,
Tianjin 300071, China
e-mail: chenzq@nankai.edu.cn

Z. Chen · Z. Liu · C. Zhang
Tianjin Key Laboratory of Intelligent Robotics, Nankai University,
Tianjin 300071, China

Q. Zhang
College of Science, Civil Aviation University of China,
Tianjin 300300, China

© Springer Nature Singapore Pte Ltd. 2018
Y. Jia et al. (eds.), *Proceedings of 2017 Chinese Intelligent Systems Conference*,
Lecture Notes in Electrical Engineering 460,
https://doi.org/10.1007/978-981-10-6499-9_19

Consensus is a fundamental problem to distributed coordinate control of multi-agent systems, and it requires the state of agents in the multi-agent systems reach an agreement by coordinating and negotiating with their neighbors [5]. A long line of research in this field has contributed a bunch of works focusing on the solution of the fascinating consensus problem. For example, in [6], a classical research framework of consensus problem is proposed. Based on algebraic graph theory, matrix theory and control theory, three cases were analyzed when the communication topology among the agents was directed and fixed, directed and switched, undirected and fixed, respectively. Inspired by the prominent work [5], many consensus algorithms were introduced for first-order integrator multi-agent systems in the face of various conditions, such as with time-delay [7], with nonlinear input constraints [8], in noisy environment [9], etc. With the deepening of researching, there has been a tremendous amount of renewed interest in multi-agent systems with second-order integrator agents [10–12]. In [10], the consensus algorithms were proposed under 4 different scenarios. As a extended case, in [12],the leader-following tracking control problem for second-order multi-agent systems under measurement noises and directed communication channels was investigated. Based on a novel velocity decomposition technique, the proposed consensus protocols could guarantee all followers track the active leader. Furthermore, in [13, 14], the consensus of multi-agent systems with inherent nonlinear dynamics and singular systems were discussed, respectively.

In the [1–14], the agents in the multi-agent systems shared the same dynamics (first-order integrator dynamics or second-order integrator dynamics). However, in the practical engineering applications, the agents are characterised as different dynamics due to the cooperation of different groups of vehicles, robots or space vehicles. It is well known that the heterogeneous multi-agent system is a more general case, and has a broader range of applications. In [15], the heterogeneous multi-agent system was investigated under undirected topology for the first time. Based on the graph theory, Lyapunov directed method and Lasalle's invariance principle, the consensus problems were discussed with linear consensus protocol and saturated consensus protocol, respectively. As a extended case, the consensus of leader-following network was also solved. In the following, under the assumption of without velocity measurements, the consensus of heterogeneous multi-agent systems was achieved [16]. In [17], due to the limitation of actuators, a special case was discussed. Based on the distributed observers and auxiliary systems methods, the consensus problem, aggregating problem and tracking problem were solved, respectively.

In this context, motivated by a realistic scenario, the mixed-order multi-agent system is considered. For example, due to the difference of controllers (the control input of some agents are velocity, the others are acceleration), it is more appropriate to formulate the multi-agent system as the mixed-order equation. And more remarkably, the directed communication topology is more realistic and more challenging. At these points, by the aid of state transformation, two different methods are proposed to solve the consensus. Finally, two meaningful examples are provided to verify the effectiveness of the gained theoretical results.

2 Preliminaries

2.1 Graph Theory

Let $\mathcal{G} = (\mathcal{V}, \mathcal{E}, \mathcal{A})$ be a weighted directed graph of order n with the set of nodes $\mathcal{V} = \{v_1, v_2, \ldots, v_n\}$, set of edges $\mathcal{E} \subseteq \mathcal{V} \times \mathcal{V}$, and an adjacency matrix $\mathcal{A} = [a_{ij}] \in \mathbb{R}^{N \times N}$ with nonnegative elements a_{ij}. $e_{ij} = (v_i, v_j)$ denotes an edge from node v_j to v_i. The set of neighbors of node v_i is denoted by $\mathcal{N}_i = \{v_j \in \mathcal{V} : (v_i, v_j) \in \mathcal{E}\}$. The graph $\mathcal{G}$ has a directed spanning tree if there exists a node called the root such that there exist directed paths from this node to every other node. The degree matrix $D = \{d_1, d_2, \ldots, d_n\} \in \mathbb{R}^{n \times n}$ of graph $\mathcal{G}$ is a diagonal matrix, where diagonal $d_i = \sum_{j \in N_i} a_{ij}$ for $i = 1, 2, \ldots, N$. Then the Laplacian matrix of $\mathcal{G}$ is defined as $\mathcal{L} = D - \mathcal{A}$.

2.2 Problem Description

Consider a mixed-order multi-agent system of size N (agents), and the number of second-order agents is $M(M < N)$, the rest are first-order agents. Without loss of generality, the dynamics of second-order agents are described by

$$\begin{cases} \dot{x}_i = v_i, \ i = 1, 2, \ldots, M \\ \dot{v}_i = u_i, \ i = 1, 2, \ldots, M \end{cases} \tag{1}$$

where $x_i, v_i, u_i \in \mathbb{R}^p$ represent the position, velocity and control input vectors of agent i, respectively. The dynamics of first-order agents are described by

$$\dot{x}_i = u_i, i = M + 1, M + 2, \ldots, N \tag{2}$$

where $x_i, u_i \in \mathbb{R}^p$ represent the position and control input vectors of agent i, respectively. Moreover, the definition of consensus of mixed-order multi-agent systems is given as follows.

Definition 1 (*see* [15]) The mixed-order multi-agent system (1)–(2) is said to reach consensus if for any initial condition we have

$$\lim_{t \to \infty} \|x_i(t) - x_j(t)\| = 0, \forall i, j \in \{1, 2, \ldots, N\}$$
$$\lim_{t \to \infty} \|v_i(t) - v_j(t)\| = 0, \forall i, j \in \{1, 2, \ldots, M\}$$

3   Main Results

3.1   Quasi First-Order System Approach

Motivated by literature [1, 10], the consensus protocol is proposed as follows.

$$\begin{cases} u_i = k_1 \sum_{j=1}^{N} a_{ij}(x_j - x_i) - k_2 v_i, & i = 1, 2, \ldots, M \\ u_i = k_3 \sum_{j=1}^{N} a_{ij}(x_j - x_i), & i = M+1, M+2, \ldots, N \end{cases} \tag{3}$$

where $k_1 > 0, k_2 > 0, k_3 > 0$ are the feedback gain. So the mixed-order multi-agent system can be written as follows.

$$\begin{bmatrix} \dot{x}_s \\ \dot{v}_s \\ \dot{x}_f \end{bmatrix} = \begin{bmatrix} 0_M & I_M & 0_{M\times(N-M)} \\ k_1(-L_s - D_{sf}) & -k_2 I_M & k_1 A_{sf} \\ k_3 A_{fs} & 0_{(N-M)\times M} & k_3(-L_f - D_{fs}) \end{bmatrix} \begin{bmatrix} x_s \\ v_s \\ x_f \end{bmatrix} \tag{4}$$

where $x_s^T = [x_1, x_2, \ldots, x_M]$, $v_s^T = [v_1, v_2, \ldots, v_M]$, $x_f^T = [x_{M+1}, x_{M+2}, \ldots, x_N]$, and L_s, $L_f, A_{sf}, A_{fs}, D_{sf}$ and D_{fs} are the sub-block of laplacian matrix L of mixed-order system.

Similar to [18, 19], the variable transformation $\bar{v}_i = \frac{v_i}{k_1} + x_i, i = \{1, 2, \ldots, M\}$ is introduced. So the system (4) can be rewritten as

$$\begin{bmatrix} \dot{x}_s \\ \dot{\bar{v}}_s \\ \dot{x}_f \end{bmatrix} = \begin{bmatrix} -k_1 I_M & k_1 I_M & 0_{M\times(N-M)} \\ (k_2 - k_1)I_M - (L_s + D_{sf}) & (k_1 - k_2)I_M & A_{sf} \\ k_3 A_{fs} & 0_{(N-M)\times M} & k_3(-L_f - D_{fs}) \end{bmatrix} \begin{bmatrix} x_s \\ \bar{v}_s \\ x_f \end{bmatrix} \tag{5}$$

Theorem 1 *Suppose the directed communication topology of mixed-order multi-agent system composed of* (1) *and* (2) *is fixed and has a spanning tree, the consensus problem can be solve if the feedback k_1 and k_2 satisfy the following condition*

$$k_2 - k_1 \geq d_i \tag{6}$$

where d_i denotes the element of degree matrix D, $i \in \{1, 2, \ldots, M\}$.

Proof After the variable transformation, the consensus problem of system (4) is converted to consensus problem of system (5). When the directed communication topology of system (4) has a spanning tree, the system (5) also has a spanning tree. Next, the system (5) is our focus, and it can be rewritten as

$$\dot{X} = HX \tag{7}$$

where $X^T = [x_s^T, \bar{v}_s^T, x_f^T]$, H is the block matrix. To investigate system (7), a star tranformation is introduced as follows.

$$Y = SX \tag{8}$$

where $S \in \mathbb{R}^{(N+M) \times (N+M)}$ is defined by

$$S = \begin{bmatrix} 1 & 0 & 0 & \cdots & 0 \\ 1 & -1 & 0 & \cdots & 0 \\ \vdots & \vdots & \vdots & \ddots & \vdots \\ 1 & 0 & 0 & \cdots & -1 \end{bmatrix}$$

It is easy to see that $S = S^{-1}$, The system (7) can be rewrriten as

$$\dot{Y} = SHS^{-1}Y \tag{9}$$

let $Y = [Y_1, Y_e^T]^T$, where $Y_1 = x_{s1}$, and $Y_e = [x_{s1} - x_{s2}, x_{s1} - x_{s3}, \ldots, x_{s1} - x_{sM}, x_{s1} - \bar{v}_{s1}, x_{s1} - \bar{v}_{s2}, \ldots, x_{s1} - \bar{v}_{sM}, x_{s1} - x_{f1}, x_{s1} - x_{f2}, \ldots, x_{s1} - x_{f(N-M)}]^T$. The system (9) can be divided into the following two subsystems.

$$\dot{Y}_1 = h_1 F Y_e \tag{10}$$

$$\dot{Y}_e = EHF Y_e \tag{11}$$

where h_1 is the first row of H, $E = \begin{bmatrix} \mathbf{1}_{N+M-1} & -I_{N+M-1} \end{bmatrix}$ and $F = [\mathbf{0}_{N+M-1} \ -I_{N+M-1}]^T$

It is easy to see that the consensus of system (7) is achieved if and only if the subsystem (11) is asymptotically stable with respect to Y_e. So the consensus problem of system (7) is converted to the asymptotically stable problem of system (11). Choose a Lyapunov function

$$V = Y_e^T P Y_e \tag{12}$$

where $P \in \mathbb{R}^{(N+M-1) \times (N+M-1)}$ is positive definite. Since the condition (6) holds, the matrix $-H$ can be considered as the Laplacian matrix of a directed topology with $N + M - 1$ agents. Assume Y_1 represents the root node in the network with $N + M$ agents that has a directed spanning tree, it is easy to see that the eigenvalues of matrix EHF are same as the eigenvalues of matrix H expect the single eigenvalue 0, i.e., EHF is Hurwitz stable. So $\lim_{t \to \infty} \|x_i(t) - x_j(t)\| = 0, \forall i, j \in \{1, 2, \ldots, N\}, \lim_{t \to \infty} \|v_i(t) - v_j(t)\| = 0, \forall i, j \in \{1, 2, \ldots, M\}$. Based on the proposed consensus protocol (3), the consensus of mixed-order system (4) is also solved.

3.2   Quasi Second-Order Systems Approach

Inspired by [14], the consensus protocols with auxiliary system for first-order agent are proposed as follows.

$$
\begin{cases}
u_i = k_1 \sum_{j=1}^{N} a_{ij}(x_j - x_i) + k_2 \sum_{j=1}^{M} a_{ij}(v_j - v_i) + k_2 \sum_{j=M+1}^{M} a_{ij}(\bar{v}_j - v_i), \; i \in \varGamma_1 \\
x_i = \bar{v}_i, \hspace{6.5cm} i \in \varGamma_2 \\
\dot{\bar{v}}_i = k_1 \sum_{j=1}^{N} a_{ij}(x_j - x_i) + k_2 \sum_{j=1}^{M} a_{ij}(v_j - \bar{v}_i) + k_2 \sum_{j=M+1}^{M} a_{ij}(\bar{v}_j - \bar{v}_i), \; i \in \varGamma_2
\end{cases}
\tag{13}
$$

where $k_1 > 0, k_2 > 0$ are the feedback gain, and $\bar{v}_0$ can be chosen arbitrarily. $\varGamma_1 = \{1, \ldots, M\}, \varGamma_2 = \{M+1, \ldots, N\}$. So the mixed-order multi-agent system with the proposed consensus protocols can be written as follows.

$$
\dot{\xi} = \begin{bmatrix} 0_{N \times N} & I_N \\ -k_1 L & -k_2 L \end{bmatrix} \xi
\tag{14}
$$

where $\xi = [x^T, v^T]^T, x = [x_1, x_2, \ldots, x_N]^T, v = [v_1, v_2, \ldots, v_M, \bar{v}_{M+1}, \ldots, \bar{v}_N]^T$. Via the star transformation, the system (14) can be divided into the following two subsystems.

$$
\dot{\eta}_1 = \begin{bmatrix} 0 & 1 \\ 0 & 0 \end{bmatrix} \eta_1 + \begin{bmatrix} 0_{N-1}^T & 0_{N-1}^T \\ -k_1 l_1 F & -k_2 l_1 F \end{bmatrix} \eta_e
\tag{15}
$$

$$
\dot{\eta}_e = \begin{bmatrix} 0_{(N-1) \times (N-1)} & I_{N-1} \\ -k_1 ELF & -k_2 ELF \end{bmatrix} \eta_e
\tag{16}
$$

where the notations are the same as those in (10) and (11). It is easy to see that mixed-order multi-agent system with consensus protocols (13) is converted to quasi second-order agent system (14), and the consensus problem is converted to the asymptotic stable problem of system (16). So the following result can be obtained.

Theorem 2 *Suppose the directed communication topology of mixed-order multi-agent system composed of (1) and (2) is fixed and has a spanning tree, the consensus problem can be solve if the feedback k_1 and k_2 satisfy the following condition*

$$
\begin{cases}
k_1 > 2\bar{\lambda} \\
k_2 > \frac{k_1^2}{2} \\
2k_1 k_2 - k_1^3 - 2\bar{\lambda} > 0
\end{cases}
$$

where $\bar{\lambda}(P) = max(\lambda(P))$, P is a positive matrix and satisfies $(ELF)^T P + P(ELF) = I_{(N-1) \times (N-1)}$.

Proof To investigate system (16), choose a Lyapunov function candidate as follows.

$$V = \eta_e^T \begin{bmatrix} \mu P & \nu P \\ \nu P & \gamma P \end{bmatrix} \eta_e = \eta_e^T \bar{P} \eta_e \tag{17}$$

where μ, ν, γ are positive constants with $\mu\gamma > \nu^2$. Differentiating V along (15) and (16), we have

$$\dot{V} = \eta_e^T \left(\bar{P} \begin{bmatrix} 0_{(N-1)\times(N-1)} & I_{N-1} \\ -k_1 ELF & -k_2 ELF \end{bmatrix} + \begin{bmatrix} 0_{(N-1)\times(N-1)} & I_{N-1} \\ -k_1 ELF & -k_2 ELF \end{bmatrix}^T \bar{P} \right) \eta_e \tag{18}$$

From the proof of Theorem 1, the matrix EHF is Hurwitz stable. It is not difficult to get that

$$\dot{V} = \eta_e^T \begin{bmatrix} -k_1^2 I_{N-1} & \mu P - k_1 k_2 I_{N-1} \\ \mu P - k_1 k_2 I_{N-1} & 2k_1 P - k_2^2 I_{N-1} \end{bmatrix} \eta_e = \eta_e^T \Xi \eta_e \tag{19}$$

Since P is positive definite, there exists an orthogonal matrix U such that $U^T P U = diag\{\lambda_1, \ldots, \lambda_{N-1}\} = \Lambda$. Using $diag\{U^T, U^T\}$ and $diag\{U, U\}$ to pre- and post-multiply Ξ. So the $-\Xi$ is positive definite if and only is the following matrix are positive definite.

$$\begin{bmatrix} k_1^2 & -\mu\lambda_i + k_1 k_2 \\ -\mu\lambda_i + k_1 k_2 & -2k_1 \lambda_i + k_2^2 \end{bmatrix} \quad i = \{1, 2, \ldots, N-1\} \tag{20}$$

Based on Vieta's Theorem, the following inequation must be hold.

$$k_1^2 - 2k_1 \lambda_i + k_2^2 > 0 \tag{21}$$

$$k_1^2(-2k_1 \lambda_i + k_2^2) - (-\mu\lambda_i + k_1 k_2)^2 > 0 \tag{22}$$

Let $\mu = 2$, based on the conditions of Theorem 2, (21) and (22) holds. So the system (16) is asymptotic stable, i.e. So $\lim_{t\to\infty} \|x_i(t) - x_j(t)\| = 0, \forall i, j \in \{1, 2, \ldots, N\}$, $\lim_{t\to\infty} \|v_i(t) - \bar{v}_j(t)\| = 0, \forall i \in \{1, 2, \ldots, M\}, \forall j \in \{M+1, M+2, \ldots, N\}$. Based on the proposed consensus protocol (13), the consensus of mixed-order system (4) is also solved.

4 Examples

In this section, two examples illustrates the effectiveness of the gained theoretical results.

Consider a mixed-order multi-agent system with 4 agents. Agent 1 and 2 are described by the second-order linear dynamics, while agent 3 and 4 are formulated as the first-order linear dynamics. The communication topology is given as $A = [0, 0, 0, 0; 1, 0, 0, 0; 0, 1, 0, 0; 0, 0, 1, 0]$.

Example 1 To verify the Sect. 3.1, suppose that $k_1 = 1$, $k_2 = 3$ and $k_3 = 1$, so the condition of Theorem 1 holds. The initial value of all agents are chosen as $x^T = [x_1, v_1, x_2, v_2, x_3, x_4] = [6, 2, -5, -4, 3, -2]$. From Fig. 1, it is known that the state trajectories of quasi first-order multi-agent system will tend to a same value.

Example 2 By calculating the positive definite P, we know that $\bar{\lambda} = 1.71$. Let $k_1 = 4$, $k_2 = 10$, so the conditions of Theorem 2 hold. The initial value of all agents are chosen as $x^T = [x_1, v_1, x_2, v_2, x_3, \bar{v}_3, x_4, \bar{v}_4] = [6, 2, -5, -4, 3, -2 - 1, 0]$. From Fig. 2, it is known that the position states of all agents will reach an agreement. Moreover, once the system reach an agreement, the states of all agents will be same as the agent 1.

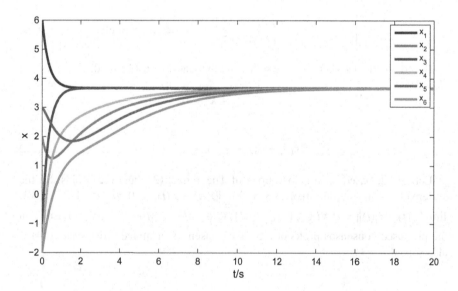

Fig. 1 The state trajectories of quasi first-order multi-agent system

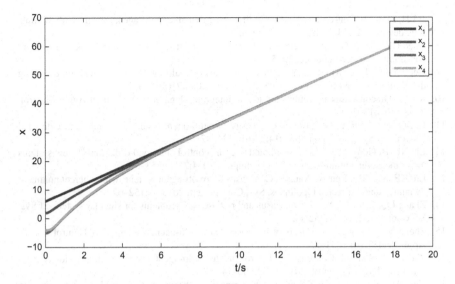

Fig. 2 The position trajectories of quasi second-order multi-agent system

5 Conclusion

This paper posed and addressed the consensus problem of mixed-order multi-agent system. Under the directed communication topology, two different methods were introduced to solve the consensus. By the variable transformation method, the mixed-order multi-agent system was converted to quasi first-order multi-agent system and second-order multi-agent system, respectively. By the analysis of the partial stability, the sufficient conditions of consensus were obtained. Finally, two examples illustrated the effectiveness of the gained theoretical results.

References

1. Ren W, Beard RW. Consensus seeking in multi-agent systems under dynamically changing interaction topologies. IEEE Trans Autom Control. 2005;50(5):655–61.
2. Olfati-Saber R. Flocking for multi-agent dynamic systems: algorithms and theory. IEEE Trans Autom Control. 2006;51(3):401–20.
3. Egerstedt M, XiaoMing H. Formation constrained multi-agent control. IEEE Trans Robot Autom. 2001;17(6):947–51.
4. Olfati-Saber R, Dunbar WB, Murray RM. Cooperative control of multivehicle systems using cost graphs and optimization. In: Proceedings of the American control conference. 2003. p. 2217–2222.
5. Keviczky T, Borelli F, Balas GJ. A study on decentralized receding horizon control for decoupled systems. In: Proceedings of the American control conference. 2004. p. 4921–4926.
6. Olfati-Saber R, Murray RM. Consensus problems in networks of agents with switching topology and time-delays. IEEE Trans Autom Control. 2004;49(9):1520–33.

7. Ji LH, Liao LF. Consensus problems of first-order dynamic multi-agent systems with multiple time delays. Chin Phys B. 2013;22(4):040203.
8. Miao GY, Ma Q. Group consensus of the first-order multi-agent systems with nonlinear input constraints. Neurocomputing. 2015;161:113–9.
9. Chen Y, JinHu L, XingHuo Y. Robust consensus of multi-agent systems with time-varying delays in noisy environment. J Syst Sci Complex. 2008;21(3):406–15.
10. Ren W. On consensus algorithms for double-integrator dynamics. IEEE Trans Autom Control. 2008;53(6):1503–9.
11. Li DQ, Wang XF, Yin ZX. Robust consensus for multi-agent systems over unbalanced directed networks. J Syst Sci Complex. 2014;27(6):1121–37.
12. Liu XL, BuGong X, Xie LH. Distributed tracking control of second-order multi-agent systems under measurement noises. J Syst Sci Complex. 2014;27(5):853–65.
13. Liu KE, Xie GM, Ren W, Wang L. Consensus for multi-agent systems with inherent nonlinear dynamics under directed topologies. Syst Control Lett. 2013;62:152–62.
14. Zhang LQ, Feng JE, Yao J. Consensus and r-consensus problems for singular systems. J Syst Sci Complex. 2014;27:252–62.
15. Zheng YS, Wang L. Consensus of heterogeneous multi-agent systems. IET Control Theory Appl. 2011;5(16):1881–8.
16. Zheng YS, Wang L. Consensus of heterogeneous multi-agent systems without velocity measurement. Intern J Control. 2012;85(7):906–14.
17. Geng H, Chen ZQ, Liu ZX, Zhang Q. Consensus of a heterogeneous multi-agent system with input saturation. Neurocomputing. 2015;166:382–8.
18. Gu MQ, Song YZ: Quasi-average consensus of directed hybrid swarm agents. In: Proceedings of the 27st Chinese control conference, Kunming. 2008.
19. Feng YZ, ShengYuan X, Lewis FL, Zhang BY. Consensus of heterogeneous first and second-order multi-agent systems with directed communication topologies. Intern J Robust Nonlinear Control. 2015;25:362–75.

Link Prediction from Partial Observation in Scale-Free Networks

Xuecheng Yu and Tianguang Chu

Abstract We study the link prediction problem in scale-free networks by using node similarity measure method to estimate the probabilities of potential links from a partial observation of links. Specifically, we give estimates of scaling parameter of the power-law distribution based on the observed node degrees and approximate solutions to a transcendental equation with Hurwitz-zeta function, whereby to obtain a local sub-network similarity measure of node pairs that do not have available observation information. Experiments on synthetic scale-free networks verify the effectiveness of our method.

Keywords Link prediction · Partial observation · Scale-free network Zeta function

1 Introduction

Recently, the study of link prediction in social networks has received considerable interest because its wide range of applications. Examples include prediction of the collaborations in co-authorship networks [1], detection of underground relationships between terrorists [2], and reconstruction of protein-protein interaction network [3].

Link prediction problem considers predicting new links which are likely to emerge in networks in future or links have existed in networks but not been observed. In general, it can be categorized into static link prediction problem and dynamic prediction problem. The former considers prediction of links using only a snapshot of the network, whereas the latter addresses that based on a stream of snapshots of the network over time [4]. In this paper, we focus on the static link prediction problem. To date, there have been several approaches to such a problem. For instance, Ref. [5] proposed a method to estimate probabilities of potential links using node similarity

X. Yu · T. Chu(✉)
College of Engineering, Peking University, Beijing 100871, China
e-mail: chutg@pku.edu.cn

X. Yu
e-mail: yuxuecheng@pku.edu.cn

© Springer Nature Singapore Pte Ltd. 2018   199
Y. Jia et al. (eds.), *Proceedings of 2017 Chinese Intelligent Systems Conference*,
Lecture Notes in Electrical Engineering 460,
https://doi.org/10.1007/978-981-10-6499-9_20

measures, which may be divided into local similarity measures and global similarity measures, according to features of the network used in similarity calculation [6]. Local similarity measures compute similarity between any two nodes using local features including the node degree, neighborhood, and number of common neighbors [4], and can be represented in terms of Salton index [7], Jaccard index, and Sørensen index [8], etc. Global similarity measures calculate the similarity between two nodes using global features such as the number of paths, information flow, etc. More details can be found in [8]. Besides, there have been other methods that make use of different information other than node similarity measures, e.g., unsupervised learning approaches based on heterogeneous nodes in network, and supervised learning approach using both network structure and node attributes [3].

In this paper, we consider the problem of link prediction in scale-free networks given a partially observed snapshot of the network, using the node similarity measure as an estimate of the probability of a potential link. To do so, we assume that the network structure is static and present estimates of scaling parameter of the power-law distribution based on observed node degrees. Then, we introduce a local sub-network similarity measure using the scaling parameter to estimate probabilities of potential links for the node pairs without observed links. Since similar nodes tend to connect to each other, links can be predicted in accordance with the rank of probabilities. We note that out work deals with the case of discrete power-law distribution, about which very fewer results have been available in literature. The existing methods for estimating scaling parameters of discrete power-law distributions usually require certain limits on the minimum value of observed node degrees, e.g., larger than six as pointed in [9]. Our methods does not have such restriction and can work well even in the case where the minimum observed node degree is one.

2 Problem Statement

Social networks are usually described by a directed graph, which can be denoted as $G(V, E)$, where V is the node set, E is the link set, and the number of nodes is $|V| = n$. For a pair of nodes i and j with link (i, j), if the information transmits from node i to node j not vice versa, node j is called an incoming neighbor of node i.

The aim of this paper is to study the link prediction problem in scale-free networks with a snapshot, which provides the knowledge of all nodes and partial links in a network.

3 Theory and Algorithm

Power-law distributions come in two expressions: continuous distributions and discrete distributions, where the quantity in distributions takes continuous real numbers or discrete set of values, respectively. Let x represent the random variable corresponding to the node degree. In this paper, we only consider the case of integer values with a probability distribution of the form

$$p(x) = \Pr(x = X) = Cx^{-\alpha} \tag{1}$$

where X is the observed value, C is a normalization constant, and α is a positive constant parameter of the distribution known as the scaling parameter.

In many cases, the value of α is obtained empirically. In [9] the scaling parameter α is in $(2,3)$, and in [10] it lies in $[2,4]$. In addition, it is pointed out in [9] that the distribution with $\alpha \leq 1$ is not normalizable and cannot occur in application. In view of this, we will estimate the value of α by fitting power-law distribution to observed node degrees under the assumption that $\alpha > 1$. Notice that given a snapshot of a network, the degrees of all nodes can be observed from a moralized graph of the network, which is obtained by adding links from each node to its all incoming neighbors in the snapshot. Here, we assume that the observed degree of each node is at least one, i.e.,

$$x_{\min} \geq 1,$$

where $x_{\min}$ is the minimum degree of observed nodes in a snapshot.

As pointed in [9], the distribution (1) is calculated by

$$p(x) = \frac{x^{-\alpha}}{\zeta(\alpha, x_{\min})}, \tag{2}$$

where

$$\zeta(\alpha, x_{\min}) = \sum_{n=0}^{\infty} (n + x_{\min})^{-\alpha}$$

is Hurwitz zeta function. So, the normalization constant C in (1) is $1/\zeta(\alpha, x_{\min})$.

To estimate the scaling parameter α, the power-law distribution (2) can be calculated by using maximum likelihood method. Given a data set containing n i.i.d. observed node degrees $x_i \geq x_{\min}$, $i \in [1, n]$, the probability that the observed node degrees drawn from the distribution (2) is proportional to

$$p(x|\alpha) = \prod_{i=1}^{n} p(x_i|\alpha),$$

which is also the likelihood of the observation. The observed node degrees are most likely to be generated by the distribution (2) with scaling parameter α that maximizes the likelihood function

$$L(\alpha) = \ln p(x|\alpha) = -n \ln(\zeta(\alpha, x_{\min})) - \alpha \sum_{i=1}^{n} \ln x_i.$$

The maximum likelihood estimate (MLE) for parameter α is written as

$$\hat{\alpha} = \arg \max_{\alpha} L(\alpha).$$

Taking $\partial L/\partial \alpha = 0$, and denoting $\frac{\partial}{\partial \alpha} \zeta(\alpha, x_{\min})$ as $\zeta'(\alpha, x_{\min})$ for simplicity, we have

$$\frac{\zeta'(\hat{\alpha}, x_{\min})}{\zeta(\hat{\alpha}, x_{\min})} = -\frac{1}{n} \sum_{i=1}^{n} \ln x_i,$$

which is difficult to solve because of Hurwitz zeta function. According to $x_{\min}$, the maximum likelihood estimate of α can be obtained in two cases.

In the case of $x_{\min} = 1$, Hurwitz zeta function $\zeta(\alpha, 1) = \zeta(\alpha)$, which is a special form known as Riemann zeta function [11]. A list of values of $\zeta'(\alpha)/\zeta(\alpha)$ is given in [12], by which one can evaluate α numerically. Here, we give an analytical evaluation of $\zeta'(\alpha)/\zeta(\alpha)$.

According to the Euler product formula for the Riemann zeta function, we have

$$-\frac{\zeta'(\alpha)}{\zeta(\alpha)} = \sum_{p \in P} \frac{\ln p \, e^{-\alpha \ln p}}{1 - e^{-\alpha \ln p}},$$

where $P = \{2, 3, 5, 7, \ldots\}$ is the set of all prime numbers in increasing order. By calculation, it can be obtained that

$$\sum_{k \geq 2} \Lambda(k) k^{-\alpha} = \frac{1}{n} \sum_{i=1}^{n} \ln x_i, \tag{3}$$

where

$$\Lambda(k) = \begin{cases} \ln p & \text{if } k = p^m, \ p \in P, \ m \in \mathbb{N} \\ 0 & \text{otherwise} \end{cases},$$

and $k \in \mathbb{N}$, $\mathbb{N}$ is the set of natural number.

Let $a_k = \Lambda(k) k^{-\alpha}$, then $b_k = \ln k \cdot k^{-\alpha} \geq a_k$. It can be verified that $\sum_{k=2}^{\infty} b_k$ is convergent. Hence, $\sum_{k=2}^{\infty} a_k$ is convergent. So, we can use a partial sum of $\sum_{k=2}^{\infty} a_k$ to approximate it. Actually, we can have

$$\sum_{k=2}^{\infty} a_k = \sum_{k=2}^{N} a_k + O\left(\frac{\ln N}{N^{\alpha-1}}\right).$$

Then, the estimate of scaling parameter $\hat{\alpha}$ can be obtained by solving (3) with the left side being approximated by the sum of its first finite terms.

In the case of $x_{min} > 1$, we have

$$\int_{x_{min}-\frac{1}{2}}^{\infty} t^{-\alpha} dt = \zeta(\alpha, x_{min}) \left(1 + \sum_{k \geq 1} d_k\right),$$

where

$$d_k = \frac{\prod_{i=0}^{2k-1}(\alpha + i)}{4^k(2k+1)!} x^{-2k}.$$

It is clear that $\{d_k\}$ is convergent based on the ratio test [13]. Hence,

$$\int_{x_{min}-\frac{1}{2}}^{\infty} t^{-\alpha} = \zeta(\alpha, x_{min})\left(1 + O(x_{min}^2)\right).$$

We have,

$$\frac{\zeta'}{\zeta} = -\frac{1}{n}\sum_{i=1}^{n}\ln x_i \approx -\left[\frac{1}{\alpha - 1} + \ln(x_{min} - \frac{1}{2})\right].$$

Therefore, the scaling parameter α can be approximated by

$$\hat{\alpha} = 1 + n\left[\sum_{i=1}^{n}\ln\frac{x_i}{x_{min} - \frac{1}{2}}\right]^{-1}. \tag{4}$$

Using $\hat{\alpha}$, we give a local sub-network (LS) similarity measure to estimate probabilities of potential links for node pairs without observed links. For a node pair (a, b), we use observed node degrees to estimate the scaling parameters of a sub-network centering on nodes a and b and the global network respectively, then calculate similarity of a and b by

$$LS(a, b) = \sum_{z \in N(a,b)}\left(x_z^{\alpha_l/\alpha_g}\right)^{-1}, \tag{5}$$

where x_z is the degree of node z, $N(a, b)$ is a sub-network centering on nodes a and b, α_g is the scaling parameter of global network, and α_l is the scaling parameter of sub-network. In $N(a, b)$, the shortest path length from any node to a or b is equal or less than a setting radius of the sub-network.

We outline the link prediction in Algorithm 1.

Algorithm 1 Link Prediction Algorithm in scale-free network
___
Input: Snapshot of a social network $G(V, E)$, the number of links to predict n_p
Output: Predicted links based on LS similarity measure
  1: **for** $(a, b) \notin E$ **do**
  2:     Calculate α_l and α_g which respectively represent the scaling parameters of sub-network and global network using (3) and (4)
  3:     Calculate the similarity score for node pair (a, b) using (5)
  4:     Sort the scores of nodes pairs in descending order
  5:     Output the top n_p links
  6: **end for**

---

4 Experiments

To verify the effectiveness of our method, experiments are preformed in synthetic scale-free networks, which are generated according to Barabasi-Albert generative procedure [10]. After randomly deleting a given percentage of links, the residual network is considered as a partially observed snapshot. Then, links are predicted using Algorithm 1. For comparison, we also use other similarity measures to predict links. The measures we used in experiments are listed as follow.

$$LA(a, b) = \sum_{z \in N(a) \cap N(b)} \left(x_z^{\alpha_l / \alpha_g} \right)^{-1},$$

$$AA(a, b) = \sum_{z \in N(a) \cap N(b)} \ln(x_z)^{-1},$$

$$RA(a, b) = \sum_{z \in N(a) \cap N(b)} x_z^{-1},$$

$$CN(a, b) = |N(a) \cap N(b)|,$$

$$PA(a, b) = x_a x_b,$$

where $N(a)$ is set of neighboring nodes of a, x_z is the degree of node z. The experiment results are given in Table 1.

Table 1 Experimental results. Numbers in the left column are percentages of deleting links in synthetic scale-free networks. Numbers in other columns are the accuracy of link prediction results

	LS%	LA%	AA%	RA%	CN%	PA%
10	66.7	50.0	50.0	50.0	50.0	16.7
20	53.8	23.1	46.1	46.1	30.8	15.4
30	30.0	20.0	25.0	25.0	40.0	15.0

5 Conclusions

In this paper, we gave estimates of scaling parameter of the discrete power-law distribution based on observed node degrees. In the case that the minimum observed node degree is one, the scaling parameter can be estimated by solving an equation with Riemann zeta function numerically. In the case that the minimum observed degree is larger than one, an approximating method was developed to solve a transcendental equation with Hurwitz zeta function. Based on the scaling parameter, a local subnetwork similarity measure was proposed to estimate probabilities of potential link for a pair of node, using both local and global information of the network structure. We further gave an algorithm to predict missing links according to the rank of probabilities. Experiments on synthetic networks show that our method gives better results compared with other existing similarity measures.

Acknowledgements This work was supported by National Natural Science Foundation of China under Grant No. 61673027, and partly by National Basic Research Program of China (973 Program, No. 2012CB821203).

References

1. Liben-Nowell D, Kleinberg J. The link-prediction problem for social networks. J. Assoc. Inf. Sci. Technol. 2007;58(7):1019–31.
2. Clauset A, Moore C, Newman ME. Hierarchical structure and the prediction of missing links in networks. Nature. 2008;453:98–101.
3. Zhao Y, Wu YJ, Levina E, Zhu J. Link prediction for partially observed networks. 2013. arXiv: 1301.7047.
4. Srinivas V, Mitra P. Link prediction in social networks. Springer International Publishing; 2016.
5. Liben-Nowell D, Kleinberg J. The link prediction problem for social networks. In: Proceedings of the 2003 ACM CIKM international conference on information and knowledge management; 2003. p. 556–559.
6. Scholz M. Node similarity as a basic principle behind connectivity in complex networks. 2010. arXiv.1010.0803.
7. Salton G, McGill MJ. Introduction to modern information retrieval. Auckland: McGraw-Hill; 1983.
8. Lu L, Zhou T. Link prediction in complex networks: a survey. Phys Stat Mech Appl. 2010;390(6):1150–70.
9. Clauset A, Shalizi CR, Newman MEJ. Power-law distributions in empirical data. Siam Rev. 2009;51(4):661–703.
10. Barabasi AL, Albert R. Emergence of scaling in random networks. Science. 1999;286(5439):509–12.
11. Borwein JM, Bradley DM, Crandall RE. Computational strategies for the Riemann zeta function. J Comput Appl Math. 2000;121(1):247–96.
12. Walther A. Anschauliches zur Riemannschen Zetafunktion. Acta Mathematica. 1926;48(3):393–400.
13. Rudin W. Principles of mathematical analysis. McGraw-Hill Book Co. Inc; 1953.

Apnea Detection with Microbend Fiber-Optic Sensor

Li Liu, Tingfeng Ye, Xuemei Guo, Ruixun Kong, Lei Bo
and Guoli Wang

Abstract As a common chronic disease, sleep apnea syndrome (SAS) seriously threatens patients' health, so it is imperative to find an effective way of apnea detection. In this paper, we present a new method that explores of using the high sensitivity characteristic of the micro-bending effect of the gradient multimode fiber to detect human breathing movement, and thereby collecting the respiratory signals. With characterizing central apnea, 8 different features in time-domain and frequency-domain are extracted from the respiratory signals, which are used to train the classifiers. In the feature modeling, we present a peak calculation method based on moving average curve (MAC) to increase the accuracy of estimating the respiratory frequency and amplitude. In our experimental studies, the forward sequence selection method (SFS) is employed to combine these features for training SVM classifier, and our approach can reach an accuracy rate of 94.6% in apnea discrimination.

Keywords Fiber-optic sensor · Frequency estimation · SAS · Apnea detection

L. Liu · T. Ye
School of Electronics and Information Technology, Sun Yat-sen University,
Guangzhou 510006, China

X. Guo · G. Wang (✉)
School of Data and Computer Science, Sun Yat-sen University,
Guangzhou 510006, China
e-mail: isswgl@mail.sysu.edu.cn

R. Kong · L. Bo
Vkan Certification & Testing Co., Ltd, Guangzhou, China

X. Guo · G. Wang
Key Laboratory of Machine Intelligence and Advanced Computing,
Ministry of Education, Guangzhou, China

© Springer Nature Singapore Pte Ltd. 2018
Y. Jia et al. (eds.), *Proceedings of 2017 Chinese Intelligent Systems Conference*,
Lecture Notes in Electrical Engineering 460,
https://doi.org/10.1007/978-981-10-6499-9_21

1 Introduction

Sleep apnea syndrome (SAS) is a common chronic disease, which seriously threatens patients' health [1]. SAS includes: central type, obstructive and mixed type. According to AASM guidelines [2], it is determined that an apnea event happens if the respiration pauses more than 10 s. Frequent apnea would cause oxygen shortage of human body, which thereby lead to the brain hypoxia and bring a bad quality of sleep. Hence, the patient would be sleepy or a drowsiness symptom occurs. For car drivers, it can cause serious traffic accidents [3] that severely threaten people's lives. Furthermore, SAS is an independent risk factor for hypertension, coronary heart disease, arrhythmia, stroke and other diseases [4]. Thus, it is of great significance to put forward a feasible method to monitor the human respiration and recognize SAS, and thereby evaluate the human health condition.

Currently, the overnight in-laboratory Polysomnography (PSG) recording is the gold standard method for the diagnosis of apnea [2], which continuously records several different physiological parameters such as respiratory intensity, air flow, oxygen saturation, and heart rate. However, a typical PSG is a costly, time-consuming, labor-intensive procedure, which would cause discomfort and is intrusive to the patients. Thus, various techniques have been proposed to simplify SAS detection. For example, air mattress sensor [5], acoustic sensor [6] and liquid pressure sensor [7] are used in the research. Some other methods with reduced number of needed parameters based on air flow [8, 9], snoring [10], photo-plethysmography (PPG) [11] and electrocardiography (ECG) [12, 13] are also used in detecting SAS.

In existing works, a variety of breath monitoring methods based on fiber optical sensor has been proposed. Plasma modified fiber sensor [14], for example, which is based on a plastic optical fiber (POF), is used to monitor patients' breath by contacting with the patient chest, but the manufacturing process is too complex to be applied. Interferometric fiber sensor [15] is developed working with other expensive instruments. In [16], micro-bending loss effect theory of multimode optic fiber was introduced. Using the theory, a high sensitivity micro-bending fiber sensor is studied by Chen et al. [17–19], which is simple and low-cost, and was used in heart rate and respiration detection, but the accuracy and stability of their method need to be improved. In this paper, the microbend fiber-optic sensor is used to build our respiratory signal acquisition system. More specifically, we use the high sensitivity characteristic of the micro-bending effect of the gradient multimode fiber to detect human breathing movement, and thereby collecting the respiratory signals.

Respiration frequency is an important parameter in breath detection. Each breathing cycle contains a crest (or trough) because of its sine characteristic, so the number of peaks in the statistical period can reflect the frequency of the signal. However, for most cases, the respiratory signal is not a standard sine wave and pseudo-peak would occur because of interference or other factors, thus, with the general peak detection method, it would easily judge a pseudo-peak as an actual peak, which may cause error. Zero-crossing detection method can, to a certain extent, avoid wrong judgment of pseudo-peak, but because the respiratory signal of the reference value may be up and down the value of zero, and the method cannot count the peak under zero, so the same, it may cause error. In [18], they simply use the peak calculation method to estimate breath frequency, which brings error due to the pseudo-peak. In order to increase the accuracy of estimating the respiratory frequency and amplitude, we present a MAC-based [20] peak calculation method.

In this study, SVM is used as classifier to discriminate apnea event. As the performance of a classifier depends primarily on the selection of good features [21, 22], we find out 8 different features from both time-domain and frequency-domain, and use sequential forward selection method (SFS) [23] to choose different combination of features to train the classifiers, and compare the classification results between them. Finally, we reach a discrimination accuracy of 94.6% by using SVM classifier.

2 Acquisition of Respiration Signal

2.1 System Description

There are three parts in our system: the cushions with embedded fiber, the peripheral circuit and the display terminal, as shown in Fig. 1. The cushion, with a length of about 70 cm, width of 10 cm, thickness of about 3 mm, is consists of outer layer, inner micro-bend deformer, and the embedded optical fiber. The transceiver circuit integrates the transmitter (HFBR1419, Avago), the receiver (HFBR 1429, Avago), the amplifier circuit and the wireless transmission module (CC2530) together. The transmitting and receiving ends respectively realize the mutual conversion between the electrical signal and the optical signal. The amplifying circuit amplifies the changing signal from the receiving end and filters part of the high frequency noise, and then send the voltage signal to the display terminal (PC) through the CC2530 wireless transmitting module.

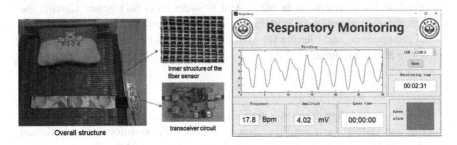

Fig. 1 The cushions with embedded fiber, the peripheral circuit and the display terminal

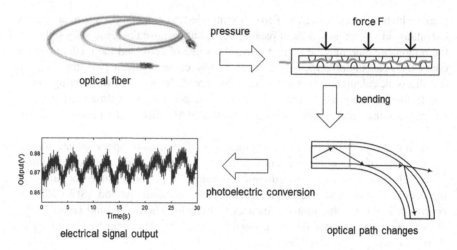

Fig. 2 Schematic diagram of monitoring breath movement using the microbend fiber-optic sensor

The optical fiber used in this paper is a graded multimode fiber with a core radius of 62.5 μm, a core refractive index of about 1.5 and a numerical aperture of about 0.0275, and the optimal period Λ_c of the optical fiber is about 6.8 mm. We make use of the high sensitivity characteristics of gradient multimode fiber micro-bending effect to measure the small breathing amplitude of the human body. Figure 2 shows the schematic diagram of monitoring breath movement using the microbend fiber-optic sensor.

2.2 Signal Acquisition

Human breathing state include: normal breathing when quiet, deep breathing, rapid breathing and apnea.

During the process of collecting respiratory signals, different subjects are asked to lie down quietly on the bed shown in Fig. 1, the body lies on the cushion to ensure contact with the sensor. For collecting normal respiratory signals, subjects are asked to breath normally in 90 s. And in order to collect apnea signals, the subjects simulate apnea in the middle 30 s of a 90 s testing period. After removing the DC component, and filtering the signal using a Butterworth low-pass filter, we get our final output signal

$$s(t) = A \, \cos(2\pi f t + \varphi) + e(t) \tag{1}$$

Those collected signals form a training set and a test set, which are used for training and testing the classifiers. Figure 3 shows the waveform of a collected signal, from which we can see that apnea happens in the middle 30 s.

Fig. 3 The waveform of
simulated apnea

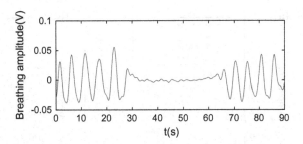

3   Respiratory Frequency Estimation

Since the irregularities of breath frequency and amplitude may occur when
breathing, a peak calculation method based on MAC is proposed to calculate the
respiratory rate and amplitude, in order to improve the accuracy of respiratory
monitoring results. First of all, we use the MAC method [20] to process the signal,
the formula is expressed as

$$
MAC(t) = \begin{cases} \overline{s(\tau)}\Big|_{0}^{2T}, & 0 \leq t \leq T \\[2mm] \overline{s(\tau)}\Big|_{t-T}^{t+T}, & T \leq t \leq L-T \\[2mm] \overline{s(\tau)}\Big|_{L-2T}^{L}, & L-T \leq t \leq L \end{cases} \tag{2}
$$

where s is the respiratory signal, L is the signal length; T is the average time
window, with the length of about a respiratory cycle. We use the maximum likelihood estimation method to get the maximum likelihood frequency f^*, and let
$T = 1/f^*$. $\overline{s(\tau)}\big|_{t1}^{t2}$ represents the mean value of signals during $[t_1, t_2]$. Using the
method above, we can get the moving average curve of the signal. Then we define
the rising and falling time flag of the signal as

$$
t_{rise} = \{s(t-1) \leq MAC(t-1)\} \cap \{s(t) \leq MAC(t)\} \tag{3}
$$

$$
t_{fall} = \{s(t-1) \geq MAC(t-1)\} \cap \{s(t) < MAC(t)\} \tag{4}
$$

The peak is defined as the maximum value between a rising flag and an adjacent
falling flag, while the valley is defined as the minimum value between a falling flag
and an adjacent rising flag. The formula is given as

$$
peak = max\{s(t_{rise}: t_{fall})\} \tag{5}
$$

$$
valley = min\{s(t'_{fall}: t'_{rise})\} \tag{6}
$$

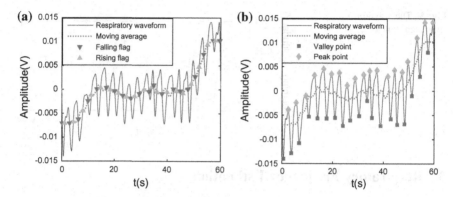

Fig. 4 **a** Find the rising and falling flags based on MAC. **b** Detecting the peak point based on the rising and falling flags

The peak point detected by the above method is shown in Fig. 4. It can be seen that the method can effectively find the peak of the signal and remove the pseudo-peak interference, and thereby improve the accuracy of the frequency estimation.

According to the number of peak points in the statistical time window, the formulas to calculate the respiratory rate and the breathing amplitude is given as

$$f = n_{peak} \times \frac{60}{L} \tag{7}$$

$$A = \frac{\sum_{1}^{n_{peak}} |peak(i) - MAC(i)|}{n_{peak}} \tag{8}$$

where n_{peak} is the number of peak points in the time window, and L is the signal length in seconds. $peak(i)$ and $MAC(t_i)$ represent the value of the i-th peak and its MAC value of the corresponding time, respectively.

In order to compare the errors between the MAC-based peak calculation method and the maximum likelihood frequency estimation method for the estimation of the respiratory frequency, we take the mean error of each method at different frequencies, and compare with the error of general peak calculation method. The result is shown in Fig. 5. It can be seen from the figure that the error of the MAC based method is significantly reduced compared with the maximum likelihood estimation method, and in most cases it can be controlled within 10%. Therefore, the use of the MAC based method for short time respiratory signal frequency estimation has a high reliability and accuracy.

Fig. 5 Frequency error of
different frequency using
maximum likelihood
estimation method

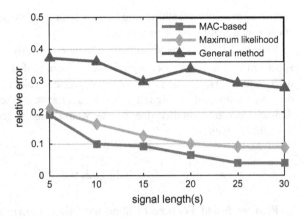

4 Apnea Detection

In this section, we will study the respiratory signals collected by the fiber-optic
sensor, and to detect apnea event using machine learning algorithms. Particularly,
we will focus on determining the central type apnea.

4.1 Feature Extraction of the Central Apnea

Compared with the normal respiratory signals, there are many different characteristics of the central apnea signals. In this paper, the following 8 apnea features are
extracted from time-domain and frequency-domain.

Feature 1: The upper 90th percentile of the signal sequence. Here we assume
that the respiratory signal sequence is $s = [s(1), s(2), \ldots, s(N)]^T$, where N is the
length of the sequence. Each value in the sequence is arranged in ascending order to
obtain the ordered sequence Z, the upper 90th percentile is equal to $Z(r)$, where
$r = \lceil 0.9N \rceil$. Obviously when the apnea occurs, the upper 90th percentile of the
respiratory signal is much smaller than the normal breath, and its size would reflect
the intensity of the breathing movement for a period of time.

Feature 2: Short-term time-domain variance, denote as σ_s^2. The variance reflects,
to a large extent, the strength of the breathing movement.

$$\sigma_s^2 = \frac{1}{N} \sum_{i=1}^{N} (s(i) - \bar{s})^2 \tag{9}$$

Feature 3: Maximum spectral value f_{max}. For the normal sine-like breathing
signal, the maximum spectral value of the spectrum is significantly greater than the

central apnea with breathing amplitude of zero, so the maximum spectral value is chosen as a feature of central apnea.

Feature 4: Frequency domain variance σ_f^2. After obtaining the short-term spectrum of the respiratory signal, the sequence of the spectrum values between 0.167 and 0.5 Hz is taken out and the variance is calculated to obtain the frequency domain variance of the signal. When apnea happens, the variance in the frequency domain becomes smaller due to the weakening of respiratory movement. The formula is given as

$$\sigma_f^2 = \frac{1}{N} \sum_{i=1}^{N} (f(i) - \bar{f})^2 \tag{10}$$

Feature 5 and **Feature 6**: Short-term time-domain maximum value S_{max} and minimum value S_{min}. Similar to the short-term time-domain variance, the stronger the respiratory motion, the more the maximum and the minimum value would deviate from the mean line. Oppositely, when apnea happens, the signal would be near the zero mean line.

Feature 7: The mean amplitude value of the short-term respiratory signal, denote as $\bar{A}$. $\bar{A}$ would be at a certain level when breathing normally, while it would be near zero when apnea happens.

Feature 8: The area of the short-term time-domain, denote as $Area(s)$, and it can be expressed as

$$Area(s) = \sum_{1}^{n_{peak}} |s(i) - MAC(i)| \tag{11}$$

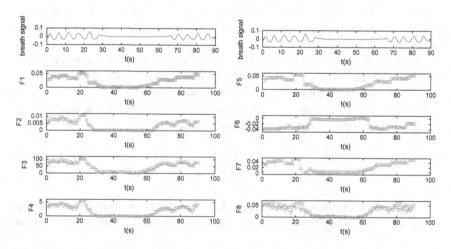

Fig. 6 The original breathing signal and its features waveform. F_i represents feature i

Table 1 Classification results of SVM

Features	Feature set	SE (%)	SP (%)	AC (%)
2	F1, F8	82.7	98.5	93.1
3	F1, F6, F8	82.6	99.0	93.8
4	F1, F6, F7, F8	85.5	99.1	94.6
5	F1, F2, F6, F7, F8	85.5	99.3	94.6
6	F1, F2, F5, F6, F7, F8	82.6	99.2	93.5
7	F1, F2, F4, F5, F6, F7, F8	76.8	99.3	91.6
8	F1, F2, F3, F4, F5, F6, F7, F8	69.1	98.7	90.7

Figure 6 shows a section containing the original breathing signal and its features waveform, where F1 represents Feature1, and so forth. Set the sliding window length to be 8 s and slides every 1 s. It can be seen from Fig. 6 that, under normal breathing and apnea, there are obvious differences among them. In this paper, we combine different features to form an eigenvector, and try to determine apnea using machine learning algorithms.

4.2 *Experimental Results and Analysis*

During the experiment, 300 normal breathing samples and 300 central apnea samples from male and female respectively were collected to form a training set, while another 240 respiratory samples were used as testing set.

For feature selection, we use sequential forward selection method (SFS), which is a kind of greedy search algorithm, to select the combination of features. Specifically, SFS starts from an empty feature set, and sequentially add a new feature that maximizes the classification accuracy when combined with the features that have already been selected. Sensitivity (SE), specificity (SP) and accuracy (AC) are used to evaluate the discrimination effect of central apnea. The discrimination results are shown in Table 1.

As can be seen from Table 1, the SVM classifier works best when the number of features is 5, wherein the sensitivity and the specificity are 85.5% and 99.3% respectively. At this time, the approach reaches an accuracy rate of 94.6% in apnea discrimination.

5 Conclusions and Discussion

In order to achieve satisfactory apnea detection performance, this work presents a new method that uses microbend fiber-optic sensor to collect respiratory signals, and we present a peak calculation method based on MAC to increase the accuracy of estimating the respiratory frequency and amplitude. After studying the features of the respiratory signal, 8 features from time-domain and frequency-domain are selected to train the SVM classifier and make a discrimination of the collected signals. Finally, our approach reaches a satisfactory accuracy rate of 94.6% in apnea detection. However, it can be seen from the experimental results that it's possible to increase the sensitivity and accuracy rate to a higher level, and these are what we will study in next step.

Acknowledgements This work was supported by the National Natural Science Foundation of P. R. China under Grant Nos. 61375080 and 61772574, and the Key Program of Natural Science Foundation of Guangdong, China under Grant No. 2015A030311049. The Guangzhou science and technology project under Grant Nos. 201510010017, 201604010101.

References

1. Young T, Peppard PE, Gottlieb DJ. Epidemiology of obstructive sleep apnea: a population health perspective. Am J Respir Crit Care Med. 2002;165(9):1217–39.
2. American Academy of Sleep Medicine (AASM) Task Force. Sleep-related breathing disorders in adults: recommendations for syndrome definition and measurement techniques in clinical research. Sleep. 1999;22:667–89.
3. Radun I, Radun JE. Convicted of fatigued driving: who, why and how? Accid Anal Prev. 2009;41(4):869–75.
4. Klar H, Yaggi MD. Obstructive sleep apnea as a risk factor for stroke and death. New England J Med. 2005;353:2034–41.
5. Chee Y, Han J, Youn J, Park K. Air mattress sensor system with balancing tube for unconstrained measurement of respiration and heart beat movements. Physiol Meas. 2005;26 (4):413–22.
6. Scanlon MV. Acoustically monitor physiology during sleep and activity. In: IEEE conference on engineering in medicine and biology; 1999. p. 787.
7. Zhu X, Chen W, Nemoto T, Kanemitsu Y, Kitamura K, Yamakoshi K, Wei D. Real-time monitoring of respiration rhythm and pulse rate during sleep. IEEE Trans Biomed Eng. 53 (12):2553–63.
8. Rathnayake SI, Wood IA, Abeyratne UR, Hukins C. Nonlinear features for single-channel diagnosis of sleep-disordered breathing diseases. IEEE Trans Biomed Eng. 2010;57(8): 1973–81.
9. Nakano H, Tanigawa T, Furukawa T, Nishima S. Automatic detection of sleep-disordered breathing from a single-channel airflow record. Eur. Respiration, J. 2007;29:728–36.
10. Abeyratne UR, Wakwella AS, Hukins C. Pitch jump probability measure for the analysis of snoring sound in apnea. Physiol Meas. 2005;26:779–98.
11. Nakajima K, Tamura T, Miike H. Monitoring of heart and respiratory rates by photoplethysmography using a digital filtering technique. Med Eng Phys. 1996;18(5):365–72.
12. Bsoul M, Minn H, Tamil L. Apnea MedAssist: real-time sleep apneamonitor using single-lead ECG. IEEE Trans Inf Technol Biomed. 2011;15(3):416–27.
13. Khandoker AH, Palaniswami M, Karmakar CK. Support vector machines for automated recognition of obstructive sleep apnea syndrome from ECG recordings. IEEE Trans Inf Technol Biomed. 2009;13(1):37–48.
14. Vallan A, Carullo A, Casalicchio ML, et al. A plasma modified fiber sensor for breath rate monitoring. In: 2014 IEEE international symposium on medical measurements and applications (MeMeA). IEEE; 2014. p. 1–5.

15. Sprager S, Zazula D. Detection of heartbeat and respiration from optical interferometric signal by using wavelet transform. Comput Methods Programs Biomed. 2013;111:41–51.
16. Berthold JWI. Historical review of microbend fiber optic sensors. J Lightwave Technol. 1995;13(7):1193–9.
17. Chen Z, Teo JT, Ng SH, et al. Portable fiber optic ballistocardiogram sensor for home use. In: SPIE BiOS. International Society for Optics and Photonics; 2012. 82180X-82180X -7.
18. Chen Z, Teo JT, Ng SH, et al. Monitoring respiration and cardiac activity during sleep using microbend fiber sensor: a clinical study and new algorithm. In: 2014 36th annual international conference of the IEEE engineering in medicine and biology society (EMBC). IEEE; 2014. p. 5377–80.
19. Deepu CJ, Chen Z, Teo JT, et al. A smart cushion for real-time heart rate monitoring. In: Biomedical circuits and systems conference (BioCAS), 2012 IEEE. IEEE; 2012. p. 53–6.
20. Lu W, Nystrom MM, Parikh PJ, et al. A semi-automatic method for peak and valley detection in free-breathing respiratory waveforms. Med Phys. 2006;33(10):3634–6.
21. Bishop CM. Neural networks for pattern recognition. NewYork, NY, USA: Oxford University Press; 1995.
22. Yu L, Liu H. Feature selection for high-dimensional data: a fast correlation-based filter solution. In: International conferences on machine learning; 2003. p. 856–63.
23. Pal M, Foody GM. Feature selection for classification of hyperspectral data by SVM. IEEE Trans Geosci Remote Sens. 2010;48(5):2297–307.

Improved Biogeography-Based Optimization Algorithm for Mobile Robot Path Planning

Jianfei Yang and Lin Li

Abstract In view of biogeography-based optimization algorithm has the disadvantages of application limitations and slow convergence speed when in solving the problem of mobile robot path planning. This paper proposes an improved biogeography-based optimization algorithm, which is used to solve the global path planning of mobile robot in static environment. In the proposed algorithm, the navigation point model is selected as the working area model of mobile robot, and the nonlinear migration model and mutation mechanism with the elite retention mechanism are introduced to the biogeography-based optimization algorithm to improve its performance. Simulation proves the feasibility and the effectiveness of the proposed path planning algorithm.

Keywords Mobile robot · Path planning · Biogeography-based optimization algorithm · The navigation point model

1 Introduction

Path Planning problem is one of the most basic and critical issues in mobile robot research [1–3]. It is to solve how mobile robots can walk in the environment. According to the degree of mastery of the information range of the robot, the path planning of mobile robot can be divided into two categories: one is the global path planning; the other is local path planning.

Over the years, experts and scholars have proposed a lot of algorithms for mobile robot path planning. One is the traditional methods, which more representative of the methods are: visual method, artificial potential field method [4–6], free space

J. Yang · L. Li (✉)
Department of Control Science and Engineering, University of Shanghai
for Science and Technology, Shanghai, China
e-mail: lilin0211@163.com

J. Yang
e-mail: 343442147@qq.com

© Springer Nature Singapore Pte Ltd. 2018   219
Y. Jia et al. (eds.), *Proceedings of 2017 Chinese Intelligent Systems Conference*,
Lecture Notes in Electrical Engineering 460,
https://doi.org/10.1007/978-981-10-6499-9_22

method; one is the intelligent methods, such as A-star algorithm [7], neural network algorithm, genetic algorithm, simulated annealing algorithm. Rajat et al. [8] adopted an efficient genetic algorithm to solve the problem of mobile robot path planning in dynamic environment. In [9], Wang Jianguo et al. proposed genetic algorithm based on grid map model which is used to solve the problem of mobile robot path planning. However, the swarm intelligent optimization algorithms have more wider application and stability, and stronger parallel search ability and faster convergence speed than the general intelligent algorithms [8, 9]. The commonly used swarm intelligent algorithms are: ant colony algorithm, particle swarm algorithm, artificial fish algorithm. For example, in [10], a combination of genetic algorithm and ant colony algorithm is proposed by He Juan et al., which was used in solving the mobile robot path planning in grid map situation. Qin Yuanqin et al. investigated a relatively new particle swarm algorithm which combined with the mutation operator to solve the problem of mobile robot path planning based on MAKLINK graph theory [11]. In [12], Yongkuo Liu et al. introduced a particle swarm optimization algorithm in solving the problem of mobile robot path planning in special working area. In [13], a hybrid particle swarm algorithm is investigated by Purcaru et al., which combined with a gravitational search algorithm based on grid map in two-dimensional static environment.

However, the above mentioned intelligent algorithms still have the shortcomings of large computational complexity, easy to fall into local optimum, low adaptability and low convergence speed. Therefore, it is necessary to find a more effective intelligent algorithm. In 2008, Scholar Simon proposed biogeography-based optimization (BBO) algorithm which is a bionic optimization. It is a more effective swarm intelligent algorithm for global optimization studies. The BBO algorithm is to achieve optimization by simulating the migration process of nature species between habitats. It has the advantages of strong global search ability, fast convergence speed and effective use of current information. At present, there are few relevant literatures on this algorithm both at home and abroad. For example, Li Xiangtao et al. introduced a BBO algorithm in solving global numerical optimization [14]. An improved BBO algorithm in solving mobile robot path planning problem is proposed by Mo Hongwei et al., which combined with the particle swarm algorithm [15]. In order to make up application limitation of basic BBO algorithm and improve its convergence rate, an improved BBO algorithm is proposed in this paper, which can be used to solve the problem of mobile robot path planning in static environment by changing the migration model and using elite retention strategy. Simulation proves the validity and feasibility of the proposed algorithm.

2 Build the Environment Model

The problem of mobile robot path planning mainly includes two subproblems: environment modeling and path search. At present, the commonly used environment models are: the undirected network graph model based on MAKLINK theory, topology map model, feature map model, grid map model, navigation point model and so on. Among them, the grid map model has been widely used because it is easy to be stored and analyzed in the computer and also convenient to updated the map. However, the combination of grid map model and BBO algorithm is not ideal in dealing with obstacle and boundary constraint problems. The navigation point model is an environment model which has been adopted gradually in recent years. In this paper, the navigation point method is used to model the environment model of mobile robot path planning. The concrete description is as follows:

Assuming that the working area range of mobile robot is a finite area of two-dimensional plane, there are a limited number of known size static obstacles in this area, and the motion of mobile robot can be regarded as a mass movement. In a $60 * 100$ two-dimensional plane space, we use different radius of the circles to represent the static obstacle, the radius represent the size of obstacle itself.

As shown in Fig. 1, create a coordinate system with the coordinates of the lower left corner of the map as the origin of the coordinates, so that the distribution of obstacles in this area can be expressed by a circle equation, such as (1):

$$(x-a)^2 + (x-b)^2 = R^2 \tag{1}$$

where (a, b) is the center coordinate of the obstacle, R is the radius of the obstacle, the starting point of the path (yellow box) and the target point (red box) are preset. Path planning is to get a best route from starting point to target point in trying to avoid all of the obstacles as far as possible. The path of mobile robot is determined

Fig. 1 Robot working area map

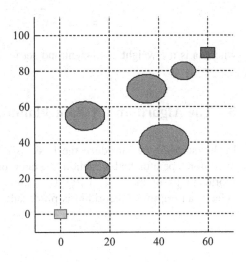

by a series of navigation point, this path is composed of short straight lines connecting these navigation points, so a path can be described by a series of sequential navigation point coordinates. This article sets up four rough navigation points and fifty careful navigation points, which constitute the path of the robot together.

In the path planning, the length of path is often used as a index. Therefore, the length function can be established according to this length. In this paper, the Euclidean distance between the two path points is taken as the path length function:

$$l = \sqrt{(x_s - x_t)^2 + (y_s - y_t)^2} \tag{2}$$

where l is the path length function, x_s, x_t, y_s, y_t is the point of path.

Owing to the traditional model-based method does not do the effective external interference constraint of the robot path planning, it is possible that the planned path collide with obstacles in practical application. In order to solve the above problem, the need to set up a safe distance, therefore, the following penalty function is set in the selected environment model:

$$v = \max(1 - d/R(k), 0) \tag{3}$$

where d is the distance between the path node and the center point of the obstacle; R (k) is the radius of the kth obstacle.

$$d = \sqrt{(x_i - a)^2 + (y_i - b)^2} \tag{4}$$

where x_i and y_i are the coordinates of the ith path point; a and b are the center coordinates of the obstacle, where the smaller the value of v, the safety coefficient of the final path is higher. The fitness functions can be obtained by combining formula (2) and (3):

$$f = \sqrt{(x_s - x_t)^2 + (y_s - y_t)^2} + \omega \cdot v \tag{5}$$

where ω is the weight coefficient and set to 100 here.

3   The Algorithm for Path Planning

In this section, we introduce the basic BBO algorithm firstly, and introduce its migration operator and mutation operator briefly. Then, we analyze briefly the shortcomings of the basic BBO algorithm. In order to solve the corresponding defects, an improved algorithm is proposed.

3.1 The Basic Biogeography-Based Optimization Algorithm

Dan Simon proposed the biogeography-based optimization algorithm in 2008. This algorithm is based on the principle of species migration in biogeography, combined with the framework structure of other swarm intelligent optimization algorithms. It has the advantages of strong global search ability and fast convergence rate. Since this algorithm has been put forward, it has been widely concerned by scholars both at home and abroad, and has become a hotspot in the path planning algorithm. It not only achieves better performance than other intelligent algorithms in solving the problems of nonlinear and non-different or other complex conditions, and also achieves good application effect in many practical optimization projects.

In the BBO algorithm, each habitat individual corresponds to the possible solution of the optimization problem and the characteristics of habitat habitable are called suitability index variables (SIV), which are consists of temperature, humidity, rainfall and other natural conditions and the habitat itself. Whether a habitat is suitable for species survival is reflected by the habitat suitability index (HSI). The idea of BBO algorithm is to improve the species diversity of habitats and improve HSI of habitat by means of species exchange between habitats according to the basic principle of species migration, and obtain HSI optimal habitat individual. A habitat is suitable for species to survive when it has a higher HSI. On the country, it has a lower HSI. The HSI is an important factor affecting population distribution and migration, because it can be used as an evaluation standard of a habitat for habitable. Often, a habitat with a higher HSI is suit for more species living here, with less HSI habitat accommodating less species. However, the habitat with lower HSI would accept new biological population from a habitat with higher HSI, to improve the diversity of the habitat with lower HSI and improve the HSI value of the habitat. In BBO, the candidate population is represented by integer vectors. Each integer in the solution vector corresponds to an SIV.

The main feature of BBO differs from the other intelligent algorithm lies in its migration operator. In BBO, each individual has its own migration rate λs and exit rate μs, which are functions in the habitat when the number of species is S, (s = 1, 2, 3 ... p). Represented by the Eqs. (6) and (7):

$$\mu_s = \frac{E \cdot S}{p} \tag{6}$$

$$\lambda_s = I \cdot \left(1 - \frac{S}{p}\right) \tag{7}$$

where E is the maximum migration rate, I is the maximum exit rate, p is the maximum number of species. Generally, E and I are unit matrix.

Assume the probability p_s that the habitat is just to accommodate the number of species is s. The probabilities of each species count can be calculated using the differential formula (8) within a certain period of time.

$$\dot{p}_s = \begin{cases} -(\mu_s + \lambda_s) \cdot p_s + \mu_{s+1} \cdot p_{s+1}, s = 0 \\ -(\mu_s + \lambda_s) \cdot p_s + \lambda_{s-1} \cdot p_{s-1} + \mu_{s+1} \cdot p_{s+1}, 1 \le s < s_{\max} - 1 \\ -(\mu_s + \lambda_s) \cdot p_s + \lambda_{s-1} \cdot p_{s-1}, s = s_{\max} \end{cases} \qquad (8)$$

where $s_{\max}$ is the largest number of species of the habitat.

During migration process, the information exchange between habitats is depended on λs and μs. The migration idea is: for each individual x_i, first determine whether x_i needs to be migrated based on the λs. If it is determined that x_i is the individual to be migrated, for each dimension SIV of x_i, in the remaining habitat selected a individual x_k which needs to be exited in the way of roulette based on μs, instead a SIV of x_i with migration operator which are combined with x_k randomly.

After migration process, the mutation is used to improve the diversity of the population to obtain better solutions. The mutation operator randomly changes a habitat's SIV based on mutation rate m_s. The mutation rate m_s is expressed by the Eq. (9):

$$m_s = m_{\max} \cdot \left(1 - \frac{p_s}{p_{\max}}\right) \qquad (9)$$

where $m_{\max}$ is the maximum mutation rate of the habitat, $p_{\max}$ is the maximum probabilities of species. The mutation strategy is: For the habitat individual x_i, determine whether it needs to be mutated based on m_s. If it is determined that x_i is the individual to be mutated, a SIV of x_i is replaced with a randomly generated value in its search space.

Finally, the habitat individual after migration and mutation is compared with the original individual. If the HSI value of this individual is higher than the originals', replace the original individual with this individual in next evolution. Otherwise, the original individual as a better individual was preserved. By continuously iterating evolution, the algorithm retains the better individual outs and evolves the poor individual and guides the search process to the global optimal individual approximation.

3.2 Improved Biogeography-Based Optimization Algorithm

The BBO algorithm is a simple and efficient swarm intelligent optimization algorithm which is proposed by Simon in 2008. There are a simple linear migration mechanism and random mutation operator. This algorithm was originally used for theoretical research and some simple experimental or engineering applications. However, it is found that the BBO algorithm based on this migration and mutation mechanism can not reflect the more complicated natural migration process and can not solve the more complicated experimental or engineering problems. It also has the lower later convergence speed in solving some simple optimization problems.

So, an improved BBO algorithm is investigated. And it is applied to the global path planning for mobile robots in static environments.

In the improved algorithm, the linear migration mechanism is replaced with nonlinear migration mechanism, so as to change the calculation method of the migration rate and the exit rate, realize the dynamic change of the migration rate according to the quantity. On the basis of this migration method, the SIV of the habitat individual x_k who will be moved out is combined with the weight of the SIV of the habitat individual x_i that will be moved into to replace the individual x_i. This migration operator also enhances the information utilization ability of the individual and the evolutionary ability of the individual itself. This nonlinear migration process can be used in more complex and realistic optimization environment and path planning environment, and we can get more real experimental data by using this migration process. However, the mutation process of the basic BBO would make the better individual become better individual by mutate its SIV, simultaneously, would bring its mutation into poor individual possibly. Therefore, in order to prevent this from happening, the elite retention mechanism is adopted in the proposed algorithm, so that the information of the individuals with better habitat is effectively protected. The nonlinear cosline migration model is described by Eqs. (10) and (11), the weighting-based migration operator is described by the Eq. (12):

$$\mu_s = \frac{E}{2} \cdot (\cos(\frac{s \cdot \pi}{s_{max}}) + 1) \tag{10}$$

$$\lambda_s = \frac{I}{2} \cdot (-\cos(\frac{s \cdot \pi}{s_{max}}) + 1) \tag{11}$$

$$x_{ij} = \alpha \cdot x_{ij} + (1 - \alpha) \cdot x_{kj} \tag{12}$$

where α is the weight coefficient, and the x_{kj} is the j-th dimension of SIV of the to be exited individual.

4   The Algorithm Flow

First, generate a navigation point model diagram under the known obstacle distribution, use the radius of the circle to represent the obstacles, preset the starting point and the target point. The maximum number of iterations of the algorithm is T, the current number of iterations is 0.

Step 1   Initialize algorithm parameters: m_{max}, T, p, and others;
Step 2   Calculate the fitness function values for each habitat;
Step 3   Determine whether meet the termination condition;
Step 4   Keep the individuals with the highest values in a temporary array;
Step 5   Calculate the migration rate λs and exit rate μs of each individual;

Step 6   Chose a habitat which is to be moved into based on the λs;
Step 7   Chose a habitat which is to be moved out in the remaining habitats based
          on μs;
Step 8   Do the migration process on the selected habitats according to the above
          two steps;
Step 9   Mutate the habitat which is selected to mutate by (9);
Step 10   Calculate the fitness values for each habitat again, and sort the habitat in
          best calculated result to the worst mode;
Step 11   Determine whether to meet the termination conditions, if satisfied, then
          continue to the next step, otherwise to go to the second step;
Step 12   Output the optimal habitat individual and stop the algorithm.

5   The Simulation and Analysis

In order to verify the performance of the proposed algorithm, this algorithm is
simulated by MATLAB 2014a in the Windows7 operating system. For comparison,
we compare the proposed algorithm with the basic BBO algorithm in the same
environment for solving same problem.

In this simulation, the maximum number of iterations is set to 250; the maximum
number of populations is set to 20; the probability of mutation is set to 0.1 and the
migration probability is set to 0.2.

Fig. 2 Path graph of
proposed algorithm

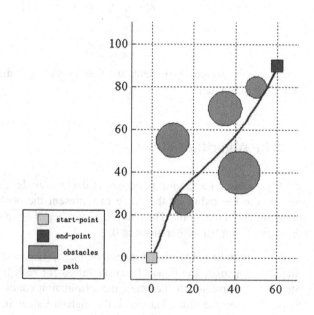

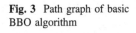

Fig. 3 Path graph of basic BBO algorithm

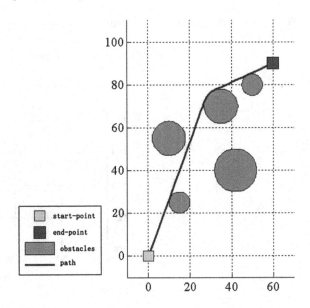

Figure 2 is a path map of proposed algorithm in the navigation point model; Fig. 3 is also a path map of basic BBO algorithm in same environment; the path length of Fig. 2 is 108.6902 and the path length of Fig. 3 is 129.3689.

Figure 4 is the convergence graph of basic BBO algorithm for mobile robot path planning in two-dimension static environment; Fig. 5 is the convergence map of proposed algorithm in solving same problem in same conditions. According to the

Fig. 4 Convergent graph of basic method

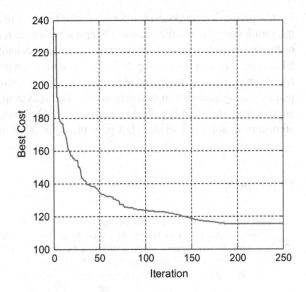

Fig. 5 Convergent graph of proposed method

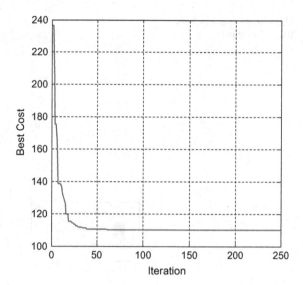

simulation results, although the two algorithms can finally get the solution but the improved algorithm has better performance, and the length of ultimate planned path is also shorter, the convergence rate has improved significantly too.

6 Conclusion

In this paper, we propose an improved BBO algorithm, which has the nonlinear migration operator and the mutation operator with elite retention mechanism instead of the linear migration and the random mutation mechanism. We use the improved BBO algorithm to find a best path based on navigation point model. The simulation results show that the proposed algorithm is effective and efficient for mobile robot path planning problem in two-dimension static environment. It has faster convergence speed and shorter path than the basic BBO algorithm. It is a new intelligent algorithm in solving mobile robot path planning problems.

References

1. Xin-Chun L, Dong-Bin Z, Jian-Qiang Y, Zuo-Shi S. A coordinated and hierarchical path planning approach for mobile manipulator. Manuf Autom. 2005;27(5):28–32.
2. Xue-Ying M, Lan-Fang L. Applications of the artificial intelligence in the robotic motion planning problem. Comput Appl Res. 2004;21(4):135–8.
3. Bo D, Xiao-Ming X, Zi-Xing C. Current status and future development of mobile robot path planning technology. Control Eng. 2005;12(3):198–202.

4. Dan W, Huang X-H, Zhou W. Path planning of mobile robot based on improved artificial potential field. J Comput Appl. 2010;30(8):2021–3.
5. Huang B, Cao G. Path planning of mobile robot based on fuzzy artificial potential field. J Shanghai Univ Sci Technol. 2006;28(4):347–50.
6. Yang Y, Wang C. Robot obstacle avoidance control based on improved artificial potential field method and its MATLAB implemention. J Shanghai Univ Sci Technol. 2013;35(4):496–500.
7. Duchon F, Babinec A, Kajan M. Path Planning with modified A-star algorithm for a mobile robot. Proc Eng. 2014;96(96):59–69.
8. Rajat KP, Choudhury BB. An effective path planning of mobile robot using genetic algorithm. In: IEEE international conference on computational intelligence, vol. 145; 2015. p. 287–91.
9. Wang JG, Ding B, Miao G-J, Bao J-W, Yang X. Path planning of mobile robot based on improving genetic algorithm, vol. 112. Berlin: Springer; 2011. p. 535–42.
10. Juan H, Zhong-Ying N-G. A method of mobile robotic path planning based on integrating of GA and ACO. Comput Simul. 2010;27(3):170–4.
11. Qin Y-Q, Sun D-B, Li N. Path Planning for mobile robot using the particle swarm optimization with mutation operator. In: Proceedings of the third international conference on machine learning and cybernetics, vol. 4; 2004. p. 2473–78.
12. Liu Y-K, Li M-K, Xie C-L, Peng M-J, Xie F. Path planning research in radioactive environment based on particle swarm algorithm. Prog Nucl Energy. 2014;74(74):184–92.
13. Constantin P, Radu-Emil P, Daniel I. Hybrid PSO-GSA robot path planning algorithm in static environments with danger zones. In: System theory, control and computing (ICSTCC), 2013 17th international conference; 2013. p. 434–439.
14. Li X-T, Wang J-Y, Zhou J-P, Yin M-H. A perturb biogeography-based optimization with mutation for global numerical optimization. Appl Math Comput. 2011;218(2):598–609.
15. Mo H-W, Li-Fang X. Research of biogeography particle swarm optimization for robot path planning. Neurocomputing. 2015;148(148):91–9.

An Improved Particle Filter Algorithm Based on Swarm Intelligence Optimization

You Lin and Lin Li

Abstract Due to the problem of particle degeneracy and loss of particle diversity in particle filter algorithm, this paper proposes an improved particle filter algorithm referring to some ideas of optimized algorithm like particle swarm and firefly group. The proposed algorithm utilizes the firefly algorithm to optimize particle filter and avoid re-sampling process; makes the particles move towards the location of better weight particles and prevents the small weight particles from disappearing after several iterations. Meanwhile, this algorithm sets a transition threshold and iteration times in order to improve the real-time property of the algorithm. Experimental results show that the improved algorithm possesses higher estimation accuracy and keeps good diversity of particle.

Keywords Particle filter · Swarm intelligence optimization · Particle impoverishment · State estimation

1 Introduction

Particle filter is a recursive Bayesian estimation method based on Monte-Carlo simulation. As it possesses the advantages of applying to nonlinear and non-Gaussian noise, particle filter is widely used in various fields including visual tracking, robot localization, aerial navigation and process fault diagnosis. Its main idea is to get a set of random sample spreading in the state space to make approximation of probability density function and replace integral operation with sample means, in order to get the minimum variance distribution of state.

Y. Lin · L. Li (✉)
Department of Control Science and Engineering, University of Shanghai
for Science and Technology, Shanghai, China
e-mail: lilin0211@163.com

Y. Lin
e-mail: linyou.shanghai@qq.com

© Springer Nature Singapore Pte Ltd. 2018
Y. Jia et al. (eds.), *Proceedings of 2017 Chinese Intelligent Systems Conference*,
Lecture Notes in Electrical Engineering 460,
https://doi.org/10.1007/978-981-10-6499-9_23

In the process of practical application, the major problem of particle filter is that after several iterations, the importance weight may focus on minority particles. But these particles could not express the posterior probability density function accurately, namely particle degeneracy. To the problem of particle degeneracy, re-sampling method could be adopted. However, re-sampling algorithm duplicate large weight samples and abandon small ones, which could leads to the impoverishment of particles. And to the problem of particle impoverishment, scholars form in and abroad have made a great deal of researches. The method of local re-sampling was proposed in [1]. Its basic idea is separating the particles into three particle assemblies—high, mid, low according to weight degrees, abandoning the lower weights, maintain the middle weights and duplicate the higher ones. The particle filters chosen based on weights was also mentioned in [2]. This method selects a certain number of particles with higher weight form a whole collection of particles to be used in state estimation at net time. Also, deterministic re-sampling particle filter algorithm was proposed in [3] which could avoid abandoning those lower weight ones directly and maintain particle diversity to a degree. However, all these methods mentioned above are still based on traditional frame of re-sampling. The problem of particle impoverishment is still not solved thoroughly.

At present, particle filter based on swarm intelligence optimization [4] is an important developing orientation in the area of particle filter theory. It could regard the particles as individuals of biotic community and simulate their acting regulations [5] to make particle distribution more reasonable. These kinds of method improve the re-sampling steps and do not abandon low weights particles, thus being able to solve impoverishment problems. Particle filter based on swarm optimization [6] is a representative of intelligence optimized particle filters. Particle filter algorithm based on self-adapted particle swarm optimization was proposed in [7]. It could control the quantity of particles from neighborhood self-adaptively, thus improving the reasonability of sample distribution and accuracy of filter. The method of optimizing particle filters based on ant group algorithm was put forward in [8]. It could make the lower weight particles move toward higher likelihood areas and make the particles have better posterior probability distribution. Particle filter algorithm based on artificial firefly swarm optimization was proposed in [9], which only uses its particle transfer formula to recombine the sample instead of iterations.

Considering the problems mentioned above, this research makes improvements on the location of firefly algorithm [10] and brightness update formula. With reference to particle swarm algorithm [11], the study utilizes global optimum to guide the integral movements of particle swarms, comes up with a simplified brightness formula to find best particle, pre-treats the particle groups, successfully combines firefly group optimization method with particle filer. This method avoids more operation complexity brought by particle interaction and meanwhile, improves the precision of particle filter greatly.

2 Basic Particle Filter

Particle filter is a kind of estimation method of Monte-Carlo simulation based on Bayesian estimation. It focuses on solving state estimation problems of dynamic random system. Assume that the dynamic process of nonlinear system is as follows:

$$x_t = f(x_{t-1}, u_{t-1}) \tag{1}$$

$$y_t = h(x_t, n_t) \tag{2}$$

x_t is status value, y_t is observed value, $f(*)$ is state transition function, $h(*)$ is observation function, u_{t-1} is system noise, n_t is observational noise.

State prediction equation is

$$p(x_t|y_{1:t-1}) = \int p(x_t|x_{t-1})p(x_{t-1}|y_{1:t-1})|dx_{t-1} \tag{3}$$

Update equation is

$$p(x_t|y_{1:t}) = \frac{p(y_t|x_t)p(x_t|y_{1:t-1})}{p(y_t|y_{1:t-1})} \tag{4}$$

$p(y_t|y_{1:t-1}) = \int p(y_t|x_t)p(x_t|y_{1:t-1})dx_t$ is normalization constant.

General steps are as follows:

Step 1: Initialization of Particles. Particle assembly $\{x_0^i, i=1,2,\ldots,N\}$ is generated by prior probability $p(x_0|y_0) = p(x_0)$, weight of all particles is 1/N.

Step 2: Importance function known and easy to be sampled is

$$q(x_{0:t}|y_{1:t}) = q(x_0) \prod_{j=1}^{t} q(x_j|x_{0:j-1}, y_{1:j}) \tag{5}$$

At this time, particle assembly is $\{\tilde{x}_t^i\}_{i=1}^{N}$

Step 3: Calculating Particle Weight

$$w_t^i = \frac{p(y_{1:t}|x_{0:t})p(x_{0:t})}{q(\tilde{x}_t|x_{0:t-1}, y_{1:t})q(x_{0:t-1}, y_{1:t})} = w_{t-1}^i \frac{p(y_t|\tilde{x}_t)p(\tilde{x}_t|x_{t-1})}{q(\tilde{x}_t|x_{t-1}, y_t)} \tag{6}$$

Step 4: Normalization of Weight

$$w_t^i = w_t^i / \sum_{i=1}^{N} w_t^i \tag{7}$$

Step 5: Re-sampling
Do re-sampling of particle assembly $\{\tilde{x}_t^i, w_t^i\}$ at this moment, then the particle assembly turns into $\{x_t^i, 1/N\}$; The weight of each particle is 1/N, preparing for the next cycling calculation.
Step 6. Output. The state value of time t is

$$\tilde{x}_t = \sum_{i=1}^{N} \tilde{x}_t^i w_t^i \tag{8}$$

Then the algorithm turns back to Step 2 and does cycling operation in order to realize cycling tracking of systematic state.

3 Particle Swarm Optimization Firefly Algorithm -PF Algorithm (PSOFA-PF)

In order to overcome the defects of common particle filter, this paper introduces the idea of particle swarm and firefly algorithm into particle filter to improve the optimization process. As there exist similarities between these two algorithms and particle filter, they could be combined to get more effective particle filter algorithm.

3.1 Particle Swarm Optimization and Firefly Algorithm

The basic principle of PSO: Initialize a particle swarm randomly which contains particles with the quantity of m, dimension of n. The location of particle I is $X_i = (x_{i1}, x_{i2}, \ldots, x_{in})$, speed $V_i = (v_{i1}, v_{i2}, \ldots, v_{in})$. In the process of iteration, particles unswervingly update their own speed and location through individual pole value p_i and global value G in order to realize particle optimization. The update formula is as follows:

$$V_i = \omega * V_i + c_1 * Rand * (p_i - X_i) + c_2 * Rand * (G - X_i) \tag{9}$$

$$X_{i+1} = X_i + V_i \tag{10}$$

Rand is a random number in $(0, 1)$; ω is inertia coefficient. The bigger ω is, the stronger the global searching ability of the algorithm is; the smaller ω is, the stronger the local searching ability is; c_1 and c_2 are learning parameters.

Firefly algorithm [12] simulates the actions of firefly groups. It has better rate and precision of convergence, and is easier for projects to realize. The rules of firefly algorithm are as follows: (1) The firefly will be attracted to move towards those brighter ones regardless of gender factors. (2) The attractive force of fireflies is in direct proportion to their brightness. Taking two fireflies in random, one will fly toward the other which is brighter than itself, and the brightness will turn down steadily as the distance increases. (3) If there are no other brighter fireflies, the firefly will move randomly. Major parameters of firefly algorithm are as follows:

Major steps of FA are as follows:

(1) Calculate the relative fluorescence brightness of fireflies:

$$I = I_0 * e^{-\gamma r_{ij}}$$ (11)

I_0 represents the maximum fluorescence brightness of fireflies i and j is relevant to objective function value; the higher the objectivity function value is, the brighter the firefly is γ represents the light absorption coefficient, Fluorescence will be weakened steadily with the increase of distance and absorption of spread medium; r_{ij} represents the spacial distance between firefly i and j.

(2) Firefly attractiveness is:

$$\beta = \beta_0 * e^{-\gamma r_{ij}^2}$$ (12)

β_0 represents the maximum attractiveness.

(3) The update formula of location is:

$$x_i = x_i + \beta * (x_j - x_i) + \alpha * \left(Rand - \frac{1}{2} \right)$$ (13)

x_i, x_j are the spacial location of firefly i and j; α represents step size, value in $[0, 1]$; $Rand$ means the random figures complying with uniform distribution within $[0, 1]$.

3.2 Particle Swarm Optimization Firefly Algorithm-PF Algorithm

Standard particle filter adopts method of re-sampling to prevent particle weight degeneracy, which eliminates particles with lower weights and duplicates those with higher weights. The advantage of this method lies in its easy operation. However, after several iterations, the diversity of particles will be reduced. The characteristics of firefly algorithm should be utilized at this time. Firstly, the weight of particles is compared as the fluorescent brightness of fireflies. By imagining the method of firefly optimization, low weight particles will move towards the location of high weights instead of abandoning low weight particles. Theoretically, after several times of movement, all particles will gather at the location of the particle of the highest weight, thus realizing final optimization. This process can not only maintain the location of high weight particles, make the particle groups distribute better, but also keep particle diversity.

In order to realize this process, firstly we should refer to the particle swam algorithm and guide the particle swarm by utilizing optimal value. Then location update formula and brightness formula of firefly algorithm should be improved to avoid interactive operation of particles and reduce complexity. Take the best weight as the comparison object of all the particles, then find the individual with the highest lightness. So we come up with a simplified formula $I = abs(y_t - \tilde{x}_t)$, the best value is

$$G = \{x_t^n | \min [I(x_t^1), I(x_t^2), \ldots, I(x_t^N)]\} \tag{14}$$

Detailed realization and steps of the algorithm are as follows:

Step 1. Initializing particles. Take N particles and generate particle assembly $\{x_0^i, i = 1, 2, \ldots, N\}$, Importance function is

$$q(\tilde{x}_t | x_{t-1}, y_{1:t}) = p(\tilde{x}_t | x_{t-1}) \tag{15}$$

Step 2. Eliminate particles of false weight in order to cooperate with firefly algorithm to track the objectives faster and more accurate.
When $x_t^i * \tilde{x}_t^i < 0$, weight should be zero, $w_t^i = 0$ on the contrary, weight maintains the same and then normalize particle weight.

$$w_t^i = w_t^i / \sum_{i=1}^{N} w_t^i \tag{16}$$

Step 3. Calculate the attractiveness degree between particle i and global optimum $\beta = \beta_0 * e^{-\gamma r_i^2}$, r_i is the distance between i and G.
Step 4. Update the location of particles

$$x_i = x_i + \beta * (x_j - x_i) + \alpha * \left(Rand - \frac{1}{2} \right) \tag{17}$$

Step 5. Update global optimum $G = \{x_t^n | \min [I(x_t^1), I(x_t^2), \ldots, I(x_t^N)]\}$
Step 6. Set iteration termination threshold
When $I > 0.01$, do not calculate Step 4.
When $I < 0.01$, calculate Step 4 until the iteration times reach the highest.
Step 7. Update weight. As firefly algorithm changes the spacial locations of each particle after an iteration. Therefore, while updating the location of particles, particle weights should also be updated. Detailed methods are as follows:

$$w_t^i = \frac{p(y_t^i|\bar{x}_t^i)p(\bar{x}_t^i|y_{1:t-1})}{q(\bar{x}_t^i)} \quad (18)$$

Step 8. Normalization.

$$w_t^i = w_t^i / \sum_{i=1}^{N} w_t^i \quad (19)$$

Step 9. Output.

$$\tilde{x}_t = \sum_{i=1}^{N} \tilde{x}_t^i w_t^i \quad (20)$$

From the above, we could see that through Step 2, the quality of the whole particle swarm could be improved; resources can be avoided from wasting; real-time aspects can be better promoted. As the firefly algorithm has strong convergence ability, it could make the particle swarms move toward the true range; high weight particles will further move to optimal value. This could greatly promotes the estimation accuracy of particle swarms and meanwhile, increase efficiency by setting number of iterations and avoid wasting calculating resources on those already optimized particles. Avoiding low weight particles getting into the process of iteration by setting termination threshold can help maintain particle diversity and promote algorithm real-time qualities. The algorithm fully utilizes the optimized information in the entire particle swarm, avoids interactive operation among each individuals, greatly increases the speed of operation and is conducive for particles to get rid of local extremum.

4 Simulation Experiment

This paper adopts a basic and common model (UNGM model) to check the effectiveness of the algorithm and compare it with standard particle filter algorithm and particle swarm optimized particle filter algorithm. The experimental environment is inter core2 processor with 4G internal storage and matlab2014a softwares.

The systematic state transfer model is:

$$x_t = \frac{1}{2}x_{t-1} + \frac{25x_{t-1}}{1+x_{t-1}^2} + 8\cos\frac{6(t-1)}{5} + u_t \quad (21)$$

The observation model is:

$$y_t = \frac{x_t^2}{20} + n_t \quad (22)$$

Root-mean-square error formula is:

$$RMSE = \left[\frac{1}{T}\sum_{t=1}^{T}(x_t - \tilde{x}_t)^2\right]^{\frac{1}{2}} \tag{23}$$

t stands for time, x_t stands for quantity of state, y_t stands for quantity of observation, u_t, n_t stands for zero-mean Guassian noise. Particle filter step is set as 50. Quantity of particles is 100. According to the universal parameters set by present references [13], maximum attractiveness degree in firefly algorithm is set as 0.9, step index 0.4, light intensity absorption coefficient 1. Generally speaking, the bigger the step index is, the stronger the global optimization ability is, but it will affect convergence precision. Otherwise, the smaller the step index is, the stronger the local optimization ability is, but it will affect convergence speed. This should be adjusted according to specific circumstances and applications (Chart 1).

As shown in Figs. 1 and 2, PSOFA-PF algorithm is better than the other two algorithms in accuracy while tracking particle. The estimated value based on Firefly algorithm has smaller error fluctuation than the other two, which means the improved algorithm possesses not only higher accuracy, but also better stability.

Chart 1 Experimental result

Algorithm	Error	Time
PF	4.6347	0.1210
PSO-PF	3.9501	0.1886
PSOFA-PF	3.8762	0.1593

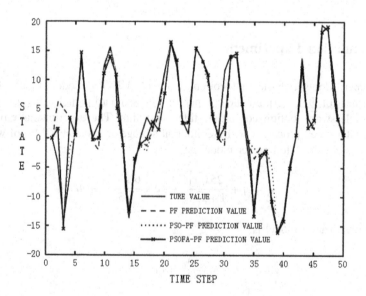

Fig. 1 Filter state estimation

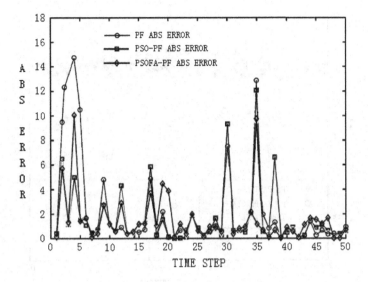

Fig. 2 Absolute error

Compared in terms of time, it uses less than PSO-PF and keeps nearly the same as basic particle filter.

To test the diversity of samples, filter step is set to 100 in order to compare particle distribution circumstances while the time of the filter is 45 and 90. From Figs. 3 and 4, at the later period of filtering by basic particle filter, particles are

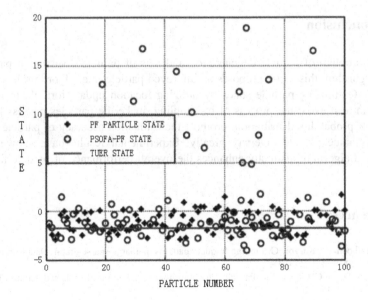

Fig. 3 Particle distribution when t = 45

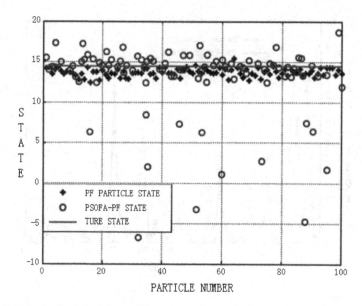

Fig. 4 Particle distribution when t = 90

mainly concentrated on minority state value and lack diversity, which is not beneficial for state estimation. While in PSOFA-PF, not only state estimation can be processed precisely, but particle diversity can also be greatly maintained.

5 Conclusion

To the problem of particle degeneration and loss of particle diversity in particle filter algorithm, this paper proposes an improved particle filter algorithm based on PSOFA Optimizing particle states by utilizing location update formula in firefly algorithm can avoid the process of re-sampling, make the particles possess better posterior probability distribution, promote the estimation accuracy of particle filter and guarantee particle diversity greatly. Experimental results have shown that improved algorithm distinctly enhances the overall performance of particle filter.

References

1. Kuramochi M, Karypis G. Finding frequent patterns in a large sparse graph. Data Min Knowl Disc. 2005;11(3):243–71.
2. Zhang Q, Hu CH, Qiao YK. Particle filter algorithm based on weight selected. Control Decis. 2008;23(1):117–20.

3. Li T, Sattar TP, Sun SD. Deterministic resampling: unbiased sampling to avoid sample impoverishment in particle filters. Sig Process. 2012;92(7):1637–45.
4. Yu Y, Zheng X. Particle filter with ant colony optimization for frequency offset estimation in OFDM systems with unknown noise distribution. Sig Process. 2011;91(5):1339–42.
5. Yuan GH, Fan CJ, Zhang HZ, Wang B, Qin TG. Hybrid algorithm integrating new particle swarm optimization and artificial fish school algorithm. J Univ Shanghai Sci Technol. 2014;36(3):223–6.
6. Fang Z, Tong GF, Xu XH. Particle swarm optimized particle filter. Control Decis. 2007;22 (3):273–7.
7. Chen ZM, Bo YM, Wu PL, Duan WY, Liu ZF. Novel particle filter algorithm based on adaptive particle swarm optimization and its application to radar target tracking. Control Decis. 2013;28(2):193–200.
8. Yq CAO, Zhong T, Huang XS. Improved particle filter algorithm based on ant colony optimization. Appl Res Comput. 2013;30(8):2402–4.
9. Zhu WC, Xu DZ. Improved particle filter algorithm based on artificial glowworm swarm optimization. Appl Res Comput. 2014;31(10):2920–4.
10. Tian MC, Bo YM, Chen ZM, Wu PL, Zhao GP. Firefly algorithm intelligence optimized particle filter. Acta Autom Sin. 2016;42(1):89–97.
11. Xian WM, Long B, Li M, Wang HJ. Prognostics of lithium-lon batteries based on the verhulst model, particle swarm optimization and particle filter. IEEE Trans Instrum Meas. 2013;63 (1):2–17.
12. Yang XS. Firefly algorithm, stochastic test functions and design optimization. Int J Bio-Inspired Comput. 2010;2(2):78–84.
13. Park S, Hwang JP, Kim E, Kang HJ. A new evolutionary particle filter for the prevention of sample impoverishment. IEEE Trans Evol Comput. 2009;13(4):801–9.

Facial Expression Recognition Using Histogram Sequence of Local Gabor Gradient Code-Horizontal Diagonal and Oriented Gradient Descriptor

Huanhuan Zhang and Lin Li

Abstract This paper present a original method for facial expression recognition, which fused with the Gabor filter and Local Gradient Code-Horizontal Diagonal (LGC-HD) as well as Histogram of Oriented Gradient (HOG). This approach firstly is used Viola-Jones algorithm to resize the facial expression image and convolve the facial expression image with Gabor filters to extract the Gabor Coefficients Maps (GCM). Then, we obtain Average Gabor Maps (AGM) by folding GCM of four orientations in each scale to reduce dimensions. The LGC-HD and HOG is applied on each AGM to obtain the LGGC-HD-HOG descriptor. At last, the Support Vector Machine (SVM) is adopted as classifier. We conclude that the method in this paper is better in recognition rate than other similar methods by analyzing the experimental result.

Keywords Gabor filters · Local gradient code-horizontal diagonal · Histogram of oriented gradient · Support vector machine

1 Introduction

The facial expression embodys wealthy information about human behaviors. Synthesizing various factors, experts divided the human emotion into six types, including happy, sad, disgust, surprise, fear and anger [1]. The Facial Recognition has been applied in many fields, such as human-computer interface, virtual reality, surveillance systems. Therefore, the Facial Recognition has a good academic value and development prospect.

H. Zhang · L. Li (✉)
Department of Control Science and Engineering, University of Shanghai
for Science and Technology, Shanghai 200093, China
e-mail: lilin0211@163.com

H. Zhang
e-mail: zhang1528163991@126.com

© Springer Nature Singapore Pte Ltd. 2018   243
Y. Jia et al. (eds.), *Proceedings of 2017 Chinese Intelligent Systems Conference*,
Lecture Notes in Electrical Engineering 460,
https://doi.org/10.1007/978-981-10-6499-9_24

Based on the two kinds of operation objects (static image and dynamic image), the facial expression recognition skill can be divided into two types [2]. According to [3], the technique based on static image cost lower computation and can meet real-time request, compared the technique based on dynamic image. So this paper focused on static images to build the facial expression recognition.

Zhu et al. [4] demonstrated that the recognition accuracy is high using the Gabor filter and SVM. But real-time is not good for the high computational time. Shan et al. [5] performed a research that combined the local binary patterns (LBP) and some classifiers like SVM. The result showed that it reduce the computational time, but it also lose 20% important information compared to the Gabor method [6]. The traditional LBP operator is by comparing the central gray value and ambient pixels to get the local texture information, which cannot better describe facial expression of facial wrinkles, muscles and other local deformations. So Tong et al. [7] proposed the local gradient code (LGC). And later, the LGC-HD of optimization of the LGC is proposed.

In order to extract more texture information at different resolutions as well as decreasing the computational time, [8] proposed a method to extract facial expression features using the Gabor filters and LGC-HD. Due to the similarities between anger, fear and surprise, it is not easy to distinguish among these facial expressions. To cover the shortage, Xu et al. [9] proposed a approach using Gabor filter and HOG.

In this paper, we decide to combine LGC-HD and HOG after convolving the facial images with the Gabor kernels. The arrangement of rest paper as follows: Sect. 2 introduces the preprocessing for facial expression images. And Sect. 3 introduces the method of LGGC-HD-HOG. Section 4 discuss the experiment result. At last, Sect. 5 concludes this paper.

2 Image Preprocessing

The preprocessing of the facial expression picture includes size normalization and extracting the facial expressions only. Therefore, Viola-Jones algorithm is applied to detect each pair eyes [10]. Cropping the interest region of input facial image is depended on the distance between the two eyes [11]. Assuming that the distance between the left eye and right eye is β. The height of cropping region is 3.4 times β (the height of the eye above is 1 times β, and the following height of the eye is 2.4 times β,) applied to β. The horizontal cropping region is demarcated by a factor of 2.2 applied to β.

It can not only remove the background information and non-face information interference, but also improve the precision of facial expression recognition. After the cropping, we resize the cropping image automatically to 112×96 pixels. Figure 1a, b show that the input and the preprocessed image.

Fig. 1 **a** The input image
b The preprocessed image

(a)   **(b)**

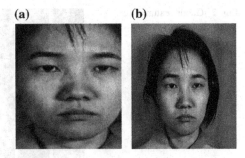

3 Feature Extraction

3.1 Gabor Wavelet Transform

The human visual system is regard as the best system for distinguishing expression, because it can get greater precision under the different distances and resolutions. Given that, Gabor filter is applied to the first step of image feature extraction for its the multi-direction and multi-scale performance [12]. Gabor wavelet transform can be seen as Fourier transform with the window function of the Gaussian function [13]. The two-dimension Gabor kernel can be expressed as:

$$G_{u,v}(z) = \frac{\|k_{u,v}\|}{\sigma^2} \exp\left(-\frac{\|k_{u,v}\|^2}{2\sigma^2}\right)[\exp(ik_{u,v}z) - \exp\left(-\frac{\sigma^2}{2}\right)] \tag{1}$$

where z represents the pixel position (x, y), u represents the orientation of Gabor wavelets, v is the scale of Gabor wavelets, σ is the sigma deviation of the Gaussian envelope [14], $\| \ \|$ is modular arithmetic [15] and k_u represents the center frequency of the filter [16]. It can be defined like the Eq. (2):

$$k_{u,v} = \left(\begin{array}{c} k_v \cos \phi_u \\ k_v \sin \phi_u \end{array}\right) \tag{2}$$

where: $k_v = 2^{\frac{v+2}{2}\pi}$, $\phi_u = u\frac{\pi}{k}$. Here, the Gabor wavelets can be expressed in four scales $u(u = 0, 1, 2, 3)$ and eight orientation v ($v = 0, 1, 2, ..., 7$). So the number of Gabor kernels is 32.

In our work, we firstly do convolution operation with a set of Gabor kernels to get GCM. The corresponding convolution operation can be defined by Eq. (3):

$$F_{x,y,u,v} = I(z) \, {}^*G_{u,v}(z) \tag{3}$$

After convolving the facial expression image with Gabor kernels, we can get Gabor feature maps as shown in the Fig. 2.

Although we used Gabor filters only with 4 scales and 8 orientations to process the images, the dimension size of the Gabor feature is still 344064 (4 × 8 112 × 96). The too high dimension size will increase the computational time.

Fig. 2 Gabor feature maps

Fig. 3 Folding images

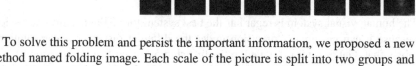

To solve this problem and persist the important information, we proposed a new method named folding image. Each scale of the picture is split into two groups and each group of four directions of the image superposition. On the one hand, it's better to retain the structural characteristics of each direction; On the other hand, it also can decrease the feature dimension. The image folding is defined by Eq. (4):

$$F_{u,v} = \sum_{i=0}^{i=4} I_{u,i}(z)\, u(u=0,1,2,3),\, v(v=0,1,2,3,4,5,6,7) \cdot \qquad (4)$$

After folding GCM, feature maps are decreased from the 32 Gabor Coefficients Maps to 8 images which are shown in Fig. 3.

3.2 LGC and LGC-HD Operator Feature

LGC is different from the traditional LBP algorithm in terms of extracting the facial expressions. That can well describe the deformation, wrinkles, facial muscles of different ages and gender features by using the LGC algorithm. It can be described using the 3 × 3 template show in the Fig. 4 by computing the pixels through the formula (5) and converting the binary coding to decimal numbers.

Fig. 4 3 × 3 template of
LGC operator

g_1	g_2	g_3
g_4	g_c	g_5
g_6	g_7	g_8

6	1	1
5	8	1
5	5	1

3	4	5
3	8	4
3	5	3

8	4	3
3	9	1
2	5	7

LBP=00000000 LBP=00000000 LBP=00000000
LGC=11110010 LGC=00000101 LGC=11010110
LGC-HD=11110 LGC-HD=00001 LGC-HD=11010

Fig. 5 LBP, LGC and LGC-HD operators

The LGC algorithm formula can be expressed as:

$$LGC = s(g_1 - g_3)2^7 + s(g_4 - g_5)2^6 + s(g_6 - g_8)2^5 + s(g_1 - g_6)2^4 + s(g_2 - g_7)2^3$$
$$+ s(g_3 - g_8)2^2 + s(g_1 - g_8)2^1 + s(g_3 - g_6)2^0$$

$$(5)$$

The vertical direction of the pixel contains less texture information. So a further optimization on the LGC method is proposed. LGC-HD can decrease the redundancy feature vectors and be described with five bits as shown in Fig. 5. So LGC-HD can be a good descriptor of texture information to retain robustness and stability in performance.

The LGC-HD operator can be expressed as:

$$LGC - HD = s(g_1 - g_3)2^4 + s(g_4 - g_5)2^3 + s(g_6 - g_8)2^2$$
$$+ s(g_1 - g_8)2^1 + s(g_3 - g_6)2^0$$

$$(6)$$

Due to compute the histogram of the whole image, it is possible to lose some important details. In this point, firstly LGC-HD is applied to each AGM. Then we divide the feature image into 42 (7×6) blocks and each block is 16×16 in pixels as shown in Fig. 6.

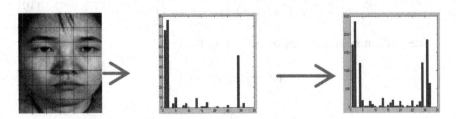

Fig. 6 The computation of LGC-HD

Calculating the histogram of the image h(x, y) is expressed as.

$$H_m = \sum_{x,y} I\{h(x,y) = m\}, \quad m = 0 \ldots k-1 \tag{7}$$

where H_m represent the number of pixels that have a label is k (Here, the maximum of k is equal to 32),

$$I\{Q\} = \begin{cases} 1, & Q \text{ is ture} \\ 0, & Q \text{ is false} \end{cases} \tag{8}$$

and we calculate each block's histogram (The number of block is 42). Here, 42 blocks are defined as B_j where $j = 0, 1, \ldots, 41$.

$$H_{u,v,B_j,i} = \sum_{(x,y \in B_j)} I\{LGC - HD(u,v,x,y)\} \quad I\{x,y\} \in B_j \tag{9}$$

where $i = 0, 1, \ldots, k\text{-}1$.

3.3 HOG Operator Feature

HOG [17] is a pretty method to extract some detailed information. To improve the stability of the algorithm and to provide wealthy texture information, HOG and LGC-HD are combined. When the texture information are abundant, the recognition is also increase.

HOG is a feature descriptor used to perform object detection in computer vision and image processing. The appearance and shape of an expression can be well depicted by the gradient or edge direction density distribution. First step is calculating the gradient value and gradient direction and then accumulating the values in the same gradient direction. The vertical gradient and horizontal gradient can be expressed by Eq. (10):

$$\begin{cases} G_x = I(x+1,y) - I(x-1,y) \\ G_y = I(x,y+1) - I(x,y-1) \end{cases} \tag{10}$$

The gradient magnitude can be defined by Eq. (11):

$$M(x,y) = \sqrt{G_x^2 + G_y^2} \tag{11}$$

And the gradient direction can be expressed by Eq. (12):

$$\theta = \arctan\left(\frac{G_y}{G_x}\right) \tag{12}$$

Using the HOG descriptor for a whole image may lose some important information, so we divided each AGM into 42 patches, in order to persist important information.

In this paper, the process to extract HOG feature vector can be split into the following three steps:

a. At first, we can do convolution to obtain the phase and magnitude of each pixel.
b. Then, we divide the circumference $(2 \times pi)$ into 9 bins [9], with each bin containing $30°$ $(2 \times pi/9)$.
c. Finally, each patch can get a 9-dimensional eigenvector. Forming a block with four patches, you can get 30 blocks. In each block, we connect each patch's eigenvector by end to end. The feature dimension is $8 \times 9 \times 4 \times 30$.

At last we combine the HOG ($9 \times 8 \times 30 \times 4 = 88640$) and LGC-HD ($32 \times 7 \times 6 \times 8 = 10752$) feature vectors to get a 19392 dimension vector.

4 Result Discussion

Experiments are performed on Japanese Female Facial Expression (JAFFE) database [18], which include 213 images of ten people. There are three to four images for each expression of one person. Since the JAFFE database is relatively small, the approach of fivefold cross validation is chosen. We randomly divide the JAFFE into five equal-sized parts, form which 171 images (80%) are trained a model and the other 42 images (20%) used to test the training model.

SVM is chosen to do our experiment for its advantages in solving small sample, nonlinear and high dimensional pattern recognition. Radial Basis Function (RBF) is a strong local kernel function, whether large or small samples have better performance. Here, the RBF as the SVM kernel function is to perform classification.

To be better described the facial expression recognition, the JAFFE database was divided into six classes and seven classes to be applied to experiment. Here we will do a comparison for Tables 1 and 2, we can find that whether the JAFFE database is divided into seven categories or six classes, we can see that the experimental accuracy impact is relatively small. This is because we extracted the well characteristics that is representative and distinguished.

In Table 3, we compared the recognition results among different algorithms based on the same database. By comparing LBP [6], LGC-HD [8], Gabor +LGC-HD [9] and LPQ-es-LBP-S [19], we could see clearly the recognition accuracy of the proposed method in this paper is best, which is 95.24% in seven types of expressions and 97.56% in six types of expressions.

Table 1 Recognition rate in seven types of expressions (%)

	Anger	Disgust	Fear	Happy	Neutral	Sad	Surprise
Anger	100	0	0	0	0	0	0
Disgust	0	100	0	0	0	0	0
Fear	0	0	85.7	0	0	14.3	0
Happy	0	0	0	100	0	0	0
Neutral	0	0	0	0	100	0	0
Sad	0	0	0	0	0	100	0
Surprise	0	0	0	0	16.7	0	83.3

Table 2 Recognition rate in seven types of expressions (%)

	Anger	Disgust	Fear	Happy	Sad	Surprise
Anger	100	0	0	0	0	0
Disgust	0	100	0	0	0	0
Fear	0	0	85.7	0	14.3	0
Happy	0	0	0	100	0	0
Sad	0	0	0	0	100	0
Surprise	0	0	0	0	0	100

Table 3 Recognition with other facial expression recognition methods (%)

Reference	Feature extraction	Classifier	Recognition rate	
			6-class (%)	7-class (%)
Ojala et al. [6]	LBP	Template matching	83.7	79.1
Al-Sumaidasee et al. [8]	LGC-HD	Euclidean distance	90	–
Xu et al. [9]	Gabor+LGC-HD	Template matching	93.3	90.0
Chao et al. [19]	LPQ+es-LBP-s	SVM	–	94.88
Proposed	Gabor+LGC-HD +HOG	SVM	97.56	95.24

5 Conclusion

The new approach proposed in this paper combine the Gabor filter with the LGC-HD and HOG. And to keep the strong features, each feature map is divided into some blocks to compute the histogram sequence. LGC-HD calculation speed is relatively low, but in the continuous area of cutting, it will lose some of the details of the texture information. The novelty is mainly based on the richness of LGC-HD and the integration of HOG auxiliary information. And the results demonstrate that the method in this paper is not only fast but also achieve better recognition performance.

References

1. Kumari J, Rajesh R, Pooja KM. Facial expression recognition: a survey. Procedia Comput Sci. 2015;58:486–91.
2. Fasel B, Luettin J. Automatic facial expression analysis: a survey. Pattern Recogn. 2003;36 (1):259–75.
3. Sun X, Xu H, Zhao C, Yang J. Facial expression recognition based on histogram sequence of local Gabor binary patterns. 2008:158–63.
4. Zhu J, Su G, Li Y. Facial expression recognition based on Gabor feature and Adaboost. J Optoelectron Laser. 2011;17(8):993–8.
5. Shan C, Gong S, McOwan PW. Robust facial expression recognition using local binary patterns. IEEE Int Conf Image Process. 2005;2:II-370–3.
6. Ojala T, Pietikainen M, Maenpaa T. Multiresolution gray-scale and rotation invariant texture classification with local binary patterns. IEEE Trans Pattern Anal Mach Intell. 2002;24:971–87.
7. Tong Y, Chen R, Cheng Y. Facial expression recognition algorithm using LGC based on horizontal and diagonal prior principle. Optik-Int J Light Electron Optics. 2014;125 (16):4186–9.
8. Al-Sumaidaee SAM, Dlay SS, Woo WL, Chambers JA. Facial expression recognition using local Gabor gradient code-horizontal diagonal descriptor. Intell Sig Process. 2015:1–6.
9. Xu X, Quan C, Ren F. Facial expression recognition based on Gabor wavelet transform and histogram of oriented gradients. Mechatron Autom. 2015:2117–22.
10. Jones P, Viola P, Jones M. Rapid object detection using a boosted cascade of simple features. Pattern Recogn. 2001;1:511–8.
11. Lopes AT, Aguiar ED, Souza AFD, Oliveira-Santos T. Facial expression recognition with convolutional neural networks: coping with few data and the training sample order. Pattern Recogn. 2017;61:610–28.
12. Elsayed RS, Kholy PAE, Elnahas PMY. Robust facial expression recognition via sparse representation and multiple Gabor filters. Int J Adv Comput Sci Appl. 2013;4(3):S95.
13. He C, Zheng YF, Ahalt SC. Object tracking using the Gabor wavelet transform and the golden section algorithm. IEEE Trans Multimedia. 2002;4(4):528–38.
14. Samad R, Sawada H. Extraction of the minimum number of Gabor wavelet parameters for the recognition of natural facial expressions. Artif Life Robot. 2011;16(1):21–31.
15. Chang JH, Ibarra OH, Pong T, et al. Two-dimensional convolution on a pyramid computer. IEEE Trans Pattern Anal Mach Intell. 1988;10(4):590–3.
16. Liu C, Wechsler H. Gabor feature based classification using the enhanced Fisher liner discriminant model for face recognition. Image Process. 2002;11(4):467–76.
17. Radman A, Zainal N, Suandi SA. Automated segmentation of iris images acquired in an unconstrained environment using HOG-SVM and GrowCut. Digit Sig Proc. 2017;64:60–70.
18. Lyons MJ, Akemastu S, Kamachi M, Gyoba J. Coding facial expressions with gabor wavelets. In: 3rd IEEE international conference on automatic face and gesture recognition; 1998. p. 200–5.
19. Chao WL, Ding JJ, Liu JZ. Facial expression recognition based on improved local binary pattern and class-regularized locality preserving projection. Sig Process. 2015;117:1–10.

Event-Triggered Consensus Control for Linear Multi-agent Systems Using Output Feedback

Xiaohui Hou and Yang Liu

Abstract This paper studies the event-triggered consensus problem of multi-agent systems with general linear dynamics. To be specific, an observer-based event-triggered control protocol is proposed using the local output feedback information. It is proved that the multi-agent system reaches consensus under the proposed protocol, and the gain matrices of the protocol can be determined by solving a group of LMIs. In the end, a simulation example is given to verify the theoretical results.

Keywords Multi-agent system · Consensus · Event-triggered control · Output feedback

1 Introduction

A multi-agent system consists of multiple agents which can operate independently and communicate to each other through a communication network. It has broad potential applications in many areas, such as sensor networks, flight formation, robotics teams and so on. Some states of interests are required to reach and maintain the same value in lots of the applications above. So the consensus problem, as a fundamental issue in the research of multi-agent system, has been widely studied during the past decades.

Some early results on consensus of multi-agent systems can be found in [1–6], to name a few. However, these early works were all based on continuous information exchange of agents, and the protocols were designed to update unremittingly, which

X. Hou (✉) · Y. Liu
School of Automation Science and Electrical Engineering,
Beihang University, Beijing 100191, China
e-mail: houxiaohui1993@163.com

Y. Liu (✉)
The Seventh Research Division, Beihang University, Beijing 100191, China
e-mail: ylbuaa@163.com

© Springer Nature Singapore Pte Ltd. 2018
Y. Jia et al. (eds.), *Proceedings of 2017 Chinese Intelligent Systems Conference*,
Lecture Notes in Electrical Engineering 460,
https://doi.org/10.1007/978-981-10-6499-9_25

is often hard to realize due to the limits on energy source and computing power. So, the event-triggered strategy has been adopted to solve this problem. References [7–11] developed the event-triggered consensus protocols for the multi-agent systems with single- or double-integrator dynamics. In addition, [12–14] considered the event-triggered consensus control for multi-agent system with general linear dynamics by the state feedback control method. But, the internal states are usually impossible to be obtained in most real systems so that the output feedback control is further considered in the consensus literature. There has already been some studies on designing continuous output feedback control laws to solve the consensus problem such as [4, 5], but rarely in event-triggered way.

Motivated by the above works, the event-triggered consensus problem is considered for multi-agent systems with general linear dynamics in this paper. Correspondingly, the contribution of our research can be summarized as follows. First, a novel output feedback event-triggered protocol is proposed with an observer form, using only local output feedback information at discrete-time instants. Second, it's proved theoretically that the consensus can be achieved. Meanwhile, the sufficient conditions are given, from which the undetermined matrices in the protocol can be also determined. Third, the protocol is designed in a distributed way, which ensures that each agent triggers independently only using its own and its neighbors' information.

The rest of the article is as follows. Section 2 introduces some preliminaries and the problem formulation. In Sect. 3, a distributed event-triggered control protocol is proposed with an observer form, and it's proved that the consensus of multi-agent system can be achieved. Then, a simulation example is given in Sects. 4 and 5 concludes the article.

2 Preliminaries and Problem Formulation

2.1 Preliminaries

The communication topology among agents is described by an undirected graph $\mathcal{G} = (\mathcal{V}, \mathcal{E}, \mathcal{A})$. $\mathcal{V} = \{v_1, v_2, \ldots, v_n\}$ is the set of nodes, where node v_i represents the i-th agent. $\mathcal{E} \subseteq \mathcal{V} \times \mathcal{V}$ is the set of undirected edges, in which edge (v_i, v_j) represents the information flow between agents i and j. If $(v_i, v_j) \in \mathcal{E}$, node v_j is called a neighbor of v_i, and then the set of neighbors of node v_i is denoted by $\mathcal{N}_i = \{v_j \in \mathcal{V} | (v_i, v_j) \in \mathcal{E}\}$. $\mathcal{A} = [a_{ij}]$ is a symmetric adjacency matrix with weighing factors $a_{ij} \geq 0$. It is stimulated that $(v_i, v_j) \in \mathcal{E}$ if and only if $a_{ij} = a_{ji} > 0$, and $a_{ii} = 0$. Then the Laplacian matrix is defined as $\mathcal{L} = \mathcal{D} - \mathcal{A}$, where $\mathcal{D} = diag\{d_1, \ldots, d_n\}$ with $d_i = \sum_{j=1}^{n} a_{ij}$ is called the degree matrix of $\mathcal{G}$. A sequence of edges of the form $(v_{i_k}, v_{i_{k+1}}), k = 1, \ldots, l-1$ is named a path. And a graph is called connected if there is a path between every two nodes.

Lemma 1 [3] *Let L be the Laplacian associated with an undirected graph $\mathcal{G}$. Then L has at least one zero eigenvalue and all of the nonzero eigenvalues are positive. Furthermore, matrix L has exactly one zero eigenvalue if and only if the undirected graph $\mathcal{G}$ is connected, and the eigenvector associated with zero is $\mathbf{1}$.*

Lemma 2 [4] *Let $L_c = [L_{cij}] \in \mathbb{R}^{n \times n}$ be a symmetric matrix with*

$$L_{cij} = \begin{cases} \frac{n-1}{n}, & i=j \\ -\frac{1}{n}, & i \neq j \end{cases}, \tag{1}$$

then the following statements hold:

(1) *The eigenvalues of L_c are 1 with multiplicity n−1 and 0 with multiplicity 1. The vectors $\mathbf{1}_n^T$ and $\mathbf{1}_n$ are the left and the right eigenvectors of L_c associated with the zero eigenvalue respectively.*
(2) *There exists an orthogonal matrix $U \in \mathbb{R}^{n \times n}$ such that*

$$U^T L_c U = \begin{bmatrix} I_{n-1} & \mathbf{0}_{n-1} \\ * & 0 \end{bmatrix} \tag{2}$$

and the last column of U is $\frac{\mathbf{1}_n}{\sqrt{n}}$. Let $L \in \mathbb{R}^{n \times n}$ be the Laplacian of a given undirected graph, then

$$U^T L U = \begin{bmatrix} L_1 & \mathbf{0}_{n-1} \\ * & 0 \end{bmatrix} \tag{3}$$

where $L_1 \in \mathbb{R}^{(n-1) \times (n-1)}$ is positive definite if and only if the graph is connected.

Lemma 3 (Schur complement formula) *Let S be a symmetric matrix of the partitioned form $S = [S_{ij}]$ with $S_{11} \in \mathbb{R}^{r \times r}$, $S_{12} \in \mathbb{R}^{r \times (n-r)}$ and $S_{22} \in \mathbb{R}^{(n-r) \times (n-r)}$. Then $S < 0$ if and only if*

$$S_{11} < 0, S_{22} - S_{21} S_{11}^{-1} S_{12} < 0 \tag{4}$$

or equivalently

$$S_{22} < 0, S_{11} - S_{12} S_{22}^{-1} S_{21} < 0 \tag{5}$$

2.2 Problem Formulation

Consider the multi-agent system consisting of n identical agents that can be modelled by the following linear dynamic system:

$$\dot{x}_i(t) = Ax_i(t) + Bu_i(t)$$
$$y_i(t) = Cx_i(t), \quad i = 1, 2, \ldots, n \tag{6}$$

where $x_i(t) \in \mathbb{R}^m$ is the state, $u_i(t) \in \mathbb{R}^{m_1}$ is the control input, $y_i(t) \in \mathbb{R}^p$ is the measured output. Without loss of generality, it's assumed that (A, B) is stabilized, (A, C) is observable and C is of full row rank. And the communication topology is described by an undirected graph $\mathcal{G}$ which is assumed to be connected. Protocol $u_i(t)$ is said to solve the consensus problem if and only if the states of agents satisfy

$$\lim_{t \to \infty} (x_i(t) - x_j(t)) = 0, \forall i, j \in \mathcal{N} \tag{7}$$

In this paper, $u_i(t)$ is designed based on the event-triggered scheme using observer-based control approach.

3 Protocol Design and Consensus Analysis

3.1 Output Feedback Event-Triggered Protocol

The event times for agent i are denoted as $t_k^i, k = 0, 1, \ldots$. Since full states are hardly obtained or detected in many systems, the distributed event-triggered protocol is designed based on an observer controller as following:

$$\begin{cases} u_i(t) = K \sum_{j \in N_i} a_{ij}(e^{A(t-t_k^i)}\tilde{x}_i(t_k^i) - e^{A(t-t_{k'}^j)}\tilde{x}_j(t_{k'}^j)) \\ \dot{\tilde{x}}_i(t) = A\tilde{x}_i(t) + Bu_i(t) + G(\tilde{y}_i(t) - y_i(t)) \\ \tilde{y}_i(t) = C\tilde{x}_i(t) \end{cases} \tag{8}$$

where $\tilde{x}_i(t) \in R^m$ and $\tilde{y}_i(t) \in \mathbb{R}^p$ are the state and output of observer respectively, $K \in \mathbb{R}^{m_1 \times m}$ and $G \in \mathbb{R}^{m \times p}$ are gain matrices to be determined, and a_{ij} is the interaction strength between agents i and j. And the triggering time of agent i in (8) is determined by the following event-triggered condition:

$$\|e_i(t)\| = \sqrt{\sigma} \left\| \sum_{j \in N_i} a_{ij}(\tilde{x}_i(t) - \tilde{x}_j(t)) \right\| \tag{9}$$

Denote the observation error as $h_i(t) = \tilde{x}_i(t) - x_i(t)$ and state measurement error as $e_i(t) = e^{A(t-t_k^i)}\tilde{x}_i(t_k^i) - \tilde{x}_i(t), t \in [t_k^i, t_{k+1}^i)$. Then substituting protocol (8) into system (6) yields the following closed loop system in compact form:

$$\dot{\tilde{x}}(t) = (I_n \otimes A + L \otimes BK)\tilde{x}(t) + (L \otimes BK)e(t) + (I_n \otimes GC)h(t) \tag{10}$$

where $\tilde{x}(t) = [\tilde{x}_1^T(t), \ldots, \tilde{x}_n^T(t)]^T \in \mathbb{R}^{mn}$, $e(t) = [e_1^T(t), \ldots, e_n^T(t)]^T \in \mathbb{R}^{mn}$, $h(t) = [h_1^T(t), \ldots, h_n^T(t)]^T \in \mathbb{R}^{mn}$, and L is the Laplacian matrix of $\mathcal{G}$.

In addition, by the definition of $h_i(t)$, we have $\dot{h}_i(t) = (A + GC)h_i(t)$, from which it is obtained

$$\dot{h}(t) = (I_n \otimes (A + GC))h(t) \tag{11}$$

Obviously, for $i = 1, \ldots, n$, $x_i(t) = \tilde{x}_i(t) - h_1(t)$ reaches consensus asymptotically if $\tilde{x}_i(t)$ reaches consensus asymptotically and $h_i(t)$ also reaches zero asymptotically. Equivalently, if both the consensus of (10) and the stability of (11) are ensured, then the state $x_i(t)$ achieves consensus asymptotically.

3.2 Model Transformation

Since the consensus trajectory of the above system is nonzero, the nonzero solution is firstly transformed into the origin to make it easier for the performance analysis.

Let $\bar{x}_i(t) = \tilde{x}_i(t) - \frac{1}{n} \sum_{j=1}^n \tilde{x}_j(t)$, which can also be written as $\bar{x}(t) = \tilde{x}(t) - \frac{1_n}{n} \otimes (\sum_{j=1}^n \tilde{x}_j(t)) = (L_c \otimes I_m)\tilde{x}(t)$. And by Eq. (10), for $L_c 1_n = 0_n$ and $L 1_n = 0_n$, it is derived that

$$\begin{aligned}
\dot{\bar{x}}(t) &= (L_c \otimes I_m)\dot{\tilde{x}}(t) \\
&= (L_c \otimes A + L_c L \otimes BK)\bar{x}(t) + (L_c L \otimes BK)e(t) + (L_c \otimes GC)h(t)
\end{aligned} \tag{12}$$

Then, based on Lemma 2, denote an orthogonal matrix $U = [U_1 \quad U_2]$ with $U_2 = \frac{1_n}{\sqrt{n}}$ being the last column such that

$$U^T L_c U = \begin{bmatrix} I_{n-1} & 0_{n-1} \\ * & 0 \end{bmatrix} \triangleq \bar{L}_c, \quad U^T L U = \begin{bmatrix} L_1 & 0_{n-1} \\ * & 0 \end{bmatrix} \triangleq \bar{L} \tag{13}$$

Let

$$\begin{aligned}
\hat{x}(t) &= (U^T \otimes I_m)\bar{x}(t) \triangleq [\hat{x}^{1T}(t) \quad \hat{x}^{2T}(t)]^T \\
\hat{e}(t) &= (U^T \otimes I_m)e(t) \triangleq [\hat{e}^{1T}(t) \quad \hat{e}^{2T}(t)]^T \\
\hat{h}(t) &= (U^T \otimes I_m)h(t) \triangleq [\hat{h}^{1T}(t) \quad \hat{h}^{2T}(t)]^T
\end{aligned} \tag{14}$$

Equation (12) is transformed into

$$
\begin{aligned}
\dot{\hat{x}}(t) &= (U^T \otimes I_m)\dot{\tilde{x}}(t) \\
&= (\bar{L}_c \otimes A + \bar{L}_c \bar{L} \otimes BK)\hat{x}(t) + (\bar{L}_c \bar{L} \otimes BK)\hat{e}(t) + (\bar{L}_c \otimes GC)\hat{h}(t)
\end{aligned}
\tag{15}
$$

which can be decoupled into the following two subsystems

$$
\dot{\hat{x}}^1(t) = (I_{n-1} \otimes A + L_1 \otimes BK)\hat{x}^1(t) + (L_1 \otimes BK)\hat{e}^1(t) + (I_{n-1} \otimes GC)\hat{h}^1(t) \tag{16}
$$

and

$$
\dot{\hat{x}}^2(t) = 0 \tag{17}
$$

which indicates that the state of the second subsystem remains unchanged.

In addition,

$$
\begin{aligned}
\dot{\hat{h}}^1(t) &= (U_1^T \otimes I_m)\dot{h}(t) = (I_{n-1} \otimes (A + GC))\hat{h}^1(t) \\
\dot{\hat{h}}^2(t) &= (U_2^T \otimes I_m)\dot{h}(t) = (A + GC)\hat{h}^2(t)
\end{aligned}
\tag{18}
$$

can be derived, which indicates that the stability of $\hat{h}^2(t)$ can be ensured if $\hat{h}^1(t)$ is asymptotically stable.

Denote $\xi^{1T}(t) = \begin{bmatrix} \hat{x}^{1T}(t) & \hat{h}^{1T}(t) \end{bmatrix}^T$ and $\hat{e}^{1T}(t) = \begin{bmatrix} \hat{e}^{1T}(t) & \mathbf{0}_{1 \times m} \end{bmatrix}^T$, so

$$
\dot{\xi}^1(t) = \hat{A}_1 \xi^1(t) + \hat{E}_1 \hat{e}^1(t) \tag{19}
$$

in which

$$
\hat{A}_1 = \begin{bmatrix} I_{n-1} \otimes A + L_1 \otimes BK & I_{n-1} \otimes GC \\ 0 & I_{n-1} \otimes (A + GC) \end{bmatrix}, \hat{E}_1 = \begin{bmatrix} L_1 \otimes BK & 0 \\ 0 & 0 \end{bmatrix} \tag{20}
$$

And from the definition of U, it's obvious that $U_1^T \eta = 0$ if and only if $\eta = \eta_0 \mathbf{1}_n$. Then combining with

$$
\hat{x}^1(t) = (U_1^T \otimes I_m)\tilde{x}(t) = (U_1^T \otimes I_m)[\tilde{x}(t) - \frac{\mathbf{1}_n}{n} \otimes (\sum_{j=1}^{n} \tilde{x}_j(t))] = (U_1^T \otimes I_m)\tilde{x}(t) \tag{21}
$$

we know that the consensus of system (10) and stability of (11) are equivalent to the stability of system (19).

3.3 Consensus Analysis

Theorem 1 *Consider system* (19) *and set the event-triggered condition as* (9). *The system is asymptotically stable if there exists a positive definite matrix* $\hat{P}$, *such that*

$$\begin{bmatrix} \hat{A}_1^T \hat{P} + \hat{P}\hat{A}_1 + \alpha\sigma\hat{L}_1^T\hat{L}_1 & \hat{P}\hat{E}_1 \\ \hat{E}_1^T \hat{P} & -\alpha I \end{bmatrix} < 0 \tag{22}$$

holds, where α *is a positive scalar and* $\hat{L}_1 = [\, L_1 \otimes I_m \quad 0_{n\times n} \otimes 0_{m\times m}\,]$. *And there is no Zeno behavior.*

Proof From (9), it yields that $\|e(t)\| \le \sqrt{\sigma}\|(L\otimes I_m)x(t)\| = \sqrt{\sigma}\|(L\otimes I_m)\bar{x}(t)\|$. By the property of orthogonal transformation, $\|\hat{e}(t)\| = \|e(t)\|$ holds obviously. And $\|\hat{e}^1(t)\| = \|\hat{e}^1(t)\| \le \|\hat{e}(t)\| = \|e(t)\|$ holds as

$$\|\hat{e}(t)\|^2 = \hat{e}^T(t)\hat{e}(t) = \hat{e}^{1T}(t)\hat{e}^1(t) + \hat{e}^{2T}(t)\hat{e}^2(t) = \|\hat{e}^1(t)\|^2 + \|\hat{e}^2(t)\|^2 \tag{23}$$

Meanwhile it holds that

$$\begin{aligned} \|(L\otimes I_m)\bar{x}(t)\|^2 &= \bar{x}^T(t)((L^T L)\otimes I_m)\bar{x}(t) \\ &= \hat{x}^T(t)((\bar{L}^T\bar{L})\otimes I_m)\hat{x}(t) \\ &= \|(L_1\otimes I_m)\hat{x}^1(t)\|^2 \\ &= \|[\,L_1\otimes I_m \quad 0_{n\times n}\otimes 0_{m\times m}\,]\xi^1(t)\|^2 \triangleq \|\hat{L}_1\xi^1(t)\|^2 \end{aligned} \tag{24}$$

and then $\|\hat{e}^1(t)\| \le \sqrt{\sigma}\|\hat{L}_1\xi^1(t)\|$ is obtained.

Define the candidate Lyapunov function as $V(t) = \xi^{1T}(t)\hat{P}\xi^1(t)$, with positive definite matrices $\hat{P}$. Then

$$\begin{aligned} \dot{V}(t) &= \dot{\xi}^{1T}(t)\hat{P}\xi^1(t) + \xi^{1T}(t)\hat{P}\dot{\xi}^1(t) \\ &= (\hat{A}_1\xi^1(t) + \hat{E}_1\hat{e}^1(t))^T\hat{P}\xi^1(t) + \xi^{1T}(t)\hat{P}(\hat{A}_1\xi^1(t) + \hat{E}_1\hat{e}^1(t)) \\ &= \xi^{1T}(t)(\hat{A}_1^T\hat{P} + \hat{P}\hat{A}_1)\xi^1(t) + 2\xi^{1T}(t)\hat{P}\hat{E}_1\hat{e}^1(t) \\ &\le \xi^{1T}(t)(\hat{A}_1^T\hat{P} + \hat{P}\hat{A}_1 + \alpha^{-1}\hat{P}\hat{E}_1\hat{E}_1^T\hat{P})\xi^1(t) + \alpha\hat{e}^{1T}(t)\hat{e}^1(t) \\ &\le \xi^{1T}(t)(\hat{A}_1^T\hat{P} + \hat{P}\hat{A}_1 + \alpha^{-1}\hat{P}\hat{E}_1\hat{E}_1^T\hat{P} + \alpha\sigma\hat{L}_1^T\hat{L}_1)\xi^1(t) \end{aligned} \tag{25}$$

holds obviously. Then by (22), using Lemma 3, we get $\dot{V}(t) < 0$ which indicates that system (19) is asymptotically stable.

The inexistence of Zeno behavior is proved as follows. By the definition of event-triggering function, $\|e_i(t)\| \leq \sqrt{\sigma} \left\| \sum_{j \in N_i} a_{ij}(\tilde{x}_i(t) - \tilde{x}_j(t)) \right\|, t \in [t_k^i, t_{k+1}^i)$ holds. The derivative of $\|e_i(t)\|$ satisfies

$$
\begin{aligned}
\frac{d}{dt}\|e_i(t)\| &\leq \|\dot{e}_i(t)\| \\
&= \left\| Ae^{A(t-t_k^i)}\tilde{x}_i(t_k^i) - \dot{\tilde{x}}_i(t) \right\| \\
&= \|Ae_i(t) - Bu_i(t) - GCh_i(t)\| \\
&\leq \|A\|\|e_i(t)\| + \|B\|\|u_i(t)\| + \|GC\|\|h_i(t)\|
\end{aligned}
\tag{26}
$$

From $\dot{h}_i(t) = (A + GC)h_i(t)$, it's derived $h_i(t) = e^{(A+GC)(t-t_k^i)}h_i(t_k^i)$. And $\|h_i(t)\| \leq \|h_i(t_k^i)\|$ as $(A + GC)$ is Hurwitz. With $u_i(t)$ designed in (8), $\|u_i(t)\| \leq \|u\|_{mi}$ holds where $\|u\|_{mi} = \max\limits_{t \in [t_k^i, t_{k+1}^i]} \left\| K \sum_{j \in N_i} a_{ij}(e^{A(t-t_k^i)}\tilde{x}_i(t_k^i) - e^{A(t-t_k^i)}\tilde{x}_j(t_k^i)) \right\|$ for the property of continuous function. Thus the derivative of $\|e_i(t)\|$ satisfies

$$
\frac{d}{dt}\|e_i(t)\| \leq \|A\|\|e_i(t)\| + \varphi_k^i
\tag{27}
$$

with $\varphi_k^i = \|GC\|\|h_i(t_k^i)\| + \|u\|_{mi}$. And from (27) it derives that

$$
\|e_i(t)\| \leq \frac{\varphi_k^i}{\|A\|}(e^{\|A\|(t-t_k^i)} - 1), t \in [t_k^i, t_{k+1}^i)
\tag{28}
$$

When consensus is not achieved, $\left\| \sum_{j \in N_i} a_{ij}(\tilde{x}_i(t) - \tilde{x}_j(t)) \right\| > 0$ holds between two neighbor event-triggered times. And there must be an infimum η_k such that

$$
\sigma \left\| \sum_{j \in N_i} a_{ij}(\tilde{x}_i(t) - \tilde{x}_j(t)) \right\| \geq \eta_k > 0, t \in [t_k^i, t_{k+1}^i)
\tag{29}
$$

holds. In addition, it's obvious that there exists one and only one solution t^* to $\frac{\varphi_k^i}{\|A\|}(e^{\|A\|(t-t_k^i)} - 1) = \eta_k$ and it satisfies that $t^* - t_k^i > 0$. Then, combining with (28) and (29) we get $t_{k+1} - t_k \geq t^* - t_k^i > 0$, which means the event-triggering time interval is strictly positive.

Theorem 2 *Under the event-triggered protocol (8) and (9), the multi-agent system (6) achieves consensus if there exist a positive definite matrix*

$$\bar{P} = \begin{bmatrix} \bar{P}_1 & 0_{m \times p} & 0 \\ * & \bar{P}_2 & 0_{p \times (m-p)} \\ * & * & \bar{P}_3 \end{bmatrix} \tag{30}$$

and a matrix

$$\bar{Q} = \begin{bmatrix} \bar{Q}_0 & 0_{(m_1+m) \times (m-p)} \end{bmatrix}, \bar{Q}_0 = \begin{bmatrix} \bar{Q}_1 & 0_{m_1 \times p} \\ 0_{m \times m} & \bar{Q}_2 \end{bmatrix} \tag{31}$$

such that

$$\begin{bmatrix} (\bar{S}\bar{P} + \bar{Y}_i\bar{Q}))^T + (\bar{S}\bar{P} + \bar{Y}_i\bar{Q}) & \bar{Z}_i\bar{Q} & \bar{P}\bar{M}^T \\ (\bar{Z}_i\bar{Q})^T & \alpha(\bar{V} - \bar{V}\bar{P} - \bar{P}\bar{V}) & 0 \\ \bar{M}\bar{P} & 0 & -(\alpha\sigma\lambda_i^2 I_m)^{-1} \end{bmatrix} < 0 \tag{32}$$

holds with a positive scalar α for $i = 1$ and $i = n - 1$, where

$$\bar{S} = V^{-1}SV, \ \bar{Y}_i = V^{-1}Y_i, \ \bar{Z}_i = V^{-1}Z_i, \ \bar{M} = MV, \ \bar{V} = V^TV \tag{33}$$

while

$$S = \begin{bmatrix} A & 0 \\ 0 & A \end{bmatrix}, \ Y_i = \begin{bmatrix} \lambda_i B & I \\ 0 & I \end{bmatrix}, \ Z_i = \begin{bmatrix} \lambda_i B & 0 \\ 0 & 0 \end{bmatrix}, \ M = \begin{bmatrix} I_m & 0_{m \times m} \end{bmatrix} \tag{34}$$

and $V = diag\{I_m, V_0\}$ is a nonsingular matrix with

$$CV = \begin{bmatrix} I_{m+p} & 0 \end{bmatrix} \tag{35}$$

and $\lambda_i(i = 1, \ldots, n-1)$ is denoted as the positive eigenvalues of Laplacian matrix L and $\lambda_1 < \cdots < \lambda_{n-1}$. Then the gain matrices G and K can be determined by

$$K = \bar{Q}_1\bar{P}_1^{-1}, \ G = \bar{Q}_2\bar{P}_2^{-1} \tag{36}$$

Proof Without generality, $\hat{P}$ is taken as $\hat{P} = I_{n-1} \otimes P$ for convenience. Pre- and post-multiply (15) with $\hat{U}_1^T = diag\{\bar{U}_1^T \otimes I_m, \bar{U}_1^T \otimes I_m, \bar{U}_1^T \otimes I_m, \bar{U}_1^T \otimes I_m\}$ and $\hat{U}_1$ respectively, in which $\bar{U}_1$ is a matrix that satisfies $\bar{U}_1^T L_1 \bar{U}_1 = diag\{\lambda_1, \ldots, \lambda_{n-1}\}$. Then the inequality can be divided into a series of inequalities as following

$$\begin{bmatrix} \bar{A}_1^T P + P\bar{A}_1 + \alpha\sigma\lambda_i^2 M^T M & P\bar{E}_1 \\ \bar{E}_1^T P & -\alpha I \end{bmatrix} < 0, \ i = 1, \ldots, n-1 \tag{37}$$

where

$$\bar{A}_1 = \begin{bmatrix} A + \lambda_i BK & GC \\ 0 & A + GC \end{bmatrix}, \bar{E}_1 = \begin{bmatrix} \lambda_i BK & 0 \\ 0 & 0 \end{bmatrix} \tag{38}$$

Then $\bar{A}_1$ and $\bar{E}_1$ can be decomposed as

$$\bar{A}_1 = \begin{bmatrix} A + \lambda_i BK & GC \\ 0 & A + GC \end{bmatrix} = \begin{bmatrix} A & 0 \\ 0 & A \end{bmatrix} + \begin{bmatrix} \lambda_i B & I \\ 0 & I \end{bmatrix} \begin{bmatrix} K & 0 \\ 0 & G \end{bmatrix} \begin{bmatrix} I & 0 \\ 0 & C \end{bmatrix} \triangleq S + Y_i \hat{K} R$$

$$\bar{E}_1 = \begin{bmatrix} L_1 \otimes BK & 0 \\ 0 & 0 \end{bmatrix} = \begin{bmatrix} \lambda_i B & 0 \\ 0 & 0 \end{bmatrix} \begin{bmatrix} K & 0 \\ 0 & G \end{bmatrix} \begin{bmatrix} I & 0 \\ 0 & C \end{bmatrix} \triangleq Z_i \hat{K} R \tag{39}$$

So (37) can be transformed into

$$\begin{bmatrix} (S + Y_i \hat{K} R)^T P + P(S + Y_i \hat{K} R) + \alpha \sigma \lambda_i^2 M^T M & P Z_i \hat{K} R \\ (Z_i \hat{K} R)^T P & -\alpha I \end{bmatrix} < 0 \tag{40}$$

By Lemma 3, (40) is equivalent to

$$\begin{bmatrix} (S + Y_i \hat{K} R)^T P + P(S + Y_i \hat{K} R) & P Z_i \hat{K} R & M^T \\ (Z_i \hat{K} R)^T P & -\alpha I & 0 \\ M & 0 & -(\alpha \sigma \lambda_i^2 I_m)^{-1} \end{bmatrix} < 0 \tag{41}$$

Since C is of full row rank, there exists a nonsingular matrix V such that (35) holds. Pre- and post-multiplying (41) with $\tilde{V}^T$, $\tilde{V} = diag\{V, V, I_m\}$ yields

$$\begin{bmatrix} (\bar{S} + \bar{Y}_i \hat{K} \bar{R})^T \tilde{P} + \tilde{P}(\bar{S} + \bar{Y}_i \hat{K} \bar{R}) & \tilde{P} Z_i \hat{K} R & \bar{M}^T \\ (\bar{Z}_i \hat{K} \bar{R})^T \tilde{P} & -\alpha V^T V & 0 \\ \bar{M} & 0 & -(\alpha \sigma \lambda_i^2 I_m)^{-1} \end{bmatrix} < 0 \tag{42}$$

in which $\tilde{P} = V^T P V$. Subsequently, pre- and post-multiplying (42) with $diag\{\tilde{P}^{-1}, \tilde{P}^{-1}, I\}$ and denoting $\bar{P} = \tilde{P}^{-1}$ and $\bar{Q} = \hat{K} \bar{R} \bar{P}$ yields

$$\begin{bmatrix} (\bar{S}\bar{P} + \bar{Y}_i \bar{Q}))^T + (\bar{S}\bar{P} + \bar{Y}_i \bar{Q}) & \bar{Z}_i \bar{Q} & \bar{P} \bar{M}^T \\ (\bar{Z}_i \bar{Q})^T & -\alpha \bar{P} V^T V \bar{P} & 0 \\ \bar{M} \bar{P} & 0 & -(\alpha \sigma \lambda_i^2 I_m)^{-1} \end{bmatrix} < 0 \tag{43}$$

Then (32) holds for $V^T V - V^T V \bar{P} - \bar{P} V^T V \geq -\bar{P} V^T V \bar{P}$. According to the convex property of linear matrix inequalities, if (32) holds when λ_i takes its extreme values λ_1 and λ_{n-1}, then it holds for $i = 1, \ldots, n-1$. So, combining with Theorem 1, the proof is completed.

4 Simulation

In this section, a simulation example is provided to illustrate the performance of the control protocol proposed in the previous section. A network of four agents is considered, with the system matrices in (6) are set as

$$A = \begin{bmatrix} 0 & -1 \\ 2 & 0.1 \end{bmatrix}, \quad B = \begin{bmatrix} 1 & 0 \\ 0 & 2 \end{bmatrix}, \quad C = \begin{bmatrix} 0 & 1 \\ 1 & 0 \end{bmatrix}$$

And the initial values are set as $x_1(0) = [-0.3 \quad -0.6]^T$, $x_1(0) = [-0.6 \quad 0.4]^T$, $x_1(0) = [0.3 \quad 0.3]^T$, $x_1(0) = [0.5 \quad 0.5]^T$. The communication topology is set as following with all weighing factors defined as 0.5 for convenience.

From Fig. 1, the Laplacian matrix is defined as

$$L = \begin{bmatrix} 1.5 & -0.5 & -0.5 & -0.5 \\ -0.5 & 0.5 & 0 & 0 \\ -0.5 & 0 & 0.5 & 0 \\ -0.5 & 0 & 0 & 0.5 \end{bmatrix}$$

and L_c is defined as (1). Besides, we choose $\sigma = 0.1$, $\alpha = 10$. Then according to Theorem 2, the gain matrices are determined as

$$K = \begin{bmatrix} -2.6855 & -1.0018 \\ -0.5064 & -1.5030 \end{bmatrix}, \quad G = \begin{bmatrix} -0.3713 & -1.9166 \\ -1.8234 & -0.3889 \end{bmatrix}$$

by solving the LMIs related to $\lambda_i = 0.5$ and 2. Then Fig. 2 shows that the four agents reach consensus after the initial disagreement is set. That is, the theorems obtained are validated correct by the simulation example.

Fig. 1 Communication topology of multi-agent system

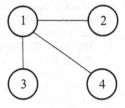

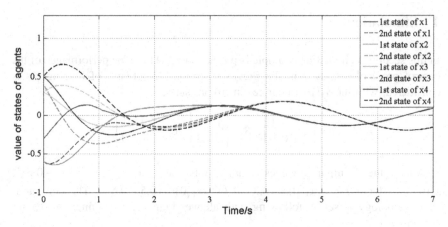

Fig. 2 States of four agents with the given initial values

5 Conclusion

In this paper, using the observer-based output feedback control method, an event-triggered protocol is proposed to solve the consensus problem of multi-agent systems with general linear dynamics. It's proved that the multi-agent system reaches consensus, and exhibits no Zeno behavior. Further research will focus on multi-agent systems subject to external disturbances.

Acknowledgements This work was supported by the National Basic Research Program of China (973 Program: 2012CB821200, 2012CB821201), and the NSFC (61473015, 61104147, 61134005, 61327807, 61520106010).

References

1. Vicsek T, Czirók A, Ben-Jacob E, et al. Novel type of phase transitions in a system of self-driven particles. Phys Rev Lett. 1995;75(6):1226–9.
2. Jadbabaie A, Lin J, Morse AS. Coordination of groups of mobile autonomous agents using nearest neighbor rules. IEEE Trans Autom Control. 2003;48(6):988–1001.
3. Olfati-Saber R, Murray RM. Consensus problems in networks of agents with switching topology and time-delays. IEEE Trans Autom Control. 2004;49(9):1520–33.
4. Liu Y, Jia Y. H_∞ consensus control of multi-agent systems with switching topology: a dynamic output feedback protocol. Int J Control. 2010;83(3):527–37.
5. Li Z, Duan Z, Chen G. Consensus of multi-agent systems and synchronization of complex networks: a unified viewpoint. IEEE Trans Circuits Syst. 2010;57(1):213–24.
6. Ren W. On consensus algorithms for double-integrator dynamics. IEEE Trans Autom Control. 2008;53(6):1503–9.
7. Noorbakhsh SM, Ghaisari J. Distributed event-triggered average consensus protocol for multi-agent systems. In: 2015 23rd Iranian conference on electrical engineering. Tehran: IEEE; 2015. p. 840–5.

8. Dimarogonas DV, Frazzoli E, Johansson KH. Distributed event-triggered control for multi-agent systems. IEEE Trans Autom Control. 2012;57(5):1291–7.

9. Liu Z, Chen Z. Event-triggered average-consensus for multi-agent systems. In: Proceedings of the 29th Chinese control conference. Beijing: IEEE; 2010. p. 4506–11.

10. Seyboth GS, Dimarogonas DV, Johansson KH. Event-based broadcasting for multi-agent average consensus. Automatica. 2013;49:245–52.

11. Han J, Zhang H, Jiang H. Event-based H_∞ consensus control for second-order leader-following multi-agent systems. J Franklin Inst. 2016;353:5081–98.

12. Hu W, Liu L. Consensus of linear multi-agent systems by distributed event-triggered strategy. IEEE Trans Cybern. 2016;46(1):148–57.

13. Hu W, Liu L, Feng G. Consensus of multi-agent systems by distributed event-triggered control. In: The international federation of automatic control. Cape Town: IFAC; 2014. p. 9768–73.

14. Zhang H, Feng G, Yan H. Consensus of multi-agent systems with linear dynamics using event-triggered control. IET Control Theory Appl. 2014;8:2275–81.

Quantitative Inspection of Shear Mark Based on Lyapunov Dimension

Bingcheng Wang and Chang Jing

Abstract Base on the analysis of the Lyapunov dimension that can be used to characterize the system dynamic status and the irregularity degree, and according to the nonlinear dynamic characteristics on surface profile of shearing marks, the analysis method on characteristics of shearing marks based on Lyapunov dimension is proposed. The collected surface profile curve of shearing marks is treated as the time series. The time series are reconstructed by phase space reconstruction theory. In order to make the reconstructed phase space fully reflect the system dynamic characteristics, the determination problem of time delay and embedding dimension are discussed. On above basis, the Lyapunov dimension is calculated, and the ratio of embedding dimension and the time delay multiplied by the Lyapunov dimension are taken as the characteristic quantity of the mark surface. It can be seen that the Lyapunov dimension and the characteristic quantity of the mark surface are different according to the analysis and calculation of the Lyapunov dimension of the surface profile curve of the shearing marks. It is proved that the characteristic quantity is an effective parameter to characterize the surface profile of different shear marks. The characteristics of marks can be extracted and recognized effectively. Therefore the theoretical basis and technical method are provided for studying the surface characteristics of shearing marks.

Keywords Shearing marks · Delay time · Embedding dimension · Lyapunov dimension · Quantitative inspection

B. Wang (✉)
Center Laboratory for Forensic Science,
Shenzhen University, Shenzhen 518060, China
e-mail: wbc8631@163.com

C. Jing
Department of Technology, Guangdong Police College, Guangzhou 510232, China

© Springer Nature Singapore Pte Ltd. 2018
Y. Jia et al. (eds.), *Proceedings of 2017 Chinese Intelligent Systems Conference*,
Lecture Notes in Electrical Engineering 460,
https://doi.org/10.1007/978-981-10-6499-9_26

1 Introduction

The surface topography has unsteady stochastic characteristics. This fact has been discovered by Sayles et al. in the 1970s. The problem of surface topography recognition is studied by time series analysis [1] on the basis of stochastic process theory. The analysis of the two-dimensional profile curve and the three-dimensional surface are treated as one-dimensional signals and three-dimensional signals [2]. In the research of past 20 years, some progress is made in this field. Wang Anliang and other researchers wrote the paper [3], etc. results. The achievements show that the time series have been applied in the study of surface topography.

The surface profile curve of shearing marks is treated as one-dimensional signals. The Lyapunov dimension is applied to the study of surface characteristics analysis of shearing marks. The analysis method of surface characteristics of shearing marks based on Lyapunov dimension is proposed. The basis of the reasonable calculation of the Lyapunov dimension is the determination of delay time τ and the embedding dimension d in the phase space reconstruction of time series. Therefore in this paper the determination of these two quantities is analyzed emphatically. The delay time of different time series is determined by mutual information function method. The minimum embedding dimension d is determined according to the improved false neighbor method which is proposed by Cao [4]. The correlation dimension is calculated for the surface profile curve of shearing marks [5–7] on the basis. The research provides a method of analysis for surface characteristics recognition of shearing marks.

2 Experimental Conditions and Data Acquisition

Scissors, wire cutters and wire clippers are used as shearing tools to shear the lead wires. Several samples of shearing marks are made. The samples are observed under the stereo microscope. The samples which can reflect the stability characteristics of the shearing marks are chosen to be collected digitally.

In the experiment, the shearing marks are collected digitally by using Austria Infinite Focus auto-zoom three-dimensional surface topography measurement device. A three-dimensional shearing mark is stored in computer. In this experiment, the objective magnification is 10 times and the sampling resolution is 1.1 μm. The stable part of the mark characteristic is marked by the application software. The profile curve which is perpendicular to the marks surface is obtained. Profile curve which is perpendicular to the mark surface formed by scissors, wire cutters, wire clippers are shown in Fig. 1.

(a)   (b)   (c)

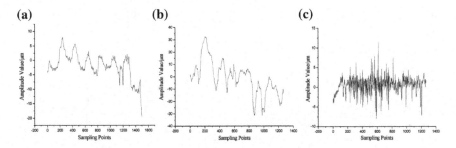

Fig. 1 Three kinds of profile curves

3 Two Parameter Determination in Restructuring of Phase Space

The delay time τ is one of the two important parameters of phase space reconstruction. In order to make the reconstructed phase space fully reflect the system dynamic characteristics, two parameters must be properly selected. When the selection of delay time τ is too big or too small, the dynamic characteristic of system is unable to reflect and the calculated chaotic characteristics may be inaccurate or even incorrect. In this paper, the delay time τ is determined by using mutual information function method.

The basic philosophy of mutual information function method is to determine the delay time τ by using the mutual information function to calculate the position of first minimum. The mutual information function is the general random connection measure between two random variables. It is the reasonable analysis method of the nonlinear analysis. Supposes the observation time series is $\{x_n\}$, and then observes the mutual information of variables between n and $n+1$ time, the definition is:

$$I(\tau) = \sum_{n=1}^{N} p(x_n, x_{n+\tau}) \ln[p(x_n, x_{n+\tau})/p(x_n)p(x_{n+\tau})] \tag{1}$$

In the formula, $p(x_n)$, $p(x_{n+\tau})$ and $p(x_n, x_{n+\tau})$ are the probability. The function value of the time series $\{x_n\}$ is obtained. Draw the curve under the coordinate system of $I(\tau) - \tau$. The value of τ, which corresponding to the first minimum point, is the time delay.

The delay time τ of phase space reconstruction of three kinds of profile curves is calculated by the formula (1) according to determine principles of delay time τ—the mutual information function method.

It can be seen from the Fig. 2 the delay time τ of the phase space reconstruction is different with different profile curves.

Embedding dimension d is one of the important parameters of the phase space reconstruction. The improved false neighbour method which is proposed by Cao has no subjective parameters in determining the embedding dimension d. It is

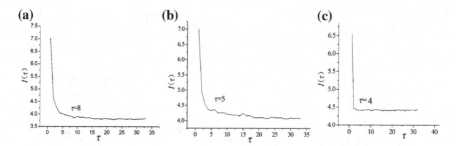

Fig. 2 Delay time τ of different profile curves

suitable the high-dimensional dynamic system and the small data quantity. The random signals can be distinguished and the signals can be determined. The principle is:

Suppose the time series is $\{x_n\}n = 1, 2, \ldots, N$, the delay time τ is determined by the formula (1), and then a time vector is constructed, such as $X_i(d) = \{x_i, x_{i+\tau}, \ldots, x_{i+(d-1)\tau}\}$ $i = 1, 2, \ldots, N - (d-1)\tau$. The d is embedding dimension. A quantity $\beta(i, d)$ is introduced refers to the false neighbour method, and to order:

$$\beta(i, d) = \frac{\|X_i(d+1) - X_{n(i,d)}(d+1)\|}{\|X_i(d) - X_{n(i,d)}(d)\|} \quad i = 1, 2, \ldots, N - d\tau \qquad (2)$$

In formula (2) $\|X_k(d) - X_l(d)\| = \max_{0 \le j \le m-1} \|x_{k+j\tau} - x_{l+j\tau}\|$, $n(i, d)$ is an integer, the size is $1 \le n(i, d) \le N - d\tau$, $X_{n(i,d)}(d)$, is the nearest point of $X_i(d)$ in the reconstruction d space.

In order to avoid that different threshold of distance of two neighbor points have to select the different threshold of distance of two neighbor points, an average value of all parameters $n(i, d)$ is defined, it is:

$$E(d) = \frac{1}{N - d\tau} \sum_{i=1}^{N-d\tau} \beta(i.d) \qquad (3)$$

It can be seen by the formula (3), the function value $E(d)$ is only related to the embedded dimension d and the delay time τ, obviously, when the delay time τ is given, the function changes is changed only with the embedding dimension d, therefore $E_1(d)$ is defined:

$$E_1(d) = E(d+1)/E(d) \qquad (4)$$

Draw the curve under $E_1(d) - d$ coordinate system; the $E_1(d)$ value does not change with the increase of d, the d corresponding to is the minimum embedding dimension.

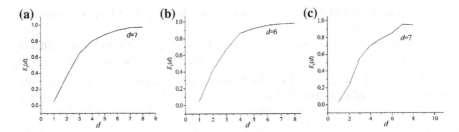

Fig. 3 The minimum embedding dimension of three profile curves

The minimum embedding dimension d is determined according to the improved false neighbor method which is proposed by Cao. The minimum embedding dimension of phase space reconstruction of three kinds of profile curves are calculated by formula (4). Figure 3 shows the minimum embedding dimension of three kinds of profile curves using the method of Cao.

It can be seen from the Fig. 3, the embedding dimension of phase space reconstruction is different with different profile curves.

4   Calculating of Lyapunov Exponent Spectrum of Profile Curves

The basic feature of chaotic motion is that it is very sensitive to initial conditions. The orbits generated by two very close initial values would separate by way of index with time changes. Lyapunov exponent is the amount of quantitative description of this phenomenon. In the Lyapunov exponent spectrum, the smallest Lyapunov exponent determines the speed of orbital convergence; the largest Lyapunov exponent determines the divergence of the orbital that is to say the speed of covering the entire attractor. The largest Lyapunov exponent reflects the change level in the phase space orbit caused by movement system's small changes in initial conditions, which is a very important parameter used to characterize the "strange" level of non-linear chaotic attractor. Therefore it is widely used to characterize of the behavior of nonlinear systems.

In fact, Lyapunov exponent is an average volume related to shrinkage and expansion characteristics of track in a different direction of the phase space. Each Lyapunov exponent λ can be seen as the average volume of local deformation for relative movement in all directions of phase space, at the same time they are determined by the evolution of the system during a long time. As a result, in terms of space and time, Lyapunov exponent is not a local content but that a characteristic of overall.

Although currently there are many methods of calculating volume of Lyapunov exponent, they generally fall into two categories: Wolf method and Jacobian method. Calculation of Lyapunov exponent spectrum based on time series usually use Jacobian method.

For dynamical system of one-dimensional phase space $X = F(X)$, the tangent vector β at $X(t)$ can be expressed as linear in its tangent space, where $\dot{\beta} = K(X(t)) \times \beta$, and $K = \partial F / \partial X$ is the Jacobian matrix for F. To define the average divergence rate of tangent vector β:

$$\lambda = \lim_{t \to \infty} \frac{1}{t} Ln \frac{|\beta(t)|}{|\beta(0)|} \tag{5}$$

Extend it to the m-dimensional space and we have:

$$\lambda_i = \lambda(X(0), \zeta_i) \tag{6}$$

In the above formula, ζ_i (i = 1, 2, ..., m) is the standard orthogonal basis vectors, while λ_i (i = 1, 2, ..., m) is called Lyapunov exponent spectrum.

To determine the time series of $X(t)$, according to Takens delay embedding theorem, m is selected as the embedding dimension for the reconstruction of phase space and τ as delay time. Phase space of reconstruction is $\{x(t), x(t+\tau), \cdots, x(t+[m-1]\tau)\}$., after tracing evolution of M times, until the time series end N, the Lyapunov exponent spectra of the system can be obtained. That is,

$$\lambda_i = \frac{1}{N_m} \sum_{k=0}^{N_m} \ln|B_k \zeta_{ki}| \tag{7}$$

where B_k is a linear operator. According to the above formula Lyapunov exponent spectrum of profile curves can be calculated. Table 1 shows three different profile curves for Lyapunov exponent spectrums

As can be seen from Table 1, when the profile curves are different, the delay time τ and embedding dimension d are different; the corresponding value of Lyapunov exponent spectrum distribution is different; The largest and the smallest Lyapunov exponent values are different. This shows Lyapunov exponent's sensitivity to the marks surface. Therefore, the value of Lyapunov exponent spectrum can be used as characteristics value to identify the profile curves.

Table 1 Lyapunov exponent spectrums of profile curves

Types	Sample a	Sample b	Sample c
λ_1	0.26583	0.71168	0.28398
λ_2	0.067919	0.080965	0.071236
λ_3	0.018113	0.023152	0.030029
λ_4	0.0045946	0.0050028	0.013824
λ_5	−0.0082603	−0.026241	−0.01595
λ_6	−0.054925	−2.047	−0.064381
λ_7	−1.2276		−1.3947

5 Characteristics Analysis of the Profile Curves Based on Lyapunov Dimension

Inspects an attractor A of dynamic system of M dimension, its Lyapunov exponent is $\lambda_1 \geq \lambda_2 \geq \cdots \lambda_m$, supposing with the positive integer $n < M$, so $\lambda_1 + \lambda_2 + \cdots + \lambda_n \geq 0$, then attracts the Lyapunov dimension of attractor A is defined as:

$$D_\lambda = n + \frac{\lambda_1 + \lambda_2 + \cdots + \lambda_n}{|\lambda_{n+1}|} \tag{8}$$

According to the definition of Lyapunov dimension, it can be seen that when the system is under different working conditions, the exponent λ of contraction or the expansion in different directions of track in phase space are different; then Lyapunov dimension of system is different. This dimension value can be used to identify the surface topography of the shearing marks. According to formula (8), Lyapunov dimension value may be calculated, the specific values shown in Table 2.

It can be seen from the calculation results that the Lyapunov dimensions of different profile curves are different. The dimension value of profile curve formed by scissors is the biggest, while the dimension value of profile curve formed by wire cutters is the smallest.

Taking into account the influence of the delay time τ and the embedding dimension d on the system characteristics, the formula (8) is modified by introduce a weighting factor. The feature quantity of profile curve is defined as:

$$D_\lambda^* = \beta \cdot D_\lambda \tag{9}$$

where the weighted coefficient β is the ratio of embedding dimension d and delay time τ, that is $\beta = d/\tau$, The quantity of characteristic calculated by formula (9) is shown in Table 3.

It can be seen from the calculation results that the characteristic quantity of different profile curves are obviously different. The characteristic quantity of profile curve formed by scissors is the biggest, while the characteristic quantity of profile curve formed by wire cutters is the smallest. Because considering the influence of the delay time τ and the embedded dimension d on the system characteristics, the distinguish of characteristic quantity calculated by the formula (9) is good and to help identify shearing marks type

Table 2 Lyapunoy dimension of different profile curves

Types	Sample a	Sample b	Sample c
D_λ	6.2389	5.3882	6.2285

Table 3 Characteristic quantity of different profile curves

Curves	Sample a	Sample b	Sample c
D_λ^*	10.9181	3.5921	8.7199

The Lyapunov dimension and the quantitative index of different mark surface profile curves are shown in Table 3. It can be seen from the calculation results, the Lyapunov dimensions and the quantitative indexes are different with different profile curves.

6 Conclusions

The embedding dimension d and the delay time τ are two important parameters of phase space reconstruction. In order to make the reconstructed phase space fully reflect the chaotic characteristics of the system, these two parameters must be properly selected. It is also the prerequisite and basis for the accurate calculation of chaotic characteristic quantities such as Lyapunov dimension. It can be seen that the embedding dimension and delay time are different with different profile curves according to the analysis of three kinds of profile curves.

The Lyapunov dimension of three profile curves is calculated based on the appropriate selection of embedding dimension and delay time. The parameter can reflect the dynamic characteristics of the nonlinear system, and it can be used to analyze and identify the surface characteristics of shearing marks. The Lyapunov dimension and characteristic quantity are different with different surface of marks. It can be quantitatively marked surface topography characteristics.

Therefore, the Lyapunov dimension and characteristic quantity can be used to describe the characteristics of marks surface. It can be used to analyze and identify the surface characteristics of shearing marks. It provides a method to analyze the surface features of shearing marks.

Acknowledgements This work is supported by the Nature Science Fund of China (NSFC), No.: 61571307

References

1. Lv J, Lu J, Chen S. Analysis and application of chaotic time series. China: Wuhan University Press; 2002.
2. Wang A, Yang C. The calculation methods for the fractal characteristics of surface topography. China Mech Eng. 2002;13(8):714–8 (in Chinese).
3. Wang A, Yang C. Application of wavelet transform to evaluate the mechanical surface topography. Chin J Mech Eng. 2001;37(8):65–74 (in Chinese).

4. Cao LY. Practical method for determining embedding dimension of a scalar time series. Physica D. 1977;110(5):43–50.
5. Li YX, Gao MZ, Ma K. Developments in calculation theory of fractal dimension of rough surface. Adv Math. 2014;41(4):397–408 (in Chinese).
6. Wang BC, Wang ZY, Jing C. Study on examination method of striation marks based on fractal theory. Appl Mech Mater. 2015;740:553–6.
7. Su YW, Chen W, Zhu AB, et al. Contact and. wear simulation fractal surfaces. J Xi Jiaotong Univ. 2014;47(7):52–5 (in Chinese).

Iron Ore Sintering Subsection Temperature Model on the Airflow Rate by PID Control

Zhifeng Ding, Qinglin Sun, Shengfei Liu and Zengqiang Chen

Abstract Sintering is a complicated process in which the sintered ores are produced for the blast furnace. In this paper, a subsection temperature model on airflow rate is proposed to explain the effect of the airflow rate on the sinter bed temperature. Furthermore, the sintering process control strategy is put forward applying PID algorithm based on the optimization of airflow rate. The sintering process is divided into 24 stages to track the sinter bed temperature. Experimental results show that the airflow quantity is sufficient for the coke combustion and remarkable effects can be achieved on the electricity consumption saving of the main exhaust fan which is up to 32.6%.

Keywords Iron ore sintering · Subsection airflow rate temperature model · PID control · Electricity consumption saving

1 Introduction

The iron ore sintering process is one of the most significant step in the steelmaking industry, which influences the production, quality and the energy consumption of the blast furnace. With the rapid development of the steelmaking technology in recent 20 years, sintering process consumes 10–15% of China's total energy consumption by far, among which the electricity consumption accounts for 13–20%. In

Z. Ding · Q. Sun (✉) · S. Liu · Z. Chen
College of Computer and Control Engineering, Nankai University, Tianjin 300350, China
e-mail: sunql@nankai.edu.cn

Z. Ding
e-mail: dingzf@mail.nankai.edu.cn

S. Liu
e-mail: liushengfei2001@163.com

Z. Chen
e-mail: chenzq@nankai.edu.cn

© Springer Nature Singapore Pte Ltd. 2018　　　　　　　　　　　　　　277
Y. Jia et al. (eds.), *Proceedings of 2017 Chinese Intelligent Systems Conference*,
Lecture Notes in Electrical Engineering 460,
https://doi.org/10.1007/978-981-10-6499-9_27

addition, the power consumption of the main exhaust fan accounts for nearly 50% of the total sintering process consumption [1, 2]. Therefore, it is necessary to reduce the energy consumption of the main exhaust fan to meet the demand of the energy-saving and emission-reduction.

The sintering is a complex physical chemistry process and many scholars have been investigated in this context. In [3–8], mechanism models of the sintering process based on thermodynamics are built to describe the effects of the various parameters on the heat change.

In addition, artificial intelligence technology has been widely used in the sintering process, such as Expert System, Artificial Neural Networks and Fuzzy Control [9, 10]. The aforementioned models emphasize the correlation between input and output variables. A large amount of actual production data is needed to train the model.

In most of domestic sintering plants, the opening of air door keeps constant to guarantee the quantity of airflow rate for sintering. But in this way, the main exhaust fan runs in operating frequency all the time which leads to the great waste of power consumption. In recent years, the manufacturing technology of high-voltage frequency converters develops rapidly and the cost decreases obviously. Some sintering plants used high-voltage frequency converters in the main exhaust fan to control the motor speed [11, 12], in order that negative pressure and airflow rate of sintering can be adjusted with the frequency control of the main exhaust fan.

2 Sintering Process

The sinter mix consists of iron ore, coke, limestone and returned ore. Then the mixture are blended with some water for granulation. The sintering process on the sinter strand in the sintering plant is illustrated in Fig. 1. There are 24 wind boxes connected with the main exhaust fan. The height of the sinter bed is 900 mm and the length of the sinter strand is 60 m. The sinter time is about 40 min. After ignition, the electric motor runs and the main exhaust fan rotates. Then the heat supplied by the coke combustion is transferred from top to bottom which forms a flame front. With the sinter strand moving, the flame front descends down and reaches the bottom at the 23rd wind box ensuring the full combustion of the coke. The whole sintering process can be divided into three zones: sinter mix zone, combustion zone and sintered zone.

Generally, the sintering process is tested on the sintering pot for simulation as is shown in Fig. 2.

After ignition, air is suctioned down through the sinter bed which supplies oxygen for coke combustion. When the flame front reaches the bottom, the sintering process ends and the sinter mix transfers to the sintered ore completely.

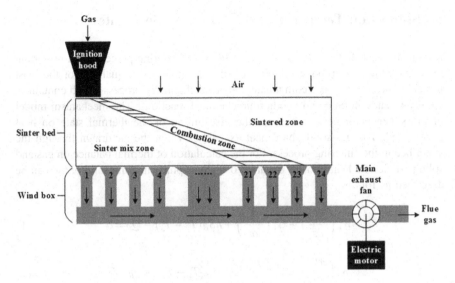

Fig. 1 Sintering process on the sinter strand

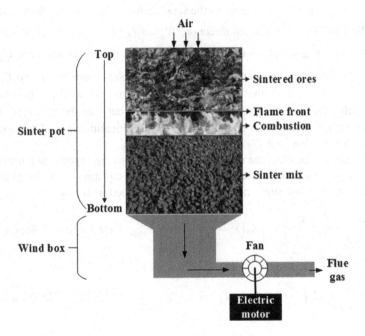

Fig. 2 Sintering process on the sinter pot after ignition

3   Subsection Temperature Model on Airflow Rate

Reasonable distribution of the thermal state in the sintering process is an important indicator which can produce sinter ores of high quality and quantity for the blast furnace. The sintering is a complicated physical chemistry process which contains a variety of endothermic and exothermic chemical reactions. The mechanism model of the sintering process is built as the distribution of the thermal state on heat transfer [6]. The gas-solid phase heat exchanges when the air drawn through the sinter bed in the sintering process. For the calculation of thermal balance on gas and solid particle, the heat transfer equation of the gas phase and the solid phase can be described as follows:

$$W_g\left(\frac{\partial T_g}{\partial z}\right) + G_g\left(\frac{\partial T_g}{\partial t}\right) + hS_B(T_g - T_s) - Q_R = 0 \tag{1}$$

$$W_s\left(\frac{\partial T_s}{\partial z}\right) + G_s\left(\frac{\partial T_s}{\partial t}\right) + hS_B(T_g - T_s) - Q_R = 0 \tag{2}$$

where $W_g\left(\frac{\partial T_g}{\partial z}\right)$ and $W_s\left(\frac{\partial T_s}{\partial z}\right)$ denotes the sensible heat caused by the flow of the gas and solid particle in the vertical direction, $W_g = v_g G_g/\varepsilon$ and $W_s = v_s G_s$ denotes the heat flow of the gas and solid per unit cross-sectional area respectively; $G_g\left(\frac{\partial T_g}{\partial t}\right)$ and $G_s\left(\frac{\partial T_s}{\partial t}\right)$ denotes the accumulation of the gas and solid particle, $G_g = \rho_g C_g \varepsilon$ and $G_s = \rho_s C_s(1 - \varepsilon)$ denotes the heat capacity of the gas and solid per unit volume respectively; $hS_B(T_g - T_s)$ denotes the heat exchange between the gas phase and the solid phase; Q_R denotes the reaction heat. In the different zones of the sintering process, the reaction heat Q_R has different meanings.

In the sinter mix zone, the reaction heat Q_R refers to the vaporization heat of the water and $Q_R = r_{H_2O}\Delta H_v$. Therefore, the heat transfer equation of the gas phase and the solid phase in the sinter mix zone can be described as follows:

$$W_g\left(\frac{\partial T_g}{\partial z}\right) + G_g\left(\frac{\partial T_g}{\partial t}\right) + h_d S_B(T_g - T_s) - (1-a)r_{H_2O}\left(3.1563\times10^6 - 2369.6T_W\right) = 0 \tag{3}$$

$$W_s\left(\frac{\partial T_s}{\partial z}\right) + G_s\left(\frac{\partial T_s}{\partial t}\right) + h_d S_B(T_g - T_s) - ar_{H_2O}\left(3.1563\times10^6 - 2369.6T_W\right) = 0 \tag{4}$$

In the combustion zone, the reaction heat Q_R refers to the exothermic carbon combustion reaction and the endothermic decomposition of limestone. Therefore, the heat transfer equation of the gas phase and the solid phase in the combustion zone can be described as follows:

$$W_g\left(\frac{\partial T_g}{\partial z}\right) + G_g\left(\frac{\partial T_g}{\partial t}\right) + h_m S_B\left(T_g - T_s\right) - \left(1 - a_1\right)\left(-279.5\right)\frac{\pi d_c^2 n_c \rho_c c_{O_2}}{k_f + k_c} \tag{5}$$
$$- \left(1 - a_2\right)h_p a_l U_l\left(T_g - T_{l1}\right) = 0$$

$$W_s\left(\frac{\partial T_s}{\partial z}\right) + G_s\left(\frac{\partial T_s}{\partial t}\right) + h_m S_B\left(T_g - T_s\right) - a_1\left(-279.5\right)\frac{\pi d_c^2 n_c \rho_c c_{O_2}}{k_f + k_c} \tag{6}$$
$$- a_2 h_p a_l U_l\left(T_g - T_{l1}\right) = 0$$

In the sintered zone, it is assumed that there is no exothermic and endothermic reactions so the reaction heat $Q_R = 0$. Therefore, the heat transfer equation of the gas phase and the solid phase in the sintered zone can be described as follows:

$$W_g\left(\frac{\partial T_g}{\partial z}\right) + G_g\left(\frac{\partial T_g}{\partial t}\right) + h_u S_B\left(T_g - T_s\right) = 0 \tag{7}$$

$$W_s\left(\frac{\partial T_s}{\partial z}\right) + G_s\left(\frac{\partial T_s}{\partial t}\right) + h_u S_B\left(T_g - T_s\right) = 0 \tag{8}$$

Considering the different bed structure in these three zones, the porosity of the sintering bed ε is equal to 0.417, 0.2, 0.8 in the sinter mix zone, combustion zone and sintered zone respectively.

The thermal state of the whole sintering process can be described by the heat transfer Eqs. (3–8). After setting the initial condition of parameters and the initial temperature field, the heat transfer Eqs. (3–8) can be solved and the whole temperature field can be obtained.

In this paper, the airflow rate temperature model is proposed that the sintering process is divided into 24 stages. Each stage lasts 100 s and has a suitable temperature target. The estimated temperature can be calculated by Eq. (9).

$$y(k) = a_1 y(k-1) + a_2 y(k-2) + bu(k) + e(k) \tag{9}$$

where $y(k)$, $y(k-1)$, $y(k-2)$ denotes the sinter bed temperature T_s at time k, $k-1$ and $k-2$ respectively; $u(k)$ denotes the airflow rate v_g at time k; $e(k)$ denotes the random error at time k; the coefficients a_1, a_2 and b can be calculated by the recursive least squares method (RLS) using the data from the thermal state model. The estimated sinter bed temperature tracks the actual sinter bed temperature accurately as illustrated in Fig. 3.

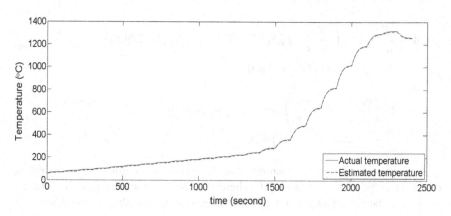

Fig. 3 The estimated sinter bed temperature tracks the actual sinter bed temperature

4 Temperature Tracking by PID Control

The PID controller has been developed for nearly 80 years, and it has become one of the most widely used technologies in industrial control because of its simple structure, good stability, reliable operation and convenient adjustment [13].

Proportional (P) control is one of the simplest control method. The output of the controller is proportional to the input error signal and the system has a steady state error.

In the integral (I) control, the output of the controller is proportional to the integral of the input error signal. The integral component is used to eliminate static error.

In the differential (D) control, the output of the controller is proportional to the differential of the input error signal. The differential component is used to improve the response speed and reduce the adjustment time.

The control error $e(t)$ can be calculated according to the reference $r_{in}(t)$ and the actual output $y_{out}(t)$ by Eq. (10).

$$e(t) = r_{in}(t) - y_{out}(t) \tag{10}$$

The control law is the linear combination of proportional (P), integral (I) and differential (D) of the control error which can be calculated by Eq. (11).

$$u(t) = K_P \left(e(t) + \frac{1}{T_I} \int_0^t e(t)dt + T_D \frac{de(t)}{dt} \right) \tag{11}$$

Sinter bed temperature tracking in the sintering process by PID control is illustrated in Fig. 4. The tracking error at the end of each stage is within $\pm 3\ ^{\circ}C$.

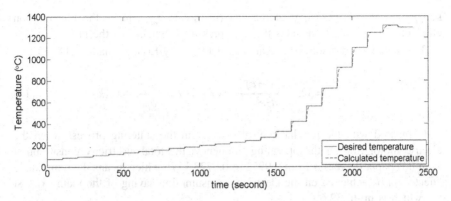

Fig. 4 Sinter bed temperature tracking in the sintering process by PID control

5 Analysis of Airflow Quantity and Power Consumption

The quantity of the airflow can be calculated by Eq. (12).

$$Q(k) = v_g(k)AT, k = 1, 2, \ldots, 2400 \tag{12}$$

where $Q(k)$ denotes the quantity of the airflow at time k; $v_g(k)$ denotes the airflow rate at time k; A denotes the cross-sectional area of the sintering pot; T denotes the sampling period and $T = 1$ s. Therefore, the total quantity of the airflow of the sintering process can be calculated by Eq. (13).

$$Q = \sum_{k=0}^{2400} Q(k) \tag{13}$$

The total electricity consumption of the electric motor can be calculated by Eq. (14).

$$W = \sum_{k=0}^{2400} p(k)T \tag{14}$$

where W denotes the total electricity consumption of the electric motor; $p(k)$ denotes the power of the electric motor at time k and $p(k)$ can be calculated by Eq. (15).

$$p(k) = \frac{c\Delta P(k)Q(k)}{\eta_d \eta_m \eta_f} \tag{15}$$

where c denotes the coefficient of reserve power; $\Delta P(k)$ denotes the negative pressure in the wind box at time k; $Q(k)$ denotes the quantity of the airflow at time k; η_d denotes the efficiency of the electric motor; η_m denotes the transmission efficiency of the fan; η_f denotes the total pressure efficiency of the fan.

The negative pressure $\Delta P(k)$ can be calculated by Ergun equation [14, 15].

$$\frac{\Delta P(k)}{H} = 323 \frac{\mu_g (1-\varepsilon)^2}{d_p^2 \varepsilon^3} v_g(k) + 3 \cdot 78 \frac{\rho_g (1-\varepsilon)}{d_p \varepsilon^3} v_g^2(k) \qquad (16)$$

Compared with the regular control strategy in the sintering process when the electric motor runs in the operating frequency, the total electricity consumption W in this paper decreases from 0.9972 to 0.6723 kWh which means the remarkable effects can be achieved on the electricity consumption saving of the main exhaust fan which is up to 32.6%.

6 Conclusion

Sintering process provides sintered ores for the blast furnace. The flame front descends down the sinter bed with the drawn of the airflow by the main exhaust fan. Therefore, airflow rate is a critical indicator in the sintering process on the sinter bed temperature. In this paper, the heat transfer in the sintering process has been studied. The effect of the airflow rate on the sinter bed temperature is explained in the thermal state model. 24 stages are combined together for modeling the thermal state of the sintering process in which the estimated sinter bed temperature can be calculated to track the actual sinter bed temperature accurately. The experimental results show that the tracking error is within ±3 °C at the end of each stage. In this way, the quantity of airflow is sufficient for the coke combustion and the electricity consumed by the main exhaust fan can be reduced by 32.6%.

Acknowledgements This work is supported by National Natural Science Foundation of China under Grant (61273138, 61573197), National Key Technology R&D Program under Grant (2015BAK06B04), and the key Technologies R&D Program of Tianjin under Grant (14JCZDJC39300).

References

1. Zhao JP. Numerical modelling of the iron ore sintering process and its experimental validation. PhD thesis. Zhejiang University; 2012.
2. Hongyi Jia. Main ventilation frequency transformation practice of sintering machine in steel plant. China Metall. 2017;27(2):57–61.
3. Toda H, Kato K. Theoretical investigation of sintering process. ISIJ Int. 1984;24:178–86.
4. Cumming MJ, Rankin WJ. Modelling and simulation of iron ore sintering; 1985.
5. Yuan X, Zhou Q. Research on the basic theory of sintering process of iron ore. Metallurgical Industry Press; 1997. p. 203–21.

6. Long HM. Research and application on sintering thermal state model of iron ore. PhD thesis. Central South University; 2007.
7. Eng LC, Carlson PG. Effect of iron ores and sintering conditions on flame front properties. ISIJ Int. 2012;52(6):967–76.
8. Yang W, Ryu C, Choi S, et al. Mathematical model of thermal processes in an iron ore sintering bed. Met Mater Int. 2004;10(5):493–500.
9. Fan X, Wang H. Mathematical model and artificial intelligence of sintering process. Central South University Press; 2002.
10. Fan XH, Huang XX, Chen XL, et al. Research and development of the intelligent control of iron ore sintering process based on fan frequency conversion. Ironmaking Steelmaking. 2016;7:1–6.
11. Qing HE. Case study on variable frequency speed control and energy saving of the main exhauster in sintering plant. Sintering Pelletizing. 2012.
12. Li S, Pu X, Yang W, et al. Energy-saving principles and effects of frequency conversion control on main sintering exhauster. Metall Equip. 2013.
13. Ang KH, Chong G, Li Y. PID control system analysis, design, and technology. IEEE Trans Control Syst Technol. 2005;13(4):559–76.
14. Ergun S. Fluid flow through packed columns. J Mater Sci Chem Eng. 1952;48(2):89–94.
15. Wang G, Wen Z, Lou G, et al. Mathematical modeling and combustion characteristic evaluation of a flue gas recirculation iron ore sintering process. Int J Heat Mass Transf. 2016;97:964–74.

A Robust Fast Type-2 Fuzzy Induction Control System

Jinghui Pan, Chengwei Wang, Yuanfeng Lan and Weicun Zhang

Abstract Vector control system for induction motor decouples current by coordinate transformation, in order to gain a wide range of speed regulation. This flux orientation control system is sensitive to the parameters of rotor in induction motor, results in poor robustness. While the Type-2 fuzzy logic, developed from type-1 fuzzy logic, is preferable for resolving the problem about uncertainty of parameters. The adoption of type-2 fuzzy logic in the field of Vector control system for induction motor is a promising approach to resolve the problem about the poor robust result from the change of rotor parameters in induction motor control system. However, due to its high complexity and the need for large amount of calculation, type-2 fuzzy logic is not applied well in real-time system like induction motor control system. In this article, a type-2 fuzzy induction motor is designed through a novel IEKM type reduction algorithm. Besides, a simplified vector control system for fast type-2 induction motor is designed based on the Column pivot element SVD-QR algorithm. The effectiveness of this system is confirmed by simulation and the test result.

Keywords Type-2 fuzzy · Induction motor · Vector control · SVD · QR with column pivoting · EKM

J. Pan (✉) · C. Wang · Y. Lan
Key Laboratory of Advanced Control of Iron and Steel Process
(Ministry of Education) Beijing, Beijing, China
e-mail: pjhpjh000@163.com

C. Wang
e-mail: 13622125719@163.com

Y. Lan
e-mail: lanyuanfeng2010@163.com

W. Zhang
School of Automation and Electrical Engineering, University of Science
and Technology Beijing, Beijing 100083, China
e-mail: weicunzhang@263.net

© Springer Nature Singapore Pte Ltd. 2018
Y. Jia et al. (eds.), *Proceedings of 2017 Chinese Intelligent Systems Conference*,
Lecture Notes in Electrical Engineering 460,
https://doi.org/10.1007/978-981-10-6499-9_28

1 Introduction

The vector control system of asynchronous motor is easily affected by the motor parameters especially the motor resistance, for the reason of the adoption of the flux linkage tracing algorithm, results in the low robustness of the system. In this article, the control algorithm of the outer loop speed regulator is improved to promote the robustness and anti-interference performance of the induction motor vector control system.

The common controller of speed controller for induction motor vector control system is PID controller, fuzzy controller and etc. Since the PID controller needs accurate mathematical model of the system, therefore, the robustness of the vector control system of induction motor using PID controller is weak. Compared with the type-2 fuzzy controller in this article, the fuzzy controller is also called as type-1 fuzzy controller [1].

Type-1 fuzzy controller has robustness to solve the change of motor parameters to some certain degree, however, due to the low robustness of vector control, type-1 fuzzy is often difficult to meet the performance requirements of high performance induction motor control system [2]. Since type-1 fuzzy controller cannot resolve the problem of uncertainty [3–5]: the first is about measuring the uncertainty of the noise, that is, the sensor measurement is affected by a variety of high noise and measurement conditions, results in the noise in the parameters of the system training; the second is about the uncertainty of actuator characteristics, that is, the change of control input caused by the change of actuator features caused by the use of time and environment; the third is about the uncertainty of language, that is, the difference between the meaning of language to different people and the difference between the results obtained by experts and questionnaire in the uniform rules; the fourth is about the uncertainty of the operating environment of the control system, which means the change of input and output caused by the change of system or its perating environment.

Among them, the asynchronous motor vector control system often has the uncertainty of actuator features.

Based on the problems existing in the type-1 fuzzy, Zadeh proposed type-2 fuzzy set in 1975. The type-2 fuzzy set is a preferable approach to the above uncertainties. The type-2 fuzzy set is characterized by the fuzzy degree of membership of type-1 fuzzy set, the membership degree of fuzzy set is a fuzzy set of type-1. Type-2 fuzzy system can model the uncertainty of language and data at the same time, so as to deal with the uncertainty of fuzzy rules directly [3]. The test shows that the type-2 fuzzy controller has better performance than that of type-1 fuzzy controller in the high uncertainty case where the control objects change significantly. Moreover, type-2 fuzzy has distinct advantages on reducing the number of rules, smoothing the output of control and optimizing the response performance of control [5]. Those features make the type-2 fuzzy theory the most preferable approach for vector control algorithm of asynchronous motor.

2 Vector Model of Asynchronous Motor

Vector control, based on coordinate transformation, converters the three-phase asynchronous motor model into two-phase DC motor model through 3/2 transform. This process simplifies the three-phase asynchronous motor model and makes motion control easier. Through the 3/2 transformation, the three-phase physical model is transformed into two physical models.

Among them, the number of turns for each phase of the three-phase winding is N3, the number of turns for each phase of the two-phase winding is N2.

The current conversion formula is as follows:

$$\begin{bmatrix} i_\alpha \\ i_\beta \end{bmatrix} = \frac{N_3}{N_2} \begin{bmatrix} 1 & -\frac{1}{2} & -\frac{1}{2} \\ 0 & -\frac{\sqrt{3}}{2} & \frac{\sqrt{3}}{2} \end{bmatrix} \begin{bmatrix} i_A \\ i_B \\ i_C \end{bmatrix} \tag{1}$$

After rotating transformation, the asynchronous motor is operated according to the rotor flux oriented vector mode. α-β coordinate system is transformed into m-t coordinate system according to Eq. (2).

ψ_r is the rotation vector of rotor flux, and the space angle of the rotor is φ, rotating angular velocity $\omega_1 = \frac{d\varphi}{dt}$.

When $\frac{d\psi_{rt}}{dt} = \frac{d\psi_{rq}}{dt} = 0$, the state equation of asynchronous motor rotor flux linkage in m-t coordinate system is:

$$\left. \begin{array}{l} \frac{di_{sm}}{dt} = \frac{L_m}{\sigma L_s L_r T_r} \psi_r - \frac{R_s L_r^2 + R_r L_m^2}{\sigma L_s L_r^2} i_{sm} + \omega_1 i_{st} + \frac{u_{sm}}{\sigma L_s} \\ \frac{di_{st}}{dt} = -\frac{L_m}{\sigma L_s L_r} \omega \psi_r - \frac{R_s L_r^2 + R_r L_m^2}{\sigma L_s L_r^2} i_{st} + \omega_1 i_{sm} + \frac{u_{st}}{\sigma L_s} \\ \frac{d\psi_r}{dt} = -\frac{1}{T_r} \psi_r + \frac{L_m}{T_r} i_{sm} \\ \frac{d\omega}{dt} = \frac{n_p^2 L_m}{J L_r} i_{st} \psi_r - \frac{n_p}{J} T_L \end{array} \right\} \tag{2}$$

Among them, J is rotational inertia of the unit, T_L is load torque, ω is angular velocity, L_m is the mutual inductance of stator and rotor windings of the coaxial equivalent, L_s and L_r are respectively the self-inductance of stator and rotor, T_r is the electromagnetic time constant of the rotor, $T_r = L_r/R_r$, R_r and R_s are respectively rotor and stator winding resistance, σ is the magnetic flux leakage coefficient of motor, u_{sm} and u_{st} are respectively the excitation component and torque component of stator voltage, i_{sm} and i_{st} are respectively the excitation component and torque component of stator current.

After the transformation, the asynchronous motor can be controlled by DC motor control strategy. The control system structure is shown in Fig. 1.

Here, AΨR is the flux regulator, ATR is the torque regulator. VR—is the reverse synchronous coordinate transformation. The inverter uses IGBT device. SPWM control method is used to construct SPWM type asynchronous motor vector control

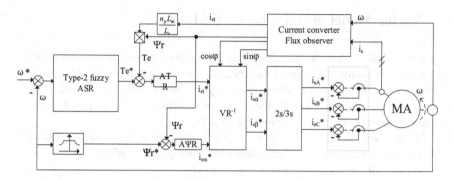

Fig. 1 Type-2 fuzzy vector control system for induction motor

system. After the current hysteresis PWM control, the motor three-phase input current can quickly track the reference current.

In the figure, the speed regulator ASR is a fast type-2 fuzzy speed regulator, which is composed of the fast type-2 fuzzy algorithm.

3 Type-2 Fuzzy Theory

Interval type-2 fuzzy systems are represented by interval valued fuzzy sets. When the membership values of the type-2 fuzzy sets are equal, the fuzzy set of the type-2 is called interval fuzzy set of type-2 which belongs to a subset of type-2 fuzzy sets. This kind of interval fuzzy sets of type-2 preserves the ability of the type-2 fuzzy for dealing with the uncertainty of the complex system, and simplifies the type-2 fuzzy system at the same time, which makes the type-2 fuzzy system easier to understand and calculate. The type-2 fuzzy system can describe the change of the rotor parameters of the asynchronous motor more accurately than the type-1 fuzzy system, so as to improve the robustness of the vector control system of induction motor. Since the interval type-2 fuzzy is more ambiguous than the traditional type-1 fuzzy, as well as less time-consuming while maintaining the features of type-2 fuzzy, it has become a practical and commonly used type-2 fuzzy system. Interval valued fuzzy sets are defined as follows.

Definition 1 Assuming that there are domains E, the membership function values of interval valued fuzzy sets as $\tilde{\mu}_{\tilde{A}}(e) = [\mu_{\tilde{A}}^{L}(e), \mu_{\tilde{A}}^{H}(e)]$, where, in $\mu_{\tilde{A}}^{L}(e), \mu_{\tilde{A}}^{H}(e)$, E [0, 1], and $\mu_{\tilde{A}}^{L}(e) < \mu_{\tilde{A}}^{H}(e), \forall e \in E$.

Definition 2 Interval fuzzy set $\tilde{A}$ is defined as $\tilde{A} = \{e, [\mu_{\tilde{A}}^{L}(e), \mu_{\tilde{A}}^{H}(e)] | e \in E\}$. From the above, the interval fuzzy set can be defined by $\mu_{\tilde{A}}^{L}(e), \mu_{\tilde{A}}^{H}(e), \mu_{\tilde{A}}^{L}(e)$ is a lower bound membership function, denoted $LMF(A)$, $\mu_{\tilde{A}}^{H}(e)$ is the upper membership

Fig. 2 Membership function
in type-2 fuzzy system

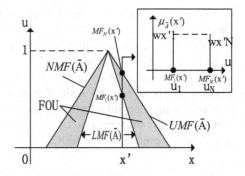

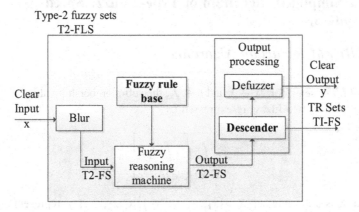

Fig. 3 Type-2 fuzzy control system

function, the upper left side is represented as $NMF(A)$, and the upper right side is represented as $UMF(A)$.

In this paper, the interval type-2 fuzzy sets are represented by triangular fuzzy membership functions, as shown in Fig. 2.

Type-2 fuzzy controller structure diagram as shown in Fig. 3. The biggest difference between type-2 fuzzy system and type-1 fuzzy system is the addition of a reducing device. The function of the reducing device is to transform the type-2 fuzzy set of fuzzy inference machine into type-1 fuzzy set. Because of its large amount of computation, the performance of the algorithm will directly affect the performance of the entire type-2 fuzzy controller. At present, how to solve the complexity of the type-2 fuzzy sets has become a hot topic in the study of type-2 fuzzy sets.

In this paper, the reducing device and fuzzy rule base are simplified.

There exist three basic methods to simplify type-2 fuzzy in order to solve its complexity, namely, by reducing the number of TR reduction iteration, avoiding iteration, and the adoption of streamline fuzzy rules. In this article, NIEKM algorithm is used to reduce the number of iterations of TR reduction [6], besides, the column pivoting SVD-QR algorithm is used to simplify the fuzzy rules [7].

NIEKM reduction algorithm is a new improved EKM reduction algorithm (Novel Improved EKM, NIEKM). The algorithm simplifies the structure of the reducer and reduces the computation time, so as to simplify the structure of the type-2 fuzzy speed regulator ASR. The column pivoting SVD-QR algorithm is used to select the most important rules in the fuzzy rules, in order to reduce the number of rules, simplify the type-2 fuzzy structure and reduce the computation time of the type-2 fuzzy speed regulator ASR.

4 The Simplified Algorithm of Type-2 Fuzzy Speed Regulator

4.1 NIEKM Reduction Algorithm

In type-2 fuzzy set, if for $\forall x \in X$ and $u \in J_x$, the sub-membership value is 1, then $\tilde{A}$ is the interval of two fuzzy, re-expressed as

$$\tilde{A} = \int_{x \in X} \left[\int_{u \in Jx} 1/u \right] /x \tag{3}$$

When X is a discrete set, $X = \{x_1, x_2, \ldots, x_N\}$, $x_1, x_2, \ldots, x_N$ represents $[\underline{x}_1, \overline{x}_1]$, $[\underline{x}_2, \overline{x}_2], \ldots, [\underline{x}_N, \overline{x}_N]$, $u = \{u_{\tilde{A}}(x_1), u_{\tilde{A}}(x_2), \ldots, u_{\tilde{A}}(x_N)\}$ represents $[\underline{f}(x_1), \overline{f}(x_1)]$, $[\underline{f}(x_2), \overline{f}(x_2)], \ldots, [\underline{f}(x_N), \overline{f}(x_N)]$.

The descent centroid of the interval type-2 fuzzy set $\tilde{A}$ is

$$y = \frac{yl + yr}{2} \tag{4}$$

where

$$yl = \frac{\sum_{i=1}^{L} \underline{x}_i \overline{f}(x_i) + \sum_{i=L+1}^{N} \underline{x}_i \underline{f}(x_i)}{\sum_{i=1}^{L} \overline{f}(x_i) + \sum_{i=L+1}^{N} \underline{f}(x_i)}$$

$$yr = \frac{\sum_{i=1}^{R} \overline{x}_i \underline{f}(x_i) + \sum_{i=R+1}^{N} \overline{x}_i \overline{f}(x_i)}{\sum_{i=1}^{R} \underline{f}(x_i) + \sum_{i=R+1}^{N} \overline{f}(x_i)}$$

N is the rule number, L and R are called the left and right switching points of yl and yr, respectively. The core of the descent algorithm is to find L and R.

yl and yr have the same calculation process, here on the yl calculation as an example to introduce the NIEKM reduction method.

(1) Put $\underline{x}_i$ (i = 1, 2, ..., N) in ascending order. And adjust the order of $f(x_i)$, corresponding to $\underline{x}_i$.

(2) Let $L = Round(N/2.4)$, calculate

$$a = \sum_{i=1}^{L} \underline{x}_i \overline{f}(x_i) + \sum_{i=L+1}^{N} \underline{x}_i \underline{f}(x_i)$$

$$b = \sum_{i=1}^{L} \overline{f}(x_i) + \sum_{i=L+1}^{N} \underline{f}(x_i)$$

$$y = a/b$$

(3) Category search $L' \in [1, N-1]$ satisfies

$$\underline{x}_{L'} < y \le \underline{x}_{L'+1}$$

(4) If $L' = L$, stop the calculation, otherwise continue

(5) Calculate s = sign $(L' - L)$

$$a' = a + s \sum_{i=\min(L,L')+1}^{\max(L,L')} \underline{x}_i(\overline{f}(x_i) - \underline{f}(x_i))$$

$$b' = b + s \sum_{i=\min(L,L')+1}^{\max(L,L')} (\overline{f}(x_i) - \underline{f}(x_i))$$

$$y = a'/b'$$

(6) Let $y = y', a = a', b = b', L = L'$, return to step (3)

In the calculation of yr process, only need to step (2) the function $L = Round(N/2.4)$ modified to $R = Round(N/1.7)$, and make the appropriate changes to the subsequent x_i and $f(x_i)$.

4.2 Column Pivoting SVD-QR Simplification Algorithm

The SVD decomposition of a matrix is to decompose a $N \times M$ matrix into three matrices U, S, and V, which satisfy the following relations:

$$\phi_r = USV^T \tag{5}$$

Among them, U is a $N \times N$ orthogonal matrix, S is a diagonal matrix, and V is a $M \times M$ orthogonal matrix.

When the rank $(\phi_r) < M$, there are redundant rules in matrix ϕ_r. As a result, we can use the method of ignoring the small singular value to extract the fuzzy rules with large singular values, and form a small number of ϕ_r, so that ϕ_r can satisfy the rank $(\phi_r) = M$, the simplest fuzzy rule matrix is obtained. The number of rules can be obtained by singular value.

This paper simplifies the algorithm by simplifying the rules in column pivoting QR method.

Column pivoting QR decomposition definition: the matrix A can be decomposed into a permutation matrix E, an orthogonal matrix Q and an upper triangular matrix R, which satisfy the following relations:

$$AE = QR \tag{6}$$

In the case of rank deficiency, the SVD decomposition of $P \in R^{N \times M}$ is obtained by (3), define the matrix $P^{(r)} \in R^{N \times r}$ satisfies

$$PE = \begin{bmatrix} P^r & P^{(M-r)} \end{bmatrix} \tag{7}$$

where, $E \in R^{M \times M}$ is a permutation matrix. It was proofed by Golub that if

$$E^T V = \begin{bmatrix} V'_{11} & V'_{21} \\ V'_{12} & V'_{22} \end{bmatrix} \tag{8}$$

And V'_{11} is nonsingular, then

$$\frac{\sigma_r(P)}{\left\| (V'_{11})^{-1} \right\|_2} \leq \sigma_r(P^{(r)}) \leq \sigma_r(P) \tag{9}$$

where, $\sigma_r(P^{(r)})$ and $\sigma_r(P)$ are the first r singular values of matrix $P^{(r)}$ and P, respectively. If the selected permutation matrix E is as likely as possible to get a well-conditioned V'_{11}, it should make $\left\| (V'_{11})^{-1} \right\|_2$ as small as possible so as to maximize the singular value of $P^{(r)}$.

By using column pivoting QR, we can get a well-conditioned V'_{11}, and get a sufficient linear independent ϕ_r subset.

Specific steps are as follows:

(1) Calculate the activation matrix ϕ_r, ϕ_l.
(2) After obtaining the matrix ϕ_r, perform SVD decomposition on it to obtain the matrix S and V.
(3) According to the singular value of the diagonal matrix S, the normalized difference of each singular value is calculated by the following below:

$$a_i = \frac{\sigma_i - \sigma_{i+1}}{\max(\sigma) - \min(\sigma)}, i = 1, 2, \ldots, r - 1 \tag{10}$$

where, r = rank(P).

Determine the threshold value, according to the number of diagonal matrix elements greater than the threshold, the number of the reduced model rules r1 is determined.

(4) Divide the orthogonal matrix V down as the following below:

$$V = \begin{bmatrix} V_{11} & V_{12} \\ V_{21} & V_{22} \end{bmatrix} \tag{11}$$

where, V_{11} is r row r column matrix.

Performed QR with column pivoting on matrix $\begin{bmatrix} V_{11}^T & V_{21}^T \end{bmatrix}$, obtain the permutation matrix E. In the first R1 column, the row of 1 elements corresponding to the position of the ϕ_r column can be obtained by simplifying the rule set T1.

(5) In the same way, the simplified rule matrix T2 can be obtained from ϕ_l.

(6) Rule $T = T1 \cup T2$ is the final streamlining rule matrix.

In this paper, we use the NIEKM reduction algorithm to reconstruct the type-2 fuzzy controller, and simplify the fuzzy rules by using SVD-QR with column pivoting.

In the asynchronous motor control system in this paper, the input of type-2 fuzzy controller is speed error E and speed error rate of change CE, the output is torque Te*. The input E and CE are divided into 7 fuzzy sets, and 49 fuzzy rules are obtained.

By using the algorithm of SVD-QR with column pivoting, the 49 fuzzy rules are reduced to 15 fuzzy rules. The serial numbers are 10, 15, 16, 17, 19, 22, 23, 24, 25, 27, 28, 38, 39, 43, 44.

Using the above two algorithms, this paper completes the design work of the vector control system of type-2 fuzzy induction motor.

The NIEKM descent algorithm used in this paper, as well as the simplification of the fuzzy rules, are designed to reduce the computational time of type-2 fuzzy system.

The commonly used descending algorithms are EKM, EIASC and NIEKM, of which the calculation time before and after the simplification of rules are shown in Table 1.

Table 1 Execute time of several type reduction algorithms with all rules and simple rules

Reduction algorithms	EKM	EIASC	NIEKM
With all rules (ms)	9.26	11.48	7.62
With simple rules (ms)	2.73	2.85	2.76

It can be seen from Table 1: first, the algorithm used in this article costs the minimum time before the fuzzy rule is simplified; second, after simplification of the fuzzy rules, the calculation time of each descent algorithm is greatly reduced, and the calculation time of NIEKM descent algorithm is reduced by about 64%.

The above results confirm the effectiveness of the simplified method used in this article to reduce computational time.

5   Simulation and Experiment

The main parameters of asynchronous motor used in simulation are shown in Table 2. The given speed is 120 rad/s, the maximum torque limit is 700 Nm.

Figure 4 shows the speed response error rate curve of the PID speed regulator and the fast type-2 fuzzy speed regulator, as well as the response results of the two algorithms of the steady state speed controller of induction motor vector control system using PID and fast type-2 fuzzy. It can be seen that, in steady state, the error rate of PID control system is greater than or equal to 0.45%, type-2 fuzzy control system error rate is less than or equal to 0.1%, the error rate of type-2 fuzzy control is less than PID. That is because type-2 fuzzy system parameters are obtained by using the data of the induction motor control system of the 20,001 groups, therefore

Table 2 Parameters of induction motor

Parameter	Numerical value	Parameter	Numerical value
Stator resistance Rs (Ω)	0.087	Rated power (kW)	37.3
Rotor resistance Rr (Ω)	0.456	Rated voltage (V)	380
Stator inductance (mH)	0.8	Rated frequency (Hz)	50
Rotor inductance (mH)	0.68	Rated current (A)	70

Fig. 4 Error rates of fast type-2 fuzzy controller and PID controller

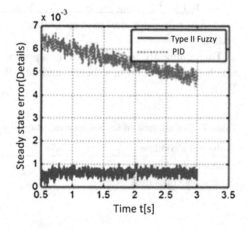

type-2 fuzzy has a strong ability to self-learn and approximate the state function of asynchronous motor, so as to improve the control precision. Figure 5 shows the robustness difference between fast type-2 fuzzy control and ordinary fuzzy control. According to the case where the rotor resistance of the asynchronous motor is about 1.5 times the maximum variable in the actual production run. The stator resistance and rotor resistance of the asynchronous motor are changed and simulated to verify the robust performance of the asynchronous motor type-2 fuzzy vector control system for the change of the parameters of the asynchronous motor.

When the motor parameters in Fig. 5a graph remain unchanged, we can compare the difference between type-1 fuzzy and type-2 fuzzy performance. It can be seen from Fig. 5a that the regulation time of type-2 fuzzy is about 0.07 s, the regulation time of type-1 fuzzy is about 0.2 s, and the regulation time of type-2 is less than that of type-1. From Fig. 5a to Fig. 5b, and to Fig. 5c, the rotor resistance and stator resistance of induction motor were increased to 1.25 times and 1.5 times, respectively. It shows that in the process of increasing resistance, the type-2 fuzzy system can always remain stable and keep steady state error less than 0.01; however, with the change of the resistance of the motor, the steady-state error of type-1 fuzzy system becomes too large to keep system following the given speed, the steady state error is increased to 0.23 when the resistance is changed to 1.5 times, meanwhile, the steady state error of type-2 is only about 0.01. It can be seen that type-2 fuzzy has a stronger robustness to the motor parameter variation than type-1 fuzzy.

The most commonly used controller is the PID controller. However, PID control requires an accurate system model, resulting in its poor robustness. As can be seen from Fig. 6, in the same asynchronous motor vector control system, when the asynchronous motor parameters increase to 1.5 times, the PID control speed output cannot follow the given value; however, the type-2 fuzzy still remains stable, with the steady-state error about 0.01. It shows that when the asynchronous motor resistance change to 1.5 times, the PID control system is unable to have enough robustness to ensure the normal operation of the system. However, the type-2 fuzzy in can still ensure the normal operation of the system in the same case.

The results above show that the type-2 fuzzy algorithm is more robust than the conventional PID and type-1 fuzzy control algorithm. This makes up for the weak robustness of vector control of induction motor.

In addition to strong robustness, the type-2 fuzzy system also has strong adaptability to the external disturbance, which makes the type-2 fuzzy induction motor control system having anti-interference ability.

Figure 7 shows the anti-jamming performance of the fast type-2 fuzzy algorithm. In the simulation in this article, the sudden load disturbance is applied to the system in 1 s steady state. When 200 Nm load torque is added to the system in Fig. 7a suddenly, the speed change rate of the type-2 fuzzy control system is less than that of the fuzzy control system; When the 600 Nm load torque is added to the system in Fig. 7b suddenly, type-2 fuzzy control system can still run stably with a high steady-state precision, while the fuzzy control system lost its stability. It is shown that the nonlinear adaptive ability of type-2 fuzzy systems is stronger than that of type-1 fuzzy systems.

Fig. 5 Robust of type-1
fuzzy and fast type-2 fuzzy
control

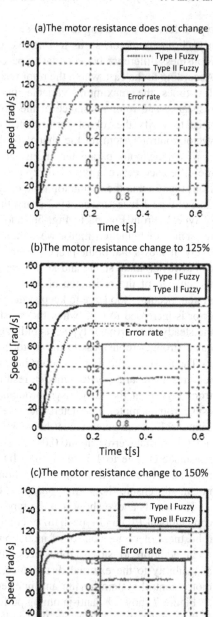

(a)The motor resistance does not change

(b)The motor resistance change to 125%

(c)The motor resistance change to 150%

Fig. 6 Robust between
fuzzy-2 and PID control with
150% changed Rr and Rs

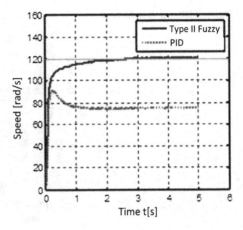

Fig. 7 Speed compared with
disturbed load between type-1
fuzzy controller and type-2
fuzzy controller

(a) Suddenly increase load torque 200Nm

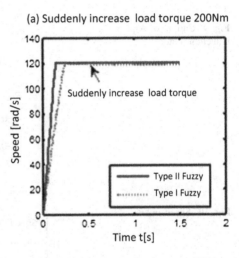

(b) Suddenly increase load torque 600Nm

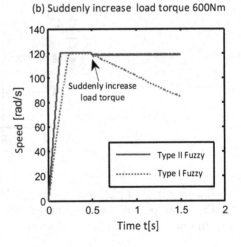

Fig. 8 Platform of induction motor

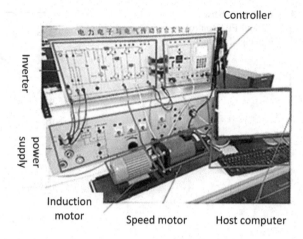

Fig. 9 Current of A-phase in induction motor

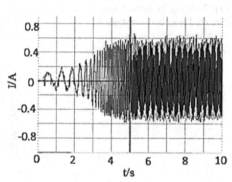

Fig. 10 Speed in test

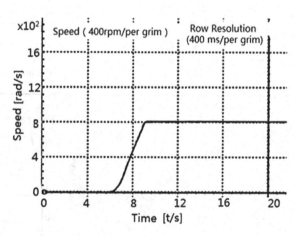

The experimental platform is shown in Fig. 8.

The given speed of the experiment is 800 r/min. The current waveforms and the experimental speed waveforms are shown in Fig. 9 and Fig. 10 respectively. With a small current ripple, fast rising of rotating speed, and the stable speed, it can meet the performance indicators in general, which proves that the type-2 fuzzy vector control system can run stably with steady state performance. In a given load disturbance experiment, the system remains stable before and after the given load disturbance, which confirms the feasibility and effectiveness of the adoption of type-2 fuzzy control as a real-time control algorithm.

6 Conclusions

In this article, a high robustness vector control system of type-2 fuzzy induction motor with is designed based on the weak robustness of vector control system of induction motor, and type-2 fuzzy theory is adopted to realize the speed regulator. The structure of the type-2 fuzzy controller is simplified together with its algorithm, which makes the calculation time of type-2 fuzzy system satisfies the need of the induction motor control system. Simulation and experimental results show that the vector control system of the type-2 fuzzy induction motor has strong robustness to the variation of motor parameters, as well as the steady state performance and superior anti-jamming performance, which confirm the validity of the type-2 fuzzy algorithm.

Besides, this article also shows the promising application of type-2 fuzzy in the field of the vector control system of induction motor. The further improvement of the performance of asynchronous motor speed control system will be achieved by a more mature type-2 fuzzy theory.

References

1. Hojabri H, Mokhtari H, Chang L. A generalized technique of modeling, analysis, and control of a matrix converter using SVD. IEEE Trans Ind Electron. 2011.
2. Pan Y, Chen J, Xu J, Guo K. State feedback linearization control for current of input side in the matrix convertor. High Volt Eng. 2014;08:2497–503.
3. Zhu Y. Household automatic cooking machine. Househ Appl Mag. 2002;(3):20–1.
4. Pan Y, Huang D, Sun Z. Overview of type-2 fuzzy logic control. Control Theory Appl. 2011;28(1):13–23 (in Chinese).
5. Coupland S, John RI. Type-2 fuzzy logic and the modelling of uncertainty. In: Fuzzy sets and extensions: representation aggregation and models. Berlin: Springer; 2008. p. 3–22.
6. Chafaa K, Laamari Y, Barkati S, Chaouch S. Adaptive type-2 fuzzy control for induction motor. In: Proceedings of the IEEE 5th international multi-conference on systems, signals and devices. Amman, India: IEEE; 2008.
7. Chen W, Yu Y, Xu D, Xu Z. Parameters estimation of induction motors at standstill with adaptive nonlinearity compensation. Proc CSEE. 2012;32(6):156–62 (in Chinese).

Research on Rotor Flux Observer for Extended Complex Kalman Filter of Asynchronous Motor

Yuedou Pan, Chengwei Wang, Jinghui Pan and Zhengming Zhao

Abstract In order to improve the observation accuracy of the rotor flux of the asynchronous motor, a flux observation method based on the extended complex Kalman filter is proposed and applied to the vector control of the asynchronous motor. The complex mathematical model of asynchronous motor is established by choosing the state variables appropriately. On this basis, a rotor flux observer based on the extended complex Kalman filter algorithm is designed to realize the accurate estimation of the rotor flux. The results show that the method can simplify the motor state equation effectively and reduce the order of the mathematical model of the asynchronous motor, it can also reduce the amount of calculation. Simulation results verify the effectiveness and feasibility of the proposed method.

Keywords Asynchronous motor · Extended complex kalman filter · Vector control · Rotor flux observation

1 Introduction

Asynchronous motor is widely used in industrial engineering, defense technology, transportation, metallurgy, food and other industries because of its simple structure, low cost and its excellent performance. Real-time tracking of motor status is critical to improving the motor control performance. The current control method of the

Y. Pan (✉)
School of Automation and Electrical Engineering, University of Science and Technology Beijing, 100083 Beijing, China
e-mail: ydpan@ustb.edu.cn

C. Wang · J. Pan
Key Laboratory of Advanced Control of Iron and Steel Process, (Ministry of Education) Beijing, 100083 Beijing, China
e-mail: 13622125719@163.com

Z. Zhao
Department of Electrical Engineering, Tsinghua University, 100084 Beijing, China

© Springer Nature Singapore Pte Ltd. 2018
Y. Jia et al. (eds.), *Proceedings of 2017 Chinese Intelligent Systems Conference*,
Lecture Notes in Electrical Engineering 460,
https://doi.org/10.1007/978-981-10-6499-9_29

Asynchronous motor is mainly vector control and direct torque control. The basic idea of the vector control is to decouple the stator current into the current excitation component and the torque component, so that the AC motor is equivalent to the DC motor to control and improve the dynamic performance of the AC motor. The vector control needs to obtain the rotor flux of the motor. However, the rotor flux can not be directly detected and can only be estimated by measuring the stator current and voltage. Therefore, the accuracy of the flux estimation is particularly critical to the vector control system.

The basic method of rotor flux observation is voltage model method and current model method. As the voltage model method contains pure integral term, DC offset error and initial error will lead to integral saturation. The current model method can solve the integral saturation problem of the voltage model method, but its observation accuracy is greatly affected by the motor parameters. In order to observe the flux better, domestic and foreign researchers have done a lot of research, put forward a number of Asynchronous motor state estimation method, there are the following categories: adaptive sliding mode method [1], model reference adaptive side method [2], speed adaptive method [3], neural network state estimation method [4]. The adaptive sliding mode method has low accuracy to the mathematical model of the system and has good robustness to the external disturbance, but the buffering phenomenon exists in the sliding mode observer can not be removed. Model reference adaptive method uses the rotor flux voltage equation as the reference model, the current model as an adjustable model, using the proportional integral adaptive law to estimate, but the introduction of the voltage model will lead to integral saturation problem, low error. In paper [4], a RBF neural network flux observer is proposed, which has strong self-learning ability and generalization ability, but its theoretical research is not mature and is only used for simulation verification.

The extended Kalman filter algorithm is the best predictive estimation method in the sense of minimum variance. In recent years, the Kalman filter has been widely concerned and applied to the Asynchronous motor vector control system, and achieved good results. Because it is a stochastic process estimation method, it has strong anti-interference ability to deal with the process noise and measurement noise in the system, so it has a strong robustness to the change of motor parameters.

As the mathematical model of Asynchronous motor is a high-order, nonlinear, strong coupling system, so it is very difficult to observe it directly. In recent years, many scholars have been trying to use various methods to reduce the order of the model in order to observe the flux. Wu Mingzhu et al. proposed a reduced-order flux linkage observer based on EKF in 2012 [5], although the order of the system state equation is reduced, but the observation error is large. Song Wenxiang et al. proposed a reduced-order flux observer based on the motor back electromotive force deviation in 2013 [6]. It has closed-loop observation characteristics, but the robustness to the system is poor. Wang Bin and others proposed a stator flux reduction order observer in 2014 [7], but the observer structure is complex, it does not apply to the fast changing speed of the occasion. Montanari et al. proposed an adaptive flux observer in 2007 [8]. Kojabadi H.M in 2005 proposed a pseudo-reduced order flux observer [9], rotor flux observation did not introduce feedback correction term, whether the error between the estimated value and the actual value need to be ignored should be further studied. In this paper, the extended complex Kalman filter algorithm (ECKF) is proposed on the basis of the traditional Kalman filter algorithm. By selecting the state variables appropriately, the state equation of the Asynchronous motor is written as a complex form, which reduces the state equation order, it not only inherits the advantages of traditional Kalman filter, but also simplifies the algorithm, accelerates the running time of the system and improves the observation accuracy of the flux linkage.

2 The Complex Mathematical Model of Asynchronous Motor [10]

In order to facilitate the calculation and analysis, suppose the Asynchronous motor magnetic circuit is linear, ignoring the effect of iron loss, Asynchronous motor mathematical model of the state equation is:

$$
\begin{cases}
i_{s\alpha} = -\delta i_{s\alpha} + \lambda\theta\psi_{r\alpha} + n_p\lambda\omega_r\psi_{r\beta} + \eta u_{s\alpha} \\
i_{s\beta} = -\delta i_{s\beta} - n_p\lambda\omega_r\psi_{r\alpha} + \lambda\theta\psi_{r\beta} + \eta u_{s\beta} \\
\dot{\psi}_{r\alpha} = \theta L_m i_{s\alpha} - \theta\psi_{r\alpha} - n_p\omega_r\psi_{r\beta} \\
\dot{\psi}_{r\beta} = \theta L_m i_{s\beta} + n_p\omega_r\psi_{r\alpha} - \theta\psi_{r\beta} \\
\dot{\omega}_r = b_1(\psi_{r\alpha}i_{s\beta} - \psi_{r\beta}i_{s\alpha}) - b_2 T_L
\end{cases}
\tag{1}
$$

where $\theta = \frac{R_r}{L_r}$, $\sigma = 1 - \frac{L_m^2}{L_sL_r}$, $\eta = \frac{1}{\sigma L_s}$, $\lambda = \frac{L_m}{\sigma L_sL_r}$, $\delta = \eta R_s + L_m\lambda\theta$, $b_1 = \frac{n_p^2 L_m}{JL_r}$, $b_2 = \frac{n_p}{J}$, $i_{s\alpha}$ and $i_{s\beta}$ are stator currents, $u_{s\alpha}$ and $u_{s\beta}$ are stator voltages, $\psi_{r\alpha}$ and $\psi_{r\beta}$ are the rotor flux, R_r denotes rotor resistance, and R_s denotes stator resistance, L_s is stator self-inductance, L_r is rotor self-inductance, L_m is mutual inductance, and ω_r is rotor electrical angular velocity.

In order to establish the reduced-order model of Asynchronous motor, this paper defines the following state variable complex form:

$$
\begin{aligned}
x &= (x_1, x_2, x_3) = (i_s, \psi_r, \omega_r)^T \\
&= (i_{s\alpha} + ji_{s\beta}, \psi_{r\alpha} + j\psi_{r\beta}, \omega_r)^T \\
u_s &= u_{s\alpha} + ju_{s\beta}
\end{aligned}
\tag{2}
$$

where j is the imaginary unit.

With the stator voltage and the stator current as input, it is assumed that the sampling frequency of the closed-loop control of the rotational speed is sufficiently large relative to the time constant of the motor, and it can be approximated that the rotational speed is constant at two adjacent sampling points k and k + 1. So $\dot{\omega}_r = 0$. This approximation error estimate does not affect the accuracy of the estimation error, but also simplifies the establishment of the motor complex model.

This method does not affect the accuracy of the estimation is mainly due to the expansion of Kalman filter excellent anti-jamming capability, this error can be attributed to the system of disturbance which is filtered out by the filter to get accurate estimates.

In Eq. (2), since the new state variable is defined, the stator voltage is selected as the input and the stator current is used as the output, the Eq. (1) becomes:

$$\begin{cases} i_{s\alpha} + ji_{s\beta} = -\delta(i_{s\alpha} + ji_{s\beta}) + (\lambda\theta - jn_p\lambda\omega_r)(\psi_{r\alpha} + j\psi_{r\beta}) + \eta(u_{s\alpha} + ju_{s\beta}) \\ \dot{\psi}_{r\alpha} + j\dot{\psi}_{r\beta} = \theta L_m(i_{s\alpha} + ji_{s\beta}) + (-\theta + jn_p\omega_r)(\psi_{r\alpha} + j\psi_{r\beta}) \\ \dot{\omega}_r = 0 \end{cases} \quad (3)$$

Put the defined state variables into the Eq. (3), we can get:

$$\begin{cases} \dot{x}_1 = -\delta x_1 + (\lambda\theta - jn_p\lambda\omega_r)x_2 + \eta u_s \\ \dot{x}_2 = \theta L_m x_1 + (-\theta + jn_p\omega_r)x_2 \\ \dot{x}_3 = 0 \end{cases} \quad (4)$$

The output of the system is:

$$y = h(x) = x_1 \quad (5)$$

Equation. (4) is the complex mathematical model of Asynchronous motor. Compared with the Eq. (4) and (1), it is not difficult to find that the state equation of the Asynchronous motor mathematical model becomes a complex form due to the introduction of the complex variable j, and the order of the state equation is reduced from 5 to 3.

3 Observerbility Analysis of Complex Mathematical Model of Asynchronous Motor

In order to facilitate the analysis of some of the parameter values in the following Kalman filter algorithm, we should discretize the complex mathematical model first. From the Eq. (4), we can conclude the system coefficient matrix is:

$$A = \begin{bmatrix} -\delta & \lambda\theta - jn_p\lambda\omega_r & \eta \\ \theta L_m & -\theta + jn_p\omega_r & 0 \\ 0 & 0 & 0 \end{bmatrix}$$

Let the sampling period be T_s, then according to the system equation discretization formula: $A_s = E + AT_s + \frac{A^2 T_s^2}{2!} + \frac{A^3 T_s^3}{3!} + \dots$ (E is a unit array), ignore the high order term, the above can approximated to $A_s = E + AT_s$, then the above matrix changes into:

$$A_s = \begin{bmatrix} 1-\delta T_s & (\lambda\theta - jn_p\lambda\omega_r)T_s & \eta T_s \\ \theta L_m T_s & 1+(-\theta + jn_p\omega_r)T_s & 0 \\ 0 & 0 & 1 \end{bmatrix}$$

The discrete equation of the Asynchronous motor is:

$$\begin{cases} x_1(k+1) = (1-\delta T_s)x_1(k) + (\lambda\theta - jn_p\lambda\omega_r)T_s x_2(k) + \eta u_s(k) \\ x_2(k+1) = \theta L_m T_s x_1(k) + [1+(-\theta + jn_p\omega_r)T_s]x_2(k) \\ x_3(k+1) = x_3(k) \end{cases} \qquad (6)$$

$$y(k) = h(x(k)) = x_1(k) \qquad (7)$$

The following is a comparative analysis of the discrete complex model, let $x(k+1) = f_s(x(k), u(k))$, then the Eq. (6) change into a matrix form:

$$f_s = \begin{bmatrix} (1-\delta T_s)x_1(k) + (\lambda\theta - jn_p\lambda x_3(k))T_s x_2(k) + \eta T_s u_s(k) \\ \theta L_m T_s x_1(k) + [1+(-\theta + jn_p x_3(k))T_s]x_2(k) \\ x_3(k) \end{bmatrix} \qquad (8)$$

The Jacobian matrix of $f_s(x(k), u(k))$ at $x(k)$ is:

$$\left.\frac{\partial f_s}{\partial x}\right|_{x(k)} = \begin{bmatrix} (1-\delta T_s) & (\lambda\theta - jn_p\lambda x_3(k))T_s & -jn_p\lambda T_s x_2(k) \\ \theta L_m T_s & [1+(-\theta + jn_p x_3(k))T_s] & jn_p T_s x_2(k) \\ 0 & 0 & 1 \end{bmatrix}$$

To facilitate the following of the matrix representation, we set the observability matrix as:

$$M = \begin{bmatrix} M_1 & M_2 & M_3 \\ M_4 & M_5 & M_6 \\ M_7 & M_8 & M_9 \end{bmatrix}$$

where the first row of the matrix:

$$dh_1 = \left.\frac{\partial h(x)}{\partial x}\right|_{x(k)} = [1 \quad 0 \quad 0] = [M_1 \quad M_2 \quad M_3]$$

the second row of the matrix:

$$\begin{aligned} dh_2 &= \left.\frac{\partial h(x)}{\partial x}\right|_{x(k+1)} \left.\frac{\partial f_s}{\partial x}\right|_{x(k)} \\ &= [1-\delta T_s \quad (\lambda\theta - jn_p\lambda x_3(k))T_s \quad -jn_p\lambda T_s x_2(k)] \\ &= [M_4 \quad M_5 \quad M_6] \end{aligned}$$

the third row of the matrix

$$dh_3 = \frac{\partial h(x)}{\partial x}\bigg|_{x(k+2)} \frac{\partial f_s}{\partial x}\bigg|_{x(k+1)} \frac{\partial f}{\partial x}\bigg|_{x(k)}$$

$$= [I_1 \quad I_2 \quad I_3]\frac{\partial f}{\partial x}\bigg|_{x(k)}$$

where $\begin{cases} I_1 = 1 - \delta T_s \\ I_2 = (\lambda\theta - jn_p\lambda x_3(k))T_s \\ I_3 = -jn_p\lambda T_s x_2(k+1) \end{cases}$

Put the above into the matrix, the matrix as follows:

$$dh_3 = [M_4^2 + M_5\theta L_m T_s \quad M_4 M_5 + M_5[1 + (-\theta + jn_p x_3(k))T_s] \quad M_4 M_6 + jn_p T_s(M_5 x_2(k) - \lambda x_2(k+1))]$$
$$= [M_7 \quad M_8 \quad M_9]$$

The determinant of the matrix M is:

$$\det(M) = M_5 M_9 - M_6 M_8$$
$$= (n_p^2\lambda^2 x_3(k)T_s^2 + jn_p\lambda^2\theta T_s^2)(x_2(k) - x_2(k+1))$$

The observability condition is that the determinant of the matrix is not zero, from the above formula, we can see the first item of the equation $n_p^2\lambda^2 x_3(k)T_s^2 + jn_p\lambda^2\theta T_s^2$ is greater than 0, and since the rotor flux in the two adjacent sampling time values are not equal, equivalent to $x_2(k) \neq x_2(k+1)$, then the observability of the Asynchronous motor in the Eq. (7) is established.

4   Design of Extended Complex Kalman Rotor Flux Observer

In this paper, the discrete complex model of Asynchronous motor has been analyzed extensively, we will deduce the expression of the extended complex Kalman filter on the basis of complex model of Asynchronous motor,so as to estimate the rotor flux of the Asynchronous motor.

4.1   ECKF Discrete Model of Asynchronous Motor

According to the above analysis, we give the discrete state equation of the Asynchronous motor for the extended complex Kalman filter (ECKF) flux estimation:

$$\begin{cases} x_{k+1} = f_s(x(k), u(k)) + w_k \\ y_k = h(x(k)) + u_k \end{cases} \tag{9}$$

The expression $f_s(x(k), u(k))$ is shown in Eq. (8), $h(x(k))$ is shown in Eq. (9), w_k and v_k are the system noise sequences and measured noise sequences with unrelated mean values of zero respectively, the covariance matrix is represented by Q_k and R_k.

4.2 ECKF Algorithm

In this paper, the design of extended complex Kalman filter algorithm and the traditional one is different from the introduction of complex variables, so the initial error covariance marix is Hermitian matrix, $P_k = P_k^H$.

The Kalman filter algotithm is an estimate of the minimum variance, it is based on the physical quantity of the system of using the recursive way, it can be estimate that the corresponding state in the noise measurement signal. It includes the forcast phase and the update phase. The prediction phase carries out a priori estimate of the next state based on the input $u(k)$ and the current state $\hat{x}_k$ at time t_k, then we can get the state estimate $\hat{x}_{k+1/k}$, and calculate the covariance matrix $P_{k+1/k}$.

$$\begin{cases} \hat{x}_{k+1/k} = f_s(\hat{x}_k, u_k) \\ P_{k+1/k} = F_k P_k F_k^H + Q_k \end{cases} \tag{10}$$

where $F_k = \frac{\partial f}{\partial x}\big|_{x(k)}$, $\hat{x}_{k+1/k}$ is a priori estimate of the state variable at time $k+1$. $P_{k+1/k}$ is state estimation error covariance matrix. The update phase needs to calculate posteriori estimates based on the output sample $y(k)$, the state priori estimate $\hat{x}_{k+1/k}$, and the Kalman filter gain K_{k+1}.

$$\begin{cases} \hat{x}_{k+1} = \hat{x}_{k+1/k} + K_{k+1} e_k \\ K_{k+1} = P_{k+1/k} H_k^T (H_{k+1} P_{k+1/k} H_{k+1}^T + R_k)^{-1} \\ P_{k+1} = (I - K_{k+1} H_k) P_{k+1/k} \end{cases} \tag{11}$$

where $e_k = y_k - H_k \hat{x}_{k+1/k}$, $H_k = \frac{\partial h(x)}{\partial x}\big|_{x(k)} = \begin{bmatrix} 1 & 0 & 0 \end{bmatrix}$

Equations (10) and (11) are the recursive expressions of extended complex Kalman filter.

The process of the ECKF is initialize it first, the initial value of the state $x(0) = \begin{pmatrix} 0 & 0 & 0 \end{pmatrix}$, sampling time $k = 0$, and noise covariance matrix Q_k and R_k. The setting of these two parameters is critical to the performance of the filter. However, there is no consistent method for setting of the two covariance matrix. After reviewing a large amount of papers, this article used trial and error method, give R_k first, and then try to get the ideal value of the Q_k. In this simulation,

$Q_k = diag(10^3, 3 \times 10^{-3}, 25)$, $R = 1$ are obtained by this method. The priori estimate value $\hat{x}_{1/0}$ and $P_{1/0}$ at time $k = 0$ is obtained from Eq. (9). And then posteriori estimates $\hat{x}_1$ and P_1 from Eq. (11). At the same time, to complete the calculation of the gain matrix, so that we finish a step size calculation. After that the obtained value is taken as the initial value recurse to the Eq. (10), and the state estimation at the next time is performed. In this paper, the complex model can reduce the amount of calculation effectively is mainly due to the following three reasons:

① Reduce the order of the filter. EKF algorithm is based on the fifth-order model to calculate, and this paper introduces the complex variable, the order becomes 3 order, so it reduce the calculation of the amount of calculation.

② No matrix inversion of the operation. The gain matrix in ECKF is a constant value compared to the traditional Kalman algorithm in the process of solving the gain matrix K.

③ Zero element reduction in the output matrix. The output matrix in the traditional Kalman filter is a matrix of 2×5, and there are eight zero elements. In this paper, the output matrix H_k is a matrix of 1×3, where the zero elements are two, which reduces the number of unnecessary calculation.

5 Simulation Studies

Simulation model of rotor flux observer for ECKF of Asynchronous motor is established to verify the effectiveness of the rotor flux observer based on Simulink/Matlab. The actual value of the rotor flux is given by the asynchronous mathematical model. The parameter of the asynchronous motor are as follows: $U_N = 380\,V$, $f_N = 50\,Hz$, $R_r = 1.395\,\Omega$, $R_s = 1.405\,\Omega$, $L_{sd} = 0.1780\,H$, $L_{rd} = 0.1780\,H$, $L_{md} = 0.1722\,H$, $p = 2$, $J = 0.511\,kg\,m^2$.

In the simulation, the asynchronous motor runs under the no-load and open-loop operation with three-phase AC(380 V, 50 Hz), and the load torque of $30\,N\,m$ is applied at 3 s. Simulation results are shown in Figs. 1, 2, 3 and 4.

Fig. 1 Speed of motor

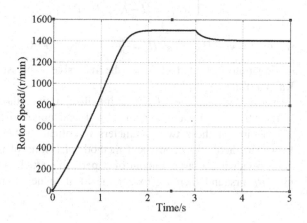

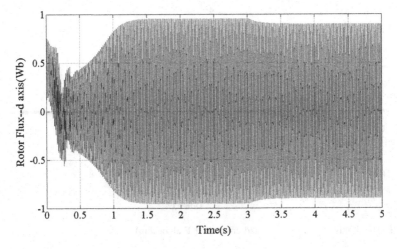

Fig. 2 Estimated (red line) and actual (blue line) d axis rotor flux

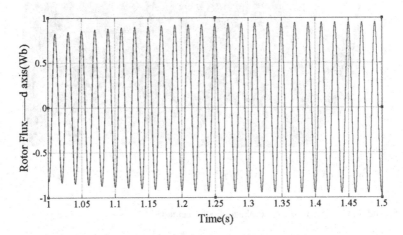

Fig. 3 Estimated (red line) and actual (blue line) d axis rotor flux (patial magnification)

From Fig. 4, the ECKF algorithm rotor fluxes error are estimated to be stable within 0.02 Wb. However, in Fig. 5, the voltage model method rotor fluxes error are estimated to be stable within 1 Wb. So we can conclude that the error of the flux observer (ECKF) is less than that of the voltage model method.

The rotor flux which is estimated by the voltage model method will be drifted when the stator resistance changes, however, the ECKF algorithm can suppress the flux drift effectively. The parameter of the observer model is constant, change the stator resistance into 2.81 Ω, other parameters of the motor is unchanged. The asynchronous motor runs under the no-load and open-loop operation with

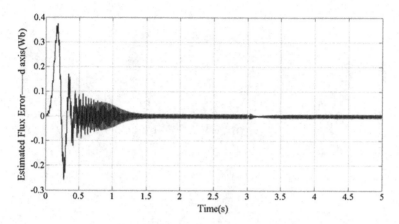

Fig. 4 The d axis rotor flux estimated error (ECKF algorithm)

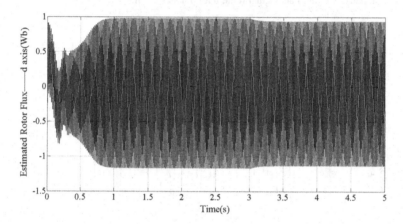

Fig. 5 The d axis rotor flux error (voltage model method)

three-phase AC(380 V, 50 Hz),and the load torque of 30 N m is applied at 3 s. The estimated error is shown in Fig. 6.

As is shown in Fig. 6a, the ECKF algorithm estimates that the fluxes are stable over a period of time and the peak-to-peak value is about 0.16 Wb. While the voltage model method in the stable peak-to-peak value is about 2.1 Wb, resulting a large flux drift. So the ECKF can suppress the flux drift well.

The ECKF algorithm is applied to the Asynchronous motor vector control system. The parameters of the motor and the observer are the same as those in the open loop. Given rotor flux $\psi = 1$ Wb, given the speed of the motor is 800 r / min. The speed regulator is controlled by PID, and $K_P = 100, K_I = 3, K_D = 0.5$. The flux regulator is controlled by PI, and $K_P = 50, K_I = 4$, simulation results are shown in

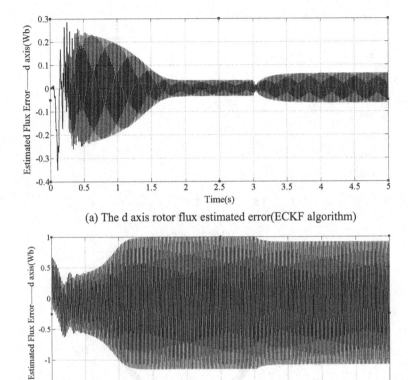

(a) The d axis rotor flux estimated error(ECKF algorithm)

(b) The d axis rotor flux error (voltage model method)

Fig. 6 The error of the estimated d axis flux $(R_s = 2.81\,\Omega)$

Fig. 7. The Asynchronous motor vector control system based on ECKF flux observer is shown in Fig. 8.

From the simulation results, we can see the ECKF rotor flux observer can observe the flux accurately, and the speed of motor also eventually stabilized at the given speed 800 r / min, which demonstrates the feasibility of the designed rotor flux observer.

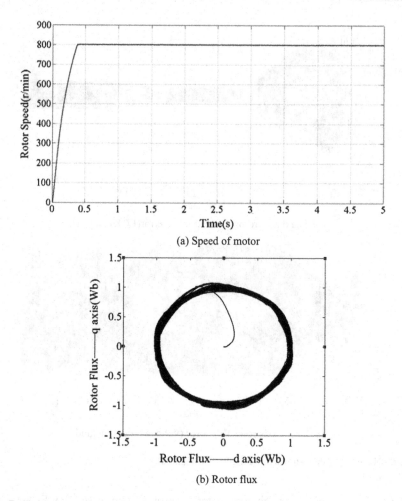

(a) Speed of motor

(b) Rotor flux

Fig. 7 Simulation results

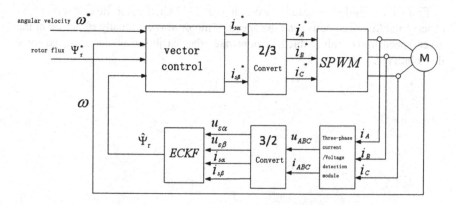

Fig. 8 Vector control system of asynchronous motor based on ECKF flux observer

6 Conclusion

This paper presents ECKF rotor flux observer, and use this observer into the vector control system. As a result of the introduction of the complex variables, so the order of the Asynchronous motor state equation is decreased and reduce the calculation of the amount of DSP. Compared with the voltage model method, this method improves the observation accuracy and produces a smaller flux drift than voltage model method when the stator resistance changes. The simulation results show that this method is effective and feasible.

References

1. Gang H, Changfan Z. Research on a new sliding mode flux observer for asynchronous motor vector control. J Dyn Control. 2011;01:75–8.
2. Trabelsi R, Khedher A, Mimouni MF, et al. Back stepping control for an asynchronous motor using an adaptive sliding rotor-flux observer. Electr Power Syst Res. 2012;93:1–15.
3. Yingjun W, Shouzhen W. Reaearch and simulation of rotor flux observer for asynchronous motor. J Guizhou Univ Technol. 2000;04:41–5.
4. Guorong LIU, Daolu Z. Research on neural network observer of motor flux. J Elect Mach Control. 2011;08:81–7.
5. Mingzhu W, Hong L, Lei W, Jiapeng X. Phase and flux observer for PMSM based on reduced order EKF. Measur Control Technol. 2012;08:59–62.
6. Wenxiang S, Jie Z, Zhiyong R. Methods and implementation of reduced-order flux observer for speed sensorless asynchronous motor. J Elect Mach Control. 2013;6:42–50.
7. Bin W, Yue W, Wei G. Fault-tolerant direct torque control system of permanent magnet synchronous motor based on stator flux reduction. J Eletrotech Soc. 2014;03:160–71.
8. Montanari M, Peresada SM, Rossi C, et al. Speed sensorless control of asynchronous motors based on a reduced-order adaptive observer. IEEE Trans Control Syst Technol. 2007;15 (6):1049–64.
9. Kojabadi HM, Chen CL. MRAS-basedadaptive pseudoreduced-order flux observer for sensorless asynchronous motor drives. IEEE Trans Power Eletron. 2005;20(4):930–8.
10. Yuedou P, Huade Li. Automatic control system for electric traction (2nd ed.). Beijing: Mechanical Industry Press;2014. p. 153–6.

Linear Active Disturbance Rejection Control for Piezo-Fexural Nanopositioning Stage

Wei Wei, Jiali Wu, Jingyu Wu and Min Zuo

Abstract Piezoelectric actuators are commonly used in nanopositioning stages. Due to the hysteresis of piezoelectric actuator, accurate position control of a nanopositioning system is always a challenging task. In this paper, linear active disturbance rejection control (LADRC) is designed to address hysteresis. With the help of extended state observer, hysteresis can be estimated, and it can be compensated by control signal in real time. Simulation results confirm LADRC is able to eliminate hysteresis and improve positioning precision.

Keywords Piezoelectric actuators · Nanopositioning · Hysteresis · LADRC

1 Introduction

Nowadays, nanopositioning stage has been extensively used [1–5], such as scanning probe microscopes, optical alignments and bio-operation devices. Piezoelectric actuator has the advantage of nanometer resolution, fast response and low power consumption, so it is commonly utilized in nanopositioning [1]. However, a distinct disadvantage is the hysteresis of piezoelectric material. Hysteresis will result in positioning errors, which is about $10 \sim 15\%$ of the whole stroke [6], or even make system be unstable [7]. In order to eliminate the influence of hysteresis and improve the positioning accuracy, different control approaches have been discussed. Based on hysteresis inverse model, feed-forward and feedback control has been proposed [8–11]. By putting the hysteretic inverse model in feed-forward channel, one can overcome hysteresis effects. At the same time, feedback control is designed to regulate positioning errors.

To realize such control approach, one needs to establish hysteretic model first. So far, different kinds of hysteresis model have been proposed, such as Preisach

W. Wei · J. Wu · J. Wu · M. Zuo (✉)
School of Computer & Information Engineering, Beijing Technology and Business
University, Beijing, China
e-mail: zuomin@btbu.edu.cn

© Springer Nature Singapore Pte Ltd. 2018
Y. Jia et al. (eds.), *Proceedings of 2017 Chinese Intelligent Systems Conference*,
Lecture Notes in Electrical Engineering 460,
https://doi.org/10.1007/978-981-10-6499-9_30

model [12], Prandtl-Ishlinskii (PI) model [13] and Bouc-Wen model [14]. However hysteresis is relatively complex and its model cannot be modelled exactly. Therefore, it is difficult to get an exact hysteresis inverse model.

To avoid such problem, other approaches have been discussed. Rather than modeling the hysteresis inverse model directly, researchers have designed control algorithms to address hysteresis directly. A novel observer is proposed and it is integrated with PID-based sliding mode controller for the tracking control of a nanopositioning system [15]. Model predictive discrete-time sliding mode control is designed to obtain desired performance for nanopositioning stage [16].

In recent years, a special kind of control technique, i.e. active disturbance rejection control (ADRC), has been proposed and applied in many industrial sectors, for its simpler structure and better performance. Furthermore, for simplifying parameters tuning, linear active disturbance rejection control (LADRC) has been proposed by Gao [17].

In this paper, we also consider the nanopositioning problem and take hysteresis as disturbance. By converting the control problem into a disturbance rejection problem, we introduce LADRC to the control of nanopositioning. Different reference signals are taken to verify LADRC.

2   Nanopositioning Stage Description

In this section, model of nanopositioning stage is presented. Structure of a nanopositioning stage is given in Fig. 1.

Here, hysteresis model is constructed by a series of Backlash model [18] (Fig. 2).

In this paper, we take 43 Backlash modules and the dead-band width for each Backlash module is $i/43$, $i = 1, 2, \ldots, 43$. When $v(t) = \sin(t), K = 1/6$, hysteresis can be drawn in Fig. 3.

For the linear dynamic system, we take the same model as [14],

$$G(s) = \frac{3}{s^2 + 1810\,s + 1000000} \tag{1}$$

In this paper, hysteresis model and linear system model presented will be utilized to simulate dynamics of a nanopositioning stage.

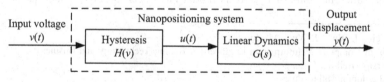

Fig. 1 Structure of a nanopositioning stage

Fig. 2 Structure of hysteresis model

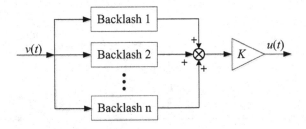

Fig. 3 Hysteresis curve

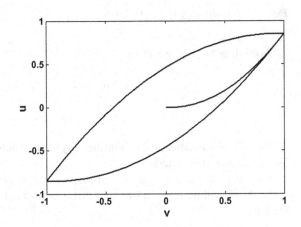

3 Linear Active Disturbance Rejection Control

For a second order nonlinear system,

$$\ddot{y} = f(y, \dot{y}, d) + b_0 u \tag{2}$$

where y is the system output, $f(\bullet)$ includes internal uncertainties and external disturbance d, b_0 represents the control gain, u is the control law.

For LADRC, control law can be designed as

$$\begin{cases} u_0 = k_p(y_r - z_1) - k_d z_2 \\ u = (u_0 - z_3)/b_0 \end{cases} \tag{3}$$

where k_p is the proportional gain, k_d is the differential gain, y_r is the reference signal, z_1, z_2, z_3 are outputs of extended state observer (ESO). ESO can be designed as

$$\begin{cases} \dot{z}_1 = z_2 + \beta_1(y - z_1) \\ \dot{z}_2 = z_3 + \beta_2(y - z_1) + b_0 u \\ \dot{z}_3 = \beta_3(y - z_1) \end{cases} \tag{4}$$

where $\beta_1, \beta_2, \beta_3$ are gains of ESO, and y is the system output.

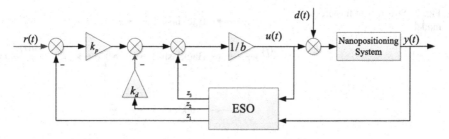

Fig. 4 Nanopositioning system by LADRC

According to [17], we have

$$\beta_1 = 3\omega_o, \beta_2 = 3\omega_o^2, \beta_3 = \omega_o^3 \tag{5}$$

$$k_p = \omega_c^2, \, k_d = 2\omega_c \tag{6}$$

where ω_c is the bandwidth of controller, ω_o is the bandwidth of ESO. Both ω_c and ω_o are tunable parameters.

For nanopositioning stage, closed-loop system is given below.

Next, we will try to get a structure that is equivalent to Fig. 4. From (5) and (8), we have

$$\begin{cases} z_1(s) = \dfrac{3\omega_o s^2 + 3\omega_o^2 s + \omega_o^3}{(s + \omega_o)^3} y(s) + \dfrac{b_0 s}{(s + \omega_o)^3} u(s) \\[2mm] z_2(s) = \dfrac{(3\omega_o^2 s + \omega_o^3)s}{(s + \omega_o)^3} y(s) + \dfrac{b_0(s + 3\omega_o)s}{(s + \omega_o)^3} u(s) \\[2mm] z_3(s) = \dfrac{\omega_o^3 s^2}{(s + \omega_o)^3} y(s) - \dfrac{b_0 \omega_o^3}{(s + \omega_o)^3} u(s) \end{cases} \tag{7}$$

According to (3), (6) and (7), we have

$$\begin{aligned} u = & \frac{1}{b_0} \frac{(s+\omega_o)^3}{(s+\omega_o)^3 + 2\omega_c s^2 + (\omega_c^2 + 6\omega_o\omega_c)s - \omega_o^3} (\omega_c^2 y_r \\ & - \frac{(3\omega_c^2\omega_o + 6\omega_c\omega_o^2 + \omega_o^3)s^2 + (3\omega_c^2\omega_o^2 + 2\omega_c\omega_o^3)s + \omega_c^2\omega_o^3}{(s+\omega_o)^3} y) \end{aligned} \tag{8}$$

In this paper, LADRC is utilized. Therefore, any discrepancies between cascade of two integrators and dynamics of nanopositioning stage are total disturbance.

If we define $G_c(s) = \dfrac{(s+\omega_o)^3}{(s+\omega_o)^3 + 2\omega_c s^2 + (\omega_c^2 + 6\omega_o\omega_c)s - \omega_o^3}$, $G_0(s) = \dfrac{1}{s^2}$, and

$H(s) = \dfrac{(3\omega_c^2\omega_o + 6\omega_c\omega_o^2 + \omega_o^3)s^2 + (3\omega_c^2\omega_o^2 + 2\omega_c\omega_o^3)s + \omega_c^2\omega_o^3}{(s+\omega_o)^3}$, structure given in

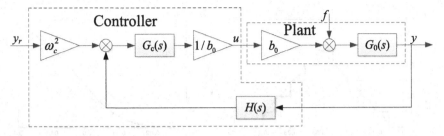

Fig. 5 An equivalent structure for second order nonlinear system by LADRC

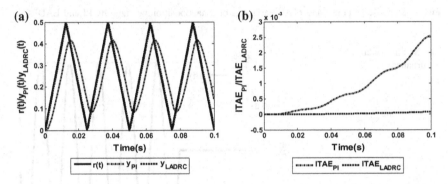

Fig. 6 Tracking response and ITAE variations of a nanopositioning stage by PI and LADRC

Fig. 5 is equivalent to the structure shown in Fig. 4. From Fig. 5, we can see that closed-loop system is stable by choosing proper ω_c, ω_o and b_0.

4 Numerical Simulations

In order to verify LADRC, simulations have been performed. PI controllers are taken to make comparisons. In all cases, let $\omega_c = 5000, \omega_o = 400000, b_0 = 270$, $K_P = 25200, K_I = 100$. Saw tooth and ladder signals are taken to be reference signals. Two cases are considered.

Case I. No external disturbance exists.

When reference signal is a saw tooth signal, system responses are shown in Fig. 6.

From Fig. 6a, we can see that the output of LADRC coincides with the reference signal, while the one of PI is inferior. Variations of ITAE presented in Fig. 6b also confirm LADRC is superior to PI.

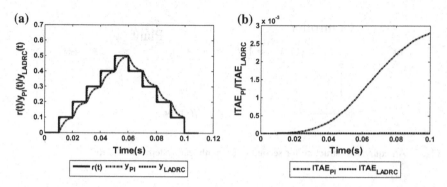

Fig. 7 Stepwise response and ITAE variations of a nanopositioning stage by PI and LADRC

Fig. 8 Sinusoidal disturbance

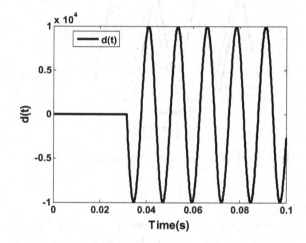

When reference signal is a ladder signal, responses of the nanopositioning system is shown in Fig. 7. Response shown in Fig. 7 also confirms that LADRC is capable of achieving desired performance. Next, external disturbance is taken into consideration.

Case II. External disturbance exists.

Sinusoidal disturbance d is introduced. Its place is shown in Fig. 4. Disturbance and system responses are given in Figs. 8 and 9, respectively.

From Fig. 9, we can see that system response of LADRC is much better than the one of PI, which confirms that LADRC is superior to PI even if disturbance exists.

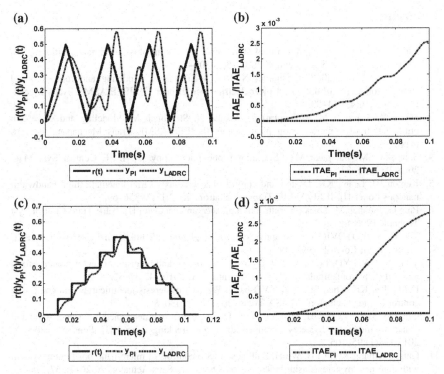

Fig. 9 System response and ITAE variations of a nanopositioning stage by PI and LADRC (in presence of sinusoidal disturbance)

5 Conclusion

Different piezoelectric materials have different hysteresis properties. Moreover, it is hard to model. It is necessary to address it properly. In this paper, from the point of ADRC, hysteresis is regarded as disturbance. It can be estimated and eliminated by LADRC to improve the positioning accuracy. By comparison with PI, we can see that LADRC provides a more practical and effective solution to the control of nanopositioning stage.

Acknowledgements This work is supported by National Natural Science Foundation of China (61403006).

References

1. Devasia S, Eleftheriou E, Moheimani SOR. A survey of control issues in nanopositioning [J]. IEEE Trans Control Syst Technol. 2007;15(5):802–23.
2. Grossard M, Boukallel M, Chaillet N, Rotinat-Libersa C. Modeling and robust control strategy for a control-optimized piezoelectric microgripper [J]. IEEE/ASME Trans Mechatron. 2011;16(4):674–83.
3. Merry R, Maassen M, Molengraft M, Wouw N, Steinbuch M. Modeling and waveform optimization of a nano-motion piezo stage [J]. IEEE/ASME Trans Mechatron. 2011;16 (4):615–26.
4. Salapaka SM, Salapaka MV. Scanning probe microscopy [J]. IEEE Control Syst Mag. 2008;28(2):65–83.
5. Kenton BJ, Leang KK. Design and control of a three-axis serial kinematic high-bandwidth nanopositioner [J]. IEEE/ASME Trans Mechatron. 2012;17(2):356–69.
6. Ping G, Jouaneh M. Tracking control of a piezoceramic actuator [J]. IEEE Trans Control Syst Technol. 1996;4(3):209–16.
7. Tao G, KOLOTOVIC PV. Adaptive control of plants with unknown hysteresis [J]. IEEE Trans Autom Control. 1995;40(2):200–12.
8. Yang YL, Wei YD, Lou JQ, et al. Hysteresis modeling and precision trajectory control for a new MFC micromanipulator [J]. Sens Actuators, A. 2016;247:37–52.
9. Lei L, Tan KK, Chen SL, et al. SVD-based Preisach hysteresis identification and composite control of piezo actuators [J]. ISA Trans. 2012;51(3):430–8.
10. Qin YD, Shirin Zadeh BJ, Tian YL, et al. Design issues in a decoupled XY stage: static and dynamics modeling, hysteresis compensation, and tracking control [J]. Sens Actuators A. 2013;194(1):95–105.
11. Gan JQ, Zhang XM, Li H, Wu H. Full closed-loop controls of micro/nano positioning system with nonlinear hysteresis using micro-vision system [J]. Sens Actuators A. 2017;257:125–33.
12. Tiberiu-Gabriel Z, Michael AEA, Zhang Z, Nils AA. Preisach model of hysteresis for the piezoelectric actuator drive [C]. In: IEEE Industrial Electronics Society;2015. p. 2788–93.
13. Liu SN, Su CY, Li Z. Robust adaptive inverse control of a class of nonlinear systems with Prandtl-Ishlinskii hysteresis model [J]. IEEE Trans Autom Control. 2014;59(8):2170–5.
14. Zhao XL, Wanh JL. Backstepping control with error transformation for Bouc-Wen hysteresis nonlinear system [J]. Control Theory Appl. 2014;31(8):1094–8.
15. Peng JY, Chen XB. Integrated PID-based sliding mode state estimation and control for piezoelectric actuators [J]. IEEE/ASME Trans Mech. 2014;19(1):88–99.
16. Xu QS, Cao ZW. Piezoelectric positioning control with output-based discrete-time terminal sliding mode control [J]. IET Control Theory Appl. 2017;11(5):694–702.
17. Gao ZQ. Scaling and bandwidth-parameterization based controller tuning [J]. Denver, CO, United States;2003. p. 4989–96.
18. Zhao T, Zhai YW, Yu XJ. Neural network hysteresis model based on the describing function of backlash [J]. J Qiongdao Univ Sci Technol (Nat Sci Ed). 2012;33(3):315–20.

Linear Active Disturbance Rejection Control for Nanopositioning System by ITAE Optimal Tuning Approach

Wei Wei, XuDong Shen, YanJie Shao and Min Zuo

Abstract Nano-positioning systems with piezoelectric actuators are widely used. However, hysteresis, a common phenomenon in piezoelectric material, makes the control of nano-positioning system be a challenge. A simple and effective approach to deal with hysteresis is necessary. In this paper, both hysteresis and other disturbances are regarded as total disturbance, and linear active disturbance rejection control (LADRC) is utilized. With the help of extended state observer (ESO), total disturbance can be estimated and it can be compensated by control signal in real time. Integral of time-multiplied absolute-value of error (ITAE) optimal bandwidth parameterization approach is utilized to determine the parameters of LADRC. Numerical results are presented to confirm LADRC and its tuning approach.

Keywords Nanopositioning · LADRC · ITAE optimal

1 Introduction

The technique of studies and applications of materials with structure sizes ranging from 1 to 100 nm is called nanotechnology [1]. It not only refers to nano-material manufacturing technology, but also contains nano-level measurement, processing, control and so on [2]. Recently, it has been widely used in many kinds of precision equipments, such as nano-positioning stages [1], electronic transformers [3], petroleum exploration and developments [4]. Since nano-positioning requires actuators with excellent operating bandwidth and low power output with large mechanical force, piezoelectric actuator is a preferable choice [1]. However, as hysteresis is a typical nonlinear phenomenon existing in piezoelectric materials, it will decrease accuracy greatly and even make the closed-loop system be unstable [5]. Therefore,

W. Wei · X. Shen · Y. Shao · M. Zuo (✉)
School of Computer & Information Engineering, Beijing Technology and Business University, Beijing, China
e-mail: zuomin@btbu.edu.cn

© Springer Nature Singapore Pte Ltd. 2018
Y. Jia et al. (eds.), *Proceedings of 2017 Chinese Intelligent Systems Conference*,
Lecture Notes in Electrical Engineering 460,
https://doi.org/10.1007/978-981-10-6499-9_31

325

it is necessary to design an effective controller to reduce negative effects of hysteresis.

In order to deal with the problem, numerous approaches have been discussed, such as fuzzy control [6], neural network control [7], and adaptive control [8]. Active disturbance rejection control (ADRC), proposed by Han in 1998 [9], takes systems' nonlinearities, uncertainties and un-modeled dynamics as total disturbances. By estimating and compensating those disturbances in real time, ADRC is able to guarantee desired system performance. Linear active disturbance rejection control (LADRC) inherits the essence of ADRC, but reduces the number of tunable parameters and accelerates the applications of ADRC. So far, ADRC/LADRC has been applied in numerous industrial processes successfully, such as superconducting cavity control [10], flywheel energy storage system [11], tracking problem of IPSRU [12], hysteresis compensation [13], piezoelectric beam control [14], internal permanent-magnet synchronous motor control [15]. Generally, second order ADRC with position, velocity and total disturbance is enough for a given controlled system. Such issue has been discussed in [16].

Integral of time-multiplied absolute-value of error (ITAE) was firstly proposed by Graham and Lathrop in 1953 [17]. It is an index that depicts system performance from the points of fast and accuracy, and a set of normalized transfer function coefficients is provided to minimize ITAE values. In order to minimize ITAE value, Maurya et al. proposed a tuning approach for fractional order PID controller [18].

In this paper, we also introduce ITAE index. Based on the bandwidth parameterization method [19] and ITAE optimal transfer function provided in [17], ITAE optimal bandwidth parameterization approach is proposed to improve the performance of nanopositioning systems.

2   Linear Active Disturbance Rejection Control Design

2.1   Second Order Linear Active Disturbance Rejection Control

A closed-loop system by second order LADRC is shown in Fig. 1.

where r is the set value and y is system output, u is the control signal, k_p and k_d are controller parameters. Nanopositioning system is represented by a cascaded system of H and M. H is the hysteresis, and M is the linear dynamics. The Preisach model can be expressed as [20]

$$y(t) = \iint_{\alpha \geq \beta} \mu(\alpha, \beta) \gamma_{\alpha\beta}[u(t)] d\alpha d\beta \tag{1}$$

where $\mu(\alpha, \beta)$ is the weight function of $\gamma_{\alpha\beta}[u]$, $\gamma_{\alpha\beta}[u]$ is the hysteresis unit. Transfer function of the linear model of nanopositioning is [21]

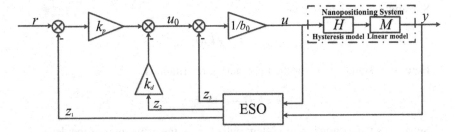

Fig. 1 Nanopositioning system by second order LADRC

$$G(s) = \frac{3.391 \times 10^{10}}{s^3 + 3759s^2 + 2.063 \times 10^7 s + 7.514 \times 10^{10}} \tag{2}$$

Extended state observer (ESO) takes the form

$$\begin{cases} \dot{z}_1 = z_2 + \beta_1(y - z_1) \\ \dot{z}_2 = z_3 + \beta_2(y - z_1) + b_0 u \\ \dot{z}_3 = \beta_3(y - z_1) \end{cases} \tag{3}$$

where $\beta_1, \beta_2, \beta_3$ are observer gains, z_1 is the estimation of output, z_2 and z_3 are estimations of the derivative of output and total disturbance, respectively.

Control law can be described as

$$\begin{cases} u = (u_0 - z_3)/b_0 \\ u_0 = k_p(r - z_1) + k_d(\dot{r} - z_2) \end{cases} \tag{4}$$

where b_0 is the control coefficient, $\dot{r}$ is the derivative of set value.

2.2 Second Order LADRC Based on ITAE Optimal Bandwidth Parameterization Approach

A second order nonlinear system can be written as

$$\ddot{y} = f(y, \dot{y}, u) + bu \tag{5}$$

where $f(y, \dot{y}, u)$ is the total disturbance of system. If ESO works well, the second order nonlinear closed-loop system can be transformed to cascaded integrators, i.e.

$$\ddot{y} \approx u_0 \tag{6}$$

Substituting (4) into (6), we have

$$\ddot{y} \approx k_p(r - z_1) + k_d(\dot{r} - z_2) \tag{7}$$

Here, step signal is taken, then (7) can be rewritten as

$$\ddot{y} \approx k_p(r - z_1) - k_d z_2 \tag{8}$$

Since z_1 is the estimation of system output, z_2 is the estimation of the derivative of output, (8) can be written as

$$\ddot{y} \approx k_p(r - y) - k_d \dot{y} \tag{9}$$

Laplace transform is performed for both sides of (9), we have

$$G_d(s) = \frac{y(s)}{r(s)} = \frac{k_p}{s^2 + k_d s + k_p} \tag{10}$$

Let $k_p = \omega_n^2, S = \frac{s}{\omega_n}$, then

$$G_{do}(S) = \frac{\omega_n^2}{\omega_n^2 S^2 + k_d \omega_n S + \omega_n^2} = \frac{1}{S^2 + \frac{k_d}{\omega_n} S + 1} \tag{11}$$

According to second order ITAE optimal transfer function [17],

$$G_0(s) = \frac{1}{s^2 + 1.41s + 1} \tag{12}$$

we have

$$k_p = \omega_n^2, \ k_d = 1.41 \omega_n \tag{13}$$

here ω_n is the natural frequency.

For bandwidth parameterization approach [19],

$$\beta_1 = 3\omega_o, \beta_2 = 3\omega_o^2, \beta_3 = \omega_o^3 \tag{14}$$

where ω_o is the bandwidth of ESO, ω_c represents bandwidth of controller.

Then, ITAE optimal bandwidth parameterization tuning approach for second order LADRC is according to (13) and (14). Adjustable parameters are b_0, ω_n and ω_o.

3 Numerical Simulations

There are two parts in simulation experiments, i.e. time domain response and frequency domain response. The parameters determined by the bandwidth parameterization tuning method are $b_0 = 300000000$, $\omega_c = 10000$, $\omega_o = 16000$. While parameters of ITAE optimal bandwidth parameterization approach are $b_0 = 100000000$, $\omega_n = 7000$, $\omega_o = 12140$.

Frequency domain response of system's linear model is given in Fig. 2. From Fig. 2, we can see clearly that there is a peak in amplitude-frequency response.

Time domain response and frequency domain response obtained by the bandwidth parameterization method and ITAE optimal bandwidth parameterization method are shown in Table 1.

Table 2 gives out the transfer functions of linearized closed-loop systems, cut-off frequencies, magnitude margins, phase margins and ITAE values for different tuning methods.

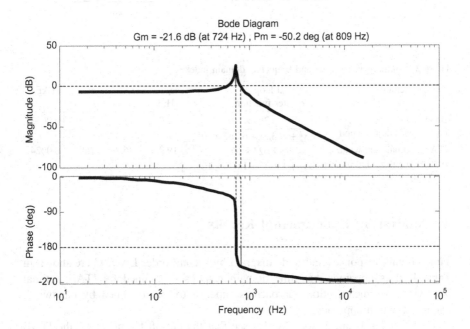

Fig. 2 Bode diagram of the linear model

Table 1 Comparisons of responses by two approaches

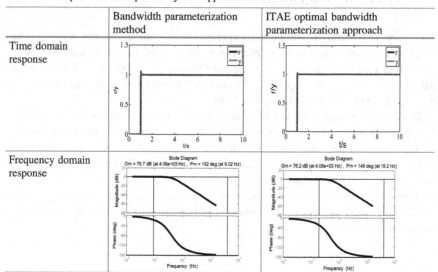

	Bandwidth parameterization method	ITAE optimal bandwidth parameterization approach
Time domain response		
Frequency domain response		

Table 2 Comparisons of time and frequency domain indexes

Tuning method	Closed loop transfer function	ω_c (Hz)	G_m (dB)	P_m (°)	ITAE
Bandwidth parameterization method	$\dfrac{6.767 \times 10^4}{s^2 + 363.3s + 6.767 \times 10^4}$	9.02	79.7	162	0.0033
ITAE optimal bandwidth parameterization approach	$\dfrac{1.024 \times 10^5}{s^2 + 436.1s + 1.024 \times 10^5}$	19.2	76.2	149	0.0024

4 Analysis of Experimental Results

Time domain responses and bode diagram by second order LADRC are shown in Table 1. It is apparent that time domain responses obtained by ITAE optimal bandwidth parameterization approach is superior to that obtained by bandwidth parameterization approach.

From Tables 1 and 2, we can also see that the cut-off frequency of the ITAE optimal tuning method is much larger than that of bandwidth parameterization method, which reveals that bandwidth obtained by ITAE optimal tuning method is larger than that by bandwidth parameterization method. However, magnitude margin and phase margin obtained by ITAE optimal tuning method are smaller than that obtained by bandwidth parameterization method. Therefore, ITAE optimal parameterization approach is capable of improving the response speed of the system, but some stability margin is sacrificed when the response speed is improved. In addition, smaller ITAE value can be obtained by ITAE optimal tuning method.

Therefore, by ITAE optimal bandwidth parameterization approach, we can make system response be faster and more accurate.

5 Conclusion

In this paper, LADRC is utilized to control the nanopositioning system, and a new tuning method, i.e. ITAE optimal bandwidth parameterization method, is taken to determine the parameters of LADRC. Time domain response, frequency domain response and ITAE values have been obtained and compared. It can be learned from the experimental results that ITAE optimal tuning method is able to improve the dynamic performance of the nanopositioning system.

Acknowledgements This work is supported by National Natural Science Foundation of China (61403006).

References

1. Devasia S, Moheimani SOR. A survey of control issues in nanopositioning. IEEE Trans Control Syst Technol. 2007;15(5):802–23.
2. Bai C. The state of the art and thinking of nanotechnology in China. Physics. 2002;31(2):65–70.
3. Contreras JE, Rodriguez EA, Taha-Tijerina J. Nanotechnology applications for electrical transformers—a review. Electr Power Syst Res. 2017;573–84.
4. Liu H, Jin X, Ding B. Application of nanotechnology in petroleum exploration and development. Pet Explor Dev. 2016;43(6):1107–15.
5. Tao G, Kokotovic PV. Adaptive control of plants with unknown hystereses. IEEE Trans Autom Control. 1992;39(1):59–68.
6. Hwang Chih-Lyang. Microprocessor-based fuzzy decentralized control of 2-D piezo-driven systems. IEEE Trans Ind Electron. 2008;55(3):1411–20.
7. Liu J, Zhou K. Neural networks based modeling and robust control of hysteresis. In: Chinese control conference;2016. p. 3051–56.
8. Chaoui H. Observer-based adaptive control of piezoelectric actuators without velocity measurement. In: 2016 IEEE 25th international symposium on industrial electronics (ISIE), Santa Clara;2016. p. 399–404.
9. Jingqing H. Auto-disturbances-rejection controller and its applications. Control Decis. 1998;13(1):19–23.
10. Geng Z. Superconducting cavity control and model identification based on active disturbance rejection control. IEEE Trans Nuclear Sci. 2017;64(3):951–8.
11. Chang X, Li Y, Zhang W, et al. Active disturbance rejection control for a flywheel energy storage system. IEEE Trans Ind Electron. 2015;62(2):991–1000.

12. Li H, Yang Y, Jia P et al. Application of active disturbance rejection control in tracking problem of IPSRU. In: International conference on control & automation (ICCA);2016. p. 616–20.

13. Goforth Frank J, Qing Zheng, Zhiqiang Gao. A novel practical for rate independent hysteretic systems. ISA Trans. 2012;51(3):477–84.

14. Zheng Q, Richter H, Gao Z. On active vibration suppression of a piezoelectric beam. In: American control conference;2013. p. 6613–18.

15. Du B, Wu S. Application of linear active disturbance rejection controller for sensorless control of internal permanent-magnet synchronous motor. IEEE Trans Ind Electron. 2016;63 (5):3019–27.

16. Yao X, Wang Q, Liu W, et al. Two-order ADRC control for general industrial plants. Control Eng China. 2002;9(05):59–62.

17. Graham D, Lathrop RC. The synthesis of 'Optimum' transient response: criteria and standard forms. AIEE Trans. 1953;72(5):273–88.

18. Maurya AK, Bongulwar MR, Patre BM. Tuning of fractional order PID controller for higher order process based on ITAE minimization. In: IEEE India conference;2015. p. 1–5.

19. Gao Z. Scaling and bandwidth-parameterization based controller tuning. Denver: CO, United States; 2003. p. 4989–96.

20. Hu H, Mrad RB. On the classical Preisach model for hysteresis in piezoceramic actuators. Mechatronics. 2003;13(2):85–94.

21. Shan Y, Leang KK. Design and control for high-speed nanopositioning: serial-kinematic nanopositioners and repetitive control for nanofabrication. IEEE Control Syst. 2013;33(6):86–105.

Nonlinear Tracking Control for Image-Based Visual Servoing with Uncalibrated Stereo Cameras

Siqi Li, Xuemei Ren, Yuan Li and Haoxuan Qiu

Abstract This paper considers the problem of uncertain camera pose and camera parameters for a 3-degree-of-freedom (DOF) robot manipulator in nonlinear visual servoing tracking control. To solve this problem, the typical Kalman filter (KF) algorithm is designed to estimate the image Jacobian matrix online, which can reduce the system noises to improve the robustness of the control system. Visual optimal feedback controller is developed to precisely track the desired position of the robot manipulator. In addition, stereo cameras are incorporated into the robot manipulator system such that the tracking errors in both camera image frame and robot base frame can simultaneously converge to zero. Experimental results are included to illustrate the effectiveness of the proposed approach.

Keywords Visual servoing · Robot manipulator · Kalman filter · Image jacobian

1 Introduction

Visual servoing has been generally used in robot manipulator nonlinear tracking control. The traditional robot hand-eye coordination control theories rely on the study of camera calibration technique. However, the calibration accuracy restricts the accuracy of system in practice. The camera pose can be changed and sometimes its working condition can be difficult to realize calibration. Compared with known camera parameters, uncalibrated visual servo system performs higher flexibility and adaptability.

Several visual servoing control methods have been developed to improve the control performance of robot manipulator. Chaumette presented a summary of visual servo control approaches [1]. Farrokh investigated the comparison of the position-based and image-based robot visual servoing methods to improve the

S. Li · X. Ren (✉) · Y. Li · H. Qiu
School of Automation, Beijing Institute of Technology, Beijing 100081, China
e-mail: xmren@bit.edu.cn

© Springer Nature Singapore Pte Ltd. 2018
Y. Jia et al. (eds.), *Proceedings of 2017 Chinese Intelligent Systems Conference*,
Lecture Notes in Electrical Engineering 460,
https://doi.org/10.1007/978-981-10-6499-9_32

dynamics performance and robustness of visual servoing (RVS) system [2]. In [3], it was mentioned that image-based visual servoing (IBVS) is a properer way under the condition of unknown camera parameters. Cai presented a 2-1/2-D visual servoing choosing orthogonal image features to enhance the behavior of tracking system [4]. The image Jocabian matrix, first presented by Weiss, is universally utilized to define the relationship between camera image frame and robot base frame. Zergerogluprovided the analytic solution of the image Jacobian matrix under the condition of ideal camera pose for the problem of position tracking control of a planar robot manipulator with uncertain robot-camera parameters [5]. While in practice, it is impossible to guarantee that the camera imaging plane is parallel to the task plane, that is, the relationship between the two coordinate systems cannot only be defined by one rotation angle. Hence, the control methods based on the direct estimation for the numerical solution of the image Jacobian matrix are widely used in robot hand-eye coordination research field. In [6], discrete-time estimator was developed by using the least squares algorithm. Yoshimi utilized geometric effect and rotational invariance to accomplish alignment [7]. Piepmeier performed online estimation using a dynamic recursive least-squares algorithm [8], while the work [9] addressed robust Jacobian estimation.

In this paper, the nonlinear visual servo controller is developed for a 3-degree-of-freedom (DOF) robot manipulator with fixed uncalibrated stereo cameras configuration. Only one-step robot movement is used to estimate the current image Jacobian matrix completely using Kalman filter (KF) algorithm. Furthermore, the filtering process adopts recursive computing without storing historical data. Then, a simple visual optimal feedback control is designed to achieve nonlinear trajectory tracking.

2   Problem Formulation

The visual sensor is used to measure the information in robot task space, and intuitively reflects the relationship between the robot end-effector and the target to be tracked. According to [10], the mapping relation between camera image frame and robot base frame is defined as follows:

$$\dot{s} = J_I(p) \cdot \dot{p} \tag{1}$$

where $p(t) = (x_b(t) \ \ y_b(t))^{\mathrm{T}}$ denotes the end-effector position in robot base frame, $s(t) = (u_l(t) \ \ v_l(t) \ \ u_r(t) \ \ v_r(t))^{\mathrm{T}}$ represents the projected position of the end-effector in stereo camera image frame. The image Jacobian matrix, expressed by $J_I(q) \in \mathbb{R}^{4 \times 2}$, is defined as

$$J_I(p) = \frac{\partial s}{\partial p} = \begin{pmatrix} \dfrac{\partial u_l}{\partial x_b} & \dfrac{\partial u_l}{\partial y_b} \\[1ex] \dfrac{\partial v_l}{\partial x_b} & \dfrac{\partial v_l}{\partial y_b} \\[1ex] \dfrac{\partial u_r}{\partial x_b} & \dfrac{\partial u_r}{\partial y_b} \\[1ex] \dfrac{\partial v_r}{\partial x_b} & \dfrac{\partial v_r}{\partial y_b} \end{pmatrix} \tag{2}$$

where $J_I(p)$ is the interaction matrix involving the end-effector position and the intrinsic and extrinsic parameters of cameras. Specifically, the corresponding spatial position can be uniquely determined from the image feature, because the dimension of image feature space is greater than the dimension of robot motion space, which means that the tracking of image space is equivalent to the tracking of robot motion space with stereo cameras configuration.

3 Online Estimation and Visual Optimal Feedback Controller

3.1 Online Estimation

To estimate the image Jacobian matrix online via KF, transforming the image Jacobian matrix into a state vector which can be expressed as the following form:

$$x = \left(\frac{\partial s_1}{\partial p} \quad \frac{\partial s_2}{\partial p} \quad \frac{\partial s_3}{\partial p} \quad \frac{\partial s_4}{\partial p} \right)^{\mathrm{T}} \tag{3}$$

where $\dfrac{\partial s_i}{\partial p} = \left(\dfrac{\partial s_i}{\partial x_b} \quad \dfrac{\partial s_i}{\partial y_b} \right)^{\mathrm{T}}, i = 1, 2, 3, 4$, denotes the transposition of i th row vector of the image Jacobian matrix $J_I(p)$.

Based on the definition of $J_I(p)$ in (1), s can be written as

$$s(k+1) \approx s(k) + J_I(k) \cdot \Delta p(k) = s(k) + J_I(k) \cdot u(k) \tag{4}$$

We define x in (3) as the system state, and the image feature variation $\Delta s(k) = s(k+1) - s(k)$ as the system output $z(k)$. Hence, the state equation can be derived as follows:

$$x(k+1) = x(k) + \eta(k) \tag{5}$$

$$z(k) = C(k) \cdot x(k) + w(k) \tag{6}$$

where $\eta(k) \in \mathbb{R}^8$ denotes the state noise and $w(k) \in \mathbb{R}^4$ is the observation noise, which are both assumed to be white Gaussian noises shown in (7).

$$E(\eta(k)) = 0, \quad \text{cov}\{\eta(k), \eta(j)\} = R_\eta \cdot \delta_{kj}$$
$$E(w(k)) = 0, \quad \text{cov}\{w(k), w(j)\} = R_w \cdot \delta_{kj} \qquad (7)$$
$$\text{cov}\{\eta(k), w(j)\} = 0 \cdots \forall k, j$$

The system input of the state Eqs. (5) and (6) is contained in $C(k)$:

$$C(k) = \begin{pmatrix} u^\mathrm{T}(k) & & 0 \\ & \ddots & \\ 0 & & u^\mathrm{T}(k) \end{pmatrix}_{4 \times 8} \qquad (8)$$

Five steps of recursive estimation based on KF algorithm are listed as follows:

Step 1. Predict the next state based on current state

$$\hat{x}(k) = x(k+1|k) = x(k|k) \qquad (9)$$

Step 2. Update the uncertainty with respect to the state estimate

$$P(k+1|k) = P(k|k) + R_\eta \qquad (10)$$

Step 3. The Kalman gain can be computed as

$$K(k+1) = P(k+1|k) \cdot C^\mathrm{T}(k) \cdot [C(k) \cdot P(k+1|k) \cdot C^\mathrm{T}(k) + R_w]^{-1} \qquad (11)$$

Step 4. The optimal state estimate value is given by

$$\hat{x}(k+1) = x(k+1|k+1) = \hat{x}(k) + K(k+1) \cdot [z(k+1) - C(k) \cdot \hat{x}(k)] \qquad (12)$$

Step 5. Update the state covariance matrix

$$P(k+1|k+1) = [I - K(k+1) \cdot C(k)] \cdot P(k+1|k) \qquad (13)$$

Given any two-step linearly independent test movement $\Delta p_1, \Delta p_2$, the corresponding image feature variations $\Delta s_1, \Delta s_2$ are gained. Thus, the initial value of $\hat{x}(0)$ (i.e. $\hat{J}_I(0)$) can be obtained:

$$\hat{J}_I(0) = (\Delta s_1 \ \ \Delta s_2) \cdot (\Delta p_1 \ \ \Delta p_2)^{-1} \qquad (14)$$

where $\hat{x}(0)$ can be remodeled by $\hat{J}_I(0)$ using the form in (3).

3.2 Visual Optimal Feedback Controller Design

After online estimation of the image Jacobian matrix, the design of the direct visual feedback controller can be easily established. Define the system error in camera image frame as:

$$e(t) = s^*(t) - s(t) \tag{15}$$

where $s^*(t) = s_o(t)$ denotes the desired image feature describing the image feature of the moving target.

The control objective of this paper is to design a control signal $u = \dot{p}$, which minimize the following objective function

$$H = \frac{1}{2} e^T e \tag{16}$$

Discretizing the control signal u, the optimal control signal at time k is given by

$$u(k) = J_I(k)^+ \cdot \left(\hat{s}_o(k+1) - s(k) \right) \tag{17}$$

where $J_I(k)^+$ is Moore–Penrose pseudoinverse of $J_I(k)$. When $J_I(k)$ is of full rank 2, its inverse can be expressed as

$$J_I(k)^+ = \left(J_I(k)^T J_I(k) \right)^{-1} J_I(k)^T \tag{18}$$

The estimation of the target image feature at time $k+1$, denoted by $\hat{s}_o(k+1)$, can be obtained by

$$\hat{s}_o(k+1) = s_o(k) + (s_o(k) - s_o(k-1)) \tag{19}$$

Theorem 1 *With the configuration of stereo cameras, the control law (17) guarantees global asymptotic position tracking at any time k.*

$$e_p(k) = p^*(k) - p(k) \to 0 \tag{20}$$

Proof Substituting (4), (17) and (19) into (16), the objective function is rewritten as the following form:

$$H(k) = \frac{1}{2} \left(s^*(k) - s(k) \right)^{\mathrm{T}} \left(s^*(k) - s(k) \right)$$

$$= \frac{1}{2} \left(s_o(k) - \widehat{s}_o(k) \right)^{\mathrm{T}} \left(s_o(k) - \widehat{s}_o(k) \right) \qquad (21)$$

$$= \frac{1}{2} \sum_{i=1}^{4} \left(s_{oi}(k) - \widehat{s}_{oi}(k) \right)^2 \to 0$$

From (21), the difference between the target image feature and its estimation tends to zero via KF algorithm. Under stereo cameras configuration, the convergence of image space is equivalent to the convergence of robot motion space with stereo cameras configuration. Hence, the conclusion of (20) is straightforward. In Eye-to-hand system, the control signal $u(k)$ is able to guarantee the asymptotic convergence of tracking error.

4   Experimental Verification

In this section, a 3-DOF robot manipulator system is employed as the test-rig to test the proposed control scheme. The configuration of the whole experimental setup is depicted in Fig. 1. The experimental setup is composed of a YASKAWA 6-DOF robot manipulator (rotation part is locked), two fixed pinhole cameras and an upper computer. The control method is written by using Visual C ++ program. The camera frame rate is set as 30 fps.

Fig. 1 A 3-DOF visual robot manipulator test-rig

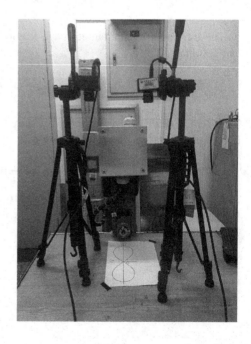

The robot end-effector performs two-dimensional translation movement, tracking the curve track target on task plane, and uncalibrated stereo cameras are fixed right above the task plane.

The linearly independent test movements lead to the initial state vector is given by

$$\hat{J}_I(0) = \begin{pmatrix} 3.7608 & 0.0081 & 2.4344 & -0.0035 \\ -4.0446 & -2.3378 & -0.4403 & -2.4993 \end{pmatrix}^{\mathrm{T}} \tag{22}$$

The variance of the each noise in the state equation and the covariance matrix is set to

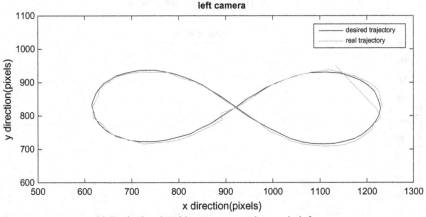

(a) Desired and real image-space trajectory in left camera

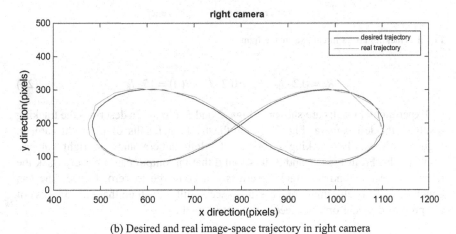

(b) Desired and real image-space trajectory in right camera

Fig. 2 Desired trajectory and end-effector trajectory in camera image frame

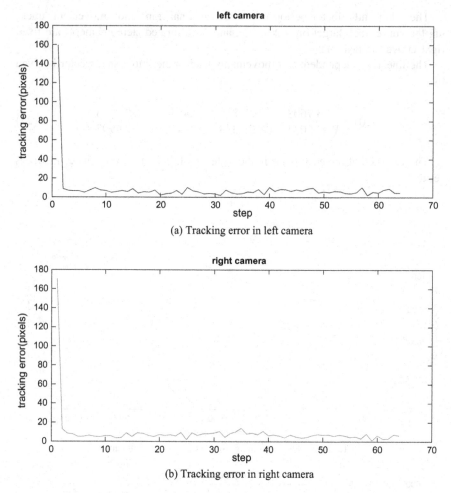

Fig. 3 Tracking errors in camera image frame

$$R_\eta = 0.2 \cdot I_8 \quad R_w = 0.2 \cdot I_4 \quad P(0) = 15 \cdot I_8 \tag{23}$$

Experimental results are shown in Figs. 2 and 3. Figure 2a describes the tracking result of the left camera, Fig. 2b gives the tracking results of the right camera. Figure 3 shows the tracking errors of the left camera and the right camera, respectively. From Figs. 2 and 3, it is found that the output can precisely track the desired trajectory and the tracking errors can converge to zero. Hence, one can conclude that the aforementioned results successfully illustrate that effectiveness of the proposed visual optimal feedback control method.

5　Conclusions

In this paper, online estimation of the image Jacobian matrix has been investigated using uncalibrated stereo cameras. This method, which combines KF algorithm and visual optimal feedback control, is presented for a 3-DOF robot manipulator with system noises. KF is utilized to estimate the image Jacobian matrix, and the visual optimal feedback control is designed based on the estimate values. Stereo cameras system is applied to guarantee tracking error converging to zero. Experimental results verify the effectiveness and robustness of the proposed algorithm.

Acknowledgements This work was supported by the National Natural Science Foundation of China under Grants 61433003, 61472037.

References

1. Chaumette F, Hutchinson S. Visual servo control. II. Advanced approaches [Tutorial]. IEEE Robot Autom Mag. 2007;14(1):109–18.
2. Janabi-Sharifi F, Deng L, Wilson WJ. Comparison of basic visual servoing methods. IEEE/ASME Trans Mechatron. 2011;16(5):967–83.
3. Tao B, Gong Z, Han D. Survey on uncalibrated robot visual servoing control. Chin J Theor Appl Mech. 2016;48(4):767–83.
4. Cai CX, Somani N, Knoll A. Orthogonal image features for visual servoing of a 6-DOF manipulator with uncalibrated stereo cameras. IEEE Trans Robot. 2016;32(2):452–61.
5. Zergeroglu E, Dawson D, Querioz M, et al. Vision-based nonlinear tracking controllers with uncertain robot-camera parameters. IEEE/ASME Trans Mechatron. 2001;6(3):322–37.
6. Hosoda K, Asada M. Versatile visual servoing without knowledge of true jacobian. In: IEEE/RSJ/GI international conference on intelligent robots and systems '94. Advanced robotic systems and the real world;1994. p. 186–93.
7. Yoshimi B, Allen P. Alignment using an uncalibrated camera system. IEEE Trans Robot Autom. 1995;11(4):516–21.
8. Piepmeier J, McMurray G, Lipkin H. Uncalibrated dynamic visual servoing. IEEE Trans Robot Autom. 2004;20(1):143–7.
9. Azad S, Farahmand AM, Jagersand M. Robust Jacobian estimation for uncalibrated visual servoing. In: IEEE international conference on robotics and automation;2010. p. 5564–69.
10. Su J, Zhang Y, Luo Z. Online estimation of Image Jacobian Matrix for uncalibrated dynamic hand-eye coordination. Int J Syst Control Commun. 2008;1(1):31–52.

Object Detection from Images Based on MFF-RPN and Multi-scale CNN

Jiayi Zhou, Hengyi Zheng, Hongpeng Yin and Yi Chai

Abstract In this paper, an object detection model based on MFF-RPN and Multi-scale CNN is proposed. Firstly, the region proposal network based on multi-level feature fusion (MFF-RPN) is presented to extract the candidate proposals. Secondly, a convolutional neural network (CNN) with different scale convolution kernels is conducted to extract features adaptively. Finally, multi-task loss is employed to establish a complex mapping between image object features and object detection mode. The experimental results prove that the proposed algorithm gets better classification performance and higher detection accuracy.

Keywords Object detection · MFF-PRN · CNN · Multi-task loss

1 Introduction

Object detection from images aims at locating the objects in the image and determining the category of each object. It is a fundamental computer vision problem that has attracted an immense amount of attention. It can be applied to many advanced visual tasks, including autonomous driving, home care for elderly people,

Key Laboratory of Dependable Service Computing in Cyber Physical Society,
Ministry of Education, Chongqing 400030, China

J. Zhou · Y. Chai
School of Automation, Chongqing University, Chongqing 400044, China

Y. Chai
Key Laboratory of Power Transmission Equipment and System Security,
Chongqing 400044, China

H. Yin
College of Mechanical Engineering, Chongqing University,
Chongqing 400044, China
e-mail: yinhongpeng@gmail.com

© Springer Nature Singapore Pte Ltd. 2018   343
Y. Jia et al. (eds.), *Proceedings of 2017 Chinese Intelligent Systems Conference*,
Lecture Notes in Electrical Engineering 460,
https://doi.org/10.1007/978-981-10-6499-9_33

intelligent robot navigation. Accurate object detection algorithms would allow computers to perform these complex tasks effectively.

Detecting an object reliably requires not only the object's fine-grained details, but also global information surrounding it. Current detection models [1–4] only use global features formed by higher convolution layers. The results of these methods reveal that the deeper layer of CNN can classify the object with high accurate but cannot find the localization effectively. Because the feature maps became coarser while layers deeper. In the other hand, the results of [5] demonstrates better localization performance benefits of the features of low layers (color, edge and texture features). But this method produce a reduced recall.

To combine both of features of low layers and high layers, a good object detection approach is proposed. For region proposal generation task, a region proposal network (RPN) based on multi-level feature fusion is proposed to combine shallow but high-resolution and deep but semantic CNN features. It decreases the number of object proposals and guarantee high recalls. For object detection task, a multi-scale CNN is presented to extract complex abstract feature of the image adaptively. Small size of convolution kernel is set to find more details of the underlying location information in the bottom of the convolution layer. A larger convolution kernel is set in the high-level convolution layer to detect the global features of category. Finally, multi-task loss is employed to establish a complex mapping between image object features and object detection mode. It generates less region proposals and improves the detection and classification accurate by elaborated multi-level convolution features.

2 The Proposed Object Detection Algorithm

2.1 Region Proposal Generation Using MFF-RPN

In the field of semantic segmentation, the Fully Convolution Network (FCN) has quite outstanding performance [6, 7]. In [6], the author divides the rough high-level information with the fine low-level information to semantic segmentation. Hyper-Net aggregates hierarchical feature maps to produce Hyper Feature. According to the above works, this paper develops a novel region proposal network that combines low-level with high-level information to produce better region proposals.

Figure 1 shows the design details of MFF-RPN. We take an image (with size 110 by 110 and 3 channels) as the input to illustrate. In conv1 layer, we set 256 different filters to convolve the entire raw image. Every filter is 7 by 7 with a stride of 2. The output feature maps is 55 by 55. Then these 256 convolved feature maps is fed into max pool layer. They are maxed with 3 by 3 regions and a stride of 2. This layer outputs 256 feature maps with size of 26 by 26. In conv2 layer, filters with size 5 by 5, 1 stride are adopt to produce 256 feature maps. Similar pooling operation is repeated in pool2 layer. In con3, conv4, conv5 layers, same size 3 by 3 of filters are employed to produce high convolution feature maps. Convolution

layers and pooling layers have different size of feature maps due to pooling operation. To make sure these feature maps at the same resolution, a deconvolutional
operation (Deconv) is adopted for conv3 layer to conduct upsampling. Max pooling
and Deconv operation can leave the image with more details of objects. Then conv4
and conv5 layers are applied to take these three feature maps as input to combine
these three feature maps (conv1, conv2 and conv3). These two layers produce more
abstract semantic features and condense these three feature maps into a uniform
space. Then, local response normalization (LRN) is utilized to normalize
multi-level feature maps. These feature maps are fed into a ROI pooling layer to
produce global features. The last two fully connected layers take features from ROI
pooling layer as input and output a 4096 dimensions vector. Finally, a classification
layer and a detection layer are connected to produce 300 proposals. The classification layer outputs the softmax probability of the objects in image. The softmax
probability is just estimated over "background" class and "non-background" class.
The detection layer produces final proposal position with four values. However,
these region proposals may highly overlap others. This can increase the complexity
of detection and reduce the accurate of our detection model. To reduce the number
of redundant proposals, non-maximum suppression (NMS) [8] is adopted according
to the probability of the proposals. In this paper, the intersection-over-union
(IoU) threshold is fixed at 0.7. NMS operation abandons the candidate proposal if
its IoU higher than 0.7. After NMS, the top-300 ranked region proposals are
selected for detection. The top-300 region proposals is used to train the detection
network, but different numbers of proposals are used to evaluate the detection
network.

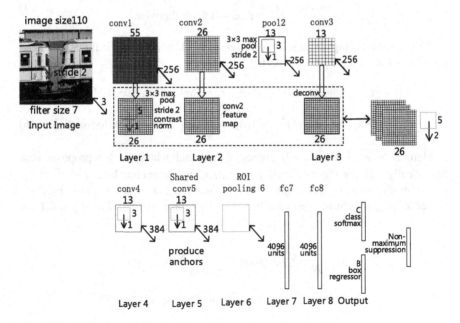

Fig. 1 The architecture of MFF-RPN

To train MFF-RPN, the proposed algorithm take the same objective loss function with SSD [9] model. In this work, IoU of candidate proposal is higher than 0.7, it is signed positive sample, while lower than 0.3, it is signed negative sample. The softmax probability is just estimated over the binary class label, "background" class and "non-background" class. For an anchor box i, its loss function is defined as:

$$L(p_i, t_i) = L_{cls}(p_i, p_i^*) + \lambda p_i^* L_{reg}(t_i, t_i^*) \tag{1}$$

Here p_i is the predicted softmax probability of the proposal i being "non-background" class. $p_i^* = 1$ if the proposal is a positive sample, and $p_i^* = 0$ if negative. $t_i^* = \{t_x^*, t_y^*, t_w^*, t_h^*\}_i$ represents the coordinates of the predicted proposal with 4 values, and $t_i = \{t_x, t_y, t_w, t_h\}_i$ represents the coordinates of the ground-truth proposal. The classification loss L_{cls} represents the softmax loss of two classes (non-background vs. background), it is defined as:

$$L_{cls}(p_i, p_i^*) = -\log[p_i p_i^* + (1 - p_i)(1 - p_i^*)] \tag{2}$$

Here, when the proposal is positive, the regression loss, $L_{reg}(t_i, t_i^*)$ is activated and is disabled otherwise. It is defined as:

$$L_{reg}(t_i, t_i^*) = R(t_i - t_i^*) \tag{3}$$

Here R is smooth-L1 function defined in [3].

$$smooth_{L_1}(x) = \begin{cases} 0.5x^2 & \text{if } |x| < 1 \\ |x| - 0.5 & \text{otherwise} \end{cases} \tag{4}$$

The loss-balancing parameter λ is set to 1. The parameterizations of the 4 coordinates are adopted following [1]:

$$t_x = (x - x_a)/w_a, \ t_y = (y - y_a)/h_a, \ t_w = \log(w/w_a), \ t_h = \log(h/h_a) \tag{5}$$

$$t_x^* = (x^* - x_a)/w_a, \ t_y^* = (y^* - y_a)/h_a, \ t_w^* = \log(w^*/w_a), \ t_h^* = \log(h^*/h_a) \tag{6}$$

Here x, y, w, h are top-left corner, height and width for the proposal box. Specifically,x is for the predicted proposal, x_a is for anchor box, and x^* is for ground-truth proposal. The regression loss function make the predicted proposal closer to the ground-truth proposal box. The loss function of an image j is defined as:

$$L(j) = \frac{1}{N_{cls}} \sum_i L_{cls}(p_i, p_i^*) + \lambda \frac{1}{N_{reg}} \sum_i p_i^* L_{reg}(t_i, t_i^*) \tag{7}$$

where N_{cls}, N_{reg} is the size of a mini-batch, in this paper, it is 256.

2.2 Object Detection Using Multi-scale CNN

After region proposals are generated by MFF-RPN, a CNN with 8 layers (5 conv layers, 1 ROI pooling layer and 2 fc layers) is employed to extract features of objects. Differ from normal CNNs, our CNN set a smaller convolution kernel to extract more details of the underlying location information in the bottom of the convolution layer and set a larger convolution kernel to detect the global features of category in the high-level convolution layer. The convolution kernel size of each convolutional layer is 3×3, 3×3, 5×5, 5×5, 7×7. The motivation of this operation is to obtain the adaptive complex abstract feature of the image. Same as MFF-RPN, the final convolution feature maps are classified and regressed based on the multi-task loss function.

The training of this CNN is a little bite different from MFF-RPN. Inspired by the MultiBox [10] and SSD model, we derive the detection objective. Let $g = \{g_0, \ldots, g_m\}$ be a discrete probability distribution over $m + 1$ categories. $x_{ij}^m = \{0, 1\}$ indicates that the i-th default box matches the j-th ground truth box of category m. It has the same meaning with p_i in MFF-RPN. The match strategy is IoU higher than 0.5 for positives and lower than 0.3 for negatives. So, the paper take the multi-task loss instead of the binary loss in the multi-task loss function. It is following as:

$$L = \frac{1}{N_{cls}} \sum_i L_{cls}^i(x, g) + \lambda \frac{1}{N_{reg}} \sum_i L_{reg}^i(x, d^m, d^t) \tag{8}$$

$d^m = \{d_x^m, d_y^m, d_w^m, d_h^m\}$, $d^t = \{d_x^t, d_y^t, d_w^t, d_h^t\}$ are for the predicted location, and ground-truth box (he coordinates in the annotation) respectively.

2.3 Joint Training

In this section, a joint training strategy is employed to share convolution features in both MFF-RPN and CNN. In practice, a 4-step training process same with Faster RCNN for joint optimization are developed.

Firstly, the MFF-RPN is trained as described above. The first three layers are initialized by the Zeiler and Fergus [11] pre-trained model. The ROI pooling layer and two fully connected layers are initialize by the Fast RCNN pre-trained model. Then MFF-RPN is fine-tuned for the region proposal generation task with dataset VOC2007 and 2012. The momentum term weight is fixed at 0.9 and the weight decay factor is 0.0005. MFF-RPN is training through 80 k mini-batches. For the first 60 k mini-batches, the learning rate is set at 0.005. At the last 20 k mini-batches, the learning rate reduce ten times than first 60 k mini-batches.

Secondly, the multi-scale CNN take the proposals by the step-1 MFF-RPN as training set. The initialized criterion is same as MFF-RPN. However, multi-scale CNN is initialized by different layers (all convolution layers and two fully connected layers, ROI pooling layer) of these two models. The training parameters are same as step1.

Thirdly, the convolution layers and the ROI pooling layer of the multi-scale CNN are utilized to initialize MFF-RPN. The convolution layers are fixed in step3 and the other layers of MFF-RPN are fine-tuned. MFF-RPN and multi-scale CNN can share convolution feature by this operation.

At the last step, the convolution layers of multi-scale CNN are fixed, the ROI pooling layer and two fully connected layers are fine-tuned. The proposals generated by the step-3 MFF-RPN is utilized to fine-tune in this step. Thus, these two networks share the same convolution features which benefits efficient performance.

3 Results and Analysis

In this section, both object proposal and object detection are evaluated comprehensively on the PASCAL VOC 2007 and VOC2012 detection datasets [12] with other popular detection methods. The proposed approach is implemented in MATLAB based on the wrapper of Caffe framework. It runs on a PC with a 4.0 GHz CPU and a GeForce GTX 1080 GPU. Our detection model are initialized with ZF model, and the Faster RCNN model. Recall is utilized to evaluate our MFF-RPN efficiency. This paper evaluates the whole detection model by Mean Average Precision (mAP).

3.1 Experimental Results on PASCAL VOC2007

As shown in Table 1, the conclusions are drawn by comparing Selective Search with MFF-RPN and the current popular RPN method. Recall represents the percentage of all targets being correctly detected. All models in the table are trained on the VOC2007 train + val sets and tested on test set. All models produce 300 proposals per image except Selective Search ("SS"). Each image can produce about 2000 proposals using Selective Search method.

Table 1 Results (Recall) on PASCAL VOC 2007 test set with MFF-RPN, SS, RPN

Method	IoUpos_thr	IoUneg_thr	Recall (%)
SS	–	–	57.79
RPN	0.7	0.3	76.85
MFF-RPN	0.7	0.3	**88.01**
	0.5	0.5	86.75
	0.5	0.3	87.46

Table 2 Results (mAP) on PASCAL VOC 2007 test set

Method	MAP	Aero	Bike	Bird	Boat	Bottle	Bus	Car	Cat	Chair	Cow
RCNN	58.5	68.1	72.8	56.8	43.0	36.8	66.3	74.2	67.6	34.4	63.5
Fast-R	66.9	**74.5**	78.3	69.2	53.2	36.6	**77.3**	78.2	**82.0**	40.7	**72.7**
SPP-N	59.2	68.6	69.7	57.1	41.2	40.5	66.3	71.3	72.5	34.4	67.3
Proposed	**70.7**	69.9	**80.6**	**70.1**	**57.3**	**49.9**	76.2	**79.5**	80.9	**50.1**	71.3

Method	Table	Dog	Horse	Mbike	Person	Plant	Sheep	Sofa	Train	Tv
RCNN	54.5	61.2	69.1	68.6	58.7	33.4	62.9	51.1	62.5	64.8
Fast-R	**67.9**	79.6	**79.2**	73.0	69.1	30.1	65.4	**70.2**	75.8	65.8
SPP-N	61.7	63.1	71.0	69.8	57.6	29.7	59.0	50.2	65.2	**68.0**
Proposed	66.2	**80.1**	78.5	**74.7**	**74.3**	**35.6**	**66.9**	65.5	**78.1**	65.6

The table shows that the proposed MFF-RPN performs quite well on PASCAL VOC 2007 benchmarks than other proposal generation methods. Specifically, Selective Search (57.79%) is inferior to the RPN's performance (76.85%). With 300 region proposals, MFF-RPN gets 88.01% recall, 11 points higher than RPN and 30 points higher than Selective Search. The table also shows MFF-RPN gets better performance when the IoU threshold of positive sample is 0.7, negative sample is 0.3 than the other sampling criterions (0.5IoU threshold for positives and 0.5IoU threshold for negatives, 0.5IoU threshold for positives and 0.3IoU threshold for negatives). This is because 0.5 IoU threshold for an candidate proposal is slightly small to recognize objects and backgrounds correctly. Thus, lower recall and bad localization performance of later detection network are produced.

For object detection, our approach is compared with R-CNN, Fast R-CNN and SPP-Net. The results are shown in Table 2. These models use bounding box regression to realize object detection. R-CNN gets 58.5% mAP and Fast R-CNN achieves 66.9%. SPP-Net gets a mAP of 59.2%. Our approach achieves a mAP of 70.7%, outperforming R-CNN by 12.2 points, SPP-Net by 11.5 points and Fast R-CNN by 3.8 points. As the results of Tables 1 and 2 shown, the proposed model has better detection performance due to MFF-RPN generates more accurate proposals than the other two methods. MFF-RPN benefits from lower and higher features, which makes for better object localization. As for "bird", "bottle", "chair", "dog", our network outperforms others because features of these objects have more details than other methods.

3.2 Experimental Results on PASCAL VOC2012

Our approach is compared with R-CNN, FAST R-CNN on PASCAL VOC 2012. The proposed detection model is initialized as described in Sect. 2.3. The other models are initialized by ZF model. The detection results are shown in Table 3. Our approach gets a mAP of 72.8% which is the top result on VOC 2012. It

Table 3 Results (mAP) on PASCAL VOC 2012 test set

Method	MAP	Aero	Bike	Bird	Boat	Bottle	Bus	Car	Cat	Chair	Cow
RCNN	62.4	79.6	72.7	61.9	41.2	41.9	65.9	66.4	84.6	38.5	67.2
Fast-R	68.4	80.3	74.7	66.9	46.9	37.7	**73.9**	68.6	**87.7**	41.7	71.1
Proposed	**72.8**	**81.3**	**76.0**	**69.8**	**47.6**	**44.0**	71.7	**72.3**	85.2	**42.2**	**73.7**
Method	Table	Dog	Horse	Mbike	Person	Plant	Sheep	Sofa	Train	Tv
RCNN	46.7	82.0	74.8	76.0	65.2	**35.6**	65.4	54.2	67.4	60.3
Fast-R	**51.1**	86.0	**77.8**	**79.8**	69.8	32.1	65.5	**63.8**	76.4	**61.7**
Proposed	48.5	**88.0**	76.9	77.5	**74.2**	34.5	**68.5**	58.4	**78.8**	59.3

outperforming R-CNN by 10.4 points and Fast R-CNN by 4.4 points. These data demonstrate that our object detection method has an excellent detection performance because MFF-RPN and multi-scale CNN can extract more details of the objects in image again. These experimental results indicate our network can detection object from images with high efficiency.

Some detection examples on VOC2007 and VOC2012 test sets are shown in Fig. 2. Our method takes the house behind the track as a train in pic.1. It also takes a doll and a human sculpture as real person in pic.2, pic.3. This is because our approach ignores the similarity between different categories. We can take similarity constraint into our future research work. In other pictures, our approach achieves excellent performance.

Fig. 2 Examples on VOC 2007 and VOC2012 test sets

4   Conclusions

A novel object detection method is presented to achieve higher object detection accuracy in this paper. MFF-RPN combines higher but abstract features which benefit object classification and lower but fine, high-resolution features which benefit object location. This network guarantees high recalls and decrease the number of region proposals that lead to higher detection efficiency. Multi-scale CNN extracts more details of the underlying location information in the bottom of the convolution layer and detects the global features of category in the high-level convolution layer. It can obtain the adaptive complex abstract feature of the image. The experimental results proves that the proposed algorithm achieves higher object detection accuracy on standard benchmarks due to these major improvements.

Acknowledgements   We would like to thank the supports by Chongqing Nature Science Foundation for Fundamental science and frontier technologies (cstc2015jcyjB0569), China Central Universities Foundation (106112016CDJZR175511).

References

1. Girshick R, Donahue J, Darrell T, et al. Rich feature hierarchies for accurate object detection and semantic segmentation. In: Computer vision and pattern recognition. IEEE;2014. p. 580 −87.
2. He K, Zhang X, Ren S, et al. Spatial pyramid pooling in deep convolutional networks for visual recognition. IEEE Trans Pattern Anal Mach Intell. 2015;37(9):1904.
3. Girshick R, Fast R-CNN. In: IEEE international conference on computer vision (CVPR). IEEE;2015. p. 1440−48.
4. Ren S, He K, Girshick R, et al. Faster R-CNN. Towards real-time object detection with region proposal networks. IEEE Trans Pattern Anal Mach Intell. (PAMI) 2015;1−1.
5. Ghodrati A, Diba A, Pedersoli M, et al. Deepproposal: Hunting objects by cascading deep convolutional layers. In: IEEE International conference on computer vision (CVPR);2015. p. 2578−86.
6. Long J, Shelhamer E, Darrell T. Fully convolutional networks for semantic segmentation. In: IEEE conference on computer vision and pattern recognition (CVPR);2015. p. 3431−40.
7. Hariharan B, Arbeláez P, Girshick R, et al. Hypercolumns for object segmentation and fine-grained localization. In: IEEE conference on computer vision and pattern recognition;2015. p. 447−56.
8. Neubeck A, Van Gool L. Efficient non-maximum suppression. In: International conference on pattern recognition. IEEE computer society;2006. p. 850−55.
9. Liu W, Anguelov D, Erhan D, et al. SSD: Single shot multibox detector. In: European conference on computer vision (ECCV). Springer International Publishing;2016. p. 21−37.
10. Erhan D, Szegedy C, Toshev A, et al. Scalable object detection using deep neural networks. In: IEEE conference on computer vision and pattern recognition (CVPR);2014. p. 2147−54.
11. Zeiler MD, Fergus R. Visualizing and understanding convolutional networks. In: European conference on computer vision. Springer International Publishing;2014. p. 818−33.
12. Everingham M, Eslami SMA, Van Gool L, et al. The pascal visual object classes challenge: a retrospective. Int J Comput Vis. 2015;111(1):98−136.

Following Consensus in Multi-vehicle Systems with Chain and Ring Coupling

Yiyi Liu, Lan Xiang and Jin Zhou

Abstract This paper investigates coordinated control problem of multi-vehicle systems by the use of the vehicle-following models with optimal velocity function. A concept of following consensus for multi-vehicle systems is first introduced, and two following consensus protocols for two topology cases of chain and ring coupling are then proposed, respectively. It is shown that the multi-vehicle systems with chain coupling are more easy to achieve the following consensus compared with the case of ring coupling. Subsequently, simulation examples illustrate and visualize the effectiveness and feasibility of the theoretical results.

Keywords Following consensus · Multi-vehicle systems · Vehicle-following model

1   Introduction

Traffic flow problems have received significant attention in both fundamental research and practical applications over the past few decades. One of the most impressive issues on traffic dynamics is traffic congestion or traffic jam [1]. As a result, many traffic flow models have been developed to describe various physical mechanisms of complex traffic phenomena, including the formation mechanism and propagating properties of various traffic waves and jams, the mechanisms of lane-changing and overtaking, the interactions among multiple vehicles [2]. Among existing traffic flow models, the developed vehicle-following model (or car-following

Y. Liu · J. Zhou(✉)
Shanghai Institute of Applied Mathematics and Mechanics,
Shanghai University, Shanghai 200072, China
e-mail: jzhou@shu.edu.cn

L. Xiang
Department of Physics, School of Science, Shanghai University,
Shanghai 200444, China

© Springer Nature Singapore Pte Ltd. 2018   353
Y. Jia et al. (eds.), *Proceedings of 2017 Chinese Intelligent Systems Conference*,
Lecture Notes in Electrical Engineering 460,
https://doi.org/10.1007/978-981-10-6499-9_34

model) with optimal velocity function is usually viewed as a fundamental model for the study of traffic flow dynamics. Accordingly, a wide variety of works related to the vehicle-following model have been presented from various perspectives in recent years [3–5].

In the past decades, coordinated behavior of multi-agent systems has been extensively investigated in various fields of science and engineering due to its many potential practical applications [6–9]. In particular, consensus problem has recently become a rather significant topic for the study of cooperative behavior of multi-agent systems, and a large variety of control strategies have been proposed for various kinds of consensus problems, including single-integrator and double-integrator consensus, and many others [9–12]. An important task undertaken in coordinated control is to design an appropriate consensus protocol or algorithm such that the group of agents can achieve a certain desired global cooperative behavior of common interest through the information exchange between agents. In addition, the vehicle-following model is usually formulated by the position and velocity between adjacent vehicles, which can well describe the traffic flow characteristics of vehicles following up one by one along single lane. Therefore, it is applicable to deal with coordinated behavior of multi-vehicle systems based on the vehicle-following models. However, so far, no relevant work fully addresses the cooperative behavior of multi-vehicle systems based on vehicle-following model. For practical reasons, it is still desirable to propose some simple yet effective consensus protocols for multi-vehicle systems. These observations motivate our present research in this paper.

This paper considers the coordinated behavior of multi-vehicle systems by the use of the vehicle-following models with optimal velocity function. Firstly, we present a concept of following consensus for multi-vehicle systems, i.e.,the position error between adjacent vehicles converge to a specific constant, while their velocity realizes complete consensus. Furthermore, we propose two following consensus protocols of multi-vehicle systems for two topology cases of chain and ring coupling, and then provide convergence analysis for such protocols, respectively. We also show that the multi-vehicle systems with chain coupling are more easy to achieve following consensus compared with the case of ring coupling. Finally, simulation examples illustrate and visualize the effectiveness and feasibility of the theoretical results.

The remainder of this paper is organized as follows. Section 2 presents problem formulation. Sections 3 and 4 give the main results of the following consensus of multi-vehicle systems for both cases with chain and ring coupling, respectively. In Sect. 5, illustrative examples are given to demonstrate the effectiveness of the proposed protocols. Finally, concluding remarks are made in Sect. 6.

2 Problem Formulation

Considering n vehicles moving in a one-dimensional Euclidean space, the dynamics behavior of each vehicle can be described by

$$\begin{cases} \dot{x}_i(t) = v_i(t), \\ \dot{v}_i(t) = u_i(t), \quad i = 1, 2, \cdots, n, \end{cases} \tag{1}$$

where $x_i(t)$, $v_i(t) \in R$ are the position and velocity of the ith vehicle at time t respectively, and $u_i(t)$ is the control input or protocol to be designed based on the information obtained by the ith vehicle from its neighbors and itself.

Let $\Delta x_i = x_{i-1}(t) - x_i(t)(i = 2, 3, \cdots, n)$ denote the relative position error between two adjacent vehicles $i - 1$ and i, and the control input $u_i(t)$ for the ith vehicle is given by [13]

$$u_i(t) = \alpha \left[V(\Delta x_i) - v_i(t) \right] + \beta(\Delta \dot{x}_i(t)), \tag{2}$$

where $\alpha > 0$ and $\beta \geq 0$ are the sensitivity coefficient, and the optimal velocity function can be chosen as follows:

$$V(x) = \begin{cases} 0, & \text{if} \quad x \in [0, h_{\text{stop}}), \\ v_{\text{max}} \dfrac{(x - h_{\text{stop}})^3}{h_{\text{stop}}^3 + (x - h_{\text{stop}})^3}, & \text{if} \quad x \in [h_{\text{stop}}, +\infty), \end{cases}$$

in which v_{max} is the maximum velocity, and h_{stop} is the minimum safe headway. It is easy to know that the optimal velocity function $V(x)$ possesses the following properties [13]:

(1) $V(x)$ is a non-negative continuous monotonically increasing function.

(2) $\lim_{x \to \infty} V(x) = v_{\text{max}}$, and $V(x)$ has an upper bound v_{max}.

(3) $V(x)$ is globally Lipschitz continuous, that is, assume that there exists a constant $\rho > 0$, such that for any two different numbers $x_1, x_2 \in R^+$,

$$|V(x_1) - V(x_2)| \leq \rho |x_1 - x_2|.$$

In this case, the multi-vehicle systems (1) with the control input (2) are the well-known full velocity difference (FVD) model.

By considering the aforementioned background, in this paper we investigate the following consensus problem of multi-vehicle systems by taking into account the vehicle-following models with optimal velocity function. Throughout this paper, the following definition will be used.

Definition 1 For the multi-vehicle systems (1), the control protocol u_i of the ith vehicle is said to solve asymptotically following consensus problem if

$$\lim_{t \to \infty} |x_{i-1}(t) - x_i(t)| = d, \quad \lim_{t \to \infty} |v_{i-1}(t) - v_i(t)| = 0, \quad i = 2, 3, \cdots, n,$$

where $d > 0$ is a specific constant.

The goal of following consensus scheme here is to implement some suitable control input protocols u_i ($i = 1, 2, \cdots, n$) based on the vehicle-following models with optimal velocity function, such that the following consensus of the multi-vehicle systems (1) can be realized in the sense of Definition 1.

3 Following Consensus with Chain Coupling

For the topology case of chain coupling, it is easy to see that all the vehicles in systems (1) move along a straight line. Using the proposed protocol (2), the dynamics of the multi-vehicle systems (1) can be described as

$$
\begin{cases}
\dot{x}_1(t) = v_1(t), \\
\dot{v}_1(t) = \alpha\left[V(d) - v_1(t)\right], \\
\dot{x}_i(t) = v_i(t), \\
\dot{v}_i(t) = \alpha\left[V(\Delta x_i) - v_i(t)\right] + \beta(\Delta \dot{x}_i(t)), \quad i = 2, 3, \cdots, n,
\end{cases}
\tag{3}
$$

where $V(d)$ is a given desired velocity with respect to the first vehicle with $d > h_{\text{stop}}$ for the purpose of practical control strategy. Actually, the first vehicle can be viewed as a virtual leader for the multi-vehicle systems (1).

By introducing the transformations $h_i(t) = \Delta x_i - d$, and $s_i(t) = \Delta \dot{x}_i$, the corresponding error systems can be written as

$$
\begin{cases}
\dot{h}_2(t) = s_2(t), \\
\dot{s}_2(t) = \alpha\left[V(d) - V(h_2(t) + d)\right] - (\alpha + \beta)s_2(t), \\
\dot{h}_i(t) = s_i(t), \\
\dot{s}_i(t) = \alpha\left[V(h_{i-1}(t) + d) - V(h_i(t) + d)\right] - (\alpha + \beta)s_i(t) + \beta s_{i-1}(t), \\
\qquad\qquad\qquad\qquad\qquad\qquad\qquad\qquad i = 3, 4, \cdots, n.
\end{cases}
\tag{4}
$$

Theorem 1 *Under the chain coupling topology, the multi-vehicle systems (1) with the control protocol (2) can always solve the following consensus problem asymptotically if $\alpha + \beta > 0$.*

Proof Consider the following Lyapunov function candidate

$$U_i(t) = \frac{1}{2}s_i^2(t) + \int_0^{h_i(t)} g(\xi)d\xi, \qquad i = 2, 3, \cdots, n, \tag{5}$$

where $g(\xi) = \alpha[V(\xi + d) - V(d)]$.

First for $i = 2$, taking derivative of U_2 with respect to t gives

$$\dot{U}_2(t) = -(\alpha + \beta)s_2^2(t) \le 0.$$

By using the well-known Krasovskii-Barlbasin's Theorem, it follows from (4) that $h_2(t) \to 0$ and $s_2(t) \to 0$ as $t \to \infty$.

Next for $i = 3$, the corresponding error systems (4) can be rewritten as

$$\begin{cases} \dot{h}_3(t) = s_3(t), \\ \dot{s}_3(t) = \alpha\left[V(d) - V(h_3(t) + d)\right] - (\alpha + \beta)s_3(t) + \delta_3(t), \end{cases}$$

where $\delta_3(t) = \alpha[V(h_2(t) + d) - V(d)] + \beta s_2(t) \to 0$ as $t \to \infty$.

Accordingly, by employing the Young's inequality, and taking derivative of U_3 with respect to t gives again

$$\dot{U}_3(t) = -(\alpha + \beta - \varepsilon)s_3^2(t) + \frac{1}{\varepsilon}\delta_3^2(t),$$

where ε is a small enough constant satisfying $0 < \varepsilon < \alpha + \beta$. It then follows from (4) that $h_3(t) \to 0$ and $s_3(t) \to 0$ as $t \to \infty$.

Finally, similarly to the above analysis procedure, it can be concluded that $h_i(t) \to 0$ and $s_i(t) \to 0$ for $i = 2, 3, \cdots, n$, as $t \to \infty$. This completes the proof.

Remark 1 Theorem 1 shows that the multi-vehicle systems (1) with chain coupling under the control protocol (2) can always achieve the following consensus provided that $\alpha + \beta > 0$ and $d > h_{\text{stop}}$.

4 Following Consensus with Ring Coupling

For the case of ring coupling topology, it is easy to see that all the vehicles in the multi-vehicle systems (1) move along a ring lane with length of L. Different from the case of chain coupling, it require to consider the following consensus problem of multi-vehicle systems with the leaderless.

In order to realize the following consensus of the multi-vehicle systems (1), a prerequisite constraint condition must be given by $\sum_{i=1}^{n} \Delta x_i(t) = L$. To do so, we introduce another following consensus protocol below,

$$u_i(t) = \alpha\left[V(\Delta x_i) - v_i(t)\right] + \beta(\Delta \dot{x}_i(t)) + k(\Delta x_i(t) - d), \ i = 1, 2, \cdots, n, \qquad (6)$$

where k is a positive feedback constant, and $d = L/n$ with $\Delta x_1(t) = L - \sum_{i=2}^{n} \Delta x_i(t)$.

Thus, the corresponding dynamics for the multi-vehicle systems (1) can be written as

$$\begin{cases} \dot{x}_i(t) = v_i(t), \\ \dot{v}_i(t) = \alpha\left[V(\Delta x_i) - v_i(t)\right] + \beta(\Delta \dot{x}_i(t)) + k(\Delta x_i(t) - d), i = 1, 2, \cdots, n. \end{cases} \qquad (7)$$

By employing the same transformations $h_i(t)$ and $s_i(t)$ ($i = 1, 2, \cdots, n$) as above, the error systems with respect to (7) can be given by

$$\begin{cases} \dot{h}_1(t) = s_1(t), \\ \dot{s}_1(t) = \alpha\left[V(h_n(t) + d) - V(h_1(t) + d)\right] - (\alpha + \beta)s_1(t) + \beta s_n(t) \\ \qquad\qquad + k(h_n(t) - h_1(t)), \\ \dot{h}_i(t) = s_i(t), \\ \dot{s}_i(t) = \alpha\left[V(h_{i-1}(t) + d) - V(h_i(t) + d)\right] - (\alpha + \beta)s_i(t) + \beta s_{i-1}(t) \\ \qquad\qquad + k(h_{i-1}(t) - h_i(t)), \qquad\qquad\qquad i = 2, 3, \cdots, n. \end{cases} \qquad (8)$$

Theorem 2 *Under the ring coupling topology, the multi-vehicle systems* (1) *with the control protocol* (6) *can always solve the following consensus problem asymptotically if the matrices $A \in R^{(n-1)\times(n-1)}$ and $C - BA^{-1}B^T \in R^{(n-1)\times(n-1)}$ are negative definite, where $A = b_5 E_{n-1}$ and*

$$B = \begin{bmatrix} \dfrac{b_3}{2} - \dfrac{b_4}{2} & \dfrac{b_4}{2} & \cdots & 0 & 0 \\ -\dfrac{b_4}{2} & \dfrac{b_3}{2} & \cdots & 0 & 0 \\ \vdots & \vdots & \ddots & \vdots & \vdots \\ -\dfrac{b_4}{2} & 0 & \cdots & \dfrac{b_3}{2} & \dfrac{b_4}{2} \\ -\dfrac{b_4}{2} & 0 & \cdots & 0 & \dfrac{b_3}{2} \end{bmatrix}, \quad C = \begin{bmatrix} b_1 - b_2 & 0 & \cdots & -\dfrac{b_2}{2} & -\dfrac{b_2}{2} \\ 0 & b_1 & \cdots & 0 & 0 \\ \vdots & \vdots & \ddots & \vdots & \vdots \\ -\dfrac{b_2}{2} & 0 & \cdots & b_1 & \dfrac{b_2}{2} \\ -\dfrac{b_2}{2} & 0 & \cdots & \dfrac{b_2}{2} & b_1 \end{bmatrix},$$

in which $b_1 = -(\alpha + \beta - \omega_1 - \alpha\rho\frac{1}{\varepsilon}), b_2 = \beta - \omega_1, b_3 = k + \omega_1(\alpha + \beta) - \omega_2,$
$b_4 = -\omega_1\beta + \omega_2$ *and* $b_5 = -\omega_1 k + \alpha\rho\varepsilon$ *with three suitable chosen positive constant* ε, ω_1 *and* ω_2 *satisfying* $\omega_2 \geq \omega_1^2$.

Proof Consider the following Lyapunov function candidate

$$U(t) = \sum_{i=2}^{n} U_i(t), \tag{9}$$

where

$$U_i(t) = \frac{1}{2}s_i^2(t) - \omega_1 s_i(t)(h_{i-1}(t) - h_i(t)) + \frac{1}{2}\omega_2(h_{i-1}(t) - h_i(t))^2.$$

Noting that $U_i(t)$ is a positive definite function with respect to $s_i(t)$ and $(h_{i-1}(t) - h_i(t))$ if $\omega_2 \geq \omega_1^2$.

In view of the property of the velocity function $V(x)$, and by combining the Young's inequality, taking derivative of U with respect to t gives

$$\dot{U}(t) = \sum_{i=2}^{n} \dot{U}_i(t) \leq \mathbf{r}^T \begin{bmatrix} A & B^T \\ B & C \end{bmatrix} \mathbf{r},$$

where $\mathbf{r} = (h_1 - h_2, h_2 - h_3, \ldots, h_{n-1} - h_n, s_2, s_3, \ldots, s_n)^T$.

It follows that $h_{i-1}(t) - h_i(t) (i = 2, 3, \cdots, n) \to 0$ and $s_i(t) \to 0 (i = 2, 3, \cdots, n)$ as $t \to \infty$. Since $\sum_{i=1}^{n} h_i(t) = 0$ and $\sum_{i=1}^{n} s_i(t) = 0$, it is easy to see that $h_i(t) \to 0$ and $s_i(t) \to 0$ for $i = 1, 2, \cdots, n$ as $t \to \infty$. Clearly, this completes the proof.

Remark 2 It should be noted that the control protocol (6) is an extension of the following consensus protocol (2) from chain coupling to ring topology. It is obvious that Theorem 1 is less restrictive than Theorem 2, which implies that the multi-vehicle systems with chain coupling are more easy to achieve following consensus than the case of ring coupling. This point will be further illustrated through numerical simulation section.

5   Simulation Examples

Two simulation examples are worked out to demonstrate the effectiveness of the theoretical results. For our convenience, we here consider a team of five vehicles moving along single lane for two topology cases both chain and ring coupling, respectively. For the reason of convenient comparison, the following consensus performance between chain and ring coupling, in the case of chain coupling we choose the relative parameters $\alpha = 0.9$, $\beta = 0.2$, $d = 30$, $v_{max} = 11$, $h_{stop} = 14$. While in

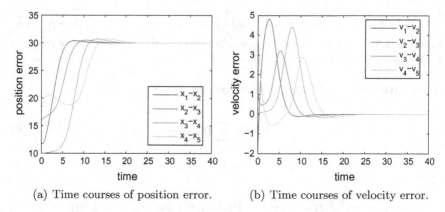

(a) Time courses of position error.   (b) Time courses of velocity error.

Fig. 1 The following consensus process in the chain coupling

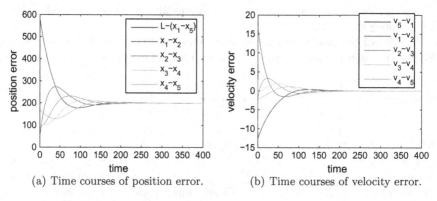

(a) Time courses of position error.   (b) Time courses of velocity error.

Fig. 2 The following consensus process in the ring coupling

the case of ring coupling we select the parameters $\alpha = 13$, $\beta = 0.03$, $k = 0.4$, $L = 1000$, $v_{\max} = 11$, $h_{\text{stop}} = 14$. It is easy to verify that all the conditions of Theorem 2 are satisfied. Figures 1 and 2 visualize the following consensus process of the multi-vehicle systems (1) of five vehicles for two cases chain and ring coupling, respectively. It can be seen that the multi-vehicle systems with chain coupling are more easy to realize the following consensus in contrast to the case of ring coupling. This conclusion is obviously in consistence with the above theoretical results.

6 Conclusions

In this paper, we have studied the following consensus issues of multi-vehicle systems based on the vehicle-following models with optimal velocity function. A concept of following consensus for multi-vehicle systems is first presented to characterize the coordinated behavior of the traffic flow for multi-vehicle systems along single lane. Two control protocols have been proposed to solve the following consensus problems for the topology cases both chain and ring coupling, respectively. It is shown that the multi-vehicle systems with chain coupling are more easy to achieve the following consensus in contrast to the case of ring coupling. Finally, simulation examples are also presented to verify the effectiveness of the theoretical results.

Acknowledgments This work is supported by the National Science Foundation of China (Grant Nos. 11672169 and 11272191).

References

1. Bando M, Hasebe K, Nakayama A, Shibata A, Sugiyama Y. Dynamical model of traffic congestion and numerical simulation. Phys Rev E. 1995;51(2):1035–42.
2. Tang TQ, Huang HJ, Xu G. A new macro model with consideration of the traffic interruption probability. Phys A Stat Mech Appl. 2008;387(27):6845–56.
3. Peng GH, Cheng RJ. A new car-following model with the consideration of anticipation optimal velocity. Physi A Stat Mech Appl. 2013;392(17):3563–9.
4. ShaoWei Yu, Liu QL, Li XH. Full velocity difference and acceleration model for a car-following theory. Commun Nonlinear Sci Numer Simul. 2013;18(5):1229–34.
5. Sipahi R, Niculescu S, Abdallah CT, Michiels W, Gu K. Stability and stabilization of systems with time delay. IEEE Control Syst. 2011;31(1):38–65.
6. Cao YC, Yu WW, Ren W, Chen GR. An overview of recent progress in the study of distributed multi-agent coordination. IEEE Trans Ind Informatics. 2013;9(1):427–38.
7. Qin JH, ChangBin Yu. Cluster consensus control of generic linear multi-agent systems under directed topology with acyclic partition. Automatica. 2013;49(9):2898–905.
8. Yu J, Wang L. Group consensus of multi-agent systems with undirected communication graphs, Proceedings of 2009 7th Asian Control Conference 2009; p. 105–10.
9. Liu J, Zhou J. Distributed impulsive group consensus in second-order multi-agent systems under directed topology. Int J Control. 2014;59(6):1–15.
10. Mei J, Ren W, Ma GF. Distributed containment control for lagrangian networks with parametric uncertainties under a directed graph. Automatica. 2012;48(4):653–9.
11. Liu J, Ji JC, Zhou J, Xiang L, Zhao LiYun. Adaptive group consensus in uncertain networked Euler-Lagrange systems under directed topology. Nonlinear Dyn. 2015;82(3):1145–57.
12. WenWu Yu, Chen GR, Cao M, Kurths J. Second-order consensus for multiagent systems with directed topologies and nonlinear dynamics. IEEE Trans Syst Man Cybern Part B. 2010;40(3):881–91.
13. Jiang R, Wu QS, Zhu ZJ. Full velocity difference moldel for a car-following theory. Phys Rev E. 2001;64(1):7101–4.

Tracking Control of Quad-Rotor Based on Non-regular Feedback Linearization

Yan Lixia and Ma Baoli

Abstract This paper addresses the tracking control problem of quad-rotor moving in three-dimensional space. The tracking controller is based upon the use of suitable input transformation and non-regular feedback linearization technique. Two steps are involved in control design. In the first step, two control inputs are respectively designed for height and yaw angle, guaranteeing their tracking errors asymptotically converge to zero. In the second step, the control inputs for roll and pitch angles are employed to govern the rest two position tracking errors by substituting designed height and yaw control inputs into the remaining position tracking errors, with the remaining position tracking errors holding a feedback linearized form in their fourth derivatives. Finally, simulations are provided to validate the proposed control scheme.

Keywords Quad-rotor · Non-regular feedback linearization · Tracking control

1 Introduction

Research on quad-rotor, being one kind of VTOL unmanned aerial vehicles, has attracted worldwide attentions due to its potential applications such as aerial photo, surveillance and geographical mappings [1, 2].

Many techniques can be used to design flight control algorithms for quad-rotor. Considering a simplified linearized model of quad-rotor, a classical PID and an LQ stabilization controller are proposed in [3]. Control designs for quad-rotor derived from backstepping methods can be seen in [4, 5]. In [6], an adaptive tracking controller based on immersion and invariance methodology is proposed. Two different adaptive sliding-mode configurations are respectively shown in [7, 8]. Via solving required roll and pitch angles to regulate position errors, controllers proposed in

Y. Lixia · M. Baoli(✉)
The Seventh Research Division, School of Automation Science
and Electrical Engineering, Beihang University, Beijing 100191, China
e-mail: mabaoli@buaa.edu.cn

Y. Lixia
e-mail: yanlixia@buaa.edu.cn

© Springer Nature Singapore Pte Ltd. 2018
Y. Jia et al. (eds.), *Proceedings of 2017 Chinese Intelligent Systems Conference*,
Lecture Notes in Electrical Engineering 460,
https://doi.org/10.1007/978-981-10-6499-9_35

[9, 10] transform trajectory tracking task into an attitude tracking task. Besides, the feedback linearization is another popular and useful technique in quad-rotor control. Ignoring most nonlinear terms, a feedback linearization controller is introduced in [8]. By designing a nested structure with attitude control being inner loop and position control being outer loop, a feedback linearization control scheme is proposed in [11]. Through calculation of Lie bracket, a feedback linearization control law is reported in [12]. However, normal feedback linearization method adopted in quad-rotor controls either omits many nonlinear terms or increases numbers of states, which may contribute to bad performance of quad-rotor. Motivated by problems exist in normal feedback linearization, we use non-regular feedback linearization method to achieve tracking control of quad-rotor, simultaneously achieving full-state feedback control and avoiding increasing state numbers. The detail discussion and some illustrations about non-regular feedback linearization can be seen in [13].

The paper is organized as follows. Section 2 contains problem formation, controller design is included in Sect. 3, simulation results and conclusion are presented in Sects. 4 and 5 respectively.

2 Problem Formation

This section considers quad-rotor configuration, some assumptions and problem formation.

Let (x, y, z) denote the position in 3-D space and z represent height, then the dynamic model of quad-rotor can be described by [4]

$$\begin{cases} \ddot{\phi} = p_{11}\dot{\theta}\dot{\psi} - p_{12}\dot{\theta}\Omega + p_{13}U_2 \overset{\Delta}{=} \bar{U}_2 \\ \ddot{\theta} = p_{21}\dot{\phi}\dot{\psi} + p_{22}\dot{\phi}\Omega + p_{23}U_3 \overset{\Delta}{=} \bar{U}_3 \\ \ddot{\psi} = p_{31}\dot{\phi}\dot{\theta} + p_{33}U_4 \overset{\Delta}{=} \bar{U}_4 \\ \ddot{z} = -g + \cos\phi\cos\theta\dfrac{1}{m}U_1 \\ \ddot{x} = (\cos\phi\sin\theta\cos\psi + \sin\phi\sin\psi)U_1/m \\ \ddot{y} = (\cos\phi\sin\theta\sin\psi - \sin\phi\cos\psi)U_1/m \end{cases} \tag{1}$$

where

$$\begin{aligned} &p_{11} = (I_y - I_z)/I_x, p_{12} = J_r/I_x, p_{13} = l/I_x, p_{21} = (I_z - I_x)/I_y, p_{22} = J_r/I_y, \\ &p_{23} = l/I_y, p_{31} = (I_x - I_y)/I_z, p_{33} = 1/I_z, \Omega = \Omega_2 + \Omega_4 - \Omega_1 - \Omega_3 \end{aligned} \tag{2}$$

and symbol definitions in (2) are
    The posed control inputs are U_1, U_2, U_3, U_4, derived from

$$\begin{aligned} U_1 &= b(\Omega_1^2 + \Omega_2^2 + \Omega_3^2 + \Omega_4^2), U_2 = b(\Omega_4^2 - \Omega_2^2) \\ U_3 &= b(\Omega_3^2 - \Omega_1^2), U_4 = d(\Omega_2^2 + \Omega_4^2 - \Omega_1^2 - \Omega_3^2) \end{aligned} \tag{3}$$

Symbol	Definition	Symbol	Definition
ϕ	Roll angle	b	Thrust factor
θ	Pitch angle	d	Drag factor
ψ	Yaw angle	l	Lever
Ω_i	Rotor speed of motor $i, i = 1, 2, 3, 4$	m	Mass
$I_{x,y,z}$	Body inertia	g	Gravitational acceleration
J_r	Rotor inertia		

Firstly, choose $\bar{U}_2, \bar{U}_3, \bar{U}_4$ as to-be-designed new inputs by input transformation and the inverse transformation is

$$U_2 = [\bar{U}_2 - (p_{11}\dot{\theta}\dot{\psi} - p_{12}\dot{\theta}\Omega)]/p_{13}, U_3 = [\bar{U}_3 - (p_{21}\dot{\phi}\dot{\psi} + p_{22}\dot{\phi}\Omega)]/p_{23}$$
$$U_4 = [\bar{U}_4 - p_{31}\dot{\phi}\dot{\theta}]/p_{33} \tag{4}$$

The input transformation for U_1 will be emphasized in the sequel. Let reference states be (ψ_r, z_r, x_r, y_r) with appropriate order of derivatives as stated in the following assumption.

Assumption 1 Reference states and derivatives $\psi_r, \dot{\psi}_r, \ddot{\psi}_r, z_r, \dot{z}_r, z_r^{(3)}, z_r^{(4)}, x_r,$ $\dot{x}_r, \ddot{x}_r, x_r^{(3)}, x_r^{(4)}, y_r, \dot{y}_r, \ddot{y}_r, y_r^{(3)}, y_r^{(4)}$ are bounded and $|\ddot{z}_r|$ satisfies $|\ddot{z}_r| < g$.

Define tracking errors by $e_\psi = \psi - \psi_r, e_z = z - z_r, e_x = x - x_r, e_y = y - y_r$. The first-order derivatives of tracking errors are

$$\dot{e}_\psi = \dot{\psi} - \dot{\psi}_r, \dot{e}_z = \dot{z} - \dot{z}_r, \dot{e}_x = \dot{x} - \dot{x}_r, \dot{e}_y = \dot{y} - \dot{y}_r \tag{5}$$

Assumption 2 Roll angle ϕ and pitch angle θ satisfy $|\phi| < \dfrac{\pi}{2}, |\theta| < \dfrac{\pi}{2}$.

Suppose that Assumptions 1 and 2 hold, the objective of this paper is to regulate $(e_\psi, \dot{e}_\psi, e_z, \dot{e}_z, e_x, \dot{e}_x, e_y, \dot{e}_y)$ asymptotically converge to zero.

3 Controller Design

Tracking controller for quad-rotor is proposed in this section. Under Assumption 2, let $\bar{U}_1 \overset{\Delta}{=} \ddot{z}$, then we can adopt $\bar{U}_1$ as one input since $U_1 = m(g + \bar{U}_1)/(\cos\phi\cos\theta)$. By (5), we obtain $(\ddot{e}_\psi, \ddot{e}_z, \ddot{e}_x, \ddot{e}_y)$ by

$$\ddot{e}_\psi = \bar{U}_4 - \ddot{\psi}_r$$
$$\ddot{e}_z = \bar{U}_1 - \ddot{z}_r$$
$$\begin{bmatrix} \ddot{e}_x \\ \ddot{e}_y \end{bmatrix} = C(\psi) \begin{bmatrix} \tan\theta \\ \tan\phi/\cos\theta \end{bmatrix} f_z - \begin{bmatrix} \ddot{x}_r \\ \ddot{y}_r \end{bmatrix} \tag{6}$$

where $f_z = g + \bar{U}_1 + \ddot{z}_r$. Shown as (1), the six states $(\phi, \theta, \psi, z, x, y)$ are regulated only by four control inputs (U_1, U_2, U_3, U_4). This is a underactuated control problem. To solve this problem, we adopt non-regular feedback linearization method to design controller. More precisely, we firstly design $\bar{U}_1$ and $\bar{U}_4$ to regulate $(e_z, \dot{e}_z, e_\psi, \dot{e}_\psi)$, and substitute $(\bar{U}_1, \bar{U}_4)$ into $(\ddot{e}_x, \ddot{e}_y)$. Then, by calculating $(e_x^{(3)}, e_y^{(3)})$ and $(e_x^{(4)}, e_y^{(4)})$, whose expressions contains inputs$(\bar{U}_2, \bar{U}_3)$, we can construct appropriate $(\bar{U}_2, \bar{U}_3)$ to ensure (e_x, e_y)-dynamics asymptotically stable. We devote to divide control design into following two steps.

Step 1. *Design* $\bar{U}_1, \bar{U}_4$ *to stabilize* $(e_\psi, \dot{e}_\psi, e_z, \dot{e}_z)$.

Suppose that Assumptions 1 and 2 hold, we firstly propose controllers $\bar{U}_1, \bar{U}_4$ by

$$\begin{aligned}
\bar{U}_1 &= \ddot{z}_r - k_{z1} \arctan e_z - k_{z2} \arctan \dot{e}_z \\
\bar{U}_4 &= \ddot{\psi}_r - k_{\psi 1} \arctan e_\psi - k_{\psi 2} \arctan \dot{e}_\psi
\end{aligned} \tag{7}$$

where control gains $k_{\psi 1}, k_{\psi 2} > 0$ and $k_{z1}, k_{z2} > 0$, satisfying $k_{z1} + k_{z2} < 2(g - |\ddot{z}_r|)/\pi$. Error dynamics $\ddot{e}_z, \ddot{e}_\psi$ become

$$\ddot{e}_z = -k_{z1} \arctan e_z - k_{z2} \arctan \dot{e}_z, \ddot{e}_\psi = -k_{\psi 1} \arctan e_\psi - k_{\psi 2} \arctan \dot{e}_\psi \tag{8}$$

Choose a Lyapunov candidate $V_\rho = k_{\varepsilon 1} \rho \arctan \rho - 0.5 k_{\rho 1} \ln\left(1 + \rho^2\right) + 0.5 \dot{\rho}^2$ and get its derivative by $\dot{V}_\rho = -k_{\varepsilon 2} \dot{\rho} \arctan \dot{\rho}$, where symbol ρ denotes e_z or e_ψ, symbol ε denotes z or ψ. LaSalle's theorem can then be used to show that $\rho, \dot{\rho}$ asymptotically converge to zero. Thus, we have

$$\lim_{t \to \infty} \left(e_z, \dot{e}_z\right) = 0, \lim_{t \to \infty} \left(e_\psi, \dot{e}_\psi\right) = 0 \tag{9}$$

Furthermore, since $\bar{U}_1 = \ddot{z}_r - k_{z1} \arctan e_z - k_{z2} \arctan \dot{e}_z$ and $k_{z1} + k_{z2} < 2(g - |\ddot{z}_r|)/\pi$, we know that $f_z = g + \bar{U}_1 > 0, \forall t \geq 0$. Also, $\bar{U}_1$ is continuous and at least twice differentiable with respect to t. So $\dot{f}_z, \ddot{f}_z$ can be given by

$$\begin{aligned}
\dot{f}_z &= z_r^{(3)} - k_{z1} \frac{\dot{e}_z}{1 + e_z^2} + k_{z1} k_{z2} \frac{\arctan e_z}{1 + e_z^2} + k_{z2}^2 \frac{\arctan \dot{e}_z}{1 + \dot{e}_z^2} \\
\ddot{f}_z &= z_r^{(4)} - k_{z1} \frac{\ddot{e}_z \left(1 + e_z^2\right) - 2 e_z \dot{e}_z^2}{\left(1 + e_z^2\right)^2} + k_{z1} k_{z2} \frac{\dot{e}_z}{\left(1 + e_z^2\right)\left(1 + \dot{e}_z^2\right)} \\
&\quad - k_{z1} k_{z2} \frac{2 \dot{e}_z \ddot{e}_z \arctan e_z}{\left(1 + \dot{e}_z^2\right)^2} + k_{z2}^2 \frac{\ddot{e}_z - 2 \dot{e}_z \ddot{e}_z \arctan \dot{e}_z}{\left(1 + \dot{e}_z^2\right)^2}
\end{aligned} \tag{10}$$

Note that f_z is a function of $(e_z, \dot{e}_z)$, we omit substituting $\ddot{e}_z = -k_{z1} \arctan(e_z) - k_{z2} \arctan(\dot{e}_z)$ for brevity.

Step 2. *Design* $\bar{U}_2, \bar{U}_3$ *to stabilize* $(e_x, \dot{e}_x, e_y, \dot{e}_y)$.

From (6), $(e_x^{(3)}, e_y^{(3)})$ are given by

$$
\begin{bmatrix} e_x^{(3)} \\ e_y^{(3)} \end{bmatrix} = - \begin{bmatrix} x_r^{(3)} \\ y_r^{(3)} \end{bmatrix} + \dot\psi \frac{\partial C}{\partial \psi} \begin{bmatrix} \tan\theta \\ \tan\phi/\cos\theta \end{bmatrix} f_z + C(\psi) \begin{bmatrix} \tan\theta \\ \tan\phi/\cos\theta \end{bmatrix} \dot f_z
$$
$$
+ C(\psi) \begin{bmatrix} 0 \\ \frac{\dot\phi}{\cos^2\phi\cos\theta} \end{bmatrix} f_z + C(\psi) \begin{bmatrix} \dot\theta/\cos^2\theta \\ \frac{\dot\theta\tan\theta\tan\phi}{\cos\theta} \end{bmatrix} f_z
$$
(11)

where $\dfrac{\partial C}{\partial \psi} = \begin{bmatrix} -\sin\psi & \cos\psi \\ \cos\psi & \sin\psi \end{bmatrix}$. A further computation yields

$$
\begin{bmatrix} e_x^{(4)} \\ e_y^{(4)} \end{bmatrix} = - \begin{bmatrix} x_r^{(4)} \\ y_r^{(4)} \end{bmatrix} + \ddot\psi \frac{\partial C}{\partial \psi} \begin{bmatrix} \tan\theta \\ \tan\phi/\cos\theta \end{bmatrix} f_z - \dot\psi^2 C(\psi) \begin{bmatrix} \tan\theta \\ \tan\phi/\cos\theta \end{bmatrix} f_z
$$
$$
+ 2\dot\psi \frac{\partial C}{\partial \psi} \begin{bmatrix} \tan\theta \\ \tan\phi/\cos\theta \end{bmatrix} \dot f_z + 2\dot\psi \frac{\partial C}{\partial \psi} \begin{bmatrix} \dot\theta/\cos^2\theta \\ \frac{\dot\phi}{\cos^2\phi\cos\theta} + \frac{\dot\theta\tan\theta\tan\phi}{\cos\theta} \end{bmatrix} f_z
$$
$$
+ 2C(\psi) \begin{bmatrix} \dot\theta/\cos^2\theta \\ \frac{\dot\phi}{\cos^2\phi\cos\theta} + \frac{\dot\theta\tan\theta\tan\phi}{\cos\theta} \end{bmatrix} \dot f_z + C(\psi) \begin{bmatrix} \tan\theta \\ \tan\phi/\cos\theta \end{bmatrix} \ddot f_z
$$
$$
+ C(\psi) \begin{bmatrix} 0 \\ \frac{2\dot\phi\tan\phi+\dot\theta\tan\theta}{\cos^2\phi\cos\theta} \end{bmatrix} f_z \dot\phi
$$
$$
\underbrace{+ C(\psi) \begin{bmatrix} 2\dot\theta\tan\theta/\cos^2\theta \\ \frac{\dot\phi\cos\theta\sin\theta+\dot\theta\sin\phi\cos\phi(1+\sin^2\theta)}{\cos^3\theta\cos^2\phi} \end{bmatrix} f_z \dot\theta}_{\Xi_1}
$$
(12)

$$
+ C(\psi) \begin{bmatrix} 0 \\ 1 \\ \cos^2\phi\cos\theta \end{bmatrix} f_z \ddot\phi + C(\psi) \begin{bmatrix} 1/\cos^2\theta \\ \frac{\tan\theta\tan\phi}{\cos\theta} \end{bmatrix} f_z \ddot\theta
$$
$$
= - \begin{bmatrix} x_r^{(4)} \\ y_r^{(4)} \end{bmatrix} + \Xi_1 + \begin{bmatrix} \cos\psi & \sin\psi \\ \sin\psi & -\cos\psi \end{bmatrix} \left\{ \begin{bmatrix} 0 \\ \frac{1}{\cos^2\phi\cos\theta} \end{bmatrix} \ddot\phi + \begin{bmatrix} 1/\cos^2\theta \\ \frac{\tan\theta\tan\phi}{\cos\theta} \end{bmatrix} \ddot\theta \right\} f_z
$$
$$
= - \begin{bmatrix} x_r^{(4)} \\ y_r^{(4)} \end{bmatrix} + \Xi_1 + A f_z \begin{bmatrix} \bar U_2 \\ \bar U_3 \end{bmatrix}
$$

where

$$
A = \begin{bmatrix} \dfrac{\sin\psi}{\cos^2\phi\cos\theta} & \dfrac{\cos\psi}{\cos^2\theta} + \dfrac{\sin\psi\tan\theta\tan\phi}{\cos\theta} \\[2mm] \dfrac{-\cos\psi}{\cos^2\phi\cos\theta} & \dfrac{\sin\psi}{\cos^2\theta} - \dfrac{\cos\psi\tan\theta\tan\phi}{\cos\theta} \end{bmatrix}, \quad \frac{\partial^2 C}{\partial\psi^2} = -C(\psi)
$$
(13)

We propose the following feedback linearization controller

$$
\begin{bmatrix} \bar U_2 \\ \bar U_3 \end{bmatrix} = A^{-1} f_z^{-1} \left(-\Xi_1 + \begin{bmatrix} -k_1 e_x - k_2 \dot e_x - k_3 \ddot e_x - k_4 e_x^{(3)} \\ -k_1 e_y - k_2 \dot e_y - k_3 \ddot e_y - k_4 e_y^{(3)} \end{bmatrix} + \begin{bmatrix} x_r^{(4)} \\ y_r^{(4)} \end{bmatrix} \right)
$$
(14)

where

$$A^{-1} = \begin{bmatrix} (\cos^2\phi \cos\theta \sin\psi & (-\cos^2\phi \cos\theta \cos\psi \\ -0.25 \sin 2\phi \sin 2\theta \cos\psi) & -0.25 \sin 2\phi \sin 2\theta \sin\psi) \\ \cos^2\theta \cos\psi & \cos^2\theta \sin\psi \end{bmatrix} \quad (15)$$

Additionally, the combination of (14) and (12) implies

$$e_x^{(4)} = -k_1 e_x - k_2 \dot{e}_x - k_3 \ddot{e}_x - k_4 e_x^{(3)}, e_y^{(4)} = -k_1 e_y - k_2 \dot{e}_y - k_3 \ddot{e}_y - k_4 e_y^{(3)} \quad (16)$$

where control gains k_1, k_2, k_3, k_4 should be chosen such that the above linear systems are asymptotically stable. Then we know that $e_X = [e_x, \dot{e}_x, \ddot{e}_x, e_x^{(3)}]^T, e_Y = [e_y, \dot{e}_y, \ddot{e}_y, e_y^{(3)}]^T$ vanish asymptotically to zero. Summarizing the above results in two steps, the following Theorem can be obtained.

Theorem 1 *Consider the quad-rotor system (1) under control of (7) and (14), suppose that Assumptions 1 and 2 hold. Then, error states $(e_z, \dot{e}_z, e_\psi, \dot{e}_\psi, e_X, e_Y)$ asymptotically converge to zero.*

Remark 1 Assumption 2 is reasonable because most quad-rotors work with roll and pitch angles no more than $\pi/2$. Actually, we know that $\ddot{e}_x, \ddot{e}_y$ stay bounded under control of (7) and (14). Considering (6), the boundedness of $\ddot{e}_x, \ddot{e}_y$ ensures $(\tan\phi, \tan\theta)$ are bounded resulting with $|\phi| < \pi/2$ and $|\theta| < \pi/2$ respectively. Also, the boundednesses of $(\dot{\phi}, \dot{\theta})$ can be drawn from (11). Thus, the closed-loop system is stable. A more detailed discussion will be included in our future work.

4   Simulation Results

Numerical simulations are given in this section to illustrate the previous results. Adopt the model parameters in [7] as $b = 2.9842 \times 10^{-5}$(N.m/rad/s), $d = 3.232 \times 10^{-7}$(N.m/rad/s), $I_x = 3.8278 \times 10^{-3}$(N.m/rad/s^2), $I_y = 3.8278 \times 10^{-3}$ (N.m/rad/s^2), $I_z = 7,6566 \times 10^{-3}$(N.m/rad/s^2), $m = 0.486$(kg), $g = 9.81$(m/s^2), $l = 0.25$(m), $J_r = 2.8385 \times 10^{-5}$(N.m/rad/s^2). Set control gains by $k_{z1} = 1, k_{z2} = 1, k_{\psi 1} = 0.5$, $k_{\psi 2} = 0.5, k_1 = 0.96, k_2 = 3.88, k_3 = 5.88, k_4 = 3.96$. Two cases are taken into consideration.

Case 1 Tracking a constant reference.

Without loss of generality, we set reference signals and their derivatives by zero in this case. Initial states of stabilization control are set by $(\phi(0), \dot{\phi}(0), \theta(0), \dot{\theta}(0), \psi(0), \dot{\psi}(0)) = (\pi/8, 0, \pi/8, 0, \pi/8, 0), (z(0), \dot{z}(0), x(0), \dot{x}(0), y(0), \dot{y}(0)) = (1, 0, -6, 0, 3, 0)$. Simulation time is 30 s. Simulation results in this case are shown in Figs. 1 and 2.

Case 2 Tracking a time-varying trajectory.

Choose a time-varying reference ellipse curve generated by $\dot{x}_r = v\cos\psi_r, \dot{y}_r = v\sin\psi_r, \dot{\psi}_r = w, \dot{z} = v_z\cos(\varrho), \dot{\rho} = c_\varrho$ with initial values $(x_r(0), y_r(0), \psi_r(0), z_r(0), \varrho(0)) = (0, -25, 0, 0, 0)$ and $v = 5, w = 0.2, v_z = 1, c_\varrho = 0.1$. Set initial states of tracking control by$(\phi(0), \dot{\phi}(0), \theta(0), \dot{\theta}(0), \psi(0), \dot{\psi}(0)) = (0, 0, 0, 0, 0, 0)$, $(z(0), \dot{z}(0), x(0), \dot{x}(0), y(0), \dot{y}(0)) = (-5, 0, -10, 0, -10, 0)$. Simulation time is 80 s Simulation results of tracking a time-varying reference trajectory are pictured in Figs. 1, 2, 3 and 4.

As shown in Fig. 1, states asymptotically converge to zero when references are zero. Tracking errors asymptotically converge to zero when tracking time-varying reference signals as posed in Fig. 3. Besides, controls U_1, U_2, U_3, U_4 in both cases are smooth and bounded, which can be drawn from Figs. 2 and 4.

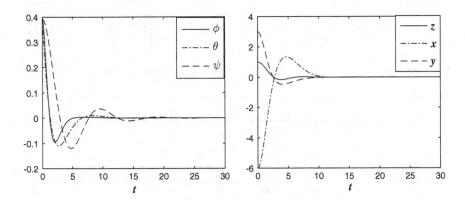

Fig. 1 State trajectories of $(\phi, \theta, \psi, z, x, y)$

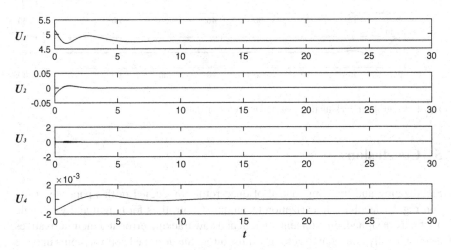

Fig. 2 Controls in tracking constant reference signals

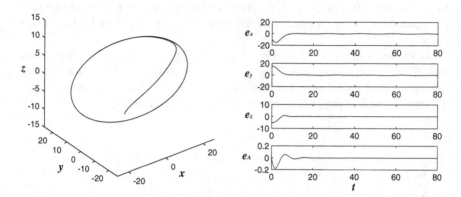

Fig. 3 Quad-rotor trajectory and tracking errors in tracking time-varying reference signals

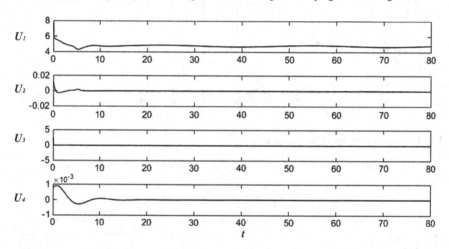

Fig. 4 Controls in tracking time-varying reference signals

5   Conclusion

Stabilization and tracking control of quad-rotor are studied in this paper by using non-regular feedback linearization technique. Controllers for height and yaw angle are firstly designed, guaranteeing height and yaw tracking error and their derivatives asymptotically converge to zero. By substituting the designed control inputs into the rest position tracking error dynamics and computing the fourth order derivatives of position tracking error, we propose a feedback linearization controller for the rest control inputs. Stability analysis shows that the tracking errors asymptotically decay to zero. In future work, we will handle the remaining issues in this paper and consider parameter uncertainties of quad-rotor.

Acknowledgements This work is supported by National Natural Science Foundation (No.61573034,No.61327807).

References

1. Hua MD, Hamel T, Morin P, Samson C. Introduction to feedback control of underactuated VTOL vehicles: a review of basic control design ideas and principles. IEEE Control Syst. 2013;33(1):61–75.
2. Chen Y, Ma B, Xie W. Robust stabilization of nonlinear PVTOL aircraft with parameter uncertainties. Asian J Control. 2016;19(3):1239–49.
3. Bouabdallah S, Noth A, Siegwart R. PID vs LQ control techniques applied to an indoor micro quadrotor. IEEE Int Conf Intell Robot Syst. 2004;3:2451–6.
4. Bouabdallah S, Siegwart R. Backstepping and sliding-mode techniques applied to an indoor micro quadrotor. In: Proceedings of the 2005 IEEE international conference on robotics and automation;2005. p. 2247–52.
5. Das A, Lewis F, Subbarao K. Backstepping approach for controlling a quadrotor using lagrange form dynamics. J Intell Robot Syst. 2009;56(1–2):127–51.
6. Zhao B, Xian B, Zhang Y. Nonlinear robust adaptive tracking control of a quadrotor UAV via immersion and invariance methodology. IEEE Trans Ind Electron. 2015;62(5):2891–902.
7. Bouadi H, Cunha SS, Drouin A. Adaptive sliding mode control for quadrotor attitude stabilization and altitude tracking. IEEE 12th international symposium on computational intelligence and informatics;2011. p. 449–55.
8. Lee D, Jin Kim H, Sastry S. Feedback linearization vs. adaptive sliding mode control for a quadrotor helicopter. Int J Control Autom Syst. 2009;7(3):419–28.
9. Zuo Z. Trajectory tracking control design with command-filtered compensation for a quadrotor. IET Control Theory Appl. 2010;4(11):2343–55.
10. Bouabdallah S, Siegwart R. Full control of a quadrotor. In: IEEE/RSJ international conference on intelligent robots and systems;2007. p. 153–58.
11. Voos H. Nonlinear control of a quadrotor micro-UAV using feedback-linearization. In: IEEE international conference on mechatronics;2009. p. 1–6.
12. Benallegue A, Mokhtari A, Fridman L. Feedback linearization and high order sliding mode observer for a quadrotor UAV. In: IEEE international workshop on variable structure systems;2006. p. 365–72.
13. Sun Z, Ge SS, Lee TH. Stabilization of underactuated mechanical systems: a non-regular backstepping approach. Int J Control. 2001;74(11):1045–51.

The Civil Aviation Crew Recovery Time-Space Network Model Based on a Tabu Search Algorithm

Qing Zhang, Yongxiu Ma, Zhengquan Yang and Zengqiang Chen

Abstract In order to reduce cost of airlines when disruption of crew scheduling happens, a mathematical programming model of crew scheduling recovery is constructed under the usage of airline crew. In the proposed model, the objective is to minimize the usage of airline crew,moreover, the estimated delay costs are considered as chance constraints. To solve the model, a tabu search algorithm ia adopted. Finally, a numerical example is carried out to illustrate the efficiency of the model and algorithm. The computed results show that the tabu search algorithm designed in this paper to solve the crew scheduling recovery problem not only achieves good optimization results, but also has a higher computational efficiency, a faster convergence rate and a more stable computing result. By this, the airlines cost waste can be better avoided.

Keywords Airline delay · Crew scheduling recovery problem · Tabu search algorithm · Optimization

1 Introduction

The crew scheduling recovery problem is always a cutting-edge and hot issue in operational research and combination optimization. In production and life, several percent points may be reduced for crew cost by the optimized crew scheduling, and then ten thousands of dollars can be saved for airlines every year [1]. This may influence their profits directly. Thus, the crew scheduling problem is concerned widely by academic and civil aviation circles.The overtime crew should be replaced

Q. Zhang · Y. Ma · Z. Yang (✉) · Z. Chen
College of Science, Civil Aviation University of China, Tianjin 300300, China
e-mail: zquanyang@163.com

Z. Chen
Department of Automation, Nankai University, Tianjin 300071, China

Z. Yang
Air Traffic Management Research Base Civil Aviation University of China,
Tianjin 300300, China

© Springer Nature Singapore Pte Ltd. 2018   373
Y. Jia et al. (eds.), *Proceedings of 2017 Chinese Intelligent Systems Conference*,
Lecture Notes in Electrical Engineering 460,
https://doi.org/10.1007/978-981-10-6499-9_36

by the other one, so the normal flight operation can be restored as soon as possible, that is,the crew scheduling is restored, The crew scheduling recovery is a comprehensive NP hard problem requiring massive calculation [2].

In the last several decades, lots of academics research the problem of crew scheduling recovery. Ellis.L.Johnson (1994) is the earlier researcher, discussing the single crew delay cannot make a normal connection to the next flight tasks [3]. Teodorovic and Stojkvic (1995) proposed a dynamic programming based method of crew recovery [4]. M.Clarke (1995) developed an airline operating control real-time decision support system, which includes the airline crew recovery module [5]. Stojkovic, Soumis et al. (1998) described the crew recovery problem which is described as the subset partition problem, the objective function is to find an execution cost minimum and on the original plan to change the smallest crew of assignment scheme [6]. Gao and Yu (1997) proposed a multi-commodity integer network flow model and a heuristic search algorithm, but they could not consider the reserve crews and deadhead crews [7]. Lettovsky, Johnson et al. (2000) proposed a so-called real-time crew recovery scheme to solve the frame [8]. Medard (2004) proposed recovery from the perspective of the crew members [9]. Rudiger Nissen and Knut Haase (2006) proposed the needs of European airlines, a model based on duty period is proposed, which is based on the model of the crew recovery [10]. Zhao (2010) presented a crew recovery model to minimum the cost considering deadhead crews and reserve crews, and promoted ant colony algorithm to solve it [11]. But Zhao constructed the model under the determining condition. Therefore, there are three ways to solve the problem of generating crews, the first is using column generation and branch and bound method; the second is the use of heuristic method; the third is the use of ant colony algorithm.

The main contribution of this paper is proposing a mathematical model to solve this problem. The objective function is to minimize the crew loss in the disrupted fight schedule recovery, the requisite constraints have been taken into account in the model to describe the crew recovery problem.moreover, under the conditions to meet the minimum cost, all the choices of the crews are taken into consideration and it avoids the influences of the aircraft types. The mathematical model describes clearly and the constraints rigorously. A tabu search algorithm [12–14] is developed for solving the model, and a memory search concept of tabu search is proposed to fit the crew problem. It can quickly obtain near optimal solution, and the solution is better than that given by dispatchers. Computational experiment has shown that the designed algorithm can meet the practical requirements for crew recovery.

In our paper, this major research work includes the following sections. First, it introduces the mathematical model for crew recovery problem. Next, proposed the tabu search algorithm, and we also use a numerical example to confirm the model. In the end, we provide some future directions.

2 Mathematical Model for Crew Recovery Problem

In this paper, we have made a further improvement on the basis of the research model proposed by Zhao et al. [11]. For introducing original model. Our improved model, firstly we list some definitions as follows.

Superscript/subscript indications:

z: Normal crew indication; j: Backup crew indication; b: Add crew indication; s: Crew base indication;

Sets:

S: Crew base set; Z: Normal crew set; J: Deadhead set; B: Backup crew set;

Parameters:

c_z^f: Delay cost for Crew z to execute Flight f; c_j^f: Cost for the deadhead j to execute Flight f; c_b^f: Cost for Backup Crew b to execute Flight f; t_z^o: Ready time for Crew z; t_f^o: Departure time for Crew f; t_z^e: Task termination time for Crew z; t_f^e: Arrival time for Flight f; d_f^s: Flight f terminating at the crew base is written as 1. Otherwise, it is 0.

Variables:

x_z^f: Crew z executing Flight f is written as 1. Otherwise, it is written as 0;
x_j^f: Crew j executing Flight f is written as 1. Otherwise, it is written as 0;
x_b^f: Crew b executing Flight f is written as 1. Otherwise, it is written as 0;
x_i^l: Crew l executing Mode i is written as 1. Otherwise, it is written as 0.

Mathematical model(we proposed in this paper):

$$Min\{\sum_{z\in Z}\sum_{f\in F} c_z^f x_z^f + \sum_{j\in J}\sum_{f\in F} c_j^f x_j^f + \sum_{b\in B}\sum_{f\in F} c_b^f x_b^f\} \tag{1}$$

subject to

$$\sum_{z\in Z} x_z^f + \sum_{j\in J} x_j^f + \sum_{b\in B} x_b^f = 1, (\forall f \in F) \tag{2}$$

$$\sum_{z\in Z}\sum_{f\in F} d_f^s x_z^f + \sum_{j\in J}\sum_{f\in F} d_f^s x_j^f + \sum_{b\in B}\sum_{f\in F} d_f^s x_b^f = A_s, (\forall s \in S) \tag{3}$$

$$x_z^f t_z^o \le t_f^o, (\forall f \in F, \forall z \in Z) \tag{4}$$

$$x_z^f t_z^e \ge t_f^e, (\forall f \in F, \forall z \in Z) \tag{5}$$

$$x_z^f \in \{0, 1\}, x_j^f \in \{0, 1\}, x_b^f \in \{0, 1\}, x_i^l \in \{0, 1\}. \tag{6}$$

In the model, Eq. (1) is the objective function which makes crew resource waste minimization. where, the first term is crew idle cost, the second term is deadhead cost, and the third term is reserve crew cost. And Eq. (1) emphasises on the use of the crew and the standby crew. Eq. (2) is the flight cover restraint, which can ensure either crew or deadhead or reserve crew is in one flight. Eq. (3) is the restraint for crew returning to base, which requires all crews go back the crew base after fulfilling flight task on that day. Eq. (4) is the Flight which are not bound by delay, the Flight f operation by crew is ready before departure. Equation (5) is the restraint that the crew does not work overtime, and the duty-fulfilling time is greater than the Flight arrival time. Equation (6) is 0 and 1 integer constraint. This model is proposed to solve the crew recovery problem for multi-crew. So it can be applied in practice widely.

3   Solving the Crew Recovery Problem by the Tabu Search Algorithm

3.1   Tabu Search Algorithm

The tabu search algorithm [12] is a global neighborhood algorithm, simulating the humans optimizing feature with memory function. It can avoid the detour searching by local neighborhood searching mechanism and trustful tabu criterion, and releases some tabooed excellent states by tabu-breaking level to ensure multiple effective exploration and realize the global optimization finally.

This paper employs an example of five-city Multi Traveling Salesman (TSP) problem to show the feature of tabu searching algorithm. Assuming the tabu length as 3, and SWAP as the neighbor operation (The starting point of solution is 1), then it is stipulated that $x \notin N(x)$, and 4 candidate solutions are selected in present neighbor set.

The distance matrix (d_{ij}) for this problem is:

$$\begin{pmatrix} 0 & 10 & 15 & 6 & 2 \\ 10 & 0 & 8 & 13 & 9 \\ 15 & 8 & 0 & 20 & 15 \\ 6 & 13 & 20 & 0 & 5 \\ 2 & 9 & 15 & 50 & \end{pmatrix}.$$

Step 1: Current state is (1,2,3,4,5;45), where 45 is a target value. By this time,H={ },then the candidate solutions are:
(1,3,2,4,5;43),(1,4,3,2,5;45),(1,2,5,4,3;59),(1,2,3,5,4;44)},

then (1, 3, 2, 4, 5; 43) is a new current solution,
and let H={(1,3,2,4,5;43)}.

Step 2: The current solution is (1,3,2,4,5;43),H={(1,3,2,4,5;43)}, the candidate solutions are:
{(1, 3, 2, 5, 4; 43), (1, 4, 2, 3, 5; 44), (1, 2, 3, 4, 5; 45), (1, 3, 5, 4, 2; 48)},
then (1, 3, 2, 5, 4; 43) is selected as the new current solution,
and let:H={(1,3,2,4,5;43),(1,3,2,5,4;43)}.

Step 3: The current solution is (1,3,2,4,5, 43), H={(1,3,4,5; 43), (1,3,2,5,4; 43)}, the candidate solution is {(1, 3, 2, 5, 4; 43), (1, 2, 3, 5, 4; 44), (1, 5, 2, 3, 4; 45), (1, 4, 2, 5, 3; 58)}, then select (1, 2, 3, 5, 4; 44) as the new current solution,
and let:H={(1,3,2,4,5;43),(1,3,2,5,4;43),(1,2,3,5,4;44)}.

Step 4: The current solution is (1, 2, 3, 5, 4; 44), H={(1, 3, 2, 4, 5; 43), (1, 3, 2, 5, 4; 43), (1, 2, 3, 5, 4; 44)}, the candidate solutions are: {(1, 3, 2, 5, 4; 43), (1, 5, 3, 2, 4; 44), (1, 2, 3, 5, 4; 45) 1, 2, 4, 5, 358}, then select (1, 5, 3, 2, 4; 44) as the new current solution,
and let:H={(1,3,2,4,5;43),(1,3,2,5,4;43),(1,2,3,5,4;44)}.

Step 5: Current solution is (1, 5, 3, 2, 4; 44), H={(1, 5, 3, 2, 4; 43), (1, 2, 3, 5, 4; 44)}, the candidate solutions are: {(1, 5, 4, 2, 3; 43), (1, 2, 3, 5, 4; 44),(1, 5, 3, 4, 2; 44), (1, 5, 2, 3, 4;58)},
then select (1, 5, 4, 2, 3; 43) as the new current solution.

4   Case Calculation and Result Analysis

The crew recovery is done after the aircraft route recovery is completed. Here the aircraft route optimized output results from one airline which are used as input data. There are five aircrafts to execute 12 flights in this case, and the flight is delayed due to partial aircraft failures in execution. After the aircraft route is recovered, four flights are still delayed. The new aircraft implementation scheme disorganizes original crew duty plan of Table 1, where the crew deadhead operates by 2,000RMB yuan once, the reserve crew is called by 10,000RMB yuan once, and the crew idle goes by 50RMB yuan per minute.

After the aircraft route is recovered, the crew on-duty scheduling is disorganized as shown in Table 2. The duty ending time for Crew C2 is 19:30, and it is 20:40 after Flight xx1338 is completed; Flight xx1338 can be implemented by Crew C2 as

Table 1 Original crew on-duty scheduling

Crew	Work	Close	Number	Take-off	Startpoint	Arrival time	Arrivalpoint
C1	7.5	17.5	xx3108	8	Shang Hai102	10.5	Guang Zhou103
C1	7.5	17.5	xx3109	11	Guang Zhou103	13.5	Shang Hai102
C1	7.5	17.5	xx3110	14	Shang Hai102	15.5	Bei Jing101
C1	7.5	17.5	xx3111	16	Bei Jing101	17.5	Shang Hai102
C2	9.5	19.5	xx1337	9.83	Guang Zhou103	14.33	Bei Jing101
C2	9.5	19.5	xx1338	15	Bei Jing101	19.5	Guang Zhou103
C3	5	15	xx1209	11	Guang Zhou103	13.5	Shang Hai102
C3	5	15	xx1210	14	Shang Hai102	16.5	Guang Zhou103
C4	11.5	21.5	xx1223	14	Bei Jing101	16.5	Guang Zhou103
C4	11.5	21.5	xx1224	17	Guang Zhou103	21.5	Bei Jing101
C5	7	17	xx1518	8	Bei Jing101	12.5	Guang Zhou103
C5	7	17	xx1519	12.5	Guang Zhou103	15.33	Bei Jing101

Table 2 Crew on-duty scheduling

Crew	Work	Close	Number	Take-off	Startpoint	Arrival	Arrivalpoint
C1	7.5	17.5	xx3108	8	Shang Hai102	10.5	Guang Zhou103
C1	7.5	17.5	xx3109	11	Guang Zhou103	13.5	Shang Hai102
C1	7.5	17.5	xx3110	14	Shang Hai102	15.5	Bei Jing101
C1	7.5	17.5	xx3111	16	Bei Jing101	17.5	Shang Hai102
C2	9.5	19.5	xx1337	11	Guang Zhou103	15.5	Bei Jing101
C2	9.5	19.5	xx1338	16.17	Bei Jing101	20.67	Guang Zhou103
C3	10.67	20.67	xx1209	11	Guang Zhou103	13.5	Shang Hai102
C3	10.67	20.67	xx1210	14	Shang Hai102	16.5	Guang Zhou103
C4	11.5	21.5	xx1223	14	Bei Jing101	16.5	Guang Zhou103
C4	11.5	21.5	xx1224	17	Guang Zhou103	21.5	Bei Jing101
C5	7	17	xx1518	8	Bei Jing101	12.5	Guang Zhou103
C5	7	17	xx1519	15	Guang Zhou103	15.33	Bei Jing101

it violates the stipulation that a crew shall not be overtime. Crew C5s duty ending time is 17:00, and it is 17:30 after C5 completes Flight xx1519 in the original flight scheduling; Crew C5 cannot execute the Flight xx1519 because of overtime. If the crew task is rearranged, Flight xx1337 and Flight xx1519 cannot be executed because of lack of crew.

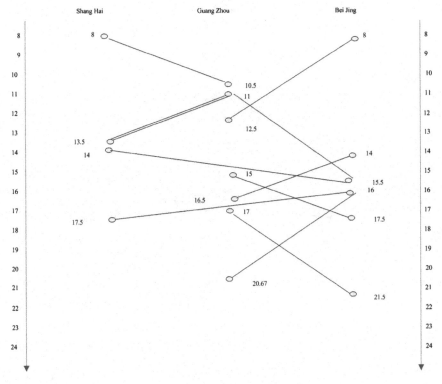

Fig. 1 The time-space network diagram for the crew time airport before recovery

The MATLABR2014a coding operating procedure is used in this study. First, read the flight timetable and the crew on-duty timetable; then all flights constitute the memory matrix by the tree divergence in time order to gain the recovery result which can meet all restraint conditions. Meanwhile, the time-space network diagram for the crew time airport is given as shown in Figs. 1 and 2.

Solution Algorithm:

Step 1. Firstly read flight schedules locations, then read crews work time and place table. Sequence the flight according to time. And random initial value is given.

Step 2. Each crew carried out all flights they can fly, select the minimum cost,and store the crew. The others repeat the cycle. When crews search the minimum cost, they can avoid the crew which has been stored, because they know it has been found.

Step 3. That crew has been stored cant stay there generally, and after a while it will rejoin the team to find out the least minimum. This is because there are more messages by then. Eventually, the memory matrix is constituted based on a tree divergence.

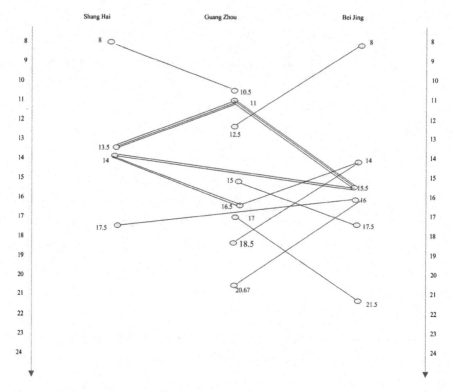

Fig. 2 The time-apace network diagram for the crew time airport after adjusting

Step 4. According to the above calculation, the local optimum crews are released. And ultimately by the local optimum to global optimum, making the cost to a minimum.

Finally, the optimized crew scheduling in Table 3 is obtained, the optimized crew space-time network is given as shown in Fig. 2. From the Table 3 we can see, we used five normal crews and two deadhead crews. It gave eight schemes and guaranteed over time which did not need too long. At the same time to enhance the degree of satisfaction of passengers to avoid the two delay of the aircraft. The operation result is that we need to spend eighty thousand yuan and 2.7 s. We can make optimization for the unexpected situation, making the application more general. The convergence rate is also increased. But the final cost is not beyond the budget.

Table 3 Operating result

0	xx1519	15	0	xx1519	15	0	xx1519	15	0	xx1519	15
0	xx1338	16	0	xx1338	16	0	xx1338	16	0	xx1338	16
1	xx3108	8	1	xx3108	8	1	xx3108	8	1	xx3108	8
1	xx3109	11	1	xx3109	11	1	xx1337	11	1	xx1337	11
1	xx3110	14	1	xx3110	14	1	xx3111	16	1	xx3111	16
1	xx3111	16	1	xx3111	16	2	xx1209	11	2	xx1209	11
2	xx1209	11	2	xx1337	11	2	xx3110	14	2	xx1210	14
2	xx1210	14	3	xx1209	11	3	xx3109	11	3	xx3109	11
3	xx1337	11	3	xx1210	14	3	xx1210	14	3	xx3110	14
4	xx1223	14	4	xx1223	14	4	xx1223	14	4	xx1223	14
4	xx1224	17	4	xx1224	17	4	xx1224	17	4	xx1224	17
5	xx1518	8	5	xx1518	8	5	xx1518	8	5	xx1518	8
0	xx1519	15	0	xx1519	15	0	xx1519	15	0	xx1519	15
0	xx1338	16	0	xx1338	16	0	xx1338	16	0	xx1338	16
1	xx3108	8	1	xx3108	8	1	xx3108	8	1	xx3108	8
1	xx1209	11	1	xx1337	11	1	xx1337	11	1	xx1209	11
1	xx3110	14	1	xx3111	16	1	xx3111	16	1	xx3110	14
1	xx3111	16	2	xx3109	11	2	xx3109	11	1	xx3111	16
2	xx1337	11	2	xx1210	14	2	xx3110	14	2	xx3109	11
3	xx3109	11	3	xx1209	11	3	xx1209	11	2	xx1210	14
3	xx1210	14	3	xx3110	14	3	xx1210	14	3	xx1337	11
4	xx1223	14	4	xx1223	14	4	xx1223	14	4	xx1223	14
4	xx1224	17	4	xx1224	17	4	xx1224	17	4	xx1224	17
5	xx1518	8	5	xx1518	8	5	xx1518	8	5	xx1518	8

5 Conclusions

It is seen from computing results in this paper that the relatively optimizing results can be gained in a shorter term similarly for the tabu search algorithm, and the solution is stable in convergence when the appropriate parameter combination is set. This further demonstrates the stability of this algorithm. In this paper, any crew can meet the requirement of all aircraft licenses in the solving process, and this is different from some crews can not execute some aircraft types to a certain extent. Therefore, our model and algorithm consolidate and strengthen the robustness of crew scheduling recovery, and delay recovery quickness.

Acknowledgements This work is supported by National Natural Science Foundation of China (61573199), National Natural Science Foundation of Tianjin (14JCYBJC18700), Basic Research Projects of High Education (3122015C025).

References

1. Jarrah A, Diamond JT. The problem of gengerating ctew bidlines, interfaces. vol. 27, 1997. p. 49–64.
2. Gang Yu. Operations research in the airline industry. Boston, MA: Kluwer Academic Publishers; 1997.
3. Johnson E, Lettovsky L, Nemhauser G, Pandit R, Querido S. Final report to northwest airlines on the crew recovery problem. The Logistic Institute, Georgia Institute of Technology, Atlanta, GA. 1994.
4. Stojkovic G. Model to reduce airline schedule disturbances. J Transp Eng. 1995;121:324–31.
5. Clarke DM. Irregular Airline Operations: A Review of the state-of-the-practice in Airline Operations Control Centers, J Air Transp Manag. 1995; 4:67–76,
6. Stojkovic G, Soumis F, Desrosiers J. The operational airline crew scheduling problem. Transp Sci. 1998; 32: 232–245
7. Wei G, Gang Y. Optimization model and algorithm for crew management during airline irregular operations. J Comb Optim. 1997;3:305–21.
8. Lettovsky, Johnson Proposed a so called real time crew recovery scheme to solve the frame, vol. 7, 2000 p. 303–312, .
9. Medard c.p., N. Sawhney. Airline crew scheduling: from planning to operations,in Carmen research and technology report. Garmen Systems AB: Goteborg, Sweden. 2004.
10. Haase Knut, Nissen Rudiger. "Duty-Period-Based Netword Model for Crew Rescheduling in European Air- line"s, Springer Science + Business. Media. 2006;9:255–78.
11. Xiuli Zhao. Research on modeling and algorithm of airline irregular recovery. Nanjing China: Nanjing University of aeronautics and astronautics; 2010.
12. Sclim SZ, Lsmail MA. K-Means "Type Algorithms: A Generalized Convergence Theorem and Characteristics of Local Optimization", IEEE Trans. Pattern Anal Mach Intell 1984. 6:81–7.
13. Chiu-Hung Chen. Liu, Tung-Kuan, Chou Jyh-Horng. Integrated short-haul airline crew scheduling using multi-objective optimization genetic algorithms. IEEE Trans Syst Man Cybern Syst. 2013. 42:1077–90.
14. Ellis L. Johnson, George L. Nemhauser. Airline crew recovery. Transp Sci. 2000. 34:337–48.

Design an Induction Motor Rotor Fux Ob-Server Based on Orthogonal Compensation of the Stator Flux and Back EMF

Bo Yu, Jiaxing Zhao and Yuedou Pan

Abstract To address the problems of error accumulation and drift caused by pure integral part in the traditional voltage model flux observer and amplitude and phase error when the first-order low-pass filter is introduced into the traditional voltage model flux observe and saturation threshold selection in the first-order low-pass filter saturated feedback link, this paper applied a compensation method based on orthogonal principle of the stator flux and back EMF in observing Induction motor rotor flux. We analyzed the principle of this new voltage model and derived the compensation algorithm due to the introduction of the amplitude and phase error with the low-pass filter. Simulation experiments verified the correction of the compensation algorithm, the application of this method in observing Induction motor rotor flux is right and effective. The induction motor vector control system had a good dynamic and steady state performance.

Keywords Voltage model · Rotor flux · Vector control system

1 Introduction

Conventional voltage model flux observer of the rotor is a pure integrator, The accumulation error and drift of pure integral parts can cause system instability [1]. Especially at low speeds, the stator voltage drop effect is so obvious that the back electromotive force is submerged by the measurement error, which causes the

B. Yu
Zhonghui Vision Energy Technology (Beijing) Co., Ltd, Beijing 100010, China
e-mail: boyu@sinowestar.com

J. Zhao (✉) · Y. Pan
School of Automation, University of Science and Technology Beijing,
Beijing 100083, China
e-mail: zhaojiaxingPJ@163.com

Y. Pan
e-mail: ydpan@ustb.edu.cn

© Springer Nature Singapore Pte Ltd. 2018
Y. Jia et al. (eds.), *Proceedings of 2017 Chinese Intelligent Systems Conference*,
Lecture Notes in Electrical Engineering 460,
https://doi.org/10.1007/978-981-10-6499-9_37

accuracy of the flux value calculated by the voltage model is reduced [2, 3]. The current common solution is to use a first-order low-pass filter to replace the pure integrator to solve the problem of error accumulation. But this method also leads to additional phase shift and amplitude errors [4]. In [1], a method of compensating the low-pass filtered flux in the forward channel with a certain change in the reference flux is proposed. However, the model in this method is an open-loop structure, which is liable to cause excess flux compensation, and the flux observation deviates from the reference value. In [5], a first-order low-pass filter with feedback clipping is proposed to replace the pure integrator. This method eliminates the cumulative DC error inherent in the accumulator, and because the method is based on feedback, the additional phase and amplitude errors lead by the first-order low-pass filter in the channel are compensated. However, the first-order low-pass filter in this method has the problem that the threshold of the feedback limit is fixed and can not be adjusted with the threshold value of the rotor flux reference threshold, which leads to poor dynamicity of the whole system and limited scope of application.

In order to solve the problems in the above-mentioned literature, this paper presents a design method of rotor-based magnetic flux observer for asynchronous motor voltage model based on orthogonality compensation of stator flux and back electromotive force. The method of designing the voltage model rotor flux observer has the advantages of the first-order low-pass filter in [5], and at the same time, it solves the problem of system dynamic poor and applicable range limited due to the fixed threshold of feedback clipping in [5]. This paper designs a new voltage model of the rotor flux observer observed ψ_{ra}, $\psi_{r\beta}$ curves sinusoidal good, dynamic response fast.

In this paper, the whole control system is modeled by Matlab/Simulink, and the rotor flux observer of the asynchronous motor based on the orthogonality compensation of stator flux and back electromotive force is simulated. The simulation results show that the observer can accurately observe the rotor flux, and the asynchronous motor vector control system with the observer is better dynamic. The whole asynchronous motor vector control system has good dynamic and steady state performance in the whole speed range.

2   Traditional Voltage Model Rotor Flux Observer

Two-phase stationary shaft $(\alpha - \beta)$, the traditional voltage model rotor flux observer is:

$$\psi_{ra} = \frac{L_{rd}}{L_{md}} [\frac{1}{s + \omega_c} e_{sa} - \sigma L_{sd} i_{sa}] \tag{1}$$

$$\psi_{r\beta} = \frac{L_{rd}}{L_{md}} [\frac{1}{s + \omega_c} e_{s\beta} - \sigma L_{sd} i_{s\beta}] \tag{2}$$

In formula (1) and (2) ψ_{ra}, $\psi_{r\beta}$ respectively said the rotor and shaft flux; u_{sa}, $u_{s\beta}$ respectively said the stator and shaft voltage; i_{sa}, $i_{s\beta}$ respectively said the stator and shaft current; R_s, L_s, σ respectively said stator resistance, stator inductance and magnetic flux leakage coefficient, among them, $\sigma = 1 - \frac{L_{md}^2}{L_{sd}L_{rd}}$, L_{sd}, L_{md}, L_{rd} respectively said the stator inductance, mutual inductance and rotor inductance.

The above two formulas contain pure integral links, the existence of the error accumulation and drift problems will lead to system instability.

3   Design of Rotor Flux Observer for Voltage Model Based on Orthogonal Compensation of Stator Flux and Back EMF

Pure integral link is $y = \frac{1}{s}x$, The first order low pass filter is $y = \frac{1}{s+\omega_c}x$, use a first order low pass filter $\frac{1}{s+\omega_c}x$ to replace pure integrator $\frac{1}{s}x$, formula (1) ,(2) rewrite as follows:

$$\psi_{ra} = \frac{L_{rd}}{L_{md}}[\frac{1}{s+\omega_c}e_{sa} - \sigma L_{sd}i_{s_a}] \tag{3}$$

$$\psi_{r\beta} = \frac{L_{rd}}{L_{md}}[\frac{1}{s+\omega_c}e_{s\beta} - \sigma L_{sd}i_{s\beta}] \tag{4}$$

In formula (3), (4), e_{sa}, $e_{s\beta}$ said the stator winding back electromotive force. In formula (3), (4) the use of a first-order low-pass filter will lead to the observed rotor flux linkage amplitude and phase a certain error [5]. To eliminate the error, in the formula (3), (4) use the compensation link: $\frac{\omega_c}{s+\omega_c}y$ to compensate the output of a first-order low-pass filter by feedback. Formula (3), (4) rewrite formula (5), (6) to get new voltage model:

$$\psi_{ra} = \frac{L_{rd}}{L_{md}}[\frac{1}{s+\omega_c}e_{sa} + \frac{\omega_c}{s+\omega_c}\psi_{sacmp} - \sigma L_{sd}i_{s_a}] \tag{5}$$

$$\psi_{r\beta} = \frac{L_{rd}}{L_{md}}[\frac{1}{s+\omega_c}e_{s\beta} + \frac{\omega_c}{s+\omega_c}\psi_{s\beta cmp} - \sigma L_{sd}i_{s\beta}] \tag{6}$$

In formula (5) ,(6) ψ_{sacmp}, $\psi_{s\beta cmp}$, are $\psi_{s\beta cmp}$ the components on α, β axis, ψ_{scmp} is stator flux compensation value.

In order to prevent excessive compensation, the magnitude of the ψ_{scmp} feedback should be limited. The usual way is to add a limiter to the feedback link, however, the disadvantage of this method is that the limit of the feedback limit is fixed, and in the actual environment, the rotor flux is constantly changing, which requires the threshold of the limiter to be constantly adjusted.

In order to solve the problem of fixed limit value of feedback limit, this paper designs an asynchronous motor voltage model rotor flux observer based on orthogonality compensation of stator flux and back electromotive force, The feedback compensation link of this observer utilizes the stator flux ψ_s linkage to the stator back electromotive force e_s, and designs the orthogonalizer replace the limiter in the feedback link. Orthogonalizer detects the orthogonality of ψ_s and e_s, outputs deviation e, The compensation amount of ψ_{scmp} outputted by the PI regulator.

$$e = \psi_{sa}e_{sa} + \psi_{s\beta}e_{s\beta} \tag{7}$$

$$\psi_{scmp} = (k_{p1} + \frac{k_{i1}}{s}) \frac{\psi_{sa}e_{sa} + \psi_{s\beta}e_{s\beta}}{|\psi_s|} \tag{8}$$

In formula (7), (8), ψ_{sa}, $\psi_{s\beta}$ said the components of ψ_s in α, β axis, e_{sa}, $e_{s\beta}$ said the components of stator back electromotive force e_s in α, β axis, k_{p1}, k_{i1} said PI regulator constants. Thus, the stator flux ψ_s in the new asynchronous motor voltage model is composed of two parts: forward link output ψ_{s1} and The amount of feedback ψ_{s2}. When ψ_s orthogonal to e_s, The angle θ between ψ_s and e_s is zero; When $\theta > 90°$, the amount of feedback ψ_{s2} is too large, the PI controller outputs a negative value to reduce ψ_{s2}, When $\theta < 90°$, the amount of feedback ψ_{s2} is too small, the PI controller outputs a positive value to increase ψ_s. The new model is shown in Fig. 1.

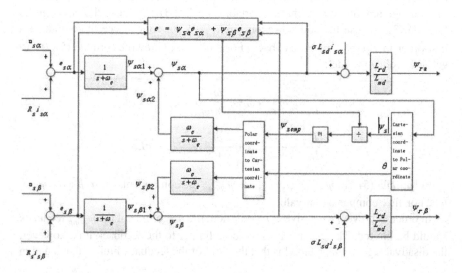

Fig. 1 New induction motor voltage model rotor flux observer

4   Vector Control System of Rotor Flux Observer with New Asynchronous Motor Voltage Model

In this paper, the design of the stator flux and back electromotive force orthogonal compensation of the asynchronous motor voltage model rotor flux observer vector control system shown in Fig. 2, In this paper, the rotor flux $\hat{\psi}_r$ observation and calculation are based on the stator flux and back electromotive force orthogonal compensation method, Motor output speed $\hat{\omega}_r$ and asynchronous motor voltage model flux observer measured $\hat{\psi}_r$ difference with the reference speed and reference flux to get the deviation control signal. The speed closed loop and the magnetic chain closed loop control are performed with the deviation control signal. The simulation model is shown in Fig. 3.

In order to precisely control the motor, this paper uses the diode midpoint clamped three-level inverter SVPWM to control the asynchronous motor. This three-level inverter has a total of 27 kinds of switch combinations, you can output 19 switch vectors, constitute the vector space of the inverter, and the vector space is divided into six large sectors and 24 small sectors. SVPWM technology combines the inverter and the motor as a whole, by combining the three-phase voltage space

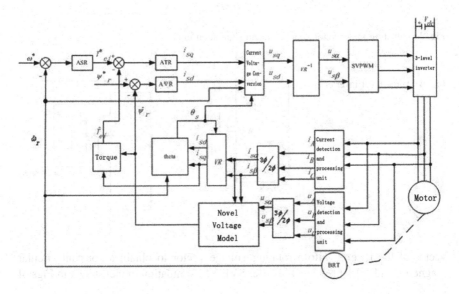

Fig. 2 Induction motor vetor contorl system

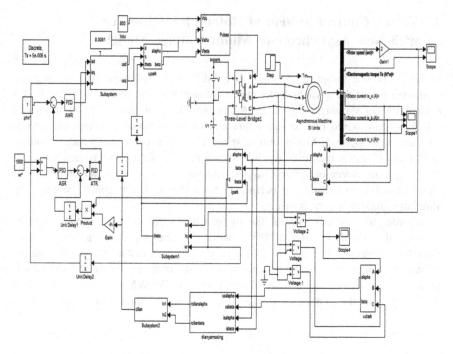

Fig. 3 Induction motor vetor contorl system simulation model

Fig. 4 Three-level inverter

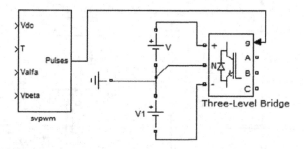

vector of the inverter into a rotating voltage vector to obtain a constant circular magnetic field [6]. Three-level inverter SVPWM simulation model shown in Figs. 4 and 5.

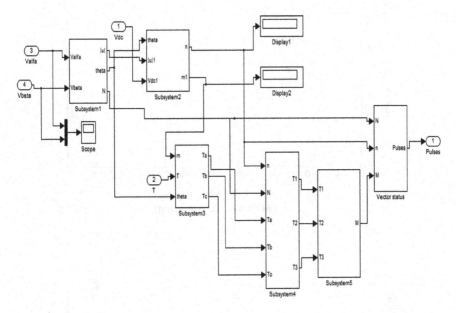

Fig. 5 SVPWM model

5 Simulation Results and Analysis

The parameters of the winding type asynchronous motor for simulation are 3.73 kW, 400 V, 50 Hz. Given the rotor flux amplitude is 1Wb, Rs = 0.7384 Ω, Rr = 0.7403 Ω, Ls = 0.003045H, Lr = 0.003045H, np = 2. Flux, speed and torque regulators are AΨR, ASR and ATR, Three-level inverter DC bus voltage is 800 V. And the triangular wave period is 0.0001 s.

Simulation results and analysis: Figs. 6, 7, 8 and 9, respectively are in very low speed 1 Hz (30r/min, low speed 10 Hz (300 r/min), 25 Hz (750r/min) and rated speed 50 Hz (1500r/min)running under the simulation results. As can be seen from the figure, the dynamic response speed is fast, and system stable within a short time in the reference value. Figure 10 is the motor at low speed 10 Hz operation, the new asynchronous motor voltage model rotor flux observer measured by the rotor flux. It can be seen from the figure that the curve of the flux $\psi_{r\alpha}$, $\psi_{r\beta}$ of the rotor is smooth, the peak is neat, the sine is better, the integral drift and the integral saturation do not occur, and the two curves are orthogonal. Figure 11 shows the rotor flux ψ_r, which is composed of multiple concentric circles. And the curve is smooth and the final flux linkage is stable at the reference value. Figure 12 is the

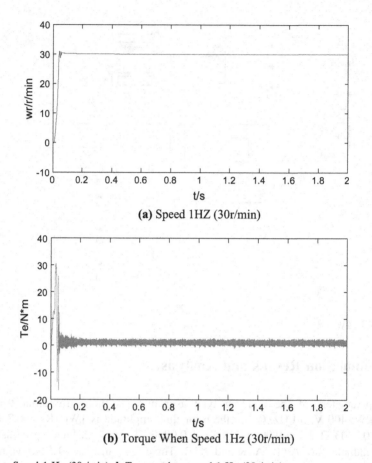

(a) Speed 1HZ (30r/min)

(b) Torque When Speed 1Hz (30r/min)

Fig. 6 a Speed 1 Hz (30r/min), **b** Torque when speed 1 Hz (30r/min)

rotor flux waveform when the motor from the constant excitation field into the weak magnetic field(50 Hz into 70 Hz). From the figure can be seen weak magnetic speed characteristics: magnetic flux amplitude was significantly reduced.

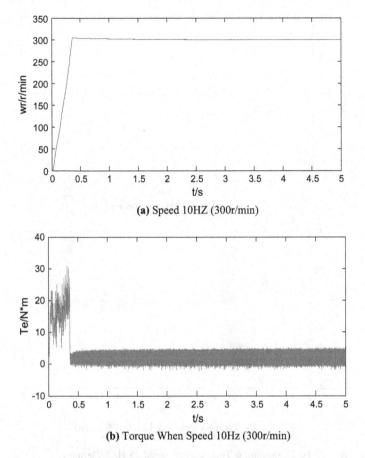

(a) Speed 10HZ (300r/min)

(b) Torque When Speed 10Hz (300r/min)

Fig. 7 a Speed 10 Hz (300r/min), **b** Torque when speed 10 Hz (300r/min)

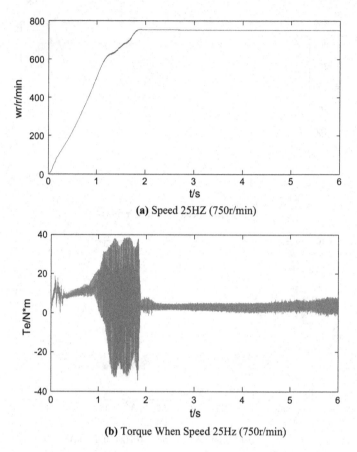

(a) Speed 25HZ (750r/min)

(b) Torque When Speed 25Hz (750r/min)

Fig. 8 **a** Speed 25 Hz (750r/min), **b** Torque when speed 25 Hz (750r/min)

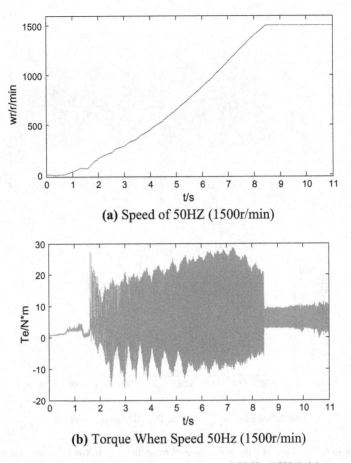

(a) Speed of 50HZ (1500r/min)

(b) Torque When Speed 50Hz (1500r/min)

Fig. 9 **a** Speed of 50 Hz (1500r/min), **b** Torque when speed 50 Hz (1500r/min)

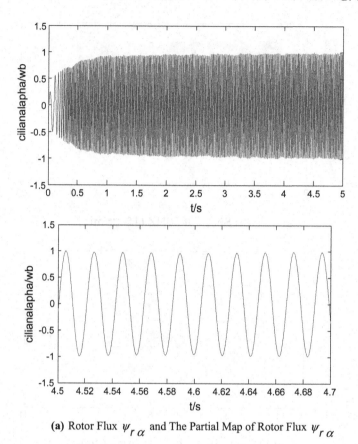

(a) Rotor Flux $\psi_{r\alpha}$ and The Partial Map of Rotor Flux $\psi_{r\alpha}$

Fig. 10 **a** Rotor Flux ψ_{ra} and the partial map of rotor flux ψ_{ra}, **b** Rotor flux $\psi_{r\beta}$ and the partial map of rotor flux $\psi_{r\beta}$, **c** Rotor flux ψ_{ra} orthogonal to $\psi_{r\beta}$

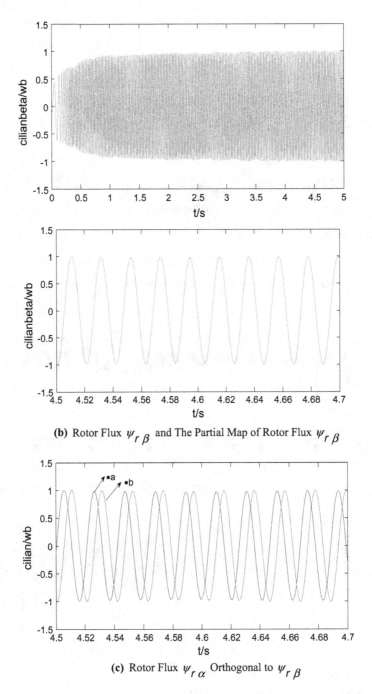

(b) Rotor Flux $\psi_{r\beta}$ and The Partial Map of Rotor Flux $\psi_{r\beta}$

(c) Rotor Flux $\psi_{r\alpha}$ Orthogonal to $\psi_{r\beta}$

Fig. 10 (continued)

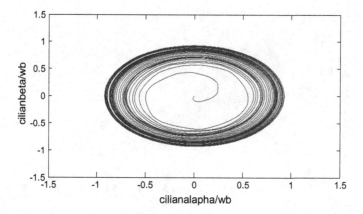

Fig. 11 Rotor flux round

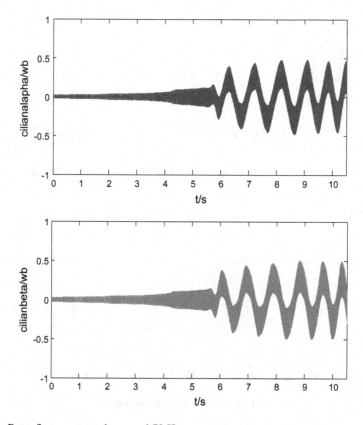

Fig. 12 Rotor flux $\psi_{r\alpha}$, $\psi_{r\beta}$ when speed 70 Hz

6   Conclusion

In this paper, the design of the stator flux and back electromotive force orthogonal compensation of the voltage model rotor flux observer used in the induction motor rotor chain observation eliminates the cumulative DC error inherent in the pure integrator, compensates the additional phase and amplitude errors, improves the dynamic performance of the system, and enlarges the variation range of the flux amplitude; Using the new asynchronous motor voltage model, the vector control system of the rotor flux observer has good dynamic and steady state performance at full speed range.

References

1. Yali Zhou, Yongdong Li, Zedong Zhen. Speed sensorless vector control of induction motor based on MRAS. Electric Air Drive. 2009;39(4):3–8.
2. Wu R. Slemon G R. A permanent magnet motror drive without a shaft sensors. IEEE Trans Ind Appl. 1991;27(5):1005–11.
3. Tajima H, Hori Y. Speed sensorles field oriented control of the induction machine. IEEE IAS Conf Rec. 1991:385–391.
4. Jin H, Huang J. Self—adaptive observation and parameter identification of induction motor flux chain based on model reference method. J Electrotech Soc. 2006;21(1):66–69.
5. Jianqiang Zhong, Lingru You, Qinwen Xun. A speed sensorless vector control method for asynchronous motor with improved voltage model. J Micro Mot. 2009;42(5):16–8.
6. Jin Y. Analysis and research on SVPWM method of three—level inverter. Hunan University;2007.

Fast Robust PCA on Background Modeling

Huini Fu, Zhihui Gao and HengZhu Liu

Abstract This paper extends the subspace learning method Robust principal component analysis (RPCA) for background modeling to recover the background scene from video sequence with static camera. We propose a novel matrix reformulation and optimization process for RPCA method to solve background modeling problem. The experiments are conducted among our proposed method and other statistical methods including RPCA algorithm and its variants under wallflower datasets and LRSLibrary benchmarks separately. The results of experiments show that our method exceeds the existing RPCA in time complexity in a great manner, while keeps and even improves the modeling performance over other modeling algorithms.

Keywords Fast RPCA · Convex function · LRSLibrary · Matrix restoration

1 Introduction

Equipping security surveillance cameras with computer vision method, which could save millions in reducing manpower and improve efficiency in detecting abnormal activities or even pre-detect dangerous events. Background model generation is the basic computer vision method for video analysis. Recovering background scene from video sequence is essential to video analysis. Nowadays background modeling is still considered a major research field in video understanding and analyzing, with proposed state-of-the-art algorithms [1–9], benchmarks [10–12], and real-time implementations [13–15]. Specific background recovering methods are able to deal with specific issues including sudden and gradual light changes, ill video conditions introduced by shadows and highlighted regions, dynamic background with period jittering, and etc. Constrained by problems like perspective and illumination issues

H. Fu (✉) · Z. Gao · H. Liu
School of Computer Science and Technology, National University of Defense Technology, Changsha 410000, China
e-mail: 798768548@qq.com

© Springer Nature Singapore Pte Ltd. 2018
Y. Jia et al. (eds.), *Proceedings of 2017 Chinese Intelligent Systems Conference*,
Lecture Notes in Electrical Engineering 460,
https://doi.org/10.1007/978-981-10-6499-9_38

and many other difficulties, background modeling method constantly lead to poor estimations.

With static camera, the background scene is assumed to be unchanged. Under this surveillance situation, the primal idea that popped-up in brain is to acquire a pure background image without any moving foreground objects, so as to subtract the background from current video image to get the foregrounds. Considering complex surveillance conditions, it is hard to get pure background. With regard to more realistic applications, the background does not remain static because of changing lights or dynamic background objects. Instead, building background models is then proposed to deal with those complex application environments. Numerous methods for background model have been proposed as referred in [16]. Background model methods can be divided into four categories: (i) basic background model. (ii) Statistical background model.

Basic background model computes each pixel's averages [17], median [18] or histogram [19] in time sequence. Then subtract the model from current frame and get the difference value. Finally segment foreground through threshold comparison. This kind of method computes fast and good with simple surveillance environment.

After reading the literature, the statistical methods are the mostly used because of the robustness to different situations. Mixture of Gaussian (MOG) and Kernel Density Estimation (KDE), and other variants are supposed to be the typical statistical methods. MOG was proposed in 1999 by Stauffer and Grimson [6] and it still remains one of the most popular background model which adapts complex dynamic surveillance environment. Considering that each background pixel has a relatively concentrated distribution, it could be described with the Gaussian distribution. Mixed Gaussians will maintain multiple models for each pixel so that the background model could effectively estimate the complex background scenarios such as leaves, the water surface ripples, etc. Since the algorithm assumes pixels are independent, it could not guarantee the integrity of the foreground.

The RPCA (Robust Principal Component Analysis) problem, also named low rank matrix recovery problem, was firstly proposed in compressed sensing technology and was used to sample and reconstruct signals. John Wright [20] showed the great improvements in dimension reduction and face recognition using RPCA.

RPCA algorithm makes good use of subspace projection and expresses video data well and reduces dimensionality greatly. Oliver [8] proposed a method of building background model with Principal Component Analysis (PCA) technique. With RPCA, frame data are sorted from a $w*h$ matrix into a $[w*h\ 1]$ vector and N frames are sorted in the same way. And finally these frames are composed into $w*h*N$ matrix. Using RPCA in background modeling for static camera, the background sequences are basically the same. If each image is saved as a column vector of the data matrix, then the matrix is low rank, and the background can be recovered by restoring the low rank matrix through convex optimization. This algorithm shows great advantage over classic Single Gaussian (SG), MOG, and KDE algorithms in Wallflower datasets in reference [21].

This paper proposes a fast RPCA method by improving the optimization computation process of existing RPCA. In this work, our approach focuses on the utilization of the fast optimization method on the existing RPCA framework. The experiments show great improvements in background recovering and show better segmentation between background and foreground, which is especially efficient in time complexity.

The rest of the paper is organized as follows. In Sect. 2, we briefly describe the existing RPCA algorithm for background recovering. In Sect. 3, the new fast RPCA algorithm is introduced. In Sect. 4, we present the improved optimization process. In Sect. 5, experiments and discussions are presented.

2   The RPCA Problem

Assuming that the video is composed of n frames of size *width*height*. Let $X = [x_1, \ldots, x_n] \in R^{d \times n}$ be a rectangular matrix, each column is a vectorized video frame. d denotes the *width*height* image dimension and n is the total number of the images. RPCA seeks a low rank data representation matrix Z and sparse matrix E considering the contained outliers, the RPCA can be formulated as,

$$\min_{Z,E} \ rank(Z) + \lambda \|E\|_0 \qquad (1)$$

where λ is a balance parameter, and E is the representation error which must be sparse matrix [22]. While the $rank(Z)$ and L0-norm computation is NP hard, then the problem can be relaxed to compute nuclear norm and L1-norm, which is convex and can be optimized. (L0-norm is the numbers of none-zero elements in the vector. L1-norm is the sum of the absolute values of the elements in the vector, and is also called sparse rule operator—lasso regularization)

$$\min_{Z,E} \ \|Z\|_* + \lambda \|E\|_1 \qquad (2)$$

3   Fast RPCA Via Reformulation

In this section, we introduce fast RPCA method, which we proposed to efficiently solve RPCA problem and background model problem.

In machine learning, compared with L1-norm, which only considers the minimization of sparse matrix, L2-norm could prevent overfitting and improve model generalization ability. On the basis of the L2-norm, L2, 1-norm is defined as follows, and it shares similar sparse patterns from multiple different tasks.

$$\|E\|_{2,1} = \sum_{i=1}^{n} \|e_i\| \tag{3}$$

where e_i is the i-th column of matrix E.

L2, 1-norm regularization is important in foreground detection when each column vector in the matrix share the same pattern after L2, 1-norm regularization. Then an improved formulation of RPCA is presented as follows,

$$\min_{L,E} \|Z\|_* + \lambda \|E\|_{2,1}$$
$$s.t., \ X = Z + E \tag{4}$$

r is the rank of X. Through SVD decomposition, X can be represented by

$$X = U_r S_r V_r^T \tag{5}$$

where $U_r \in R^{d*r}$ and $V_r \in R^{n*r}$ are two column-wise orthogonal matrices. $U_r' U_r = V_r' V_r = I_r$, and $S \in R^{r*r}$ is a diagonal matrix.

According to the theory in reference [23], let W^* denote the optimal solution of the following problem, the inference can be conducted as,

$$\min_{W,E} \|U_r W\|_* + \lambda \|E\|_{2,1}$$
$$s.t., \ U_r S_r V_r^T = U_r W + E \tag{6}$$

Then $\{Z^*, E^*\}$ can be defined as

$$\begin{cases} Z^* = V_r W^* \\ E^* = X - V_r W^* \end{cases} \tag{7}$$

which is proved to share the same optimal solution of the problem in (4).

Replacing E with $E = U_r S_r V_r^T - U_r W$ in (6), the RPCA problem can be rewrite as follows,

$$\min \|U_r W\|_* + \lambda \|U_r S_r V_r^T - U_r W\|_{2,1} \tag{8}$$

Which is equivalent to

$$\min_W \|W\|_* + \lambda \|S_r V_r^T - W\|_{2,1} \tag{9}$$

then the objective function can be rewritten as,

$$\min_{W,Q} \|W\|_* + \lambda\|Q\|_{2,1}$$

$$\text{s.t., } S_r V_r^T = W + Q \tag{10}$$

4 Optimization

To efficiently solve the problem, ALM (Augmented Lagrangian Method) is adopted. ALM is an improvement of quadratic penalty method, which can ensure the convergence of algorithm and also increase its adaptability. ALM is widely used in the optimization of RPCA and its variants. Then the corresponding augmented Lagrangian of formulation (10) is

$$\ell(W,Q,\Lambda) = \|W\|_* + \lambda\|Q\|_{2,1} + \langle L, P - W - Q\rangle + \frac{\mu}{2}\|P - W - Q\|_F^2 \tag{11}$$

where $\mu > 0$ is penalty parameter, and L is the lagrangian multiplier.

In this section, iteratively updating W, Q as a two-variable optimization problem is conducted as follows.

4.1 Updating W

The sub-problem $\min_{W} \ell(W,Q,L)$ for updating W is given by,

$$
\begin{aligned}
\min_{W} \ell(W,Q,L) &= \|W\|_* + \langle L, P - W - Q\rangle + \frac{\mu}{2}\|P - W - Q\|_F^2 \\
&= \|W\|_* + \frac{\mu}{2}\left\|P - Q - W + \frac{L}{\mu}\right\|_F^2 \\
&= \|W\|_* + \frac{\mu}{2}\|W - G\|_F^2
\end{aligned}
\tag{12}
$$

where G is defined as $G = P - Q_t + \frac{L}{\mu}$.

Apply the singular values shrinkage operator to G and we can get the optimal solution, i.e.,

$$D_{1/\mu}(G) = U_G\, diag\left(\left[S_G - \frac{1}{\mu}\right]_+\right)V_G^T \tag{13}$$

Algorithm 1. Solving the problem in formulation 9

Input : S_r, V_r, λ.

Initialize $W_0 = Q_0 = L_0 = 0_{r*n}, t = 0$.

Parameters $\mu, \mu_{max}, \gamma, \varepsilon$.

While not converged **do**

　1. $W_{t+1} = \arg\min_{W} \ell(W, Q_t, L_t)$.

　2. $Q_{t+1} = \arg\min_{W} \ell(W_{t+1}, Q, L_t)$.

　3. $L_{t+1} = L_t + \mu(P - W - Q)$.

　4. $\mu = \min(\gamma * \mu, \mu_{max})$.

　5. $i\,t = t + 1$.

　6. Check whether $\left\|P - W_t - Q_t\right\|_{max} \leq \varepsilon$(the convergence condition) is satisfied.

End while

Output: $W^* = W_t$

---

where $U_G \, diag\,(S_G)V_G'$ is the skinny SVD of G, and $\lfloor \cdot \rfloor_+$ is a threshold operator which sets negative elements to zeros.

4.2　Updating Q

The subproblem $\min_{Q} \ell(W, Q, L)$ for updating Q is given by,

$$\min_{Q} \ell(W, Q, L) = \min_{Q} \lambda\|Q\|_{2,1} + \frac{\mu}{2}\|Q - C\|_F^2 \qquad (14)$$

where $C = P - W_{t+1} + \dfrac{L_t}{\mu}$. Denote the i-th column of Q and C as q_i and c_i respectively. Then problem 14 can be rewritten as

$$\min_{q} \lambda \sum_{i=1}^{m} \|q_i\| + \frac{\mu}{2} \sum_{i=1}^{m} \|q_i - c_i\|^2 \qquad (15)$$

which is separable with regard to q_i. Therefore it can be solved by optimized m subproblems. The optimal solution is

$$q_i = \begin{cases} \frac{\|c_i\| - \lambda/\mu}{\|c_i\|} c_i, & if \ \|c_i\| > \lambda/\mu \\ 0, & otherwise \end{cases} \quad (16)$$

where c_i denotes the i-th element of C.

4.3 Time Complexity

Empirically, the number of iteration is relatively small. In Algorithm 1, the most computational cost is in updating $\{W, Q\}$. So the time complexity of each iteration is $O(rm\min(r,m) + rm)$.

5 Experiments and Analysis

There are three parameters λ, γ, μ in the implementation of fast RPCA. γ and μ are the growth parameter and penalty parameters in the ALM algorithm which monitor the convergence rate of algorithm. For they make no difference to the result if they are set in an appropriate range, so we set it to $\lambda = 0.01, \gamma = 1.01, \mu = 0.03$ empirically.

5.1 Time Expense

Experiment of time expense is conducted under MAC OS X Yosemite (2.2 GHz Intel Core i7 16 GB 1600 MHz DDR3) with matlabR2014b. In this experiment, we compete our fast RPCA algorithm on the benchmark—**LRSLibrary** [21]. The

Table 1 The expense of time (sec)

DataSets [24]	ALM [25]	RPCA [26]	GoDec [27]	FPCP [28]	Fast RPCA
Car (51)	6.91	**6.04**	0.01	0.01	**0.005**
Escalator (197)	176	**109.89**	0.601	0.41	**0.21**
Highway (1698)	>2550	2550	37.2	6.78	**16.2**
Shop (156)	281	**98**	0.59	0.152	**0.164**

Note Numbers in parentheses is the total number of frames in the video

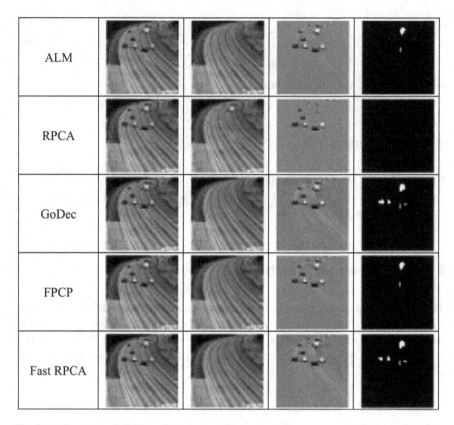

Fig. 1 Performance of RPCA variants on car video sequence

result is encouraging. Table 1 shows that fast RPCA is super efficient than RPCA and is roughly the same as FPCP.

After reading the literature, the speeds of RPCA algorithm and its variants are sorted as shown in Table 1 above conventionally.

Figures 1, 2, 3 and 4 shows the performance of Fast RPCA.

Results show that our algorithm is much faster than RPCA while have better performance than RPCA through directly observation.

5.2 Performance

Precision and recall experiments are conducted under windows (Intel Core i7-4790 CPU 3.6 GHz 16 GB RAM) with matlabR2011a 64bit (Table 2).

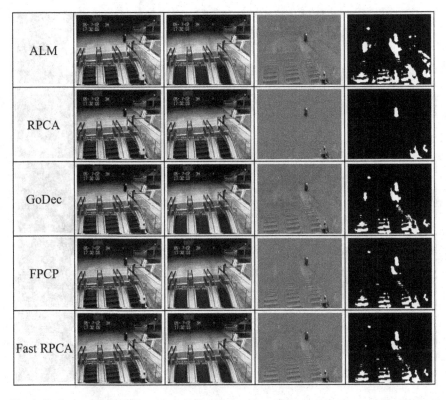

Fig. 2 Performance of RPCA variants on escalator video sequence

Fig. 3 Performance of RPCA variants on highway video sequence

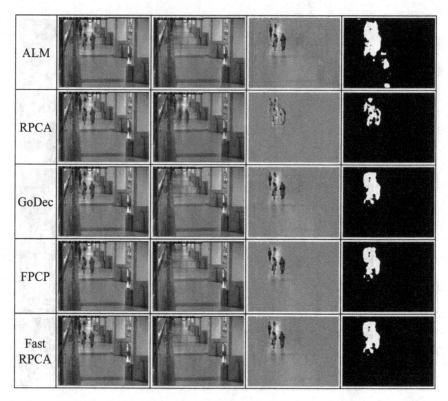

Fig. 4 Performance of RPCA variants on shop video sequence

Table 2 Results comparison with [21] using Wallflower Datasets [29]

Datasets / Algorithms		Bootstrap (3055)	Camouflage (353)	Foreground aperture (2113)	Light switch (2715)	Moved object (1745)	Time of day (5890)	Waving tree (287)
SG	FN	2215	4101	3464	1857	0	949	3110
	FP	92	2040	1290	15123	0	535	357
MOG	FN	1874	398	2442	1633	0	1008	1323
	FP	217	3098	530	14169	0	20	341
KDE	FN	1755	238	2413	760	0	1298	170
	FP	933	3392	624	14153	0	125	589
SL_PCA	FN	304	350	2441	962	0	879	1027
	FP	6129	1548	537	362	1065	16	2057
Fast RPCA	FN	1180	1840	3340	998	0	1144	**336**
	FP	**463**	7423	623	776	**630**	144	6673

FN	GroundTruth = 1	Test Result = 0	Meaning: missing
FP	GroundTruth = 0	Test Result = 1	Meaning: wrong

Note Numbers in parentheses is the total number of images in dataset

5.3   Conclusion

Experiment 1.5.1 shows that time expense of fast RPCA is greatly improved than RPCA and other variants, which represents that the novel optimization process is valid and efficient. Experiment 1.5.2 is conducted only the initialization process of background modeling. The results show no obvious improvement because that the background model is built without updating procedure, which is good enough than others in initialization process at first sight.

This paper showed that our proposed method exceeds the existing RPCA in time complexity in a great manner, while keeps and even improves the modeling performance over other statistical modeling algorithms and other RPCA variants.

References

1. Chiranjeevi P, Sengupta S. Spatially correlated background subtraction, based on adaptive background maintenance. J Vis Commun Image Represent. 2012;23(23):948–57.
2. Marghes C, Bouwman T. Background modeling via incremental maximum margin criterion. In: International conference on computer vision. Springer;2010. P. 394–03.
3. Huang DY, Chen CH, Hu WC, et al. Reliable moving vehicle detection based on the filtering of swinging tree leaves and raindrops. J Vis Commun Image Represent. 2012;23(4):648–64.
4. Wren CR, Azarbayejani Ali, Darrell Trevor, et al. Real-time tracking of the human body. IEEE Trans Pattern Anal Mach Intell. 1997;19(7):780–5.
5. Elgammal, A, Harwood D, Davis L. Non-parametric model for background subtraction. Lect Notes Comput Sci.
6. Stauffer, Chris, Grimson, W.E.L. Adaptive background mixture models for real-time tracking. CVPR. IEEE Comput Soc. 1999;2246.
7. Barnich O, Van DM. ViBe: a universal background subtraction algorithm for video sequenc-es. IEEE Trans Image Process Publ IEEE Signal Process Soc. 2011;20(6):1709–24.
8. Oliver N, Rosario B, Pentland A. A Bayesian computer vision system for modeling human interactions. IEEE Trans Pattern Anal Mach Intell. 2000;22(8):831–43.
9. Pham V Q, Takahashi K, Naemura T. Foreground-background segmentation using iterated distribution matching. In: CVPR, IEEE computer society conference on computer vision and pattern recognition. IEEE computer society conference on computer vision and pat-tern recognition;2011. P. 2113–20.
10. Brutzer S, Hoferlin B, Heidemann G. Evaluation of background subtraction techniques for video surveillance. In: IEEE conference on computer vision & pattern recognition. IEEE;2011, 1937–44.
11. Karaman M, Goldmann L, Yu D, et al. Comparison of static background segmentation methods. In: Proceedings of SPIE—The international society for optical engineering, vol. 5960(4);2005. p. 2140–51 2005.
12. Menze M, Geiger A. Object scene flow for autonomous vehicles. Computer vision and pattern recognition. IEEE;2015.
13. Appiah K, Hunter AA. Single-Chip FPGA implementation of real-time adaptive background model. In: IEEE international conference on field-programmable technology;2005. P. 95–02.
14. Kryjak T, Komorkiewicz M, Gorgon M. Real-time background generation and fore-ground object segmentation for high-definition colour video stream in FPGA device. J Real-time Image Process. 2014;9(1):61–77.

15. Rodriguez-Gomez R, Fernandez-Sanchez EJ, Diaz J, et al. FPGA implementation for real-time background subtraction based on horprasert model. Sensors. 2012;12(1):585–611.
16. Bouwmans T. Traditional and recent approaches in background modeling for foreground detection: an overview. Comput Sci Rev. 2014;11–12:31–66.
17. Lee B, Hedley M. Background estimation for video surveillance. IVCNZ. 2002;2002:315–20.
18. Mcfarlane NJB, Schofield CP. Segmentation and tracking of piglets in images. Mach Vis Appl. 1995;8(3):187–93.
19. Wang N, Zheng J, Wang Y, et al. Extracting roadway background image: mode-based approach. Trans Res Rec J Trans Res Board. 2006;1944(1).
20. Wright J, Ganesh A, Rao S, et al. Robust principal component analysis: exact recovery of corrupted low-rank matrices. J ACM. 2009;58(3):11.
21. Bouwmans T, Baf FE, Vachon B. Statistical background modeling for foreground detection: a survey. Handbook of pattern recognition and computer vision;2009. P. 181–99.
22. Guyon C, Bouwmans T, Zahzah E H. Robust principal component analysis for background subtraction: systematic evaluation and comparative analysis. Principal component analysis. InTech;2012.
23. Xiao S, Li W, Xu D, et al. FaLRR: a fast low rank representation solver. In: IEEE conference on computer vision and pattern recognition. IEEE;2015. P. 4612–20.
24. Sobral A, Bouwmans T, Zahzah EH. LRSLibrary: Low-Rank and sparse tools for background modeling and subtraction in videos. Handbook on "robust low-rank and sparse matrix decomposition: applications in image and video processing"; 2016.
25. Tang G, Nehorai A. Robust principal component analysis based on low-rank and block-sparse matrix decomposition. Information sciences and systems. IEEE;2011. P. 1–5.
26. Torre FDL, Black MJ. Robust principal component analysis for computer vision. In: Proceedings of the eighth IEEE international conference on computer vision, 2001. ICCV 2001, vol. 1, IEEE;2001. P. 362–9.
27. Zhou T, Tao D. GoDec: Randomized low rank & sparse matrix decomposition in noisy case. In: International conference on Machine learning, ICML 2011, Bellevue, Washington, USA, June 28–July. DBLP;2011. P. 33–40.
28. Rodriguez P, Wohlberg B. Fast principal component pursuit via alternating minimization. In: IEEE international conference on image processing. IEEE;2014. P. 69–73.
29. Toyama K, Krumm J, Brumitt B, et al. Wallflower: principles and practice of background maintenance. In: IEEE international conference on computer vision, vol. 1, IEEE Xplore;1999. p. 255–61.

Esophagus Tumor Segmentation Using Fully Convolutional Neural Network and Graph Cut

Zhaojun Hao, Jiwei Liu and Jianfei Liu

Abstract The development of Esophagus radiation treatment plan demands accurate Esophagus tumor segmentation. However, such task was often prevented by random distribution and weak boundaries of Esophagus tumors on CT images. To address these challenges, we develop a novel framework based on the combination of Fully Convolutional Neural Network (FCN) and graph cut algorithms. FCN is utilized to establish an Esophagus tumor classifier on the training dataset with expert-labeled tumor regions. When segmenting Esophagus tumors on the test dataset, the tumor probability maps are first estimated. Graph cut is next used to extract the actual tumor regions by enforcing the spatial constraints. 87 CT sequences were selected as the validation dataset, and 3-fold cross-validation was performed to evaluate the segmentation accuracy. Tumor volume overlap between ground-truth and segmentation results was only 71% by exploiting FCN alone, while it was improved to 80% by combining graph cut algorithm. These promising results suggest that the combination of FCN and graph cut can accurately segment Esophagus tumors, which has a great potential to reduce human burden in contouring tumor regions as well as improve the accuracy of radiation treatment planning.

Keywords Fully convolutional neural network · Graph cut · Esophagus tumor · Segmentation

Z. Hao · J. Liu (✉)
School of Automation and Electrical Engineering, University of Science and Technology Beijing, 100083 Beijing, China
e-mail: liujiwei@ustb.edu.cn

J. Liu
National Eye Institute, Nation Institutes of Health, Bethesda, MD 20892, USA

© Springer Nature Singapore Pte Ltd. 2018
Y. Jia et al. (eds.), *Proceedings of 2017 Chinese Intelligent Systems Conference*,
Lecture Notes in Electrical Engineering 460,
https://doi.org/10.1007/978-981-10-6499-9_39

413

1 Introduction

The Esophagus tumors always come from the Esophagus. It is between the throat and the stomach [1]. Symptoms often manifest as swallowing difficult, weight loss, swallowing pain, hoarseness, expansion of lymph nodes, dry cough or hematemesis [2].

All the time, segmentation of Esophagus tumor is an important step to improve the accuracy of radiation treatment planning and computer-aided decision support systems (CAD). And contouring Esophagus tumor regions in CT are important biomarkers of tumor disease progression. However, manual or even semi-manual techniques are usually operator-dependent, time-consuming and non-objective in clinical practice [3].

In order to improve the productivity of radiologists, it's significant to explore an efficient method of automatic Esophagus tumor segmentation. As we know, one cut algorithm effectively replaces approximate iterative optimization techniques based on the descent of block coordinate. One cut term is good-fitting for interactive segmentation [4] (with user seeds interfaces or bounding box). However, in Esophagus ct volumes, it's impossible to sign the bounding box or seeds for low-contrast between neighbor organs (as is shown in Fig. 2). Furthermore, the tumor has variation of shape and size. So, traditional segment method alone is no longer valid to Esophagus tumor segmentation. There is rarely successful method of automatic Esophagus tumor segmentation for now.

In recent years, Deep Convolutional Neural Networks CNN has gained high focus in field of science. CNN has shown excellent performance at solving computer vision tasks including object recognition [5], classification and also object segmentation. These improvement motivate us to explore a machine learning solution that not only robust but also effective. Our proposed method extract the feature of Esophagus tumor and then complete the segmentation with graph cut, for the reason that FCN is only able to localize segment boundaries at a low level of accuracy [6], then graph cut is able to cover the shortage of FCN. Our method exploits both local features as well as global contextual features simultaneously.

2 FCN Architecture

Fully Convolutional Neural Network is a rich class of deep learning model [7]. Our FCN architecture is first fine-tuned based on voc-fcn8 s caffemodel weights, which fuse additional predictions from pool3 and pool4, provides further precision. It combines rough, high layer information and fine, low layer information [6], then trained and tested with Caffe [8]. Further more, we use structure image as the ground truth and then sample the bounding boxes of the largest region which has Esophagus tumors as the input images.

During training step, in order to generate more data for training, we apply non-rigid deformations first. The degree of rotate is 30°, 60°, 90° and 110°. This

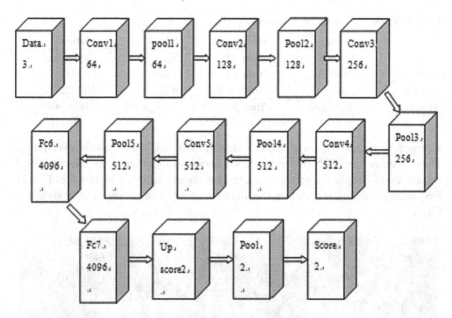

Fig. 1 Architecture of our FCN method

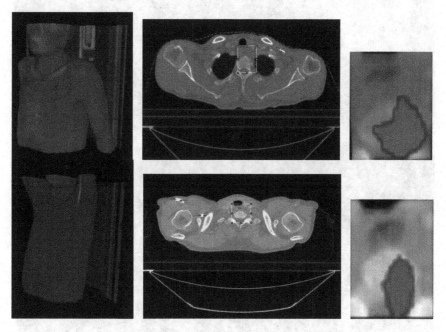

Fig. 2 First column is a ct series of two patients. Second column is one of the slices of CT volumes and the yellow box is the largest possible region of tumor. Third column is the largest possible region with Esophagus (labeled in red color)

Table 1 Result of 3-fold cross validation network

	Patient1	Patient2	Patient3	Patient4
Weights	Voc-fcn8 s	Train_iter	Train_iter1	Train_iter2
Accuracy	0.8809	0.8913	0.8913	0.8887
Mean IoU	0.6853	0.6951	0.7210	0.7375
Model	Train_iter	Train_iter1	Train_iter2	Train_iter3

approach is usually referred to augment data and can help to avoid over-fitting. Then we trained the FCN network in a 3-fold cross-validation on the Esophagus CT dataset, as 87 Esophagus ground-truth images are available (as is shown in Fig. 2). Fine-tuning once takes more or less than one hour. The architecture is shown in Fig. 1.

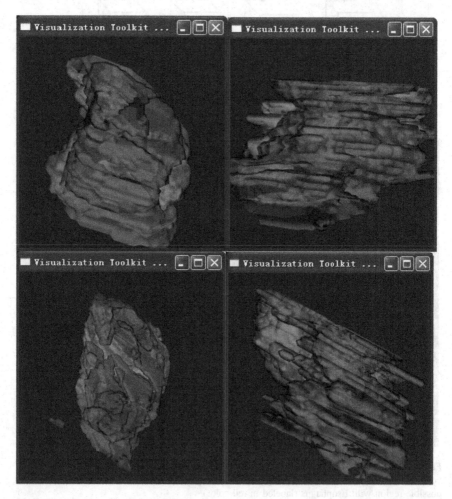

Fig. 3 Esophagus tumor of each patient

Table 1 shows the training result of our Fully Convolutional Neural Network, we fine-tuned the network step by step to get our final caffemodel. IOU means intersection over union. Now we get the segmentation result of FCN and possibility of each pixel belonging to background as well as tumor region (Fig. 3).

3 Method

In the image segmentation algorithm, there are two main groups. One of the methods assumes the appearance model is known and the other method estimates the appearance model jointly with the segmentation result [9]. Recently, algorithms combining graph cut have already been widely concerned in the field of vision and graphics for the reason that graph cuts combine not only boundary regularization but also the spatial properties. Approaches combining graph cut have some interesting connections with segmentation methods such as snakes, geodesic active contours, and level-sets, and so on. The segmentation energy is optimized by max-flow methods.

In order to segment a given image, first of all, we create a graph with every nodes corresponding to every pixel of the given image [10]. And there are two extra nodes: a "foreground or object" terminal (called a source S) and a "background" terminal (called a sink T). Then we assign weights to every edge in the graph. The set of edges E consists of two types of edges: n-links (neighborhood links) and t-links (terminal links). Each pixel has two links s-link and t-link connecting it to each terminal source S and sink T. Each pair of neighboring pixels in N (neighbor pixels sets) is connected by an n-link, it is related to the size of the neighborhood.

The weights assigned to the edges are shown in Table 2, where C and K are Const, N{p, q} is sets of neighbor pixels, prob (bg) is the possibility of each pixel belonging to background which is get from FCN segmentation result, prob (fg) is the possibility of every pixel belonging to foreground or object.

Many segment tasks contain assigning a label (such as disparity) to every pixel in the given image [11]. Hard constraint is that the labels should vary smoothly almost everywhere as sharp discontinuities may exist, for example, at object boundaries. These tasks are naturally stated in terms of energy minimization, the

Table 2 Weight assign of the graph

Edges	Weight	When
{p, q}	N{p, q}	{p, q} $\in$ N
{p, S}	0	P $\in$ bg
	C	P $\in$ fg
	K*pr(bg)	else
{p, T}	C	P $\in$ bg
	0	P $\in$ fg
	K*pr(fg)	else

first item is determined by data properties and the second item is determined by region properties.

Our energy function is defined as below function.

$$E(x) = \sum_i pr(x_i) + \sum_{ij} \theta_{ij}(x_i, x_j)$$

where

$$
\begin{aligned}
pr(x_i) &= pr(bg) = -\log p(x_i) & \text{when } x_i \in \text{background} \\
pr(x_i) &= pr(fg) = -\log p(x_i) & \text{when } x_i \in \text{foreground}
\end{aligned}
\tag{1}
$$

$$\sum_{ij} \theta_{ij}(x_i, x_j) \propto \exp\left(-\frac{(I_p - I_q)^2}{2\sigma^2}\right) \cdot \frac{1}{dist(p, q)} \tag{2}$$

Function (1) reflect on how the intensity of pixel fits into given intensity models (e.g. heatmap) of the object and background, function (2) penalizes a lot for discontinuities between pixels of similar intensities.

Our algorithms is shown below

1. FCN segment result, including image I(x, y) and heatmap which shows prob(fg) and prob(bg).
2. Erode I(x, y) as foreground seed fg, dilate I(x, y) as background seed bg.
3. Initialize graph

   3.1 if $\{p, q\} \in N$, then cost = N{p, q}
   3.2 if $p \in fg$, s-link = C, t-link = 0

   else if $p \in bg$, s-link = 0, t-link = C

   else s-link = K*pr(bg), t-link = K*pr(fg)

4. Max-flow [12], minimize E(x)

Erode FCN segmentation result as the region seed and dilate result as the background seed.

4   Experimental Result and Analysis

Different from other kind of tumor, low-contrast between Esophagus and neighbor organs (as is shown in Fig. 4), as well as variation of shape and size (as is shown in Fig. 2) result to traditional segment method, such as watershed, level set, threshold and so on, rarely perform good. Here we evaluate the result using VO (volume overlap), DC (dice coefficient), RA (relative absolute volume difference), MSD

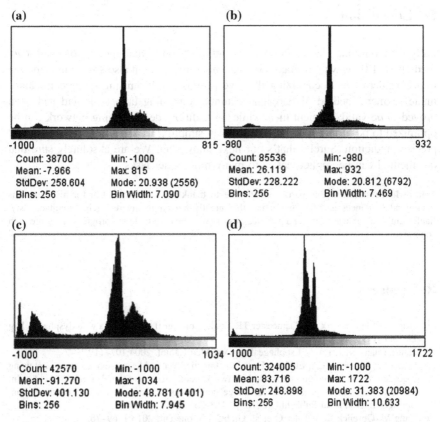

Fig. 4 First column is the pixel value analyze of the patient1. **a** is the region with tumor, **b** is region without tumor. Second column is the pixel value analyze of the patient2. **c** is the region with tumor, **d** is region without tumor

Table 3 Results evaluation with FCN-Graph cut segment methods

	DC (%)	VO (%)	RA (%)	MSD (mm)	ASD (mm)
Patient1	78	83	16	12	7
Patient2	74	80	6	18	9
Patient3	70	76	18	24	15
Patient4	81	81	10	21	13
Average	76	80	12	18	11

(maximum symmetric absolute surface distance), ASD (average symmetric absolute surface distance). The final experimental result comparison is shown in Table 3. Figure 3 shows the segmentation result of each patient, the blue region is the ground truth and the red region is our segment tumor region.

5 Conclusion

Fully Convolutional Neural Network is definitely up-to-date method of pixel-level prediction [13], as well as graph cut takes not only data but also spatial information into consideration. Recognizing this, we proposed a promising method for automatic segment method of Esophagus tumor combining both tools and had been proved to be valid, efficient meanwhile. In addition, deep learning network can be optimized to be further more simultaneously simplifying, speeding up learning and precise prediction. Surely, that's our next study plan. We are absolutely sure that our method can be applied easily to many more other tasks.

Acknowledgements In this part, I would like to thank my tutor Jiwei Liu for his numerous scientifical guidances and for motivation that greatly helped in my research. Simultaneously, Jianfei Liu's valuable advising and patience of supervise means a lot to my progress. They are both honorable.

References

1. Attwood SEA, Smyrk TC, Demeester TR, et al. Esophageal eosinophilia with dysphagia. Dig Dis Sci. 1993;38(1):109.
2. Haustermans K, Lerut A. Esophageal tumors. Med Radiol. 2004:107–119.
3. Christ PF, Elshaer MEA, Ettlinger F et al. Automatic liver and lesion segmentation in ct using cascaded fully convolutional neural networks and 3D conditional random fields. In: International conference on medical image computing and computer-assisted intervention. Springer International Publishing;2016. p. 415–23.
4. Tang M, Gorelick L, Veksler O et al. GrabCut in one cut. 2013:1769–76.
5. Krizhevsky A, Sutskever I, Hinton GE. ImageNet classification with deep convolutional neural networks. In: International conference on neural information processing systems. Curran Associates Inc;2012. p. 1097–105.
6. Shelhamer E, Long J, Darrell T. Fully Convolutional Networks for Semantic Segmentation. IEEE Trans Pattern Anal Mach Intell. 2017;39(4):640.
7. Chen LC, Papandreou G, Kokkinos I, et al. Semantic image segmentation with deep convolutional nets and fully connected CRFs. Comput Sci. 2014;4:357–61.
8. Jia Y, Shelhamer E, Donahue J et al. Caffe: convolutional architecture for fast feature embedding. 2014:675–78.
9. Boykov Y, Veksler O, Zabih R. Fast approximate energy minimization via graph cuts. IEEE Trans Pattern Anal Mach Intell. 2002;23(11):1222–39.
10. Boykov Y, Funka-Lea G. Graph cuts and efficient N-D image segmentation. Int J Comput Vis. 2006;(2):109–31.
11. Rother C, Kolmogorov V, Blake A. "GrabCut": interactive foreground extraction using iterated graph cuts. In: ACM SIGGRAPH. ACM;2004. p. 309–14.
12. Boykov Y, Kolmogorov V. An experimental comparison of min-cut/max-flow algorithms for energy minimization in vision. IEEE Trans Pattern Anal Mach Intell. 2004;26(9):1124–37.
13. Havaei M, Davy A, Wardefarley D, et al. Brain tumor segmentation with Deep Neural Networks. Med Image Anal. 2017;35:18–31.

Hierarchical Consensus Algorithm of Multi-agent System Based on Node-Contribution-Based Community Decomposition

Siyu Ye and Chen Wei

Abstract To improve the convergence speed of multi-agent system, a hierarchical consensus algorithm based on community decomposition is proposed. Considering converting the single-layer consensus problem to multi-layers consensus problem, the topology graph is divided into several sub-graphs by utilizing community decomposition algorithm based on node contribution firstly, and the sub-graphs achieve consensus respectively. And then apply the hierarchical decomposition consistency algorithm to the system. The convergence speed of the multi-agent system is improved significantly by optimizing the topology on the premise of maintaining the original topology constraints. For the first-order linear system, the effectiveness of this algorithm is demonstrated by simulations compared with the standard model.

Keywords Multi-agent system · Convergence speed · Community decomposition · Hierarchical decomposition consensus algorithm

1 Introduction

In recent years, consensus problem for multi agent has been greatly developed. Saber [1] first put forward the theoretical framework of the consistency problem, and design the most general consistency algorithm for first-order system. Inspired by the swarm model, Ren [2] introduced consensus problem for second-order system and derive necessary and sufficient conditions, through the analysis of eigenvalues.

The convergence rate of multi-agent is one of the most important problem of consensus problem. Saber et al. proposed that the algebraic connectivity is an important indicator to measure the convergence rate of the system. Xiao et al. [3]

S. Ye (✉) · C. Wei
School of Automation Science and Electrical Engineering, Beihang University,
Beijing 100191, China
e-mail: jill0707@buaa.edu.cn

© Springer Nature Singapore Pte Ltd. 2018   421
Y. Jia et al. (eds.), *Proceedings of 2017 Chinese Intelligent Systems Conference*,
Lecture Notes in Electrical Engineering 460,
https://doi.org/10.1007/978-981-10-6499-9_40

designed fast linear iterations for distributed averaging to design the network weights and reallocate the weights. Aysal et al. [4] proposed a method based on local node state prediction and She et al. [5] proposed a fast consensus protocol using both current states and outdated states for second-order directed network systems. Saber et al. [6] made the complex network into a small world network, the algebraic connectivity can be sharply increased. These consistency algorithms are generally in the single-layer topology, which are limited in the practical application so that scholars try to enhance the convergence rate based on multi-layers topology.

Newman et al. first proposed the concept of a complex network of community structure [7], the quality of its classification by the modularity Q [8] to measure. Based on this, Girvan and Newman [9] proposed a classic community detection algorithm, called the GN algorithm. For large-scale network with millions of nodes, Neman proposed NF algorithm. Qi et al. [10] proposed a community division framework based on the connection strength of nodes. Li et al. [11] proposed a hierarchical decomposition algorithm that transforms a single-layer topology into a multi-layer structure, effectively increasing the convergence rate. Tang Man et al. [12] introduced the spectral decomposition method in the community structure discovery, and utilize a hierarchical algorithm for improving the convergence speed.

The main contribution in this work is to introduce a new hierarchical consensus based on community decomposition of node contribution, and the main idea is to convert the single-layer consensus problem to multi-layers consensus problem without changing the original topology. For the single-layer topology, the topology graph is divided into several sub-graphs by employing a less complex community decomposition algorithm firstly, and the sub-graphs achieve consensus respectively, which the whole process speed up the convergence rate significantly.

The remainder of this paper is organized as follows. In Sect. 2, some background and necessary preliminaries are presented. In Sect. 3, the hierarchical consensus of Multi-agent Systems based on node contribution community decomposition algorithm is introduced. A simulation example is provided in Sect. 4 to demonstrate the effectiveness of this algorithm. Finally, the study is summarized in Sect. 5.

2   Background and Preliminaries

2.1   Graph Theory Notations

In this paper, undirected graphs are used to model the information exchange between agents. Let $G = (V, E)$ be an undirected graph with a set of n nodes $V = \{v_1, v_2, \ldots, v_n\}$ and an edge set $E = \{\varepsilon_1, \varepsilon_2, \ldots, \varepsilon_m\} \subseteq V^2$. The set of neighbors of agent i is denoted by $N_i = \{v_j \in V | (v_i, v_j) \in E\}$. The adjacency matrix $A = [a_{ij}]$ with nonnegative elements a_{ij} is a weighted adjacency matrix. In the following, it is assumed that $\varepsilon_{ij} \in E \Leftrightarrow a_{ij} > 0$ (for simplicity, choosing $a_{ij} = 1$), otherwise $a_{ij} = 0$. Moreover, it is assumed $a_{ii} = 0$. Thus, the adjacency matrix A is a symmetric nonnegative matrix. The notion D $(D = diag\{d_1, d_2, \ldots, d_n\})$ represents a degree matrix with d_i $\left(d_i = \sum_{j \neq i} a_{ij} \right)$ denoting the degree of node v_i. The graph Laplacian matrix associated with the undirected graph G is defined as $L(G) = L = D - A$.

2.2 Consensus Protocols

Considering that the multi-agent system consists of n agents. Each agent can be considered as a node in an undirected graph, and the information transfer between two agents can be regarded as an edge. The dynamics model of each agent is described in terms of the continuous-time first-order integrator:

$$\dot{x}_i(t) = u_i(t) \tag{1}$$

where $x_i(t)$ and $u_i(t)$ denotes the state value and the control input of agent i at time t respectively.

The linear standard consensus protocol proposed by Saber is written as:

$$u_i(t) = \sum_{j \in N_i} a_{ij}(x_i(t) - x_j(t)) \tag{2}$$

and the consensus process with fixed topology is presented by

$$\dot{x}(t) = -Lx(t) \tag{3}$$

where $x(t) = [x_1(t), x_2(t), \ldots, x_n(t)]^T$ and L is the corresponding Laplacian matrix of graph G.

Definition 1 The second small eigenvalue [1] $\lambda_2(L)$ of the Laplacian matrix of the multi-agent system is defined as algebraic connectivity.

The connectivity of a multi-agent system is often presented as an important index to measure the convergence rate and performance of the system consistency algorithm. Following reference 11, the time of convergence on a connected graph G can be approximately proportional to $\frac{1}{\lambda_2(L)}$. The larger λ_2 means that the information exchange between agents is better and the convergence rate is faster. For a hierarchical system, the overall convergence time can be assessed $\sum_{i=1}^{r} \frac{1}{\lambda_2^i}$, where r is the number of layers in the hierarchy.

3 The Hierarchical Consensus Algorithm

Community structure is one of the important characteristics of complex networks. Nodes belonging to the same community are tightly connected, and the connection between the different communities are sparse. The community decomposition of complex networks has great significance in real life. Therefore, the idea of community decomposition is considered to solve the problem improving the convergence rate.

Considering the utilization of hierarchical decomposition method to optimize the topology of multi-agent system, without changing the original topology structure, the multi-agent single-layer structure is converted into multi-layer structure. If each layer sub-graph convergence speed is faster than the convergence speed of the original topology, the overall convergence time is expected to improve than the original graph. The whole algorithm can be divided into two parts as following.

3.1 Single Layer Decomposition Algorithm

In this part, a community decomposition algorithm based on node contribution is proposed so that the topology is divided into a number of internally closely connected subgraphs. The community is a group that nodes belong to it are connected tightly and the connection between groups are often sparse. In order to describe the closeness of nodes, according to the idea of community attraction [13], the concept of community density is introduced firstly.

Definition 2 The density of community C is the ratio of the number of edges e and the number of edges that may exist in community C. The expression is as follows:

$$D(C) = \frac{2|e|}{|V|(|V|-1)} \tag{4}$$

where $|e|$ and $|V|$ denotes the number of edges and nodes in community C respectively. In addition, the density of the community $D(C) \in [0, 1]$. When $D(C) = 1$, indicating that all nodes in the community C are connected to each other, the nodes in the community formed a global coupling network. In the process of community formation, the nodes will be assigned to the community whose D is higher; On the other hand, the node will choose the community which have more number of connected edges with it.

It can be considered that the contribution of the node to the community is determined by the number of edges between the node and the community and the density of the connected community. The density of community is greater, the possibility of the node belonging to the community is higher. Inspired by Qi et al. algorithm [10], the concept of node contribution is proposed.

Definition 3 The contribution of node i to community is $Con(i)$:

$$Con(i, C_k) = \alpha D(C_k) + \beta \frac{\sum\limits_{j \in C_k} A_{ij}}{d(i)} \tag{5}$$

where α, β are adjustment parameter, which satisfied $\alpha + \beta = 1$. $D(C_k)$ denotes the density of community C_k, and $\sum\limits_{j \in C_k} A_{ij}$ denotes the number of edges connected node i with community C_k; $d(i)$ represents the degree of node i.

In the network, any community C is a subgraph of network G, that is $C \subset G$. And for each node i in C, whose contribution to community C is required to be more than to any other community.

The main idea of the algorithm is as follows: Assuming that the initial division of the network G is $C_1, C_2, \ldots, C_k$, for each node i in the network, calculate the contribution of the node i to each neighborhood community, and divide the node i to the community corresponding to the maximum contribution, all nodes in the graph repeat the process until the community structure is stable.

The specific process of the algorithm is described as following. The multi-agent system has n nodes, the topology graph is G.

(1) Supposed that all nodes are in the same community and calculate the density of it. If $D(C) \neq 1$, the community will be divided.
(2) The nodes in the graph are initially divided by the rule that each node and its half of the neighbor nodes are divided into a community until each node in the graph is divided into a single community $C_1, C_2, \ldots, C_k$.
(3) Calculate the contribution $Con(i)$ according to (5) of each node i to all neighborhood communities that denotes all the communities connected with node i.
(4) If the new maximum node contribution is greater than the contribution of node i to the original community, the node i will be moved into the new community k, otherwise it will remain in the original community;
(5) Repeat steps (3)–(4) until the structure of the community in the network is stable.

This method ensure that each layer can be effectively divided into several subgraphs which internal connections are close, the convergence rate of each subgraph is faster than the original topology of the convergence rate. And the single layer decomposition is finished through the progress.

3.2 Hierarchical Decomposition Consistency Algorithm

After the single layer decomposition algorithm, all nodes are decomposed into different communities that means the original graph is decomposed into several subgraphs which are tightly connected. So we can get a new internal connection graph consists of these subgraphs, each of whose convergence time is faster than the original graph. And these decomposed subgraphs are considered as new nodes in the next layer. Then construct the information transfer relationship between the higher-layer topology and the lower-layer topology, and convert the topology to the multi-layers structure. In the process of higher-layer nodes changing the state, the state of the lower-layer node is always kept consistent with the state of the node that is considered as a new node in the next layer belongs to the community where the lower-layer node is in. The specific process of hierarchical decomposition consistency algorithm is introduced as following:

The multi-agent system has n nodes, the topology graph is G. r^i denotes the number of decomposed communities in ith-layer.

(1) Initialization.
(2) Decomposition. Decompose the original graph using single layer decomposition algorithm proposed in part 1.
(3) Getting r^i subgraphs after community decomposition, which are $G_1, G_2, \ldots G_{r^i}$. Applying the standard consensus protocol to r^i subgraphs make them achieve convergence respectively.
(4) Structuring new graph. Considering decomposed subgraphs as new nodes, repeat the step (2) to (3) in the next layer until the graph cannot be divided any more.
(5) Applying the standard consensus protocol to the last layer graph, the whole system achieve convergence eventually.

4 Simulation

Considering an agent system consists of 10 nodes, whose topology is depicted in Fig. 1a by assuming the weight of each edge to be equal to 1. The density of this graph is $D(G_a) = 0.29$. For this system, the hierarchical decomposition algorithm is used for community decomposition. The hierarchical structure is shown in Fig. 1. Apply single layer decomposition algorithm to obtain the first layer structure, which consists of three subgraphs. The second layer structure can also be obtained by hierarchical decomposition consistency algorithm. The second smallest eigenvalues of the original graph Laplacian is $\lambda_2(L) = 0.176$. The convergence time of the original topology graph utilizing standard consensus protocol (Fig. 2a) is 16 s. And the second smallest eigenvalues of the first layer subgraphs are $\lambda_2^1 = 3, \lambda_2^2 = 3, \lambda_2^3 = 2$, the reciprocal sum of which $\sum_{i=1}^{r} 1/\lambda_2^i = 1.167$ is larger than $\lambda_2(L) = 0.176$. The convergence time (Fig. 2b) after applying hierarchical decomposition algorithm is 6.2 s faster than utilizing standard consensus protocol (2) directly.

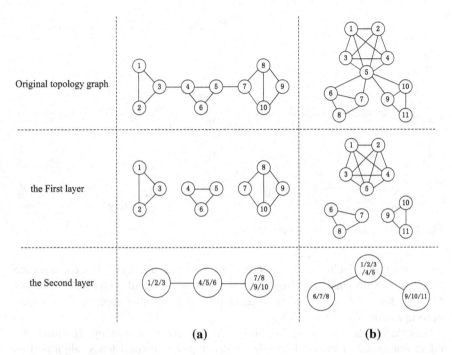

(a)   **(b)**

Fig. 1 Hierarchical structure

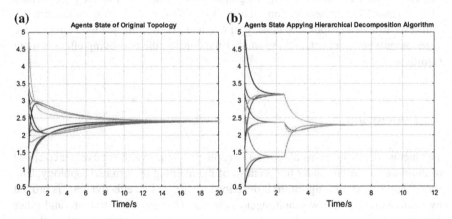

Fig. 2 Multi-agents states of topology (**a**) in Fig. 1

Considering an agent system consists of 11 nodes, whose topology is depicted in Fig. 1b. The density of this graph is $D(G_b) = 0.36$. The hierarchical structure is also shown in Fig. 1. Apply single layer decomposition algorithm to obtain the first

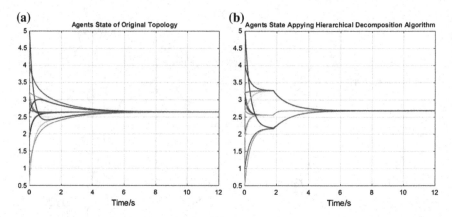

Fig. 3 Multi-agents states of topology (**b**) in Fig. 1

layer structure, which consists of three fully coupled subgraphs. The convergence time of the original topology graph utilizing standard consensus protocol (2) (Fig. 3a) is 7.2 s, and the convergence time (Fig. 3b) after applying hierarchical decomposition algorithm is 5.8 s.

Comparing the time assuming of applying different consistency algorithm, the effectiveness of proposed hierarchical decomposition consistency algorithm is demonstrated. However, different topology of different graph has different convergence time. The density of original graph is not the same so that the effectiveness of the hierarchical decomposition algorithm is different. According to the simulation results of two graphs, the density of original graph influence the convergence speed improvement effect, and the smaller the density, the improving effect is more significant.

5   Conclusion

The hierarchical consensus algorithm is applied to improve the convergence speed of multi-agent system under the premise of maintaining the original topology. The specific process of this algorithm is described clearly and its effectiveness, is proved by simulation through two multi-agent systems. The topology structure and other factors' influences on the effectiveness of the algorithm, whether this algorithm is suitable for second-order system and time delay system will be further studied in the future.

References

1. Olfati-Saber R, Murray RM. Consensus problems in networks of agents with switching topology and time-delays. IEEE Trans Autom Control. 2004;49(9):1520–33.
2. Ren W, Atkins E. Second-order consensus protocols in multiple vehicle systems with local interactions. In: AIAA guidance, navigation, and control conference and exhibit;2005. p. 15–18.
3. Xiao L, Boyd S. Fast linear iterations for distributed averaging. Syst Control Lett. 2004;53:65–78.
4. Aysal TC, Oreshkin BN, Coates MJ. Accelerated distributed average consensus via localized node state prediction. IEEE Trans Signal Process. 2009;57(4):1563–76.
5. She Y, Fang H, Fast consensus for multi-agent systems in directed networks. Control Decis. 2010; 25(7):1026–30.
6. Olfati-Saber R. Ultrafast consensus in small-world networks. In: Proceeding of American control conference, Portland, OR;2005. p. 2371–78.
7. Girvan M, Newman MEJ. Community structure in social and biological networks. Proc Nat Acad Sci. 2002; 99(12):7821–26.
8. Newman MEJ. Finding community structure in networks using the eigenvectors of matrices. Phys Rev E. 2006; 74:036104.
9. Girvan M, Newman MEJ. Finding and evaluating community structure in networks. Phys Rev E. 2004; 69.
10. Qi Y, Bin W et al. Detecting communities in massive networks based on local community attractive force optimization. In: Proceedings of the international conference on advances in social networks analysis and mining, Odense, Denmark. 2010. p. 291–95.
11. Li X, Xi Y. Hierarchically decomposing multi-agent system to accelerate group consensus. In: Proceedings of 2011 8th Asian control conference, Kaohsiung;2011. p. 347–52.
12. Tang M, Li X, Zhao S, Liu Y. Hierarchical consensus of multi-agent systems based on community decomposition. IFAC Proc Vol. 2013; 46(13).
13. Hu YQ, Chen HB et al. Comparative definition of community and corresponding identifying algorithm. Phys Rev E. 2008; 78(026121):1–11.

An Immersive Roaming Method Based on Panoramic Video

Quanfa Xiu, Xiaorong Shen, Tianlong Zhang and Luodi Zhao

Abstract The idea for panoramic video immersive roaming is from the current mature street view, such as Google Street View, and Baidu Street View. The method of video's immersive roaming proposed in this paper will overcome the disadvantages of the Google Street View which based on panoramic pictures. The method mainly involves the adaptive tuning of the camera algorithm, the immersive viewing angle forward and camera switching algorithm. Finally, the method is proved to achieve panoramic video immersive roaming through panoramic video, and there is high practical value.

Keywords Panoramic video immersive roaming · Street view · Adaptive tuning of the camera · Camera switching

1 Introduction

As we all know, Google Street View is a kind of service in Google maps, Google Street View is synthetic based on a series of continuous panorama pictures, which were taken from Google Street View cars along the road. The user can view panoramic images in any position on the road and use the mouse to control the arrows on the road to move forward and backward, given a feeling of immersion.

With the development of street view technology,the use of Google Street View is improving. Street view can help people to know unfamiliar environment, and get people to recognize them before they go to a strange scene, which make people outside lives easier. Meanwhile, as street view has grown, more and more people have made innovations in their own research by using the Google Street View API.

Chang Ting-Tzu [1] make long-scene panoramic visualization for Google Street View images, Firstly, User input addresses of the starting and goal point. Then the

Q. Xiu (✉) · X. Shen · T. Zhang · L. Zhao
School of Automation Science and Electrical Engineering, Beihang University (BUAA),
Beijing 100191, China
e-mail: xiuquanfa@163.com

© Springer Nature Singapore Pte Ltd. 2018
Y. Jia et al. (eds.), *Proceedings of 2017 Chinese Intelligent Systems Conference*,
Lecture Notes in Electrical Engineering 460,
https://doi.org/10.1007/978-981-10-6499-9_41

system fetch data from Google Street View automatically, and establish models through SFM [2] framework. Finally, using Graph-Cut [3] to produce muli-viewpoint panorama from sparse omnidirectional consecutive images, which give user a visual summary of all route.

Chi Peng [4] also fetch data from Google Street View by inputting addresses of the starting and goal point by users. Then, creating a continuous video with visually real quality by combining each picture with image alignment [5] and poisson [6] blending technique.

Chris Clews [7] investigated the use of Google Street View (GSV) as a novel research method for collecting alcohol-related data in the urban environment. Billy Chen [8] present a user interface with video to help user familiar strange route.

However, Google Street View also has its own limits. The street view application is based on images so sometimes the pictures we see in Google Street View may be taken a few months ago. The Google Street View is neither real-time nor dynamic monitoring.

In this paper, the immersive roaming method of panoramic video is proposed, which use the panoramic video from the location of the roadside. The adaptive tuning of the camera, the immersive forward viewing angle and camera switching algorithm were mainly involved in this method.

The rest of this paper is organized as follows. The adaptive tuning of the camera algorithm is introduced in Sect. 2. In Sect. 3, we show the immersive forward viewing angle and camera switching algorithm. The results are shown in Sect. 4. Finally, conclusions are drawn in Sect. 5.

2 The Adaptive Tuning of the Camera Algorithm

Marnix Kraus [9] has ever used the distance, visibility and angle three factors to determine the PTZ camera, which to monitor people in video surveillance. In this paper, we also need to consider the influence of distance and angle.

The overall process framework of algorithm as follow (Fig. 1).

In the process of panoramic video roaming implementation, users need to input the Two-dimensional picture whose camera position was marked firstly. The coordinate of the camera i is (X_i, Y_i). The view position is (X, Y) which was got by click the mouse on the picture. Then choose the direction A and A has four choices those are east $(1, 0)$, west $(-1, 0)$, south $(0, -1)$, north $(0, 1)$.

Calculate the distance D_i between the view position and camera i:

$$D_i = \sqrt{(X_i - X)^2 + (Y_i - Y)^2} \tag{1}$$

We can get the distance array D

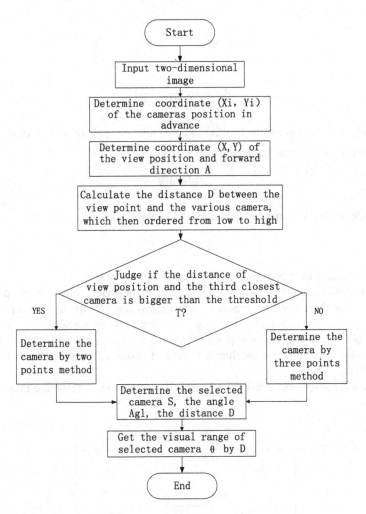

Fig. 1 The adaptive tuning of the camera algorithm

$$D = [D_1, D_2 \ldots D_n] \tag{2}$$

Then ordered the data of D by quick sort method and get the array H that was from small to big.

$$H = [H_1, H_2 \ldots H_n] \tag{3}$$

Agl is the intersection angle between the vector $\vec{V}$ and direction $\vec{A}$. $\vec{V}$ is the vector from camera position to view position.

$$\cos (Agl) = \frac{\overrightarrow{V} * \overrightarrow{A}}{|V| * |A|} \tag{4a}$$

$$Agl = \text{arc} \left(\frac{\overrightarrow{V} * \overrightarrow{A}}{|V| * |A|}\right) \tag{4b}$$

In order to distinguish the positive and negative of angle relative to direction A:
While vector $\overrightarrow{V}$ can be obtained by A clockwise

$$Agl = Agl \tag{5}$$

While vector $\overrightarrow{V}$ can be obtained by A counterclockwise

$$Agl = -Agl \tag{6}$$

The cameras distribution in reality usually as follow (Fig. 2).

As is shown in the two-dimensional picture, the best view camera in any position is one of the three closest cameras, so that we just consider the first three points of H. The picture show that when the view point in the corner, we need to consider the three point, but while the view point is close to the four sides we just consider the two smaller one. So we set the threshold value T (usually, T equal to the distance of two adjacent camera):

When $H_3 > T$, we use two points method, but while $H_3 \leq T$, we use three points method.

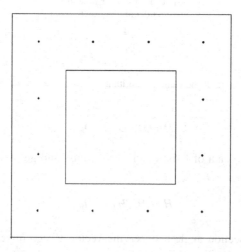

Fig. 2 The cameras distribution

Two points method:

Because two points method just consider the two points around the view point, angle is the main factor. The intersection angle between the view point and two cameras position is Agl1 and Agl2.

While

$$|Agl1| < 90 < |Agl2| \tag{7}$$

The corresponding camera point of $Agl1$ was chosen.

But when:

$$|Agl2| < 90 < |Agl1| \tag{8}$$

The corresponding camera point of $Agl2$ was chosen.

Three points method:

Because the view position located in the corner of the street, we should consider three point. The angle is not the only factor, we should also need to consider the influence of D_i. Do the normalization for distance and angle respectively. We know that the distance closer, the angle smaller and the influence bigger. So that,

For distance:

$$F_D = 1 - \frac{D_i}{D_T} \tag{9}$$

D The distance between view position and camera;
D_T The distance between two adjacent cameras.

For angle:

$$F_A = 1 - \frac{|Agl_i|}{180} \tag{10}$$

Agl The angle between the vector from camera to view position and A.

Finally we get F:

$$F_i = W_D * F_D + W_{Agl} * F_A \tag{11}$$

W_D The weight of distance parameter
W_{Agl} The weight of angle parameter.

Two weight parameters meet:

$$W_D + W_{Agl} = 1 \tag{12}$$

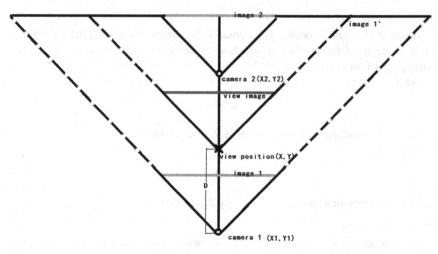

Fig. 3 The visual range of two adjacent cameras and view position

In three points method, we choose the camera by calculating the F_i, which is biggest in the three.

After that, we will get the chosen camera and its Agl_i and D_i.

The visual range θ of two adjacent cameras and view position as follow (Fig. 3).

Firstly, we should determine the center line L and the visual range θ of initial video. The center line L is related to A and Agl_i. In this paper, we set that dividing one picture into 360 parts on average in horizontal direction from left to right based on pixel. The line marked $0° \sim 360°$. And the line marked $0° \sim 180°$ in vertical direction. The center angle line L' of North A(0, 1) is $90°$ line; The center angle line L' of South A(0, −1) is $270°$ line; The center angle line L' of East A(1, 0) is $180°$ line; The center angle line L' of West A(−1, 0) is coincidence of $0°$ line and $360°$ line.

So that

$$w : h = 2 : 1 \tag{13}$$

At the same time, we set the initial visual range θ: $90°$ in horizontal direction and $90°$ in vertical direction. For panoramic video, the ratio of horizontal length w and vertical length h is $w : h = 2 : 1$. But, our visual range is $w : h = 1 : 1$. It is a square. So that the initial visual range:

$$\theta_0 = \left(\frac{w}{4}, \frac{h}{2}\right) \tag{14}$$

When distance of view position and chosen camera is D, Its visual range is:

$$\theta_D = \left(\frac{w}{4} - 2D, \frac{h}{2} - 2D\right) \tag{15}$$

3 Immersive Viewing Angle Forward and Camera Switching

We know that Zooming can give people a feeling of forward and point of interest (POI) can be used to switch camera [10]. Some other algorithm also be used, such as Mosic [11], Tunnel [12], Transitional [13]. In this paper, we refer to the Zooming algorithm because the way was straight.

The Overall process framework of algorithm as follow (Fig. 4).

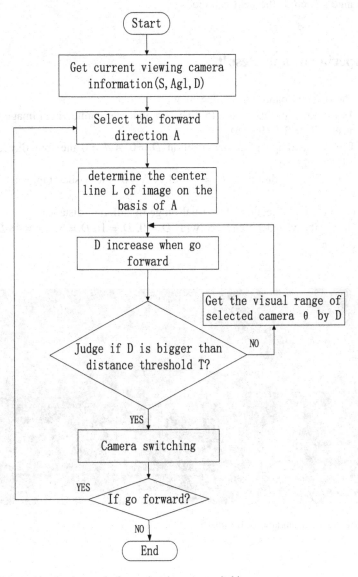

Fig. 4 Immersive viewing angle forward and camera switching

In the process of forward, user can change the direction, then get the vector A. we can determine the center line L and the visual range θ_D according to the formula (15). When forward, the D_i increase at the speed of k pixels (in this paper, k = 2):

$$D_i = D_i + k \tag{16}$$

Then calculate the visual range θ_D by formula (15). As the D_i increase, the visual range in the changes and give people the feeling of forward. When the D_i increase to the threshold T (usually, T equal to the distance of two adjacent camera), the visual range switch to the next camera.

4 Experiment and Result

1.4.1 The original panoramic video image
       As is shown in the picture (5). The original panoramic video image should meet w:h = 2:1 (Fig. 5).
1.4.2 The visual range of initial position $D_i = 0$, $Agl_i = 0$ the four direction as follow (Fig. 6).
       When the condition is $D_i \neq 0$ and $Agl_i \neq 0$, the visual range as follow (Fig. 7).
1.4.3 On the same direction, the visual range of different distance
       North: the visual range vary with $D_i = 0$, $D_i = 16$ $D_i = 32$ $D_i = 48$ $D_i = 64$ (Fig. 8).

Fig. 5 The original panoramic video image

(a) North   (b) East

(c) South   (d) West

Fig. 6 The visual range of initial position

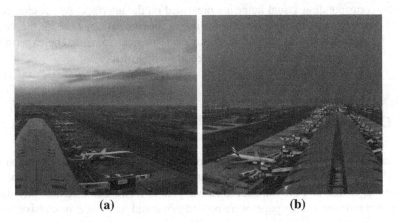

(a)   (b)

Fig. 7 The visual range of other position

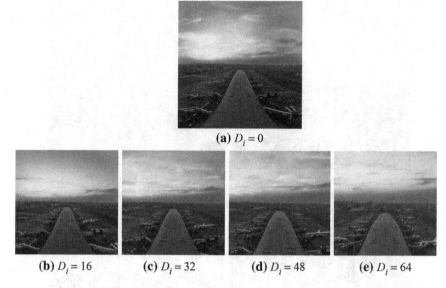

(a) $D_i = 0$

(b) $D_i = 16$ **(c)** $D_i = 32$ **(d)** $D_i = 48$ **(e)** $D_i = 64$

Fig. 8 On the same direction, the visual range of different distance

We use Two-dimensional picture, and input the corresponding panoramic video at each point. We set the distance of two adjacent camera 75, $W_D = 0.3$ and $W_{Agl} = 0.7$. As is shown in the picture of experiment result, in any view position, user choose the direction, and then get the corresponding visual range. It has realized that the camera adaptive tuning and the algorithm is proved to be accurate. In the process of testing video forward, for the same direction, Starting with $D_i = 0$, as D_i increases, different D's corresponding visual range are presented on the interface. As the change of visual range, the user has the feeling of forward. When $D_i = 75$, the visual range switches to the initial visual range of next camera. The experiment prove that the algorithm gives a good feeling of forward.

5 Conclusion

The immersive video roaming based on the panoramic video is an optimization for Google Street View based panoramic image roaming. The adaptive tuning of the camera algorithm and the immersive viewing angle forward and camera switching algorithm proposed in the paper were proved that could realize the immersive video roaming based on the panoramic video. The adaptive tuning of the camera algorithm can tune the responding visual range in any position with random direction.

The view forward algorithm give user a good feeling of forward. In future, we will do more work on camera switches smoothly to ensure the best experience of immersive roaming.

References

1. Tingtzu C. Long-scene panoramic visualization for google street view image. National Taiwan University Department of Computer Science & Information Engineering;2010.
2. Torii A, Havlena M, Pajdla T. From Google street view to 3D city models. In: IEEE international conference on computer vision workshops. IEEE;2009. p. 2188–95.
3. Rav-Acha A, Engel G, Peleg S. Minimal aspect distortion (MAD) mosaicing of long scenes. Int J Comput Vis. 2008;78(2):187–206.
4. Chi P. Integrated smooth Google street view videos and maps for route planning. National Taiwan University Department of Networking and Multimedia;2010.
5. Szeliski R. Image alignment and stitching: a tutorial. Found Trends® Comput Graph Vis. 2004;2(1):1–104.
6. Rez P, Gangnet M, Blake A. Poisson image editing. ACM Trans Graph. 2003;22(3):313–8.
7. Clews C, Brajkovich-Payne R, Dwight E, et al. Alcohol in urban streetscapes: a comparison of the use of Google street view and on-street observation. BMC Public Health. 2016;16(1):1–8.
8. Chen B, Neubert B, Ofek E et al. Integrated videos and maps for driving directions. In: ACM symposium on user interface software and technology, Victoria, Bc, Canada, October. DBLP;2009. p. 223–32.
9. Kraus M. Active multi-camera navigation in video surveillance systems. 2012.
10. De HG, Piguillet H, Post FH. Spatial navigation for context-aware video surveillance. IEEE Comput Graph Appl. 2010;30(5):20.
11. Biocca F, Tang A, Owen C et al. Attention funnel:omnidirectional 3D cursor for mobile augmented reality platforms. In: Conference on human factors in computing systems, CHI 2006, Montréal, Québec, Canada, April. DBLP;2006. p. 1115–22.
12. Girgensohn A, Kimber D, Vaughan J et al. DOTS: support for effective video surveillance. In: International conference on multimedia 2007, Augsburg, Germany, September. DBLP;2007. p. 423–32.
13. Grasset R, Looser J, Billinghurst M. Transitional interface: concept, issues and framework. In: IEEE/ACM international symposium on mixed and augmented reality, ISMAR 2006, October 22–25, 2006, Santa Barbara, CA, USA. DBLP;2006.

Dual Radio Tomographic Imaging with Bayesian Compressive Sensing

Baolin Shang, Jiaju Tan, Xuemei Guo and Guoli Wang

Abstract A common difficulty in device-free localization (DFL) using received signal strength (RSS) is that the uninformative and redundant RSS data in wireless sensor network (WSN) usually degrades the localization performance. This paper develops a dual radio tomographic imaging (RTI) to keep data efficiency in sparse signal recovery as well as eliminating the redundant measurements. In addition to Tikhonov regularization, we also utilize a robust loss function in sparse Bayesian learning for RTI to handle with the outlier data. Moreover, by taking advantage of proposed spatial continuous filter in position estimation, our dual RTI achieves lower localization errors in DFL system.

Keywords Radio tomographic imaging · Data efficiency · Sparse Bayesian learning · Spatial continuous filter

1 Introduction

Target location is an emerging requirement of elder care, security, and many other location based service (LBS) applications [1, 2]. Radio frequency (RF) WSN based DFL becomes more attractive as no more electronic devices or tags need to be attached to users. RTI makes localization more desirable since it can obtain satisfactory results in obstructed environments and chaotic offices. When RF waves travel through some obstructions like walls with numbers of peer-to-peer sensor nodes, waves will experience shadowing losses, leading the RSS measurement variation.

Baolin Shang and Jiaju Tan contribute equally to this work.

B. Shang · X. Guo · G. Wang(✉)
Key Laboratory of Machine Intelligence and Advanced Computing, (Sun Yat-sen University), Ministry of Education, Guangzhou, China
e-mail: isswgl@mail.sysu.edu.cn

J. Tan
Institute of Robotics and Automatic Information System, Nankai University (Tianjin Key Laboratory of Intelligent Robotics), Tianjin, China

© Springer Nature Singapore Pte Ltd. 2018
Y. Jia et al. (eds.), *Proceedings of 2017 Chinese Intelligent Systems Conference*,
Lecture Notes in Electrical Engineering 460,
https://doi.org/10.1007/978-981-10-6499-9_42

And these RSS can be obtained by inexpensive wireless devices, which facilitates the low-cost deployment of commercial promotion.

The task of a RTI based localization system is to localize the target by exploring the shadowing losses on links between two nodes within the network area and reconstructing the shadow fading image of the interested environment [3]. In order to realize the efficient area coverage, the monitoring area should be covered by many dense radio links. However, in some specific location cases, only a few links directly pass through the target providing valuable RSS data, while the other links are mostly uninformative and redundant. To handle with those futile RSS data, Wilson et al. [4] creatively uses the variance of signal strength(VRTI) over a window to capture the motion in spatial voxels of a wireless sensor network. By locating the largest variance of the windowed RSS, the method excludes uninformative links. Meanwhile, the travelling ways in WSN may be affected by diffracting, reflecting, or scattering a subset of their multipath component [5]. And various models are explored to explain the uncertainty RSS measurement level of each radio link due to the multipath fading interferences [6, 7]. These links with high uncertainties become less informative or even destructive to localization issues. In [8], a novel data-efficient strategy is developed to select the informative yet non-redundant radio links with the mutual information criterion. However, this work only operates off-line thus fails to accommodate some specific localization task with extra redundancy included. In this paper, we aim to develop a redundancy exclusive yet algorithm-efficient dual RTI for performance improvement in real-life localization issues.

The sparse distribution of target induced shadow fading makes Bayesian compressed sensing (BCS) [9] desirable for RTI. Instead of point estimation, the posterior distribution of signal is obtained for efficiency design. More specifically, the probabilistic prediction can be improved by a fully probabilistic framework which introduces a prior over the model weights governed by a set of hyperparameters. And comparable generalization performance with higher sparse solution is used in many applications like [10]. Here, we focus on a robust redundancy-exclusive RTI scheme under the BCS framework. Besides, the highest attenuation pixel can be determined as the estimated location in RTI. While this position is usually surrounded by a few attenuation pixels in the region. Then a spatial continuous filter should be introduced for the location estimation of the target. As a result of our works, valuable radio links are extracted while the uninformative links with high uncertainties can be excluded, leading to location improvement.

The paper is organized as follows. Section 2 presents the RTI based localization problem and formulates our dual RTI. In Sect. 3, the Tikhonov regularization and HBCS learning rules are given. The spatial continuous filter strategy is introduced in Sect. 4. Finally, experimental results are reported to demonstrate the effectiveness of proposed method.

2 Dual Radio Tomographic Imaging

This section discusses the proposed dual RTI in details. Firstly, the traditional RTI is provided. Then we discuss the destructive contribution of redundant measurements. At last, the implementation of proposed dual RTI are given.

2.1 Radio Tomographic Imaging

Suppose that a wireless sensor network is deployed to cover the sensing area. The task is to image the target-induced attenuation from the RSS measurement on links in the network. If P is the number of sensors deployed along the perimeter, there are $N = P(P-1)/2$ unidirectional links that cover the sensing area. In RSS-based human detection, the RSS on a single link $i \in \{1, 2, \ldots, N\}$ at time t can be modelled in logarithmic(dB) scale as [5]

$$s_{it}(\varpi) = s_i(o) + \Delta s_{it}(\varpi)$$

where ϖ is the current state of the link, and $s_i(o) = E[s_{it}(\varpi = o)]$ when o is the human-free state. The additive deviation $\Delta s_{it}(\varpi)$ models the body-induced perturbation. In RTI scheme, the body-induced perturbation is mostly ascribed to the shadow fading of a target, which is approximated as a spatial integral of the sensing area. Thus, the RSS variation for link i can be described as

$$\Delta s_{it}(\varpi) = \sum_{j=1}^{N} \Phi_{ij} w_j + \varepsilon_i \tag{1}$$

where $w = [w_1, w_2, \ldots, w_M]^T$ denotes the fading distribution of pixels, Φ_{ij} is the weight of pixel j for link i and ε_i represents observation noise.

To all links covered the monitoring area, the observation model is given by

$$s = \Phi w + \varepsilon \tag{2}$$

where $s \in R^N$ is the changes in RSS measurement due to the presence of the target, $\Phi \in R^{N \times M}$ is the measurement matrix depending on the weights model. To specify a proper Φ, various parametrization method of RSS variations is used to explore the relationship between the excess path length and the shadow fading effect in our prior work [7, 11], and for brevity, the elliptical model is adopted to depict the contribution of each pixel to the fading of a link.

Fig. 1 **a** The shadow fading image with 30 selected links and **b** The shadow fading image with 10 more links, 40 links. The RTI artifact increases with more used links

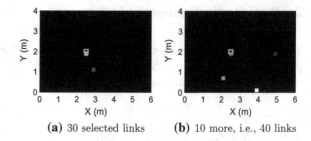

(**a**) 30 selected links   (**b**) 10 more, i.e., 40 links

2.2 Problem Statement

This section discuss the performance degradation of RTI localization technic when the redundant and high uncertainty links are involved in the reconstruction step. Without loss of generality, consider a RTI task of a rectangular sensing area. Firstly, we manually select 30 of 120 links in the area for the shadow fading image reconstruction. Measurements of selected links drop notably and to this end, most valuable data is utilized for image recovery. As shown in Fig. 1a, the location could be estimated roughly from the shadow fading image when a pixel with the largest attenuation is simply considered as the estimated location. Secondly, as a contrast, ten more links are randomly chosen for target localization. The result is shown in Fig. 1b. We can see that more RTI artifacts are occurred in shadow fading image. Thus, localization performance is degraded with the number of used links increases.

As shown, the 30 selected links with dramatic dropping provide amounts of informative information while ten more randomly chosen links introduce more redundant and uncertain information. Therefore, it is the key to finding a way of eliminating redundancy and uncertainty in radio links and achieving sparse measurements, which leads to better localization performance.

2.3 Dual Radio Tomographic Imaging

As we discussed, it's a crucial requirement to keep data efficiency in signal recovery to avoid the use of those uninformative data. To this end, it motivates the exploration of dual representation of w, given by

$$w = \Phi^T z \tag{3}$$

where z is a $N \times 1$ vector, w is a $M \times 1$ vector. Substitute (3) into (2), the dual representation of RTI is reformulated in the form of

$$s = \Psi z + \varepsilon \tag{4}$$

where $\Psi = \Phi^T \Phi$ with the element of $\Psi_{ij} = \Phi_i^T \Phi_j$. Φ is designed in previous section. We can see that z is a lower dimension vector which excludes the redundant links with high uncertainties due to destructive multipath components. Moreover, the lower dimension of our algorithm facilitates the iteration step and consumes less time in our real-time localization system.

3 Image Reconstruction

In this section, we firstly give the Tikhonov Regularization method to reconstruct the image signal and then the robust loss function based sparse Bayesian learning algorithm is derived for the recovery.

3.1 Tikhonov Regularization

An energy term is added to the least-squares formulation [3], resulting in the objective function

$$f(z) = \frac{1}{2}\|\Psi z - s\|^2 + \alpha(\|D_X z\|^2 + \|D_Y z\|^2) \tag{5}$$

where the last part is the Tikhonov matrix enforcing the solution with some certain desired properties. By setting the derivative of (5) to zero, we have

$$\hat{z} = (\Psi^T \Psi + \alpha(\|D_X z\|^2 + \|D_Y z\|^2))^{-1}\Psi^T s \tag{6}$$

The solution is simply a linear transformation of the measurement data and very appealing for our real-time localization system. However, as a special case of sparse Bayesian learning (SBL), Tikhonov regularization doesn't provide any prior information on image signal or noise model for recovery, thus fails to handle with outlier data in more complicated environments and can not keep z sparse.

3.2 Robust Bayesian Method

Different from the compressive step in traditional SBL, our works focus on the non-compressed sparse signal recovery and try to explore the heterogenous model under SBL to reconstruct the dual vector of attenuation image. The prior distribution for signal z and noise ε are both assumed to be gaussian as

$$p(z|\alpha) = \prod_{i=1}^{N} N(z_i|\alpha_i), p(\varepsilon|\sigma) = \prod_{i=1}^{N} N(\varepsilon_i|\sigma_i^2)$$

The likelihood of RSS measurements s is given by

$$p(s|z,\sigma) = N(s|\Psi z,\sigma) \tag{7}$$

By applying the Bayesian inference, the posterior distribution over z is

$$p(z|s,\alpha,\sigma) = \frac{p(s|z,\sigma)p(z|\alpha)}{p(s|\alpha,\sigma)} = N(\mu,\Sigma) \tag{8}$$

with the mean μ and covariance Σ

$$\mu = \Sigma\Psi^T Bs \tag{9}$$

$$\Sigma = \left[\Psi^T B\Psi + A\right]^{-1} \tag{10}$$

where $A = diag(\alpha)$ and $B = diag(\sigma^{-2})$. By setting the derivative of type Π maximum likelihood of equation (7) to zero, the α and σ can be estimated as

$$\alpha_i = \frac{1}{\mu_i^2 + \Sigma_{ii}} \tag{11}$$

$$\sigma_i^2 = (s_i - \Psi_i^T \mu) + tr(\Psi_i \Psi_i^T \Sigma) \tag{12}$$

It can be seen that when the iteration goes convergence, hyperparameters and the sparse signal z both are estimated. And then substitute the sparse z into Eq. (3), the attenuation imaging is sparsely reconstructed.

4   Spatial Continuous Filter

We introduce a spatial continuous filter to enforce position estimation. Let $x_{largest}$ denote the largest component in the attenuation image X and Λ denote the coordinates corresponding to X. The estimated position is given by

$$\hat{\Lambda}_E = \sum_{i=1}^{K} p_i \hat{\Lambda}_i \tag{13}$$

with $\hat{\Lambda}_i$ is

$$\hat{\Lambda}_i \in \Lambda\{X|x_i > \alpha x_{largest}\}$$

where α is a user-defined parameter and $i \in [1, \ldots, K]$. p_i is the individual probability of object placement in space. Generally, p_i is set to be $\frac{1}{M}$, implying a random placement of the target.

5   Experimental Results

5.1   Experimental Setting

Scene 1: The cluttered indoor environment locates in room 415 of the Information
Science and Technology building in SYSU. The sensors are placed along
the perimeter of a 4m×5m rectangle area with 16 nodes of 1 m height. As
shown in Fig. 2a, the sensing area is surrounded by walls and equipments.

Scene 2: The heavily obstructed indoor environment locates in laboratory 421 in
the same building above. As Fig. 2b shown, amounts of chairs, desks and
computers are blocking radio links, leading to extensively rich multipath.
16 RF nodes of 1 m height are deployed to form a 4m × 4m area.

Scene 3: The cluttered through-wall area in Fig. 3a, divided by a wooden wall (the
thickness is about 8 cm) into two part, is located in the same office 415. A
24 peer-to-peer network was placed in a rectangular perimeter of 7m×8m,
and there are many other obstructions inside and outside the sensing area.

Scene 4: This more obstructed one is in our laboratory 421 with 30 nodes deployed
around the perimeter of the whole room to form a 8m × 8m area. As
shown in Fig. 3b, the height of each node is much less than the height of
tables, and RF waves travel through many dense objects from transmitter
to receivers.

Besides, the optimal mass transfer (OMAT) metric [12] is applied to evaluate the
localization performance, expressed as:

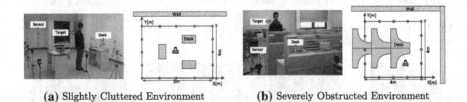

(a) Slightly Cluttered Environment   **(b)** Severely Obstructed Environment

Fig. 2   Scenes (left) and Sketches (right) of Scene 1 and Scene 2 respectively

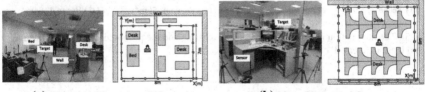

(a) Cluttered Through-wall Scene   **(b)** More Obstructed Scene

Fig. 3   Scenes (left) and Sketches (right) of Scene 3 and Scene 4 respectively

$$d_O(\mathbf{X}, \hat{\mathbf{X}}) = \left(\frac{1}{N_t} \min_{\pi \in \Pi} \sum_{i=1}^{N_t} d(x_i, \hat{x}_{\pi(i)})^p \right)^{\frac{1}{p}} \tag{14}$$

where Π is the set of possible permutations π between the set of the estimated targets $\hat{\mathbf{X}} = \{\hat{x}_1, \ldots, \hat{x}_{N_t}\}$ and the set of the real targets $\mathbf{X} = \{x_1, \ldots, x_{N_t}\}$. We set the order of metric $p = 2$ and the width of pixel in RTI to be 0.2 m. The estimated location of targets is considered be the center of a pixel. And the OMAT error is averaged from above 50 trails at a specific position.

5.2 Benefit from Dual Space

The reconstructed image by dual RTI and existing advanced method is shown in Figs. 4 and 5. As Dual RTI eliminates redundant measurements it recoveries the image with much less artifacts and performs remarkably better in Scene 3 and Scene 4, which are more complicate and larger than other 2 scenes.

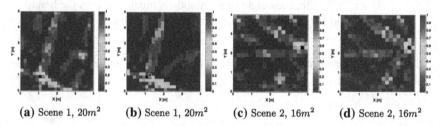

(a) Scene 1, $20m^2$ (b) Scene 1, $20m^2$ (c) Scene 2, $16m^2$ (d) Scene 2, $16m^2$

Fig. 4 The attenuation images reconstructed by existing and the proposed Dual RTI respectively. **a** and **b** are in Scene 1. **c** and **d** are in Scene 2

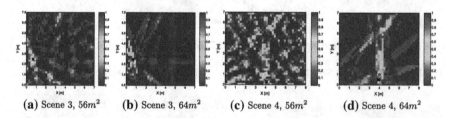

(a) Scene 3, $56m^2$ (b) Scene 3, $64m^2$ (c) Scene 4, $56m^2$ (d) Scene 4, $64m^2$

Fig. 5 The attenuation images reconstructed by existing and the proposed Dual RTI respectively. **a** and **b** are in Scene 3. **c** and **d** are in Scene 4

5.3 Localization Result

The error cumulative probability results with existing advanced RTI algorithm and the Dual RTI are both reported in Fig. 6. In Scene 1 and Scene 2, dual RTI outperforms others with all OMAT errors are under 1.2 m or less. This performance is mostly comparable due to the relatively simple environments.

As shown in Fig. 6c, d, all performance drop notably by objects induced multipath components, while dual RTI has better adaptability and more robustness to data redundancy. For instance, in office through-wall experiment, 90% OMAT errors of proposed method are under 2 m while existing one is under 80% and Tikhonov regularization is about 70% of 2 m. These large errors are due to the through-wall attenuation, particularly links have to cross the metallic obstruction to complete the communication cycle in the office scene in scene 4.

The multi-target localization is more challenging, summarized in Table 1.

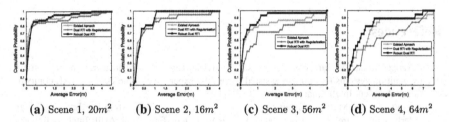

(a) Scene 1, $20m^2$ (b) Scene 2, $16m^2$ (c) Scene 3, $56m^2$ (d) Scene 4, $64m^2$

Fig. 6 The localization error cumulative probability

Table 1 Localization performances in meter: average OMAT errors and 90% values

Targets	Scene	Existing approach		Regularization method		Dual RTI with Bayesian	
		Mean	90%	Mean	90%	Mean	90%
1	Scene1	0.36	0.51	0.50	0.45	0.34	0.45
1	Scene2	0.54	1.06	0.65	1.28	0.60	1.10
1	Scene3	0.91	3.75	1.28	4.20	0.62	1.20
1	Scene4	1.95	7.36	1.88	4.61	1.50	3.26
2–3	Scene1	0.98	1.38	1.13	1.62	1.06	1.28
2–3	Scene2	0.95	1.20	0.70	1.40	1.02	1.40
2–3	Scene3	1.62	2.32	1.43	2.18	1.44	3.00
2–3	Scene4	2.12	3.48	2.07	0.928	2.82	2.28

6   Conclusion

In this paper, a dual radio tomographic imaging strategy is introduced. The advantage of our approach is to keep data efficiency by eliminating redundant data as well as the invalid measurements. To reconstruct the attenuation image, both Tikhonov regularization and a robust loss function based sparse Bayesian learning are experimentally explored to handle with outlier data for performance improvement and localization enhancement. The device-free localization experimental results conducted in four diverse scenarios are reported, which demonstrate the superior performance of proposed approach, especially in the severely obstructed indoor environments.

Acknowledgements This work was supported by the Special Project of Sharing Large Scientific Instruments and Equipments with the Public under Grant No. 2015B030304001, the National Natural Science Foundation of P.R. China under Grant Nos. 61772574 and 61375080, the Key Program of Natural Science Foundation of Guangdong, China under Grant No. 2015A030311049.

References

1. Álvarez-García JA, Barsocchi P, Chessa S, Salvi D. Evaluation of localization and activity recognition systems for ambient assisted living: the experience of the 2012 EvAAL competition. J Ambient Intell Smart Environ. 2013;5(1):119–32.
2. Suryadevara N, Mukhopadhyay SC, Wang R, Rayudu R. Forecasting the behavior of an elderly using wireless sensors data in a smart home. Eng Appl Artif Intell (Elsevier). 2013;26(10):2641–52.
3. Wilson J, Patwari N. Radio tomographic imaging with wireless networks. IEEE Trans Mob Comput. 2010;9(5):621–32.
4. Wilson J, Patwari N. See-through walls: motion tracking using variance-based radio tomography networks. IEEE Trans Mob Comput. 2011;10(5):612–21.
5. Savazzi S, Sigg S, Nicoli M, Rampa V, Kianoush S, Spagnolini U. Device-free radio vision for assisted living: leveraging wireless channel quality information for human sensing. IEEE Signal Process Mag. 2016;33(2):45–58.
6. Wilson J, Patwari N. A fade-level skew-laplace signal strength model for device-free localization with wireless networks. IEEE Trans Mob Comput. 2012;11(6):947–58.
7. Luo Y, Huang K, Guo X, Wang G. A hierarchical RSS model for RF-based device-free localization. Pervasive Mob Comput (Elsevier). 2016;31(C):124–136.
8. Huang K, Guo Y, Yang L, Guo X. Optimal information based adaptive compressed radio tomographic imaging. In: Control conference, 2013. pp. 7438–7444.
9. Donoho DL. Compressed sensing. IEEE Trans Inf Theory. 2006;52(4):1289–306.
10. Yuan J, Wang K, Yu T, Fang M. Integrating relevance vector machines and genetic algorithms for optimization of seed-separating process. Eng Appl Artif Intell (Elsevier). 2007;20(7): 970–9.
11. Guo Y, Huang K, Jiang N, Guo X, Li Y, Wang G. An exponential-rayleigh model for RSS-based device-free localization and tracking. IEEE Trans Mob Comput. 2015;14(3):484–94.
12. Schuhmacher D, Vo B-T, Vo B-N. A consistent metric for performance evaluation of multi-object filters. IEEE Trans Signal Process. 2008;56(8):3447–57.

Simulation and Study of the Incinerator's Combustion Control System

Yu Mu, Lingyu Fang and Jiwei Liu

Abstract This paper introduces two modeling methods of the incinerator's combustion control system. Aiming at the features of the control system which apply to the incinerator's combustion system, this paper uses the techniques of system identification which are based on the least square method and the neural network, introduces two ways of model identification in the incinerator's combustion control system, then they can be helpful for the following design of the control system. By simulating the model of the incinerator's combustion control system, desired results can be obtained for following the track of the main control target-temperature.

Keywords Incinerator · System identification · The least square method · BP neural network · Temperature

1 Introduction

As the social economy and modern industry develop, more waste will be produced, which become a pressing issue around the world. The waste incinerator can make enough effort on waste processing. However, the existing incinerator will still give up a lot of harmful substance during the combustion process, which is a risky factor of human health and natural environment. This will require us to continue to improve the incineration plant combustion control system so that it can limit the fuel excess and air pollution caused by improper operation under the premise of ensuring the combustion temperature. The most important step in this is to propose a systematic identification method for the combustion process of the incinerator and to verify the reliability of the model after identification. It provides the guarantee for

Y. Mu
The 101 Research Institute of Ministry of Civil Affairs, 100071 Beijing, China

L. Fang (✉) · J. Liu
School of Automation and Electrical Engineering, University of Science
and Technology Beijing, 100083 Beijing, China
e-mail: fang_ly1993@163.com

© Springer Nature Singapore Pte Ltd. 2018   453
Y. Jia et al. (eds.), *Proceedings of 2017 Chinese Intelligent Systems Conference*,
Lecture Notes in Electrical Engineering 460,
https://doi.org/10.1007/978-981-10-6499-9_43

the safety, energy saving and high-efficiency operation of the incinerator, and will establish the foundation for further research and design of control system.

2 Control Strategy of Incinerator Combustion System

In most incinerator main circuit temperature control system, the general use of the control strategy is the PID-based self-tuning fuzzy control. This control method not only keeps the advantages of conventional PID control system such as simple, easy to use, strong robustness [1], but also has more other characteristics for flexibility, adaptability, accuracy.

In order to achieve better economic combustion, we need to provide a reasonable flow of air and gas for combustion process in accordance with the output flow of main loop temperature controller [2, 3]. The best way is to make a certain proportion of the amount of air and fuel, to achieve a state of mutual coordination between the two. For the whole combustion process, the furnace negative pressure is also an important index of combustion control [4]. Under ideal operating conditions, the negative pressure of the furnace in the incinerator should not be too large or too small.

In the whole process of the work, we need to continue to supply the reasonable amount of air and fuel, and ensure the reasonable state in the joint action of the two. On this basis, we should ensure the stability of the furnace negative pressure within a reasonable range according to the relationship between the air supply volume and the negative pressure so that the system can complete the work flow. It is not difficult to see that we need to build a model around the relationship among the amount of air supply, fuel quantity, and temperature value in the next step.

3 Model Identification of Incinerator Control System

According to the variables that control system needs to use, we observe and record the two groups of the incinerator working process data. The data recorded the parameters of the combustion process in detail, such as fuel quantity, fan rotation frequency, temperature and negative pressure, which play a decisive role in the identification of the system. Due to the lack of the system identification in the existing combustion control system, we can use the least square method and neural network method to identify the system around these parameters and the characteristics of the control system. And we should verify the reliability of the model in the simulation after identification, and lay a foundation for the design of the following control system.

3.1 Identification Model of Incinerator Based on Least Square Method

We can consider the fuel and the fan frequency that both are used for combustion control system of incinerator as the system input, and consider the variation of temperature and negative pressure in the process of combustion as the system output, so a simple double input double output system is established [5–7]. The range of the air supply, fuel quantity and temperature value is larger, and conforms to a certain change rule. However, the absolute value of the negative pressure state is basically stable within a very small range, and it is not convenient to introduce it to the system model. Therefore, we may wish to put the system as a dual-input single-output system to study the effect of air volume and fuel volume on the temperature change. On this basis, it can be concluded that the negative pressure of the furnace is in a stable state.

For the whole combustion process, we record the data per minute to get the system-related parameters. According to the data characteristics, we may wish to temporarily see the system as a discrete system. Then according to the desired establishment of the dual input single output system model, the difference equation is in the form of [8]:

$$y(k+n)+a_{n-1}y(k+n-1)+\cdots+a_1y(k+1)+a_0y(k)=b_0u(k)+c_0f(k) \quad (1)$$

where k is the kT time; T is the sampling period; y(k) is the output at time kT; u(k) and f(k) are the input quantities at time kT. In the incinerator system, y(k) is the kT time furnace temperature, the unit is degree; u(k) and f(k) are the fuel quantity and fan frequencies of the furnace at time kT, respectively, in units of elevations and Hz; $a_i(i=0,1,2,\cdots,n; a_n=1)$, b_0 represents the constant coefficient of system characteristics.

Set the state variables as follows, we can get the dynamic equation:

$$
\begin{aligned}
x_1(k+1) &= x_2(k) \\
x_2(k+1) &= x_3(k) \\
&\vdots \\
x_{n-1}(k+1) &= x_n(k) \\
x_n(k+1) &= -a_0x_1(k)-a_1x_2(k)-\cdots-a_{n-1}x_n(k)+b_0u(k)+c_0f(k) \\
y(k) &= x_1(k)
\end{aligned}
\quad (2)
$$

The state equation expressed in vector-matrix form is:

$$
\begin{bmatrix} x_1(k+1) \\ x_2(k+1) \\ \vdots \\ x_{n-1}(k+1) \\ x_n(k+1) \end{bmatrix} = \begin{bmatrix} 0 & 1 & 0 & \cdots & 0 \\ 0 & 0 & 1 & \cdots & 0 \\ \vdots & \vdots & \vdots & & \vdots \\ 0 & 0 & 0 & \cdots & 1 \\ -a_0 & -a_1 & -a_2 & \cdots & -a_{n-1} \end{bmatrix} \cdot \begin{bmatrix} x_1(k) \\ x_2(k) \\ \vdots \\ x_{n-1}(k) \\ x_n(k) \end{bmatrix} + \begin{bmatrix} 0 & 0 \\ 0 & 0 \\ \vdots & \vdots \\ 0 & 0 \\ b_0 & c_0 \end{bmatrix} \cdot \begin{bmatrix} u(k) \\ f(k) \end{bmatrix}
$$

$$
y(k) = \begin{bmatrix} 1 & 0 & \cdots & 0 \end{bmatrix} \cdot x(k) \tag{3}
$$

For the state space model, the system order refers to the minimum number of states. In the identification of the incinerator model, the order of the model can not be determined by theoretical deduction, so we need to first identify the order of the model.

We might as well use the classic F-test method [9]. By introducing a hypothesis test, the problem of determining the order of the model is attributed to the question of whether the decrease of $J_2(n_2)$ is more significant than $J_1(n_1)$ when n_1 is increased to n_2. J refers to the sum of squared residuals of the model output and the observed output. In general, J decreases with the increase of model order. In the case of actual system identification, the number of input and output signal samples is far more than the number of parameters to be identified. In this case, as the model order increases, the J value decreases significantly. When the order of the model is greater than the real order of the system, the significant decline of J will cease. We could use this principle to determine the order of the order. It can be shown that when N is large enough, if $n_2 > n_1 \geq n$ is established, then:

$$
\frac{N}{\sigma_{\varepsilon1}^2} J_2(n_1) \sim \chi^2(N - 2n_1)
$$

$$
\frac{N}{\sigma_{\varepsilon1}^2} J_2(n_2) \sim \chi^2(N - 2n_2) \tag{4}
$$

$$
\frac{N}{\sigma_{\varepsilon1}^2} J_2(n_1) - J_2(n_2) \sim \chi^2(2n_2 - 2n_1)
$$

$J_2(n_2)$ and $J_2(n_1) - J_2(n_2)$ are independent random variables. Introduce the statistics t:

$$
t = \frac{J_2(n_1) - J_2(n_2)}{J_2(n_2)} \cdot \frac{N - 2n_2}{2n_2 - 2n_1} \tag{5}
$$

$$
\sim F(2n_2 - 2n_1, N - 2n_2)
$$

That is, when $n_2 > n_1 \geq n$, the statistic t obeys the F distribution that the degrees of freedom are $2n_2 - 2n_1$ and $N - 2n_2$, respectively. Then suppose there is a significant level of α, when $t < t_\alpha$, $n_2 > n_1 \geq n$ is established, otherwise n is larger than

Table 1 Object functions and t statistics in the different orders

System order	Correlation coefficient				
n	a_i	b_i	c_i	J	t(n−1, n)
1	$a_1 = 1.0171$	$b = 0.0348$	$c = -0.0203$	1.563	
2	$a_1 = 1.7657$ $a_2 = -0.7642$	$b = 0.0250$	$c = -0.0281$	1.209	10.32
3	$a_1 = 1.5659$ $a_2 = -0.2932$ $a_3 = -0.2735$	$b = 0.0275$	$c = -0.0359$	1.106	3.61
4	$a_1 = 1.6352$ $a_2 = -0.2199$ $a_3 = -0.6888$ $a_4 = 0.2750$	$b = 0.0273$	$c = -0.0315$	1.051	0.92

n_1 and n_2. Here take $\hat{n} = n_1$ and $\hat{n} + 1 = n_2$, then make $t_a = F_a(2, N - 2\hat{n} - 2)$. When the order of $n = 1, 2, \cdots$, increases gradually, the F test is used to determine whether $\hat{n} \geq n_1$ is established.

Then, on the basis of the method of order identification, we can identify the parameters of the model. We use the recursive least squares method, which is a kind of on-line real-time estimation of the model parameters identification method. The basic idea can be summarized as follows: new estimate $\theta(k) =$ old estimate $\theta(k-1)$ + correction item.

We adopt the first set of experimental data, assuming the system order values of n were 1, 2, 3 and 4. And the recursive least square method is used to identify the parameters of the system. The J values are compared according to the calculated parameter values, and Table 1 is the parameter values and statistics identified under different orders:

There are about 60 sets of experimental data. Look-up table can get F (2, 60) = 3.15. From the table, we can get the conclusion that when the order of the system is 4, t (3,4) = 0.92 < 3.15. This shows that the decrease of J value has not been significant, thus the order of 3 can be determined. The system model parameter is the value of a, b and c when n = 3.

The final mathematical model is:

$$y(k) - 1.5659y(k-1) + 0.2932y(k-2) + 0.2735y(k-3)$$
$$= 0.0275u(k-1) - 0.0359f(k-1) \tag{6}$$

The state equation expressed in vector-matrix form is:

$$
\begin{bmatrix} x_1(k+1) \\ x_2(k+1) \\ x_3(k+1) \\ x_4(k+1) \end{bmatrix} = \begin{bmatrix} 0 & 1 & 0 & 0 \\ 0 & 0 & 1 & 0 \\ 0 & 0 & 0 & 1 \\ 0.2735 & 0.2932 & 1.5659 & 1 \end{bmatrix} \cdot \begin{bmatrix} x_1(k) \\ x_2(k) \\ x_3(k) \\ x_4(k) \end{bmatrix} + \begin{bmatrix} 0 & 0 \\ 0 & 0 \\ 0 & 0 \\ 0.0275 & -0.0359 \end{bmatrix} \cdot \begin{bmatrix} u(k) \\ f(k) \end{bmatrix}
$$

$$
y(k) = \begin{bmatrix} 1 & 0 & \cdots & 0 \end{bmatrix} \cdot x(k) \tag{7}
$$

3.2 Identification Model of Incinerator Based on BP Neural Network

Although we can use the least square method to identify the combustion system of the incinerator, but due to the system with strong coupling and nonlinear characteristics [7], it is difficult for us to use the model to accurately describe the mechanism. Because neural network has very strong nonlinear mapping ability, it has a good effect on describing complex systems. Through the actual input and output data obtained from the experiment, we can get a relatively accurate system model. Therefore, it is also a suitable identification method to establish the model of incinerator by using neural network.

Neural network is a nonlinear system which is composed of a large number of neurons connected by a certain structure to accomplish different information (including intelligent information) processing tasks. Different neurons express different connections through their respective synaptic weights. In the course of learning, the synaptic weights are adjusted continuously, and the actual output of the network is approaching the desired output. BP neural network is a kind of neural network. BP algorithm is called error back propagation [10], using gradient search method, so that the error of the actual output value of the network and the expected output value is the smallest.

There are many ways to adjust the weights of the neural network system [11, 12]. The method we use is to compute the total error of the network after all samples are input. Then we calculate the error signals of each layer according to the total error and adjust the weights. The data used is still the actual data collected during the operation of the incinerator, including the furnace temperature, the fan frequency and the instantaneous value of the oil intake. At the same time, the negative pressure term not used in the least square method is introduced to make the modeling process of the whole system more perfect.

In conclusion, the use of neural network on the incinerator system modeling can give full play to the advantages of neural networks to deal with nonlinear problems [13]. At the same time, it can avoid the disadvantages such as too large amount of calculation, too long time consuming. The model can be shown in Fig. 1.

Fig. 1 Modeling of the incinerator's combustion system

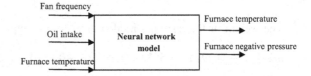

The incinerator combustion system is a time lag system, and the temperature and pressure changes at a certain time are related to the frequency of the fan and the amount of oil in the previous time. While the change has the closest proximity to the influence of the input a moment before. The frequency of fan f(t−1), the oil inflow u(t−1) and furnace temperature v(t−1) are selected as the input. In this paper, we set up a BP neural network model with three inputs and two outputs with momentum item. The number of nodes in the hidden layer is set to be 6. The model structure is shown in Fig. 2, in which x_1, x_2 and x_3 are three inputs of the model: fan frequency f(t−1), oil intake u(t−1) and Furnace temperature v(t−1); y_1 and y_2 are two outputs of the model: the furnace temperature and furnace pressure.

Then, with the basic structure of the neural network model, we can carry out simulation research based on the existing data. As with the least squares method, we also use the first set of data to simulate and select the learning rate as $\eta = 0.1$. The initial value of the network is determined by the actual training data collected.

Finally, the weight of the input layer to the hidden layer is:

$$\omega 1k1 = \begin{bmatrix} -0.5693 & -0.0492 & -0.4663 \\ -0.4947 & -0.7425 & 1.2459 \\ -0.3014 & -0.5754 & 0.0126 \\ -1.5681 & -0.3941 & -0.4213 \\ 0.0145 & -0.6130 & -0.8645 \\ -0.5498 & -0.6386 & -1.7884 \end{bmatrix} \tag{8}$$

Fig. 2 BP neural network of the incinerator's combustion system

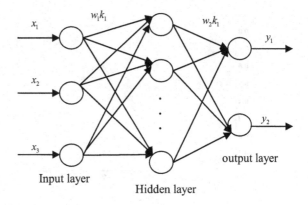

The weight of the resulting hidden layer to the output layer is:

$$\omega 2k1 = [\,-0.0974 \quad 1.5034 \quad 0.4826 \quad -0.2530 \quad -0.2664 \quad -1.3435\,] \quad (9)$$

3.3 Comparison of Two Kinds of System Identification Methods

Train the first set of data, and then test the training effect with the second sets of data. Compare the two models obtained by the two methods mentioned above (Fig. 3).

For the data, the cumulative error of the neural network identification method is $egt1 = \sum \dfrac{egt(k)^2}{2} = 0.0357$, and the cumulative error of the least square method is $egt2 = \sum \dfrac{eg(k)^2}{2} = 0.0376$.

In summary, in the two sets of data testing, the two can be better track the output data—temperature, to achieve the ultimate identification purposes. The cumulative error of the neural network is smaller than that of the least squares identification method, and the neural network model does not require the establishment of the actual system identification format, which eliminates the need for the system structure modeling. For most non-linear, the system can achieve a better identification. In the process of identification, the system parameters such as learning rate and error index will affect the convergence rate of identification, but once the recognition is over, the obtained intermediate weights can be applied directly so that

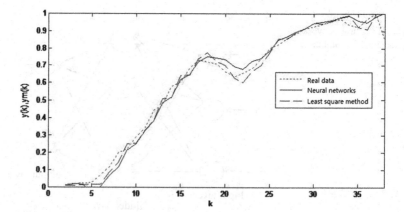

Fig. 3 The temperature curve of the real data, system identification by the neural network and system identification by the least square method

the whole system can be replaced by a neural network and further into the follow-up of the relevant control design to achieve the incineration equipment safety and environmental protection and efficient design of protection.

4 Conclusion

(1) In this paper, based on the least square method, the state equation of the combustion control system is established based on the relationship among the fuel quantity, the frequency of the fan and the temperature. Based on the relevant data, we use F test method to identify the order and the parameters of the system, e.g. Eq. (7). The simulation results verified the effectiveness of the model to the relevant data.

(2) At the same time, using the method of system identification based on BP neural network, we build three layer BP neural network model on the basis of the combustion process lag characteristics. The network is based on the relationship between two inputs and two outputs. By referring to the relevant data and taking reasonable learning rate, we demonstrate that this model realizes the tracking process of temperature and furnace negative pressure.

(3) Finally, both of the two identification methods can track the output data—temperature and achieve the final identification purposes. The cumulative error of the neural network is slightly smaller than that of the least squares method. Under the premise of achieving the ideal identification effect, the model completes the simulation research of the incinerator system, laying the foundation for realizing the safety, environmental protection and high efficiency of the incinerator.

References

1. Lu GM. Research on fuzzy PID control system of temperature for electric boiler [Dissertation]. Heilongjiang: Harbin University of Science and Technology; 2007.
2. Sheng SB, Yuan LH. Cross control by the air-fuel in the burning control of the boiler. Therm Power Gen. 1998;4:49.
3. Zhou W, Lu HP, Zhang C, et al. The application of the double-cross-limiting ratio control in the temperature control of the reheating furnace. Metal Autom. 2011;2:434.
4. Li JT, Gao X, Zhang Q. The use of fuzzy control technique in water-coal starch boiler negative pressure of boiler's chamber control system. China Instrument. 2006;6:73.
5. Qin P, Zhu J. Data hierarchical modeling algorithm for multi-input multi-output system. Mini-Micro Syst. 2003;24(1):76.
6. Yuan P. Comparisons and studies of identification methods for multivariable systems [Dissertation]. Jiangsu: Jiangnan University; 2008.
7. Wu B. Research of identification and pid control methods in multivariable system [Dissertation]. Beijing: Beijing University of Chemical Technology; 2008.

8. Yu KP, Dong HZ. Fast recursive least square scheme for time-varying nonlinear system identification base on feed forward neural network. In: The 9th vibration theory and application of academic conference. Zhejiang; 2007. p. 143.

9. Chen GC, Xie PF. Recognition and examine methods in the slither of Hydrology's variation. J China Hydrol. 2006;26(2):57.

10. Chen CP, Cha YX, Qian P, et al. MapReduce-based BP Neural Network Genetic Algorithm to Study Nonlinear System Identification//The 18th National Youth Communication Conference. Beijing; 2013. p. 242.

11. Liu YM. The research of PID controller based on improved bp neural network [Dissertation]. Beijing: University of Chinese Academy of Sciences; 2012.

12. Yuan MH. System identification method study based on neural networks [Dissertation]. Beijing: Beijing University of Technology; 2008.

13. Wu JH, Wu JF, Li L. Nonlinear system identification based on BP neural network. In: The 12th annual conference of control and application, CSAA. Beijing; 2006 p. 73.

View-Based 3D Model Retrieval
via Convolutional Neural Networks

Bowen Du, Haisheng Li and Qiang Cai

Abstract In this paper, we propose a multi-view fusion 3D model retrieval using convolutional neural network to solve the problem of the local perception in feature descriptor. By view pooling, we combine information from multiple views of a 3D model to eliminate the position correlation caused by the viewing angle of camera. In addition, integrating pre-processed RGB view-feature with Binary view-feature in the same model is used to generate a single model descriptor. Experiments on ETH dataset demonstrate the superiority of the proposed method.

Keywords Convolutional neural networks · Multi-view · 3D object retrieval · Feature fusion

1 Introduction

Three-dimensional (3D) models are now increasingly widely used in various fields such as games, computer-aided design, and augmented reality. Because a single 3D model has a large number of space occupied and the complexity of extracting features, it becomes an important research direction that how to achieve accurate, efficient and fast retrieval of the required model in the massive 3D model dataset [1].

View-based 3D model retrieval [2, 3] first capture a series of two-dimensional (2D) views from the original model, then regard these view images as the representation of the model. The most primitive method is the Light Field Descriptor proposed by Passalis et al. [3], which projects a 3D model from a dodecahedron and selects 10 representative representations as a description of the 3D model. Adaptive

B. Du · H. Li (✉) · Q. Cai
School of Computer and Information Engineering, Beijing Technology and Business University, Beijing 100048, China
e-mail: lihsh@th.btbu.edu.cn

B. Du · H. Li · Q. Cai
Beijing Key Laboratory of Big Data Technology for Food Safety, Beijing 100048, China

© Springer Nature Singapore Pte Ltd. 2018   463
Y. Jia et al. (eds.), *Proceedings of 2017 Chinese Intelligent Systems Conference*,
Lecture Notes in Electrical Engineering 460,
https://doi.org/10.1007/978-981-10-6499-9_44

Views Clustering [2] uses two units of positive icosahedra to obtain a 3D model of the views, from 320 view images to select representative images. In addition to focusing on global features, it is also focused on extracting the local features of the view [4, 5]. Gao et al. [5] proposed a Bag of Region Words algorithm locating the mapping regions of local SIFT feature-encoded visual words, and clustering as a model feature according to a certain weight. But in fact, most view-based retrieval methods highly rely on camera location and environment setting. More specifically, when projecting multiple views of 3D model, there is a strict constraint of camera arrays. Improper capturing position or difference of view points will lead to poor performance.

The use of convolutional neural network (CNN) is an important application of deep learning for image retrieval, classification. In particular, by being trained through large databases such as the ImageNet database, CNNs have long been used in the field of object detection, scene recognition, text recognition and fine-grained classification for image feature learning [6]. Krizhevsky et al. [7] achieved top-1 and top-5 error rates of 39.7 and 18.9% which is considerably better than the previous state-of-the-art results. CNN is a local 2D connection of the neural network structure. By limiting hidden area of the receiving area, our method focuses on local feature extraction.

In view of the above problems, we employ CNN and multi-view method to extract feature from multi-view 3D model. Our method is based on Multi-View Convolutional Neural Networks (MVCNN) [8]. We employ MVCNN of the view-pooling layer and improve the local perception accuracy, when extracting features. And we propose feature fusion method using element-wise average pooling operation based on multi-view. Compared with other algorithms, the advantages of our algorithm are as follows:

(1) By using CNN, our method improves the description of the local feature to retrieve 3D model;
(2) Multiple views are fused by element-wise average pooling, using a novel convolutional neural network structure. Our method is better to retain the characteristics of each view and the interference of abnormal values;
(3) Using of binary images and RGB images, we integrate different kinds of view-feature under the same model to enhance the search results.

The architecture of the paper is organized as follows. In Sect. 2, we propose the multi-view convolutional neural network using average pooling framework (MVCNN-AVG) on 3D model retrieval. Section 3 shows the experimental results on ETH dataset. Finally, Sect. 4 draws the conclusion.

2   The Proposed Method

As discussed above, we focus on developing view-based descriptors for 3D object to search the optimal retrieval for the given query model in the dataset. The multi-view convolutional neural network using average pooling framework (MVCNN-AVG) is to learn to combine information from multiple views (Fig. 1). Our method is employed to produce a single compact descriptor for the 3D model.

2.1   Model Pre-processing

The background information of views disturbs the analysis of images. In order to avoid noise interference when extracting features, we pre-process the 3D model view first. In this paper, the binary image is used to remove the background in the original RGB image, and the pre-processed image is used as the input of the neural network. Figure 1 shows the contrast before and after the pre-processing.

2.2   View-Based CNN Feature Extraction

AlexNet [7] is used as the neural network to extract features. AlexNet mainly contains five convolutional layers $conv_{1,\dots,5}$, three fully-connected layers $fc_{6,\dots,8}$ with a final 1000-way Softmax. The convolution process is as follows:

$$X^l = f\left(\delta\left(X^{l-1}\right) + b^l\right) \tag{1}$$

where l represents the number of layers, f is the activation function of the layer, $\delta(X)$ is the convolution operation of the upper layer input image, b^l is the bias of the

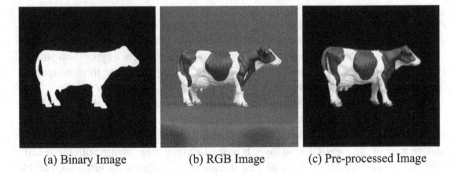

(a) Binary Image   (b) RGB Image   (c) Pre-processed Image

Fig. 1 The contrast before and after the pre-processing by binary image

layer, and x^l is the characteristic map obtained after the convolutional layer processing.

The pooling formula is as follows:

$$X_j^l = pooling\left(X_j^{l-1}\right) \tag{2}$$

where $pooling()$ represents a pooling function pooling the input. A typical operation sums all the pixels of the different $n \times n$ blocks of the input image. The output image is reduced by n times in both dimensions.

In this section, the fc_7 layer output (4096 dimensions) is used as the image feature. Since the single 3D model is represented by multiple views, the model feature dimension is $N \times D$, where N is the number of views, D is the feature dimensions of a view. The neural network is pre-trained through the 1000 images on the ImageNet database, and then fine-tuned the network model with the data in the training dataset.

After obtaining all the 3D model features, we calculate the distance between two models. Because of the use of different angles of camera location in a 3D model, we rank the views of the single model by S_i values. The weight of each view S_i is set as follows:

$$S_i = \frac{1}{n_i} \sum_{j=1}^{n_i} (x_{i,j} - \mu_i)^2 \tag{3}$$

where μ is the average of the ith view feature. Finally, the similarity distance between the two models M_1, M_2 is calculated using Euclidean distance:

$$D(M_1, M_2) = \sum_{i=1}^{M} (V_1^i - V_2^i)^2 \tag{4}$$

2.3   MVCNN-AVG Feature Extraction

Although extracting multiple discriminatory descriptors for each 3D model retrieval can improve the effect to existing 3D descriptors, how to assign weights to multiple views has an impact on the results. We propose a multi-view aggregation to synthesize all the information to form a single 3D model descriptor to eliminate the inter-view positional interference.

Figure 2 shows the algorithmic flow designed in this paper, where CNN_f is composed of five convolutional layers. The view-pooling layer is leveraged between CNN_f and CNN_b, and is placed after the last convolutional layer (fc_5). The feature map of the view-pooling layer is a one-to-many relationship with the feature map of the previous layer. Assume that the previous feature map has a total of M, the size of feature map is $N \times N$, and the jth feature map matrix is

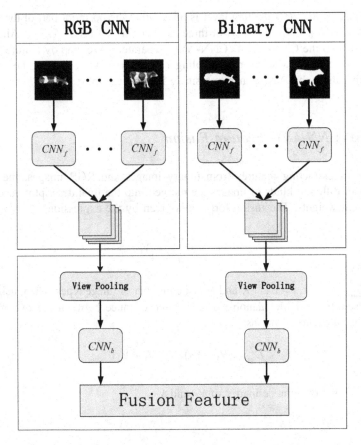

Fig. 2 Multi-view convolutional neural networks using average pooling framework

$$x_j^{l-1} = \left(a_{stj}^{l-1} \right)_{N \times N} \tag{5}$$

where $s, t \in [1, N]$, and a_{stj}^{l-1} is the jth feature map matrix elements of the convolutional layer. The filter size of the view-pooling layer is $n \times n$. The view-pooling layer output is

$$x_j^l = \left(a_{stj}^l \right)_{\frac{N}{n} \times \frac{N}{n}} \tag{6}$$

where $p, q \in \left[1, \frac{N}{n}\right]$. The elements of each pooling area are averaged in the form of:

$$a_{pqj}^l = \frac{1}{n \times n} \sum_{s=np-n+1}^{np} \sum_{t=nq-n+1}^{nq} a_{stj}^{l-1} \tag{7}$$

The data of the view-pooling layer is placed into the posterior part of the neural network (CNN_b). CNN_b consists of three fully-connected layers $fc_{6,\ldots,8}$. All views first input into the CNN_f of MVCNN-AVG separately. We employ feature fusion method by element-wise average pooling operation in view-pooling layer. And processing through CNN_b generate a single model descriptor.

2.4 MVCNN-AVG Feature Fusion

Because of extracting features from binary images and RGB images, the CNN features of different kinds of images are merged into the final descriptor according to different weights. The fusion formula is given by the expression

$$V_{\text{Fusion}} = \sum_{i=1}^{K} \omega_i V_i \tag{8}$$

where $\sum_{i=1}^{K} \omega_i = 1, i = 1 \ldots K$ and K is the number of image types. Then, using the final descriptor and Mahalanobis distance, the distance between the two models M_1, M_2 is calculated using as:

$$D_M^2 = (V_1 - V_2)\Sigma^{-1}(V_1 - V_2)^T \tag{9}$$

where Σ is the covariance matrix of V_1 and V_2.

3 Experiments

3.1 Database and Evaluation Criterion

In this paper, ETH-80 dataset is used as experimental dataset. In the evaluation index of 3D model retrieval, we use the NN, FT, ST, F, DCG, P-R curves and so on commonly used in SHREC [9] as the standard.

(1) Nearest Neighbor (NN) evaluates the retrieval accuracy of the first returned result.
(2) First Tier (FT) is defined as the recall for the first τ matches, where τ is the amount of similar objects in the same category.
(3) Second Tier (ST) is the recall for the first 2τ matches, where τ is the amount of similar objects in the same category.
(4) F-measure (F) equals to $\frac{2*PR}{P+R}$, where P is precision and R is recall of top 20 results.

(5) Discounted Cumulative Gain (DCG) is a statistic that assigns relevant results at the top ranking positions with higher weights under the assumption that a user is less likely to consider lower results.

(6) Precision-Recall Curve comprehensively demonstrates retrieval performance; it is assessed in terms of average recall and average precision, and has been widely used in multimedia applications.

3.2 Experimental Results

In Fig. 3, the method MVCNN-AVG is implemented in the ETH-80 dataset for experimental comparison. The neural network structures mentioned in Sect. 3 of this paper are compared. Each network structure uses three types of images as input to represent a 3D model. CNN-Fusion and MVCNN-AVG-Fusion are fused by Eq. (9).

All branches of CNN_f share the same weights and biases. First, the $conv_1$ layer filters the 227 * 227 * 3 input view with 96 kernels of size 11 * 11 * 3 with a stride of 4 pixels (the stride is adjacent to the neuron's sensory center distance). The $conv_2$ takes the output of $conv_1$ with 256 kernels of size 5 * 5. The $conv_3$ has 384 kernels of size 3 * 3. The $conv_4$ and $conv_5$ respectively has 384 kernels and 256 kernels of the same size of 3 * 3. The view-pooling layer leverage element-wise average operation and the pooling kernel of size 2 * 2. We also employ

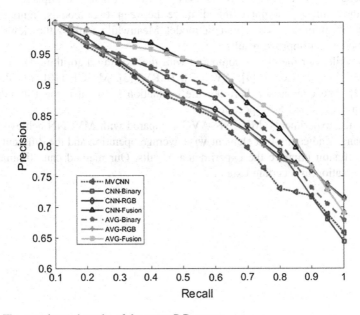

Fig. 3 The experimental results of the seven P-R curves

Table 1 The comparison of the five evaluation scores in seven ways

Method	NN	FT	ST	F	DCG
MVCNN [8]	0.9250	0.7208	0.9028	0.4366	0.9060
CNN-Binary [7]	0.9375	0.7667	0.9347	0.4384	0.9200
CNN-RGB [7]	0.9500	0.7625	0.9181	0.4390	0.9182
CNN-Fusion [7]	**0.9750**	0.8014	0.9431	0.4390	**0.9472**
MVCNN-AVG-Binary	0.9375	0.7736	0.9403	0.4390	0.9274
MVCNN-AVG-RGB	0.9375	0.7611	0.9125	0.4390	0.9166
MVCNN-AVG-Fusion	**0.9750**	**0.8042**	**0.9472**	0.4390	0.9458

element-wise maximum operation, but the experimental results not perform well. The view-pooling layer can also be placed anywhere in the entire process, but found that only after $conv_5$ the search ability to achieve the best. The view-pooling layer aggregates the features of multiple views into a single feature.

In fact, in this experiment, we extract feature from fc_6, fc_7 and fc_8 respectively for comparative experiments. Although the performances of these features are very close, features extracted from fc_7 have better accuracy. Therefore, we selected fc_7 as the model descriptor in MVCNN-AVG. Compared to MVCNN, MVCNN-AVG improve the retrieval efficiency.

Table 1 shows the comparison of the five evaluation scores in seven ways. Figure 3 shows the experimental results of the seven P-R curves. This article takes MVCNN-AVG-Fusion as the final form.

In Fig. 3 and Table 1, the results show that MVCNN-AVG achieves better performance in most cases. CNN also has good performance. Compared to CNN, our method directly compute the distance between two models, ignoring the sequence of multiple views in a single model. Meanwhile, ranking the views of the single model can improve results.

Meanwhile, our method is compared with the existing algorithm. The contrast algorithms include CCFV [4], AVC [2], BoRW [5], MMGF [10], FDDL [11], BGM [9]. The comparison between the evaluation scores and the P-R curve is shown in Figs. 4 and 5, respectively.

From the experiments, MVCNN-AVG compared with MVCNN achieve better performance, indicating that element-wise average operation and the different kinds of image fusion improve the experimental results. Our method can eliminate the position relation to a certain extent.

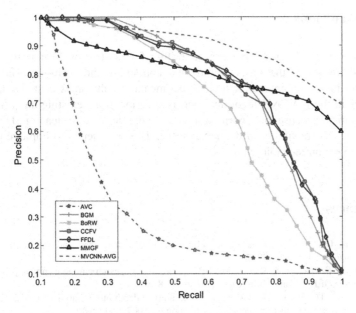

Fig. 4 The algorithm is compared with the P-R curve of the existing algorithm

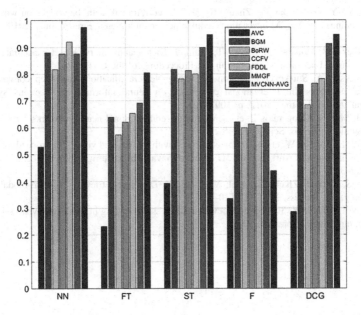

Fig. 5 This algorithm is compared with the five evaluation scores of the existing algorithm

4 Conclusions

In this paper, a 3D model retrieval algorithm for multi-view fusion structure is proposed, by using the CNN feature and combining the multi-view fusion to eliminate the viewpoint correlation. Experimental results show that our method improves in the MVCNN algorithm, and is superior to the existing algorithm. At the same time, because the algorithm input by multiple views instead of 3D model, we improve the flexibility in the actual scene. The next step is to improve the 3D model projection technology.

References

1. Gao Y, Dai QH. View-based 3D object retrieval: challenges and approaches. IEEE Multimedia. 2014;21(3):52–7.
2. Ansary T-F, Daoudi M, Vandeborre J-P. A bayesian 3D search engine using adaptive views clustering. IEEE Trans Multimedia. 2007;9(1):78–88.
3. Passalis G, Theoharis T, Kakadiaris I-A. Ptk: a novel depth buffer-based shape descriptor for three-dimensional object retrieval. Visual Comput. 2007;23(1):5–14.
4. Gao Y, Tang J, Hong R, et al. Camera constraint-free view-based 3-D object retrieval. IEEE Trans Image Process. 2012;21(4):2269–81.
5. Gao Y, Yang Y, Dai Q, Zhang N. 3D object retrieval with bag-of-region-words. In: Proceedings of the ACM international conference on multimedia, Firenze, Italy. 2010. p. 955–8.
6. Girshick RB, Donahue J, Darrell T, Malik J. Rich feature hierarchies for accurate object detection and semantic segmentation. In: Proceedings of the CVPR. 2014.
7. Krizhevsky A, Sutskever I, Hinton GE. ImageNet classification with deep convolutional neural networks. In: International conference on neural information processing systems. Curran Associates Inc.; 2012. p. 1097–105.
8. Su H, Maji S, Kalogerakis E, et al. Multi-view convolutional neural networks for 3D shape recognition. Comput Sci. 2015:945–53.
9. Gao Y, Liu A, Nie W, et al. 3D object retrieval with multimodal views. In: Proceedings of the 2015 Eurographics workshop on 3D object retrieval, 2015. Eurographics Association; 2015. p. 129–36.
10. Zhao S, Yao H, Zhang Y, et al. View-based 3D object retrieval via multi-modal graph learning. Sig Process. 2015;112:110–8.
11. Nie W, Liu A, Su Y. 3D object retrieval based on sparse coding in weak supervision. J Visual Commun Image Represent. 2015.

The Path Tracking Control Method Based on LOS Algorithm for Surface Self-propelled Model

ZHI-lin Liu, Yingkai Ma, SHOU-zheng Yuan and ZE-cai Zhou

Abstract Aiming at the surface self-propulsion model system, the paper design a path tracking control method based on the LOS (Line of sight) algorithm to achieve automatic path tracking control purposes. The algorithm uses a simple principle to obtain the desired heading angle, and through the PID control, which makes the ship closer to the expected course, finally sailing at heading, complete path tracking control. Simulation and experiment results for a Surface Self-Propulsion Vessel are provided to validate our method.

Keywords Surface self-propelled model · Path tracking control · LOS algorithm

1 Introduction

With the continuous development of surface self-propelled model, the path tracking control method of surface self-propelled model has been the hotspot of scholars. In the currently known theory of path tracking, the path tracking controller based on Backstepping is the main way [1–3], which has a higher request for the parameter of model. However, it is difficult to identify various parameters of model accurately [4], which makes it hard to carry out in many applications.

Z. Liu · Y. Ma (✉) · S. Yuan
College of Automation, Harbin Engineering University, Harbin 150001, China
e-mail: 1305634562@qq.com

Z. Liu
e-mail: liuzhilin@hrbeu.edu.cn

S. Yuan
e-mail: yuan19960628@qq.com

Z. Zhou
College of Shipbuilding Engineering, Harbin Engineering University, Harbin 150001, China
e-mail: zhouzecai@hrbeu.edu.cn

© Springer Nature Singapore Pte Ltd. 2018   473
Y. Jia et al. (eds.), *Proceedings of 2017 Chinese Intelligent Systems Conference*,
Lecture Notes in Electrical Engineering 460,
https://doi.org/10.1007/978-981-10-6499-9_45

The LOS Control Method used in this paper is the theoretical basis of surface self-propelled model [5], which derives the heading angle required by the self-propelled model to track the desired heading angle with PID control method. Then we designed a controller which converges to the aiming heading angle from present course angle which can remove the heading angle deviation. And the ship finally travels to the target heading point to complete the tracking control of the target.

The biggest advantage of the LOS algorithm is that it does not depend on the model of the controlled object, that is, the controller can be designed in an environment where the parametric model is uncertain or the external disturbance affects the ship model too much, then we complete the control of target model. Secondly, LOS algorithm is simple to design, anti-interference ability, excellent control effect is also the main factor to be used. Here we rely mainly on the LOS algorithm to identify the self-propelled model of the desired course, and then use the PID to adjust the actual course of the self-propelled model to achieve the purpose of path tracking.

2 The Introduction of LOS Algorithm

The guiding principle of the line of sight is reflected in its intuitive understanding of the operator's steering and the movement of the ship [6–8]. The principle is that if the course of the controlled model is kept at the "LOS angle" (line of sight angle), the controlled ship will be able to reach the desired position and track the track. And the LOS algorithm can reduce the traditional control from the three degrees of freedom of the ship's position and the heading angle to two degrees of freedom of the ship heading angle and speed of navigation, this feature is of vital importance for the control of under actuated ships.

2.1 Access to the Heading Angle

According to the LOS algorithm, as long as the controlled model to maintain the alignment of the "LOS angle", it will be able to achieve the desired location, so how to get "LOS angle" has become the focus of the study. As is shown in Fig. 1, assume that the desired position of the track trace is that the current position of the controlled model is recorded and assumed and can be measured. The "LOS angle" can be calculated by the following formula:

$$\psi_{los} = \arctan\left(\frac{y_{k+1} - y}{x_{k+1} - x}\right) \tag{1}$$

Among them, $\psi_{los} \in \langle -\pi, \pi \rangle$ is the LOS angle.

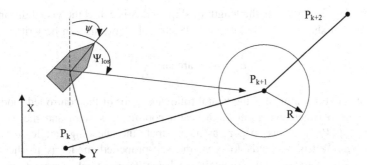

Fig. 1 LOS track tracking control algorithm schematic

Because LOS method is easy to understand and the control effect is superior to other control methods, the applicability is strong, thus LOS method has been widely used in ship track tracking control. In this paper, we will analyze the LOS method of the two navigation methods.

2.2 LOS Method Navigation Mode Analysis

The LOS control algorithm can track linear track and curve track, which is a better track tracking control algorithm. The algorithm is assisted by the introduction of a visual distance to complete the principle shown in Fig. 2 [8].

As it can be seen from the figure, the expected point $P_{los} = [x_{los}, y_{los}]^T$ is $\Delta = nL_{pp}$ from the projection point of the ship at the desired location on the desired track,

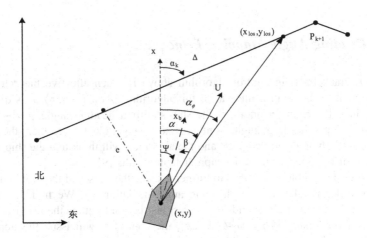

Fig. 2 LOS track tracking and navigation mode diagram

where n takes 2 to 5. L_{pp} is the length of ship, and Δ is called the visual distance of the controlled ship. At this point, the LOS control algorithm can be written as:

$$\alpha_\phi = \alpha_k + \arctan\left(\frac{-e}{\Delta}\right) \tag{2}$$

In the above formula, e is the lateral following error of the controlled model, α_k is the angle between the north of the earth coordinate system and the expected trajectory $P_k P_{k+1}$. This paper assumes that the heading angle deviation $\psi_e = \alpha_\phi - \psi$. When ψ_e tends to zero, the self-propelled model is to the target heading point of travel, and ultimately reach the desired track, complete the path tracking. Finally, through simple PID control, can effectively make ψ_e tend to zero, the current heading angle ψ will track the expected heading angle α_ϕ, through continuous control, self-aircraft model can travel along the desired track.

The ship is followed by a control algorithm to track an expectation point $P_{k+1} = [x_{k+1}, y_{k+1}]^T$. When the ship enters a range of the desired heading, it is necessary to automatically abandon the track of the desired heading, and to track the next desired heading, that is to let k = k+1. This requires a control algorithm to identify and switches the desired heading points. This paper uses the following algorithm:

Assuming that there is a circle of the current desired heading point (x_{k+1}, y_{k+1}) followed by ship, and R_0 is the radius of the circle. If the current position of the controlled model satisfies the following condition at a certain moment:

$$(x_{k+1} - x)^2 + (y_{k+1} - y)^2 \le R_0^2 \tag{3}$$

We need to convert the currently expected point to the next expected heading point. When calculating ψ_{los}, we need to switch to the next expected heading point and use the data of it to calculate.

2.3   Heading Angle Mapping Principle

The LOS track tracking control algorithm above is much effective theoretically. However, since the control interval of this algorithm is on $\langle -\pi, \pi \rangle$, it is discontinuous in both $-\pi$ and π nodes, if the ship's current heading angle $\psi = -179°$, when the ship's heading angle continues to change clockwise, then the next moment is likely to occur $\psi = 179°$ and the algorithm will think that the ship made the speed of rotation, resulting in unpredictable results [9].

To solve this problem, we need to further analyze the causes of the problem. The ship controller is related to the heading angle deviation ψ_e. We find that when $\psi_e \in (-\pi, \pi)$, the controller sends the correct instruction to guide the ship model to track the target point. When $\psi_e \notin (-\pi, \pi)$, the controller will issue the opposite instruction so that the actual flight angle of the ship model may converge from the

Fig. 3 Mapping of the
heading angle deviation

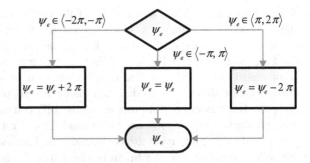

opposite direction to the expected heading angle. So we can solve this problem by
simply mapping $\psi_e \in (-2\pi, -\pi) \cup (\pi, 2\pi)$ to $\psi_e \in (-\pi, \pi)$. The specific mapping
method is shown in Fig. 3.

3 Design of PID Controller Based on Line-of-Sight Method

According to the above LOS algorithm, we can get the expected heading angle of
the self-propelled model in the course of the track tracking control experiment. The
ultimate aim of the control algorithm is to make the current heading angle of the
self-propelled model be controlled by a period of time equal to the expected course
obtained by the LOS algorithm.

$$\begin{cases} \psi_e = \psi - \psi_d \\ \psi_e \xrightarrow{t->0} 0 \end{cases}$$

Due to the characteristics of the self-propelled vehicle, the only control volume
is the steering angle of the self-propelled model. In order to achieve the purpose of
track tracking, this section uses the PID controller to converge the current heading
angle to the desired heading angle. Even if the heading angle deviation ψ obtained
based on the LOS control algorithm approaches zero, the purpose of track tracking
is achieved.

PID (Proportional—Integral—Derivative) controller has been the earliest controller that for more than 70 years and is still the most widely used controller [10].
The PID controller is easy to understand and requires a very accurate model of the
controlled object. The idea of PID controller design is to deal with the state of the
object to do the error processing, by introducing the error to do proportional control,
the output of the controlled object to eliminate the state of the required error control.

For example, the PID controller is used to control the heading of the
self-propelled model. The difference between the desired self-propelled heading and
the actual heading of the self-propelled model is the state error of the controlled

object. This error is taken as the input of the controller, and the control steering angle needed to eliminate the heading error of the self-propelled model can be obtained by the proportional control. But the proportional control cannot achieve the best control effect. By introducing the integral module, the control error accumulated in the self-propelled model control can be eliminated, and the intro-duction of the differential module can effectively suppress the overshoot in the self-propelled model control process. As the PID controller is widely used, only three parameters (K_p, K_i, K_d) need to set and the control effect is significant, so it gets a wide range of applications. In addition, because of the diversity of the controlled object, it may be in many cases that the adjustment of the integral module is not required or the adjustment of the differential module is not required, but the proportion of modules in any case are indispensable, plus the proportion of modules in the project easier to achieve, so we lost integral module or differential module for controller design.

Figure 4 is the PID control flow chart, we can see from the figure, as long as the design of a PID controller so that ψ_e tends to zero, then we achieve the goal of track tracking control, control law as shown in Eq. (4):

$$\delta = k_1 \psi_e + k_2 \psi_e + k_3 \int_0^t \psi_e dt \tag{4}$$

Since the rudder angle δ is proportional to the current heading angle ψ and the heading angle deviation is:

$$\psi_e = \psi - \psi_{los} \tag{5}$$

It can be seen from Eqs. (4 and 5) that if there is a large ψ_e in the system at some point, a rudder angle will be generated by PID control, and the steering angle will control the self-propelled flight to the desired heading, and the current heading angle of the self-propelled model approaches the desired heading angle of the self-propelled model. By constant control, as long as ψ_e exists, there will be a control of the steering angle to reduce ψ_e, until ψ_e tends to zero, that is, since the aircraft model reached the desired trajectory.

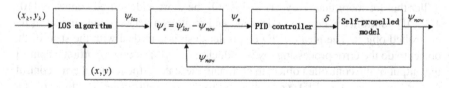

Fig. 4 PID control diagram

4   Simulation Results and Analysis of Matlab

In this simulation experiment, the main design for two corners, the first is a small corner less than 90°, the second is a larger corner more than 90°, the test results show that LOS algorithm can control the ship to achieve path tracking in different circumstance. (Figs. 5, 6, 7 and 8)

Fig. 5 Comparison of expected track and simulated track

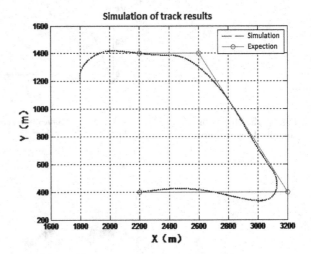

Fig. 6 Heading angular velocity curve

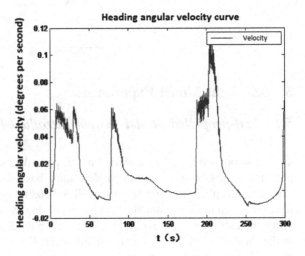

Fig. 7 Course angle change curve

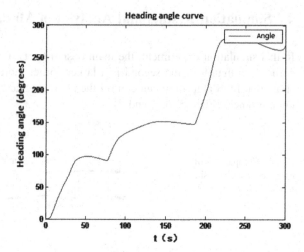

Fig. 8 The change of controlled rugger curve

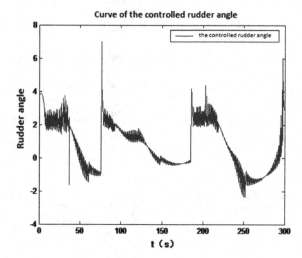

5 Self-flight Model Experiment

5.1 *Self-propelled Model System Introduced*

The self-propelled model system used in this paper is a small self-propelled aircraft platform, which can be used to complete some basic simulation experiments. The experimental platform consists of small inflatable pool, self-propelled model, monocular vision camera and the host computer. Figure 9 is a physical environment of the inflatable pool and the small self-propelled model in the simulation platform of the ship model in the laboratory environment. Figure 10 is a schematic diagram of the physical simulation platform.

Fig. 9 The small pool and
the small self-propelled
model for the
experimentation

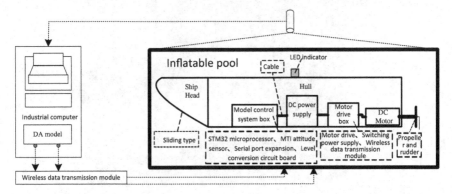

Fig. 10 The schematic diagram of the physical simulation platform

On the basis of this platform, we can study the trajectory tracking control
experiment algorithm of small self-propelled model, to achieve the purpose of
physical simulation. The desired trajectory in this experiment consists of a discontinuous set of ordered path points. A line segment is obtained between two
adjacent path points, and all the path points are connected to each other to get our
desired path.

5.2 Self-aircraft Model Overall Design

In this paper, the self-propelled model is formed by the propeller as the propulsion
device and the controllable steering angle is used to realize the heading control
function. Requiring the captain of about 0.3 m, the maximum radius of rotation is
not greater than 0.5 m, the hull is fiberglass material, and the ship is installed above
the three led lights, for self-aircraft positioning system in the camera to capture the
hull real-time location and heading information. Figure 11a is a vertical plan view,
Fig. 11b is a schematic view of the aerial model

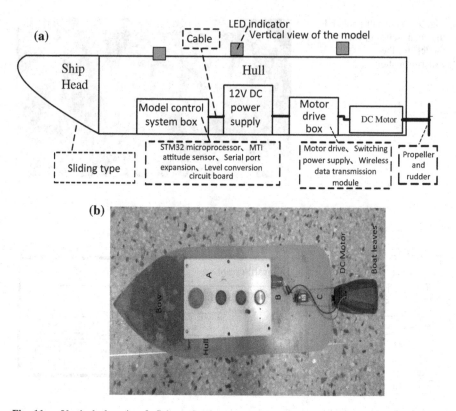

Fig. 11 **a** Vertical plan view **b** Schematic view

Self-propelled hull mainly includes control system box, DC power supply device, motor drive box, DC motor and propeller and rudder, as shown in Fig. 11a. In Fig. 11b control box is for the part A, mainly including STM32 microprocessor, MTI attitude sensor, and serial port expansion circuit and level conversion circuit board. Motor drive box is for the part B of the motor switch, motor drive and wireless data transmission module. And part C is the power supply position. In the design process, we try to meet the real ship in shape, kinematics, hydrodynamics and other aspects of the performance similar requirements (Fig. 12).

The figure shows the flow pattern of the self-propelled model control system, mainly for the camera to capture the current navigation status (position, heading angle) to the PC host computer control software, through the controller to calculate the control input of the boat, through the wireless data transmission module sent to the self-propelled model, to achieve real-time control of the self-propelled model.

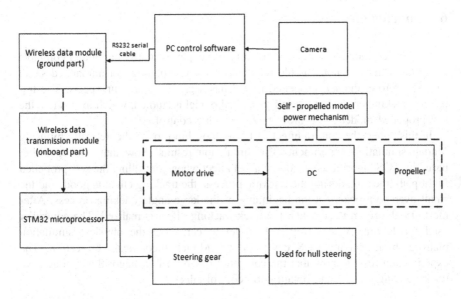

Fig. 12 Schematic diagram of the control system implementation

Fig. 13 Curve track based on
LOS algorithm

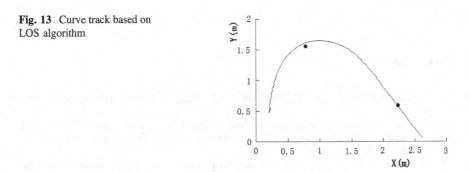

5.3 Self-propelled Model Experiment Analysis

Figure 13 is the LOS algorithm of self-propelled model curve tracking experiment results based on LOS algorithm. The path in the test to track the trajectory for the acute angle, and this curve can be seen as two straight track path [11, 12]. As shown in Fig. 13, the self-propelled model first tracks the first desired heading (0.8, 1.5). As can be seen from the figure, the tracking effect is good at this stage. When the LOS control algorithm decides to abandon the track of the current desired point and track the next desired course point, the self-propelled model is quickly turned, but due to the inertia of the self-propelled model itself, there is a certain overshoot, with the LOS algorithm and the rapid adjustment of the PID controller, the self-propelled model finally tracks the desired track quickly and reaches the second path point (2.3, 0.5) to achieve track tracking control.

6 Conclusion

Self-propelled model can be used to carry out experimental research and complete many tasks that are not suitable for the crew to complete as an unmanned small ship. Therefore, the self-propelled model has a great research prospects. The trajectory tracking control of the self-propelled model is studied, which means that the self-propelled model can achieve better tracking control effect.

In this paper, the control effect of LOS algorithm proved theoretically by using Matlab simulation experiment. The simulation results show that, the controller based on LOS algorithm has good control effect when controlling the self-propelled flight path tracking, during the tracking process, the tracking effect is good, and the small overshoot occurs at the beginning of the large angle tracking process. After the LOS algorithm is adjusted, the track tracking effect is realized. The controller used the laboratory self-propelled model to carry out the physical simulation platform based on the LOS design in the Matlab theoretical verification. The experimental results show that the controller based on LOS algorithm can achieve track tracking control and obtain good control effect.

Acknowledgements This work was supported by the National Natural Science Foundation of China (51379044) and the Fundamental Research Funds for the Central Universities (HEUCF0417).

References

1. Ping-zhong JIN. Ship water jet propulsion. Beijing: National Defense Industry Press; 1986. p. 41–136.
2. Sun CL, Wang Y-S. Computational fluid dynamics for marine waterjet design and performance analysis. Colledge of Naval Architecture and Power,Naval University of Engineering,Wuhan 430033, China.
3. Zhu H-H, YE H-k, Sun J-L, Feng D-K, WU Wen-qian. Fast analysis of axisymmetric rotator and theoretical calculation of thick boundary layer. China Water Trans. 2007; 1, 5, 1: 179–180.
4. Rangel-Huerta A, Ballinas-Hernández AL, Muñoz-Meléndez A. An entropy model to measure heterogeneity of pedestrian crowds using self-propelled agents. Physica A: Statistical Mechanics and its Applications; 2016.
5. Yi Lee. Ma YN, Song Z. Method for obtaining line-of-sight rate with a strap-down imaging guidance. J Beijing Inst Technol. 2009;04:457–62.
6. Warn-Gyu P, Hyun SY, Ho HC, Moon CK. Numerical flow simulation of flush type intake duct of waterjet. Ocean Eng. 2005;2107–2120
7. Park Warn-Gyu. Jin Ho Jang. Numeriacl flow and performance analysis of waterjet propulsion system. Ocean Eng. 2005;32:1740–61.
8. Thor IF, Morten B, Roger S. Line-of-Sight Path following of underactuated marine craft. University of Science and Technology (NTNU); p. 7491.
9. Sv. A$_A$. HARVALD. Resistance and propulsion of Ships. Wiley. 1983. p. 57–599.
10. Liu J-K. Advanced PID control MATLAB simulation. Electronic Industry Press, 2004.

11. Javad S, Araz H. Correlation between rolling noise generation and rail roughness of tangent tracks and curves in time and frequency domains. Applied Acoustics, 2016, p. 107.
12. Yiming M, Jianwei G, Huiyan C, Yanhua J, Yuwen H. Intelligent Vehicle Research Center, Beijing Institute of Technology, Beijing 100081. Design optimization of control parameters for strapdown line-of-sight stable tracking platform based on the Taguchi method. Northeastern University, China. IEEE Industrial Electronics (IE) Chapter, Singapore, Guilin University of Electronic Technology, China.

A New Localization System for Tracking Capsule Endoscope Robot Based on Digital 3-Axis Magnetic Sensors Array

Ming Xu, Deyu Kong, Linfei Ye and Jiansheng Xu

Abstract This paper proposes a new magnetic localization system for capsule endoscope robot which works wirelessly in the human body by using a 4 × 4 sensors array, composed by the latest digital magnetic sensors, QMC5883L. In order to locate a capsule effectively and non-invasively, we apply the static magnetic field method by enclosing a small cylindrical permanent magnet in the capsule. The position and orientation of the permanent magnet are related to its magnetic signals like field intensity and direction which can be measured by the sensors array outside the human body. The relations between the magnet and its signals can also be formulated by equations, as a cylindrical permanent magnet can be considered as a magnetic dipole model. In Matlab, we can approach the equations for parameters containing location information by using non-linear least squares algorithm with Levenberg-Marquardt optimization method. The experimental results show that accuracy is well controlled and the digital sensors are applicable for the tracking systems.

Keywords Localization system · Magnetic dipole · QMC5883L · Non-linear least squares · Levenberg-Marquardt optimization

M. Xu · D. Kong · L. Ye · J. Xu (✉)
Institute of Electrical Engineering, Chinese Academy of Sciences,
100190 Beijing, China
e-mail: jsxu@mail.iee.ac.cn

M. Xu
e-mail: xuming15@mail.iee.ac.cn

D. Kong
e-mail: 646284278@qq.com

L. Ye
e-mail: 865531976@qq.com

M. Xu · D. Kong · L. Ye
University of Chinese Academy of Sciences, 100190 Beijing, China

© Springer Nature Singapore Pte Ltd. 2018
Y. Jia et al. (eds.), *Proceedings of 2017 Chinese Intelligent Systems Conference*,
Lecture Notes in Electrical Engineering 460,
https://doi.org/10.1007/978-981-10-6499-9_46

487

1 Introduction

Capsule endoscope robot is an ingestible pill-like device that contains a tiny camera and an illuminating system for capturing images and a transmission module for sending the images wirelessly to external receivers. It is a great breakthrough in medical endoscopy, which helps reducing the damage to other tissues, shortening the recovery time, eliminating the side effects caused by surgery and reducing the cost of medical treatment. When the robot is moving through gastrointestinal tract, it is essential to accurately obtain the position information for controlling it actively and mastering the exact position of illness [1–4].

Magnetic localization technology is widely used in the research of robot localization due to its non radiation and high accuracy. Sufficient magnetic field data information measured by the sensor is crucial for positioning. So far, a lot of research has been done on magnetic positioning by using magnetic sensors array. For example, Vincent et al. use a 2D-array of 16 Hall sensors [5], Liao Ying et al. use a 4 × 4 sensors array of CSA-1 V sensors [6], Zheng xiao-lin et al. provide a 4 × 4 sensors array of SEN-S65 sensors [7] and Chao Hu et al. propose a cubic magnetic sensors array made of Honeywell 3-axis magnetic sensors, HMC1043 [8, 9]. However, they all pick analog sensors requiring complex support circuits to work, which may reduce accuracy and stability of the system. In this paper, we propose a localization system using a 4 × 4 array with 16 latest digital magnetic sensors, QST product QMC5883L, and a data-fitting algorithm using non-linear least squares method with Levenberg-Marquardt optimization [10] in MATLAB.

2 System Description

A cylindrical permanent magnet can be modeled as a magnetic dipole. The magnetic field of a permanent magnet is a high-order nonlinear function, which is related to the position and orientation of magnet. It can be represented by:

$$\mathbf{B}_l = B_{lx}\mathbf{i} + B_{ly}\mathbf{j} + B_{lz}\mathbf{k} = B_T\left(\frac{3(\mathbf{H}_0 \cdot \mathbf{P}_l)\mathbf{P}_l}{R_l^5} - \frac{\mathbf{H}_0}{R_l^3}\right) \quad (l = 1, 2, \ldots, N) \quad (1)$$

In (1), $\mathbf{B}_l = (B_{lx}, B_{ly}, B_{lz})^T$ is a vector defining the 3-coordinate axes magnetic flux intensity in the sensor l located at position $(x_l, y_l, z_l)^T$; B_T is a constant defining the magnetic intensity of the magnet; $\mathbf{H}_0 = (m, n, p)^T$ is a vector defining the direction of the magnet; $\mathbf{P}_l = (x_l - a, y_l - b, z_l - c)^T$ is a spatial vector from the sensor l to the magnet center $(a, b, c)^T$; R_l is a scalar defining the length of P_l.

Figure 1 shows the magnetic dipole in space rectangular coordinate system.

With three orthogonal components:

Fig. 1 Magnetic dipole

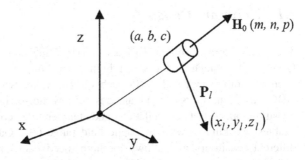

$$B_{lx} = B_T \left\{ \frac{3[m(x_l - a) + n(y_l - b) + p(z_l - c)](x_l - a)}{R_l^5} - \frac{m}{R_l^3} \right\} \tag{2}$$

$$B_{ly} = B_T \left\{ \frac{3[m(x_l - a) + n(y_l - b) + p(z_l - c)](y_l - b)}{R_l^5} - \frac{n}{R_l^3} \right\} \tag{3}$$

$$B_{lz} = B_T \left\{ \frac{3[m(x_l - a) + n(y_l - b) + p(z_l - c)](z_l - c)}{R_l^5} - \frac{p}{R_l^3} \right\} \tag{4}$$

(constraint equation: $m^2 + n^2 + p^2 = 1$)

Equations (2–4) represent the relationship between the components of B_l and the position of the sensor l with reference to the origin. We can solve the nonlinear data-fitting (curve-fitting) problems in least-squares sense by function *lsqcurvefit()*.

3 System Design

The three dimensional magnetic data measured by the 4 × 4 sensors array is transmitted to the microcontroller STC89C51 by the I2C protocol and then sent to the computer through serial port. The framework of the system is shown in Fig. 2.

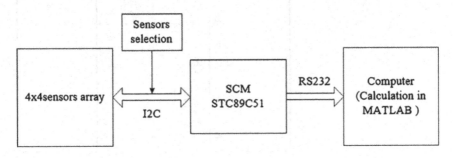

Fig. 2 System framework

3.1 Hardware Design

The sensors array contains 16 latest 3-axis digital magnet sensors, QST product QMC5883L, which is based on high resolution, magneto-resistive technology licensed from Honeywell AMR technology. Along with custom-designed 16-bit ADC ASIC, it offers the advantages of low noise, high accuracy, low power consumption, offset cancellation and temperature compensation. QMC5883L enables two kind of wide magnetic field range ($\pm$ 2 Gauss and $\pm$ 8 Gauss) with different sensitivity and allows for high-speed data communications, maximum 200 Hz data output rate. The I2C serial bus allows for easy interface.

However, all the same type digital sensors have the same I2C address, which makes it difficult to receive data separately. Meanwhile, due to the real-time requirements of the system, we need to obtain all sensors data simultaneously. Finally, we've come up with the idea of using I2C bus to transmit data in parallel, that is, to enable different sensors to share the same clock line (SCL) and to connect different data line (SDA) to different I/O ports, unlike the usual I2C bus connection. Figure 3 shows single sensor's connection with external circuits and Fig. 4 shows parallel connection of sensor modules.

3.2 Software Design

The software program of the system data transmission includes configuration of sensors and UART, data acquisition and sending. The initialization of the sensors array contains definition of ports, measurement mode, range of magnetic field, over sample ratio and output date rate; the initialization of UART contains baud rate,

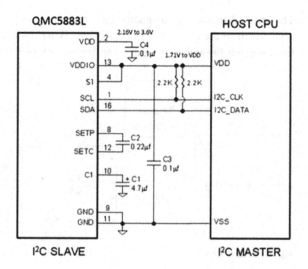

Fig. 3 Sensor connection

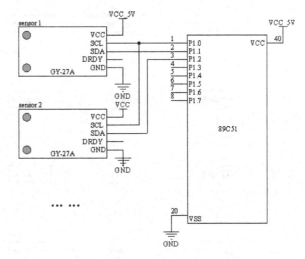

Fig. 4 Sensor modules connection

parity and other parameters. Data transmission between sensors and host CPU 89C51 follows the I2C protocol; the data stored in the 6 registers of each of the sensors are read out bit by bit simultaneously when the SCL line is high; we save the data to an intermediate 6 × 16 size data array and then send to computer by UART module. Figure 5 presents the system flow chart.

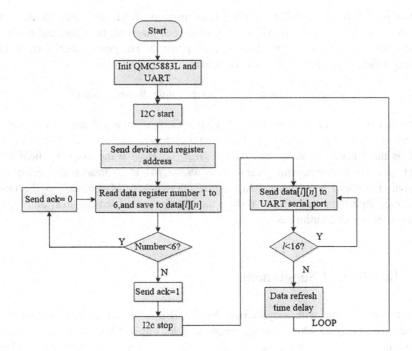

Fig. 5 The system flow chart

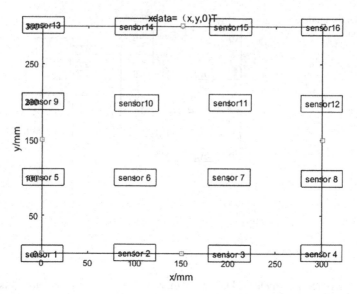

Fig. 6 Arrangement of sensors array

4 Experiment

After getting the data and converting it to magnetic field form, we calculate the magnetic data in MATLAB. The relations between magnetic field data and position coordinates are fitting with the least square approach. The grammatical form of the fitting function *lsqcurvefit( )* is shown below:

$$p = lsqcurvefit(@fun, p_0, xdata, ydata, lb, ub, options)$$

with function *fun( )* represents Eqs. (2–4); p = $(a, b, c, m, n, p)^T$ are the parameters to be solved with p_0, *lb* and *ub* as the given initial value and ranges; *xdata* = $(x_l, y_l, z_l)^T$ is the position coordinates; *ydata* = $(B_{lx}, B_{ly}, B_{lz})^T$ is the magnetic field data and we use the statement *ydata(:,4) = ones(1,16)* to represent the constraint equation; *options* = optimset ("Algorithm", "Levenberg-Marquardt") is the optimization algorithm used. Figure 6 show the arrangement of sensors array and Fig. 7 shows the exact coordinates.

5 Results and Suggestions

We test the system with a cylindrical Nd–Fe–B magnet, which is a kind of high field density magnet whose surface magnetic field is up to 5800 Gauss in the size of Ø6 mm × L8 mm. The designed test path is a line trajectory on a plane

Fig. 7 Coordinates of sensors

number	x/mm	y/mm	z/mm
1	0	0	0
2	100	0	0
3	200	0	0
4	300	0	0
5	0	100	0
6	100	100	0
7	200	100	0
8	300	100	0
9	0	200	0
10	100	200	0
11	200	200	0
12	300	200	0
13	0	300	0
14	100	300	0
15	200	300	0
16	300	300	0

perpendicular to the array. We move the magnet along the line from the x axis 0 to 30 cm and calculate positional parameters p in Matlab at the same time. The track of the magnet is shown in Fig. 8 and preliminary experimental result shows that the average z error of the localization system is 1.4 cm, which will be reduced significantly after we perfect the calibration of the sensors in the future.

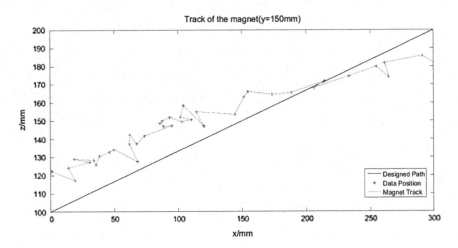

Fig. 8 Track of the magnet

The test results show that our new hardware design is applicable, simple and effective. Besides, the use of digital sensors is very convenient for a large scale use, which will lead to the improvement of accuracy due to large amount of data and many improvements can be made in the algorithm to reduce the position error too.

References

1. Xu J, Song T. Differential magnetic localization method for the microrobot with two cylindrical permanent magnets. 2012:4798–802.
2. Xu J. Research on sensors arrangement for capsule endoscope microrobot magnetic localization. In: Control and decision conference. IEEE; 2013. p. 5208–11.
3. Andrä W, Danan H, Kirmsse W, et al. A novel method for real-time magnetic marker monitoring in the gastrointestinal tract. Phys Med Biol. 2000;45(45):3081–93.
4. Than TD, Alici G, Zhou H, et al. A review of localization systems for robotic endoscopic capsules. IEEE Trans Biomed Eng. 2012;59(59):2387–99.
5. Vincent S, Predrag D. A magnetic tracking system based on highly sensitive intergrate hall sensor. JSME Int J. July 2002; No. 02–5096.
6. Ying L, Pingping J, Guozheng Y, et al. In vitro magnetic tracking and positioning system based on Holzer sensor array. Beijing Biomed Eng. 2012;5:501–6.
7. Zheng X, Li J, Hou W, et al. Localization and tracking of digestive tract diagnostic and therapeutic capsule using magnetic sensor array. Opt Precis Eng. 2009;17(3):576–82.
8. Hu C, Meng MQ, Mandal M. Efficient magnetic localization and orientation technique for capsule endoscopy. Int J Inform Acquisit. 2011;02(1):23–36.
9. Hu C, Li M, Song S, et al. A Cubic 3-axis magnetic sensor array for wirelessly tracking magnet position and orientation. IEEE Sens J. 2010;10(5):903–13.
10. Levenberg K. A method for the solution of certain non-linear problems in least squares. J Heart Lung Transplant Offic Publicat Int Soc Heart Transplant. 1944;2(4):436–8.

Analysis on Time Triggered Flexible Scheduling with Safety-Critical System

Yue Lin, Yuan-lin Zhou, Song-tao Fan and Ying-min Jia

Abstract Time triggered Network communication technology has been applied to the China's next generation multi-purpose complex Aerospace Vehicle. This vehicle is based on the technology in the system satisfy safety critical communication need, and can also take into account other large amount of data of low priority communication tasks. In this paper, combined with the actual characteristics of aerospace vehicle's mission requirements, according to the study of the key security information system level scheduling method based on this technology, a global optimal requirement decomposition and message scheduling algorithm based on automatic planning is proposed. The automatic generation algorithm of message scheduling table according to the system message transmission needs, and to meet the needs of key safety messages in the table does not meet the labeling and suggestions, providing the basis for the design of aircraft overall communication timing and strategy. Matlab/Simulink is used to modeling, simulating and verifying the algorithm. The simulation results meet the requirements of the communication design of the aircraft system.

Keywords Time triggered · Safety critical · Message scheduling · Design iteration

Y. Lin (✉) · Y. Zhou · S. Fan
Beijing Institute of Control Engineering, Beijing 201707, China
e-mail: linyue371@163.com

Y. Zhou
e-mail: YL198504@yahoo.com

S. Fan
e-mail: tigerst@yahoo.com

Y. Jia
School of BUAA 7th Research Division, Beijing 201707, China
e-mail: ymjia@buaa.edu.cn

© Springer Nature Singapore Pte Ltd. 2018
Y. Jia et al. (eds.), *Proceedings of 2017 Chinese Intelligent Systems Conference*,
Lecture Notes in Electrical Engineering 460,
https://doi.org/10.1007/978-981-10-6499-9_47

1 Introduction

The international mainstream spacecraft manufacturers have recently introduced the next generation of complex spacecraft mostly based on the next generation of time-triggered technology and networked field bus, such as TTEthernet, Arinc659, Spacewire, AFDX and so on. As the backbone of the next generation of aerospace/space-borne aircraft data bus. The security performance, transmission bandwidth, architecture and timing control capabilities are significantly better than the traditional realization, such as 1553B bus and CAN bus. It is of great help to the overall design and performance of the aircraft. Also, it could support the aircraft in the lower weight, lower power consumption, and higher security requirements, which of the case to complete a more complex system-level tasks.

By using the next generation field bus of time-triggered networked, requires the pre-planning of critical security messages and time-triggered messages (referred to as TT messages) in advance in the system is needed. The traditional planning schemes requires a great quantity of manual analysis and calculation to ensure the completeness and correctness of the system-level message scheduling strategy. At the same time, the system-level message communication demand will continue to change with the continuous design of the system. So a lot of manpower and working time is needed in the system development process.

Based on the analysis of time-triggered message scheduling strategy from literature [1, 2], this paper presents a system-level message scheduling algorithm combining the actual mission requirements of space-borne aircraft.

The algorithm is divided into three steps. First, abstraction and classify the system-level task requirements. Second, the improved genetic algorithm will be used to optimize the scheduling strategy of the whole system. Finally, on the basis of the optimization, the requirements of the task demand are analyzed automatically, and the suggestions will be given.

According to the analysis results and recommendations on the needs, the overall design staff will adjust the system design to match the goal. After several adjustments and iterations, the system design results will meet the mission requirements and overall goals.

2 System Requirement

2.1 Aircraft Information Architecture

Generally, the spacecraft would use of secondary network switch structure to fulfil the requirements of information architecture. This is mainly for two reasons, one is the length of the aircraft is long, and the second is based on the distributed system architecture considerations.

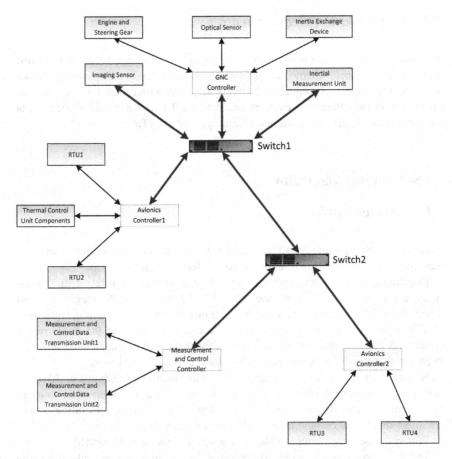

Fig. 1 Information architecture based on time trigger technology

As Fig. 1 shown below, the entire aircraft is mainly composed of flight control subsystem, avionics subsystem, measurement and control subsystem and mechanism subsystem.

The flight control system, avionics system and monitoring/manual control system using time-triggered network communication. Which part of the core devices are directly connected to the backbone network, such as the system controller, imaging sensor and inertial measurement unit. The other devices, including traditional parts, are connected to the system controller, and exchange information with other systems via the backbone of the system controller.

2.2   Task Function Requirements

According to the mission design, the system clock minimum resolution unit (time slice) is set to 8 ms. All message communication cycles are 2 ms integer times (must be greater than or equal to 8 ms). Every communication message will be transferred in the determined cycle phase. And the TT message will be received at the initial time of the next communication cycle phase (Table 1).

3   Scheduling Algorithm

3.1   Message Model

According to the mission requirements and the aircraft information architecture, a message model is established and used for system message analysis.

Depending on the safety considerations of the aircraft, all safety-critical messages are required to be implemented with TT messages. Message loss and scheduling conflict are not permitted during transmission of safety-critical messages [3]. Therefore, taking into account the clock synchronization of the nodes in the system, all TT messages should be reserved for sufficient communication intervals to ensure that the TT message does not collide during transmission.

However, the amount of reservation time will affect the feasibility of TT message static arrangement directly [4]. But on the other hand, the reservation time can be used for RC and BE messages in the system of competitive scheduling, it will not affect the global communication efficiency (Table 2).

Where the Secmsg represents the area where the message is located.

The TT message in the system is composed by global messages and partial area messages. The global message can be received (Code-named T) for all nodes in the

Table 1 TT message requirements

System	Cycle (ms)	Length (Byte)	Num
Flight control	8	64	5
Flight control	16	256	24
Flight control	128	256	40
Avionics 1	32	512	32
Avionics 1	128	512	64
Avionics 1	1024	2048	32
Avionics 2	16	64	24
Avionics 2	256	1024	64
Avionics 2	2048	4096	10
Tele command	128	1024	16
Tele command	1024	4096	8
Synchronization	1024	–	1

Table 2 Symbols and definitions tables

Symbol	Definition
T_C	Time slice period (ms)
N_{msg}	Message name
Sec_{msg}	The area where the message is located
T_{msg}	Message cycle (ms)
PT_{msg}	Expected message transfer time slice phase
PR_{msg}	Actual message transfer time slice phase
L_{msg}	Message length (Byte)
Mar_{msg}	TT message minimum interval (us)
BT_{msg}	TT message basic transmission reservation (us)
BR_{msg}	Communication rate (bps)
RT_{msg}	Communication efficiency (%)
$Trans_{msg}$	Message transfer time (us)
$Trans_{sw}$	Switch forwarding time (us)
$Trans_{ES}$	Terminal send/receive processing time (us)
$Trans_{CM}$	Compression master node compression delay (us)
T_{Syn}	System clock synchronization period (ms)

system, and some of the regional messages are transmitted only between some nodes. Will not affect other areas of TT message transmission, the scheduling can be executed in parallel (code-named A, B, C).

3.2 Offline Planning Algorithm

The problem can be summarized as a two—dimensional packing problem with boundary condition. As the packing problem is a typical NP problem, the adaptive genetic algorithm can be used to solve the problem.

A complete message cycle is 2048 ms, each single message is of individual. Such as flight control system's 32 ms cycle message, a total of 2048/32 = 64 times are sent in the full message cycle, with the same interval for each transmission cycle. So the gene parameter on the individual is the phase PRmsg of the message. The range of values for each message is:

$$PR_{msg} = \begin{cases} 0 \sim \frac{T_{msg}}{T_C}, & normal \\ \forall k \in Aera_{msg}, & Unique - requirements \end{cases} \tag{1}$$

The main parameters of each message can be abstracted as the message transmission time $Trans_{msg}$, where:

Fig. 2 Message scheduling figure

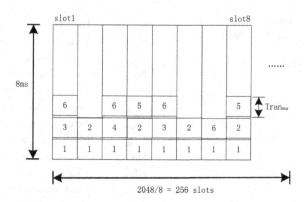

$$Trans_{msg} = Trans_{SW} + 2 \times Trans_{ES} + BT_{msg} + Mar_{msg} + \frac{L_{msg}}{BR_{msg} \times RT_{msg}} \quad (2)$$

Initialize the population, assign random parameters to each message individual, and perform adaptive analysis.

From Fig. 2 and the overall design, the total length of the TT message that can be stored in each slot must be less than 8 ms. The following steps show how to calculate the total length of the message transmission time in each slot, and definition the fitness function fitmsg:

$$SumT_k = \sum_{i=1}^{n} Trans_i \quad (3)$$

$$slotfit_k = \begin{cases} 8000 - SumT_k, & SumT_k < 8000\,us \\ 0, & SumT_k \geq 8000\,us \end{cases} \quad (4)$$

$$fit_{msg} = \min\{slotfit_k, \forall k \in Aera_{msg}\} \quad (5)$$

Calculate the length of the total message in each slot. The length of time is inversely proportional to the fitness parameter value. If the time exceeds 8 ms, the fitness become zero. Each message takes the lowest degree of fitness for all slots as its fitness value. If a slot exceeds 8 ms, all messages in this time sheet need to be reassigned.

Using the phase of the message as the result of the optimization output. Also, the phase range of the message depends on the period of the message. The cross-exchange of the genetic factors takes the same period of the message as the parent cross-swap and variation the next generation is also the group of messages in which the parent is in the group (and satisfies the special requirements of the phase selection). If the parent group is completely eliminated in the last fitness selection, the corresponding number of sets of messages are randomly initialized.

3.3   Feedback Recommendations

Set the maximum number of iterations to 5000 generations. If the number of iterations reaches the upper limit and has not reached all the individual fitness in the system greater than 0, the iteration is stopped.

For each iteration process, statistics the situation of fitness degree become zero. If this happens frequently, it is necessary to adjust the message from the system requirements, to improve the communication cycle and reduce the single communication data length.

Under the current design conditions, the parameters that do not meet the requirements of the most adaptive degree will automatically double communication cycle, and then re-enter the system algorithm analysis. If still cannot meet system requirements, the fitness does not meet the requirements of the parameters of the communication word. The number of nodes decreased by 20%, and then go into the system algorithm analysis for the third time.

4   Scheduling Algorithm

4.1   Modeling and Simulation

The algorithm is modeled and simulated in Matlab/Simulink, and the system-level communication simulation is carried out for scheduling the static scheduling results of each algorithm.

If all system messages have none-specify phase range, then only the communication cycle and communication length are specified. The scheduling result can be found with any given conditions (the total communication time of all messages does not exceed 2048 ms).

If specify the range of the phase for some messages, it is possible that to no schedule list could be got for the system constraints.

For example, take 10 messages in flight control 32 ms cycle messages and 24 messages in avionics 16 ms cycle messages as consideration. If all of these messages are specified in phase 2, and there is an unresolvable scheduling situation (Fig. 3).

The system-level communication simulation model shown in the above figure is established, and the clock of each node in the system is shifted. When the clock shift reaches 10 us, the TT message scheduling in the system still satisfies the expectation, no collision or loss, satisfies the high security of the system Requirements.

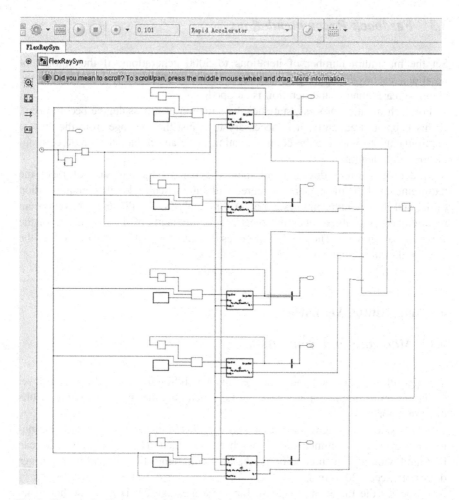

Fig. 3 System—level communication simulation model

4.2 Formal Verification

Verification boundary situation of the system by using simulation method, is not enough to fully prove the security and completeness of the system. So formal verification tools is to be used for verifying the system performance.

Based on the system-level communication model, where Simulink Test Case is joined for using. Take the phase difference between the nodes is between −10 and +10 us as system condition. Then take the system verification. First, verify that there is an overlap in the message in each slot during system-level communication. And then, verify that all message traffic in each slot exceeds the slot time range (Fig. 4).

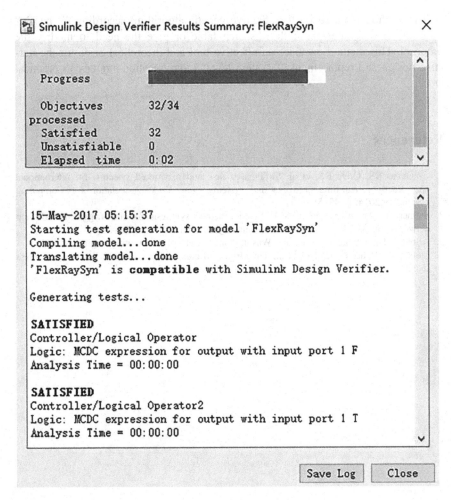

Fig. 4 Use Matlab formal verification tool to verify

The verification results show that, the system-level communication indicators meet the requirements within the given conditions, and there are no anomalies and conflicts.

5 Conclusion

Based on the actual demand of spacecraft, a method is proposed to solve the automatic generation of Ethernet triggered flexible message scheduling strategy.

The method is based on the adaptive genetic algorithm, according to the task needs to give the design of the dispatch table or give the need to modify the proposal. This method provides an effective tool and theoretical basis for the iterative design and replacement of system level communication strategy in practical work.

References

1. Craciunas SS, Olive RS, et al. SMT-based task and networked systems. In: International conference on real-time networks and systems. New York: Association for Computing Machinery;2014. p. 45–54.
2. Steiner W. An evaluation of SMT-based schedule synthesis for time-triggered multi-hop networks. In: 31st IEEE Real-Time Systems Symposium;2010. p. 375–84.
3. AS6802. Time-Triggered Ethernet. Washington: Society of Automotive Engineers;2011.
4. Steiner W, Bauer G, Hall B, et al. Time-triggered communication. Boca Raron: CRC Press Inc;2011. p. 88–9.

Research of Ore Particle Size Detection Based on Image Processing

Ruxue Wang, Weicun Zhang and Lizhen Shao

Abstract The size distribution of ore particles is an important basis for evaluating the crushing effect, and is also the main index for the optimal control of mineral processing equipment. Image processing technology is used to process the images collected by the camera on line, which can get the information of ore particle size in real time and feedback or adjust the parameters of the crushing machine in time. We first study the image pretreatment methods on ore images to separate foreground and background images. Then the improved watershed segmentation algorithm is used to segment the ore images. Finally, the distribution information of the ore particle can be obtained through the projected diameter of the ore particles. The experimental results show that the analysis results are effective in real applications, which can meet the needs of on-line detection of ore particles.

Keywords Particle size distribution · Image processing · Image pretreatment · Ore segmentation

1 Introduction

At present, image processing technology is widely used in our daily life, especially in the industrial field. Image processing has become a very common, efficient and economical means of detection [1]. However, the application of measurement and statistics of the gravel by image method is relatively unusual. In this context, we attempt to explore the method of automatic segmentation of the stacked ore particle image, and receive satisfactory results.

The ore particle size distribution is an important basis for the evaluation of crushing effect. The particle size parameters mainly include the particle area,

R. Wang · W. Zhang (✉) · L. Shao
School of Automation and Electrical Engineering, University of Science
and Technology Beijing, Beijing 100083, China
e-mail: weicunzhang@ustb.edu.cn

© Springer Nature Singapore Pte Ltd. 2018 505
Y. Jia et al. (eds.), *Proceedings of 2017 Chinese Intelligent Systems Conference*,
Lecture Notes in Electrical Engineering 460,
https://doi.org/10.1007/978-981-10-6499-9_48

perimeter, equivalent diameter, shape factor and so on [2]. In order to obtain the ore particle size distribution, we mainly study the following aspects:

(1) The characteristics of the ore particle image are analyzed, and some solutions are proposed according to the characteristics of the ore particle image.
(2) In this paper, the basic principle of mathematical morphology is introduced and the mathematical morphology is applied to the image pretreatment of ore particle to achieve a well effect.
(3) In this paper, we propose a segmentation method based on boundary information extraction, morphological reconstruction and watershed transformation. The adaptive binary algorithm and edge detection algorithm are used to extract enough boundary information. Finally, the binary image is transformed into gray image through the algorithm of distance transformations and then combined with the gradient image to realize the watershed segmentation. Experimental results show that these algorithms can be combined to form a new ore particle boundary recognition algorithm which can successfully separate the accumulated ore particles.

2 Image Pretreatment

As the ore image collection environment is complex, several problems in the ore image are collected by machine vision system:

(1) Ore images are collected in high dust, highly polluted environments so that the ore images are blurred and have severe noise.
(2) The ore image collection scene is outside, some unfavorable factors, such as natural light irradiation, occlusion from the equipment and irregular shape of the ore, which cause the great difference in the image color and brightness, produce shadow around the ore particles;
(3) Ore particles are serious stacked and adhered to each other. Meanwhile, the image background difference is small and the color information is not obvious.

According to the above problems, the ore image pretreatment needs to study the following contents.

Firstly, the image noise is filtered and the edge of the ore is preserved as much as possible. Secondly, mathematical morphology reconstruction is applied to the gray image of ore to strengthen the ore boundary in order to extract foreground objects better [3]. And at last, the foreground objects are extracted by the adaptive binary algorithm to remove the shadows and background around the ore. The specific pretreatment process is shown in Fig. 1.

Gaussian smoothing filter is a common method of image smoothing filter. Gaussian filter is a filter that selects weights based on the shape of the Gaussian function. Gaussian filter will smooth all pixels uniformly, which does not consider

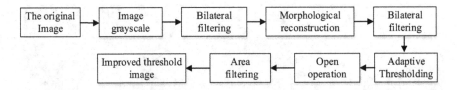

Fig. 1 The pretreatment of ore image

the characteristics of the pixels themselves, will lead to the edge of ore particles be blurred.

Bilateral filter is an edge preserving filter technique which is proposed on the basis of Gaussian filter. Bilateral filter take the Gaussian filter as a kernel function, and the weight of the pixel decreases with the distance from the center of the kernel. As a nonlinear filter, bilateral filter considers the spatial relation and the gray relation of image pixels, and through the non-linear combination of the two, to achieve adaptive filtering of the image finally. The bilateral filter is defined as follows:

$$BF[I_p] = \frac{1}{W_p} \sum_{q \in S} G_{\sigma_d}(\|p-q\|) G_{\sigma_r}(I_p - I_q) I_q \tag{1}$$

where BF is a symbol of bilateral filter, W is a scalar.

$$W_p = \sum_{q \in S} G_{\sigma_d}(\|p-q\|) G_{\sigma_r}(I_p - I_q) \tag{2}$$

$$G_{\sigma_d} = e^{-\frac{1}{2}\left(\frac{d(p,q)}{\sigma_d}\right)^2} \tag{3}$$

$$G_{\sigma_r} = e^{-\frac{1}{2}\left(\frac{\delta(I(p),I(q))}{\sigma_r}\right)^2} \tag{4}$$

where σ_d and σ_r are scalars to measure the degree of image filtering.

Formula (2) is a normalized weighted average, in which G_{σ_d} is a spatial distance function that is used to represent the influence of the pixels at different distances from the center pixel on the filtering scale. G_{σ_d} is a luminance range function that reduces the effect of the gray value different from the pixel of I_q. Because the bilateral filter can keep the edge of the ore particles in the filter, the bilateral filter is used to enhance the image filtering. Figure 2 is a comparison between the bilateral filter and the Gaussian filter. The comparison of the two figures shows that the bilateral filter can better enhance the edge of ore image compared with Gaussian filter.

Morphological open operation is the result of the first corrosion of the image, and then the result of the expansion. The processed image is called the target image, which is usually represented by a set of A. The "probe" used to collect information

(a) Gaussian filter   (b) Bilateral filter

Fig. 2 Comparison of filter results

is called the structural element, which is usually represented by a set of B. Structural element B is used to define the morphological open operation of A, the formula is (5). Morphological closed operation is the result of the first expansion of the image, and then the result of the corrosion, defined as formula (6). The open operation can eliminate the isolated points and the tiny adhesion between the targets, and it can smooth the object contour. A closed operation can fill a gap or hole which is smaller than the structural element, and it also has the function of smoothing filtering.

$$A \circ B = (A \ominus B) \oplus B \tag{5}$$

$$A \bullet B = (A \oplus B) \ominus B \tag{6}$$

Do morphological open operation in the grayscale images can filter out the shape which is less than the shape of the structure, but this will often lead to the loss of the original image information. In addition, there is also a problem in how to better select structural element, its function is similar to the "filter window" from signal processing, and its size and shape are directly related to the output of the results [4]. This paper uses the morphological reconstruction method to use the structural element of the original image to open the operation, the shape which is less than the structural element will be filtered out. And then let the result of the open operation as a mark image, let the original image as the mask image to rebuild the mark image [5]. This method not only removes the noise, but also retains the original image of the information. The reconstructed open operation of an image F with the size of n is defined as follows:

$$O_R^{(n)}(F) = R_F^D[(F \ominus nB)] \tag{7}$$

where $F \ominus nB$ represents the nth corrosion of B to F, R_F^D represents dilation to F. Similarly, the reconstruction of the closed operation is defined as follows:

$$C_R^{(n)}(F) = R_F^E[(F \oplus nB)] \tag{8}$$

where $F \oplus nB$ represents the nth dilation of $B-F$.

(a) Morphological open reconstruction

(b) Morphological open-and-closed reconstruction

(c) Morphological open operation

(d) Morphological open-and-closed operation

Fig. 3 Comparison of morphological operations

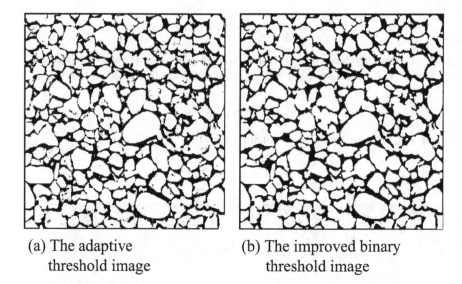

(a) The adaptive   (b) The improved binary
    threshold image   threshold image

Fig. 4 Image binarization

Compared morphological open reconstruction Fig. 3a, morphological open-and-closed reconstruction Fig. 3b with morphological open operation Fig. 3c, morphological open-and-closed operation Fig. 3d can be seen, morphological open-and-closed reconstruction can keep the contours of the original particles unchanged while highlighting the boundaries of adhesion particles.

In order to strengthen the boundary, the bilateral filtering operation is performed on the reconstructed image again, then the edge is extracted by threshold algorithm. Commonly used image threshold algorithms are OTSU, fixed threshold, adaptive threshold, bilateral threshold and single threshold operation [6]. In this paper, Adaptive threshold algorithm is adopted. The result of adaptive threshold on the basis of image preprocessing is shown in Fig. 4a. From the binary image it can be seen that some stones have black spots inside and there are burrs on the edges. Morphological operations are applied to the binary images to filter out the shape which is very small, and then use the area filtering to improve its result. And the result is shown in Fig. 4b.

3   Segmentation Method

Watershed algorithm is widely used in the field of particle image segmentation because of its fast computing speed and accurate location of boundaries [7]. The basic idea of watershed segmentation is to view the image as the topological landform of geodesy. The pixel gray value of each point represents the altitude, and many seed spots as the starting points of catchment are used to flow high to low in the surrounding areas. Each seed region expands outward, the intersection of different waters can constitute a watershed [8].

In the practical application, directly use the watershed algorithm to segment the picture is easy to be affected by noise, which leads to over-segmentation of the image. As a result of the watershed algorithm, the minimum level of gray scale is iterated to the largest gray level. The watershed transformation of the preprocessing image not only leads to over-segmentation and under-segmentation, but also reduces the efficiency of the algorithm. The direct watershed transform image is shown in Fig. 5b.

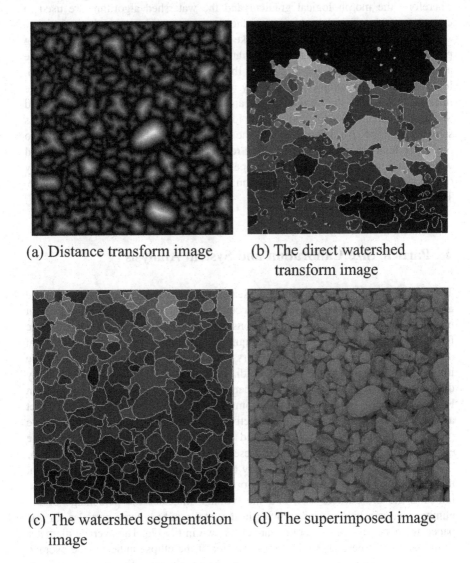

(a) Distance transform image

(b) The direct watershed transform image

(c) The watershed segmentation image

(d) The superimposed image

Fig. 5 The improved watershed segmentation images

After preprocessing the image, the improved watershed algorithm is used to segment the image. The threshold image is transformed by the distance to locate the seed region. The distance transform is used to grayscale the foreground object in the threshold image. The gray value is defined as the shortest distance from the foreground object to the background pixel, thus a three-dimensional mountain shape is formed through the chamfering template [6]. Distance transformation effect is shown in Fig. 5a.

After locating the ore's distribution position and quantity, it is necessary to find the boundary of each ore particle, especially the contour of the adhesive stack. Therefore, the morphological gradient and the watershed algorithm are used to extract the edge effectively, so as to reduce error rates of segmentation.

The ore particles of the ore image take the extracted seed region as the starting point (the white mark of Fig. 5a) for watershed segmentation, are divided in the acquired gradient image and stopped at the edge of each ore [9]. As the morphological gradient operation can effectively remove the noise information such as the internal holes of the ore image, and avoid the over-segmentation and under-segmentation, so we can get the ideal segmentation result image which is shown in Fig. 5c. And it is clear that the effect is better than Fig. 5b. In order to obtain the final gravel area segmentation results, the improved threshold image and watershed segmentation image can be operated on with the logical AND. For easy observation, the result of the segmentation is superimposed with the original image. The result image is shown in Fig. 5d.

4 Particle Size Calibration and System Analysis

The size of a particle is defined as the scale of the space occupied by the particle and is used to describe particle size. According to the purpose and measurement methods, different particle size definitions are defined. Generally, there are three forms: projection diameter, geometric equivalent diameter and physical equivalent diameter [10]. Projection diameter refers to a straight line size observed by the image. This system uses the projection diameter to calibrate the ore projection by drawing ellipses. Based on the open operation of the binary image (as shown in Fig. 6a), we carry on the connection domain mark to each independent ore target area, obtain the label of each ore objective area, and finally extract the ore size parameter which needs to be measured for the marked ore area. Each of the elliptical box in the marked image represents a separate ore, as shown in Fig. 6b.

In order to more easily and intuitively observe the ore particle size distribution, the histogram is used to represent the ore particle size distribution, the abscissa indicates the ore particle size, and the ordinate indicates the percentage of the number a certain kind of ore particles in the total number of ore particles. In this paper, the histogram of ore particle area is shown in Fig. 7a. The average diameter of the largest diameter and minimum diameter of the ellipse indicates the average particle size of the ore particles, and the histogram is shown in Fig. 7b. The analysis

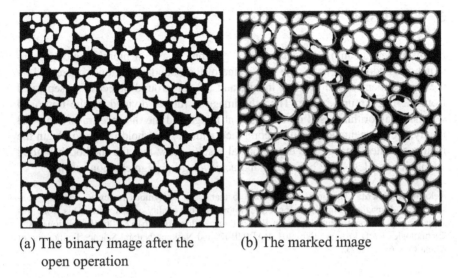

(a) The binary image after the open operation

(b) The marked image

Fig. 6 The marked image of the ore particle

of the two histograms shows that the area of the pixels in the [0, 300] and [300, 600] ranges accounts for about 55 and 38% of the total number of ore particles, and the average diameter of the ore in the [0, 30], [30, 60] ranges accounted for about 42 and 55% of the total number of ore particles. It can be seen from the Fig. 7 that most of the ore particles are distributed in the area [0, 600] and the average diameter [0, 60], a small number of ore particles which have a large particle size distribute in the area of more than 1200 and the average diameter of 120 or more areas.

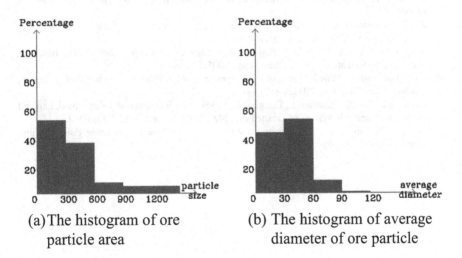

(a) The histogram of ore particle area

(b) The histogram of average diameter of ore particle

Fig. 7 Ore particle size distribution

5 Conclusion

In this paper, an improved watershed algorithm based on morphological reconstruction is adopted. Firstly, the gray value of the boundary region is increased by the pretreatment of the original image, and the foreground objective is separated from the background image. Then the image is further divided by the improved watershed algorithm based on the gradient image and the distance transform. And finally, the segmentation result is combined with the threshold image to obtain the final segmented image. The experimental results show that the method has good continuity and less false boundary, and achieves good segmentation results.

Acknowledgements The author would like to thank the anonymous reviewers for their constructive and insightful comments for further improving the quality of this work. This work was supported by National Natural Science Foundation of China (No. 61520106010), National Key Technologies R&D Program (No. 2013BAB02B07) and National Natural Science Foundation of China (No. 61603362).

References

1. Li L. Study on online detection methods of particle size based on digital image processing technology. Jiangxi University of Science and Technology;2014.
2. Gaipin C, Longmao L, Zhihong J. Design of ore particle size measurement system based on image processing. Metall Ind Autom. 2013;37(6):63–6.
3. Liu Q. Image segmentation based on grayscale morphological reconstruction. Xiangtan University;2016.
4. Yanling Z, Guixiong L, Dong C. Basic operators of mathematical morphology and application in image preprocessing. Sci Technol Eng. 2007;7(3):356–9.
5. Salembier P. Morphological multiscale segmentation for image coding. Sig Process. 1994;38(3):359–86.
6. Zhu T. Research of measurement and statistical methods for rock aggregates based on image. Xi'an University of Technology;2007.
7. Yingrong Y, Yingle F, Quan P. Separate algorithm for overlapping cell images based on distance transformation. Comput Eng Appl. 2005;20:206–8.
8. Xi L, Tianjiang W, Peng Z. An Improved Segmenting Algorithm for Touched Particle Image. J Hunan Univ (Nat Sci). 2012;39(12):84–8.
9. Guoying Z, Bo Q, Guanzhou L, Donghua L, Pengyun H. Image-based volume modeling and particle size analysis system of crushed ore. Metall Ind Autom. 2012;36(3):63–7.
10. Dong K. Research on ore granularity detection technology based on machine vision. Beijing University of Technology;2013.

The Design and Realization of Control System for a Two Degree of Freedom Parallel Manipulator

Biao Xu, Yuedou Pan and Haibo Liu

Abstract With the development of Computer numerical control (CNC) technology, motion control is widely used in many fields of our life. There are more and more demands on the accuracy, short distance transportation, and operation of small materials. Manipulator control can not only ensure the accuracy, stability and speed, but also relief the people from the harsh, tedious and hard work. Besides, it ensures the safety of people and improves the efficiency of the work simultaneously. In this paper, the control system of a two degree of freedom parallel manipulator is designed. Finally, manipulator is controlled to move arbitrarily, according to its dynamic model.

Keywords Motion control · Degree of freedom · Manipulator · CNC technology

1 Introduction

Manipulator reflects its unique advantage of high speed, stable operation and gradually replaces the tedious labor workers in electronics, medicine and other light industrial environment [1]. The parallel manipulator is more stable to realize its control function, compared with the series manipulator [2]. The positioning accuracy and stability are important reference indexes for the manipulator [3]. Therefore, the controlled precision and stability of the parallel manipulator have been the focus of the research.

B. Xu (✉) · Y. Pan · H. Liu
School of Electrical Engineering and Automation, Henan Polytechnic University,
Jiao Zuo 454000, China
e-mail: 18811329586@163.com

Y. Pan
e-mail: pyd88165@163.com

H. Liu
e-mail: liuhaibo@hpu.edu.cn

© Springer Nature Singapore Pte Ltd. 2018
Y. Jia et al. (eds.), *Proceedings of 2017 Chinese Intelligent Systems Conference*,
Lecture Notes in Electrical Engineering 460,
https://doi.org/10.1007/978-981-10-6499-9_49

This paper combined the two degree of freedom parallel manipulator [4], and established a mechanical dynamics model, realizing the positive solution algorithm from the servo motor (spindle) to the two-dimensional flat plate (virtual axis) and the reverse solution algorithm from the two-dimensional flat plate (virtual axis) to the servo motor (spindle) [5]. And the control system is equipped with Omron Corporation's latest controller NJ501-1300 controller and Sysmac Studio programming software, achieving the whole manipulator control action stably and quickly [6].

2   System Analysis

As shown in Fig. 1, the object is a two degree of freedom parallel manipulator, two servo motors installed in static platform. Firstly, servo motors control the driving arm. Secondly, the driving arm movement drives the slave arm. At last, the slave arm drives the flat plate, achieving a small material's grab ultimately [7].

As shown in Fig. 1, the code 1 refers to the active arm, the code 2 refers to the servo motor, the code 3 refers to slave arm, and the code 4 refers to the flat plate. In the Fig. 2, the control system uses Omron Corporation's NJ501-1300 intelligent controller, achieving human-computer interaction of the host computer to the lower machine connection and the EtherNet/IP network to touch screen or computer connection. The EtherCAT network is connected with the controlled object and the

Fig. 1 Two degrees of freedom parallel manipulator

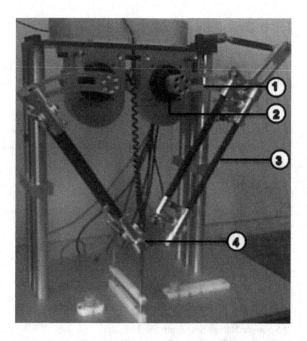

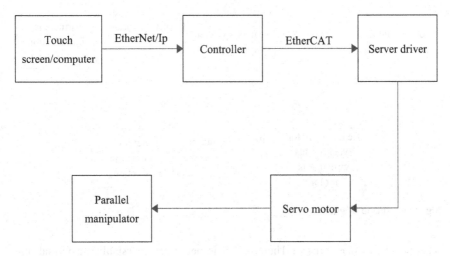

Fig. 2 System structure

servo driver to realize the control of the servo motor. The NJ controller receives the control command from the touch screen, sending the control signal to the actuator and controls the actuator to realize the movement of the manipulator.

The control system is mainly to achieve the following functions:

(1) Absolute position movement: The manipulator can run to the specified position quickly, accurately and stably.
(2) Return to original position: When the manipulator isn't on the origin position, it will return to the original position.
(3) Jog: The robot can move up, down, left and right, within the range of motion.
(3) Automatic operation: The manipulator grabs the iron sheet automatically.
(5) Manipulator simulation: The simulation and implementation of the real-time movement of the manipulator is realized by the simulation of the motion block.
(6) The trajectory tracking: The system tracks the total running path.
(7) Over limit alarm: When the manipulator moves the range, it stops and alarms signal timely.
(8) The manipulator stops expeditiously: The manipulator button can be used to stop the robot expeditiously.
(9) NS touching screen: The touch screen can control the whole motion.

3 Design Principles

According to the controlled object, it is necessary to control the rotation angle of the motor to move the position of the manipulator. And controlling motor rotation can't only determine the angle of the motor rotation but also determine the specific

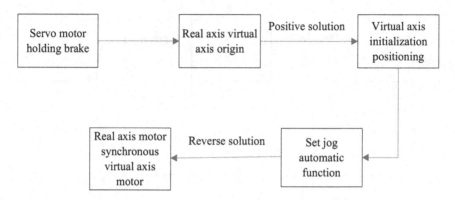

Fig. 3 Schematic diagram

position of the manipulator. Therefore, it is necessary to establish a virtual axis coordinate system to achieve a specific algorithm in the program, which corresponds to the actual movement of the manipulator. Based on this idea, the design schematic diagram is shown in Fig. 3.

Principle analysis: Firstly, the real axis and the imaginary axis are placed at the origin. Secondly, the initial position of the imaginary axis is obtained by the positive solution of the real axis. Lastly, the imaginary axis moves to the initial position. With MC_Move, MC_MoveVelocity, MC_linear interpolation, circular interpolation and other instructions to design the absolute position of movement, automatic and other functions, the imaginary axis position is obtained by reverse solution algorithm. We can complete synchronization of the real axis and the imaginary axis by MS_SyncMoveAbsolute instruction. Eventually manipulator controls the whole movement.

Hardware implementation of the system is designed. The two degree of a freedom parallel manipulator is realized by Industrial Personal Computer, PLC (controller), servo driver (two), servo motor (two) and controlled object. IPC achieves programming of the control system. PLC controller uses Omron Corporation's latest controller NJ501-1300 controller, whose CPU executes fast and includes linear-circular interpolation and other motion control module. The servo controller are two R88D-KN01H-ECT controllers. The servo motors are Panasonic motors.

Software implementation of the System is designed. The software system adopts the modular programming, including the positive solution module, the reverse module, the trajectory planning module, the manipulator trajectory tracking module, the automatic operation module and the point control module. The interface of modular programming is simple and clear as shown in the Fig. 4.

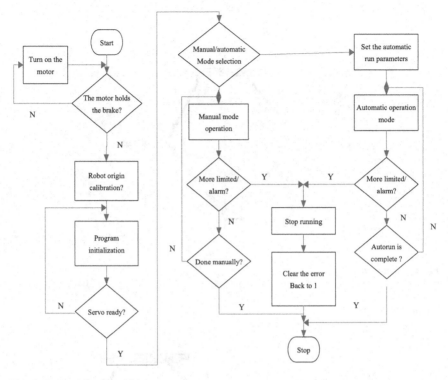

Fig. 4 Software process chart

4 System Dynamics Model

The controlled object's position movement is realized by the rotation angle of two servo motors. The movement of the flat plate is the position of the virtual (X-Y) axis located in the two-dimensional plane. So it is necessary to establish the relationship between the dynamic model of the servo motor (real axis) and the flat plate (virtual axis). Establishing positive solution algorithm and reverse solution algorithm realize the controlled object's position movement. The positive solution algorithm and reverse solution algorithm are used to establish the relationship between angle $\theta1\theta2$ and coordinate point X, Y. The coordinates can be obtained by the following formula 4.1.

$$\begin{cases} (x + 150\cos\theta1 + 38)^2 + (y + 150\sin\theta1)^2 = 300^2 \\ (x - 150\cos\theta2 - 38)^2 + (y + 150\sin\theta2)^2 = 300^2 \end{cases} \tag{4.1}$$

As shown in Fig. 5, positive solution algorithm is that the coordinates of the center point H, K are obtained from the angle $\theta1$, $\theta2$, the length of the spindle, the length of the slave shaft, and the length of the flat plate. Moving the H-point to

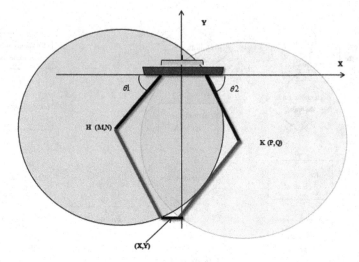

Fig. 5 Positive chart

Fig. 6 Reverse chart

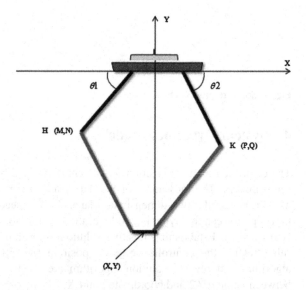

the right half of the length of the flat plate, K-point to the left half of the length of the flat plate, and making the move-point as the center, the shaft length as the radius of the circle, finally the intersection point is the end point coordinate.

The reverse solution algorithm and positive solution algorithm are similar, but not the same exactly. As known in Fig. 6, the angle of the motor's rotation is obtained by trigonometric function and the X, Y coordinates of the flat plate. Firstly, move the X coordinate to the left and right by half the flat plate. Secondly, the length of the slave axis and spindle axis are right angle triangles respectively. Lastly, combine two right-angled triangles to get X, Y coordinates and the angle of $\theta 1$ and $\theta 2$ are obtained by trigonometric function relation.

From this formula 4.2, we can get angle $\theta 1 \theta 2$ by X, Y coordinates.

$$\begin{cases} (m+80)^2 + n^2 = 150^2 \\ (m+38-x)^2 + (y-n)^2 = 300^2 \\ (p-80)^2 + q^2 = 150^2 \\ (p-38-x)^2 + (y-q)^2 = 300^2 \end{cases} \tag{4.2}$$

5 Optimal Path Planning

Reasonable trajectory planning plays an important role in reducing the inertia load and vibration effectively, improving the speed and precision of the end actuator. And it is necessary to consider the obstacle avoidance problem in the process of motion. If only considering the position from the initial point to the target point and to avoid the middle of the obstacles, the establishment of linear motion is the shortest path design, but the simple linear motion at the corner must be a large vibration problem so that this system can't guarantee the smooth operation of the manipulator. And it is necessary to ensure the velocity and the acceleration are zero at the initial and final points of each trajectory and continuous during the motion to minimize the vibration of the motion process. Under the conditions of maximum acceleration, there are polynomial rules, sine rules, trapezoidal curves and double S trapezoidal curves, which satisfy the above conditions [8]. We choose the cycloid rule in this paper.

As shown in Fig. 7, according to the height and location of obstacles, the design of straight and circular interpolation can make the minimum vibration and the fastest operation of the manipulator [9]. The interpolation algorithm is a kind of control algorithm commonly used in the machining of CNC machining tools, by comparing the actual moving position and the theoretical position to obtain the deviation, and determining the direction of the next step [10].

Fig. 7 Path planning

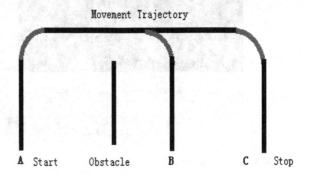

Movement Trajectory

A Start   Obstacle   B   C   Stop

The interpolation equivalent will also affect the speed and accuracy of the movement of the manipulator, if the equivalent is too small, the manipulator moves slowly, equivalent is too large, the accuracy is not enough, so selecting the appropriate equivalent will directly affect the movement of the manipulator effect. At the same time, linear and circular interpolation are specified by setting the scale factor. The manipulator will be stable, accurate and fast regardless of the height of the obstacle.

6 System Simulation Tracking Results

As shown in Fig. 8, the CX Designer function is used to detect the running of the manipulator and observe the movement of the manipulator so that the manipulator can make processing at any time.

With the CX Designer, you can observe the running status of the manipulator in real time. The manipulator can also track and keep the motion trajectory. In a separate X-axis, Y-axis direction of movement, the actual effect of tracking is better than in the implementation of circular interpolation instructions.

As shown in Fig. 9, when the manipulator moves, the trajectory of the manipulator will be tracked and saved by the Date-trace function. The Date-trace function not only implements a simulation function of the motion, but also supervises the movement of the actual manipulator. From the trajectory analysis, although the effect of arc interpolation is not very good, the manipulator moves the planning trajectory completely. So this control system is feasible.

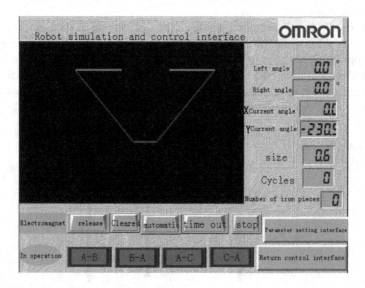

Fig. 8 Robot simulation and control interface

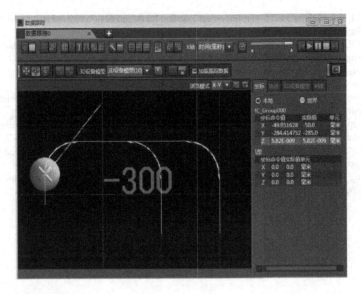

Fig. 9 Robot trajectory diagram

7 Conclusion

In this paper, the dynamic model of a two degree of freedom parallel manipulator in industrial control is studied. And its positive solution algorithm and reverse solution algorithm are deduced according to the dynamic model. According to the optimal time and the shortest path principle, the optimal trajectory of the manipulator is designed, and the data trace function of the manipulator is also tracked by CX Designer software. Finally the design and realization of the whole system are completed. In order to make the system be more efficient, faster and stable operation, our future work will be focused on achieving more accurate circular interpolation.

Acknowledgements This work was supported by National Natural Science Foundation of China (No. 61520106010), National Key Technologies R&D Program (No. 2013BAB02B07) and National Natural Science Foundation of China (No. 61603362).

References

1. Zhenxian L, Kai Y. Research on Two Degree of Freedom Parallel Manipulator Control System Based on PLC. J Xuzhou Inst Technol (Nat Sci Ed). 2014;29(1):67–71.
2. Wu P. Modeling of two DOF parallel manipulator and its fuzzy variable structure control. Harbin Institute of Technology;2007.
3. Yanan L, Chengrui Z. Continuous trajectory planning of delta parallel manipulator based on matlab image recognition. Sch Mech Eng Shandong Univ. 2014;42(15):55–6.

4. Mei J, Wu M. Diamond 600 robot control system hardware design and interface development. J Mach Des. 2004;21(32):61–3.
5. Tao S, Yi F. Algorithm of motion trajectory interpolation for five bar parallel manipulator. J Qingdao Univ. 2010;04(15):53–5.
6. Zhang X. Research on control method of high speed parallel manipulator. Tianjin University;2014.
7. Zhang Y, Zhang X, Hu J. Optimal time trajectory planning and realization of high speed parallel manipulator. Mech Electr Eng Technol. 2010;39(10):42–5.
8. Jiangping M, Panfeng W, Tian H. Trajectory planning and motion control of high-speed parallel manipulator with two degrees of freedom. Manuf Autom. 2004;26(9):30–2.
9. Yuan J, Zhang X. Scale synthesis of a new two degree of freedom high speed parallel manipulator. South China University of Technology;2011.
10. Wang H, Sun W. Research and trajectory simulation of arbitrary quadrant linear and circular interpolation. Xinjiang University of Mechanical Engineering;2012.

Corrosion Map Software System Based on Empirical Model

Dong-Mei Fu, Xing-Feng Zhao and Li-Jun Yang

Abstract Comprehensive understanding of corrosion intensity for material in atmosphere has an important significance for selecting equipments and designing material protecting schemes. In this paper, a Chinese visual corrosion map software is written by C#WinForm and SQL Server, it designs a transformation function that map corrosion categories and corrosion rates to gradient pseudo colors, which greatly improves the accuracy and sensitivity in displaying corrosion intensity. In order to establish corrosion map by a small amount of data, an empirical model that speculating the corrosion rates based on climatic and environment data is used to expand data. The software shows distribution of corrosion intensity in China with integrated database management capabilities, and will help researchers a lot who are in the field of corrosion.

Keywords Atmospheric corrosion · Visual map · Pseudo color · C#WinForm · Software

1 Introduction

Corrosion is the phenomenon that metal progressively loses it's properties in natural environment [1], atmospheric corrosion of metal material not only cause huge economic losses, but also threaten the safety of equipments, buildings, and personal safety [2, 3]. At present, 15 corrosion test sites have been established in our country, which are widely distributed in Chinese regions, so that it is difficult to obtain corrosion intensity of any region in China. Based on this property, mining the relationship between corrosion rates and atmospheric environmental factors, and mapping the corrosion intensity distribution have very important significance to carry out the work of corrosion protection.

D.-M. Fu (✉) · X.-F. Zhao · L.-J. Yang
School of Automation and Electrical Engineering, University of Science
and Technology Beijing, Beijing 100083, China
e-mail: fdm_ustb@ustb.edu.cn

© Springer Nature Singapore Pte Ltd. 2018
Y. Jia et al. (eds.), *Proceedings of 2017 Chinese Intelligent Systems Conference*,
Lecture Notes in Electrical Engineering 460,
https://doi.org/10.1007/978-981-10-6499-9_50

The relationship between corrosion rates of carbon steel and atmosphere environmental factors has been studied by many researchers [4–6]. In this paper, we use
the model in literature [6] to extend data and use these data to establish Chinese
visual corrosion map.

The Chinese visual corrosion map software is written by C#WinForm and SQL
Server. On the one hand, corrosion categories and corrosion rates are shown with
gradient and translucent Chinese map, which makes it easy for corrosion workers to
directly and completely understand the distribution of corrosion intensity. On the
other hand, it allows user to manage the database including querying, adding,
modifying, updating, deleting and downloading data, which helps user to make
better use of the data.

2 Software Architecture and Framework

The Chinese visual corrosion map software includes 5 modules: user management,
query data, database management, corrosion map and other actions. The system
framework is shown in Fig. 1.

3 Corrosion Rates Prediction Based on Empirical Model

There are many models that establish the quantitative relation between climatic,
environment factors and carbon corrosion rates [6–8]. Especially, the corrosion
rates calculated by the model in literature [8] can most fit the real carbon corrosion
rates in China, such as the formula (1) shows. Based on the empirical model,
corrosion rates of many cities besides the 15 test sites are calculated, and these data
are used to establish the Chinese visual corrosion map.

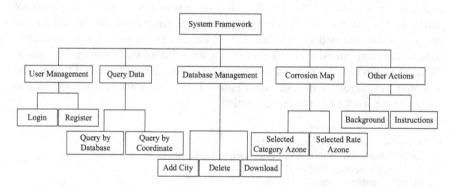

Fig. 1 System framework

$$\begin{cases} E = \sqrt{0.51 * d_1^2 + 0.29 * d_2^2 + 0.11 * d_3^2 - 0.09 * d_4^2 + 0.99 * d_5^2 + 0.51 * d_6^2 + 0.75 * d_7^2} \\ v = e^{(2.53*E + 1.97) - 20.2} \end{cases}$$

$$(1)$$

where, d_1 is annual average temperature ($^\circ C$), d_2 is annual average relative humidity (%), d_3 is annual rainfall (mm), d_4 is annual sunshine hours (h), d_5 is concentration of sulfur dioxide (ug/m^3), d_6 is concentration of nitrogen dioxide (ug/m^3), d_7 is concentration of chloride (ug/m^3), v is annual corrosion rates (um/a).

4 Visual Corrosion Map

RGB color mode is a color standard in the industry, which includes wide range of colors by the changes of red (R), green (G), blue (B) channels and their superposition between each other [9]. Pseudo color enhancement transform different gray levels of the gray image into colorful RGB encodings, so as to get the color image that includes varieties of gradient colors. And the continuous change of the colors makes the image hierarchical and containing more effective information [10–13].

Literature [8] establishes corrosion map by making every corrosion category with a single color, so that the color and corrosion category are corresponding, as is shown in Fig. 2. This map can display Chinese corrosion intensity clearly that according to ISO9223-2012 [14] corrosion standard of the six category classification. But there are obvious shortcomings for this method, that is, the sensitivity and accuracy are both lower.

Based on this property, a new gradient pseudo color transformation function is designed according to the theory of colorimetry [11]. There are two advantages for this function:

(1) The six corrosion categories are correspond to six different color systems, that is, gray, yellow, green, blue, red and purple. The color deepen along with the increase of corrosion category.
(2) Between the same corrosion category or color system, different corrosion rates are distinguished through gradient color, and the bigger the corrosion rates, the deeper the color.

The transformation function map corrosion category and corrosion rates to pseudo color coding is given by formula (2–3),

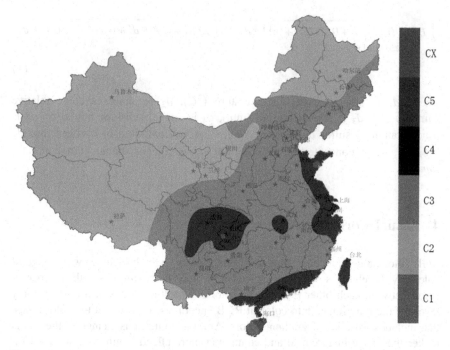

Fig. 2 The corrosion map in literature [6]

$$(R,G,B) = \begin{cases} R=G=B=rgb, & C=C1 \\ R=0, G=B=rgb, & C=C2 \\ R=B=0, G=rgb, & C=C3 \\ R=G=0, B=rgb, & C=C4 \\ R=rgb, B=G=0, & C=C5 \\ R=B=rgb, G=0 & C=CX \end{cases} \tag{2}$$

$$rgb = \begin{cases} 255 + (0-v)*100, & 0 < v \le 1.3 \text{ um/a} \\ 255 + (1.3-v)*8, & 1.3 < v \le 25 \text{ um/a} \\ 255 + (25-v)*8, & 25 < v \le 250 \text{ um/a} \\ 255 + (50-v)*8, & 50 < v \le 80 \text{ um/a} \\ 255 + (80-v)*1.5, & 80 < v \le 200 \text{ um/a} \\ 255 + (200-v)*0.5 & 200 < v \le 700 \text{ um/a} \end{cases} \tag{3}$$

where, C is the corrosion category, whose value takes $C1, C2, C3, C4, C5, CX$. v is the corrosion rates, whose value takes range $(0 \text{ um/a}, 700 \text{ um/a}]$.

Finally, establishing Chinese corrosion map as is shown in Fig. 3. Compared with the corrosion map in literature [8], the map by the proposed method has more accuracy and sensitivity in displaying corrosion category and rates, Table 1 shows the comparison results.

Fig. 3 The corrosion map established by proposed method

Table 1 The Accuracy Comparison of two corrosion map

Corrosion category	Corrosion rates (um/a)	Method in literature [6]		Proposed method	
		Color	Accuracy (um/a)	Color	Accuracy (um/a)
C1	0–1.3		0.65		0.01
C2	1.3–25		11.85		0.125
C3	25–50		12.5		0.125
C4	50–80		15		0.125
C5	80–200		60		0.67
CX	200–700		250		2

Note Accuracy represent the average accuracy between the same corrosion category

5　Software Interface Display

Among the 5 modules of Chinese visual corrosion map software, the user management module consists of login and register. The query data module consists of
query by database and query by coordinate. The database management module

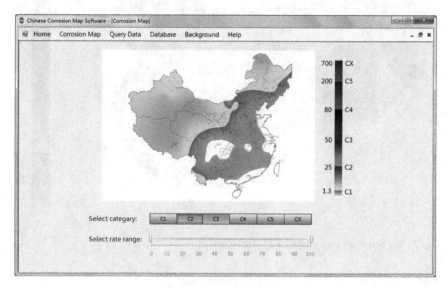

Fig. 4 Show the selected corrosion category area

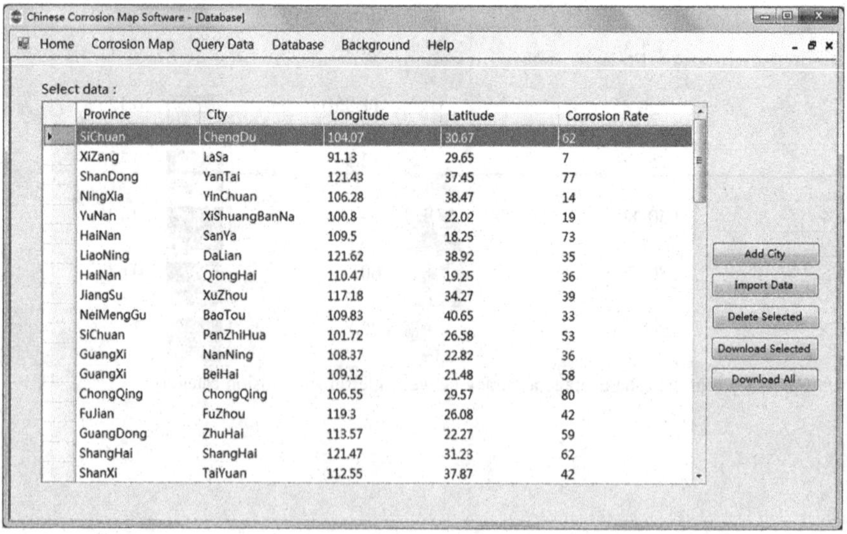

Fig. 5 Database management

consists of add cities, deleting and downloading data, and the user can select parts of the data which he or she want to download. The corrosion map module consists of mapping selected category area and mapping selected corrosion rates range area. The other actions consist of background knowledge and instructions for the software. Parts of the software interface are shown as Figs. 4 and 5.

6 Conclusions

In this paper, a new method for establishing visual corrosion map is proposed, which realizes the visual display of corrosion intensity by mapping corrosion categories to different colors. And also a transformation function is designed to map corrosion rates to pseudo colors, which greatly improves the accuracy and sensitivity for representing corrosion rates in a map. Besides, we develop a Chinese visual corrosion map software, which consists of corrosion mapping, corrosion data management, and other functions. The software make it easier for researchers to log data, modify data and download data. In brief, the software intuitively shows the corrosion intensity distribution in various regions of China, which is helpful for the research of related corrosion work.

References

1. Wan Y. Atmospheric Corrosion of metals and their experimental methods. Beijing: Chemical Industry Press; 2014. p. 45–62.
2. Li X. Research progress and prospects of materials environment corrosion. China Sci Found. 2012;(5):257–26.
3. Mingan DAI, Zhenfang LIU. Analysis of the conjunction between environment factors and atmospheric corrosion of carbon stell. Corros Prot. 2015;21(4):147–8.
4. Tidblad J. Atmospheric corrosion of metals in 2010–2039 and 2070–2099. Atmos Environ. 2012;55:1–6.
5. Surnam BY, Oleti CV. Atmospheric corrosion in Mauritius. Corros Eng Technol. 2012;47 (6):446–55.
6. Kucera V, Tidblad J, Kreislova K, et al. UN/ECE ICP materials dose–response functions for the multi-pollutants situation. Water Air Soil Pollut. 2007;7(4–3):249–58.
7. Lombardo T, Ionescu A, Chabas A, et al. Dose–response function for the soiling of silica–soda–lime glass due to dry deposition. Sci Total Environ. 2010;408:976–84.
8. Cui M. Corrosion rate speculation of carbon steel in atmospheric environment and data map establishment of carbon steel corrosion rate. Beijing: University of Science and Technology Beijing; 2016.
9. Li J. No-reference image quality assessment based on natural scene statistics in RGB color space. Acta Autom Sin. 2015;41(9):1601–15.
10. Li Z, Chen Z, Kang K. Application of color harmonic in radiation imaging. Nucl Electron Detect Technol. 2000;20(3):233–5.
11. Liu W, Liu J, Huang H. Palette's continuous coding method for sonar image pseudo-color processing. J Syst Simul. 2005;17(7):1724–31.

12. Fan Fan X, Gu G, Liu N. Design of pseudo-color coding and display for infrared digital images. Infrared Technol. 2013;35(7):398–403.
13. Gonzales. Digital image processing, 2nd ed.. Beijing: Electronic Industry Press; 2007. p. 135–40.
14. ISO9223-2012. Corrosion of metals and alloys—corrosivity of atmospheres—classification, determination and estimation.

Person Re-identification Based on Minimum Feature Using Calibrated Camera

Tianlong Zhang, Xiaorong Shen, Quanfa Xiu and Luodi Zhao

Abstract Although several approaches of person re-identification can be found in the paper, tracking people restricted in an open area is still an active research. In this paper, we propose a method assuming each pedestrian as a collection of multiple elements of the database. From the lowest point for the sector, layer by layer up to the top, we collect the minimum feature of the object. The features include statistic feature based on Bhattacharyya distance, SURF feature and Color histogram. After the pedestrian identification, and then start tracking. All the patches can be tracked individually and the vectors are calculated with the world coordinate. Not only that, the algorithm also uses the calibrated parameters of multi-camera to directly compute the pedestrian scale, and at the same time limit the region of searching, rather than using all the complex features as before. The experiment is carried out using two sets of publicly available database (PETS and CHUK). According to the experiment results, our method has a strong robustness, even for the larger changes in the scale of person can also be identified. Algorithm can also deal with many occasions, the accuracy is also high, compared to many of the existing state-of-art algorithm.

Keywords Re-identification · Multi-camera · Pattern recognition · Pedestrian tracking

1 Introduction

In video-surveillance systems, it is important to determine whether a person is currently visible in another place from a camera network that is being observed. This problem is often referred to as "object re-identification", and the related performance can be found in general such as [1]. Re-identification algorithm should be robust, even in different viewpoints and the orientations of the camera, alternative lighting conditions, changes in posture of the people, and rapidly changing clothes

T. Zhang(✉) · X. Shen · Q. Xiu · L. Zhao
School of Automation Science and Electrical Engineering, Beihang University,
Beijing 100083, China
e-mail: ztl275113335@hotmail.com

© Springer Nature Singapore Pte Ltd. 2018   533
Y. Jia et al. (eds.), *Proceedings of 2017 Chinese Intelligent Systems Conference*,
Lecture Notes in Electrical Engineering 460,
https://doi.org/10.1007/978-981-10-6499-9_51

appearance, all result in a challenging circumstance. The streets, parks, and shop-ping centers in public places such as airports and schools can create a lot of oppor-tunities for commercial and public security applications, including threat detection, monitoring parks and streets, store customer asset tracking, motion control, testing unusual events in hospitals, elderly home monitoring, employee placement, etc. The optimization of these applications requires the ability to automatically detect, iden-tify and analyze video or image data to track people and objects. Although video surveillance has been used for decades, it is still an active research area on how to automatically detect and track people. Many methods have been proposed in the last few years. The cameras that had been used varies from different types. Such as mono and stereo cameras and other differences such as the type of environment (indoor or outdoor). Most camera experiments are focusing on covering an area of a room or a hallway, corridor, a parking lot, an expressway, etc. and the position of the camera varies from whether it is with or without overlapping the field of view. There are a lot of works devoted to this area, and we will cover some of the related works below. However, the performance of most methods is still far from meeting the require-ments of practical applications. This is because of three main reasons, first of all, the performance of real video and image analysis techniques is related to the complex-ity of the problem. Face recognition technology, for example, the most commonly used method for identifying people in images, still has many very sensitive factors. Firstly, this technology has to face the environmental challenges such as luminosity changes, visual angles, and even simple camouflage. Secondly, the majority of such systems are developed and tested to control the "simulated" environment. For exam-ple, a person deliberately moves around in a room, is different than a real public area where events occur at a natural location. Finally, the current work effort is focused on the analysis of video data without considering the knowledge of the environment domain, which can fundamentally decrease the accuracy of tracking and recogni-tion. The size of useful information from a large number of surveillance videos is becoming increasingly difficult to manage. The purpose of this paper is to set up an object retrieval system for such a large collection. The detection and matching of human body appearance is a challenging problem. Change of views points leads to differences in appearance. Confliction is frequent because of the lack of lighting, streetlights and clouds, as well as the presence of other objects and traffic. Weather conditions such as rain can also cause dramatic changes in background and fore-ground.

2   Related Work

Several studies have presented this topic, such as tracking methods discussed in [2–5], and pedestrian detection methods [6]. The interest points matching is in [7], differs from [2, 5] using image matching, [4] using color matching. In this paper, we propose a new algorithm for the re-identification in camera network objects. The main contribution is to use local features to target very efficient descriptions, and

Fig. 1 Part of the CHUK dataset

to subsequently propose communication and matching algorithms. First, the system sends the image with interest targets, identifying and describing powerful local features in this image. The target is described as a very subsequent representation called the signature. This signature can be further communicated to adjacent camera nodes. Objects seen in camera nodes are again described, and specific properties are created for objects. The point correlation algorithm compares the signature of the camera signer and can accomplish effective reidentification and track the target camera network. Our method requires no knowledge of topology for the environment or training of the objects. Figure 1 is part of the CHUK dataset, this dataset is a person re-identification dataset with 3884 images of 972 pedestrians. All pedestrian images are manually cropped, and normalized to 160 * 60 pixel [8].

The advantage of our approach is that we minimize the transmission of information of adjacent camera nodes, and another feature is that every single camera view is not retrained. In addition, one of our work is to use uncalibrated settings. Calibration of multi-camera is still a difficult task. Our method is simulating a variety of traffic scenario. First, a camera is given in several cases, before simulating two camera network; secondly, a scene with synthetic pictures and Gauss noise. The result shows that our approach is useful and encourages the local features of the target to be re-identified in further studies.

3 Re-identification Method

As is mentioned, the entire object reidentification approach presented in this paper is solely based on local appearance features. Although the usage of local features has some shortcomings, i.e. computational costs, minimum requirements on image quality, etc., and highly relies on robust matching methods, it has the advantage of being partially robust to image scale, rotation, affine distortion, addition of noise and illumination changes. By using a setup such as the proposed one, it is not necessary to tediously train classifiers for each camera or add information of additional cues or sensors. In the following, we shortly introduce our choice of local features and describe the point correlation algorithm.

A. Statistical features: Statistical feature descriptors have been used in tracking algorithms in recent years [2, 9] because this descriptor requires only a small amount of memory cost and feature matching indicators, which is more convenient for implementation. In addition, different feature types can be easily mixed or embedded to improve the performance of the method. For real-time tracking and re-identification, we must achieve an appropriate balance between the discrimination ability and the computational efficiency while choosing the features and statistical descriptors. We keep the method proposed in [2] where the RGB pixel values of a given region i, is used to estimate μ_i and C_i, as a mean vector and a covariance matrix, respectively. Notice that this selection results in a compiled representation of the region. Because the covariance matrix C_i has only $d(d+1)/2$ distinct parameters, where d represents the dimension of the feature, only 9 parameters (6 for C_i and 3 for μ_i) are necessary to describe a image region. Bhattacharyya distance is a popular method based on comparing first-order and second-order parameter distributions. The region candidate value i of this distance can be derived by

$$b_i = \frac{1}{8} \left(\mu_i - \mu_m \right)^T \left[\frac{C_i + C_m}{2} \right]^{-1} \left(\mu_i - \mu_m \right) + \frac{1}{2} ln \left(\frac{|(C_i + C_m)/2|}{\sqrt{|C_i||C_m|}} \right) \quad (1)$$

where C_m is the model covariance matrix and μ_m is the mean vector. As is proposed in [2], we used $\gamma = 0.15$ in this equation.

B. Feature Description: In this stage, the interest points are described in Color histogram and SURF Description.

1. Color histograms: Histograms have been used in the past to model people appearance, such as in [1, 10]. Given a set of histograms computed at initialization, the objective is to find, at each time step, the region with the most similar histograms. To compute the histogram, first the image is converted to the CIELab space, which separates the lightness of the color (L-channel) from the color channels (a and b); only the a- and b-channels are used because of their robustness to illumination changes.

For efficiency reasons, we assume that these two channels are independent, so a given image region is described by two histograms with N_b bins, normalized such that $\sum_{j=1}^{N_b} h(j) = 1$. To evaluate a given region candidate i, with associated histograms h_i^a and h_i^b (for the a- and b-channels, respectively) we use the Bhattacharyya distance between the candidate histograms and the model histograms [10]:

$$b_i = \frac{1}{2} \left(\sqrt{1 - BC \left(h_m^a, h_i^a \right)} + \sqrt{1 - BC \left(h_m^b, h_i^b \right)} \right) \quad (2)$$

where h_m^a and h_m^b are the model histogram and BC is the Bhattacharyya coefficient defined as follows:

$$BC(h_1, h_2) = \sum_{j=1}^{N_b} \sqrt{h_1(j)h_2(j)} \qquad (3)$$

In the proceed of implementation, initializations are set with the number of bins to $N_b = 64$ and the model histograms. All of the initializations are computed and remain unchanged throughout the calculation. The γ parameter was set up to $\gamma = 0.25$. In our experiment, the use of statistical features, comparing with color histogram, can lead to a better tracking performance. In addition, it is a very efficient way to compute histograms, using integral histogram data structures presented in [11].

2. SURF Description: SURF is a feature descriptor which describes a region in a robust way, it is an invariant factor when natural environment changes. Invariant factors include scale, rotation, illumination and view variance. SURF [12] descriptor works with high matching performance, at the same time improves the speed of the SIFT [13]. The field of interest is divided into 4 * 4 square grid, such as SIFT [13]. However, we are not using the image gradient histogram to determine the feature of each subregion. A "vertical" and "horizontal" Haar wavelet response sum calculation is carried out by SURF. The actual direction and interest points are related. Each region descriptor has four values: the value of Haar wavelet response summary, and the absolute value of the response calculation of each wavelet.

4 Experiment and Results

In this section, we conduct several experimental results of proposed algorithms. Qualitative verification is carried out by visual inspection of the tracking results, and quantification is made by calculating error estimates using ground truth data for tracking estimates. To perform the experiments, we choose a dataset which is often used to evaluate state-of-art methods: the PETS datasets [15] (specifically: S2. L1). The dataset contains the camera calibration parameters and annotated pedestrian locations in the scene 1.

In the PETS dataset, seven synchronous cameras can be used for multi-camera tracking, because we are interested in monocular situations and use "view001". Sequence "S2. L1" can repeat the difficulty in real situations, such as occlusion, scale change, similar appearance, intersection of people, and so on. The sequence has a lower frame rate (FPS = 7), which can represent many surveillance cameras because of a common feature, and it is a challenge for track algorithm because the distance between pedestrians is often large. Experiment was carried out with PETS dataset, and each person in the photo was framed with different colors (Fig. 2). We then evaluate the performance differences between our tracking methods and similar approaches called FragTrack [2].

Fig. 2 proposed method performance on PETS dataset

As we can see from the comparison (Fig. 3), the loss of train and validation of our method tends to be stable when the data become enormous while other algorithms are not able to disgust the data. This is important when dealing with a high dimensional model. As for the error rate of the method, our proposed method is more accurate and the error rate is stable. This is due to our minimum feature algorithm, the less features are matching, the better performance it will get, and since our research background need no complex features and high definition, this result actually works.

We list the performance of ours on VIPeR dataset and then compare with meta descriptors in Table 1. All the descriptors are with multiple descriptors, which may reduce the local structures of interest regions. Descriptors with single layer distributions exhibit similar performance. These results confirm the validity of the minimum distribution. Our method outperforms FragTrack, because it also presents the mean information, which is lack of covariance. It is obvious that our method achieves the new state-of-art results, 40.63% of rank = 1 rate on VIPeR. It is obvious that our methods succeeded in the minimum feature and better descriptor.

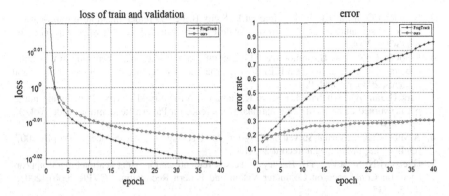

Fig. 3 The performance of two methods

Table 1 Performance comparison on VIPeR dataset

Methods	Rank = 1	Rank = 5	Rank = 10	Rank = 20
Ours	40.63	68.51	78.04	85.92
FragTrack	30.2	52	65	–
MLFL	29.1	–	65.9	70.9

5 Conclusion

In this work, we propose a re-identification tracking method based on SURF features of pedestrian and the Bhattacharyya distance of the targets. The world coordinate frame of our method combines separate tracking and estimation. The usage of calibrated camera parameters is to directly compute scale and limit search regions, and to compute translational displacement vectors in world coordinates. This system can deal with occlusion and the scale change and reduce the computational cost, compared to some state-of-art methods. There are several other ways to accomplish current and further work. First, we intend to improve the tracking performance by exploring the usage of different features in our framework. We are also interested in developing a completely automatic initialization tracking system, whose detection method is similar to what is proposed in [14], and the camera self-calibration proposed method in [13] to expand our algorithm with uncalibrated cameras.

References

1. Tu PH, Doretto G, Krahnstoever NO, Rittscher J, Sebastian TB, Yu T, Harding KG. An intelligent video framework for homeland protection. 2007.
2. Wei G, Petrushin VA, Gershman AV. Multiple-camera people localization in a cluttered environment. In: Multimedia Data Mining Workshop. 2004.

3. Hamdoun O, Moutarde F, Stanciulescu B, Steux B. Person reidentification in multi-camera system by signature based on interest point descriptors collected on short video sequences. In: ICDSC08. 2008. p. 16.

4. Morioka K, Mao X, Hashimoto H. Global color model based object matching in the multi-camera environment. In: 2006 IEEE/RSJ International Conference on Intelligent Robots and Systems, 2006. p. 2644–9.

5. Park U, Jain AK, Kitahara I, Kogure K, Hagita N. Vise: Visual search engine us-ing multi-ple networked cameras. In: International Conference on Pattern Recognition. 2006. vol. 3, pp. 1204–7.

6. A Multi-Camera Visual Surveillance System for Tracking of Reoccurrences of People, 2007. [Online]. http://dare.uva.nl/record/265108.

7. Gheissari N, Sebastian TB, Hartley R. Person reidentification using spatiotemporal appear-ance. In: Conference on Computer Vision and Pattern Recognition (CVPR 2006). 2006, p. 1528–35.

8. Zhao R, Ouyang W, Wang X. Unsupervised Salience Learning for Person Re-identification. In: CVPR. 2014.

9. Swain MJ, Ballard DH. Indexing via color histograms. In: Proceedings of the DARPA Image Understanding Workshop, Pittsburgh, PA, USA; 1990.

10. Bay H, Tuytelaars T, Van Gool L. SURF: Speeded Up Robust Features. In: Proceedings of the Ninth European Conference on Computer Vision. 2006.

11. Actions, "Actions as space-time shapes," 2005, [Online; acessado 10-Agosto-2008]. [Online]. http://www.wisdom.weizmann.ac.il/vision/SpaceTimeActions.html.

12. Cantata, "Datasets for cantata project," 2000, [Online; acessado 10-Agosto-2008]. [Online]. http://www.hitechprojects.com/euprojects/cantata/datasetscanta-ta/dataset.html.

13. Lowe D. Distinctive image features from scale-invariant keypoints. In: International Jour-nal of Computer Vision. 2003. vol. 20, p. 91110. [Online] https://citese-er.ist.psu.edu/lowe04distinctive.html.

14. Object Reacquisition and Tracking in Large-Scale Smart Camera Networks. In: Viena: First ACM/IEEE International Conference on Distributed Smart Cameras. 2007.

15. Gray D, Brennan S, Tao H. Evaluating appearance models for recognition, reacquisition, and tracking. IEEE International Workshop on Performance Evaluation for Tracking and Surveil-lance (PETS). 2007. vol. 3, no. 5.

Fusion Filter for Wireless Transmission with Random Data Packet Dropouts and Delays

Sumin Han, Fuzhong Wang and Bin Cao

Abstract This paper extends the fusion filter with random packet dropouts and random delays in the wireless transmission systems. Considering the uncertain dynamics of the factual system, a transmission output model is proposed by assuming the possible transmission delays and pockets dropouts. According to the method of the minimum cross-covariance, we derive the optimum fusion filter and the cross-covariance matrices. A simulation example is given to verify the effectiveness of the system.

Keywords Fusion filter · Packet dropouts · Delays · Wireless transmission

1 Introduction

In recent years, state estimation problem has become a hot research topic for systems with the development of Internet of things due to the partially or completely unknown parameters and environmental disturbances. The uncertainties can be approximated mathematically by the random noises accompanied by the input and measurement signals. These systems are widely used in target tracking, detection, signal processing and other areas. Especially, there will be transport errors and data delays or packet dropouts in the wireless sensors network. Hence, the filtering problem with random delays and packet dropouts is significantly challenging in information exchange over unreliable communication links [1, 2].

In the past few years, the estimation problems for systems with random delays and packet dropouts have begun to give rise to attention. For systems with one or multi-step random delays, many research results have been reported, including the suboptimal linear filters [3–5], the optimal linear filters and the robust filter [6, 7]. However, the above mentioned literatures have two issues: one is that a packet is

S. Han (✉) · F. Wang · B. Cao
School of Electrical Engineering and Automation, Henan Polytechnic University, Jiaozuo 454000, China
e-mail: hansumin@hpu.edu.cn

© Springer Nature Singapore Pte Ltd. 2018   541
Y. Jia et al. (eds.), *Proceedings of 2017 Chinese Intelligent Systems Conference*,
Lecture Notes in Electrical Engineering 460,
https://doi.org/10.1007/978-981-10-6499-9_52

sent several times to avoid packet dropouts as much as possible, which, however, can bring the network congestion; the other is that one packet at most is only used for estimation update at each moment. Nevertheless, in practice, two or multiple packets may arrive at the same time since there are possible one- or multi-step random delays. To overcome the two problems, for systems with possible one-step delays and packet dropouts, a novel simple model is presented by two Bernoulli distributed random variables, which assumes that a packet at the sensor side is only transmitted once to avoid the network congestion and the data processing center may receive one latest packet at each moment in this paper.

In this paper, the fusion filtering problem for a single sensor systems with one-step random delay, packet dropouts. The remainder of this paper is organized as follows. Section 2 gives the system model and problem statement. Section 3 derives the optimal filter. Section 4 provides an example. Section 5 draws a conclusion. Finally, we provide the relevant proofs of the equations in the Appendices A–C.

2 Problem Formulation

Consider the following stochastic sensor system with stochastic uncertainties of white noises:

$$\begin{cases} x(k+1) = \phi(k)x(k) + w(k) \\ y(k) = H(k)x(k) + v(k) \end{cases} \tag{1}$$

where $x(k) \in \mathfrak{R}^n$ is the state vector, and $y(k) \in \mathfrak{R}^m$ is the measured output for m sensors. $w(k), v(k)$ are mutually uncorrected gauss white noises. For system (1) awe make the following assumptions.

$$E[w(k)] = 0, \ E[v(k)] = 0$$
$$E[w(k)w^T(j)] = R, \ E[v(k)v^T(j)] = Q, \ E[w(k)v^T(j)] = 0$$

In the process of data wireless transmission, that is, from sensor to the local processor, random delays or packet loss may appear. In order to avoid network congestion, we assume that the packets from the sensor end will be sent only once in each moment. When the data is delayed, it is possible that the adjacent two measurement data arrives at the local processor at the same time. Hence, we assume that the local processor receives a packet, or none at this moment. If not, the latest received measurement data is used for compensation. The data received by the local processor is referred as transmission output through the wireless transmission, expressed by the following model:

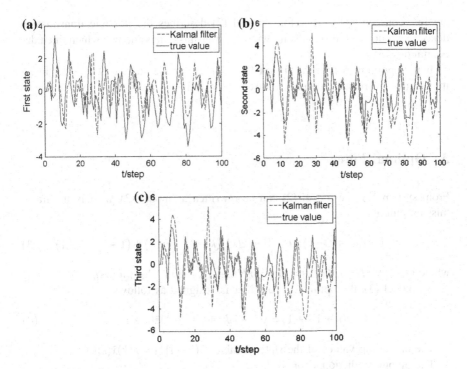

Fig. 1 Fusion filter

$$z(k) = \gamma y(k) + (1 - \gamma)\eta y(k - 1) + (1 - \gamma)(1 - \eta)z(k - 1) \qquad (2)$$

where $\gamma, \eta \in \mathfrak{R}^m$ are mutually uncorrected random variables obeying Bernoulli distributions with the probabilities $Pr[\gamma = 1] = \bar{\gamma}, Pr[\gamma = 0] = 1 - \bar{\gamma}$ and $Pr[\eta = 1] = \bar{\eta}, Pr[\eta = 0] = 1 - \bar{\eta}, \ 0 \leq \bar{\gamma} \leq 1, \ 0 \leq \bar{\eta} \leq 1$.

Model (2) shows that $z(k) = y(k)$ if $\gamma = 1$, i.e. the local processor only receives the sensor measurement at the current moment k; $z(k) = y(k - 1)$ if $\gamma = 0 \& \eta = 1$, i.e. the local processor receives the sensor measurements at the moment k − 1; there is nothing received at k if $\gamma = 0 \& \eta = 0$, then the latest measurement received is used for compensation, i.e. $z(k) = z(k - 1)$. To better explain model (2), we use Table 1 to demonstrate the transmission case.

From Table 1, we know that $y(2), y(4), y(5), y(8)$ are received on time, $y(10)$ is delayed one step, and $y(1), y(3), y(6), y(7), y(9)$ are lost. Moreover, $y(6)$ and $y(7)$

Table 1 Example of $z(k)$

k	1	2	3	4	5	6	7	8	9	10
γ	0	1	0	1	1	0	0	1	0	0
η	0	0	1	1	0	1	0	0	1	1
$z(k)$	$z(0)$	$y(2)$	$y(2)$	$y(4)$	$y(5)$	$y(5)$	$y(5)$	$y(8)$	$y(8)$	$y(8)$

are lost consecutively. Therefore, model (2) describes the possible transmission delays and pockets dropouts in wireless transmission systems, which includes continuous losses.

It is important to find the fusion filter $\hat{x}(k|k)$ for the state $x(k)$ t in the linear minimum variance sense based on the received measurements $z(k), z(k-1), \ldots, z(0)$.

3   Optimal Filter

From system (1), the model (2) can be rewritten as model (3), namely the transmission output:

$$z(k) = G_1(k)x(k) + G_2(k)x(k-1) + L(k)z(k-1) + \gamma v(k) + (1-\gamma)v(k-1) \quad (3)$$

where $G_1(k) = \gamma H(k)$, $G_2(k) = (1-\gamma)\eta H(k-1)$, $L(k) = (1-\gamma)(1-\eta)$
By model (3), the optimal fusion filter is designed as follows

$$\tilde{x}(k+1|k+1) = \tilde{x}(k+1|k) + K(k+1)\tilde{z}(k+1) \quad (4)$$

The predicting values of the state variables $\tilde{x}(k+1|k) = \phi(k)\tilde{x}(k|k)$
The output prediction error

$$\tilde{z}(k+1) = z(k+1) - \tilde{z}(k+1|k) = z(k+1) - G_1(k)\tilde{x}(k+1|k) - G_2(k)\tilde{x}(k|k) - L(k)z(k)$$

The estimation error matrices are $\tilde{x}(k+1) = x(k+1) - \tilde{x}(k+1|k+1) = \phi(k)x(k) + w(k) - \tilde{x}(k+1|k) - K(k+1)[z(k+1) - G_1(k)\tilde{x}(k+1|k) - G_2(k)\tilde{x}(k|k)$
$(k)z(k)] = \phi(k)x(k) + w(k) - [I - K(k+1)G_1(k)][\phi(k)\tilde{x}(k|k)] - K(k+1)[z(k+1)$
$sG_2(k)\tilde{x}(k|k) - L(k)z(k) = [I - K(k+1)G_1(k)][\phi(k)\tilde{x}(k) + w(k)] - K(k+1)[G_2(k)\tilde{x}$
$(k) + \gamma v(k+1) + (1-\gamma)v(k)]$

$$\begin{aligned}
\tilde{x}^T(k+1) &= [\phi(k)\tilde{x}(k) + w(k)]^T[I - K(k+1)G_1(k)]^T \\
&\quad - [G_2(k)\tilde{x}(k) + \gamma v(k) + (1-\gamma)v(k-1)]^T K^T(k+1) \\
&= [\tilde{x}^T(k)\phi^T(k) + w^T(k)][I - G_1^T(k)K^T(k+1)] \\
&\quad - [\tilde{x}^T(k)G_2^T(k) + \gamma v^T(k+1) + (1-\gamma)v^T(k)]K^T(k+1)
\end{aligned}$$

The proof is presented in Appendix A.

The cross-covariance matrices are

$$
\begin{aligned}
P(k+1) &= E\left[\tilde{x}(k+1)\tilde{x}^T(k+1)\right] = E(\{[I - K(k+1)G_1(k)][\phi(k)\tilde{x}(k) + w(k)] - K(k+1) \\
&\quad [G_2(k)\tilde{x}(k) + \gamma v(k+1) + (1 - \gamma)v(k)]\}\{[\tilde{x}^T(k)\phi^T(k) + w^T(k)][I - G_1^T(k)K^T(k+1)] \\
&\quad - [\tilde{x}^T(k)G_2^T(k) + \gamma v^T(k+1) + (1 - \gamma)v^T(k)]K^T(k+1)\}) \\
&= [I - K(k+1)G_1(k)]\{\phi(k)P(k)\phi^T(k) + R\}[I - G_1^T(k)K^T(k+1)] \\
&\quad - K(k+1)G_2(k)P(k)\phi^T(k)[I - G_1^T(k)K^T(k+1)] \\
&\quad - [I - K(k+1)G_1(k)]\phi(k)P(k)G_2^T(k)K^T(k+1) \\
&\quad + K(k+1)[G_2(k)P(k)G_2^T(k) + (2\gamma^2 - 2\gamma + 1)Q]K^T(k+1)
\end{aligned}
$$

The proof is presented in Appendix B.
We define the following equations

$$
E\left[\tilde{x}(k)\tilde{x}^T(k)\right] = P(k)
$$

$$
K(k+1) = K, \phi(k) = \phi, G_1(k) = G_1, G_2(k) = G_2
$$

Then the cross-covariance matrices can be expressed as (5)

$$
\begin{aligned}
P(k+1) &= [I - KG_1]\left[\phi P(k)\phi^T + R\right][I - KG_1]^T - KG_2P(k)\phi^T[I - KG_1]^T \\
&\quad - [I - KG_1]\phi P(k)G_2^T K^T + K\left[G_2P(k)G_2^T + (2\gamma^2 - 2\gamma + 1)Q\right]K^T
\end{aligned} \tag{5}
$$

For the sake of the minimum cross-covariance matrices, we can construct the following equation and make the linear section to the minimum.

$$
\begin{aligned}
P(k+1) + \Delta P &= [I - (K + \Delta K)G_1]\left[\phi P(k)\phi^T + R\right][I - (K + \Delta K)G_1]^T \\
&\quad - (K + \Delta K)G_2P(k)\phi^T[I - (K + \Delta K)G_1]^T \\
&\quad - [I - (K + \Delta K)G_1]\phi P(k)G_2^T(K + \Delta K)^T \\
&\quad + (K + \Delta K)\left[G_2P(k)G_2^T + (2\gamma^2 - 2\gamma + 1)Q\right](K + \Delta K)^T
\end{aligned}
$$

We can derive the gain matrix

$$
\begin{aligned}
K &= \{\left[\phi P(k)\phi^T + R\right]G_1^T + \phi P(k)G_2^T\} \\
&\quad \{G_1\left[\phi P(k)\phi^T + R\right]G_1^T + G_2P(k)\phi^T G_1^T + G_1P(k)G_2^T \\
&\quad + \left[G_2P(k)G_2^T + (2\gamma^2 - 2\gamma + 1)Q\right]\}^{-1}
\end{aligned} \tag{6}
$$

The proof is presented in Appendix C.

Fig. 2 Fusion filter error
variance

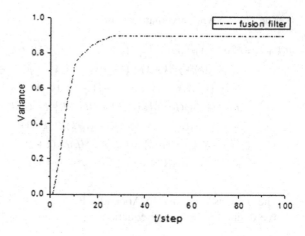

4 Simulation Example

Consider the discrete linear system including one sensor node. The
corresponding discrete-time model is given as follows:

$$\begin{cases} x(k+1) = \begin{bmatrix} 0.8761 & -0.5238 & 0 \\ 1 & 0 & 0 \\ 0 & 1 & 0 \end{bmatrix} x(k) + w(k) \\ y(k) = [\,18.760 \quad 16.542 \quad 0\,]x(k) + v(k) \end{cases}$$

Each variable is defined as above. $w(k)$, $v(k)$ are mutually uncorrected gauss
white noises with zero means and covariance matrices $E[w(k)w^T(j)] =$
$R = \begin{bmatrix} 1.160 & 0 & 0 \\ 0 & 1.022 & 0 \\ 0 & 0 & 1.211 \end{bmatrix}, E[v(k)v^T(j)] = Q = 0.222$

In the simulation, we assume $\bar{\gamma} = 0.8$ and $\bar{\eta} = 0.75$. The initial values are
$\tilde{x}(0|0) = [0 \quad 0 \quad 0]^T$ and $P(0) = 10I$, where I is the unit matrix.

The proposed fusion filter is shown in Fig. 1, where subplots (a)–(c) give the
filtering results of the three state components. The filtering error variances are
shown in Fig. 2. We see that there exists the steady-state fusion filter and the fusion
filter gives decent accuracy.

5 Conclusion

This paper studies the optimal and steady-state fusion filtering problem for
single-sensor random systems with one-step delays and packet dropouts. We first
propose a new transmission output model with a dimension output.

By the new model, the optimal linear local filter is designed in the linear minimum variance sense. The cross-covariance matrices between any two local filtering errors are derived. In the future, we will research the multi-sensor random system as well as consider the more complex conditions such as packet congestion etc.

Acknowledgements We would like to acknowledge all the co-workers for the technical support from Lab. of EMC of Henan Polytechnic University. This work was supported by National Key Research and Development Program (2016YFC0600906).

Appendix A

The estimation error matrices are

$$
\begin{aligned}
\tilde{x}(k+1) &= x(k+1) - \hat{x}(k+1|k+1) \\
&= \phi(k)x(k) + w(k) - \hat{x}(k+1|k) - K(k+1) \\
&\quad \times [z(k+1) - G_1(k)\hat{x}(k+1|k) - G_2(k)\hat{x}(k|k) - L(k)z(k)] \\
&= \phi(k)x(k) + w(k) - [I - K(k+1)G_1(k)][\phi(k)\hat{x}(k|k)] \\
&\quad - K(k+1)[z(k+1) - G_2(k)\hat{x}(k|k) - L(k)z(k) \\
&= \phi(k)\tilde{x}(k) + w(k) + K(k+1)[G_1(k)\phi(k) + G_2(k)] \\
&\quad \times \hat{x}(k|k) - K(k+1)[z(k+1) - L(k)z(k)] \\
&= \phi(k)\tilde{x}(k) + w(k) + K(k+1)[G_1(k)\phi(k) + G_2(k)]\hat{x}(k|k) \\
&\quad - K(k+1)[G_1(k)x(k+1) + G_2(k)x(k) + \gamma v(k+1) + (1-\gamma)v(k)] \\
&= \phi(k)\tilde{x}(k) + w(k) + K(k+1)[G_1(k)\phi(k) + G_2(k)]\hat{x}(k|k) - K(k+1) \\
&\quad \times \{[G_1(k)\phi(k) + G_2(k)]x(k) + G_1(k)w(k) + \gamma v(k+1) + (1-\gamma)v(k)\} \\
&= \phi(k)\tilde{x}(k) + w(k) - K(k+1)[G_1(k)\phi(k) + G_2(k)]\tilde{x}(k) \\
&\quad - K(k+1)[G_1(k)w(k) + \gamma v(k+1) + (1-\gamma)v(k)] \\
&= [I - K(k+1)G_1(k)][\phi(k)\tilde{x}(k) + w(k)] \\
&\quad - K(k+1)[G_2(k)\tilde{x}(k) + \gamma v(k+1) + (1-\gamma)v(k)]
\end{aligned}
$$

Appendix B

The cross-covariance matrices

$$
\begin{aligned}
P(k+1) &= E\left[\tilde{x}(k+1)\tilde{x}^T(k+1)\right] \\
&= E(\{[I - K(k+1)G_1(k)][\phi(k)\tilde{x}(k) + w(k)] \\
&\quad - K(k+1)[G_2(k)\tilde{x}(k) + \gamma v(k+1) + (1-\gamma)v(k)]\} \\
&\quad \times \left\{[\tilde{x}^T(k)\phi^T(k) + w^T(k)][I - G_1^T(k)K^T(k+1)]\right. \\
&\quad \left. - [\tilde{x}^T(k)G_2^T(k) + \gamma v^T(k+1) + (1-\gamma)v^T(k)]K^T(k+1)\}\right) \\
&= [I - K(k+1)G_1(k)]E\left\{[\phi(k)\tilde{x}(k) + w(k)][\tilde{x}^T(k)\phi^T(k) + w^T(k)]\right\} \\
&\quad \times [I - G_1^T(k)K^T(k+1)] - K(k+1)E\{[G_2(k)\tilde{x}(k) + \gamma v(k+1) + (1-\gamma)v(k)] \\
&\quad \times [\tilde{x}^T(k)\phi^T(k) + w^T(k)]\}[I - G_1^T(k)K^T(k+1)] - [I - K(k+1)G_1(k)] \\
&\quad \times E\left\{[\phi(k)\tilde{x}(k) + w(k)][\tilde{x}^T(k)G_2^T(k) + \gamma v^T(k+1) + (1-\gamma)v^T(k)]\right\} \\
&\quad \times K^T(k+1) + K(k+1)E\{[G_2(k)\tilde{x}(k) + \gamma v(k+1) + (1-\gamma)v(k)] \\
&\quad \times [\tilde{x}^T(k)G_2^T(k) + \gamma v^T(k+1) + (1-\gamma)v^T(k)]\}K^T(k+1) \\
&= [I - K(k+1)G_1(k)]\left\{\phi(k)E\left[\tilde{x}(k)\tilde{x}^T(k)\right]\phi^T(k) + E\left[w(k)w^T(k)\right]\right\} \\
&\quad \times [I - G_1^T(k)K^T(k+1)] - K(k+1)G_2(k)E\left[\tilde{x}(k)\tilde{x}^T(k)\right]\phi^T(k)[I - G_1^T(k)K^T(k+1)] \\
&\quad - [I - K(k+1)G_1(k)]\phi(k)E\left[\tilde{x}(k)\tilde{x}^T(k)\right]G_2^T(k)K^T(k+1) + K(k+1) \\
&\quad \times \left\{G_2(k)E\left[\tilde{x}(k)\tilde{x}^T(k)\right]G_2^T(k) + \gamma^2 E\left[v(k+1)v^T(k+1)\right] + (1-\gamma)^2 E\left[v(k)v^T(k)\right]\right\}K^T(k+1) \\
&= [I - K(k+1)G_1(k)]\left\{\phi(k)P(k)\phi^T(k) + R\right\}[I - G_1^T(k)K^T(k+1)] \\
&\quad - K(k+1)G_2(k)P(k)\phi^T(k)[I - G_1^T(k)K^T(k+1)] - [I - K(k+1)G_1(k)]\phi(k)P(k)G_2^T(k) \\
&\quad \times K^T(k+1) + K(k+1)\left[G_2(k)P(k)G_2^T(k) + (2\gamma^2 - 2\gamma + 1)Q\right]K^T(k+1)
\end{aligned}
$$

$$
E\left[\tilde{x}(k)\tilde{x}^T(k)\right] = P(k)
$$

Assume $K(k+1) = K$, $\phi(k) = \phi$, $G_1(k) = G_1$, $G_2(k) = G_2$

$$
\begin{aligned}
P(k+1) &= [I - KG_1]\left[\phi P(k)\phi^T + R\right][I - KG_1]^T - KG_2 P(k)\phi^T[I - KG_1]^T \\
&\quad - [I - KG_1]\phi P(k)G_2^T K^T + K\left[G_2 P(k)G_2^T + (2\gamma^2 - 2\gamma + 1)Q\right]K^T
\end{aligned}
$$

Appendix C

$$
\begin{aligned}
P(k+1) + \Delta P &= \left[I - (K + \Delta K)G_1\right]\left[\phi P(k)\phi^T + R\right]\left[I - (K + \Delta K)G_1\right]^T - (K + \Delta K)G_2 P(k) \\
&\quad \times \phi^T\left[I - (K + \Delta K)G_1\right]^T - \left[I - (K + \Delta K)G_1\right]\phi P(k)G_2^T(K + \Delta K)^T \\
&\quad + (K + \Delta K)\left[G_2 P(k)G_2^T + \left(2\gamma^2 - 2\gamma + 1\right)Q\right](K + \Delta K)^T \\
&= \left[I - KG_1\right]\left[\phi P(k)\phi^T + R\right]\left[I - KG_1\right]^T - KG_2 P(k)\phi^T\left[I - KG_1\right]^T \\
&\quad - \left[I - KG_1\right]\phi P(k)G_2^T K^T + K\left[G_2 P(k)G_2^T + \left(2\gamma^2 - 2\gamma + 1\right)Q\right]K^T \\
&\quad - \left[\phi P(k)\phi^T + R\right]G_1^T(\Delta K)^T + KG_1\left[\phi P(k)\phi^T + R\right]G_1^T(\Delta K)^T \\
&\quad - \Delta K G_1\left[\phi P(k)\phi^T + R\right]\Delta K G_1\left[\phi P(k)\phi^T + R\right]G_1^T K^T \\
&\quad + \Delta K G_1\left[\phi P(k)\phi^T + R\right]G_1^T(\Delta K)^T + KG_2 P(k)\phi^T G_1^T(\Delta K)^T - \Delta K G_2 P(k)\phi^T \\
&\quad + \Delta K G_2 P(k)\phi^T G_1^T K^T \Delta K G_2 P(k)\phi^T G_1^T(\Delta K)^T + \Delta K G_1 P(k)G_2^T K^T - \phi P(k)G_2^T(\Delta K)^T \\
&\quad + KG_1 P(k)G_2^T(\Delta K)^T + \Delta K G_1 P(k)G_2^T(\Delta K)^T + K\left[G_2 P(k)G_2^T + \left(2\gamma^2 - 2\gamma + 1\right)Q\right](\Delta K)^T \\
&\quad + \Delta K\left[G_2 P(k)G_2^T + \left(2\gamma^2 - 2\gamma + 1\right)Q\right]K^T + \Delta K\left[G_2 P(k)G_2^T + \left(2\gamma^2 - 2\gamma + 1\right)Q\right](\Delta K)^T \\
&= \left[I - KG_1\right]\left[\phi P(k)\phi^T + R\right]\left[I - KG_1\right]^T - KG_2 P(k)\phi^T\left[I - KG_1\right]^T \\
&\quad - \left[I - KG_1\right]^T\phi P(k)G_2^T K^T + K\left[G_2 P(k)G_2^T + \left(2\gamma^2 - 2\gamma + 1\right)Q\right]K^T \\
&\quad + \Big\{ -\left[\phi P(k)\phi^T + R\right]G_1^T + KG_1\left[\phi P(k)\phi^T + R\right]G_1^T + KG_2 P(k)\phi^T G_1^T - \phi P(k)G_2^T \\
&\quad + KG_1 P(k)G_2^T + K\left[G_2 P(k)G_2^T + \left(2\gamma^2 - 2\gamma + 1\right)Q\right]\Big\}(\Delta K)^T \\
&\quad + \Delta K\Big\{ -G_1\left[\phi P(k)\phi^T + R\right] + G_1\left[\phi P(k)\phi^T + R\right]G_1^T K^T - G_2 P(k)\phi^T \\
&\quad + G_2 P(k)\phi^T G_1^T K^T + G_1 P(k)G_2^T K^T + \left[G_2 P(k)G_2^T + \left(2\gamma^2 - 2\gamma + 1\right)Q\right]K^T\Big\} \\
&\quad + \Delta K\Big\{ G_1 P(k)G_2^T + G_1\left[\phi P(k)\phi^T + R\right]G_1^T + \left[G_2 P(k)G_2^T + \left(2\gamma^2 - 2\gamma + 1\right)Q\right]\Big\}(\Delta K)^T
\end{aligned}
$$

Make the linear section about ΔK be zero, then

$$
\begin{aligned}
&- \left[\phi P(k)\phi^T + R\right]G_1^T + KG_1\left[\phi P(k)\phi^T + R\right]G_1^T + KG_2 P(k)\phi^T G_1^T \\
&- \phi P(k)G_2^T + KG_1 P(k)G_2^T + K\left[G_2 P(k)G_2^T + \left(2\gamma^2 - 2\gamma + 1\right)Q\right] = 0
\end{aligned}
$$

References

1. Xia YQ, Fu MY, Liu GP. Analysis and synthesis of networked control systems. Heidelberg: Springer; 2011.
2. You KY, Xie LH. Survey of recent progress in networked control systems. Acta Autom. Sin. 2013;39(2):101–18.
3. Ray A, Liou LW, Shen JH. State estimation using randomly delayed measurements. J Dyn Syst Meas Control. 1993;115(1):19–26.
4. Yaz E, Ray A. Linear unbiased state estimation under randomly varying bounded sensor delay. Appl Math Lett. 1998;11(4):27–32.
5. Moayedi M, Foo YK, Soh YC. Filtering for networked control systems with single/multiple measurement packets subject to multiple-step measurement delays and multiple packet dropouts. Int J Syst Sci. 2011;42(3):335–48.

6. Ma J, Sun S. Distributed fusion filter for networked stochastic uncertain systems with transmission delays and packet dropouts. Sig Process. 2017;130:268–78.
7. Caballero-Águila R, Hermoso-Carazo A, Linares-Pérez J. Distributed fusion filters from uncertain measured outputs in sensor networks with random packet losses. Inf Fusion. 2017;34:70–9.

Extended Set-Membership Filter for Dynamic State Estimation in Power System

Haoyang He, Shanbi Wei, Yi Chai, Xiuling Sun and Jialin Deng

Abstract This paper introduces an extended set-membership filter (ESMF) method in the field of power system state estimation, when the noises is unknown but bounded (UBB), this method can enhance the reliability and accuracy of estimation result. The method is based on a dynamic state model. ESMF is applied to this system. ESMF provides 100% confidence in the reliability and safety of the power system. In this paper, ESMF is derived and demonstrated through the IEEE 30 test system.

Keywords State estimation · Extended set-membership filter · Power system

1 Introduction

In electric power networks, to make sure that the states of power networks meet the requirement, to estimate the state is necessary, and it is the basic feature of Energy Management Systems (EMS) [1]. In order to carry out a variety of key tasks, for instance bad data detection, and keep the state of the system stable and reliable, the voltage magnitude and phase angle at each bus is indispensable [2].

The dynamic model of power system reflects the change of the system state variables with time, and the dynamic model is based on the state variables of the system. In order to denote the dynamic state of the system, when assumed the behavior of the system is quasi-static, the transient process after disturbance is usually ignored. [3–5] proposed different dynamic state models, the quasi-static calculation step is usually measured in minutes, calculated every few minutes, [4] uses the Holt method, this method still has a good predictive effect even when the state variables are independent of each other. In paper [6], the actual state transfer equation is calculated based on the network equation, and the transfer of each state quantity affects the neighboring state variable. There are also the other methods

H. He (✉) · S. Wei · Y. Chai · X. Sun · J. Deng
College of Automation, Chongqing University, Chongqing 400044, China
e-mail: haoyang626@163.com

© Springer Nature Singapore Pte Ltd. 2018
Y. Jia et al. (eds.), *Proceedings of 2017 Chinese Intelligent Systems Conference*,
Lecture Notes in Electrical Engineering 460,
https://doi.org/10.1007/978-981-10-6499-9_53

which are based on the actual bus load forecasting method to calculate the state transfer method [7, 8], with load forecast instead of dynamic equation, for the load and the generator are the key factors which determine the system dynamics, besides, the load variables are independent, the change mode is easier to predict. Once the load of all the buses is predicted, the load flow calculation can provide the predicted value of the state at the next moment.

In order to solve the problem above, there are several techniques can be used. Kalman filter (KF) is a classic approach to use, however it requires that the mean is zero, and the noise belongs to the Gaussian distribution [8]. Even so, the above method always requires that all uncertainties in the random frame, so that i can provide the highest probability state estimate [9].

The noise belongs to the Gaussian distribution. Estimated probabilies require using mean and variance to describe the state distribution. However, in actual case, it is impractical to suppose that the probability of interference statistics, including the relevant bias noise, is not significant. In fact, model error constraints are usually the only information available [10].

In order to solve the problem above, we apply the Set-Membership Filter (SMF) theory. When we use the SMF method, we assume that the noise is UBB, besides, to seek for a feasible set is the goal of SMF, in which it contains all the useful information(measurement data, modeling uncertainty, and the bounds of the noise) [11]. When we deal with non-linear problems, we use the ESMF method, which is the extension of SMF, and in [11–13] has approached the issue.

When describe state estimation of nonlinear discrete-time systems, we usually use the ellipsoid set, and it is the most common mean. Based on the work of [14, 15], this paper presents a more efficient and targeted power system state estimation method.

2 Dynamic Model

The dynamic model adopted in this paper is Holt two-parameter method which is commonly used. This method also called linear extrapolation method. It can also be used as a simple short-term forecasting method with the advantages of less storage variables and faster calculation speed, and it is suitable for online operation. The representation of $f(x)$ is as follows:

$$x_{k+1|k} = S_k + b_k \tag{1}$$

$$S_k = \alpha_H x_{k|k} + (1 - \alpha_H) x_{k|k-1} \tag{2}$$

$$b_k = \beta_H (S_k - S_{k-1}) + (1 - \beta_H) b_{k-1} \tag{3}$$

where S_k and b_k are the horizontal and the tilt components respectively, α_H and β_H are the smoothing parameters, and the value is between 0 and 1. The method uses the actual values and estimated values of the state variables of the previous time to predict the next state variable by appropriately assigning the regular parameters α_H and β_H. So get:

$$\begin{cases} F_k = \alpha_H(1+\beta_H) \\ G_k = (1+\beta_H)(1-\alpha_H)x_{k|k-1} - \beta_H S_{k-1} + (1-\beta_H)b_{k-1} \end{cases} \quad (4)$$

The model of the power system can be identified as

$$x_{k+1} = F_k x_k + G_k + w_k \quad (5)$$

where x_k is the system state variable at time step k with dimension $(n \times 1)$, usually take the node's voltage amplitude and phase angle, F_k is the $n \times 1$-dimensional state transition matrix, G_k is the control matrix, w_k is the system process noise, and its variance is Q_k.

The dynamic here is a slow motion, it is the stable operation of the power system, in the case of load or generator mutation, this model is no longer applicable. In addition, there are dynamic models established according to the dynamic model of the generator. At this time, the transient process is no longer neglected, but the model based on the dynamic model of the generator is more computationally expensive for the system with more nodes.

Considering the system measurement is all PMU, in order to ensure data redundancy, the measurement is $z_k = [P_k, Q_k, V_k, \theta_k]^T$, so the measurement equation contains the nonlinear part. For node i, $h(x)$ can be expressed as follows:

$$\begin{cases} P_i = V_i \sum_{i \in I} V_j(G_{ij}(\cos\theta_i \cos\theta_j + \sin\theta_i \sin\theta_j) + B_{ij}(\sin\theta_i \cos\theta_j - \cos\theta_i \sin\theta_j)) \\ Q_i = V_i \sum_{i \in I} V_j(G_{ij}(\sin\theta_i \cos\theta_j - \cos\theta_i \sin\theta_j) - B_{ij}(\cos\theta_i \cos\theta_j + \sin\theta_i \sin\theta_j)) \\ V_i = V_i \\ \theta_i = \theta_i \end{cases}$$

$$z_k = h(x_k) + v_k \quad (6)$$

Which include all nodes that connect $i \in I$ to i, besides the case of $a = b$. Because of the PMU node measurement accuracy is high, the mixed measurement is still non-linear model.

3   Extended Set-Membership Filter

The ellipsoid E_{k+1} can be put like this:

$$E_{k+1} = \left\{ x_{k+1} : (x_{k+1} - \hat{x}_{k+1})^T P_{k+1}^{-1} (x_{k+1} - \hat{x}_{k+1}) \le 1 \right\} \tag{7}$$

where x_{k+1} is the system state at the time instant $k+1$, $\hat{x}_{k+1}$ stands for the state estimate of x_{k+1} and the shape is defined by P_{k+1}. That is to say, we seek for $\hat{x}_{k+1}$ and P_{k+1} so that

$$(x_{k+1} - \hat{x}_{k+1})^T P_{k+1}^{-1} (x_{k+1} - \hat{x}_{k+1}) \le 1 \tag{8}$$

subject to $v_k \in V_k$ suppose that the E_k is obtained.

When take the Taylor approximation of $f(x_k)$ around $\hat{x}_{k|k}$ into account, Eq. (5) can be put as:

$$x_{k+1} = \hat{x}_{k+1/k} + F_k \Delta x_k + R_2(\Delta x_k, \bar{X}_k) \tag{9}$$

where F_k is the gradient operator:

$$F_k = \left. \frac{\partial f(x)}{\partial x} \right|_{x = \hat{x}_{k|k}} \tag{10}$$

where $\Delta x_k = x_k - \hat{x}_{k|k}$, $\bar{X}_k = [\hat{x}_{k|k}^i - \sqrt{P_{k|k}^{i,i}}, \hat{x}_{k|k}^i + \sqrt{P_{k|k}^{i,i}}]$. The Lagrange remainder By using the interval algorithm, $R_2(*)$ is treated as a bounded noise, and a detailed derivation is presented [11]. $R_2(\Delta x_k, \bar{X}_k)$ is the feasible set, and it is calculated by a box Γ_k [14]:

$$\Gamma_k = \left\{ \bar{w}_k : \left| w_k^i \right| \le b_i, i = 1, .., n \right\} \tag{11}$$

$$b_i = rad(R_2^i(\Delta x_k, \bar{X}_k)) \tag{12}$$

where $R_2^i(\Delta x_k, \bar{X}_k) = \frac{1}{2} \Delta x_k^T \frac{\partial^2 f^i(\bar{X}_k)}{\partial x^2} \Delta x_k, i = 1, \ldots, n$. The radius of interval variable $\bar{X}_k$ is denoted by $rad(R_2^i(\Delta x_k, \bar{X}_k))$.

Thus, $E_{k+1|k}$ is the minimum volume ellipsoid which contains the box Γ_k and the ellipsoid $E_{k|k}$, and the ellipsoid $E_{k+1|k}$ is calculated through the time-update process. In [12, 14], they proposed a algorithm whose calculation is relatively small, through using the following method, we get $E_{k+1|k}$.

Time update:

1. $\hat{x}_{k+1/k}$ is the ellipsoid center, and it is calculated by:

$$\hat{x}_{k+1|k} = f(\hat{x}_{k|k}) \tag{13}$$

2. $P^0_{k+1|k}$ is the time-updated ellipsoid matrix, and its initialization equation is:

$$P^0_{k+1|k} = F_k P_k F_k \tag{14}$$

3. for $i = 0, \ldots, n-1$, the recursion calculation is done by:

$$P^{i+1}_{k+1|k} = (1+p_i)P^i_{k+1|k} + (1+p^i_{k|k+1})b^2_i I_i I^T_i \tag{15}$$

parameter p_i is the positive root of

$$np^2_i + (n-1)a_i p_i - a_i = 0 \tag{16}$$

where $a_i = tr(b^2_i (P^i_{k+1|k})^{-1} I_i I^T_i) = b^2_i I^T_i (P^i_{k+1|k})^{-1} I_i.$

4. Finally, $P_{k+1|k} = P^n_{k+1|k}.$

Measurement update:

We get the gradient operator by:

$$G_{k+1} = \frac{\partial g(x)}{\partial x}\bigg|_{x=\hat{x}_{k+1|k}} \tag{17}$$

As we did in Eqs. (9)–(12), we can get y_{k+1}.

Just as it showed in [14], we know that the measurement-update ellipsoid E_{k+1} is the ellipsoid that we want to get, which has the minimal-volume, and in each iteration it contains ellipsoid-strip intersection of strip set R_{k+1} and the time-update ellipsoid $E_{k+1|k}$. That is to say: $E_{k+1} \supseteq E_{k+1|k} \bigcap R_{k+1}$. Through the following algorithm, we can get the shape matrix P_{k+1} and the center matrix $\hat{x}_{k+1}$ of E_{k+1}.

Initialization:

$$\hat{x}^0_{k+1} = \hat{x}_{k+1|k} \tag{18}$$

$$P^0_{k+1} = P_{k+1|k} \tag{19}$$

The recursion calculation can be calculated as: for $i = 1, \ldots, m$,

$$h_i = G_{k+1,i}^T P_{k+1}^{i-1} G_{k+1,i} \tag{20}$$

$$\phi_i^+ = (z_{k+1}^i - G_{k+1,i}^T x_{k+1}^{i-1} + \delta_{k+1}^i)/\sqrt{h_i} \tag{21}$$

$$\phi_i^- = (z_{k+1}^i - G_{k+1,i}^T x_{k+1}^{i-1} - \delta_{k+1}^i)/\sqrt{h_i} \tag{22}$$

where $z_{k+1}^i = y_{k+1}^i - g_i(\hat{x}_{k+1|k}) + G_{k+1,i}^T \hat{x}_{k+1|k}$, $\delta_{k+1}^i = rad(R_2^i(\Delta\tilde{x}_{k+1}, \tilde{X}_{k+1})) + \varepsilon_{k+1}^i$.

$$R_2^i(\Delta\tilde{x}_{k+1}, \tilde{X}_{k+1}) = \frac{1}{2}\Delta\tilde{x}_{k+1}^T \frac{\partial^2 g^i(\tilde{X}_{k+1})}{\partial x^2} \Delta\tilde{x}_{k+1} \tag{23}$$

and $\Delta\tilde{x}_{k+1} = x_{k+1} - \tilde{x}_{k+1|k}$, $\tilde{X}_{k+1} = [\hat{x}_{k+1|k}^i - \sqrt{P_{k+1|k}^{i,i}}, \hat{x}_{k+1|k}^i + \sqrt{P_{k+1|k}^{i,i}}]$.

we set $\phi_i^+ = 1$, If $\phi_i^+ > 1$; then we set $\phi_i^- = -1$, if $\phi_i^- < -1$; and if $\phi_i^+ \phi_i^- \leq -1/n$, then $P_{k+1}^i = P_{k+1}^{i-1}$, and $\hat{x}_{k+1}^i = \hat{x}_{k+1}^{i-1}$. Otherwise for $i = 1, \ldots, m$,

$$\hat{x}_{k+1}^i = \hat{x}_{k+1}^{i-1} + q_i \frac{D_{k+1}^i G_{k+1,i} e_i}{d_i^2} \tag{24}$$

$$P_{k+1}^i = (1 + q_i - \frac{q_i e_i^2}{d_i^2 + q_i h_i})D_{k+1}^i \tag{25}$$

where

$$D_{k+1}^i = P_{k+1}^{i-1} - \frac{q_i}{d_i^2 + q_i h_i}P_{k+1}^{i-1}G_{k+1,i}G_{k+1,i}^T P_{k+1}^{i-1} \tag{26}$$

$$e_i = \sqrt{h_i}\frac{\phi_i^+ + \phi_i^-}{2} \tag{27}$$

$$d_i = \sqrt{h_i}\frac{\phi_i^+ - \phi_i^-}{2} \tag{28}$$

While q_i is the positive root of

$$(n-1)h_i^2 q_i^2 + [(2n-1)d_i^2 - h_i + e_i^2]h_i q_i + [n(d_i^2 - e_i^2) - h_i]d_i^2 = 0 \tag{29}$$

4. Finally, $P_{k+1} = P_{k+1}^m$, $\hat{x}_{k+1}^i = \hat{x}_{k+1}^m$.

4 Numerical Experiments

Taking the IEEE30 circuit as a test system,the state estimation performance of different sampling strategies is compared under the condition that the load is slowly increased by 20% in the case of 60 acquisition points in 60 min.

Assuming all nodes are configured PMU, the PMU measurement includes the node voltage amplitude, the node voltage phase angle, the node input active power and the reactive power, the standard deviation of the measurement error is 0.005, the mean value is 0. When the load increases by 30%, $\alpha_H = 0.6$, $\beta_H = 0.1$, while the load variation is in accordance with the load curve, $\alpha_H = 0.001$, $\beta_H = 0.05$.

For the comparison of performance, the performance index function—rms error (rmse) is introduced to measure the deviation between the true value and the estimated value. The root mean square error at time k is defined as

$$R_{mse}(k) = \sqrt{\frac{1}{N} \sum_{i=1}^{N} (\hat{y}_{k,i} - y_{k,i})},$$ where the ith component of the state estimate is $\hat{y}_{k,i}$,

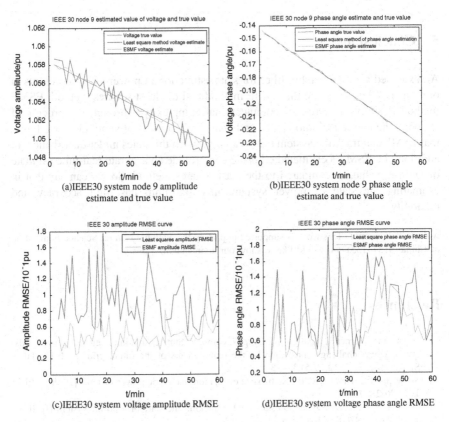

Fig. 1 The amplitude and phase angle estimation of node 9 in IEEE 30 test system and the corresponding RMSE

$y_{k,i}$ is the ith component of the true value of the state, the state dimension is N. The indicator function—the rms error represents the overall average estimation effect. The smaller the value is, the closer the estimate is to the true value. Since the state quantity contains the amplitude and phase angle of the voltage, the magnitude difference between the two is large, so in the following example, the root mean square error of the amplitude and phase angle needs to be compared separately.

(1) IEEE 30 test system, the load linear increase by 30%

In Fig. 1a, we can see that, when compared ESMF with the least squares method, ESMF is more accurate. In Fig. 1b, the estimated effect of the voltage phase angle is similar to that in graph Fig. 1c. The mean square error (MSE) of the amplitude of the ESMF method is less than that of the least squares method. As can be seen from Fig. 1d, the MSE of the phase angle of ESMF is to some extent smaller than the MSE of the phase angle of the least squares method.

In Fig. 1a, we can see that, when compared ESMF with the least squares method, ESMF is more accurate.

5 Conclusions

An extended set-membership filter for state estimation in power system presents in our paper. When compare the ESMF method and the least squares method, ESMF method is simple in terms of computational complexity. Through the process of studying the theoretical knowledge and analyze the results of simulation, we found that ESMF can track the system states rapidly, when the states under disturbance. In addition, the ESMF's estimated ellipsoid can provide an accurate real-time estimate of the true value, as compared to the least squares method. So we can say that in estimating the state of power system, this algorithm has high accuracy and reliability.

Acknowledgements This work is supported in part by the National Natural Science Foundation of China (61374135), we would like to show our appreciation.

References

1. Huang YF, Werner S, Huang J, Kashyap N, Gupta V. State estimation in electric power grids: Meeting new challenges presented by the requirements of the future grid [J]. IEEE Trans Signal Process. 2012;29(5):33–43.
2. Schweppe FC, Handschin EJ. Static state estimation in electric power systems [J]. Proc IEEE. 1974;7(62):972–82.
3. Debs A, Larson RA. Dynamic estimator for tracking the state of a power system [J]. IEEE Trans PAS. 1970;89(7):1670–8.

4. Wei Z, Sun G, Pang B. Application of unscented Kalman filter and its square root form in dynamic state estimation of power system [J]. Proc CSEE. 2011;(16):74–80.
5. da Silva AML, Do Coutto Filho MB, de Queiroz JF. State forecasting in electric power systems [C]. IEE Proc C (Generation, Transmission and Distribution). IET Digital Library, 1983;130(5):237–44.
6. Durgaprasad G, Thakur SS. Robust dynamic state estimation of power systems based on M-estimation and realistic modeling of system dynamics [J]. IEEE Trans Power Syst. 1998;13 (4):1331–6.
7. Sinha AK, Mondal JK. Dynamic state estimator using ANN based bus load prediction [J]. IEEE Trans Power Syst. 1999;14(4):1219–25.
8. Blood EA, Krogh BH, Ilic MD. Electric power system static state estimation through Kalman filtering and load forecasting [C]. In: Power and energy society general meeting-conversion and delivery of electrical energy in the 21st century, 2008 IEEE. IEEE;2008. p. 1–6.
9. Nishiyama K. Nonlinear filter for estimating a sinusoidal signal and parameters in white noise: on the case of a single sinusoid [J]. IEEE Trans Signal Process. 1997;45(2):970–81.
10. Scholte E, Campbell ME. On-line nonlinear guaranteed estimation with application to a high performance aircraft [C]. Proc Am Control Conf. 2002;184–190.
11. Kieffer M, Jaulin L, Walter E. Guaranteed recursive non-linear state bounding using interval analysis [J]. Int J Adapt Control Signal Process. 2002;16(3):193–218.
12. Maksarov D, Norton J. State bounding with ellipsoidal set description of the uncertainty [J]. Int J Control. 1996;65(5):847–66.
13. Schole E, Cambell ME. A nonlinear set-membership filter for on-line applications. Int J Robust Nonlinear Control. 2003;13(15):1337–58.
14. Bertsekas DP, Rhodes IB. Recursive state estimation for a set-membership description of uncertainty [J]. IEEE Trans Autom Control. 1971;16(2):117–28.
15. Qing X, Yang F, Wang X. Extended set-membership filter for power system dynamic state estimation [J]. Electr Power Syst Res. 2013;99:56–63.

Global Tracking Control of Underactuated Surface Vessels with Non-diagonal Matrices

Wenjing Xie and Baoli Ma

Abstract In this paper, two control laws are proposed to solve the tracking control problem of underactuated surface vessels with non-diagonal inertia and damping matrices. Based on the inherent cascaded structure of dynamics, the original tacking control problem of vessels is converted to stabilization control problem of two subsystems. Then, two tracking control laws are designed respectively for known and unknown model parameters. The first surge control law is simple without any model parameter, and hence is possible to be extended to the case of unknown model parameters. Stability analysis indicates that both the two control laws realize global $\mathcal{K}$-exponential tracking under persistently exciting condition. Effectiveness of the proposed controllers is demonstrated by numerical simulations.

Keywords Underactuated surface vessels · Tracking control · Global $\mathcal{K}$-exponential stability

1 Introduction

Control of marine surface vessels has become an active topic in the control community, due to its wide application, such as maritime rescue, self-positioning and autonomous docking. Both the full-actuated [1] and the underactuated [2–5] surface vessels are researched by control scientists. Since the underactuation is very common in practical vessels, this paper focuses on the trajectory tracking control problem of underactuated surface vessels.

In the literature, numerous of promising results related to tracking control problem of underactuated surface vessels have been presented. In [6–20], tracking control

W. Xie
School of Computer and Information Science, Southwest University,
Chongqing 400715, China

B. Ma(✉)
The Seventh Research Division, School of Automation Science and Electrical Engineering,
Beihang University, Beijing 100191, China
e-mail: mabaoli@buaa.edu.cn

© Springer Nature Singapore Pte Ltd. 2018   561
Y. Jia et al. (eds.), *Proceedings of 2017 Chinese Intelligent Systems Conference*,
Lecture Notes in Electrical Engineering 460,
https://doi.org/10.1007/978-981-10-6499-9_54

laws are proposed for the simplified vessel dynamics with diagonal system matrices. Indeed, the mass and damping matrices in practical vessels are usually non-diagonal [21–24]. In [21, 22], suitable state and input transformations were constructed to convert the original error system to a new one, whose stability could be guaranteed by the stability of chained-like subsystem, then global $\mathcal{K}$-exponential tracking control laws were proposed using Lyapunov method and Barbalat's lemma. In [23], based on the cascaded system theory, a control law were constructed to deal with the global asymptotic tracking control problem of underactuated ships with unknown model parameters. In [24], coordinate transformation, neural network, observer technique, and additional control inputs were adopted to solve the tracking problem despite the unknown model parameters and disturbances, guaranteeing local uniform boundedness. However, the control laws in the previous literature [21–24] cannot achieve global $\mathcal{K}$-exponential tracking for underactuated surface vessels with uncertain model parameters. The above review motivates us to develop a tracking control law for uncertain underactuated surface vessel with non-diagonal system matrices, which ensures the global $\mathcal{K}$-exponential convergence of tracking errors to zero.

The purpose of this paper is to develop new solutions to the global $\mathcal{K}$-exponential tracking control problem of underactuated surface vessels. From the cascaded system theory, we reveal that the original global $\mathcal{K}$-exponential tracking control problem is equivalent to global $\mathcal{K}$-exponential stabilization control problem of two subsystems. To stabilize the two subsystems, the first control law is derived in case of precise model parameters. Then, the second control law is obtained by extending the first control scheme and using the sliding mode technique, which can realize global $\mathcal{K}$-exponential tracking control of asymmetric underactuated surface vessels with unknown model parameters.

The rest of this paper is organized as follows. The stabilization problem of an underactuated surface vessels is formulated in Sect. 2. Main results are presented in Sect. 3. Simulation examples are provided in Sect. 4. Section 5 concludes the work.

2   Problem Formulation

In this paper, we consider the underactuated surface vessel with non-diagonal mass and damping matrices. The model is described by [21]

$$\dot{\eta} = \mathbf{J}(\psi)\mathbf{v},$$
$$\mathbf{M}\dot{\mathbf{v}} + \mathbf{C}(\mathbf{v})\mathbf{v} + \mathbf{D}\mathbf{v} = \tau \tag{1}$$

with $\eta = \begin{bmatrix} x, y, \psi \end{bmatrix}^{\mathrm{T}}, \mathbf{v} = [u, v, r]^{\mathrm{T}}, \tau = \begin{bmatrix} \tau_u, 0, \tau_r \end{bmatrix}^{\mathrm{T}}$ and

$$\mathbf{J}(\psi) = \begin{bmatrix} \cos\psi & -\sin\psi & 0 \\ \sin\psi & \cos\psi & 0 \\ 0 & 0 & 1 \end{bmatrix}, \mathbf{D} = \begin{bmatrix} d_{11} & 0 & 0 \\ 0 & d_{22} & d_{23} \\ 0 & d_{32} & d_{33} \end{bmatrix},$$

$$\mathbf{C}(\mathbf{v}) = \begin{bmatrix} 0 & 0 & C_{13} \\ 0 & 0 & C_{23} \\ -C_{13} & -C_{23} & 0 \end{bmatrix}, \mathbf{M} = \begin{bmatrix} m_{11} & 0 & 0 \\ 0 & m_{22} & m_{23} \\ 0 & m_{23} & m_{33} \end{bmatrix},$$

$$C_{13} = -m_{22}v - m_{23}r, C_{23} = m_{11}u$$

where (x, y) is the position of vessel, ψ is the yaw angle of vessel in the earth fixed frame, (u, v, r) denote the surge, sway and yaw velocities respectively; $(d_{11}, d_{22}, d_{33}, d_{23}, d_{32})$ and $(m_{11}, m_{22}, m_{33}, m_{23})$ are the damping and the mass parameters respectively, and the control inputs τ_u and τ_r are the surge force and the yaw moment respectively. Write the model (1) as the component form:

$$\begin{cases} \dot{x} = u\cos\psi - v\sin\psi, \dot{y} = u\sin\psi + v\cos\psi, \dot{\psi} = r, \\ \dot{u} = \dfrac{m_{22}}{m_{11}}vr - \dfrac{d_{11}}{m_{11}}u + \dfrac{m_{23}}{m_{11}}r^2 + \dfrac{1}{m_{11}}\tau_u, \\ \dot{v} = -\dfrac{m_{23}}{m_{22}}\dot{r} - \dfrac{m_{11}}{m_{22}}ur - \dfrac{d_{22}}{m_{22}}v - \dfrac{d_{23}}{m_{22}}r, \\ \dot{r} = \dfrac{1}{\Delta}(a_{uv}uv + a_{ur}ur + a_v v + a_r r + m_{22}\tau_r), \end{cases} \quad (2)$$

where $\Delta = m_{22}m_{33} - m_{23}^2, a_{uv} = m_{22}(m_{11} - m_{22}), a_{ur} = (m_{11} - m_{22})m_{23}, a_v = m_{23} d_{22} - m_{22}d_{32}$, and $a_r = m_{23}d_{23} - m_{22}d_{33}$. Since the matrix $\mathbf{M}$ is positive definite and symmetric, the constants m_{11}, m_{22}, m_{33} and Δ are all positive. Moreover, (d_{11}, d_{22}, d_{33}) are positive too.

Assume that the reference trajectory is produced by a virtual vessel:

$$\dot{x}_r = u_r\cos\psi_r - v_r\sin\psi_r, \dot{x}_r = u_r\sin\psi_r + v_r\cos\psi_r, \dot{\psi}_r = r_r,$$

$$\dot{u}_r = \dfrac{m_{22}}{m_{11}}v_r r_r - \dfrac{d_{11}}{m_{11}}u_r + \dfrac{m_{23}}{m_{11}}r_r^2 + \dfrac{1}{m_{11}}\tau_{u_r},$$

$$\dot{v}_r = -\dfrac{m_{23}}{m_{22}}\dot{r}_r - \dfrac{m_{11}}{m_{22}}u_r r_r - \dfrac{d_{22}}{m_{22}}v_r - \dfrac{d_{23}}{m_{22}}r_r, \quad (3)$$

$$\dot{r}_r = \dfrac{1}{\Delta}(a_{uv}u_r v_r + a_{ur}u_r r_r + a_v v_r + a_r r_r + m_{22}\tau_{r_r}).$$

Assumption 1 The reference trajectory satisfies the bounded condition $(u_r, v_r, r_r, \tau_{u_r}, \tau_{r_r}) \in \mathcal{L}_\infty$ and the persistently exciting (PE) condition $\lim\limits_{t\to+\infty} r_r(t) \neq 0$.

The trajectory tracking control problem of underactuated surface vessels considered in this paper is stated as: *under Assumption 1, design a control law $\tau_u(\cdot)$ and $\tau_r(\cdot)$ to steer the vessel globally $\mathcal{K}$-exponentially track the reference trajectory.*

3   Controller Development

3.1   Error Model Development

Define the tracking errors

$$z_{1e} = z_1 - z_{1r}, \; z_{2e} = z_2 - z_{2r}, \; \psi_e = \psi - \psi_r, \\ u_e = u - u_r, \; \bar{v}_e = \bar{v} - \bar{v}_r, \; r_e = r - r_r, \tag{4}$$

where

$$z_1 = x\cos\psi + y\sin\psi, \; z_2 = -x\sin\psi + y\cos\psi, \; \bar{v} = v + \frac{m_{23}}{m_{22}}r,$$

$$z_{1r} = x_r\cos\psi_r + y_r\sin\psi_r, \; z_{2r} = -x_r\sin\psi_r + y_r\cos\psi_r, \; \bar{v}_r = v_r + \frac{m_{23}}{m_{22}}r_r. \tag{5}$$

The error dynamics is computed as

$$\begin{cases}
\dot{z}_{1e} = \underbrace{u_e + z_{2e}r_r}_{\triangleq f_{z1}} + \underbrace{(z_{2e} + z_{2r})}_{\triangleq g_{z1}}r_e, \\[2mm]
\dot{z}_{2e} = \underbrace{\bar{v}_e - z_{1e}r_r}_{\triangleq f_{z2}} - \underbrace{\left(\frac{m_{23}}{m_{22}} + z_{1e} + z_{1r}\right)r_e}_{\triangleq -g_{z2}}, \\[2mm]
\dot{u}_e = \underbrace{\frac{m_{22}}{m_{11}}\bar{v}_e r_r - \frac{d_{11}}{m_{11}}u_e + \frac{1}{m_{11}}(\tau_u - \tau_{u_r})}_{\triangleq f_u} + \underbrace{\frac{m_{22}}{m_{11}}(\bar{v}_e + \bar{v}_r)r_e}_{\triangleq g_u}, \\[2mm]
\dot{\bar{v}}_e = \underbrace{-\frac{m_{11}}{m_{22}}u_e r_r - \frac{d_{22}}{m_{22}}\bar{v}_e}_{\triangleq f_v} + \underbrace{\left[\frac{m_{23}d_{22} - m_{22}d_{23}}{m_{22}^2} - \frac{m_{11}}{m_{22}}(u_e + u_r)\right]r_e}_{\triangleq g_v},
\end{cases} \tag{6a}$$

$$\begin{cases}
\dot{\psi}_e = r_e, \\[2mm]
\dot{r}_e = \frac{1}{\Delta}\left[a_{uv}(uv - u_r v_r) + a_{ur}(ur - u_r r_r) + a_v v_e + a_r r_e + m_{22}(\tau_r - \tau_{r_r})\right],
\end{cases} \tag{6b}$$

where (6a) and (6b) are respectively the position and the orientation error subsystems. Now, the global $\mathcal{K}$-exponential trajectory tracking control problem is converted to global $\mathcal{K}$-exponential stabilization control problem of (6), since the transformation in (4) is globally invertible.

3.2   Controller Design in Case of Precise Model Parameters

To employ the cascaded system theory in Lemma 2.4.5 of [25], denote

$$\mathcal{Z}_1 = [z_{1e}, z_{2e}, u_e, \bar{v}_e]^T, \quad \mathcal{Z}_2 = [\psi_e, r_e]^T, \quad g(t, \mathcal{Z}_1, \mathcal{Z}_2) = \begin{bmatrix} 0 & 0 & 0 & 0 \\ g_{z1} & g_{z2} & g_u & g_v \end{bmatrix}^T$$

for description convenience. According to

$$\|g(t, \mathcal{Z}_1, \mathcal{Z}_2)\| \leq \underbrace{|z_{2r}| + \left| \frac{m_{23}}{m_{22}} + z_{1r} \right| + \left| \frac{m_{22}}{m_{11}} \bar{v}_r \right| + \left| \frac{m_{23}d_{22} - m_{22}d_{23}}{m_{22}^2} - \frac{m_{11}}{m_{22}} u_r \right|}_{\triangleq \kappa_1(\|\mathcal{Z}_2\|)}$$

$$+ |z_{2e}| + |z_{1e}| + \left| \frac{m_{22}}{m_{11}} \bar{v}_e \right| + \left| \frac{m_{11}}{m_{22}} u_e \right| \tag{7}$$

$$\leq \kappa_1(\|\mathcal{Z}_2\|) + \underbrace{(2 + \frac{m_{11}}{m_{22}} + \frac{m_{22}}{m_{11}}) \|\mathcal{Z}_1\|}_{\triangleq \kappa_2(\|\mathcal{Z}_2\|)},$$

it follows that $g(t, \mathcal{Z}_1, \mathcal{Z}_2)$ satisfies (2.22) in [25]. Thus, the global $\mathcal{K}$-exponential stabilization of (6) can be realized if we can construct τ_r to ensure (6b) globally $\mathcal{K}$-exponentially stable, and simultaneously derive τ_u to guarantee the system

$$\begin{cases} \dot{z}_{1e} = u_e + z_{2e}r_r, \quad \dot{z}_{2e} = \bar{v}_e - z_{1e}r_r, \\ \dot{u}_e = \frac{m_{22}}{m_{11}} \bar{v}_e r_r - \frac{d_{11}}{m_{11}} u_e + \frac{1}{m_{11}}(\tau_u - \tau_{u_r}), \\ \dot{\bar{v}}_e = -\frac{m_{11}}{m_{22}} u_e r_r - \frac{d_{22}}{m_{22}} \bar{v}_e, \end{cases} \tag{8}$$

fulfil the conditions in Lemma 2.4.5 of [25]. System (8) is the system (6a) with the vanishing $g(t, \mathcal{Z}_1, \mathcal{Z}_2)$, similar to (2.17) in [25].

Step 1: Design τ_r to stabilize subsystem (6b). Construct the Lyapunov function candidate:

$$V_2 = 0.5 \left[\frac{m_{22}}{\Delta} \psi_e^2 + \left(r_e + k_\psi \psi_e \right)^2 \right], \tag{9}$$

where $k_\psi > 0$, $m_{22} > 0$ and $\Delta > 0$. Design the yaw moment control law

$$\tau_r = \tau_{r_r} - \psi_e - k_r \left(r_e + k_\psi \psi_e \right)$$

$$- \frac{1}{m_{22}} \left[a_{uv}(uv - u_r v_r) + a_{ur}(ur - u_r r_r) + a_v v_e + a_r r_e + \Delta k_\psi r_e \right], \tag{10}$$

where $k_r > 0$. Under (10), $\dot{V}_2$ is calculated as

$$\dot{V}_2 = -\frac{k_\psi m_{22}}{\Delta} \psi_e^2 - \frac{k_r m_{22}}{\Delta}\left(r_e + k_\psi \psi_e\right)^2. \tag{11}$$

This implies that the origin of system (6b) is globally $\mathcal{K}$-exponentially stable.

Step 2: Design τ_u to stabilize system (8). Let us construct

$$\tau_u = \tau_{u_r} - k_1 z_{1e}, \quad 0 < k_1 < \frac{d_{11}d_{22}}{m_{11}}. \tag{12}$$

where k_1 is a control coefficient. To probe into the stability of the closed-loop sub-system (8) and (12), define $\bar{z}_{1e} = \sqrt{\frac{k_1}{d_{22}}}z_{1e} + \frac{d_{11}}{\sqrt{k_1 d_{22}}}u_e$. Then the subsystem (8) and (12) is converted into

$$\begin{bmatrix} \dot{\bar{z}}_{1e} \\ \dot{u}_e \\ \dot{z}_{2e} \\ \dot{\bar{v}}_e \end{bmatrix} = \underbrace{\begin{bmatrix} -\frac{d_{11}}{m_{11}} & \sqrt{\frac{k_1}{d_{22}}} & \sqrt{\frac{k_1}{d_{22}}}r_r & \frac{d_{11}m_{22}}{\sqrt{k_1 d_{22}}m_{11}}r_r \\ -\frac{\sqrt{k_1 d_{22}}}{m_{11}} & 0 & 0 & \frac{m_{22}}{m_{11}}r_r \\ -\sqrt{\frac{d_{22}}{k_1}}r_r & \frac{d_{11}}{k_1}r_r & 0 & 1 \\ 0 & -\frac{m_{11}}{m_{22}}r_r & 0 & -\frac{d_{22}}{m_{22}} \end{bmatrix}}_{\triangleq A(r(t))} \begin{bmatrix} \bar{z}_{1e} \\ u_e \\ z_{2e} \\ \bar{v}_e \end{bmatrix}. \tag{13}$$

For system (13), define the Lyapunov function:

$$V_1 = 0.5[\bar{z}_{1e}, u_e] \underbrace{\begin{bmatrix} \frac{d_{22}}{k_1} & \frac{k_1 m_{11} - d_{11}d_{22}}{k_1\sqrt{k_1 d_{22}}} \\ \frac{k_1 m_{11} - d_{11}d_{22}}{k_1\sqrt{k_1 d_{22}}} & \frac{d_{11}(d_{11}d_{22} - k_1 m_{11}) + k_1 m_{11}d_{22}}{k_1^2 d_{22}} \end{bmatrix}}_{\triangleq P_1} \begin{bmatrix} \bar{z}_{1e} \\ u_e \end{bmatrix}$$

$$+ 0.5[z_{2e}, \bar{v}_e] \underbrace{\begin{bmatrix} 1 & \frac{m_{22}}{d_{22}} \\ \frac{m_{22}}{d_{22}} & \frac{m_{22}^2 d_{12}}{k_1 m_{11}d_{22}} \end{bmatrix}}_{\triangleq P_2} \begin{bmatrix} z_{2e} \\ \bar{v}_e \end{bmatrix}, \tag{14}$$

where $d_{12} = d_{11} + d_{22} > 0$. For P_1 and P_2 in (14), since $0 < k_1 < d_{11}d_{22}/m_{11}$, it follows that P_1 and P_2 are positive definite. As a result, under $0 < k_1 < d_{11}d_{22}/m_{11}$, V_1 is a positive definite function for system (13).

Taking the derivative of V_1 along (13), one has

$$\dot{V}_1 = -\frac{d_{11}d_{22}}{k_1m_{11}}\bar{z}_{1e}^2 - \frac{d_{11}d_{22} - k_1m_{11}}{k_1d_{22}}u_e^2 - \frac{m_{22}(d_{12}d_{22} - k_1m_{11})}{k_1m_{11}d_{22}}\bar{v}_e^2 \tag{15}$$

$$\triangleq -[\bar{z}_{1e}, u_e, z_{2e}, \bar{v}_e]C^T C[\bar{z}_{1e}, u_e, z_{2e}, \bar{v}_e]^T \leq 0,$$

where $C = \mathrm{diag}\left\{\sqrt{\frac{d_{11}d_{22}}{k_1m_{11}}}, \sqrt{\frac{d_{11}d_{22}-k_1m_{11}}{k_1d_{22}}}, 0, \sqrt{\frac{m_{22}(d_{12}d_{22}-k_1m_{11})}{k_1m_{11}d_{22}}}\right\} \geq 0$. Since $(A(r(t)), C))$ is uniformly completely controllable under Assumption 1, and $\dot{V}_1 \leq 0$, hence system (13) (and so (8) and (12)) is globally exponentially stable and so globally $\mathcal{K}$-exponentially stable ([26], Example 8.11).

Based on the above analysis, the conditions in Lemma 2.4.5 of [25] are satisfied with $C_1 = 0.5\min\{\lambda_{\min}(P_1), \lambda_{\min}(P_2)\} > 0$, $\alpha_2(s) = 0.5\max\{\lambda_{\max}(P_1), \lambda_{\max}(P_2)\}s^2 \in \mathcal{K}_\infty$, $W(\mathcal{Z}_1) = [\bar{z}_{1e}, u_e, z_{2e}, \bar{v}_e]C^T C[\bar{z}_{1e}, u_e, z_{2e}, \bar{v}_e]^T \geq 0$, $C_4 = 0.5\|\mathrm{diag}\{P_1, P_2\}\| > 0$, where $\lambda_{\min}(\cdot)$ and $\lambda_{\max}(\cdot)$ are the minimum and the maximum eigenvalues of matrix.

Theorem 1 *Under Assumption 1, the control law (10) and (12) guarantees the tracking errors of underactuated surface vessels in (6) globally $\mathcal{K}$-exponentially convergent to zero, provided that $0 < k_1 < \frac{d_{11}d_{22}}{m_{11}}, k_\psi > 0$ and $k_r > 0$.*

3.3 Controller Design in Case of Unknown Model Parameters

In fact, the accurate model parameters cannot be obtained even under well developed identification technologies. Thus, it is necessary to extend the control scheme (10) and (12) to the case of unknown model parameters. In this case, only the upper and lower bounds of model parameters are assumed to be known.

Since the control law (12) maintains effective for all the USVs with model parameters satisfying $d_{11}d_{22}/m_{11} > k_1$, the surge control force τ_u can still be designed as (12) with the new condition $k_1 < \underline{d}_{11}\underline{d}_{22}/\bar{m}_{11}$, and only the yaw control moment $\tau_r(\cdot)$ is needed to be redesigned in this subsection, where $\bar{\chi}$ and $\underline{\chi}$ represent the maximum and the minimum values of χ.

Let us use the sliding mode technique to modify the previous law (10) as:

$$\tau_r = \tau_{r_r} - \psi_e - k_r\left(r_e + k_\psi\psi_e\right) - \mathrm{sign}(r_e + k_\psi\psi_e)\underline{m}_{22}^{-1}$$
$$\times\left(\bar{a}_{uv}|uv - u_rv_r| + \bar{a}_{ur}|ur - u_rr_r| + \bar{a}_v|v_e| + \bar{a}_r|r_e| + \bar{\Delta}k_\psi|r_e|\right), \tag{16}$$

where $|a_{uv}| \leq \bar{a}_{uv}, |a_{ur}| \leq \bar{a}_{ur}, |a_v| \leq \bar{a}_v, |a_r| \leq \bar{a}_r$. Under the control law (16),

$$\dot{V}_2 \leq -\frac{k_\psi m_{22}}{\Delta}\psi_e^2 - \frac{k_r m_{22}}{\Delta}\left(r_e + k_\psi\psi_e\right)^2. \tag{17}$$

So, the control law (16) globally $\mathcal{K}$-exponentially regulates the subsystem (6b).

Theorem 2 *Under Assumption 1, the control law* (12) *and* (16) *ensures the error system* (6) *globally $\mathcal{K}$-exponentially stable, provided that*

$$0 < k_1 < \underline{d}_{11}\underline{d}_{22}/\bar{m}_{11}, k_\psi > 0, k_r > 0, |a_{uv}| \leq \bar{a}_{uv}, |a_{ur}| \leq \bar{a}_{ur}, |a_v| \leq \bar{a}_v, |a_r| \leq \bar{a}_r.$$

4 Simulation

In this section, effectiveness of the proposed control schemes is illustrated by simulation on the underactuated surface vessels with model parameters selected the same as [21]. Let the circular reference trajectory be generated by the model (3) with $\tau_{u_r} = -m_{22}v_r r_r + d_{11}u_r - m_{23}r_r r_r^2, \tau_{r_r} = -m_{22}^{-1}(a_{uv}u_r v_r + a_{u_r}u_r r_r + a_v v_r + a_r r_r), (x_r, y_r, \psi_r, u_r, v_r, r_r)(0) = (0, -10, 0.1703, 0.2, -0.0344, 0.0203).$

We assume that the upper and lower bounds of model parameters are $\underline{m}_{11} = 24, \bar{m}_{11} = 27, \underline{m}_{22} = 32, \bar{m}_{22} = 35, \underline{m}_{33} = 1, \bar{m}_{33} = 4, \underline{m}_{23} = 0.5, \bar{m}_{23} = 2, \underline{d}_{11} = 0.4, \bar{d}_{11} = 2, \underline{d}_{22} = 2, \bar{d}_{22} = 4, \underline{d}_{33} = 0.3, \bar{d}_{33} = 0.7, \underline{d}_{23} = \underline{d}_{32} = -0.3, \bar{d}_{23} = \bar{d}_{32} = -0.2.$ Using the above values, we have $\underline{d}_{11}\underline{d}_{22}/\bar{m}_{11} = 0.0296, \bar{a}_{uv} = 385, \bar{a}_{ur} = 22, \bar{a}_v = 18.5, \bar{a}_r = 25.1, \bar{A} = 139.75.$ Let the initial values of the vessel as $(x, y, \psi, u, v, r)(0) = (0, -13, 0, 0, 0, 0).$ Carry out the control strategy (12) and (16) with $k_1 = 0.02, k_\psi = 0.1, k_r = 0.1.$ To make the computer program run easier, we further replace the $\text{sign}(\cdot)$ by $\tanh(\cdot/0.1)$. Based on the upper and lower bounds of model parameters instead of their real values, the simulation results are shown in Fig. 1. It can be seen that the second control strategy achieves global $\mathcal{K}$-exponential tracking in case of unknown model parameters.

Fig. 1 Geometric path of the vessel under the second control scheme

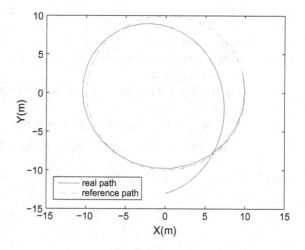

5 Conclusion

In this paper, two control laws are proposed for solving the global $\mathcal{K}$-exponential tracking control problem of underactuated surface vessels with non-diagonal matrices. The second control law is robust with respect to the model parameter. Design of control laws is based on Lyapunov direct method, and especially on the smart construction of Lyapunov function for the position tracking error subsystem. Future research includes the extension of this control scheme to path following control problem of underactuated surface vessels.

Acknowledgements This work was supported by National Nature Science Foundation of China (No. 61573034, No. 61327807), Fundamental and Frontier Research Project of Chongqing (No. cstc2016jcyjA0404), and Fundamental Research Funds for the Central Universities (SWU XDJK2016C038, SWU115046).

References

1. Zheng Z, Sun L. Path following control for marine surface vessel with uncertainties and input saturation. Neurocomputing. 2016;177:158–67.
2. Gao T, Huang J, Zhou Y, Song YD. Robust adaptive tracking control of an underactuated ship with guaranteed transient performance. Int J Syst Sci. 2017;48(2):272–9.
3. Xie W, Ma B. Robust position stabilization of underactuated surface vessels with unknown modeling parameters via simple P/D-like feedback: The center manifold approach. Asian J Control. 2015;17(4):1222–32.
4. Xie W, Ma B. Robust global uniform asymptotic stabilization of underactuated surface vessels with unknown model parameters. Int J Robust Nonlinear Control. 2015;25(7):1037–50.
5. Xie W, Ma B, Fernando T, Iu HHC. A simple robust control for global asymptotic position stabilization of underactuated surface vessels. Int J Robust Nonlinear Control. 2017.
6. Chwa D. Global tracking control of underactuated ships with input and velocity constraints using dynamic surface control method. IEEE Trans Control Syst Technol. 2011;19(6):1357–70.
7. Ghommam J, Mnif F, Derbel N. Global stabilisation and tracking control of underactuated surface vessels. IET Control Theory Appl. 2010;4(1):71–88.
8. Jiang ZP. Global tracking control of underactuated ships by lyapunov's direct method. Automatica. 2002;38(2):301–9.
9. Lefeber E, Pettersen KY, Nijmeijer H. Tracking control of an underactuated ship. IEEE Trans Control Syst Technol. 2003;11(1):52–61.
10. Yu R, Zhu Q, Xia G, Liu Z. Sliding mode tracking control of an underactuated surface vessel. IET Control Theory Appl. 2012;6(3):461–6.
11. Ashrafiuon H, Muske KR, McNinch LC, Soltan RA. Sliding-mode tracking control of surface vessels. IEEE Trans Ind Electron. 2008;55(11):4004–12.
12. Elmokadem T, Zribi M, Youcef-Toumi K. Trajectory tracking sliding mode control of underactuated AUVs. Nonlinear Dyn. 2016;84(2):1079–91.
13. Yan Z, Wang J. Model predictive control for tracking of underactuated vessels based on recurrent neural networks. IEEE J Ocean Eng. 2012;37(4):717–26.
14. Do KD, Jiang ZP, Pan J. Underactuated ship global tracking under relaxed conditions. IEEE Trans Autom Control. 2002;47(9):1529–36.
15. Pettersen KY, Nijmeijer H. Underactuated ship tracking control: theory and experiments. Int J Control. 2001;74(14):1435–46.

16. Do KD, Jiang ZP, Pan J, Nijmeijer H. A global output-feedback controller for stabilization and tracking of underactuated ODIN: a spherical underwater vehicle. Automatica. 2004;40(1):117–24.
17. Do KD, Jiang ZP, Pan J. Universal controllers for stabilization and tracking of underactuated ships. Syst Control Lett. 2002;47(4):299–317.
18. Behal A, Dawson DM, Dixon WE, Fang Y. Tracking and regulation control of an underactuated surface vessel with nonintegrable dynamics. IEEE Trans Autom Control. 2002;47(3):495–500.
19. Wang K, Liu Y, Li L. Vision-based tracking control of underactuated water surface robots without direct position measurement. IEEE Trans Control Syst Technol. 2015;23(6):2391–9.
20. Wu Y, Zhang Z, Xiao N. Global tracking controller for underactuated ship via switching design. J Dyn Syst Meas Control. 2014;136(5):054506.
21. Ma B, Xie W. Global asymptotic trajectory tracking and point stabilization of asymmetric underactuated ships with non-diagonal inertia/damping matrices. Int J Adv Robot Syst. 2013;10(9):336.
22. Zhang Z, Wu Y. Further results on global stabilisation and tracking control for underactuated surface vessels with non-diagonal inertia and damping matrices. Int J Control. 2015;88(9):1679–92.
23. Xie W, Sun H, Ma B. Global asymptotic tracking of asymmetrical underactuated surface vessels with parameter uncertainties. J Control Theory Appl. 2013;11(4):608–14.
24. Park BS, Kwon JW, Kim H. Neural network-based output feedback control for reference tracking of underactuated surface vessels. Automatica. 2017;77:353–9.
25. Lefeber AAJ. Tracking control of nonlinear mechanical systems. Universiteit Twente; 2000.
26. Khalil HK. Nonlinear systems. Upper Saddle River: Prentice Hall; 1996.

A Visualization Method for Hierarchical Structure Information of the Food Inspection Data Based on Force-Directed Model

Qiang Cai, Qiang Li, Haisheng Li and Dianhui Mao

Abstract Food safety supervision and management departments have accumulated a large number of food sampling data, and the data shows hierarchical and high-dimensional characteristics. Because of the lack of appropriate visualization methods, it is difficult to carry out efficient analysis. In this paper, we extracted a hierarchical and high-dimensional dataset from the food inspection data for visualization, and proposed a method based on the force-directed model. Besides, we applied an optimized method to design the interactive mode. Through the visual interaction, the information can be displayed much more efficiently and thus facilitate the understanding of information.

Keywords Food safety · Information visualization · Force-directed model · FR-algorithm

1 Introduction

In recent years, illegal production and management phenomenon exists in China's food enterprises. The public pay high attention to food safety issues. With the promulgation and implementation of the Food Safety Law, China's food safety supervision departments have been increasing the monitoring of food quality. They gain a lot of food sampling data each year. Through the analysis of these data, the government can realize the food safety situation in a certain region for a period of time.

Q. Cai (✉) · Q. Li · H. Li · D. Mao
School of Computer and Information Engineering,
Beijing Technology and Business University, Beijing 100048, China
e-mail: caiq@th.btbu.edu.cn

Q. Li
e-mail: 1105717904@qq.com

Q. Cai · Q. Li · H. Li · D. Mao
Beijing Key Laboratory of Big Data Technology for Food Safety,
Beijing 100048, China

© Springer Nature Singapore Pte Ltd. 2018
Y. Jia et al. (eds.), *Proceedings of 2017 Chinese Intelligent Systems Conference*,
Lecture Notes in Electrical Engineering 460,
https://doi.org/10.1007/978-981-10-6499-9_55

In the past, most of the display of the sampling data is through the form of statistical tables. The statistical tables are tables drawn from vertical and horizontal lines. The statistical data can be very detailed, but it is not conducive to intuitively analyze the problems. Therefore, the statistical chart graph or diagram is necessary, and the bar chart is the most commonly used. The chart can clearly show the quantities of the variables, intuitively reflects the differences in the quantities between variables, but the bar chart can only show two-dimensional data, only for small data sets. With the increase in data dimensions and data sets scale, the bar chart does not suitable for use any more. Due to the reasons above, some novel techniques such as tree-map, radial layout and force-oriented method appeared with the development of high-dimensional data visualization.

The tree-map can show the relationships between nodes. This layout can commonly be understood more easily to the lay-person, but rapidly use screen blank space up. When too much data are generated, the layout can be much more complicated. In the meantime, the visualization of different hierarchies usually raises an additional degree of complexity. The radial layout, which derives from tree-map, is applied to depict interactive system arranging data in an oval-shaped appearance. In comparison with the tree-map, it is more user-friendly. But because of the screen space's restrictions, the radial layout cannot visualize pretty larger hierarchies [1].

The force-oriented layout algorithm proposed by Peter Eades simulates the concept of physical mechanics. Any node in the network graph is gravitational and repulsive. The distance between the nodes will not be too close nor too far due to the gravity and the repulsive force. So the whole network map will be just appropriate. In the force-oriented layout algorithm, the physical model automatically adjusts the node position by the initial position of the given node until a stable condition is reached. In few words, in a force-directed algorithm the graph is represented as a physical system of particles with forces acting between them [2]. Due to the simplicity and the reasonable quality of the layout, this kind of algorithms is one of the most successful methods [3]. So in this paper, for the characteristics of high dimensional and hierarchical structure of food inspection data, we propose a visualization method based on force-directed model to visualize the sampling test data.

2   Visualization Pipelines

In general, the process of data visualization is divided into four phases: the original data phase, the analysis phase, the visualization phase, and the visual view phase. The four phases correspond to data preprocessing, visual mapping, visual transformation, and views (Fig. 1).

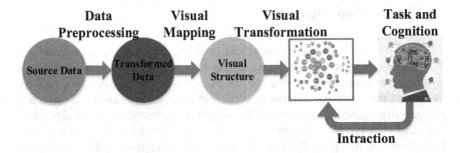

Fig. 1 The visualization pipelines and interaction flow

Data preprocessing refers to a set of activities that is done to make raw data suitable for further processing in the visualization. It is one of the first and critical step to data visualization, and the preprocessed data is directly inputted to visual mapping and obtained the final results. Since analyzing the raw data with these problems may produce misleading results, good efforts of research and development have been devoted to data preprocessing including data cleaning, data integration, data reduction, and data transformation techniques [4–6]

Our food sampling test dataset has nearly one hundred thousand records and more than 100 attributes. It is a huge and complex dataset. So firstly, we have to understand the features of the data. We found that most of the attributes in the dataset are closely related to the food classification. So if the food classification dataset is visualized properly, it can greatly promote the interpretation of the overall data. In this paper, we visualized the sampling test results of each food classification as an example. In this sub-dataset, there are some records that are illegal or unrelated to the attributes, and they were filtered in the data cleaning stage. Most of attributes in the dataset are not important to this purpose, so we firstly dropped them and retained some of the necessary attributes in the data reduction stage. The retained attributes are food classifications and sampling test results. There are many repetitive records to be integrated, and we calculated the amount of each kind of record. Then remove the duplicate records. After that, we transformed the dataset into a json format in order to visualization.

The transformed data is mapped to a hierarchical visual structure due to its character. And each record corresponds to a series node that linked to an edge. In this paper, the algorithm of force-directed model that we applied will be introduced later. However, it is not the end of the data visualization process to generate a visual image. Visual interaction helps to explore the data, resulting in a deep understanding and insight. Troy M. and Moller T. reveals how interactive functions play a unique role in information visualization. A visual view of data is generated by a certain visualization technology, then the user receives it. After the analysis, analysts may raise some questions and give feedbacks to the visualization system, and the system will retrieve the relevant data information through a series of operations.

3 Visualization Algorithm

The force-directed algorithm is one of the most popular methods to generate aesthetically pleasing layouts of graphs in 2D [7]. The objective of force directed algorithms is to start with an initial layout and achieve a final layout for a network by moving the nodes maintaining their connectivity, so as to achieve aesthetically pleasing visualization of the network [8]. At present, there are a group of algorithms that adopt the force-directed method, such as Spring Embedders, Kamada & Kawai, Fruchterman & Reingold, Davidson & Harel. After investigating their advantages and shortcomings by many experiments, G. Yuan analyzed and compared their results on time consumptions and placement styles, finding FR algorithm is better than the others [9].

The FR algorithm proposed by Fmchtenllall and Reingold introduced the force-directed model based on the classic spring algorithm. It is the most typical algorithm. The force-oriented model is based on the theory of particle physics, and the nodes in the undirected graph are simulated into atoms. The positional relationship between nodes is calculated by simulating the force field between atoms. The velocity and acceleration of the nodes are calculated by considering the interaction between atomic forces and repulsive forces. The motion of nodes is similar to that of atoms or planets. After continuous iteration, the system finally enters a dynamic equilibrium state (Fig. 2).

Each iteration of the FR algorithm is divided into three parts: first, the mutual repulsion between the nodes is calculated, and then the attraction between the nodes in the graph is calculated. Finally, the resultant force of the attraction and repulsion is calculated, and limit the movement distance by the maximum displacement. When the two nodes in the same position, FR algorithm will give the two nodes a relatively large repulsion, so that the two nodes can be separated smoothly.

For a display area with height H and width W, where any node v has two layout parameters (component): the position pos of the node and the position offset disp of the resultant force.

The basic definition of FR is as follows:

Display area:

$$area = W * H \tag{1.1}$$

Balance distance:

$$k = \sqrt{\frac{area}{|v|}} \tag{1.2}$$

|v| represents the number of nodes in the graph.

Algorithm : Fruchterman-Reingol

$area \leftarrow W * L$;

initialize $G = (V, E)$;

$k \leftarrow \sqrt{area/|V|}$;

function $f_r(x) = k^2 / x$;

for $i = 1$ to iterations do

    foreach $v \in V$ do

        $v.disp := 0$;

        for $u \in V$ do

            if $(u \neq v)$ then

                $\Delta \leftarrow v.pos - u.pos$;

                $v.disp \leftarrow v.disp + (\Delta / |\Delta|) * f_r(|\Delta|)$;

    function $f_a(x) = x^2 / k$;

    foreach $e \in E$ do

        $\Delta \leftarrow e.v.pos - e.u.pos$;

        $e.v.disp \leftarrow e.v.disp - (\Delta / |\Delta|) * f_a(|\Delta|)$;

        $e.u.disp \leftarrow e.u.disp + (\Delta / |\Delta|) * f_a(|\Delta|)$;

    foreach $v \in V$ do

        $v.pos \leftarrow v.pos + (v.disp/|v.disp|) * \min(v.disp, t)$;

        $v.pos.x \leftarrow \min(W/2, \max(-W/2, v.pos.x))$;

        $v.pos.y \leftarrow \min(L/2, \max(-L/2, v.pos.y))$;

    $t \leftarrow cool(t)$;

Fig. 2 The FR algorithm

Distance between nodes:

$$dist(u, v) = \sqrt{(u.pos_x - v.pos_x)^2 + (u.pos_y - v.pos_y)^2} \tag{1.3}$$

The attractive function between adjacent nodes u and v is:

$$f_a(u, v) = \frac{(dist(u, v))^2}{k} \tag{1.4}$$

The repulsive force function between adjacent nodes u and v is:

$$f_r(u, v) = \frac{k^2}{dist(u, v)}$$ (1.5)

The simulated annealing function involves simulated annealing algorithm. It is used to optimize the layout, but this algorithm is not the focus of this study, we don't do in-depth discussion about it.

The time complexity of the FR algorithm iteration is $O(|V^2\|E|)$.

4 Experiments

At first, we visualized the data by implementing the FR algorithm. The generated network as Fig. 3 shows, some nodes are surrounded by other nodes, so the central nodes represent sampling test results, such as qualified items, unqualified items, unchecked items, and other nodes represent the food classifications. Although the network shows some clustered characteristic of the dataset, it is too mass and confusing. As the consequence of quickly growing of the nodes quantity, how to get large-scale graphs with explicit expressions of its structures and data is a

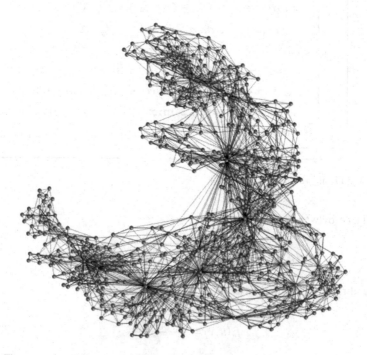

Fig. 3 The network graph generated by the FR algorithm

challenge. The central problem here is not just to provide a comprehensive presentation of large graphs for users on the screen, but also a user-friendly navigable visual structure for users scanning through the structure to discover a particular detail of the data [10], so we consider to apply an optimized algorithm based on the FR algorithm as Fig. 4 shows.

As Fig. 4 shows, the colors represent different category levels while the area of terminal nodes represent the size of data. The highest level of the category is in the center of the graph and connects to the lower level categories. The blue or white nodes represent the food sampling test results, when click them, it unfolds next level nodes. The yellow nodes represent the lowest level food classification. This kind of graph is much better to express the information of the dataset compared with Fig. 3. Based on this method and the preprocessed data, we employed a better visual layout as displayed in Fig. 5.

In Figs. 5 and 6, different food categories are distinguished by colors, and different sizes of nodes mean different amount of records in dataset. We set the result nodes as terminal nodes. Users could explore the information by clicking the nodes. For example, when users click the central node, it will unfold the highest level nodes, and they represent categories such as egg products, meat products, drink etc. Then click the drink node, it pops the lower level categories nodes. When click the fruit juice node, it pops the result nodes. Among them, the node labeled by qualified is much bigger than the other two. That means qualified fruit juice products account for the vast majority compared with the unqualified ones. By this kind of interaction, users could execute many interesting tasks, such as exploring the categories of tested food and the corresponding results, comparing the scale of tested food records, and presenting the results in a lively way etc.

Fig. 4 The network graph and interaction mode

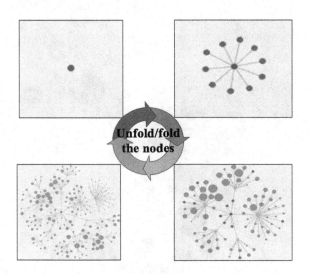

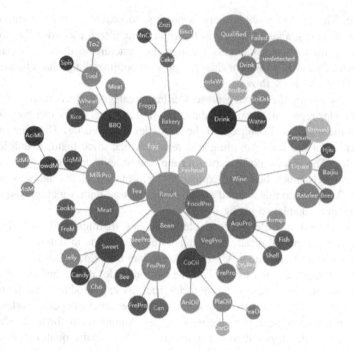

Fig. 5 The visualization generated by the improved method

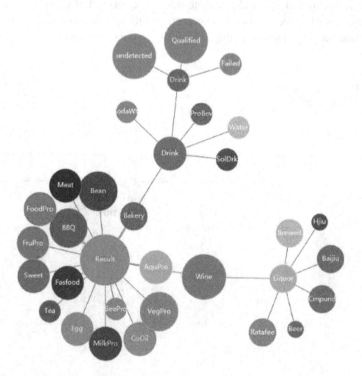

Fig. 6 The sampling test result nodes are set to be terminal nodes

5 Conclusion and Future Work

Force directed algorithms for drawing graphs have a long history. Their intuitive simplicity appeals to researchers from many different fields, and this accounts for dozens of available implementations. If we just use the normal FR algorithm, we can only get the clustered information about the sampling test results. With the hierarchical structure, this algorithm cannot visualize the dataset very well, instead, we can visualize the optimal hierarchical structure if we use multilayer method. On this basis, we designed the interaction mode; this may improve the operational experience greatly.

We have done first steps by implementing the optimized FR algorithm and designing interactive mode to visualize the subset of food sampling test dataset. Our future works might consider multiple algorithms and visualization methods to visualize the food sampling test dataset, so that multiple visualization methods can be integrated to help analyzing the dataset.

Acknowledgements This work was partially supported by Beijing Science and Technology Project (No. Z161100001616004) and Beijing Natural Science Foundation (4162019).

References

1. Gan Z, Li N, Ma Y, Lu H. Trust network visualization based on force-directed layout. In: 2013 10th Web Information System and Application Conference. Yangzhou;2013. p. 199 −204.
2. Debiasi A, Simões B, De Amicis R. Force directed flow map layout. In: 2014 International Conference on Information Visualization Theory and Applications (IVAPP). Lisbon: Portugal;2014. p. 170−7.
3. Toosi and FG, Nikolov NS. Vertex-neighboring multilevel force-directed graph drawing. In: 2016 IEEE International Conference on Systems, Man, and Cybernetics (SMC). Budapest;2016. p. 002996−3001.
4. Han J, Kamber M, Pei J. Data preprocessing in data mining concepts and techniques. Morgan: Kaufmann;2012. p. 83−124.
5. García S, Luengo J, Herrera F. Data preprocessing in data mining, vol. 72. Cham: Springer International Publishing; 2015. p. 5.
6. scikit-learndevelopers,'4.3 Preprocessing data'. http://scikit-learn.org/stable/modules/ preprocessing.html. [Accessed on 13-June-2016].
7. Kobourov GS. Force-directed drawing algorithms. In: Roberto Tamassia, editor, Handbook of Graph Drawing and Visualization, volume 81 of Discrete Mathematics and Its Applications, chapter 12. Chapman and Hall/CRC;2013. p. 383−408.
8. Teja SC, Yemula PK. Power network layout generation using force directed graph technique. In: 2014 Eighteenth National Power Systems Conference (NPSC), Guwahati;2014. p. 1−6
9. Yuan G, Dancheng L, Chunyan H, Zhiliang Z. An improved network topology auto-layout solution based on force-directed placement. In: 2009 Ninth International Conference on Hybrid Intelligent Systems, Shenyang; 2009. p. 10−14.
10. Hua J, Huang ML, Nguyen QV. Drawing large weighted graphs using clustered force-directed algorithm. In: 2014 18th International Conference on Information Visualisation, Paris;2014. p. 13−17.

Simulation of Wind Farm Scheduling Algorithm Based on Predictive Model Control

Zhixiang Luo, Shanbi Wei, Yi Chai, Yanxing Liu and Xiuling Sun

Abstract In this paper, a simulation study is carried out on the coordinated active and reactive dispatch of wind farm based on Model Predictive Control (MPC). Compared with the traditional scheduling algorithm, this method combines the active and reactive scheduling, and balances the active and reactive target by changing the weight coefficients of the cost functions. The proposed control method can not only track the scheduling instructions accurately and smoothly, but also can make fluctuations of bus' voltage in the wind farm stable. Therefore the anti-perturbation capability of wind farm system is improve. A simulation model is built in MATLAB/SIMULINK environment, and the results show that the MPC is effective for wind power dispatch.

Keywords Wind farm · Power dispatch · Coordinated active · Reactive · MPC

1 Introduction

In the case of conventional wind farm control methods, the independent control strategy of active or reactive power is mainly studied by researchers, while the coupling effect is ignored [1]. As for the scheduling strategy of active power, the power generation task of the Wind Turbine Generators (WTGs) are arranged on the condition of considering the constraints of the WTGs' own characteristics, so that the output of wind farm can track the dispatching instruction quickly and accurately. Furthermore, the strategy based on the principle of reducing the regulation frequency of the actuators is put forward in [2, 3], to alleviate the fatigue loads of the WTGs. Reactive power control is mainly researched on how to suppress the voltage fluctuation by controlling various voltage regulation devices in the wind

Z. Luo (✉) · S. Wei · Y. Chai · Y. Liu · X. Sun
State Key Laboratory of Power Transmission Equipment and System Security
and New Technology, College of Automation, Chongqing University,
Chongqing 400044, China
e-mail: 20115033@cqu.edu.cn

© Springer Nature Singapore Pte Ltd. 2018
Y. Jia et al. (eds.), *Proceedings of 2017 Chinese Intelligent Systems Conference*,
Lecture Notes in Electrical Engineering 460,
https://doi.org/10.1007/978-981-10-6499-9_56

farm [4]. Further, a three-layer reactive power allocation method is proposed in [5], which takes into account of Static Var Compensator (SVC), Doubly-fed wind turbines (DFIG) and the internal converter of DFIG comprehensively.

In recent years, the Model Predictive Control (MPC) is favored in the wind power control research due to its superior ability of solving multi-constrained problems online [6]. The authors of [7] have applied the MPC to the active control and realized the smooth output of wind power. Meanwhile, the predictive control method that link wind power reactive power compensator to output voltage management is established in [8], which realizes the rolling optimization control of system voltage. However, these approaches above listed are separate considerations of active and reactive power control, there is little literatures involved in the coordination control of active and reactive power.

However, the reactive power output capability of the WTGs is limited by the converters capacity, and the reactive output ability is determined by the terminal voltage and the active output power [9]. With the WTG running close to full load, the Var capacity will be close to zero. Additionally, the ratio of the reactance to the resistance of the medium voltage (MV) feeders in the wind farm is small, which means the influence of the active power on the voltage can not be ignored. In this paper, the MPC is applied to solve the scheduling problem of wind farm. According to comprehensively taking into account of the active/reactive power output capability of WTGs, as well as the compensation capability of SVC, this approach can improve the voltage control performance of wind farm effectively, and reduce the fatigue loads of WTGs obviously.

2   Structure of Control System

The wind farm shown in Fig. 1, which adopts the radiation topology and includes I lines. There is a total of $N_{wt} \left(N_{wt} = \sum_{i=1}^{I} J_i \right)$ WTGs with the ith line linking J_i WTGs. Each WTG is connected to the feeder line through the unit transformer, and then connected to the low voltage side of the main transformer station. After the boosting of the main transformer station, the wind power injects power grid by the transmission line.

Figure 2 shows the block diagram of the wind farm control structure based on MPC. The wind farm is regarded as a general control object which consisted of N_{wt} WTGs and a SVC. The wind farm controller receives the dispatching command of the system operator (including the active power reference P_{wf}^{ref}, and the voltage reference at the Point of Connection (POC), V_{poc}^{ref}), and the operation state information of the wind farm from sensors. Based on the above information, the MPC calculates the active power and reactive power command of each WTG ($P_{wt}^{ref}, Q_{wt}^{ref}$) and the reference voltage of the SVC (V_s^{ref}). Then those commands are transmitted

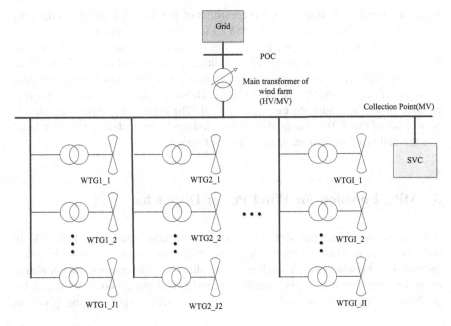

Fig. 1 Configure of a wind farm

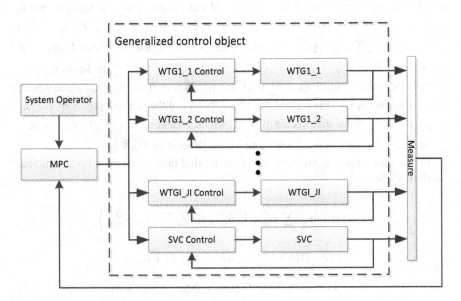

Fig. 2 The block diagram of the wind farm control structure

the controller of the bottom layer (the controller of WTGs/SVC), and the underlying controller will direct actuator (WTGs/SVC) to follow the instruction.

On the basis of the above-mentioned control frame, the power generation task of the system operator is completed by coordinating the control of WTG and SVC. Generally, SVC has two control models, constant voltage control mode and constant power control mode. The first mode is chosen so that it can quickly provide reactive power to adjust the bus voltage [10]. The control objectives include the service life of the WTGs, the quality of the wind power generated by the wind farm, and the ability of wind farm to resist the disturbance.

3 MPC Function for Wind Power Dispatch

This section will build the MPC function of wind farm scheduling. The MPC of wind power scheduling is time-varying since the prediction model depends on the operation of the wind farm. Therefore, the prediction model needs to be updated at every forecast period t_p. If the sampling period t_s is divided into n_s steps and the prediction cycle is completed by n_p steps, the relationship can be given as $t_s n_p = t_p n_s$.

Selecting the active/reactive power reference of the WTGs and voltage reference of SVC as input variables of the generalized object, therefore, the input vector is defined as $u \triangleq [P_{wt}^{ref'}, Q_{wt}^{ref'}, V_s^{ref'}]'$, where $P_{wt}^{ref} = [P_{wt\_1}^{ref}, P_{wt\_2}^{ref}, \ldots, P_{wt\_N_{wt}}^{ref}]'$ and $Q_{wt}^{ref} = [Q_{wt\_1}^{ref}, Q_{wt\_2}^{ref}, \ldots, Q_{wt\_N_{wt}}^{ref}]'$. The state vector of the generalized object is defined as $x \triangleq [x_{wt\_1}^{p'}, \ldots, x_{wt\_N_{wt}}^{p'}, Q_{wt\_1}, \ldots, Q_{wt\_N_{wt}}, x_s']'$, and x_{wt}^p is the state vector of WTG's active control loop, $x_{wt}^p \triangleq [\theta, w_r, w_f]$. Where θ, w_r, w_f, Q_{wt} are the pitch angle, the rotor speed, the filtered generator speed and the reactive power in WTGs, respectively. x_s is the state vector of SVC, which is defined as $x_s \triangleq [Q_s, V_{int}]'$, where Q_s are the reactive power of SVC and V_{int} is defined as $V_{int} \triangleq \frac{(V_s^{ref} - V_s)}{s}$.

Furthermore, the discrete MPC function of wind farm scheduling is constructed, as shown below,

$$F_{\cos t} = \min_u \left(\sum_{k=1}^{n_p} F_{\cos t}^P(k) W_{\cos t}^P + \sum_{k=1}^{n_p} F_{\cos t}^Q(k) W_{\cos t}^Q \right) \tag{1}$$

$$s.t. \quad x(k+1) = A_{t_0} x(k) + B_{t_0} u(k) + E_{t_0}; \tag{2}$$

$$x(k+1) \in [x_{\min}, x_{\max}] \cap [x(k) - \Delta x_{\lim}, x(k) + \Delta x_{\lim}]; \tag{3}$$

$$u(k) \in [u_{\min}(k), u_{\max}(k)]; \tag{4}$$

$$u(i \cdot n_s + k) = u(i \cdot n_s), i \in [0, n_p - 1], k \in [0, n_s - 1]; \tag{5}$$

$$\sum_{i=1}^{n_{wt}} P_{wt\_i}^{ref} = P_{wf}^{ref}; \tag{6}$$

A. Cost Function

As shown in Eq. (1), the cost function of MPC is divided into active control part and reactive power control part. For active power scheduling, reducing the fatigue load of WTG to extend the service life of WTG on the basis of ensuring accurately and smoothly tracking of power command. As the gearbox is the most vulnerable parts of the WTG, this paper deduces the fatigue load of the WTG by reducing the action of the shaft torque T_s. According to [11], the following expression is yielded.

$$\begin{aligned} F_{\cos t}^P(x_{wt}(k), P_{wt}^{ref}(k)) &= \left\| T_s(k) - T_s^0 \right\|^2 \\ &= \left\| A_{T_{s0}} x_{wt}(k) + B_{T_{s0}} P_{wt}^{ref}(k) + E_{T_{s0}} - T_s^0 \right\|^2 \end{aligned} \tag{7}$$

The matrixes above refer to [7].

The purpose of reactive power scheduling is to minimize the deviation between the bus voltage and the reference voltage ($\Delta V_{poc}, \Delta V_{wt}$). At the same time, in order to enhance the anti-load disturbance capability of the wind farm, The fast reactive capacity of the wind farm is achieved by minimizing the deviation between Q_s and its intermediate value Q_s^{mid}.

$$\begin{aligned} F_{\cos t}^Q(k) &= \left\| V_{poc}(k) - V_{poc}^{ref} \right\|^2 W_{\cos t\_poc}^Q \\ &+ \left\| V_{wt}(k) - V_{wt}^{ref} \right\|^2 W_{\cos t\_wt}^Q + \left\| Q_s^{ref} - Q_s^{mid} \right\| W_{\cos t\_s}^Q \end{aligned} \tag{8}$$

This paper considers the effect of the active power on the voltage. So the bus voltage ($V_{poc}(k), V_{wt}(k)$) are the function of $P_{wt}^{ref}, Q_{wt}^{ref}, Q_s^{ref}$, which can be processed approximately by First-order Taylor equation.

The active and reactive scheduling targets are balanced by selecting different weight values ($W_{\cos t}^P, W_{\cos t}^Q, W_{\cos t\_poc}^Q, W_{\cos t\_wt}^Q, W_{\cos t\_s}^Q$). Normally $W_{\cos t}^P$ is the maximum. When $\Delta V_{poc}, \Delta V_{wt}$ is too large, choosing larger $W_{\cos t\_poc}^Q, W_{\cos t\_wt}^Q$ to recover the bus voltage quickly.

B. Constraints

In the model constraints, the formula (2) represents the transfer function constraints, which will be discussed in Part C. The formula (3) represents the state constraints of the actuator, as follow,

$$\theta(k+1) \in [\theta_{\min}, \theta_{\max}] \cap [\theta(k) - \Delta\theta_{\lim}, \theta(k) + \Delta\theta_{\lim}]; \qquad (9)$$

$$\omega_r(k) \in [\omega_r^{\min}, \omega_r^{\max}]; \qquad (10)$$

$$Q_s(k) \in [Q_s^{\min}, Q_s^{\max}]; \qquad (11)$$

The formula (4) donate the constraints on the actuator input, as follow,

$$P_{wt}^{ref}(k+1) \in [0, P_{wt}^{avi}(k)] \cap [P_{wt}(k) - \Delta P_{\lim}, P_{wt}(k) + \Delta P_{\lim}]; \qquad (12)$$

$$Q_{wt}^{ref}(k) \in [Q_{wt}^{\min}(k), Q_{wt}^{\max}(k)]; \qquad (13)$$

$$V_s^{ref} \in [V_s^{\min}, V_s^{\max}]; \qquad (14)$$

The Eq. (5) donate the sampling constraint, which means the input variable should keep constant values during the sampling period. The Eq. (6) means that the wind power output should track the task.

C. Generalized Object Modeling

The generalized control object model includes WTG and SVC model. The input vector and state vector of model are defined by,

$$u(k) = [P_{wt\_1}^{ref}(k), \ldots, P_{wt\_N_{wt}}^{ref}(k), Q_{wt\_1}^{ref}(k), \ldots, Q_{wt\_N_{wt}}^{ref}(k), V_s^{ref}(k)]'; \qquad (15)$$

$$x(k+1) = \left[x_{wt\_1}^{p}(k)', \ldots, x_{wt\_N_{wt}}^{p}(k)', Q_{wt\_1}(k), \ldots, Q_{wt\_N_{wt}}(k), x_s(k)'\right]'; \qquad (16)$$

Then the discrete state space is expressed as

$$x(k+1) = A_{t_0}x(k) + B_{t_0}u(k) + E_{t_0} \qquad (17)$$

The state matrices are,

$$A_{t_0} = \begin{bmatrix} A_{wt0}^{p\_set} & 0 & 0 \\ 0 & A_{wt}^{q\_set} & 0 \\ 0 & F_{s0} & A_{s0} \end{bmatrix}, B_{t_0} = \begin{bmatrix} B_{wt0}^{p\_set} & 0 & 0 \\ 0 & B_{wt}^{q\_set} & 0 \\ E_{s0} & 0 & B_{s0} \end{bmatrix}, E_{t_0} = \begin{bmatrix} E_{wt0}^{p\_set} \\ E_{wt0}^{p\_set} \\ G_{s0} \end{bmatrix}$$

Here, $(A_{wt0}^{p\_set} \ B_{wt0}^{p\_set} \ E_{wt0}^{p\_set})$ donates the active control loop of WTG, $(A_{wt}^{q\_set} \ B_{wt}^{q\_set})$ donates the reactive control loop of WTG, $(A_{s0} \ B_{s0} \ E_{s0} \ F_{s0} \ G_{s0})$ donates the control loop of SVC.

Where $A_{wt0}^{p\_set}$ composed by all the WTGs' active control loop state transition matrices is the block diagonal matrix. And $B_{wt0}^{p\_set}, E_{wt0}^{p\_set}, A_{wt0}^{q\_set}, B_{wt0}^{q\_set}$ have the similar structure.

(1) Active control loop of WTG: In [12, 13], a simplified and linear WTG model is proposed by the National Renewable Energy Laboratory (NREL), which defines

the state vector as $x_{wt\_p}(k) \triangleq [\theta(k), \omega_r(k), \omega_f(k)]$. Then the discrete state space model are,

$$x_{wt\_p}(k+1) = A^{t0}_{wt\_p} x_{wt\_p}(k) + B^{t0}_{wt\_p} P^{ref}_{wt}(k) + E^{t0}_{wt\_p} \tag{18}$$

The state space matrices are,

$$A_{wt\_p0} = \begin{bmatrix} 0 & -\frac{\eta_g K_{p\_\theta}}{\tau_g} & \frac{K_{p\_\theta}}{\tau_g} - K_{i\_\theta} \\ \frac{K_{\theta T_r}}{J_t} & \frac{K_{w_r T_r}}{J_t} + \frac{P^0_{wt} \eta_g}{J_t \mu \omega_g^{0^2}} & 0 \\ 0 & \frac{\eta_g}{\tau_g} & -\frac{1}{\tau_g} \end{bmatrix},$$

$$B_{wt\_p0} = \begin{bmatrix} 0 \\ -\frac{\eta_g}{J_t \mu \omega_g^0} \\ 0 \end{bmatrix}, E_{wt\_p0} = \begin{bmatrix} 0 \\ \frac{K_{v_w T_r} v_w^0}{J_t} \\ 0 \end{bmatrix}$$

where $K_{\theta T_r}, K_{w_r T_r}, K_{v_w T_r}$ are the first-order Taylor expansion coefficients of the rotor torque T_r. J_r, J_g, J_t are the wind turbine inertia, the generator inertia and the equivalent inertia (where $J_t = J_r + J_g \eta_g^2$), respectively. η_g, τ_g, μ are the gearbox ratio, the time coefficient of generator speed filter and the generation efficiency, respectively. $P^0_{wt}, \omega_g^0, T_s^0, v_w^0$ mean the power output, generator speed, the shaft torque and the wind speed at the forecast time point, respectively. $K_{p\_\theta}, K_{i\_\theta}$ represent the coefficient in PI controller of the pitch angle.

(2) Reactive control loop of WTG: The authors of [14] have pointed out that the dynamic characteristics of the reactive power control loop are in the form of a first order function when the WTG operates in constant-Q mode. Defining the state vector as $x_{wt\_q}(k) \triangleq Q_{wt}(k)$, then the discrete state space model will be obtained as follow, where τ_q is the time constant of the reactive control loop.

$$\begin{aligned} x_{wt\_q}(k+1) &= -\frac{1}{\tau_q} x_{wt\_q}(k) + \frac{1}{\tau_q} Q^{ref}_{wt}(k) \\ &= A_{wt\_q} x_{wt\_q}(k) + B_{wt\_q} Q^{ref}_{wt}(k) \end{aligned} \tag{19}$$

(3) Control loop of SVC: In [15], The authors have proved that the dynamic characteristics of the SVC are described by a first order function when the SVC operates in constant-Q mode. Q^{ref}_s is obtained by V^{ref}_{poc} after a PI controller. As follows,

$$\begin{aligned} Q_s &= \frac{1}{1 + s\tau_s} Q^{ref}_s \\ &= \frac{1}{1 + s\tau_s} \left[Q^0_s + K_{p\_s}(V^{ref}_{poc} - V_{poc}) + K_{i\_s} \frac{(V^{ref}_{poc} - V_{poc})}{s} \right] \end{aligned} \tag{20}$$

where K_{p_s}, K_{i_s} donate the coefficients in PI controller of the SVC. τ_s is the time constant. Defining the state vector as $x_{wt\_q}(k) \triangleq [Q_s(k), V_{int}(k)]'$, and processing approximately V_{poc} in the forecast time point, then the discrete state space model of SVC will be obtained as follow,

$$x_s(k+1) = A_{s0}x_s(k) + B_{s0}V_{poc}^{ref}(k) + E_{s0}P_{wt}(k) + F_{s0}Q_{wt}(k) + G_{s0} \quad (21)$$

The matrices refer to [15].

4 Case Study

The experiment is carried out in the MATLAB/SIMULINK, and the model of the wind farm is built by NREL's SimWindFarm toolkit, which is a toolbox for dynamic wind farm model. The wind field model considers wake and turbulence, and the wind turbine model is full power conversion.

The simulation model is shown in Fig. 3. The voltage of the bus are calculated by the power flow calculation. Through the SIMULINK to build S-function to complete the power flow calculation. The MPC function is also accomplished by writing S-function in SIMULINK. The MPC problem of this study is a standard quadratic programming problem, which can be solved by the toolbox YALMIP quickly and accurately. In this case, the simulation time is set to 600 s. The sampling period is 1 s, with $n_s = 5$, and the prediction period is set to 5 s.

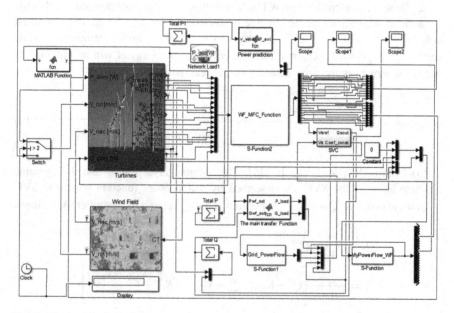

Fig. 3 The model of simulation

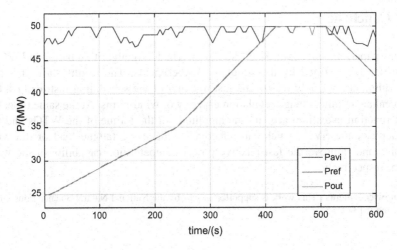

Fig. 4 Active power curve of wind farm

5 Analysis of Experimental Results

The active power reference P^{ref}, the maximum available power generation P^{avi}, and the power output P^{out} are shown in Fig. 4. It can be seen that whether the power is increased or reduced, P^{out} can track P^{ref} quickly and accurately, as long as P^{ref} is less than P^{avi}. During the periods 420–510 s, when the system operator requires that the wind farm to generate electricity at the maximum capacity, P^{out} can not completely follow P^{ref} due to the limit of the maximum capacity.

The bus voltages at POC (V_{poc}) and the 5th node (V_{wt_5}) are shown in Fig. 5. Obviously, there are deviations in both voltages, but the differences are within the acceptable range. It is obvious that the maximum of ΔV_{poc} is less than the maximum of ΔV_{wt_5}, which is because $W^{Q}_{cos\,t\_poc}$ is larger than $W^{Q}_{cos\,t\_wt}$ in the cost function.

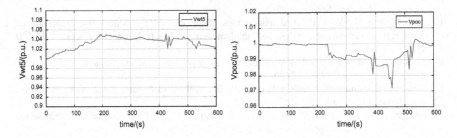

Fig. 5 Voltage at the 5th Node and the POC

6 Conclusion

In this paper, the effectiveness of the power scheduling algorithm based on MPC in wind farm is verified by the simulation experiments. The results show that this algorithm can not only realize the fast tracking of the scheduling instructions, but also enhancing the voltage regulation capacity of wind farms. At the same time, the cost function takes into account the amplitude of the torque of the WTGs and the capacity of the SVC, which can effectively reduce the fatigue load of the wind turbine and improve the fast reactive power compensation capability of the wind farm, respectively.

Acknowledgements This work is supported in part by the National Natural Science Foundation of China (61374135).

References

1. Knudsen T, Bak T, Svenstrup M. Survey of wind farm control—power and fatigue optimization. Wind Energy. 2015;18(8):1333–51.
2. Xue Y, Yu C, Zhao J, et al. A review on short-term and ultra-short-term wind power prediction. Autom Electr Power Syst. 2015;39(6):141–51.
3. Zhang B, Wu W, Zheng T, et al. Design of a multi-time scale coordinated active power dispatching system for accommodating large scale wind power penetration. Autom Electr Power Syst. 2011;35(1):1–6.
4. Yang S, Wang WS, Liu C, et al. Coordinative control strategy for reactive power and voltage of wind farms with doubly-fed induction generators. Autom Power Syst. 2013;37(12):1–6.
5. Li R, Tang F, Liu Y, et al. A new scheme of reactive power compensation and voltage for DFIG based wind farm. Proc CSEE. 2012;32(19):16–23.
6. Xi Y, Li D, Lin S. Model predictive control—status and challenges. Acta Automatica Sinica. 2013;39(3):222–36.
7. Khalid M, Savkin AV. A model predictive control approach to the problem of wind power smoothing with controlled battery storage. Renew Energy. 2010;35(7):1520–6.
8. Pathak AK, Sharma MP, Bundele M. A critical review of voltage and reactive power management of wind farms. Renew Sustain Energy Rev. 2015;51:460–71.
9. Amaris H, Alonso M, Ortega CA. Reactive power management of power networks with wind generation. London: Springer; 2013.
10. Guo Q, Sun H, Wang B, et al. Hierarchical automatic voltage control for integration of large-scale wind power: design and implementation. Electr Power Syst Res. 2015;120 (120):234–41.
11. Spudić V, Baotić M, Jelavić M. Wind turbine power references in coordinated control of wind farms. Automatika. 2011;52(2):82–94.
12. Grunnet JD, Soltani M, Knudsen T, et al. Aeolus toolbox for dynamics wind farm model. In: Simulation and control; 2010.
13. Jonkman J, Butterfield S, Musial W, et al. Definition of a 5-MW reference wind turbine for offshore system development. Off Sci Tech Inf Tech Rep. 2009.

14. Soens J, Driesen J, Belmans R, et al. Equivalent transfer function for a variable-speed wind turbine in power system dynamic simulations. Int J Distrib Energy Resour. 2005;1:111–33.
15. Peng FZ, Lai JS. Dynamic performance and control of a static VAr generator using cascade multilevel inverters. IEEE Trans Ind Appl. 1996;33(3):748–55.

A Hybrid Discrete Artificial Bee Colony Algorithm for Multi-objective Blocking Lot-Streaming Flow Shop Scheduling Problem

Dunwei Gong, Yuyan Han and Jianyong Sun

Abstract A blocking lot-streaming flow shop (BLSFS) scheduling problem involves in splitting a job into several sublots and no capacity buffers with blocking between adjacent machines. It is of popularity in real-world applications but hard to be effectively solved in light of many constrains and complexities. Thus, the research on optimization algorithms for the BLSFS scheduling problem is relatively scarce. In view of this, we proposed a hybrid discrete artificial bee colony (HDABC) algorithm to tackle the BLSFS scheduling problem with two commonly used and conflicting criteria, i.e., makespan and earliness time. We first presented three initialization strategies to enhance the quality of the initial population, and then developed two novel crossover operators by taking full of valuable information of non-dominated solutions to enhance the capabilities of HDABC in exploration. We applied the proposed algorithm to 16 instances and compared with three previous algorithms. The experimental results show that the proposed algorithm clearly outperforms these comparative algorithms.

Keywords Lot-streaming flow shop · Blocking · Multi-objective optimization
Artificial bee colony · Pareto local search

D. Gong
School of Information and Electrical Engineering, China University of Mining
and Technology, Xuzhou 221116, China
e-mail: dwgong@vip.163.com

Y. Han(✉)
School of Computer Science, Liaocheng University, Liaocheng 252000, China
e-mail: yuyanhan1023@163.com

J. Sun
School of Mathematics and Statistics, Xian Jiaotong University, Xian 710049, China
e-mail: jy.sun@xjtu.edu.cn

© Springer Nature Singapore Pte Ltd. 2018
Y. Jia et al. (eds.), *Proceedings of 2017 Chinese Intelligent Systems Conference*,
Lecture Notes in Electrical Engineering 460,
https://doi.org/10.1007/978-981-10-6499-9_57

1 Introduction

The LSFS scheduling problem can be generally classified into two categories according to whether there are buffers or not, namely, the one with infinite buffers and the one with finite buffers. The former does not result in blocking any job since there are enough intermediate buffers used to store those completed jobs; whereas, the latter maintains a limited capacity of in-process inventories, that is, either there is no buffers or the buffers with a limited capacity due to finite storage facilities. The LSFS scheduling problem with no intermediate buffers, named the blocking LSFS scheduling problem, is a special case of the latter, where a sublot has to be remained or blocked in the current machine until the downstream ones are available.

Several methods of minimizing the criterion of makespan in the LSFS scheduling problem have provided excellent results in the acceptable computation time. In [4], the authors demonstrated that utilizing the LSFS system can lead to potential benefits by evaluating such criteria as makespan and the average flow time. Zhang et al. [14] proposed two heuristic algorithms to minimize the average completion time in a multi-job lot-streaming scheduling problem with two-stage hybrid flow shops. Marimuthu et al. [5] developed the Ant Colony Optimization (ACO) algorithm and the threshold accepting (TA) algorithm to optimize makespan and the total flow time of the LSFS scheduling problem, respectively. Martin [6] presents a hybrid genetic algorithm (GA)/mathematical programming heuristic to solve the LSFS scheduling problem. Recently, estimation distribution algorithm (EDA) proposed in 1966 [7] based on the job permutation was developed to minimize makespan of the LSFS scheduling problem with the sequence-dependent setup time [8].

Both earliness and tardiness time criterion reflect strict requirements between the due date and the completion time, which embodies the users satisfaction. With the advent and development of the just-in-time manufacturing system, much progress has been made in the LSFS scheduling problem with minimizing the earliness time and tardiness time, such as [12]. In [12], a new genetic algorithm (NGA) was employed to solve the LSFS scheduling problem with a limited capacity of buffers, whose objective is to minimize the total earliness and tardiness time.

The rest of this paper is organized as follows. In Sect. 1, the literature review of related work about the basic ABC algorithm is stated. Section 2 formulates the mathematical model of the BLSFS scheduling problem. The proposed algorithm is presented in Sect. 3. Section 4 provides the experimental results. Finally, the paper ends with some conclusions in Sect. 5.

2 Mathematical Model of BLSFS Scheduling Problem

In this section, we build the mathematical model of the BLSFS scheduling problem. In the BLSFS scheduling problem, a job is split into several sublots, with each having different processing time on different machines. Due to lacking of intermediate buffer storages between adjacent machines, a sublot has to be remained in the current machine until the downstream one is available. Thus, the blocking time should be included in the start time of each sublot. In addition, any two adjacent sublots of job i allow idle time at the same machine, and both the setup and the sublot transportation time are included in the processing time. We supposed that there are n jobs represented by $\pi(1), \pi(2), ..., \pi(n)$ and m machines, and denote the sequence formed by these jobs as $\pi = \{\pi(1), \pi(2), ..., \pi(n)\}$, where each job is processed on these machines in the same order. Let $l_{\pi(i)}$ be the total number of sublots in job $\pi(i)$, and e be its eth sublot with $e = 1, 2, ..., l_{\pi(i)}$. Beside, $S_{\pi(i)t,e}$ and $C_{\pi(i)t,e}$ are the start and the completion time of the eth sublot in job i on machine t, respectively. The optimization functions of this problem can be formulated as follows:

$$\begin{cases} S_{\pi(1),1,1} = 0 \\ C_{\pi(1),1,1} = S_{\pi(1),1,1} + P_{\pi(1),1} \end{cases} \tag{1}$$

$$\begin{cases} S_{\pi(1),t,1} = C_{\pi(1),t-1,1} \\ C_{\pi(1),t,1} = S_{\pi(1),t,1} + P_{\pi(1),t} \\ t = 2, 3, ...m \end{cases} \tag{2}$$

$$\begin{cases} S_{\pi(i),1,1} = \max\{C_{\pi(i-1),1,l_{\pi(i-1)}}, S_{\pi(i-1),2,l_{\pi(i-1)}}\} \\ C_{\pi(i),1,1} = S_{\pi(i),1,1} + P_{\pi(i),1} \\ i = 2, 3, ..., n \end{cases} \tag{3}$$

$$\begin{cases} S_{\pi(i),t,1} = \max\{C_{\pi(i),t-1,1}, S_{\pi(i-1),t+1,l_{\pi(i-1)}}\} \\ C_{\pi(i),t,1} = S_{\pi(i),t,1} + P_{\pi(i),t} \\ t = 2, 3, ..., m - 1 \\ i = 2, 3, ..., n \end{cases} \tag{4}$$

$$\begin{cases} S_{\pi(i),m,1} = \max\{C_{\pi(i),m-1,1}, C_{\pi(i-1),m,l_{\pi(i-1)}}\} \\ C_{\pi(i),m,1} = S_{\pi(i),m,1} + P_{\pi(i),m} \\ e = 2, 3, ..., l_{\pi(i)} \\ i = 2, 3, ..., n \end{cases} \tag{5}$$

$$\begin{cases} S_{\pi(i),1,e} = \max\{C_{\pi(i),1,e-1}, S_{\pi(i),2,e-1}\} \\ C_{\pi(i),1,e} = S_{\pi(i),1,e} + P_{\pi(i),1} \\ e = 2, 3, ..., l_{\pi(i)} \\ i = 1, 2, 3, ..., n \end{cases} \tag{6}$$

$$\begin{cases} S_{\pi(i),t,e} = \max\{C_{\pi(i),t-1,e}, S_{\pi(i),t+1,e-1}\} \\ C_{\pi(i),t,e} = S_{\pi(i),t,e} + P_{\pi(i),t} \\ e = 2, 3, ..., l_{\pi(i)} \\ t = 2, 3, ..., m - 1 \\ i = 1, 2, 3, ..., n \end{cases} \tag{7}$$

$$\begin{cases} S_{\pi(i),m,e} = \max\{C_{\pi(i),m-1,e}, C_{\pi(i),m,e-1}\} \\ C_{\pi(i),m,e} = S_{\pi(i),m,e} + p_{\pi(i),m} \\ e = 2, 3, ..., l_{\pi(i)} \\ i = 1, 2, 3, ..., n \\ t = m \end{cases} \tag{8}$$

In the above formula, Eqs. 1, 2 give the completion time of the first sublot of the first job on m machines, respectively. Equation 3 compute the completion time of the first sublot in job i on the first machine. Equations 4, 5 states the completion time of the first sublot in job i on machine t. Equation 6 computes the completion time of the eth sublot in job i on the first machine. Equations 7, 8 gives the completion time of the eth sublot in job i on machine $t(t = 2, 3, ..., m)$. With respect to Eqs. 3–8, they reflect the blocking of adjacent sublots and adjacent jobs on machines.

$$\min f_1(\pi) = \min(C_{\max}(\pi)) = C_{\pi(n),m,l_{\pi(n)}} \tag{9}$$

$$\min f_2(\pi) = \max(0, \ d_i - C_{\pi(i),m,l_{\pi(i)}}) \tag{10}$$

where d_i is the due date of job i. Functions f_1 and f_2 are makespan and earliness time, respectively.

3 Proposed Algorithm

3.1 Initializing Population

NEH heuristic, proposed by Nawaz et al., has aroused much attention and been successfully applied to the LSFS scheduling problem [8]. Followed that Ronconi et al. [10] presented the MinMax (MM) heuristic and MME heuristic based on the criterion of makespan. The MME heuristic is the combination of MM and NEH and employs the shortest critical path to reduce the blocking time of a job on machines. In the viewpoint of good performance of the above heuristics, The following strategies give two initialization strategies based on the variants of NHE and MME heuristics, i.e., vNEH and vMME, to generate an initial population with high quality in this study.

Strategy 1: vNEH heuristic

Step1: Computer the total processing time, $T_i = \sum\limits_{t=1}^{m} C_{\pi(i),t,1_{\pi(i)}}$ $i = \{1, 2, ..., n\}$ on all machines for each job, and obtain a sequence, $\pi = \{\pi(1), \pi(2), ..., \pi(n)\}$, by sorting jobs according to the total processing time in a descendant order.

Step2: Take the firs two jobs of π, evaluate the quality of two possible subsequences, and select the one with the minimal value of f_2 as the current sequence π^*. Let $k = 2$.

Step3: Let $k = k+1$, take the k-th job of π, and obtain k subsequences by inserting it into k possible positions of the current sequence π^*, then select the subsequence with the minimal value of $f_2(\pi^*)$ as the current sequence π^*.

Step4: If $k < n$, go to Step 3; otherwise, go to Step 5.

Step5: Take the n-th job of π, and insert $\pi(n)$ into n possible positions of the current sequence π^*. Select a number of good solutions from n complete sequences according to the Pareto dominance relation.

Strategy 2: vMME heuristic

Step1: Set the job with the shortest processing time on the first machine at the first position of permutation π, and put the job with the shortest processing time on the last machine at the last position of permutation π. Let $i = 2$.

Step2: Select the i-th ($i = \{2, 3, \ldots, n-1\}$) job of permutation π according to the following smallest index function:

$$\varphi_k = \phi \times \sum_{i=1}^{m-1} \left| P_{k,j} - P_{k-1,j+1} \right| + (1 - \phi) \sum_{t=1}^{m} P_{k,j}$$

$$\phi \in [0, 1] \quad \pi(k) \neq \pi(a) \tag{11}$$

$$k = \{1, 2, \ldots, i\}$$

$$a = \{1, 2, \ldots, i\}$$

where φ ia a parameter. The above equation is composed of two components, where the first, $\sum_{i=1}^{m-1} \left| P_{k,j} - P_{k-1,j+1} \right|$, refers to the difference between the processing time of successive jobs on successive machines, and the second, $\sum_{t=1}^{m} P_{k,t}$, is the total processing time.

Step3: If $i = i + 1$, $i = n$, output a complete sequence, $\pi = \{\pi(1), \pi(2), \ldots, \pi(n)\}$, and go to Step 4, otherwise, go to Step 2.

Step4: Pick up the first two jobs of π, evaluate the quality of two possible subsequences, and select the one with the minimal value of f_1 as the current sequence π^*. Let $k = 2$.

Step5: Let $k = k + 1$, take the k-th job of π, and obtain k subsequences by inserting it into k possible positions of the current sequence π^*, then select the subsequence with the minimal value of $f_1(\pi *)$ as the current sequence π^*.

Step6: If $k < n$, go to Step 4; otherwise, go to Step 7.

Step7: Take the nth job of π, and insert $\pi(n)$ into n possible positions of the current sequence π^*. Select a number of good solutions from n complete sequences according to the Pareto dominance relation.

3.2 Updating Employ Bees

The process of ISOJX is stated as follows: (1) build an artificial chromosome and put it into array artificial_position. First, count the times that job i appears at position k

in all non-dominated solutions. Second, put the job with the most times into the k-th position of artificial_position. Repeat the above steps until the artificial solution is a legal and complete sequence; (2) randomly select two different father individuals, named *parent*1 and *parent*2, and put the common jobs at the same positions between artificial_position and *parent*1 into *parent*2. Similarly, inherit the common jobs at the same positions between artificial_position and *parent*1 to *parent*2; (3) execute OP operator; (4) fill in the rest jobs of *offspring*1 and *offspring*2. Unselected jobs of *parent*2 supplement the missing jobs of *parent*1 in turn. In this way, put unscheduled jobs of *parent*1 into the blank positions of *parent*2 in turn.

The process of ISBOX is described as follows: (1) build a block artificial chromosome and put it into array artificial_block, that is, put job j with the largest times that appears immediately after job i in all non-dominated solutions into the ith position of artificial_block. Repeat this step until the artificial solution is legal and complete; (2) randomly selected two different father individuals, named *parent*1 and *parent*2, and put the common block of jobs between artificial_block and *parent*1 into *offspring*1. Similarly, *offspring*2 inherits the identical jobs between artificial_block and *parent*2; (3) execute OP operator; (4) fill in the rest jobs of *offspring*1 and *offspring*2, and the unselected jobs of *parent*2 supplement the missing jobs of in turn. In this way, put unscheduled jobs of *offspring*1 into the blank positions of *offspring*2 in turn. It is worth noting that ISBOX considers the job blocks not only at the same position between the father and artificial_block, but also at different positions, avoiding excellent blocks being destroyed.

3.3 Modifying Selected Solutions

Onlooker bees share solutions generated by all employ bees, and modify a number of them. During solving the multi-objective optimization problem considered in this study, we applied the Pareto local search (PLS) based on the Pareto dominance relation [1] when updating the onlooker bees with a small probability, pl, to generate solutions with high quality.

4 Experiments

The proposed algorithm (HDABC, for short) is applied to 16 BLSFS scheduling problems, and compared with TA [5], INSGA [3], NGA [12], BBEDA [2], and HDABC without local search (HABCnl, for short) to evaluate the performance of the proposed algorithm in this section. All these algorithms are implemented with C++ in a PC environment with Pentium(R) Dual 2.8 GHZ and 2 GB memory. Each instance is independently run 30 times for each algorithm. Following Yoon and Ventura [13], and Tseng and Liao [11], the related data for each LSFS scheduling problem are given according to the discrete uniform distributions as below (Table 1).

Table 1 Parameter settings

Parameter	Notation	Value
Number of jobs	N	50, 70, 90, 110
Number of machines	M	5, 10, 15, 20
Due date of job j	d_i	$d_j =$ $rand()\%(15 \times m + 1) + 15 \times n$
Number of sub-lots of job j	$l_{\pi(j)}$	$l_{\pi(j)} = rand()\%6 + 1$
Processing time of job j on machine i	$P_{\pi(j),i}$	$P_{\pi(j),i} = rand()\%31 + 1$
Population size	PS	20
Crossover rate	P_c	0.6
Pareto local search rate	Pl	0.4
Archive size	AS	50
Stopping condition	CPU time	30 nm

4.1 Indicators

Distance Between Reference Set and Solution Set

The distance between the reference set and solution set $S(j)$, denoted as $D(S(j))$, was utilized to reflect the convergence of the solution set $S(j)$ to the reference set [9], and the smaller the value of $D(S(j))$, the better $S(j)$ approaches to the reference set. $D(S(j))$ was given as follows.

$$D(S(j)) = \frac{1}{|S*|} \sum_{y \in S*} d_y(S(j)) \tag{12}$$

$$d_y(S(j)) = \min_{x \in S(j)} \sqrt{\sum_{k=1}^{2} (\frac{f_k(x) - f_k(y)}{f_k^{max} - f_k^{min}})^2} \tag{13}$$

where $f_k(\bullet)$ denotes the kth objective value, and f_k^{max}, f_k^{min} are the minimal and maximal of the kth objective value in the reference set $S*$, respectively. $|S*|$ is the number of reference solutions in $S*$.

Entropy Value

Entropy is used to measure the expectation of a stochastic variable and counts the occurrence probabilities of the variable or component, also known as comentropy. In this study, we used comentropy to evaluate the diversity of the population. Denote the individual of the population as x, which is composed of $x_1, x_2, \cdots, x_n$ with their occurrence probabilities as $p(x_1), p(x_2), \cdots, p(x_n)$, respectively, and $\sum_{i=1}^{n} p(x_i) = 1$. The comentropy, denoted as $H(x)$, was calculated as follows:

Table 2 Performance of different algorithms in D(S(j)) and $H(x)$

$n \times m$	D(S(i))					H(EA(i))				
	TA	INSGA	NGA	BBEDA	HDABC	TA	INSGA	NGA	BBEDA	HDABC
505	0.21	0.14	0.34	0.16	0.01	183.83	200.36	86.11	82.68	415.42
5010	0.15	0.02	0.21	0.00	0.01	183.30	198.34	235.69	79.67	265.20
5015	0.05	0.02	0.12	0.00	0.00	181.33	192.89	221.21	342.64	139.14
5020	0.62	0.25	0.19	0.17	0.42	188.00	231.56	203.61	288.91	151.68
705	0.21	0.20	0.08	0.00	0.00	269.76	287.64	242.69	92.07	463.86
7010	0.05	0.06	0.67	0.08	0.00	267.82	294.36	412.80	213.37	433.41
7015	0.18	0.30	0.03	0.21	0.00	270.92	296.57	260.15	318.77	405.94
7020	0.16	0.08	0.20	0.09	0.00	274.77	356.11	371.11	389.01	389.28
905	0.02	0.19	0.13	0.25	0.00	358.78	469.32	391.72	522.65	747.61
9010	0.05	0.03	0.09	0.03	0.00	354.79	455.23	327.67	285.79	741.69
9015	0.06	0.11	0.04	0.05	0.00	363.19	472.36	492.97	371.97	414.02
9020	0.12	0.10	0.58	0.08	0.01	358.27	462.31	479.52	468.34	576.00
1105	0.08	0.08	0.17	0.04	0.00	452.30	500.44	543.15	523.51	789.34
11010	0.03	0.09	0.19	0.05	0.02	351.03	461.31	494.07	410.32	450.48
11015	0.05	0.05	0.24	0.08	0.01	346.82	568.67	395.90	365.23	389.78
110 20	0.16	0.07	0.20	0.12	0.03	346.74	543.95	297.31	198.23	371.67
mean	0.14	0.11	0.22	0.09	0.03	296.98	374.46	340.98	309.57	446.53

$$H(x) = \frac{1}{\text{max\_iter}} \sum_{t=1}^{\text{max\_iter}} \left(- \sum_{i=1}^{n} p(x_i)\log_2 p(x_i) \right) \tag{14}$$

where max_iter is the total number of iterations of an algorithm. It can be observed that the larger the value of $H(x)$, the more diverse the population, which reflects that different jobs occupy different positions, and no similar job blocks exist among these sequences.

4.2 Experimental Results

Performance of TA, DHS, NGA, ABC, HDABCnl and HDABC

Table 2 shows the shortest distance, D(S(j)), between the reference set and the solution set obtained by different algorithms. From this table, within the same computational time, HDABC has the distance of 0.03, far smaller than 0.14, 0.11, 0.22, and 0.09, obtained using TA, INSGA, NGA, and BBEDA, respectively. Therefore, with respect to the indicator of D(S(j)), HDABC has a better performance than the comparative ones for all these instances.

In addition to D(S(j)), Table 2 also lists the comentropy values of $H(EA(i))$ obtained by different algorithms. As shown in Table 2, the mean entropy values of

HDABC are much larger than those of the other algorithms, suggesting that the proposed algorithm has good diversity, that is, different jobs occupy different positions and few similar job blocks exist among these sequences obtained by HDABC. In terms of the values of $D(S(j))$ and $H(EA(i))$, the superior performance of the proposed algorithm in convergence and diversity can be justified.

In summary, the non-dominated solutions obtained by HDABC have good capabilities in convergence and diversity due to the proposed initialization strategy, novel crossover and mutation operators.

5 Conclusions

The BLSFS scheduling problem with more than one criterion is challenging and has various applications. In this study, we presented a HDABC algorithm to tackle the BLSFS scheduling problem with two criteria of makespan and earliness time. It is worth mentioning that the three mutation operators are randomly selected to generate the offspring population in this study. In addition, we can develop a self-adaptive learning mechanism to select the above mutation operators, so as to improve the capability of the proposed algorithm in exploration. For the Pareto local search, seeking for appropriate strategies to reduce its computational complexity is of considerable necessarily, which is helpful to enhance the efficiency of the proposed algorithm. We will further consider the above topics in the future.

Acknowledgements This research is partially supported by Natural Science Foundation of China under Grant number 61375067, 61473299.

References

1. Behnamian J, Ghomi SMTF, Zandieh M. A multi-phase covering pareto-optimal front method to multi-objective scheduling in a realistic hybrid flowshop using a hybrid metaheuristic. Expert Syst Appl. 2009;36(8):11057–69.
2. Chang PC. A pareto block-based estimation and distribution algorithm for multi-objective permutation flow shop scheduling problem. Int J Prod Res. 2015;53(3):793–834.
3. Han YY, Gong DW, Sun XY, Pan QK. An improved nsga-ii algorithm for multi-objective lot-streaming flow shop scheduling problem. Int J Prod Res. 2014;52(8):2211–31.
4. Kalir AA, Sarin SC. Evaluation of the potential benefits of lot streaming in flow-shop systems. Int J Prod Econ. 2000;66(2):131–42.
5. Marimuthu S, Ponnambalam SG, Jawahar N. Threshold accepting and ant-colony optimization algorithms for scheduling m-machine flow shops with lot streaming. J Mater Process Tech. 2009;209(2):1026–41.
6. Martin CH. A hybrid genetic algorithm/mathematical programming approach to the multi-family flowshop scheduling problem with lot streaming. Omega. 2009;37(1):126–37.
7. Mhlenbein H, Paa G. From recombination of genes to the estimation of distributions I. Binary parameters. Berlin Springer; 1996.

8. Pan QK, Ruiz R. An estimation of distribution algorithm for lot-streaming flow shop problems with setup times. Omega. 2012;40(2):166–80.

9. Pan QK, Wang L, Qian B. A novel differential evolution algorithm for bi-criteria no-wait flow shop scheduling problems. Comput Oper Res. 2009;36(8):2498–511.

10. Ronconi DP. A note on constructive heuristics for the flowshop problem with blocking. Int J Prod Econ. 2004;87(1):39–48.

11. Tseng, C.T., Liao, C.J.: A discrete particle swarm optimization for lot-streaming flowshop scheduling problem. Eur J Oper Res. 2008;191(2) 360–373

12. Ventura JA, Yoon SH. A new genetic algorithm for lot-streaming flow shop scheduling with limited capacity buffers. J Intell Manuf. 2013;24(6):1185–96.

13. Yoon SH, Ventura JA. An application of genetic algorithms to lot-streaming flow shop scheduling. IIE Trans. 2002;34(9):779–87.

14. Zhang W, Yin C, Liu J, Linn RJ. Multi-job lot streaming to minimize the mean completion time in m -1 hybrid flowshops. Intern J Prod Econ. 2005;96(2):189–200.

Cooperative Path Planning for Intelligent Vehicle Using Unmanned Air and Ground Vehicles

Qiuxia Hu, Jin Zhao and Lei Han

Abstract This paper focus on the intelligent vehicle path planning issue using cooperation between Unmanned aerial vehicle (UAV) and Unmanned ground vehicle (UGV). For the UGV's path planning problem in the cooperative system, the theory of traditional Artificial Potential Field (APF) was analyzed, and an algorithm for the intelligent vehicle was presented. The proposed algorithm allows the intelligent vehicle to navigate through obstacles, and finding a path which can reach the target without collision. The novelty of the algorithm is that directly establish the function model of repulsive and attractive force, and redefine the local minimum point to solve the problem of local minimum point. The proposed algorithm was illustrated with MATLAB, as a result, a safe and feasible path can be generated which can ensure the intelligent vehicle reach the target smoothly.

Keywords Path planning · Artificial potential field (APF) · Cooperation · Local minimum point

1 Introduction

Recently, heterogeneous robot system consisting of an Unmanned aerial vehicle (UAV) and ground Vehicle (UGV) is being studied. Unmanned aerial vehicles have a wide field of view and can cover large areas quickly. However, sensors on UAV are typically limited in their accuracy of localization of targets on the ground. Unmanned ground vehicles can be deployed to accurately locate ground targets, but they have the disadvantage of not being able to move rapidly or see through such

Q. Hu · J. Zhao (✉) · L. Han
School of Mechanical Engineering, Guizhou University, Guiyang 550000, China
e-mail: zhaojin9485@163.com

Q. Hu
e-mail: huqiuxiafind@163.com

L. Han
e-mail: ahanllei33@163.com

© Springer Nature Singapore Pte Ltd. 2018
Y. Jia et al. (eds.), *Proceedings of 2017 Chinese Intelligent Systems Conference*,
Lecture Notes in Electrical Engineering 460,
https://doi.org/10.1007/978-981-10-6499-9_58

obstacles as buildings or fences [1]. Accordingly, deploy the teams of two sorts of vehicles working cooperatively is desirable.

Many approaches to this topic have been carried out for the cooperation between UAVs and UGVs. A framework for experimental and a algorithm for coordinated control with the goal of localizing and searching for targets are presented in [1]. Vidal et al. [2] develop a probabilistic approach to pursuit-evasion games involving UAVs and UGVs. But this approach does not consider occluded vision and sensor date fusion. The influence of vision occlusions for target tracking using UAVs is considered in [3]. However, there are not UGV take into account. Yu et al. [4] present a cooperative path planning algorithm for tracking a moving target, and evaluate the performance of the algorithm with varying degree of occlusions. Tanner [5] presents a switched cooperative control scheme, to coordinate aerial and ground vehicles for the purpose of locating a moving target in a given area.

When a UGV knows the target's position, it is necessary to plan an optimal and feasible path avoiding obstacles. And minimizing a cost such as energy, time and distance. Therefore, the main work for path planning for ground vehicle is to search a collision free path.

Several approaches have been proposed. If the environment is a known static terrain, it is said to be static path planning. If there have moving obstacles in environment, it is said to be dynamic path planning. The static path planning usually use grid method, visual graph method, etc. The dynamic path planning includes artificial potential field method, particle swarm algorithm, genetic algorithm (ACO), ant colony algorithm and so on. The artificial potential field algorithm is widely used in the field of path planning [6], it has high efficiency, good real-time and safety. However, it needs to solve the problem of local minimum point. A number of potential functions have been proposed. On the one hand, some researchers aimed at improve the potential field function [7]. On the other hand, a combination of algorithms is used [8]. In improved the potential field function, Sato [9] propose the Laplace potential field method. By defining the potential field equation reasonably, the local minimum in the potential field was eliminated, but the optimal planning path was not considered. Reference [10] use the filling potential field to eliminate the influence of the local minimum. Geva et al. [11] add an adjustable dynamic repulsive force gain function to the repulsive force function. When the vehicle is near the obstacles, the repulsion increases rapidly, so that the vehicle can change direction quickly. In combination of algorithms, Reference [12] uses a random search strategy, by randomly generating a −1~1 rad angle as the next motion direction, at the same time, a secure processing strategy is proposed to ensure the vehicle will not collide with obstacles. However, it lacks of heuristic information, search efficiency is very low in complex environments.

This paper presents an algorithm for the intelligent vehicle's path planning in the UAV-UGV cooperative system. The main contribution of the proposed algorithm is to directly establish the function model of repulsive and attractive force, and redefine the local minimum point to solve the problem of local minimum point. At first, we model the planning environments use grid. Then, the repulsive and attractive force model are built. Finally, we tested the performance of the path planning algorithm under different environments by using MATLAB.

This paper is organized as follows. In Sect. 2, the structure of Unmanned Aerial Vehicle (UAV) and Unmanned Ground Vehicle (UGV) cooperative system and hierarchical hybrid architecture are presented. In Sect. 3, the path planning algorithms for the intelligent vehicle in the UAV-UGV cooperative system is introduced. Numerical results are shown in Sect. 4. Section 5 concludes this paper.

2 Overall Cooperation System

Even though this paper focuses on the path planning of the intelligent vehicle, there are many necessary modules, as shown in Fig. 1. In this paper, we proposed a hierarchical hybrid system architecture that segments the control of each agent into different layers of abstraction. By extracting the information of sensor and controlling each agent, a framework between UAVs and UGVs is built. The map building module receives the reference position from the vehicle-level sensor fusion module and computes the map. The path planner is responsible for planning a path between vehicle position and target while avoiding the obstacles. The final path is sent to the path tracking controller. Which performs the real-time control of the vehicle. The global environment's information is captured by the camera which

Fig. 1 System architecture

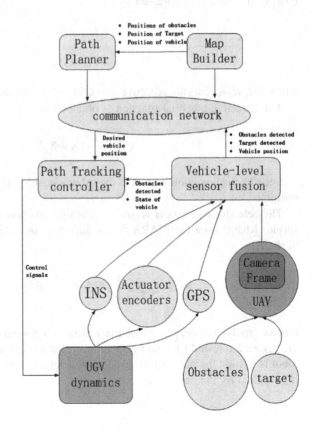

mounted on UAV. UGV gets its state using a variety of sensors for position. Sensor fusion model is used to improve the quality of the measurements. UAV and UGV share the information through the communication network.

3 Path Planning Algorithm

The artificial potential field should be designed to meet the manipulator stability condition, it also should be a nonnegative continuous and differentiable function. Two different potential function are analyzed in this part.

3.1 Classical Artificial Potential Field Approach

The classical artificial potential field approach imitates the concept of electric field in Physic. Assumed that target point and obstacles as the field source, robot's motion is determined by the force of attraction and repulsion.

The artificial potential field was proposed by Khatib in 1986. The potential energy and force are expressed by;

$$U = U_{att} + U_{rep} \tag{1}$$

$$F = F_{att} + F_{rep} \tag{2}$$

where U_{att} represents the attractive potential, U_{rep} represents the repulsive potential. The attractive potential function is obtained as;

$$U_{att} = \frac{1}{2}k(X - X_g)^2 \tag{3}$$

Let k be the attraction gain, X is the robot's position in workspace. x_g is the goal position.

The behavior of a proper repulsive potential function should be like a tightened spring, Khatib used the FIARS (Force Inducing an Artificial Repulsion from the Surface);

$$U_{rep} = \begin{cases} \frac{1}{2}\eta\left(\frac{1}{\rho} - \frac{1}{\rho_0}\right)^2, \rho \le \rho_0 \\ 0, \rho > \rho_0 \end{cases} \tag{4}$$

η is the repulsion gain. ρ is the shortest distance between robot and obstacles. ρ_0 is the distance threshold for an obstacle to create a repulsion effect on robot. If the robot is outside the range of obstacle's influence, it's not affected. On the contrary, it's repelled by the obstacles.

Derivative of the potential functions (3), (4) are the attractive force and repulsive force which are;

$$F_{att} = -grad(\mathrm{U}_{att}) = -k(\mathrm{X} - \mathrm{X}_g) = k(\mathrm{X}_g - \mathrm{X}) \tag{5}$$

$$F_{rep} = -grad(\mathrm{U}_{rep}) = \begin{cases} \eta\left(\frac{1}{\rho} - \frac{1}{\rho_0}\right)\frac{1}{\rho^2}\frac{\partial\rho}{\partial x}, \rho \leq \rho_0 \\ 0, \rho > \rho_0 \end{cases} \tag{6}$$

Because the potential field and force are all can be superimposed, when there are multi obstacles, the total potential and force are expressed as;

$$U = U_{att} + \sum_i U_{repi} \tag{7}$$

$$F = F_{att} + \sum_i F_{repi} \tag{8}$$

3.2   Improved Artificial Potential Field Approach

In this present work, we directly establish the function model of repulsive and attractive force, and redefine the local minimum point to solve the problem of local minimum point [13]. This repulsive force and attractive force can be computed by:

$$F_{rep} = \frac{w}{(J-Y)^2} + (I-X)^2 \tag{9}$$

$$F_{att} = w\sqrt{(J-GoalY)^2 + (I-GoalX)^2} \tag{10}$$

where I and J represent all points in the map, X and Y represent the center of the obstacle, w represent the weight of charge, GoalX and GoalY is the target coordinates.

Assume the environment is divided into square grid. The cell's size depends on the density of the obstacles and the final value at least equal to the length of architecture of intelligent vehicle. This is ensure the vehicle will not be stuck in the process. In the grid, each cell is assigned of three states: occupied, free, and unknown. A cell is occupied means it contain one or more obstacles, free means it is no obstacles, others are marked unknown.

In the grid environment, obstacles are encoded by rectangular coordinate system and swelled. One example is presented in the Fig. 2. Each item in the array represents one of the squares on the grid. The dark color represents unwalkable

Fig. 2 An example of obstacle

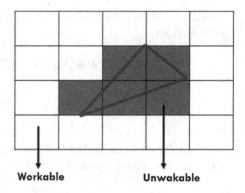

space; whereas the white color is interpreted as walkable space. Having determined the walkable space, the algorithm works and operated on the free grid.

We describe that the configuration grid is a representation of the configuration space and mark it by the resultant force. Firstly, the algorithm assume the size of the

Fig. 3 The flow chart of the algorithm

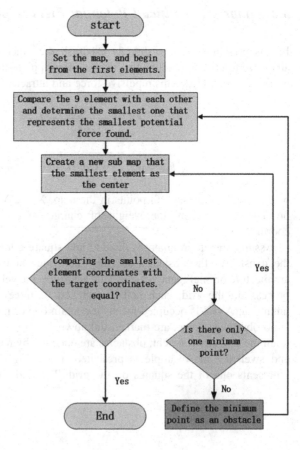

map is 100 * 100 and form a 3 * 3 sub map that contain the 9 element. After forming the sub map, it begin to compare the element with each other and the smallest one was found. Then the smallest one is used as the center to build a new sub map and repeat the previous step. Finally, comparing the smallest element coordinates with the target coordinates. If there is a unique minimum and equals to the final coordinates then stop else continue. If there is more than one minimum and not equals to the final coordinates, means the intelligent vehicle fall in a local minimum region. In order to solve this problem, an obstacle with high potential will be found at here. The algorithm's step are shown in Fig. 3.

4 Numerical Results

The path planning algorithm is tested using a simulation environment developed in MATLAB. In different environments with static obstacles, the simulation results which provide the most preferable path rather than the shortest. The snapshot of simulation environment is shown in Fig. 4a, there are two obstacle, they are placed at the coordinate A {(2.5, 5.5) (2.5, 8.5)}, B {(7, 2) (7, 5) (5.5, 5)}, the target is F(9, 7), and the parameter w was set at 10.

Figure 4 is the potential field map of the planning environment. The knobs represent the obstacle repulsion field, and the valleys represent the target point. It easy to get that each point's potential value is the superposition of the repulsive force field of the obstacle and the attractive field of the target point. The potential field value at the target point is the global minimum, and the feasible path can be planned according to the trend of potential field.

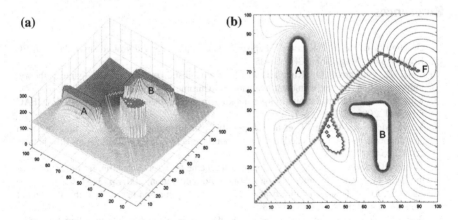

Fig. 4 Path planning results

The feasible path is given by Fig. 4b. Even local minima appear around the obstacle B, it's solved quickly by the proposed algorithm, and reach the target point F successfully.

5   Conclusion

In this paper, a new path planning approach based on the traditional artificial potential field method is proposed. Note that focusing on introducing the application of the prosed planning algorithm, we implemented the algorithm in the cooperative system between unmanned air and ground vehicles. For the proposed algorithm, the repulsion and attraction functions are modeled directly, and the functional expressions are given respectively. At the same time, the minimum point is defined as an obstacle, so that the algorithm can quickly jump out of the minimal point.

The proposed algorithm was tested by using MATLAB. It is shown that local minimum point is effectively solved, a safe and feasible path can be got. Further, the environment information can be modified in real-time. The results validate that the proposed algorithm not only greatly enhance the visibility, but also improve the applicability.

Acknowledgements   This work was supported by National Science Foundation of China (Project No.: 61164007), Science Foundation of Guizhou Province (Project No.: KY[2014]226). The Outstanding Young Talent of Guizhou Province, Postgraduate Innovation Fund of Guizhou University (Project No.: 2016034).

References

1. Grocholsky B, Keller J, Kumar V, Pappas G. Cooperative air and ground surveillance. IEEE Robot Autom Mag. 2006;13(3):16–25.
2. Vidal R, Shakernia O, Kim HJ, et al. Probabilistic pursuit-evasion games: theory, implementation, and experimental evaluation. J Chronic Dis. 2002;18(5):662–9.
3. Kim J, Kim Y. Moving ground target tracking in dense obstacle areas using UAVs. IFAC Proc Vol. 2008;41(2):8552–7.
4. Yu H, Meier K, Argyle M, et al. Cooperative path planning for target tracking in urban environments using unmanned air and ground vehicles. IEEE/ASME Trans Mechatron. 2015;20:541–52.
5. Tanner HG. Switched UAV-UGV cooperation scheme for target detection. In: Proceedings of the IEEE international conference on robotics and automation; 2007. p. 3457–62.
6. Zhang D, Liu F. Research and prospect of path planning method based on artificial potential field. Comput Eng Sci. 2013;35(6):88–95.
7. Pozna C, Troester F, Precup RE, et al. On the design of an obstacle avoiding trajectory: method and simulation. Math Comput Simul. 2009;79(7):2211–26.
8. Zhang W, Lin A, Liu T, et al. Robot path planning based on improved artificial fish swarm algorithm. Comput Simul. 2016;33(12):374–9.

9. Keisuke S. Deadlock-free motion planning using the Laplace potential field. Adv Robot. 1992;7(5):449–61.

10. Zhen-zhong Yu, Ji-hong Yan, Jie Zhao, et al. The improved artificial potential field method for mobile robot path planning. J Harbin Inst Technol. 2011;43(1):50–5.

11. Geva Y, Shapiro A. A combined potential function and graph search approach for free gait generation of quadruped robots. In: IEEE international conference on robotics and automation. IEEE; 2012. p. 5371–6.

12. Liu ZX, Yang LX, Wang JG. Soccer robot path planning based on evolutionary artificial field. Adv Mater Res. 2012;562–564:955–8.

13. Mandal P, Barai RK, Maitra M, Roy S. Path planning of autonomous mobile robot. In: 2013 7th international conference on intelligent systems and control (ISCO), vol 2, no 4; 2013. p. 178–190.

Attribute Reduction of Preference Linguistic-Valued Decision Information Systems

Tingquan Deng, Xuefang Zhang, Qingguo Lin, Dongsheng Ye and Yang Liu

Abstract Dominance-based linguistic-valued information systems have attracted much attention in practical applications in which more information can be provided and employed than real-valued information systems. This paper proposes an approach to attribute reduction of dominance-based linguistic-valued decision information systems. All linguistic-values are represented by trapezoidal fuzzy numbers and a new fuzzy dominance relation is constructed to characterize the degree of one fuzzy number dominating another. Based on the presented fuzzy dominance relation, a model of dominance fuzzy rough set is developed and its properties are investigated. A way to measure attributes significance is established by introducing new fuzzy entropy. Comparative experiments are conducted to verify the effectiveness of the proposed model.

Keywords Linguistic-valued information system · Dominance fuzzy rough set · Dominance relation · Attribute reduction

1 Introduction

As pointed out by Zadeh, granules are fuzzy rather than Boolean in human reasoning and concept formation [1, 2]. Fuzzy rough set model is proposed to deal with uncertainty [3]. Based on a fuzzy similarity relation, it is incapable of analyzing data with preference structures, such as credit approval, stock risk estimation and car evaluation. To study data with preference structures, a (fuzzy) preference relation has been introduced and widely used.

Wu et al. proposed a fuzzy rough sets model based on fuzzy relation [4]. Yeung et al. showed models capable of defining fuzzy rough sets based on arbitrary fuzzy relations [5]. Greco et al. introduced fuzzy dominance rough sets using possibility

T. Deng · X. Zhang (✉) · Q. Lin · D. Ye · Y. Liu
College of Science, Harbin Engineering University, Harbin 150001, China
e-mail: 593169413@qq.com

© Springer Nature Singapore Pte Ltd. 2018
Y. Jia et al. (eds.), *Proceedings of 2017 Chinese Intelligent Systems Conference*,
Lecture Notes in Electrical Engineering 460,
https://doi.org/10.1007/978-981-10-6499-9_59

613

and necessity measures [6]. Yang et al. introduced the α-dominance rough set to deal with interval-valued data [7].

Traditional dominance-based fuzzy rough set model always deals with some discrete data or real data [8]. However, information systems with complex data such as incomplete data [9–12], gray data [13], interval-valued data [7], mixed data [14] and linguistic-valued data are found almost everywhere in real-world applications. To fill this gap, many models were proposed [14].

The purpose of this paper is to further investigate dominance-based fuzzy rough set model to deal with linguistic-valued data. We transform linguistic values as fuzzy numbers and construct a dominance relation to compute the degree that one linguistic value dominates the other. We propose a new fuzzy rough set model and a new measure to assess the attribute significance in the linguistic-valued information system. According to the significance we can get the reduction of attribute set.

This paper is organized as follows. Firstly, we review some basic conception. New fuzzy dominance relation is proposed in Sect. 3. Based on this relation we propose a new fuzzy rough set model in Sect. 4. In Sect. 5, we present the measure of significance about the attributes and then consider the attribute reduction. Experimental analysis is provided in Sect. 6. We conclude the paper at last.

2 Preliminaries

2.1 Rough Set

Generally speaking, rough set is used to deal with information systems, in which objects or samples are characterized by attributes or features. An information system is a quadruple $IS = (U, C, V, f)$, where $U = \{x_1, x_2, \ldots, x_n\}$ is a non-empty finite set of objects, called a universe; $C = \{a_1, a_2, \ldots, a_m\}$ is a non-empty finite set of attributes; V is the range of all attributes, i.e., $V = V_C = \cup_{a \in C} V_a$; $\forall a \in C$, $\forall x \in U, f(x, a)$ is the value that x holds on a.

A decision system is an information system $IS = (U, C \cup D, V, f)$, in which D is a decision attribute set and $C \cap D = \emptyset$.

For a subset of attributes set, $B \subseteq C$, a binary relation on U is defined as $IND(B) = \{(x, y) \in U \times U : f(x, a) = f(y, a), \forall a \in B\}$, which is an equivalence relation on U dividing the objects into a family of disjoint subsets. For an arbitrary subset $X \subseteq U$, the lower and upper approximations of X is defined as

$$\underline{B}X = \cup\{[x]_B : [x]_B \subseteq X\}, \bar{B}X = \cup\{[x]_B : [x]_B \cap X \neq \phi\}$$

where $[x]_B = \{y \in U : (x, y) \in IND(B)\}$.

2.2 Dominance-Based Rough Set

Definition 2.1 Let $S = (U, C \cup D, V, f)$ be a decision information system, $B \subseteq C$, a dominance relation is defined as

$$DOM(B) = \{(x, y) \in U \times U : f(x, a) \geq f(y, a), \forall a \in B\}$$

Definition 2.2 Let $S = (U, C \cup D, V, f)$ be a decision information system, $B \subseteq C$, $\forall X \subseteq U$, the lower and upper approximations of X with respect to the dominance relation $DOM(B)$ is defined by

$$\underline{R}_B^+(X) = \{x \in U : [x]_B^+ \subseteq X\}, \bar{R}_B^+(X) = \{x \in U : [x]_B^+ \cap X \neq \phi\}$$

where $[x]_B^+ = \{y \in U : (y, x) \in DOM(B)\}$.
The dependency of D with respect to B is computed by

$$\gamma_B(D) = \frac{|\bigcup_i \underline{R}_B^+(D_i)|}{|U|}$$

where $D_i \subseteq U / DOM(D) = \{[x]_D^+ | x \in U\}$.
For $a \in B$, the significance of a based on the dependency is defined as

$$sig(a, B, D) = \gamma_B(D) - \gamma_{B\backslash\{a\}}(D) \tag{1}$$

3 Fuzzy Dominance Relation

3.1 Linguistic Information System

Let $S = (U, C \cup D, V, f)$ be a decision information system, if the elements in V are linguistic values, S is refer to as a linguistic-valued decision information system.

For example, when one describes whether a car is safe or not, one may use words like "safe" or "very safe". When one describes whether the car is expensive or not, one may use words like "cheap" or "expensive". Table 1 shows a linguistic-valued decision information system, where $U = \{x_1, \ldots, x_{10}\}$ is the set of considered cars, $C = \{$price, safety, amount of oil consumption, appearance of car, comfortable level$\}$ is the conditional attribute set, and D is a decision attribute, representing whether this car is acceptable for someone.

We can use the classical discernibility matrix method or dominance–based rough set model to obtain the attribute reduction. However, the discernibility matrix can't reflect the preference of data and dominance-based rough set use integer numbers to denote the linguistic variables, which may lose much information.

Table 1 A linguistic decision information system about cars evaluation

U	a_1	a_2	a_3	a_4	a_5	D
x_1	Middle	Low	More	Beautiful	Comfortable	Unacceptable
x_2	Low	High	Much more	Middle	Not comfortable	Good
x_3	High	High	Middle	Very beautiful	Very comfortable	Very good
x_4	High	Low	Middle	Not beautiful	Middle	Unacceptable
x_5	High	Very high	More	Beautiful	Middle	Very good
x_6	Low	Middle	Little	Beautiful	Very comfortable	Acceptable
x_7	Middle	High	Much more	Middle	Not comfortable	Acceptable
x_8	Low	Low	More	Not beautiful	Middle	Unacceptable
x_9	High	Very high	Middle	Very beautiful	Comfortable	Very good
x_{10}	Middle	High	Little	Not beautiful	Comfortable	Good

3.2 Fuzzification of Linguistic Values

Linguistic-values are usually involved in uncertainty information and can be characterized by fuzzy quantities. In this paper, we fuzzify the linguistic-valued data as trapezoidal fuzzy numbers. Firstly, we set the universe of linguistic values of an attribute as a subset of integer numbers, say $\{A, A + 1, \ldots, A + n\}$. Then each linguistic value in the linguistic-valued information system is fuzzified as a fuzzy number on $[A, A + n]$. Figure 1 shows the trapezoidal fuzzy numbers of the data with respect to attribute a_1, where $A = 1$ and $n = 3$. Three trapezoidal fuzzy numbers of 'low', 'middle' and 'high' are plotted by blue line, red line and green line, and denoted by (1,1,2,3), (1,2,3,4) and (2,3,4,4).

When fuzzifying, different A and n may be chosen for different attribute.

Fig. 1 Trapezoidal fuzzy numbers

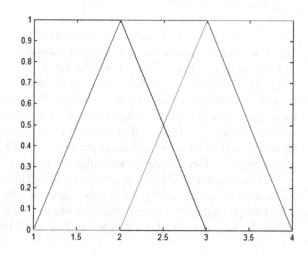

3.3 Fuzzy Dominance Relation

Let $S = (U, C, V, f)$ be an information system, if the values in V are interval-valued data, S is referred to as an interval-valued information system.

For an interval-valued information systems S, a formation of dominance relation was proposed in [7]. $\forall a \in C, \forall x, y \in U$ the degree of x dominating y on a, donated by $D_a(x, y)$, is defined by

$$DOM_a(x, y) = \begin{cases} 1, & x_a = y_a \\ 1, & x_a, y_a \in \mathbb{R}, x_a^- \geq y_a^+ \\ 0, & x_a, y_a \in \mathbb{R}, x_a^- \geq y_a^+ \\ \frac{\min\{l_{(x,a)} + l_{(y,a)}, \max\{x_a^+ - y_a^-, 0\}\}}{l_{(x,a)} + l_{(y,a)}}, & \text{otherwise} \end{cases}$$

where $\mathbb{R}$ denotes the set of real numbers, $x_a = [x_a^-, x_a^+]$ is an interval, representing the value of object x holding on attribute a, and $l_{(x,a)} = x_a^+ - x_a^-$ is the length x_a.

According to the dominance relation, $D_a(x, y) = 1$ when $x_a = y_a$. It is the same with the situation when $x_a^- \geq y_a^+$. However, the dominance degrees of the situations $x_a = y_a$ and $x_a^- \geq y_a^+$ should be distinct. So the above definition of dominance degree of one interval dominating another is irrational. In the following, we propose a new measure of one interval dominating another one to solve this problem.

Definition 3.1 For two intervals $x = [x^-, x^+], y = [y^-, y^+]$, and a small positive real number ε, let $L = \max(\min(y^-, y^+) - \max(x^-, x^+), 0)$ and

$$L^* = L + \max\left\{ \frac{\varepsilon}{|x^+ - x^-| + \varepsilon} \cdot \frac{|x^+ - y^+| - (x^+ - y^+)}{2|x^+ - y^+| + \varepsilon} \cdot (\max(x^+ - x^-, y^+ - y^-) + \min(x^+ - y^+, 0)), 0 \right\}$$

the degree of x dominating y, denoted by $IR(x, y)$, is defined by

$$IR(x, y) = \frac{L^*}{2\max(x^+ - x^-, y^+ - y^-) + \varepsilon} + \min\left\{ \frac{\max(\min(\max(x^+ - y^+, x^- - y^-), x^+ - x^-), \min(\max(x^+ - x^-, y^+ - y^-), \varepsilon)) + \varepsilon}{x^+ - x^- + \max(y^+ - x^-, -\varepsilon) + 2\varepsilon}, 1 \right\}$$

It is obvious when $x = y$, $IR(x, y) = 0.5$, meaning that any interval cannot dominate itself. When $y^+ \leq x^-$ (in this case, one says that x is larger than y), then $IR(x, y) = 1$. In this aspect, the new definition is different from the definition proposed in [7].

According to Definition 3.1, it is verified that $IR(x, y) + IR(y, x) \leq 1$.

As any cut set of a fuzzy number is an interval, we propose a measure of a fuzzy number dominating another one based on the interval dominance degree.

Definition 3.2 Let P and Q be two fuzzy numbers, the dominance degree of P dominating Q is defined by

$$R(P,Q) = \int_0^1 IR(P_\lambda, Q_\lambda)d\lambda$$

where P_λ denotes the λ-cut set of P and $IR(P_\lambda, Q_\lambda)$ denotes the dominance degree of interval P_λ dominating Q_λ, $\lambda \in [0, 1]$.

Example 1 In Fig. 1, $P(1,1,2,3)$ and Q $(1,2,3,4)$ are two trapezoidal fuzzy numbers, we can compute the dominance degree of P dominating Q.

Let $\lambda \in [0, 1]$, then $P_\lambda = [1, -\lambda + 3)]$ and $Q_\lambda = [\lambda - 1, -\lambda + 4]$, then $R(P,Q) = \int_0^1 IR(P_\lambda, Q_\lambda)d\lambda = 0.2594$ and $R(Q,P) = \int_0^1 IR(P_\lambda, Q_\lambda)d\lambda = 0.7406$.

4 Dominance-Based Fuzzy Rough Set Model

In this section, a new fuzzy rough set model based on fuzzy dominance relation is proposed to deal with linguistic-valued information system.

Definition 4.1 Let $S = (U, C, V, f)$ be a linguistic-valued information system and $B \subseteq C$, for a fuzzy set F on U, the lower approximation and upper approximation of F with respect to the fuzzy dominance relation R_B are defined by

$$\underline{R_B}(F)(x) = \frac{1}{n}\sum_{y \in U} \frac{R_B(y,x) \wedge F(y)}{R_B(y,x)}, \bar{R}_B(F)(x) = \frac{1}{n}\sum_{y \in U} R_B(y,x) \cdot F(y)$$

where $R_B(x,y) = \wedge_{a \in B} R_a(x,y)$ and R_a is the fuzzy dominance relation on U with respect to the attribute a, $R_a(x,y) = R(x_a, y_a)$, $\forall x, y \in U$.

Based on Definition 4.1, it can be derived that

(1) $\underline{R_B}(\phi) = \phi$; (2) $\underline{R_B}(U) = U$; (3) $\forall x, y \in U$, $\underline{R_B}(F_1 \cap F_2)(x) \leq \underline{R_B}(F_1)$ $(x) \wedge \underline{R_B}(F_2)(x)$; (4) $\forall x, y \in U$, $\underline{R_B}(F_1 \cup F_2)(x) \geq \underline{R_B}(F_1)(x) \vee \underline{R_B}(F_2)(x)$.

Definition 4.2 Let $IS = (U, C \cup D, V, f)$ be a linguistic-valued decision information system, the positive domain of the decision set D with respect to the conditional attribute subset B is defined as

$$Pos_B(D)(x) = \vee_{z \in U} \underline{R_B}([z]_D)(x) = \vee_{z \in U} \frac{1}{n}\sum_y \frac{R_B(y,x) \wedge R_D(y,z)}{R_B(y,x)}$$

where $[z]_D$ is the fuzzy dominance class of $z \in U$ with respect to R_D.

Theorem 4.1 (Monotonicity) *Let* $IS = (U, C \cup D, V, f)$ *be a linguistic-valued decision information system, and* $B \subseteq A \subseteq C$, *then* $\forall x \in U, Pos_B(D)(x) \leq Pos_A(D)(x)$.

Definition 4.3 The dependency degree of D with respect to B is computed by

$$\gamma_B(D) = \frac{\sum_{x \in U} Pos_B(D)(x)}{|U|}$$

Obviously, $0 \le \gamma_B(D) \le 1$ and $\gamma_B(D) \le \gamma_C(D)$.

Definition 4.4 Let $IS = (U, C \cup D, V, f)$ be a linguistic-valued decision information system, for $B \subseteq C$, if $\gamma_B(D) = \gamma_C(D)$, then B is referred to as a lower approximate consistent set in IS. Furthermore, if B is a lower approximate consistent set and $\forall a \in B$, $\gamma_B(D) > \gamma_{B \setminus \{a\}}(D)$, then B is called a lower approximate reduction in IS.

To measure the significant degree of conditional attributes, here we present a new measure based on the condition entropy.

Definition 4.5 Let $IS = (U, C \cup D, V, f)$ be a linguistic-valued decision information system, the condition entropy of the decision attribute set D relative to the conditional attribute subset B is defined as

$$H(D|B) = - \frac{e}{|U|^3} \sum_x \left(\sum_z \left(\sum_y (R_B(y, x) \wedge R_D(y, z)) \log \frac{\sum_y (R_B(y, x) \wedge R_D(y, z))}{\sum_y R_B(y, x)} \right) \right)$$

It is clear that $H(D|B) \ge 0$.

Theorem 4.2 *Let $IS = (U, C \cup D, V, f)$ be a linguistic-valued decision information system and let $B \subseteq C$, then*
$$H(D|B) \ge H(D|C).$$

Theorem 4.3 *Let $IS = (U, C \cup D, V, f)$ be a linguistic-valued decision information system. If $B \subseteq C$, then $\gamma_B(D) = \gamma_C(D) \Leftrightarrow H(D|B) = H(D|C)$.*

Proof "$\Rightarrow$" $\gamma_B(D) = \gamma_C(D) \Leftrightarrow \forall x, Pos_B(D)(x) = Pos_C(D)(x)$. Assume $x_k \in U$, and it satisfies $\underline{R_B}([x_k]_D)(x) = Pos_B(D)(x)$. According to the monotonicity of $Pos_B(D)(x)$, we have $R_D(y, x_k) = \vee_i R_D(y, x_i)$. As $Pos_C(D)(x)$ is increasing with $R_D(y, x_i)$, we have $\underline{R_C}([x_k]_D)(x) = Pos_C(D)(x)$. On the other hand, $\forall x, Pos_B(D)(x) = Pos_C(D)(x)$, $\underline{R_B}([x_k]_D)(x) = \underline{R_C}([x_k]_D)(x)$ hold. According to the monotonicity of $\underline{R_B}(F)(x)$, we have
$$\forall x \in U, Pos_B(D)(x) = Pos_C(D)(x) \Leftrightarrow \frac{R_B(y,x) \wedge R_D(y,x_i)}{R_B(y,x)} = \frac{R_C(y,x) \wedge R_D(y,x_i)}{R_C(y,x)}$$
$\Leftrightarrow R_B(y, x) = R_C(y, x)$ or $\forall y, x, x_i \in U, R_B(y, x) \le R_D(y, x_i)$ and $R_B(y, x) \le R_D(y, x_i)$ holds.

When $R_B(y, x) = R_C(y, x)$, we have $H(D|B) = H(D|C)$. When $\forall y, x, x_i \in U$, $R_B(y, x) \le R_D(y, x_i)$ and $R_C(y, x) \le R_D(y, x_i)$. Thus $H(D|B) = H(D|C)$, and so $H(D|B) = H(D|C)$.
"$\Leftarrow$": $H(D|B) = H(D|C) \Rightarrow$

$$\sum_{x_i} \left(\sum_y (R_B(y,x) \wedge R_D(y,x_i)) \log \frac{\sum_y (R_B(y,x) \wedge R_D(y,x_i))}{\sum_y R_B(y,x)} \right)$$

$$= \sum_{x_i} \left(\sum_y (R_C(y,x) \wedge R_D(y,x_i)) \log \frac{\sum_y (R_C(y,x) \wedge R_D(y,x_i))}{\sum_y R_C(y,x)} \right)$$

For the same monotonicity with respect to $R_D(y,x_i)$, $\forall x_i \in U$

$$\sum_y (R_B(y,x) \wedge R_D(y,x_i)) \log \frac{\sum_y (R_B(y,x) \wedge R_D(y,x_i))}{\sum_y R_B(y,x)}$$

$$= \sum_y (R_C(y,x) \wedge R_D(y,x_i)) \log \frac{\sum_y (R_C(y,x) \wedge R_D(y,x_i))}{\sum_y R_C(y,x)}$$

$\Leftrightarrow R_B(y,x) = R_C(y,x)$ or $\forall y, x, x_i \in U, R_B(y,x) \le R_D(y,x_i)$ and $R_B(y,x) \le R_D(y,x_i)$ holds.

It is the same with the equivalent condition in " $\Rightarrow$ ", so $H(D|B) = H(D|C) \Rightarrow \gamma_B(D) = \gamma_C(D)$. In conclusion, we have $H(D|B) = H(D|C) \Leftrightarrow \gamma_B(D) = \gamma_C(D)$.

According to Theorem 4.3, if a conditional attribute is added into a conditional attribute subset, the value of the conditional entropy decreasing monotonously. It implies that the entropy can be used to measure the significance of attributes.

Definition 4.6 Let $IS = (U, C \cup D, V, f)$ be a decision system and $B \subseteq C$. For any $a \in C \backslash B$, the significance of attribute a with respect to D is defined as

$$sig(a, B, D) = H(D|B) - H(D|B \cup \{a\}) \tag{2}$$

Based on the significance of attributes, one can obtain the reduction of linguistic-valued decision system through a heuristic method.

5 Experiments

In this section, we use the proposed method to compute the reduction of the linguistic decision information system presented in Sect. 3.1.

The degree of one object dominating another about each attribute can be computed by Definition 3.2. Table 2 lists the fuzzy dominance relation related to the attribute a_1

According to Definition 4.6, the significance of each attribute can be computed and the results are shown in the first line of Table 3. The second line of Table 3 shows the significance of each attribute computed by (1) based on the dominance-based rough set.

According to the monotonicity of the condition entropy, we first choose the attribute with the largest significance as the initial candidate reduction set and then

Table 2 The fuzzy dominance relation related to the attribute a_1

U	x_1	x_2	x_3	x_4	x_5	x_6	x_7	x_8	x_9	x_{10}
x_1	0.5000	0.2248	1.0000	0.8933	0.8933	0.2248	0.5000	0.2248	0.8933	0.5000
x_2	0.6885	0.5000	1.0000	1.0000	1.0000	0.5000	0.6885	0.5000	1.0000	0.6885
x_3	0.0000	0.0000	0.5000	0.2248	0.2248	0.0000	0.0000	0.0000	0.2248	0.0000
x_4	0.0476	0.0000	0.6291	0.5000	0.5000	0.0000	0.0476	0.0000	0.5000	0.0476
x_5	0.0476	0.0000	0.6291	0.5000	0.5000	0.0000	0.0476	0.0000	0.5000	0.0476
x_6	0.6885	0.5000	1.0000	1.0000	1.0000	0.5000	0.6885	0.5000	1.0000	0.6885
x_7	0.5000	0.2248	1.0000	0.8933	0.8933	0.2248	0.5000	0.2248	0.8933	0.5000
x_8	0.6885	0.5000	1.0000	1.0000	1.0000	0.5000	0.6885	0.5000	1.0000	0.6885
x_9	0.0476	0.0000	0.6291	0.5000	0.5000	0.0000	0.0476	0.0000	0.5000	0.0476
x_{10}	0.5000	0.2248	1.0000	0.8933	0.8933	0.2248	0.5000	0.2248	0.8933	0.5000

Table 3 The significance of each attribute with fuzzy rough set theory and rough set theory

Attributes	a_1	a_2	a_2	a_4	a_5
sig(DFR)	0.3478	0.5310	0.1243	0.0987	0.0000
sig(DR)	0	0	0.3	0	0

Table 4 The significance of each attribute with fuzzy dominance rough set model and dominance rough set model

Attributes	a_1	a_2	a_2	a_4	a_5	a_6
sig(DFR)	0.4800	0.5452	0.3856	0.4936	0.3536	0.5266
sig(DR)	0.1620	0.1620	0.0600	0.2300	0.1480	0.2920

add the conditional attribute with the second largest significance into the candidate reduction set. With this step, when the entropy of candidate reduction set is unchangeable, the candidate reduction set is our pursuit. With this procedure, we can obtain the reduction $\{a_1, a_2, a_3, a_4\}$. However, the reduction cannot be derived by using the dominance-based rough set, since all attributes but one has the significance 0.

In the following, we employ the Car Evaluation dataset from UCI machine learning data repository [15] to verify the efficiency and effectiveness of the proposed method of reduction of linguistic-valued decision information system. This dataset has 1728 objects and each object has 6 conditional attributes and one decision attribute. The values of all objects regarding attributes are linguistic variables.

With our proposed method, Table 4 lists the significance of all conditional attributes based on definition of the proposed entropy (2) and the dependency (1).

With the steps of computing reduction above, we can get the reduction of the Car Evaluation Data as $\{a_2, a_3, a_4, a_5, a_6\}$ with the proposed method. However, the reduction of this dataset can't be obtained with dominance-based rough set method.

6 Conclusions

In this paper, we proposed a new dominance-based fuzzy rough set model to deal with linguistic-valued decision information systems and a feature selection method is presented regarding such information systems. A rational dominance fuzzy relation is established. Based on this relation, a new measure to compute the significance of each attribute is developed. According to the significance together with heuristic method we get the reduction. Comparative experiments are carried out to verify the efficiency and effectiveness of the proposal. In the future, we will deal with more datasets from UCI and from real-world applications with the proposed model.

Acknowledgements This work was supported by the National Natural Science Foundation of China (grant no. 11471001).

References

1. Zadeh LA. Toward a theory of fuzzy information granulation and its centrality in human reasoning and fuzzy logic. Fuzzy Sets Syst. 1997;90(2):111–27.
2. Zadeh LA. Is there a need for fuzzy logic. In: Fuzzy Information Processing Society, 2008. Nafips 2008 Meeting of the North American. IEEE; 2008. p. 1–3.
3. Deng T, Chen Y, Xu W, et al. A novel approach to fuzzy rough sets based on a fuzzy covering. Inf Sci. 2007;177(11):2308–26.
4. Wu WZ, Mi JS, Zhang WX. Generalized fuzzy rough sets. Inf Sci—Inf Comput Sci Int J. 2003;151(3):263–82.
5. Yeung DS, Chen D, Tsang ECC, et al. On the generalization of fuzzy rough sets. IEEE Trans Fuzzy Syst. 2005;13(3):343–61.
6. Greco S, Inuiguchi M, Slowinski R. Dominance-based rough set approach using possibility and necessity measures. In: Rough sets and current trends in computing. Berlin, Heidelberg: Springer; 2002. p. 85–92.
7. Yang X, Qi Y, Yu DJ, et al. α-Dominance relation and rough sets in interval-valued information systems. Inf Sci. 2015;294(5):334–47.
8. Du WS, Hu BQ. Dominance-based rough set approach to incomplete ordered information systems. Elsevier Science Inc.; 2016.
9. Khalid A, Beg I. Incomplete interval valued fuzzy preference relations. Inf Sci. 2016;348: 15–24.
10. Du WS, Hu BQ. Dominance-based rough set approach to incomplete ordered information systems. Inf Sci. 2016;346(C):106–29.
11. Yang X, Yang J. Incomplete information system and rough set theory. In: Models and attribute reductions. Science Press: Springer; 2012.
12. Deng Ti, Yang C, Hu Q. Feature selection in decision systems based on conditional knowledge granularity. Int J Comput Intell Syst. 2011;4(4):655–71.
13. Yamaguchi D, Li GD, Nagai M. A grey-based rough approximation model for interval data processing. Inf Sci Int J. 2007;177(21):4727–44.
14. Zhang X, Mei C, Chen D, et al. Feature selection in mixed data: a method using a novel fuzzy rough set-based information entropy. Pattern Recogn. 2016;56(1):1–15.
15. Marko B, Blaz Z, et al. UCI Machine learning repository. http://archive.ics.uci.edu/ml/datasets/Car+Evaluation.

Circuit Fault Diagnosis Method of Wind Power Converter with Wavelet-DBN

Yanxing Liu, Yi Chai, Shanbi Wei and Zhixiang Luo

Abstract With the increasing wind capacity, the proportion of wind power in the grid is getting higher. Therefore, it is critical for the stable operation of the power grid to find out the location of the wind turbine failures. This paper proposes a fault diagnosis method of the wind turbine converter based on the deep belief network. Firstly, multiscale analysis of the signal is carried out by using wavelet transform to extract the characteristic vector of fault signal. DBN is used to obtain fault recognition models by supervised learning that uses the feature vector. Finally, the simulation results reveal that the method has a good ability to identify the converter fault.

Keywords Fault diagnosis · Wind power · Deep belief network · Wavelet transform

1 Introduction

In recent years, with the development of human society, the demand of energy has risen rapidly. Due to the rapid and massive consumption of traditional fossil fuels, leading to the growing global environmental pollution and greenhouse effects, as well as the non-renewable nature of fossil fuels, wind energy as a kind of green renewable energy is a priority for many countries.

As the installed capacity of the wind energy conversion system (WECS) increases rapidly, the failure rate of WECS gets higher and higher. Since WECS usually works in the harsh conditions of high temperatures, dust, oil and so forth, the wind power converters (WPCs) become one of the most vulnerable components. Therefore, quickly and accurately determining the fault type and fault

Y. Liu (✉) · Y. Chai · S. Wei · Z. Luo
State Key Laboratory of Power Transmission Equipment and System Security
and New Technology, College of Automation, Chongqing University,
Chongqing 400044, China
e-mail: 20151302015@cqu.edu.cn

© Springer Nature Singapore Pte Ltd. 2018
Y. Jia et al. (eds.), *Proceedings of 2017 Chinese Intelligent Systems Conference*,
Lecture Notes in Electrical Engineering 460,
https://doi.org/10.1007/978-981-10-6499-9_60

623

location of the converter is a necessary and significant strategy for guaranteeing the stable operation of the wind power system.

IGBT is usually used in wind power converter, hence, the converter failure mainly considered in this paper includes short-circuit and open-circuit of IGBT. In case of IGBT short-circuit fault, it can be real-time monitored through protection circuit integrated with the drive module, however, the diagnosis technology of IGBT open-circuit fault is still in the study. In [1, 2], a simulation analysis and diagnosis method is proposed. The fault current, voltage waveform obtained from the simulation of WPC with failure, are utilized to compare with the normal waveform for realizing fault diagnosis of the converter. Then, [3, 4] put forward a model-based fault diagnosis method, the failure is determined by comparing the state estimation of the state space model with the real output. Whereas, the main disadvantage is that the inaccurate mathematical model will affect the accuracy of the diagnosis. A knowledge-based fault diagnosis method is used via analyzing the current trajectory after the component of converter failures in [5]. Although the fault can be detected and diagnosed, as the outside electromagnetic interference, the current trajectory will be seriously affect, and it is not conducive to fault separation. A BP neural network fault diagnosis method based on waveform analysis is proposed in [6].

Based on the above research, the diagnosis method can be divided into two major categories: the model based and the model-free methods [7]. The model based methods mainly depends on the mathematical model of the object to be diagnosed, but the more complicated the system becomes, the more difficult to establish mathematical model, which makes the application of model-based method is very limited. While the model-free methods can overcome this drawback, therefore, based on the model-free method is more and more widely used in the field of fault diagnosis, especially with the development of deep learning [8].

Focus on the open-circuit fault of IGBT element in wind power converter, Wave-let analysis method and deep belief network (DBN) are used to diagnose the open fault in this paper. By analyzing the three-phase current signal of rectifier, the wavelet analysis is applied to extract the information at different scales as the data sample of DBN. Thus, the open-circuit fault diagnosis of WPC is realized.

2 Fault Analysis and Fault Feature Extraction

The structure diagram of wind power converter is show in Fig. 1. The method of fault diagnosis of inverter is the same as the method of rectifier. So only the rectifier is taken in consideration.

As shown in Fig. 1, there are 6 IGBTs in the rectifier so 21 kinds of failures can occur during the operation [9]. A table for faulty components and the responding fault label is illustrated in Table 1.

Part of the three-phase current signals are presented in Fig. 2 when open circuit faults occur.

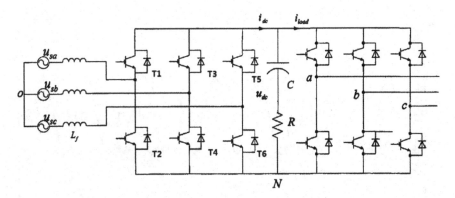

Fig. 1 The structure of converter

Table 1 The faulty components and the responding fault label

Faulty components	Fault label	Faulty component	Fault label
T1	1	T3, T5	12
T2	2	T2, T4	13
T3	3	T2, T6	14
T4	4	T4, T6	15
T5	5	T1, T4	16
T6	6	T1, T6	17
T1, T2	7	T2, T3	18
T3, T4	8	T3, T6	19
T5, T6	9	T2, T5	20
T1, T3	10	T4, T6	21
T1, T5	11		

Wind power converter fault signal as shown in Fig. 2 is a typical non-stationary signal, which can't be fully extracted its characteristic information with Fourier transform method. Wavelet analysis is a typical time-frequency analysis method with multi-resolution characteristics by which it can gradually observe the signal from coarse to fine and it can also be seen as a group of band pass filter to filter the signal. A telescopic window can be obtained by appropriately selecting the scale factor and the translation factor. With appropriate selection of the fundamental wavelet, the wavelet transform has the ability to characterize the local characteristics of the signal in both the time domain and the frequency domain. According to this feature, the multi-resolution characteristic can be applied to the extraction of the power spectrum characteristics of the signal. The Mallat algorithm, which is a fast algorithm of wavelet decomposition and reconstruction, can effectively extract the power spectrum characteristics.

In this paper, the three-phase current signal of wind power converter is decom-posed by 5 layers in db3 wavelet base function. Signal reconstruction is

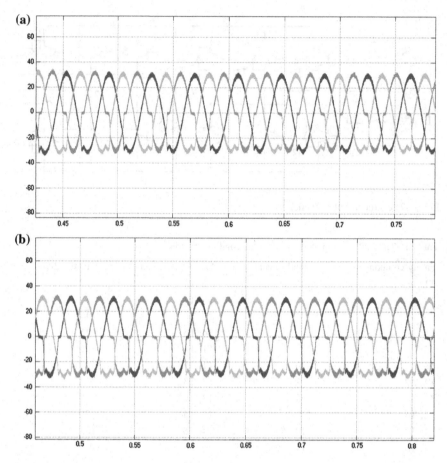

Fig. 2 Signals when open circuit fault occurs. **a** T1; **b** T2

carried out in each detail layer, and then the energy value of each layer is obtained. The magnitude of the energy value of the fault signal under different scale decomposition reflects its characteristics at the corresponding frequency of the scale, and the energy values are arranged in a row in order to form a vector. The vector can represent the characteristics of the fault signal at each frequency. Although most of the fault signals differ significantly in the characteristics at each scale, there are no differences in the frequency characteristics of the different fault signals as shown in the following figure, because the fault signal is the same, but the location is different (Fig. 3).

The time-frequency characteristic of the wavelet analysis can accurately deter-mine the position of the fault. Therefore, the position characteristic variable is introduced. When faults occur in the positive half cycle, the variable is defined to be 1, while it is equal to −1 in the negative half cycle, otherwise, it is 0.

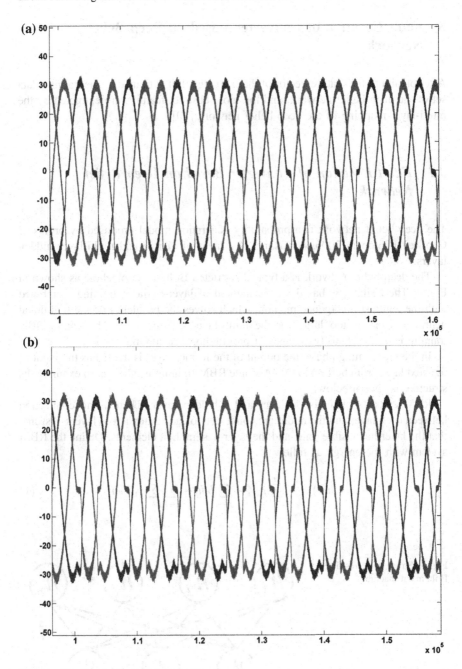

Fig. 3 Different signals with the same frequency

3 Fault Location of Converter Based on Deep Belief Network

After obtaining the feature vector of the fault signal, it is trained as the training data set of the deep belief network, and the more abstract features are extracted by the multi-layer mapping of the deep belief network [10].

3.1 Fault Location of Converter Based on Deep Belief Network

The deep belief network is a probability generation model proposed by professor Geoffrey Hinton in 2006, consisting of a layer of visible and multi-layer hidden layers.

The deep belief network is a typical restricted Boltzmann machine as shown in Fig. 4. The training is based on the method of layer-by-layer training, combined with the contrast divergence method, which solves the problem that the traditional neural network is too large for the multi-layer training [11]. The whole DBN training is divided into two stages of pre-training and tuning.

In the pre-training phase, the output of the former layer is treated as the input of the next layer from bottom to top. A single RBM training is taken as an example. Its structure is shown below.

RBM consists of a visible layer and a hidden layer. There is no connection between each layer and the nodes are connected to each other across layers. Assume that the layer has visible units and the layer has implicit elements. Define the RBM system with the energy as follow.

$$E(v, h|\theta) = - \sum_{i=1}^{n} a_i v_i - \sum_{j=1}^{m} b_j h_j - \sum_{i=1}^{n} \sum_{j=1}^{m} v_i W_{ij} h_j \qquad (1)$$

Fig. 4 Structure of restricted Boltzmann machine

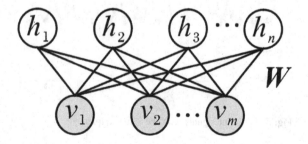

where v_i is the state of the visible unit i; h_j is the state of the hidden unit j; $\{\theta = W_{ij}, a_i, b_j\}$ is the parameter of RBM, W_{ij} is the connection weight between the visible unit i and the hidden unit j, a_i is the offset of the visible element, b_j is the Hide the offset of unit j. Based on this energy function, the joint probability distribution of RBM model is

$$P(v, h|\theta) = \frac{e^{-E(v, h|\theta)}}{Z(\theta)}, \quad Z(\theta) = \sum_{v, h} e^{-E(v, h|\theta)} \tag{2}$$

where $Z(\theta)$ is the normalization factor, also known as the partition function. Training RBM is to calculate the values of θ so that the RBM model can fit the sample data. Parameter θ can be obtained by maximizing RBM in the likelihood function on the training set (assuming T samples are included) [12].

$$\theta^* = \arg\max_{\theta} \ell(\theta) = \arg\max \sum_{t=1}^{T} \log P(v^{(t)}|\theta) \tag{3}$$

the optimal parameter θ^* can be obtained by finding the maximum value of (4) with the stochastic gradient rise method.

$$\ell(\theta) = \sum_{t=1}^{T} \log P(v^{(t)}|\theta) \tag{4}$$

The key is to calculate the partial derivative of $\log P(v^{(t)}|\theta)$ for θ.

Due to

$$\ell(\theta) = \sum_{t=1}^{T} \log P(v^{(t)}|\theta) = \sum_{t=1}^{T} \log \sum_{h} \log P(v^{(t)}, h|\theta)$$

$$= \sum_{t=1}^{T} \log \frac{\sum_{h} \exp[-E(v^{(t)}, h|\theta)]}{\sum_{v} \sum_{h} \exp[-E(v, h)|\theta]} \tag{5}$$

$$= \sum_{t=1}^{T} \left(\log \sum_{h} \exp[-E(v^{(t)}, h|\theta)] - \log \sum_{v} \sum_{h} \exp[-E(v, h)|\theta] \right)$$

The logarithmic likelihood function's partial derivative for θ is as follow.

$$\ell(\theta) = \sum_{t=1}^{T} \frac{\partial}{\partial\theta} \left(\log \sum_{h} \exp[-E(v^{(t)}, h|\theta)] - \log \sum_{v} \sum_{h} \exp[-E(v, h)|\theta] \right)$$

$$= \sum_{t=1}^{T} \left(\sum_{h} \frac{\exp[-E(v^{(t)}, h|\theta)]}{\sum_{h} \exp[-E(v^{(t)}, h|\theta)]} \times \frac{\partial(-E(v^{(t)}, h|\theta))}{\partial\theta} \right.$$

$$\left. - \sum_{v} \sum_{h} \frac{\exp[-E(v, h|\theta)]}{\sum_{v} \sum_{h} \exp[-E(v, h|\theta)]} \times \frac{\partial(-E(v, h|\theta))}{\partial\theta} \right) \tag{6}$$

$$= \sum_{t=1}^{T} \left(\left\langle \frac{\partial(-E(v^{(t)}, h|\theta)}{\partial\theta} \right\rangle_{P(h|v^{(t)}, \theta)} - \left\langle \frac{\partial(-E(v, h|\theta)}{\partial\theta} \right\rangle_{P(v, h|\theta)} \right)$$

where $\langle \cdot \rangle_p$ is the mathematical expectation of distribution P. $P(h|v^{(t)}, \theta)$ is the probability distribution of the hidden layer where the visible unit is the training sample v.

$$P(h_j = 1|v, \theta) = \sigma(b_j + \sum_i v_i W_{ij}) \tag{7}$$

where $\sigma(x) = \frac{1}{1 + \exp(-x)}$ is the sigmoid activation function. Since the structure of the RBM is symmetric, when the state of the hidden unit is given, the activation probability of the i-th visible unit is.

$$P(v_i = 1|h, \theta) = \sigma(a_i + \sum_j W_{ij} h_j) \tag{8}$$

$P(h, v|\theta)$ denotes the joint distribution of the visible unit and the hidden unit. The value of $Z(\theta)$ must be calculated to determine the distribution. However, the computational complexity is $O(2^{n+m})$. Obviously this cannot be directly calculated. Some sampling methods are usually used to obtain their approximations. The partial derivatives of W, a, b are

$$\frac{\partial \log P(v|\theta)}{\partial W_{ij}} = \langle v_i h_j \rangle_{\text{data}} - \langle v_i h_j \rangle_{\text{model}}$$

$$\frac{\partial \log P(v|\theta)}{\partial a_i} = \langle v_i \rangle_{\text{data}} - \langle v_i \rangle_{\text{model}} \tag{9}$$

$$\frac{\partial \log P(v|\theta)}{\partial b_j} = \langle h_j \rangle_{\text{data}} - \langle h_j \rangle_{\text{model}}$$

After the pre-training is completed, each RBM can get the initial parameters, constituting the initial framework of DBN [13]. Then need to adjust the DBN training, and further optimize the parameters of the network layer, so that the network can have a better discrimination. The tuning process is a supervise learning process [14], it uses label data presented in Table 1 for training using BP algorithm for fine tuning of the network parameters, and ultimately makes the network to achieve global optimal. This performance is better than the simple BP algorithm training, because it only needs to search the network parameters of a local search, compared to BP neural network, training faster, and the convergence time is short.

3.2 Realization of Fault Location Method Based on Deep Belief Network

The specific steps of fault location based on the deep belief network are shown as follows

(1) Run the simulation model under different faults condition to get the signal of the three-phase current on the machine side.
(2) Use the wavelet transform to decompose the three-phase current signal at each scale to obtain the corresponding energy coefficient at each scale
(3) Construct the characteristic vector with energy coefficient and normalize it
(4) Use the obtained characteristic vector to train the deep belief network and obtain the model parameters
(5) Use the test data set to test the classification accuracy of the deep belief network

4 Analysis of Experimental Results

In the experiments, a permanent magnet synchronous generator(PMSG) model as show in Fig. 5 is established to simulate the fault diagnosis of wind power converter in the Matlab/Simulink environment (Fig. 5).

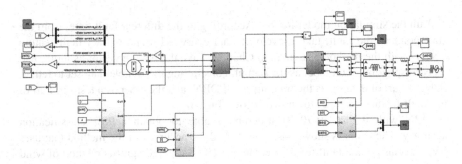

Fig. 5 Simulink model of PMSG

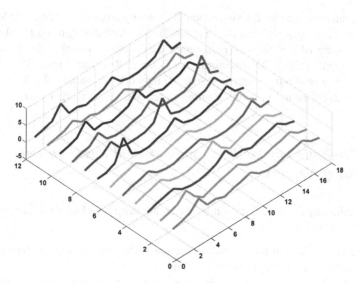

Fig. 6 Fault feature of different switches

Table 2 The diagnosis results of DBN	Faulty components	Output	Faulty component	Output
	T1	1	T3, T5	12
	T2	2	T2, T4	13
	T3	3 1	T2, T6	14
	T4	4	T4, T6	15
	T5	5	T1, T4	16
	T6	6	T1, T6	17
	T1, T2	7	T2, T3	18
	T3, T4	8	T3, T6	19
	T5, T6	9	T2, T5	20
	T1, T3	10	T4, T6	21
	T1, T5	11		

Run the simulation model for 2 s. According to the different fault time, from the start point of the fault, take a cycle of data every 0.1 s and extract fault feature vectors with the method presented in Chap. 2. Take the feature vector as input, and its corresponding fault label as the output, constituting the sample data set together. Select a part of data set as the training set of DBN, and the other as a test set. Part of the fault feature vectors are shown below (Fig. 6).

DBN with the 20-70-70-30 structure is able to obtain a 98% classification accuracy. The result is shown in Table 2. And the accuracy of the method that uses SVM to classify the fault feature vectors is 91%. So fault diagnosis method of wind power converter proposed in this paper is more accurate and effective.

5 Conclusion

With the wind power capacity keep increasing, safe and reliable operation of wind power system become more and more practical. In this paper, a fault diagnosis algorithm for the converter of wind turbines are proposed, and these techniques are verified through simulation results. The considered fault types are switching device open faults. Faulty switches can be identified by deep belief net-work which is trained by the three-phased current signal that is transformed by the Wavelet transformation first. The simulation results illustrate that the method is accurate and effective.

References

1. Wallcae AK, Spee R. The simulation of brushless dc driver failures. In: PESC record-IEEE power electronics specialists conference; 1988. p. 199–206.
2. Speed R, Wallace AK. Remedial strategies for brushless dc drive failures. IEEE Trans Ind Appl. 1990;26(2):259–66.
3. Catellani S, Saadi J, Champenois G. Monitoring of a static converter fed machine using average reference models. In: Proceedings of the conference IFAC'91 safeprocess, vol. 2; 1991. p. 205–9.
4. Berendsen CS, Champenois G, Davoine J. How to detect and to localize a fault in a DC/DC converter. In: IEEE-IECON'92; 1992.
5. Peuget R, Courtine S, Rognon JP. Fault detection and isolation on a PWM inverter by knowledge-based model. IEEE Trans Ind Appl. 1998;34(6):1318–26.
6. Hinton GE, Osindero S, Teh YW. A fast learning algorithm for deep belief nets. Neural Comput. 2006;18(7):1527–54.
7. Yang S, Li W, Wang C. The intelligent fault diagnosis of wind turbine gearbox based on artificial neural network. In: International conference on condition monitoring and diagnosis, 2008. CMD 2008. IEEE; 2008. p. 1327–30.
8. Kamel T, Biletskiy Y, Diduh CP, et al. Open circuit fault diagnoses for power electronic converters. In: 2015 IEEE 24th international symposium on industrial electronics (ISIE). IEEE; 2015. p. 361–6.
9. Karimi S, Gaillard A, Poure P, et al. FPGA-based real-time power converter failure diagnosis for wind energy conversion systems. IEEE Trans Industr Electron. 2008;55(12):4299–308.
10. Yang S, Li W, Wang C. The intelligent fault diagnosis of wind turbine gearbox based on artificial neural network. In: International conference on condition monitoring and diagnosis, 2008. CMD 2008. IEEE; 2008. p. 1327–30.
11. Ye F, Zhang Z, Chakrabarty K, et al. Board-level functional fault diagnosis using artificial neural networks, support-vector machines, and weighted-majority voting. IEEE Trans Comput Aided Des Integr Circuits Syst. 2013;32(5):723–36.
12. Lin J, Qu L. Feature extraction based on Morlet wavelet and its application for mechanical fault diagnosis. J Sound Vib. 2000;234(1):135–48.
13. Li B, Chow MY, Tipsuwan Y, et al. Neural-network-based motor rolling bearing fault diagnosis. IEEE Trans Industr Electron. 2000;47(5):1060–9.
14. Duan Q, Rong X, Zhang L. Open circuit fault diagnosis of dual PWM converter for doubly fed wind power generation system. Electr. Drive. 2010;40(04):32–5.

Adaptive Tracking a Linear System with Unknown Periodic Signal in Multi-agent Systems

Feifei Li and Ya Zhang

Abstract This paper studies the tracking control problem of multi-agent systems where each agent has homogeneous sensor and heterogeneous dynamic system, the moving target has unknown periodic input signal and the unknown periodic input can be modelled as a finite dimensional Fourier decomposition. Since some agents can not detect the target, a distributed estimation based tracking control algorithm is applied. We first design a consensus based distributed observer to estimate the state and the unknown periodic input of the system from the available measurement outputs. Leader-follower consensus protocol is applied, and the stability condition of the estimation errors is given. Then, based on the estimations, a model reference adaptive control (MRAC) algorithm is adopted to design the tracking controller. It is proved that under the proposed distributed estimation based tracking control algorithm, each agent can asymptotically track the target. A numerical simulation is given to prove the feasibility of the algorithm in this paper.

Keywords Multi-agent systems · Tracking control · Distributed estimation MRAC

1 Introduction

Tracking control is one of important problems in multi-agent systems due to its wide applications. In practical applications, some input signal of the dynamical target may be unknown to all agents, and just some agents in the sensing range can get the measurements of the target. To successfully track the target, each agent should cooperate, and not only design the tracking controller but also design the observer. This paper

F. Li · Y. Zhang(✉)
School of Automation, Southeast University, Nanjing 210096, China
e-mail: yazhang@seu.edu.cn

F. Li · Y. Zhang
Key Laboratory of Measurement and Control of Complex Systems of Engineering,
Ministry of Education, Nanjing 210096, China

© Springer Nature Singapore Pte Ltd. 2018
Y. Jia et al. (eds.), *Proceedings of 2017 Chinese Intelligent Systems Conference*,
Lecture Notes in Electrical Engineering 460,
https://doi.org/10.1007/978-981-10-6499-9_61

focuses on designing coordinated controller and observer for multi-agent systems to track a target with unknown input signal.

There have been many works on distributed tracking control of multi-agent systems. Adaptive control method is widely applied in many problems including estimation. Model Reference Adaptive Control (MRAC) is used to guarantee the stability and the accuracy of estimation and tracking. References [1, 2] studied the adaptive control for discrete-time systems with nonlinear input and switching topology through a dynamic distributed state feedback control law. Miller [3] proposed a new method of adaptive control by extending the controller to the time-varying way. Liu and Jia [4] studied the adaptive control in multi-agent systems with bounded external disturbance, the model of the dynamic was partly unknown. Some paraments of the system were needed to be determined, and a reference model was adopted to get the unknown part. Ye and Cao [5] used the adaptive control method to solve the fault-tolerant tracking problems in multi-agent systems.

In many distributed tracking problems, each agent can just obtain the measurement outputs of the target while the state information is not available. Thus an observer based tracking control is applied. Since just part of agents can obtain the target's measurement output due to limited sensing range, coordinated estimation algorithm should be applied. There have been many works on distributed estimation of multi-agent systems [6–10], while in them it is assumed that the target's system parameters are precisely known to all agents. Moreover, it is important to design the cooperation protocol of the sensor network because the stability and estimation accuracy are strictly connected to the topology and communication algorithm, which has been studied in [11–18]. Leader-follower protocol is discussed in [13, 14], while consensus algorithm is studied on in [15–18].

This paper studies the distributed tracking control problem of heterogeneous multi-agent systems in which the target has an unknown input signal and just part of agents can obtain the target's measurement outputs. A distributed estimation based tracking control is applied. In distributed estimation part, both coordinated estimation of unknown input signal and coordinated observer of target's state are applied. Unlike most of existing works on estimating the unknown signal with the method of fitting of a polynomial, in this paper we take the strategy of destructing the estimation model through its Fourier decomposition within a certain number of dimensions with the assumption that the period of the signal is known. By applying leader-follower consensus algorithm to Fourier parameter estimations and state estimations, each agent can asymptotically estimate the target's unknown input signal and target's state. Then, an adaptive state feedback controller is constructed to track the state of the target. Based on the convergence of the estimation errors of the state and unknown input, and by using the Lyapunov functional analysis, We prove the stability of the tracking errors. Our main contribution in this paper is to track the target with unknown but periodical signal input for multi-agent systems with limited sensing range. Leader-follower consensus estimation algorithm with coordinated parameter estimation is proposed, and an adaptive tracking controller based on estimations is provided.

2 Problem Formulation

Consider a linear continuous dynamic system to be tracked satisfies:

$$\dot{x}_0 = A_0 x_0(t) + B_0 r(t), \tag{1}$$

where $x_0(k) \in \mathbb{R}^n$ denotes the state of the system, $A_0 \in \mathbb{R}^{p \times p}$ denotes the system matrix of the dynamic system, $r(t) \in \mathbb{R}^{r \times 1}$ denotes an unknown periodical input signal, $B_0 \in \mathbb{R}^{p \times r}$ denotes the input matrix. A_0 is assumed to be Hurwitz stable, and A_0, B_0 are known.

It is assumed that the unknown periodic input $r(t)$ can be modelled as a finite dimensional Fourier decomposition, and assume its period T_0 is known, then the T_0-periodic signal can be described as:

$$r(t) \triangleq \sum_{s \in \ell_h} c_s e^{i s \omega_0 t} \tag{2}$$

where ℓ_h denotes the h modes, $\omega_0 = 2\pi / T_0$.

In the network, there are n agents with different dynamic equations:

$$\dot{x}_i(t) = A_i x_i(t) + B_i u_i(t) \tag{3}$$

and homogeneous measuring equations

$$y_i(t) = C x_0(t) \tag{4}$$

where $x_i(t) \in \mathbb{R}^p$ denotes the state of the ith agent of the network, $A_i \in \mathbb{R}^{p \times p}$ denotes the system matrix of agent i, $B_i \in \mathbb{R}^{p \times r}$ denotes the input matrix of agent i, $u_i(t)$ denotes its control input, $y_i(t) \in \mathbb{R}^q$ denotes the measurement output of the ith agent. Assume (A_0, C) is observable.

Due to limited sensing range, in the network there are some agents that can not detect the target system. Our goal is to design tracking controllers for all agents to track the state of the target.

3 Main Result

In this section, we propose a consensus estimation based tracking control algorithm composed of consensus based estimation algorithms and adaptive tracking controllers, and analyse the stability of the tracking error system.

3.1 Target Estimation

In this subsection, we propose a leader-follower consensus based estimation algorithm to estimate the state and the unknown periodic input of the system (1) from the available measurement outputs.

Chauvina [19] has estimated the unknown periodic signal estimation through its Fourier decomposition within a certain number of dimensions. To give our results, we firstly introduce a *lemma* from [19]:

Lemma 1 *Consider a system (1) and output equation* $y(t) = Cx_0(t)$, $Ker(A) = Ker(C^T) = \{0\}$, $Ker(A)$ *denotes the set of all vectors which map to the zero vector of matrix A. If we apply observers*

$$
\begin{cases}
\dot{\hat{x}}(t) = A_0\hat{x} + B_0 \sum_{s \in \varphi_h} \hat{c}_s e^{is\omega_0 t} - F(C\hat{x}(t) - y(t)) \\
\dot{\hat{c}}_s = -e^{-is\omega_0 t} L_s(C\hat{x}(t) - y(t)), \forall s \in \ell_h
\end{cases}
\tag{5}
$$

where the gains F, L_s *are designed as:*

$$
L_s \triangleq \varepsilon\beta_s[is\omega_0 - (A_0 - FC)^{-1}B_0]^\dagger C^T,
\tag{6}
$$

F *is the observe gain that* $A - FC$ *is Hurwitz stable,* β_s *are strictly positive reals,* $\dagger$ *denotes the Hermitian transpose. Then for small enough* $\varepsilon > 0$, *the estimation error systems*

$$
\begin{cases}
\dot{\tilde{x}} = (A_0 - FC)\tilde{x}(t) + B_0 \left(\sum_{k \in \ell_h} \tilde{c}_k e^{ik\omega_0 t} \right) \\
\dot{\tilde{c}}_k = -e^{-ik\omega_0 t} L_k C\tilde{x}, \forall k \in \ell_h
\end{cases}
$$

converge to 0, where $\tilde{x} = \hat{x} - x$, $\tilde{c}_k = \hat{c}_k - c_k$.

First, we take into account the leader-follower consensus protocol of the sensor network. In the network just part of the agents can observe the target because of the distance and other limits. We define the agents that can directly measure the target as the leaders, the agents that can not detect the target as the followers. Then, for the leaders, from Lemma 1 we can design the observers as following:

$$\begin{cases} \dot{\hat{x}}_i = A_0\hat{x}_i(t) + F(y_i(t) - C\hat{x}_i(t)) + B_0 r_i(t) \\ \dot{\hat{\alpha}}_{si} = -e^{-is\omega_0 t} L_s(C\hat{x}_i(t) - y_i(t)) \\ L_s \triangleq \epsilon\beta_s[is\omega_0 - (A_0 - FC)^{-1}B_0]^\dagger C^T, \\ r_i(t) = \sum_{s\in\ell_h} \hat{\alpha}_{si} e^{is\omega_0 t} \end{cases} \tag{7}$$

where $\hat{x}_i$ denotes the estimation of system state made by the ith node in the network, $\hat{\alpha}_{si}$ denotes the parameter estimation of α_s in the unknown periodic input $r(t)$ made by the ith node, $1 \le s \le h$, F is an observer gain stabilizing $A_0 - FC$.

For the followers, since direct measurement outputs can not be obtained, consensus protocol is applied:

$$\begin{cases} \dot{\hat{x}}_i = A_0\hat{x}_i(t) + B_0 r_i(t) + w_1 \sum_{j=1}^n a_{ij}(\hat{x}_j(t) - \hat{x}_i(t)) \\ \dot{\hat{\alpha}}_{si} = w_2 \sum_{j=1}^n a_{ij}(\hat{\alpha}_{sj} - \hat{\alpha}_{si}) \\ r_i(t) = \sum_{s\in\ell_h} \hat{\alpha}_{si} e^{is\omega_0 t} \end{cases} \tag{8}$$

Define $e_i = \hat{x}_i - x_0$, $\tilde{\alpha}_{si} = \hat{\alpha}_{si} - \alpha_s$ as the estimation error accordingly. From (1) and (7), the estimation errors of the leaders can be described as

$$\begin{cases} \dot{e}_i = (A_0 - FC)e_i + B_0 \sum_{s\in\ell_h} \tilde{\alpha}_{si} e^{is\omega_0 t} \\ \dot{\tilde{\alpha}}_{si} = -e^{-is\omega_0 t} L_s C e_i \end{cases}$$

According to Lemma 1, when $\epsilon > 0$ is small enough, the estimation errors of the leaders converge to 0.

For follower i, its estimation errors have

$$\begin{cases} \dot{e}_i = A_0 e_i + B_0 \sum_{s\in\ell_h} \tilde{\alpha}_{si} e^{is\omega_0 t} + w_1 \sum_{j=1}^n a_{ij}(e_j - e_i) \\ \dot{\tilde{\alpha}}_{si} = w_2 \sum_{j=1}^n a_{ij}(\tilde{\alpha}_{sj} - \tilde{\alpha}_{si}) \end{cases}$$

Renumber the nodes such that the leader's number is smaller than all followers', and assume there are m leaders. Define $\tilde{\alpha}_s = [\tilde{\alpha}_{s(m+1)}^T, \ldots, \tilde{\alpha}_{sn}^T]$, $e = [e_{(m+1)}^T, \ldots, e_n^T]^T$, then

$$\dot{\tilde{\alpha}}_s = -w_2(L_1 \otimes I)\tilde{\alpha}_s - w_2 L_0 \otimes I[\tilde{\alpha}_{s1}^T, \dots, \tilde{\alpha}_{sm}^T]^T,$$

$$\dot{e} = (I \otimes A_0 - w_1 L_1 \otimes I)e - w_1 L_0 \otimes I[e_1^T, \dots, e_m^T]^T + \sum_{s \in \ell_h} I \otimes B_0 \tilde{\alpha}_s e^{is\omega_0 t},$$

where $\begin{bmatrix} 0 & 0 \\ L_0 & L_1 \end{bmatrix}$ is the Laplacian matrix of the topology.

Obviously, if the target node in the extended topology is globally reachable, then all eigenvalues of L_1 have positive real parts. Since for $1 \le i \le m$, $\tilde{\alpha}_{si} \to 0$, $e_i \to 0$, in the above consensus of parameters, for any positive value $w_2 > 0$, $\tilde{\alpha}_s(t)$ converges to 0 if the target node in the extended topology is globally reachable. And sequentially, for any positive gain w_1, all eigenvalues of $I \otimes A_0 - w_1 L_1 \otimes I$ have negative real parts. Then we can obtain the following theorem.

Theorem 1 *Consider a dynamic system (1) with unknown periodic input (2). If we apply leader-follower consensus based estimation algorithm (7), (8) to the agents, then for small enough $\varepsilon > 0$, the estimation errors of the agents converge to 0 if the target node in the extended topology is globally reachable.*

3.2 Target Tracking

Tracking control of the multi-agent system will be discussed in this subsection. Mode Reference Adaptive Control (MRAC) is adopted. Each agent applies its estimations of target's state and periodic input to construct the adaptive tracking controller. The stability of the estimations is already proven in the above subsection, so it is reasonable to track the target by using the estimations of the dynamic.

To begin with, we assume that there exist constant matrices B_i^*, K_i^* such that

$$A_0 = A_i + B_i H_i^*, B_0 = B_i K_i^*. \tag{9}$$

We construct the control law:

$$u_i(t) = H_i(t)x_i(t) + K_i(t)r_i(t)$$
$$\dot{H}_i = -\Gamma_{2i} x_i(t)\tilde{x}_i^T(t)PB_i \tag{10}$$
$$\dot{K}_i = -\Gamma_{1i} r_i(t)\tilde{x}_i^T(t)PB_i$$

where $\tilde{x}_i = x_i - \hat{x}_i$, $\tilde{H}_i(t) = H_i(t) - H_i^*$, $\tilde{K}_i(t) = K_i(t) - K_i^*$. $\Gamma_{1i}, \Gamma_{12i}$ are symmetric positive definite matrices which indicate the adaptive gains of the protocol, P is a positive definite matrix satisfying $A_0^T P + PA_0 < 0$.

Define $\tilde{e}_i = x_i - x_0$ as the tracking error of agent i, $\tilde{r}_i = r_i - r$. Then we have the following tracking errors:

$$\dot{e}_i = A_i x_i + B_i(H_i(t)x_i(t) + K_i(t)r_i(t)) - A_0 x_0(t) - B_0 r(t)$$
$$= A_i x_i(t) + B_i(H_i^* x_i(t) + K_i^* r_i(t)) + B_i((H_i(t) - H_i^*)x_i(t)$$
$$+ (K_i(t) - K_i^*)r_i(t)) - A_0 x_0(t) - B_0 r(t) \tag{11}$$
$$= A_0 \tilde{e}_i(t) + B_0 \tilde{r}_i(t) + B_i \tilde{H}_i(t)x_i(t) + B_i \tilde{K}_i(t)r_i(t)$$

Theorem 2 *Consider a dynamic system satisfies (1) with unknown periodic input (2). If we apply consensus estimation based adaptive tracking controller (10) with consensus estimation algorithm (7) and (8) to the agents, then for small enough $\varepsilon > 0$, the tracking errors of the agents converge to 0 if the target node in the extended topology is globally reachable.*

Proof First, we construct a Lyapunov function with $\tilde{e}, \tilde{H}, \tilde{K}$ taken into account. The Lyapunov function for system (11) can be designed as:

$$V_i(t) = \tilde{e}_i^T(t)P\tilde{e}_i(t) + tr(\tilde{K}_i^T \Gamma_{1i}^{-1}\tilde{K}_i + \tilde{H}_i^T \Gamma_{2i}^{-1}\tilde{H}_i) \tag{12}$$

where tr denotes the trace of a matrix. According to the control laws (10), the first-order derivative of $V_i(t)$ is:

$$\dot{V}_i(t) = \tilde{e}_i^T(A_0^T P + PA_0)\tilde{e}_i$$
$$+ (B_0\tilde{r}_i + B_i\tilde{H}_i x_i + B_i\tilde{K}_i r_i)^T P\tilde{e}_i + \tilde{e}_i^T P(B_0\tilde{r}_i + B_i\tilde{H}_i x_i + B_i\tilde{K}_i r_i)$$
$$+ 2tr(\tilde{K}_i^T \Gamma_{1i}^{-1}\dot{\tilde{K}}_i + \tilde{H}_i^T \Gamma_{2i}^{-1}\dot{\tilde{H}}_i) \tag{13}$$
$$= \tilde{e}_i^T(A_0^T P + PA_0)\tilde{e}_i + 2\tilde{e}_i^T PB_0\tilde{r}_i(t)$$
$$+ 2e_i^T(t)PB_i\tilde{K}_i^T r_i + 2e_i^T(t)PB_i\tilde{H}_i^T x_i(t)$$

In the above subsection, we already prove the convergence of the estimation errors of the dynamic system, i.e., $e_i \to 0$, $\tilde{r}_i \to 0$. Moreover, since A_0 is Hurwitz, $x_i(t)$ must be bounded. We can consider the Lyapunov function by letting $\tilde{r}_i(t) = 0$ and $e_i(t) = 0$. So the first-order derivative of the Lyapunov function can be:

$$\dot{V}_i(t) = \tilde{e}_i^T(A_0^T P + PA_0)\tilde{e}_i < 0$$

From Barbalat's Lemma, we can conclude that $\lim_{t\to\infty} \tilde{e}_i(t) = 0$. The system (11) is asymptotically stable, which complete the proof of the theorem.

4 Examples

Consider a dynamic system with $A_0 = \begin{bmatrix} -1 & -2 & -2 \\ 0 & -1 & 1 \\ 1 & 0 & -1 \end{bmatrix}$, $B_0 = \begin{bmatrix} 2 \\ 0 \\ 1 \end{bmatrix}$, $C_0 = \begin{bmatrix} 1 & 0 & 0 \end{bmatrix}$. There are five agents in the network. Agent 1,3 can obtain the measurement outputs of the

Fig. 1 Topology of the
communication network

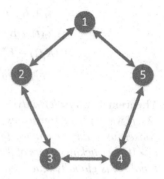

target and are considered as the leaders of the network. Agent 2, 4, 5 are the followers.
The topology of the communication network are given in Fig. 1 (Fig. 2).

We take the leader-follower consensus algorithm in this paper to design the estimation of the unknown signal, which is modeled as a triangle wave in this paper. This
given result is based on the leader-follower model, we choose the average strategy
to design the communication weights. Since only node 1, 3 can observe the target,
the other nodes are followers of these two nodes, we design estimators for these two
leaders. We choose $L = \begin{bmatrix} 1 & -1 & -1 \end{bmatrix}$, and $\varepsilon = 0, 01$. For the leaders, the estimation
errors of the three state variables and the estimation of the first state variable x_1 are
shown in Figs. 3 and 4. From Fig. 3, the estimation errors of the three state variables are small while not converge to 0. The errors are caused by the fact that the
unknown input signal triangle wave has infinite dimensional Fourier decomposition.
From Fig. 4, the leaders can nearly estimate the state of the target.

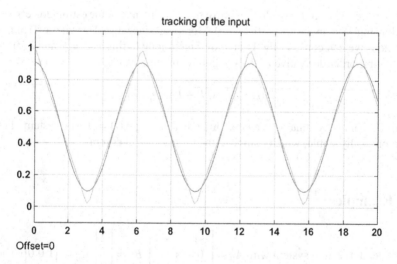

Fig. 2 Tracking of unknown input

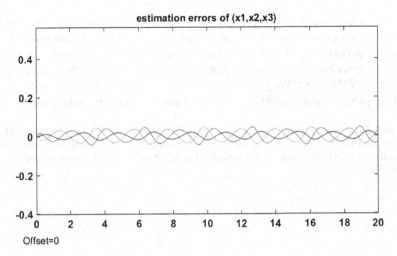

Fig. 3 Estimation errors of the three state variables

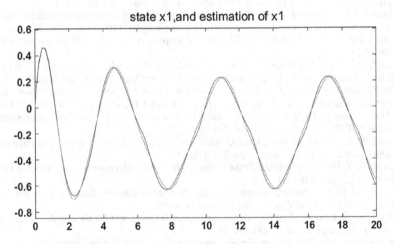

Fig. 4 Estimation of the first state variable

The tracking result of the unknown input is given in Fig. 2, The control law in this simulation is designed according to Theorem 2, and P is chosen arbitrarily as long as $A_0^T P + P A_0 < 0$.

5 Conclusion

In this paper, we study how to track a dynamic system with unknown periodic signal input in heterogeneous multi-agent systems. The controller design algorithm can be divided into two parts. One is the design of the consensus estimator, and the other is the adaptive controller. In the estimation part, we take the method of modelling the periodic signal as a finite dimensional Fourier decomposition and apply leader-follower consensus based estimation algorithm to estimate the state and periodic input of the target. We give stability conditions of the estimation errors. In the tracking part, we apply MRAC and prove the stability of the tracking errors.

Acknowledgements This work is supported by National Natural Science Foundation (NNSF) of China under Grant 61473081 and 61673106, Natural Science Foundation of Jiangsu Province under Grant BK20141341, the Fundamental Research Funds for the Central Universities under Grant 2242015R30013.

References

1. Chi RH, Han JZ. A novel periodicity-based adaptive control discrete time nonlinear systems. 27th Chinese control and decision conference (CCDC), May 2015. pp. 421–4.
2. Yuan CZ. Distributed adaptive switching consensus control of heterogeneous multi-agent systems with switched leader dynamics. 2016. https://doi.org/10.1109/ACC.2016.7525144.
3. Miller DB. A new approach to model reference adaptive control. IEEE Trans Autom Control. 2003;48(5).
4. Liu Y, Jia Y. Adaptive leader-following consensus control of multi-agent systems using model reference adaptive control approach. IET Control Theory Appl. 2012;6(13):2002–8.
5. Ye D, Zhao XG, Chao B. Distributed adaptive fault-tolerant consensus tracking of multi-agent systems against time-varying actuator faults. IET Control Theory Appl. 2016;10(5):554–63.
6. Zhou Z, Fang H, Hong Y. Distributed estimation for moving target based on state-consensus strategy. IEEE Trans Autom Control. 2013;58(8):2096–101.
7. Yang W, Yang C, Shi H, Shi L, Chen G. Stochastic link activation for distributed filtering under sensor power constraint. Automatica. 2017;75:109–18.
8. Li W, Jia Y, Du J. Distributed Kalman consensus filter with intermittent observations. J Frankl Inst. 2015;352(9):3764–81.
9. Li W, Jia Y. Distributed consensus filtering for discrete-time nonlinear systems with non-Gaussian noise. Signal Process. 2012;92(10):2464–70.
10. Battistelli G, Chisci L. Stability of consensus extended Kalman filter for distributed state estimation. Automatica. 2016;68(6):169–78.
11. Karasalo M, Hu XM. An optimization approach to adaptive Kalman filtering. Automatica. 2011;47(8):1785–93.
12. Li Z, Duan Z, Chen G, Huang L. Consensus of multiagent systems and synchronization of complex networks: a unified viewpoint. IEEE Trans Circuits Syst I: Regular Pap. 2010;57(1):213–24.
13. Chen Y, Shi Y. Leader-following consensus for multi-agent systems with switching topologies and time-varying delays: a switched system perspective. 2015 IEEE 54th annual conference on decision and control (CDC), December 2015. pp. 374–9.
14. Qiu ZR, Xie LH, Hong YG. Quantized leaderless and leader-following consensus of high-order multi-agent aystems with limited data rate. IEEE Trans Autom Control. 2016;61(9):2432–47.
15. Chen Q, Wang WC. Distributed state estimation based on cubature Kalman filtering. Chinese control and decision conference(CCDC), May 2016. pp. 3384–9.
16. Zhang S, Mourikis AI. Distributed estimation for sensor networks with arbitrary topologies. American control conference (ACC), Boston Marriott Copley Place, July 216. pp. 7048–54.
17. Chong CY, Chang KC, Mori S. Comparison of optimal distributed estimation and consensus filtering. 19th International conference on information fusion, Heidelberg, Germany, July 2016. pp. 1034–41.

18. Xiao L, Boyd S, Lall S. A Scheme for asynchronuous distributed sensor fusion based on average consensus. Networks: fourth international symposium on information processing in sensor, April 2005. pp. 63–70.
19. Chauvin J, Cordeb G, Petita N, Rouchona P. Periodic input estimation for linear periodic systems: automotive engine applications. Automatica. 2007;43(6):971–80.

System and Walking Gait Design for Hexapod Search and Rescue Robot

Chen Yang-Yang and Huang Yingying

Abstract In order to adapt to the complex disaster environment, this paper considers the system design of hexapod search and rescue robot. Such hexapod robot is suitable to different kinds of roads and obstacle, which can avoid to the shortcomings of crawler robots. This hexapod search and rescue robot includes the six foot body, voice modular, obstacle avoidance and remote monitoring function. Based on the relationship between the center of gravity and the supporting polygon, the design of the hexagon robot's walking gait is presented. The feasibility of the hexapod search and rescue robot is verified by the prototype experiment.

Keywords Hexapod search and rescue robot · The design of walking gait · Prototype experiment

1 Introduction

Robotics is an important signs to measuring a country's science and technology innovation and high level of manufacturing development. In 2016, "robot industry development plan" is promulgated by National Development and Reform Commission. As a main direction of robotics, search and rescue robot has received extensive attention in the past 10 years.

C. Yang-Yang (✉)
School of Automation, Southeast University, Nanjing 210096, China
e-mail: yychen@seu.edu.cn

C. Yang-Yang
Key Laboratory of Measurement and Control of Complex Systems of Engineering,
Ministry of Education, Southeast University, Nanjing 210096, China

H. Yingying
School of Electronic Science and Engineering, Southeast University,
Nanjing 2017514, China
e-mail: 213131866@seu.edu.cn

© Springer Nature Singapore Pte Ltd. 2018
Y. Jia et al. (eds.), *Proceedings of 2017 Chinese Intelligent Systems Conference*,
Lecture Notes in Electrical Engineering 460,
https://doi.org/10.1007/978-981-10-6499-9_62

Traditional search and rescue robots are mostly used the crawler robot due to its low unit pressure on the ground and the traction reserve index on the weak carrying ground. United States IRobot company gives Packbot [1] series of crawler robots which have finned track structure and first use the modular design. Thus such robots become one of the classic design in the world. In order to develop the robot's environmental adaptability, Inuktun's VGTV [2] changes the track shape of the crawler robot. However, crawler robots often stick and slip even worn in the practice [3]. Also, crawler robots have a larger size and thus have greater damage to the ground when turning. There is a trend to use the hexapod robot to replace the crawler type. Compared to the crawler type, the hexapod robot can cross through more rugged terrain and higher obstacles due to its discrete contact and it also has good mobility, energy consumption and independent isolation. In 1985, Robert B. McGhee et al. developed the adaptive suspension vehicle ASV [4] by using the leg structure, which can be regarded as one of the ancestors of six foot robot. NASA's Jet Propulsion Laboratory presents ATHLETE [5–7] for aerospace cargo transport. The domestic research for walking robots start relatively late. Shanghai University developed a spherical wall crawling robot [8]. Shanghai Jiao Tong University considered the design of "six-feet octopus" rescue robot [9]. Compared to the foreign application hexapod robot, there are also many technical challenges to overcome in the field of hexapod search and rescue robot at home.

In this paper, we consider the system design of hexapod search and rescue robot. In order to achieve the search and rescue mission in the complex environment, we not only give the hardware and software design of six-legged body, but also add voice modular and remote monitoring function for human-computer interaction and obstacle avoidance by using ultrasonic and infrared sensors. Note that the robot's center of gravity is unstable in the slope and step environment, we discuss the initial layout of the leg at the time of slope travel and the six-legged sequence in the step according the influence of the relationship between the polygon and the center of gravity to the robot stability. Finally, two the prototype experiments are given to verify the feasibility of the gait in slope and step.

2   System Description

Since the search and rescue robot is required to have excellent adaptability for the environment, the design of the six-legged robot should not only have a flexible six-legged body, but also include automatically avoid or cross the obstacles and communicate with the operator. Therefore, the overall design for our search and rescue robot is listed as Fig. 1. In Fig. 1, the key of system is the embedded board, which includes the gait control algorithm of 18 electric machineries for servicing the motion of six legs, obstacle avoidance algorithm according to the distance measurements by ultrasonic and infrared sensors, voice recognition module for dealing with the decision makers' demands and image compression and transmission supplying to the decision makers. From the hardware point of view to achieve

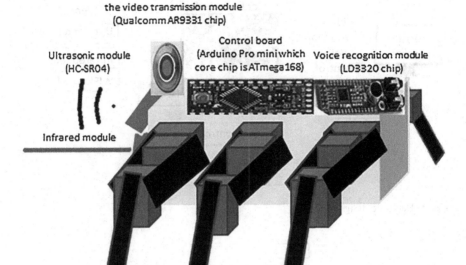

Fig. 1 Six-legged robot system

the above functions, we use the wrap-around layout for the six-legged robot, that is six legs are installed on the hexagonal vertices of octagon and each leg are controlled by three digital servos. The ultrasonic (HC-SR04) and infrared sensors are located at the front face of the robot and the WIFI camera is mounted at the top of the robot. The voice recognition module (LD3320 chip) and the video transmission module (Qualcomm AR9331 chip) line to our main control module (Arduino Pro mini which core chip is ATmega168). The details can be found in Fig. 2.

To make rational use of hardware resources, the responding software flow is listed in Fig. 3. At the beginning, the voice recognition module is in the standby state. When it receives the command issued by the decision makers, the module uses the Voice. Read function to identify whether the instruction command match with the library, if not match the module keeps standby, else sends instruction to the steering control module. Next, the gait algorithms are chosen based on the distance values coming from ultrasonic and infrared sensors and then drive the digital servo. At last, the software drives the WIFIcam to record the picture and translates the picture to the decision markers. It should be emphasized that we determine the slope or the step by using the different high position between the infrared sensor and the ultrasonic sensor and their measurements.

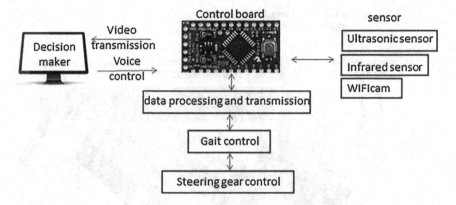

Fig. 2 Hardware design

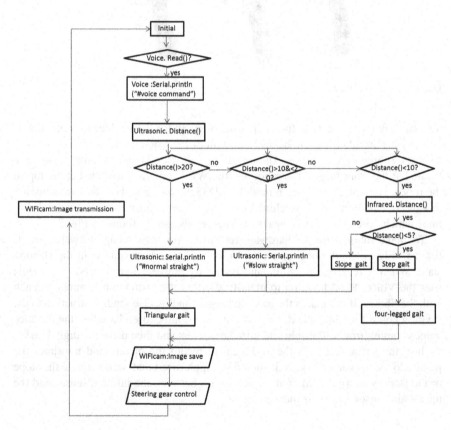

Fig. 3 Software flow

3   Gait Control

According to the difference of the supporting feet, the gait of six-legged robot can divide into the triangular, fluctuating and quadruped gait. Triangle gait is the most flexible and fastest among the three gaits. Fluctuation gait is slowest but in good stability while quadruped gait has intermediate performance between triangle gait and fluctuation gait. Considering the practicability of robot, the six-legged robot uses the triangle gait when it moves on the flat ground. Noting that the movement on the ground can also be applied to slopes with the small slope angle < 30°, we follow the triangle gait when the six-legged robot moves on some slopes with its slope angle < 30°. When the six-legged robot moves on a step, the quadruped gait is applied for the six-legged robot. In fact, the general obstacles can be reduced to the form of slopes and steps. The gait design is very important to the six-legged robot. In the following, we will introduce the gait design according to the triangle gait and the quadruped gait.

3.1   Flat Gait Design

The gait of the six-legged robot on the flat ground usually use the triangular gait. Although there are three degrees of freedom of each leg, its movement is still full of regularity. Triangular gait means there are always three adjacent legs in the support phase while the other legs are in the swing phase when the robot moves. As shown in Fig. 4, the six legs distribute at the hexagonal vertices of octagon. 1'3'5' and 2'4' 6' are the theoretical ground contact points. The solid line in Fig. 4 indicates the leg in support phase and the dotted line denotes the leg lift, which implies that there are three supporting points at any time for the purpose of the stability of the robot's barycenter.

When the six-legged robot moves along a straight line, the six-legged gait is shown in Fig. 5, that is 2',4',6' lift up and turn to the head at beginning, then put down the ground. The alternate execution of two triangular fulcrums can make the six-legged robot move forward alternately. When the robot turns right, the six-legged gait is listed as follows: Firstly, 1',3',5' lift up, 2' turns to the tail and 4' 6' turn to the head. Then 2'4'6' put down and touch the ground while 1',3',5 lift up. Finally, 2',4',6' turn to the reverse rotation in order to drive the whole body to

Fig. 4 Distribution and support triangle illustration

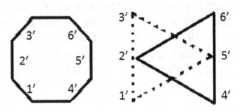

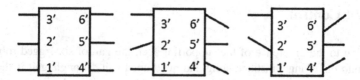

Fig. 5 Move illustration

Fig. 6 State phase of six-legged triangular gait

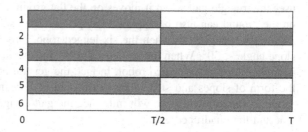

the right. A similar process turns into the Left turn and we omitted it due to the limitation of thesis space.

In the triangular gait, only the supporting legs for reverse propulsion make the six-foot robot displacement. The swing phase is to adjust the body shape and angle. Figure 6 indicates the phase diagram in the single gait cycle. From this picture, one can see that the time of support phase and swing phase are T/2, that is, the coverage factor of triangular gait is 1/2, where T is the time of legs in contact with the ground (that is the state phase).

3.2 Slope Gait Design

Although the flat gait can be extended to the slope movement, we have to consider the friction coefficient between the foot and slope when the slope angle increases. When the robot moves in a flat ground, its gravity center just locates on the center of the supporting polygon. When it travels on a slope, the projection of gravity center has a offset from the center position. It is due to the fact that the component of gravity in the direction of the slope satisfies $F = Mg * \sin\theta$, where M is the mass of the robot, g is the gravitational acceleration and θ is the slope angle. It is obvious that the gravity component increases with the increase of the slope angle. When the friction between the robot foot and the slope is smaller than the gravity component, the robot will slip. Thus, the six feet can only climb a certain angle of the slope.

In 1976, McGhee and IsWandhi proposed the concept of static stability margin SSM (Static Stability Margin), which can be used to determine whether the robot is in a stable state. SSM is defined as the minimum distance from the projection point of the center of gravity to the sides of the projection support polygon. The mathematical expression is

$$S_{SSM} = \min_{i} (\lambda_{iG})$$

where i is the number of supporting legs, λ_{iG} is the minimum distance from the projection point of the center of gravity to the sides of the projection support polygon. When SSM > 0, the robot is stable, otherwise unstable. In the following, we will show the best swing angles of front legs according to the simulation results of λ_{iG}.

The wide and high of our prototype are 12 cm and 12 cm, respectively. The gravity center of robot is 6 cm from the ground, and the distance from the toes to the center is 14 cm when it stands. The details can be found in Fig. 7.

In the slope state, the six-legged robot can improve the stability by changing the position of the support point relative to the fuselage. It is hard to analyze the relationship between the stability and the position of the support point. Since the position of the support point depends on the angle of the hip joint, we will show the relationship between the stability and the angle of the hip joint by simulations. Restricted by mechanical structure, the swing angle of each middle leg (that is 2′, 5′) is limited in $(-60°, 60°)$ while the swing angle range of each front/rear leg (that is 1′,4′/3′,6′) is $(-30°, 90°)$. Taking the triangle formed by 2′,4′,6′ as a example, Figs. 8, 9 and 10 show $\lambda_{(2',4')G}, \lambda_{(4',6')G}$ and $\lambda_{(6',2')G}$, respectively. From these pictures, one can see that the best swing angles of legs $\theta_{1w} = \theta_{6w} = 15°$, $\theta_{2w} = \theta_{5w} = 30°$ and $\theta_{3w} = \theta_{4w} = 45°$.

Fig. 7 Prototype of robot

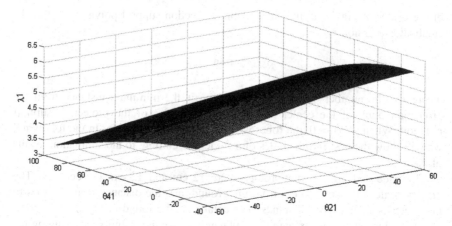

Fig. 8 Plot of $\lambda_{(2',4')G}$

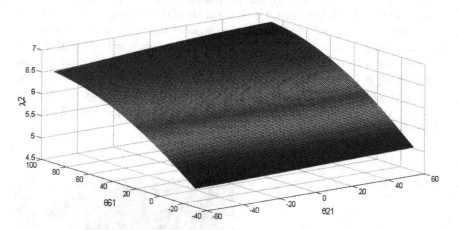

Fig. 9 Plot of $\lambda_{(4',6')G}$

3.3 Step Gait Design

When the six-legged robot crosses the step, the quadruped gait is used. In the quadruped gait, the robot divides the six feet into three groups, that is the front legs (that is $1',4'$), the middle legs (that is $2',5'$) and the rear legs (that is $3',6'$). The three groups are rotated in the order such the front legs, the middle legs and the rear legs. The phase diagram of the quadruped gait is shown in Fig. 11. From this picture, one can see that there are always four legs in the support phase at one time. During the gait cycle, the support phase occupies $2T/3$ and the swing phase occupies $T/3$, which means the coverage factor is $2/3$.

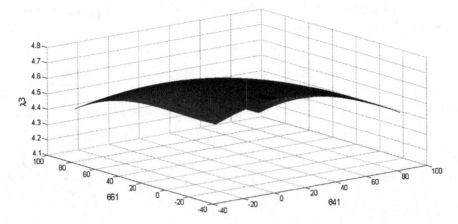

Fig. 10 Plot of $\lambda_{(6',2')G}$

Fig. 11 Quadruple gait phase diagram

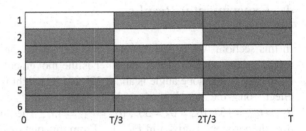

When the six-legged robot moves in the quadruped gait, the worst body stability appears in the situation that the front or hind legs lift up. Let the front legs in the swing phase as a case. If the middle and rear legs are in the support phase robot and the quadrilateral consisting of four vertices corresponding to legs cannot exceed the center line of the body, then the robot body is stable (see Fig. 12). From above discussion, the step gait design is given as follows:

(a) Front leg swings to the head, middle foot remain intact, and rear foot rotates to tail in support phase to promote the body forward;

(b) Front foot down, middle foot rotates to tail in support phase to promote the forward, and rear foot swings to the head;

(c) Front foot rotates to tail in support phase to promote the body forward, middle foot swings to the head, and rear foot down to support.

Fig. 12 Six-foot robots
support polygons

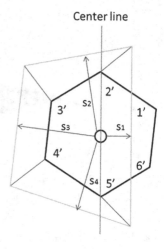

4 Experimental Results

In this section, two experimental results are given. One is the movement of prototype on the outdoor slope, the other is the movement of prototype on the steps.

In case 1, the slope angle is about 30° and its height is less than 1 cm that has no effect on the six-legged robot. The initial angles of six legs are $\theta_{1w} = \theta_{6w} = 15°$, $\theta_{2w} = \theta_5 = 30°$, $\theta_{3w} = \theta_{4w} = 45°$. The movement of prototype on the outdoor slope is given in Fig. 13. From this picture, one can obviously see that our prototype robot can move on the outdoor slope smoothly and rapidly.

In case 2, we use the table to build a step such that the step height is 3 cm and its width 5 cm. The movement of prototype on the steps is shown in Fig. 14. From this picture, it is obvious that our prototype robot can cross the steps easily.

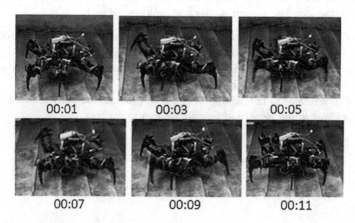

Fig. 13 The movement on the outdoor slope

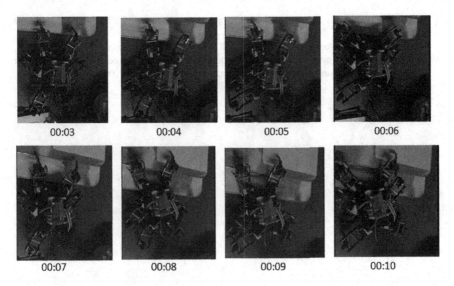

Fig. 14 The movement on the steps

5 Conclusion

This paper presents the software and hardware design of the six-legged robot. We give experimental results for proving the gait design. In ongoing research, we will devote to the feedback control of the gaits.

Acknowledgements This work is supported by National Natural Science Foundation (NNSF) of China under Grant 61673106, Natural Science Foundation of Jiangsu Province under Grant BK20171362, the Fundamental Research Funds for the Central Universities and in part by a Project Funded by the Priority Academic Program Development of Jiangsu Higher Education Institutions.

References

1. Yamauch IB. Packbot: a versatile platform for military robotics[C], Unmanned Ground Vehicle Technology VI. Proc SPIE. 2004;(5422):228–237.
2. Jeehong K, Chan GL, Gunho K. Study of machine design for a transformable shape single-tracked vehicle system [J]. Mech Mach Theory. 2010;(45):1082–1095.
3. Liu J. Mine rescue robot key technology research [D]. China Univ Min Technol. 2014;12.
4. Espenschied KS, Quinn R, Chiel HJ. Biologically-based distributed control and local reflexes improve rough terrain locomotion in a hexapod robot. Robot Auton Syst. 1996;18:59–64.
5. Wilcox BH. Athlete: an option for mobile lunar landers [C]. In: Proceedings of the aerospace conference, 2008. IEEE; 2008. p. 1–8.
6. Wilcox BH. Athlete: a cargo-handling vehicle for solar system exploration [C]. In: Proceedings of the aerospace conference, 2011. IEEE; 2011. p. 1–8.

7. Wilcox BH. Athlete: a cargo and habitat transporter for the moon [C]. In: Proceedings of the aerospace conference, 2009. IEEE; 2009. p. 1–7.
8. Tan S, Wang J. Development of spherical wall crawling robot. Robotics. 2002;24(6):517–21.
9. Shanghai Jiaotong University research and development "six paw octopus" robot disaster rescue, People's Network. http://scitech.people.com.cn/n/2013/1029/c1007-23368017.html.

Path Tracking Control for Four-Wheel Steering Vehicles by Hyperbolic Projection-Based Feedback Dominance Backstepping Method

Mingxing Li and Yingmin Jia

Abstract The path tracking problem is studied for four-wheel steering vehicles in this paper. A path tracing controller with decoupling performance between lateral velocity and yaw rate is designed by using the hyperbolic projection-based feedback dominance backstepping method to achieve the tracking goal. Simulations are shown that the lateral offset converges to zeros and the vehicle heading angle converges to the actual road direction of the desired path. These results conclude the new designing controller is very effective.

Keywords Four-wheel steering vehicle · Path tracking · Decoupling performance

1 Introduction

Automatic driving technology has attracted a growing attention in the last years with better road utilization and performances of safety and comfort. Path tracking is the basis of the automatic driving technology with the object tracking the desired path. In the path tracking process, lateral dynamics must be considered and the stability should be guaranteed [1]. There are many such control methods which have been developed to achieve the path tracking goal, such as PID control [2], LQR control [3], model predictive control [4], sliding mode control [5] and so on. However, there are conflicting between the zero sideslip angle control and the yaw rate control in the common front-wheel steering vehicles. And this conflicting makes the above mentioned path tracking methods cannot work well in high-speed driving or around high-curvature road.

M. Li · Y. Jia (✉)
The Seventh Research Division and the Center for Information and Control,
School of Automation Science and Electrical Engineering, Beihang University (BUAA),
Beijing 100191, People's Republic of China
e-mail: ymjia@buaa.edu.cn

M. Li
e-mail: lmx196@126.com

© Springer Nature Singapore Pte Ltd. 2018
Y. Jia et al. (eds.), *Proceedings of 2017 Chinese Intelligent Systems Conference*,
Lecture Notes in Electrical Engineering 460,
https://doi.org/10.1007/978-981-10-6499-9_63

To improve the tracking performance, [6] points out that the desired heading in path following of vehicles should not be the tangent direction of the desired path for the common front-wheel vehicles and the side angle need to be measured or estimated while the stable region of the vehicle may change and the new stability problem of the vehicle should be maintained. Furthermore, [7] investigates the path tracking problem using integrated control of active front-wheel steering and dynamics yaw-moment control for four-wheel independently actuated vehicles. And problems mentioned in [6] are avoided and the transient performance is improved. In this paper, the path tracking problem of active four-wheel steering vehicles (4WS) is investigated. Different from the existing works, a controller with decoupling performance is designed based on our prior works [8–12] by using the hyperbolic projection-based feedback dominance backstepping (HFDB) method, and disadvantages of path tracking mentioned in [6] are avoided.

This paper is organized as follows: modeling and problem formulation are introduced in Sect. 2. The method and designing process of the controller is shown in Sect. 3. The simulation results are described in Sect. 4 and the conclusion is made at last in Sect. 5.

2   Modeling and Problem Formulation

As shown in Fig. 1, there are three frames which are inertial frame $F_I = \{0, i, j\}$, vehicle frame $F_v = \{P_v, i_v, j_v\}$ and frame $F_s = \{P_s, i_s, j_s\}$ indexed the curve C. The origin P_v of frame F_v is the center of the vehicle mass and the origin P_s of frame F_s

Fig. 1   Path tracking figure

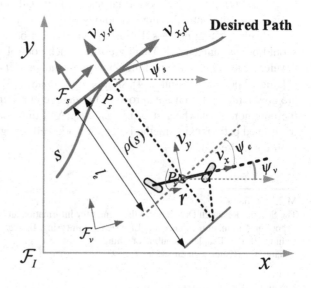

is the closest point on the desiring path from point P_v. The unit tangent vector and principal normal vector of F_s can be obtained directly as follows:

$$i_s = \begin{bmatrix} \cos(\psi_s) \\ \sin(\psi_s) \end{bmatrix}, j_s = \begin{bmatrix} -\sin(\psi_s) \\ \cos(\psi_s) \end{bmatrix} \tag{1}$$

Let $P_s = (x_s, y_s)$ and $P_v = (x_v, y_v)$ in F_I, then we have the following relationships:

$$l_e j_s = (x_s - x_v)i + (y_s - y_v)j \tag{2}$$

$$\dot{s} i_s = \dot{x}_s i + \dot{y}_s j \tag{3}$$

$$\begin{bmatrix} \dot{x}_v \\ \dot{y}_v \end{bmatrix} = \begin{bmatrix} \cos(\psi_v) & -\sin(\psi_v) \\ \sin(\psi_v) & \cos(\psi_v) \end{bmatrix} \begin{bmatrix} v_x \\ v_y \end{bmatrix} \tag{4}$$

where $\dot{s}$ is the desired linear velocity value which is often a varying parameter. Differentiating Eq. (2) with respect time and substituting $\dot{x}_i, \dot{y}_i, i = v, s$ from Eqs. (3) and (4) obtains that:

$$\begin{bmatrix} \dot{s} + \dot{\psi}_s l_e \\ \dot{l}_e \end{bmatrix} = \begin{bmatrix} \cos(\psi_s - \psi_v) & \sin(\psi_s - \psi_v) \\ -\sin(\psi_s - \psi_v) & \cos(\psi_s - \psi_v) \end{bmatrix} \begin{bmatrix} v_x \\ v_y \end{bmatrix} \tag{5}$$

and to $\dot{\psi}_s$ we have the following equation:

$$\dot{\psi}_s = \frac{\partial \psi_s}{\partial s} \dot{s} \tag{6}$$

Hence, if let $k(s) = \frac{\partial \psi_s}{\partial s}$ and $\psi_e = \psi_v - \psi_s$ then Eq. (5) can be rewritten into as follows:

$$\begin{cases} \dot{s} = \frac{1}{1+l_e k(s)} [\cos(\psi_e)v_x - \sin(\psi_e)v_y] \\ \dot{l}_e = \sin(\psi_e)v_x + \cos(\psi_e)v_y \\ \dot{\psi}_e = r - \frac{k(s)}{1+l_e k(s)} [\cos(\psi_e)v_x - \sin(\psi_e)v_y] \end{cases} \tag{7}$$

This equation is the generalization of the path tracking system. And there is a very typical situation that is $1 + l_e k(s) = 0$. In this situation, $l_e = -1/k(s)$ and point P_v is just locating on the instantaneous center of the path, thus point P_s is not uniqueness, and $\dot{s}$ is infinity or the following relationship is established:

$$\cos(\psi_e)v_x = \sin(\psi_e)v_y$$

which means there are ψ_e such that $v_x = \tan(\psi_e)v_y$. If we take the following assumption:

Assumption A1 The distance l_e is small enough such that P_s is uniqueness.

then the above mentioned typical situation can be avoided and the path tracking system can be described by:

$$\begin{cases} \dot{l}_e = \sin(\psi_e)v_x + \cos(\psi_e)v_y \\ \dot{\psi}_e = r - \frac{k(s)}{1+l_e k(s)}[\cos(\psi_e)v_x - \sin(\psi_e)v_y] \end{cases} \tag{8}$$

To the 4WS steering vehicles, the dynamics of the steering system on the flag ground can be described by:

$$\begin{cases} \dot{v}_x = \frac{1}{m}[T + mv_y r + v_x^2 c_a] \\ \dot{v}_y = \frac{1}{m}[c_f \delta_f + c_r \delta_r - (c_f + c_r)\frac{v_y}{v_x} - (l_f c_f - l_r c_r + mv_x^2)\frac{r}{v_x}] \\ \dot{r} = \frac{1}{J}[l_f c_f \delta_f - l_r c_r \delta_r - (l_f c_f - l_r c_r)\frac{v_y}{v_x} - (l_r^2 c_r + l_f^2 c_f)\frac{r}{v_x}] \end{cases} \tag{9}$$

In the above equations, to simple, the assumption v_x is constant and the braking/accelerating force is zero are taken. This model is simplified from [13].

In our prior works [8–12], we have focus on the coupling among v_x, v_y and r and how to improve the robust and safety performances of the vehicles by decoupling method. In this paper, the affections of the coupling on the path tracking results are discussed. Notice that, the longitudinal velocity is just a tunable parameter and braking/accelerating motion is neglected in the steering process, thus we just consider the following path tracking system of the vehicle:

$$\begin{cases} \dot{l}_e = \sin(\psi_e)v_x + \cos(\psi_e)v_y \\ \dot{\psi}_e = r - \frac{k(s)}{1+l_e k(s)}[\cos(\psi_e)v_x - \sin(\psi_e)v_y] \\ \begin{bmatrix} \dot{v}_y \\ \dot{r} \end{bmatrix} = \begin{bmatrix} -\frac{c_f+c_r}{mv_x} & v_x - \frac{l_f c_f - l_r c_r}{mv_x} \\ -\frac{l_f c_f - l_r c_r}{Jv_x} & -\frac{l_r^2 c_f + l_r^2 c_r}{Jv_x} \end{bmatrix} \begin{bmatrix} v_y \\ r \end{bmatrix} + \begin{bmatrix} \frac{c_f}{m} & \frac{c_r}{m} \\ \frac{l_f c_f}{J} & -\frac{l_r c_r}{J} \end{bmatrix} \begin{bmatrix} \delta_f \\ \delta_r \end{bmatrix} \end{cases} \tag{10}$$

To this system, design controller such that ψ_e and l_e both converge to zeros while t tends to infinity.

3 Controller Design

In this section, we design a controller such that v_y, l_e and ψ_e all tend zeros by using HFDB method. Let:

$$V_1 = l_e^2/2, z_1 = \psi_e - \alpha_1, \alpha_1 = -\arctan(l_e/\epsilon_1) \tag{11}$$

where $\epsilon_1 > 0$, then:

$$\dot{V}_1 = l_e v_x(\sin(z_1)\cos(\alpha_1) + \cos(z_1)\sin(\alpha_1)) + l_e v_y \cos(\psi_e).$$

Furthermore, let:

$$V_2 = V_1 + 2\sin^2(z_1/2), z_2 = r - \alpha_2 \tag{12}$$

$$\alpha_2 = \frac{k(s)}{1 + l_e k(s)}(v_x \cos(\psi_e) - \sin(\psi_e)v_y) - k_1 z_1 - v_x l_e \cos(\alpha_1) + \dot{\alpha}_1 \tag{13}$$

then:

$$\dot{V}_2 = l_e v_y \cos(\psi_e) + l_e v_x \cos(z_1)\sin(\alpha_1) + \sin(z_1)z_2 - \sin(z_1)k_1 z_1 \tag{14}$$

To z_2, we have the following differential equation:

$$\dot{z}_2 = \dot{r} - \dot{\alpha}_2 \tag{15}$$

If let $l = l_f + l_r$ and take δ_r, δ_f as following:

$$\delta_r = -\frac{m}{c_r}[(v_x - \frac{l_f c_f - l_r c_r}{m v_x})r + \frac{c_f}{m}\delta_f] \tag{16}$$

$$\delta_f = -\frac{l_r}{lc_f}(mv_x - \frac{l_f c_f - l_r c_r}{v_x})r + \alpha_2 \frac{l_f^2 c_f + l_r^2 c_r}{lc_f} + \frac{J}{l_f c_f}\dot{\alpha}_2 \tag{17}$$

then we have the following differential equation:

$$\begin{bmatrix} \dot{v}_y \\ \dot{z}_2 \\ \dot{z}_1 \end{bmatrix} = \begin{bmatrix} -\frac{c_f + c_r}{m v_x} & 0 & 0 \\ -\frac{l_f c_f - l_r c_r}{J v_x} & -\frac{l_f^2 c_f + l_r^2 c_r}{J v_x} & 0 \\ 0 & 1 & -k_1 \end{bmatrix} \begin{bmatrix} v_y \\ z_2 \\ z_1 \end{bmatrix} + \begin{bmatrix} 0 \\ 0 \\ ev_x \sin(\alpha_1) \end{bmatrix} \tag{18}$$

This system is stable and both v_y and z_2 converge to zeros even with nonzero initial conditions. As a summary of above analysis, we have:

Theorem 1 *Controller shown as follows:*

$$\begin{cases} \alpha_1 = -\arctan(\frac{l_e}{\epsilon}) \\ \alpha_2 = \frac{k(s)}{1+l_e k(s)}v_x \cos(\psi_e) - k_1 z_1 - v_x l_e \cos(\alpha_1) + \dot{\alpha}_1 \\ \delta_r = -\frac{m}{c_r}[(v_x - \frac{l_f c_f - l_r c_r}{m v_x})r + \frac{c_f}{m}\delta_f] \\ \delta_f = -\frac{l_r}{lc_f}(mv_x - \frac{l_f c_f - l_r c_r}{v_x})r + \alpha_2 \frac{l_f^2 c_f + l_r^2 c_r}{lc_f} \end{cases} \tag{19}$$

This controller can make system (10) is stable and l_e, ψ_e, v_y converge to zeros with t tending to infinity and ϵ is small enough.

Proof For the following Lyapunov function:

$$V = V_2 + x_{yz}^T P x_{yz} \tag{20}$$

where $x_{yz} = [v_y, z_2]^T$ and $P = \text{diag}\{p_1, p_2\} > 0$, then we have:

$$\dot{V} = \dot{V}_2 + 2x_{yz}^T P \dot{x}_{yz} < 0 \tag{21}$$

if there are $p_1 > 0$ and $p_2 > 0$ such that:

$$\begin{bmatrix} -v_x \cos(z_1)\cos(\alpha_1)/\epsilon & \cos(\psi_e)/2 \\ \cos(\psi_e)/2 & -2k_y p_1 + k_{yz}^2 p_1^2(k_z p_2 - k_1^{-1}\sin(z_1)/(4z_1)) \end{bmatrix} < 0 \tag{22}$$

where $k_y = \frac{c_f + c_r}{mv_x}$, $k_{yz} = \frac{l_r c_f - l_r c_r}{Jv_x}$ and $k_z = \frac{l_f^2 c_f + l_r^2 c_r}{Jv_x}$. Notice that there always exist such $p_1 > 0$ and $p_2 > 0$ which make the inequality (22) established. Thus, Theorem 1 is proved.

Remark 1 The controller described by (19) is just depend on v_x, r and vehicle parameters. Thus v_y or slide angle is not needed to be measured or observed in our controller.

Remark 2 To active front-wheel steering vehicles, this goal can be achieved and the path tracking result is well only in lower-speed driving or around lower-curvature road. That is because the coupling between lateral velocity and yaw rate is very weak in this situation and it becomes strong in high-speed driving or around high-curvature road. To the 4WS vehicles under controlled by controller (19), the rear angle δ_r is used to control v_y and the front angle δ_f is used to control r in the closed-loop system. And v_y and r are decoupled, thus the path tracking results is well not only in high-speed also around high-curvature road.

4 Simulations

In this section, simulations for the closed-loop system of system (10) and controller (19) are done. In these simulations, the parameters of the vehicle are taken as: $c_f = c_r = 160,000\,\text{N/rad}$, $l_f = 1\,\text{m}$, $l_r = 1.6\,\text{m}$, $m = 1500\,\text{mg}$ and $J = 3240\,\text{kg} \cdot \text{m}^2$. Parameters ϵ and k_1 in controller (19) are taken as $\epsilon = 0.1$ and $k_1 = 36$.

The tracking results are shown in Fig. 2. Two situations are simulated in this figure, which are the S-turn (or the lane changing) and U-turn. And the output e is defined by:

$$e = \sqrt{(x_v - x_s)^2 + (y_v - y_s)^2}$$

From Fig. 2, we know that the tracking error e is increasing with time t while $t \in [0, 0.12]\,\text{s}$ and it reaches its biggest value while time t is bigger than 1.2 s.

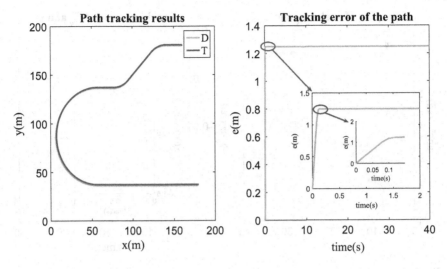

Fig. 2 Path tracking results

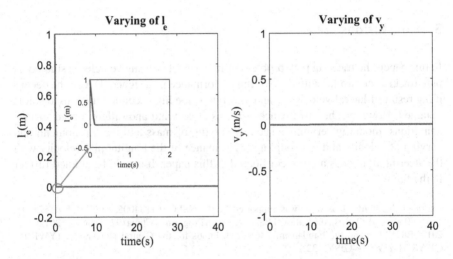

Fig. 3 Varying of lateral offset l_e and lateral velocity v_y

From Fig. 3, we know the lateral offset l_e and v_y are stable and l_e converges to zero while time t is bigger than 0.12 s. Here we can make the lateral velocity is zero while yaw rate r and angle ψ_v are varying, that because two control inputs δ_f and δ_r are taken and the controller (19) decouples lateral velocity v_y and yaw rate r. Furthermore, from Fig. 4 we know that ψ_v converges to ψ_s while t is bigger than 0.12 s.

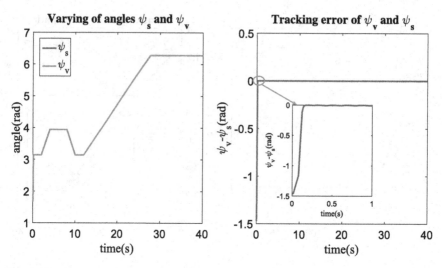

Fig. 4 Varying of angles ψ_v, ψ_s and tracking error $\psi_v - \psi_s$

5   Conclusions

In this paper, the tracking path problem of four-wheel steering vehicles is studied. A path tracking controller with decoupling performance is designed. Due to the decoupling result of lateral velocity v_y and yaw rate r, the high accurate tracking result is obtained. However, the disturbances such as cross wind and different road surface conditions, model uncertainties such as varying of mass and inertia moment, and varying of velocity and the coupling performance of the longitudinal velocity with the steering dynamics are not considered in this paper. It should be studied further in the future.

Acknowledgements This work was supported by the NSFC (61134005, 61327807, 61521091, 61520106010, 61703020), the National Basic Research Program of China (973 Program: 2012CB821200, 2012CB821201), and the Fundamental Research Funds for the Central Universities (YWF-16-GJSYS-31,YWF-16-GJSYS-32).

References

1. Naranjo JE, Gonzalez C, Garcia R, de Pedro T. Lane-change fuzzy control in autonomous ground vehicles for the overtaking maneuver. IEEE Trans Intell Trasp Syst. 2008;9(3):438–50.
2. Marino R, Scalzi S, Netto M. Nested PID steering control for lane keeping in autonomous vehicles. Control Eng Pract. 2011;19(12):1459–67.
3. Goodarzi A, Sabooteh A, Esmailzadeh E. Automatic path control based on integrated steering and external yaw-moment control. Proc Inst Mech Eng K J Multi-Body Dyn. 2008;222(2):189–200.

4. Lenain R, Thuilot B, Cariou C, Martinet P. High accuracy path tracking for vehicles in presence of sliding: application to farm vehicle automatic guidance for agricultural tasks. Auton Robot. 2006;21(1):79–97.
5. Fang H, et al. Robust anti-sliding control of autonomous vehicles in presence of lateral disturbances. Control Eng Pract. 2011;19(5):468–78.
6. Hu C, Wang R, Yan F, Chen N. Should the desired heading in path following of autonomous vehicles be the tangent dirctionn of the desired path? IEEE Trans Intell Trasp Syst. 2015;16(6):3084–94.
7. Wang R, Hu C, Yan F, Chadli M. Composite nonlinear feedback control for path following of four-wheel independently actuated autonomous ground vehicles. IEEE Trans Intell Trasp Syst. 2016;17(7):2063–74.
8. Li M, Jia Y, Du J. LPV control with decoupling performance of 4WS vehicles under velocity-varying motion. IEEE Trans Control Syst Technol. 2014;22(5):1708–24.
9. Li M, Jia Y. Decoupling control in velocity-varying four-wheel steering vehicles with H_∞ performance by longitudinal velocity and yaw rate feedback. Veh Syst Dyn. 2014;52(12):1563–83.
10. Li M, Jia Y. Precompensation decoupling control with H_∞ performance for 4WS velocity-varying vehicles. Int J Syst Sci. 2016;47(16):3864–75.
11. Li M, Jia Y, Matsuno F. Attenuating diagonal decoupling with robustness for velocity-varying 4WS vehicles. Control Eng Pract. 2016;56:49–59.
12. Li M, Jia Y. Decoupling and robust control of velocity-varying four-wheel steering vehicles with uncertainties via solving attenuating diagonal decoupling problem. J Franklin Inst. 2017;354(1):105–22.
13. Hebden RG, Edwards C, Spurgeon SK. Automotive steering control in a split-manoeuvre using an observer-based sliding mode controller. Veh. Syst. Dyn. 2004;41(3):677–94.

Singularity-Free Path Following Control for Miniature Unmanned Helicopters

Xujie Ma and Wei Huo

Abstract A singularity-free nonlinear controller is presented for the miniature unmanned helicopter to follow a reference path described by implicit expressions. Based on the time-scale separation principle, the controller is designed with hierarchical inner-outer loop structure. The outer-loop position controller is constructed with hyperbolic tangent function, and temporary-path generation method is developed to keep the control matrix invertible and obviate large control energy. The desired command attitude can be derived from position controller without singularity by choosing appropriate controller parameters. The inner-loop attitude controller is designed with the unit-quaternion attitude representation and backstepping technique to achieve attitude tracking. Numerical simulation is provided to verify the theoretical results.

Keywords Miniature unmanned helicopter · Path following · Singularity-free control · Quaternion

1 Introduction

Tracking tasks for aircrafts can be classified as two categories [1]: trajectory tracking and path following. In the first case, the aircraft is required to track a time-varying reference trajectory at every transient. While in the second case it is required to fly along a reference path at a desired speed. Unlike trajectory tracking, there is no temporal requirements on path following. Besides the path following has some control performances that can not be obtained from trajectory tracking in some specific cases [2].

X. Ma · W. Huo (✉)
The Seventh Research Division, School of Automation Science
and Electrical Engineering, Beihang University, Beijing 100191, PR China
e-mail: weihuo@buaa.edu.cn

X. Ma
e-mail: maxujie@buaa.edu.cn

© Springer Nature Singapore Pte Ltd. 2018
Y. Jia et al. (eds.), *Proceedings of 2017 Chinese Intelligent Systems Conference*,
Lecture Notes in Electrical Engineering 460,
https://doi.org/10.1007/978-981-10-6499-9_64

Various linear and nonlinear controllers have been proposed for path following control of miniature helicopters. The linear control methods are simple and reliable, such as PID [3] and LQ control [4, 5], but have the limitation to realize full envelope flight. To this end, some nonlinear control methods such as backstepping approach [6], feedback linearization technique [7] or hybrid control methods [8, 9] have been applied. Due to the under-actuated property, the helicopter model is always simplified to position outer-loop and attitude inner-loop structure [10, 11]. However, the controller designed based on the hierarchical structure may suffer from singularity when deriving the desired attitude from position controller. So far, only a few literatures [6] take into account this problem, and most reference paths are parameterized curves.

The parameterized path following [6, 8, 9] is the most commonly used problem formulation. The path is described with a time-varying parameter, and the task is to design control law and parameter timing law such that helicopter can keep up with the moving point determined by the parameter. Another problem formulation is based on implicit expressions [1, 7]. The path is given by the intersection of two manifolds. Unlike parameterized path following which turns path following problem to point-tracking problem, the task for implicit path following is to follow the entire path and the helicopter will enter an invariant set around the reference path. However, controller design for implicit path following always relates to control matrix of the closed-loop system, and it suffers from singularity when the matrix is not invertible.

In this paper a singularity-free implicit path following controller for miniature helicopters is presented. The control design is based on the hierarchical structure. The outer-loop position controller is constructed with the hyperbolic tangent function to realize path following, and a temporary path is planned to guarantee the control matrix invertible. From the position controller, the desired command attitude can be derived without singularity by choosing appropriate controller parameters. The inner-loop attitude controller is designed with the unit-quaternion representation to realize tracking for the command attitude.

2 Preliminaries

In following sections, $c(\cdot)$ and $s(\cdot)$ are shorts of $\cos(\cdot)$ and $\sin(\cdot)$. $|\cdot|$ denotes absolute value of a real number, $\|\cdot\|$ denotes Euclidean norm for a vector or induced Euclidean norm for a matrix. $\bar{\lambda}(\cdot)$ and $\underline{\lambda}(\cdot)$ denote the maximum and minimum eigenvalues, respectively. For $x = [x_1, \dots, x_n]^T \in \mathbb{R}^n$, define hyperbolic tangent function vector $\tanh(x) = [\tanh(x_1), \dots, \tanh(x_n)]^T$ and hyperbolic tangent function matrix $\text{Tanh}(x) = \text{diag}\{\tanh(x_1), \dots, \tanh(x_n)\}$. For a continuously differentiable scalar function $f(x)$, define $\partial_{x_i} f = \frac{\partial f}{\partial x_i} (i = 1, \dots n)$ and $\partial_x f = [\frac{\partial f}{\partial x_1}, \dots, \frac{\partial f}{\partial x_n}]^T$. For continuously differentiable vector function $f(x) = [f_1(x), \ldots, f_m(x)]^T$, define $\partial_x f =$

$$\begin{bmatrix} \frac{\partial f_1}{\partial x_1} & \cdots & \frac{\partial f_1}{\partial x_n} \\ \vdots & \ddots & \vdots \\ \frac{\partial f_m}{\partial x_1} & \cdots & \frac{\partial f_m}{\partial x_n} \end{bmatrix}$$ and $\partial_x^2 f = \partial_x(\partial_x f)$.

Lemma 1 ([12]) *For $x \in \mathbb{R}^n$, if $\|x\| < \bar{x} < \infty$, then there exists a constant $\chi(\bar{x}) \in (0, 1)$ satisfying $\chi\|x\| \leq \|\tanh(x)\| \leq \|x\|$.*

Lemma 2 ([13]) *For $x \in \mathbb{R}$ and $\varepsilon > 0$, $0 \leq |x| - x\tanh(x/\varepsilon) \leq k_q\varepsilon$, where $k_q = 0.2785$ satisfies $k_q = e^{-(k_q+1)}$.*

Lemma 3 ([14]) *Given a smooth function $\beta(t) : \mathbb{R}^+ \to \mathbb{R}$, its derivatives can be estimated by the command filter $\ddot{x} = -2\xi\omega_n\dot{x} - \omega_n^2(x - \beta)$ such that $\dot{\beta} \approx \dot{x}, \ddot{\beta} \approx \ddot{x}$, where $\omega_n > 0$ is chosen large enough to ensure estimation accuracy.*

3 Problem Statement

Mathematical modeling Figure 1 shows the helicopter configuration, it is modeled in two frames: the earth frame $\mathcal{I} = \{Oxyz\}$ and fuselage frame $\mathcal{B} = \{O_b x_b y_b z_b\}$. Frame $\mathcal{I}$ is fixed to the earth and its origin locates on the ground. Frame $\mathcal{B}$ is fixed to the helicopter body and its origin locates at the helicopter center of gravity (c.g.). The rotation matrix R [1], which defines the rotation from $\mathcal{B}$ to $\mathcal{I}$ around an unit vector $\hat{k} \in \mathbb{R}^3$ by an angle φ, is mostly used to describe attitude of the helicopter. The unit-quaternion $Q = [\mu, q^T]^T = \{Q \in \mathbb{R} \times \mathbb{R}^3 | \mu^2 + q^T q = 1\}$ [15] can also be used to describe attitude, where $\mu = \cos\frac{\varphi}{2}$ and $q = [q_1, q_2, q_3]^T = \sin\frac{\varphi}{2}\hat{k}$. R and Q satisfy following relation

$$R(Q) = (\mu^2 - q^T q)I_3 + 2qq^T + 2\mu S(q) \tag{1}$$

Fig. 1 Helicopter model and frames

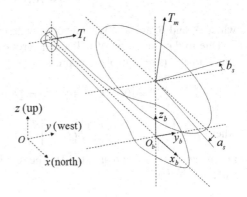

where I_3 is 3×3 identity matrix, $S(q) = \begin{bmatrix} 0 & -q_3 & q_2 \\ q_3 & 0 & -q_1 \\ -q_2 & q_1 & 0 \end{bmatrix}$. Given $Q_1 = \begin{bmatrix} \mu_1 \\ q_1 \end{bmatrix}$ and

$Q_2 = \begin{bmatrix} \mu_2 \\ q_2 \end{bmatrix}$, the quaternion multiplication $Q_1 \otimes Q_2 = \begin{bmatrix} \mu_1\mu_2 - q_1^T q_2 \\ \mu_1 q_2 + \mu_2 q_1 + S(q_1)q_2 \end{bmatrix}$. The

inverse of Q is defined as $Q^{-1} = [\mu, -q^T]^T$ and satisfies $Q^{-1} \otimes Q = [1, 0, 0, 0]^T$.

Based on the Newton–Euler equations, the dynamic model of the helicopter can be derived as follows [1]

$$\dot{p} = v, \quad m\dot{v} = -mg_3 + R(Q)f \tag{2}$$

$$\dot{Q} = \frac{1}{2}\Theta(Q)\omega, \quad J\dot{\omega} = -\omega \times J\omega + \tau \tag{3}$$

where $p = [x, y, z]^T$ and $v = [u, v, w]^T$ are position and velocity of helicopter denoted in $\mathcal{I}$, m is the helicopter mass, $g_3 = [0, 0, g]^T$ and g is gravitational acceleration, $\omega = [\omega_x, \omega_y, \omega_z]^T$ is angular velocity denoted in $\mathcal{B}$, $\Theta(Q) = \begin{bmatrix} -q^T \\ \mu I_3 + S(q) \end{bmatrix}$ is attitude transition

matrix, $J = \begin{bmatrix} J_x & 0 & -J_{xz} \\ 0 & J_y & 0 \\ -J_{xz} & 0 & J_z \end{bmatrix}$ is the inertial matrix denoted in $\mathcal{B}$. The applied force f and

torque τ denoted in $\mathcal{B}$ are given by [1]

$$f = \begin{bmatrix} T_m sa_s cb_s \\ -T_m sb_s ca_s + T_t \\ T_m ca_s cb_s \end{bmatrix}, \tau = \begin{bmatrix} T_m h_m sb_s + L_b b_s + T_t h_t + \tau_m sa_s \\ T_m l_m + T_m h_m sa_s + M_a a_s + \tau_t - \tau_m sb_s \\ -T_m l_m sb_s - T_t l_t + \tau_m ca_s cb_s \end{bmatrix} \tag{4}$$

where T_m, τ_m, T_t, τ_t are thrust and anti-torque generated by the main and tail rotors, respectively; M_a, L_b are stiffness coefficients of main rotor; h_m, h_t, l_m, l_t denote vertical and horizontal distances between the rotor centers and helicopter c.g.; a_s, b_s denote longitudinal and lateral flapping angles of the main rotor. The relationship between thrusts T_i and anti-torque τ_i ($i = m$ or t) is given by [12]

$$\tau_i = A_i |T_i|^{1.5} + B_i \tag{5}$$

where A_i and B_i are aerodynamic constants.

Due to the limitation of helicopter physical structure, a_s, b_s, T_t and τ_t are fairly small, thus it is reasonable to express (4) as [11]:

$$f = [0, 0, T_m]^T + f_\Delta, \quad \tau = Q_A\rho + \tau_B + \Delta_\tau = \tau_1 + \Delta_\tau \tag{6}$$

where $\rho = \begin{bmatrix} T_t \\ a_s \\ b_s \end{bmatrix}$, $\tau_B = \begin{bmatrix} 0 \\ T_m l_m \\ \tau_m \end{bmatrix}$, $Q_A = \begin{bmatrix} h_t & \tau_m & T_m h_m + L_b \\ 0 & T_m h_m + M_a & -\tau_m \\ -l_t & 0 & -T_m l_m \end{bmatrix}$ and $\det(Q_A) \neq 0$ [1]; f_Δ

and Δ_τ are bounded and small [12]. The real input $\rho = Q_A^{-1}(\tau_1 - \tau_B)$. Equation (2) can be rewritten as

$$\dot{v} = -g_3 + (T_m/m)r_3 + \bar{f}_\Delta \tag{7}$$

where $r_3 = Re_3$, $e_3 = [0, 0, 1]^T$, $\bar{f}_\Delta = \frac{1}{m}Rf_\Delta$.

Control objective The reference path $\mathcal{P}_r$ is a regular curve described by implicit expressions, i.e. $\mathcal{P}_r = \{[x, y, z]^T | f_1(x, y, z) = 0, f_2(x, y, z) = 0\}$, and $\partial_p f_1 \times \partial_p f_2|_{p \in \mathcal{P}_r} \neq 0$, where $f_1, f_2 \in C^\infty$; $\|\partial_p f_1\|, \|\partial_p f_2\|$ are bounded on $\mathcal{P}_r$. In this paper the manifold $f_1 = 0$ is specified as a plane $f_1 = ax + by + cz + d = 0$.

Remark 1 From $f_1, f_2 \in C^\infty$ it follows that $\partial_p f_1 \times \partial_p f_2 \in C^\infty$ and $\partial_p f_1 \times \partial_p f_2 \neq 0$ in neighbourhood of $\mathcal{P}_r$.

The control object is: (*i*) designing control inputs T_m, T_t, a_s, b_s such that the helicopter trajectory converges to the reference path ultimately and the magnitude of its velocity projection on reference path tends to desired speed $v_r > 0$, i.e. there exit small positive constants o_1, o_2, o_3 such that

$$\lim_{t\to\infty} |f_1(x, y, z)| < o_1, \quad \lim_{t\to\infty} |f_2(x, y, z)| < o_2$$

$$\lim_{t\to\infty} \left| \left(\frac{\partial_p f_1 \times \partial_p f_2}{\|\partial_p f_1 \times \partial_p f_2\|} \right)^T v - v_r \right| < o_3 \tag{8}$$

(*ii*) No singularity occurs in control process.

4 Controller Design

Define the path-following and velocity errors

$$\varsigma_1 = f_1(x, y, z), \quad \varsigma_2 = f_2(x, y, z), \quad \varsigma_3 = \eta^T v - v_r \tag{9}$$

where $\eta = \frac{\partial_p f_1 \times \partial_p f_2}{\|\partial_p f_1 \times \partial_p f_2\|}$, we can derive

$$[\dot{\varsigma}_1, \dot{\varsigma}_2, \dot{\varsigma}_3]^T = Gv - [0, 0, v_r]^T, \quad [\ddot{\varsigma}_1, \ddot{\varsigma}_2, \dot{\varsigma}_3]^T = h + G\dot{v} \tag{10}$$

where $h = [0, v^T(\partial_p^2 f_2)v, v^T(\partial_p \eta)v - \dot{v}_r]^T$, $G = [\partial_p f_1, \partial_p f_2, \eta]^T$ is the control matrix. From (10) and (7) we know

$$[\ddot{\varsigma}_1, \ddot{\varsigma}_2, \dot{\varsigma}_3]^T = h + G[-g_3 + (T_m/m)r_3] + \Delta_f \tag{11}$$

where $\Delta_f = G\bar{f}_\Delta$. Define the position loop controller $u_c = T_m r_3$ and design

$$u_c = [u_{cx}, u_{cy}, u_{cz}]^T = m(g_3 + G^{-1}(-h + v_c)) \tag{12}$$

Fig. 2 Temporary path generator

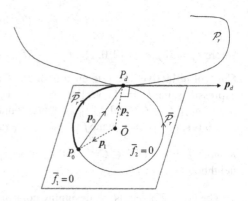

where v_c is a new control input to be determined. The singularity occurs when $\det(G) = (\partial_p f_1 \times \partial_p f_2)^T \boldsymbol{\eta} = \|\partial_p f_1 \times \partial_p f_2\| = 0$. From Remark 1, G is invertible in neighborhood of $\mathcal{P}_r$. When helicopter initial position is far from $\mathcal{P}_r$, G cannot be guaranteed to be invertible. Besides, when helicopter is far from the path, the control energy will be large. A solution for the two problems is to plan a temporary path $\bar{\mathcal{P}}_r$ from initial position to $\mathcal{P}_r$ such that corresponding control matrix $\bar{G}$ is invertible on $\bar{\mathcal{P}}_r$. In this paper define P_0 to be the initial position, when $\sqrt{|f_1(P_0)|^2 + |f_2(P_0)|^2} > 10$, it is need to plan a temporary-path.

Assumption 1 $\boldsymbol{\Delta}_f = [\Delta_{fx}, \Delta_{fy}, \Delta_{fz}]^T$ and $\boldsymbol{\Delta}_\tau = [\Delta_{\tau x}, \Delta_{\tau y}, \Delta_{\tau z}]^T$ are bounded and satisfy $|\Delta_{fi}| \leq \delta_i, |\Delta_{\tau i}| \leq \gamma_i$ ($i = x, y, z$), where $\boldsymbol{\delta} = [\delta_x, \delta_y, \delta_z]^T$ and $\boldsymbol{\gamma} = [\gamma_x, \gamma_y, \gamma_z]^T$ are known upper bounds; The coefficient c in $f_1 = 0$ satisfies $|c| > \delta_x/g$ for avoiding singularity in deriving desired unit-quaternion. The physical meaning is that $f_1 = 0$ is not perpendicular to $x - y$ plane of $\mathcal{I}$.

Path planning As illustrated in Fig. 2, the initial position $P_0 = [x_0, y_0, z_0]^T$. Choose a point $P_d = [x_r, y_r, z_r]^T$ on $\mathcal{P}_r$ and compute tangent vector $\boldsymbol{p}_d = [\bar{x}, \bar{y}, \bar{z}]^T = \partial_p f_1 \times \partial_p f_2|_{P_d}$. Define $\boldsymbol{p}_0 = \overrightarrow{P_0 P_d}$ and compute $\boldsymbol{p}_0 \times \boldsymbol{p}_d = [\bar{x}_1, \bar{y}_1, \bar{z}_1]^T$. If P_d are chosen such that $\bar{z}_1 \neq 0$ and $\boldsymbol{p}_0$ is not collinear with $\boldsymbol{p}_d$, a temporary path can be planned with following steps:

Firstly, passing through $\boldsymbol{p}_0$ and $\boldsymbol{p}_d$ we can define a plane $\bar{f}_1 = a_1 x + b_1 y + c_1 z + d_1 = 0$. Since P_d lies on $\bar{f}_1 = 0$ and its normal vector is $\boldsymbol{p}_0 \times \boldsymbol{p}_d$, we know $d_1 = -a_1 x_r - b_1 y_r - c_1 z_r$ and $[a_1, b_1, c_1] = k[\bar{x}_1, \bar{y}_1, \bar{z}_1]$, where k is a constant. Considering $\bar{z}_1 \neq 0$, choosing $|k| > \frac{\delta_x}{|\bar{z}_1|g}$ can guarantee that $c_1 > \delta_x/g$.

Then, define $\bar{f}_2 = 0$ to be a sphere, its center $\bar{O} = [a_2, b_2, c_2]^T$, $\boldsymbol{p}_1 = \overrightarrow{\bar{O}P_0}$ and $\boldsymbol{p}_2 = \overrightarrow{\bar{O}P_d}$. Since $\boldsymbol{p}_2 \perp \boldsymbol{p}_d, \bar{f}_1(\bar{O}) = 0$ and $\|\boldsymbol{p}_1\| = \|\boldsymbol{p}_2\|$, $\bar{O}$ can be determined by

$$
\begin{bmatrix} a_2 \\ b_2 \\ c_2 \end{bmatrix} = \begin{bmatrix} \bar{x} & \bar{y} & \bar{z} \\ a_1 & b_1 & c_1 \\ x_r - x_0 & y_r - y_0 & z_r - z_0 \end{bmatrix}^{-1} \begin{bmatrix} \bar{x}x_r + \bar{y}y_r + \bar{z}z_r \\ -d_1 \\ \frac{x_r^2 - x_0^2 + y_r^2 - y_0^2 + z_r^2 - z_0^2}{2} \end{bmatrix}
$$

Since p_0 is not collinear to p_d, above inverse matrix is well defined and $[a_2, b_2, c_2]^T$ can be determined uniquely. Define $\|p_1\| = r$, we have $\bar{f}_2 = \frac{1}{k_1}((x - a_2)^2 + (y - b_2)^2 + (z - c_2)^2 - r^2) = 0$, where $k_1 \neq 0$ is a constant.

Finally, temporary path $\bar{P}_r$ is the intersection of $\bar{f}_1 = 0$ and $\bar{f}_2 = 0$. Since two paths are generated (bold line and normal one in Fig. 2), the one with smaller angle from its tangent line at P_0 to helicopter head direction should be selected.

Remark 2 $\bar{P}_r$ satisfies that 1) $\partial_p \bar{f}_1 \times \partial_p \bar{f}_2$ is collinear with $\partial_p f_1 \times \partial_p f_2$ at P_d. 2) $\partial_p \bar{f}_1 \perp \partial_p \bar{f}_2$ on $\bar{P}_r$, i.e. $\partial_p \bar{f}_1 \times \partial_p \bar{f}_2 \neq 0$ on $\bar{P}_r$. Since $\bar{f}_1, \bar{f}_2 \in C^\infty$, it holds near $\bar{P}_r$, i.e. control matrix $\bar{G}$ invertible near $\bar{P}_r$.

$\bar{P}_r$ **following** The position loop controller design is divided to two steps: $\bar{P}_r$ following and then $\mathcal{P}_r$ following. Firstly define errors for the temporary path $\bar{P}_r$

$$\varsigma_1 = a_1 x + b_1 y + c_1 z + d_1, \quad \varsigma_2 = \frac{1}{k_1}((x - a_2)^2 + (y - b_2)^2 + (z - c_2)^2 - r^2)$$

$$\varsigma_3 = \bar{\eta}^T v - \bar{v}_r \tag{13}$$

where $\bar{\eta} = \frac{\bar{\eta}_1 \times \bar{\eta}_2}{\|\bar{\eta}_1 \times \bar{\eta}_2\|}$, $\bar{\eta}_1 = [a_1, b_1, c_1]^T$, $\bar{\eta}_2 = \frac{2}{k_1}[x - a_2, y - b_2, z - c_2]^T$. It yields

$$[\dot{\varsigma}_1, \dot{\varsigma}_2, \dot{\varsigma}_3]^T = \bar{G}v - [0, 0, \bar{v}_r]^T \tag{14}$$

$$[\ddot{\varsigma}_1, \ddot{\varsigma}_2, \dot{\varsigma}_3]^T = \bar{h} + \bar{G}[-g_3 + (T_m/m)r_3] + \Delta_f \tag{15}$$

where $\bar{G} = [\bar{\eta}_1, \bar{\eta}_2, \bar{\eta}]^T$ is control matrix, $\bar{h} = [\bar{h}_1, \bar{h}_2, \bar{h}_3]^T = [0, \frac{2}{k_1} v^T v, v^T (\partial_p \bar{\eta}) v - \dot{v}_r]^T$. Define $\bar{u}_c = T_m r_3$ and design it as

$$\bar{u}_c = [\bar{u}_{cx}, \bar{u}_{cy}, \bar{u}_{cz}]^T = m(g_3 + \bar{G}^{-1}(-\bar{h} + \bar{v}_c)) \tag{16}$$

$$\bar{v}_c = \begin{bmatrix} \bar{v}_1 \\ \bar{v}_2 \\ \bar{v}_3 \end{bmatrix} = \begin{bmatrix} -\bar{k}_{11}(\tanh(\varsigma_{k1}) + \tanh(\dot{\varsigma}_1)) - \tanh(\frac{\vartheta_1}{\bar{\varepsilon}_1})\delta_x \\ -\bar{k}_{21}(\tanh(\varsigma_{k2}) + \tanh(\dot{\varsigma}_2)) - \tanh(\frac{\vartheta_2}{\bar{\varepsilon}_2})\delta_y \\ -\bar{k}_{31}\tanh(\varsigma_3) - \tanh(\frac{\varsigma_3}{\bar{\varepsilon}_3})\delta_z \end{bmatrix} \tag{17}$$

where $\varsigma_{k1} = \bar{k}_{12}\varsigma_1 + \dot{\varsigma}_1$, $\varsigma_{k2} = \bar{k}_{22}\varsigma_2 + \dot{\varsigma}_2$, $\vartheta_1 = \tanh(\varsigma_{k1}) + \tanh(\dot{\varsigma}_1) + \frac{\bar{k}_{12}}{\bar{k}_{11}}\dot{\varsigma}_1$, $\vartheta_2 = \tanh(\varsigma_{k2}) + \tanh(\dot{\varsigma}_2) + \frac{\bar{k}_{22}}{\bar{k}_{21}}\dot{\varsigma}_2$; $\bar{k}_{11}, \bar{k}_{12}, \bar{k}_{21}, \bar{k}_{22}, \bar{k}_{31}, \bar{\varepsilon}_1, \bar{\varepsilon}_2, \bar{\varepsilon}_3$ are positive constants. Since $\|r_3\| = 1$, $T_m = \|\bar{u}_c\|$. If $\bar{u}_c \notin \mathcal{L} = \{[0, 0, u]^T, u \leq 0\}$, then the desired unit-quaternion $Q_c = [\mu_c, q_c^T]^T$ can be derived from $\bar{u}_c$ [12]

$$q_c = \frac{1}{2\|\bar{u}_c\|\mu_c} \begin{bmatrix} \bar{u}_{cy} \\ -\bar{u}_{cx} \\ 0 \end{bmatrix}, \quad \mu_c = \sqrt{\frac{\|\bar{u}_c\| + \bar{u}_{cz}}{2\|\bar{u}_c\|}} \tag{18}$$

Theorem 1 *If initial velocity $v(0) = 0$ and the parameters in (17) satisfy*

$$\bar{k}_{12} \le \bar{k}_{11} < \frac{|c_1|g - \delta_x}{2}, \bar{k}_{22} \le \bar{k}_{21}, \bar{\varepsilon}_1 < \frac{0.382\bar{k}_{12}}{k_q\delta_x}, \bar{\varepsilon}_2 < \frac{0.382\bar{k}_{22}}{k_q\delta_y} \quad (19)$$

Then control law (16) guarantees: (i) $\varsigma_1, \dot{\varsigma}_1, \varsigma_2, \dot{\varsigma}_2, \varsigma_3$ are bounded and converge to a small neighbourhood of origin. (ii) $\bar{G}$ is invertible during control and no singularity occurs in deriving desired unit-quaternion with (18).

Proof (i) Define $\bar{\varsigma} = [\varsigma_{k1}, \dot{\varsigma}_1]$, choose Lyapunov function

$$V = \int_0^{\varsigma_{k1}} \tanh(s)ds + \int_0^{\dot{\varsigma}_1} \tanh(s)ds + \frac{\bar{k}_{12}}{2\bar{k}_{11}}\dot{\varsigma}_1^2 \quad (20)$$

$$\dot{V} = -\bar{k}_{11}[\tanh(\varsigma_{k1}) + \tanh(\dot{\varsigma}_1)]^2 - \bar{k}_{12}\dot{\varsigma}_1\tanh(\dot{\varsigma}_1) - \vartheta_1[\tanh(\frac{\vartheta_1}{\bar{\varepsilon}_1})\delta_x - \Delta_{fx}]$$
$$\le -\tanh^T(\bar{\varsigma}) \begin{bmatrix} \bar{k}_{11} & \bar{k}_{11} \\ \bar{k}_{11} & \bar{k}_{11} + \bar{k}_{12} \end{bmatrix} \tanh(\bar{\varsigma}) - \vartheta_1\tanh(\frac{\vartheta_1}{\bar{\varepsilon}_1})\delta_x + \vartheta_1\Delta_{fx} \quad (21)$$

Based on Lemma 2, we know $\vartheta_1\Delta_{fx} \le |\vartheta_1|\delta_x \le [\vartheta_1\tanh(\frac{\vartheta_1}{\bar{\varepsilon}_1}) + k_q\bar{\varepsilon}_1]\delta_x$. Considering $\begin{bmatrix} \bar{k}_{11} & \bar{k}_{11} \\ \bar{k}_{11} & \bar{k}_{11} + \bar{k}_{12} \end{bmatrix} = \Lambda^T \begin{bmatrix} \bar{k}_{11} & 0 \\ 0 & \bar{k}_{12} \end{bmatrix} \Lambda$, where $\Lambda = \begin{bmatrix} 1 & 1 \\ 0 & 1 \end{bmatrix}$, (21) yields $\dot{V} \le k_q\bar{\varepsilon}_1\delta_x - \underline{\lambda}(\Lambda^T\Lambda)\min\{\bar{k}_{11}, \bar{k}_{12}\}\|\tanh(\bar{\varsigma})\|^2 = k_q\bar{\varepsilon}_1\delta_x - 0.382\bar{k}_{12}\|\tanh(\bar{\varsigma})\|^2$. From (19) we know $\frac{k_q\bar{\varepsilon}_1\delta_x}{0.382\bar{k}_{12}} < 1$. When $\|\tanh(\bar{\varsigma})\| > \sqrt{\frac{k_q\bar{\varepsilon}_1\delta_x}{0.382\bar{k}_{12}}}$, $\dot{V} < 0$, which means $\bar{\varsigma}$ is bounded. Then from Lemma 1 we have $\chi\|\bar{\varsigma}\| \le \|\tanh(\bar{\varsigma})\|$, where $0 < \chi < 1$, it follows

$$\dot{V} \le -0.382\bar{k}_{12}\chi^2\|\bar{\varsigma}\|^2 + k_q\bar{\varepsilon}_1\delta_x \quad (22)$$

Besides, from (20) we have $\frac{\chi}{2}\|\bar{\varsigma}\|^2 \le V \le \chi_2\|\bar{\varsigma}\|^2$, where $\chi_2 = \frac{1}{2}(1 + \frac{\bar{k}_{12}}{\bar{k}_{11}})$. Then $\dot{V} \le -\frac{0.382\bar{k}_{12}\chi^2}{\chi_2}V + k_q\bar{\varepsilon}_1\delta_x$. Integrating it yields

$$\frac{\chi}{2}\|\bar{\varsigma}\|^2 \le V \le [V(0) - \frac{k_q\bar{\varepsilon}_1\delta_x\chi_2}{0.382\bar{k}_{12}\chi^2}]e^{\frac{-0.382\bar{k}_{12}\chi^2}{\chi_2}t} + \frac{k_q\bar{\varepsilon}_1\delta_x\chi_2}{0.382\bar{k}_{12}\chi^2} \quad (23)$$

So $\bar{\varsigma}$ exponentially converges to the set $\mathbb{Z}_{v1} = \{\bar{\varsigma}|\|\bar{\varsigma}\| \le \sqrt{\frac{0.73\bar{\varepsilon}_1\delta_x}{\chi^3}(\frac{1}{\bar{k}_{12}} + \frac{1}{\bar{k}_{11}})}\}$. Due that $|\dot{\varsigma}_1| \le \|\bar{\varsigma}\|$, $\dot{\varsigma}_1$ also converges to $\mathbb{Z}_{v1}$. Considering $\dot{\varsigma}_1 = -\bar{k}_{12}\varsigma_1 + \varsigma_{k1}$ and $|\varsigma_{k1}| \le \|\bar{\varsigma}\|$, integrating $\dot{\varsigma}_1$ yields

$$|\bar{\varsigma}_1| \le e^{-\bar{k}_{12}t}(|\bar{\varsigma}_1(0)| - \sqrt{\frac{2k_q\bar{\varepsilon}_1\delta_x\chi_2}{0.382\bar{k}_{12}^3\chi^3}}) + \sqrt{\frac{2k_q\bar{\varepsilon}_1\delta_x\chi_2}{0.382\bar{k}_{12}^3\chi^3}} \tag{24}$$

Thus $\bar{\varsigma}_1$ converges to set $\mathbb{Z}_{p1} = \{|\bar{\varsigma}_1| \le \frac{1}{\bar{k}_{12}}\sqrt{\frac{0.73\bar{\varepsilon}_1\delta_x}{\chi^3}(\frac{1}{\bar{k}_{12}} + \frac{1}{\bar{k}_{11}})}\}$. Similarly, $\bar{\varsigma}_2$ and $\dot{\bar{\varsigma}}_2$

exponentially converge to sets $\mathbb{Z}_{p2} = \{\bar{\varsigma}_2||\bar{\varsigma}_2| \le \frac{1}{\bar{k}_{22}}\sqrt{\frac{0.73\bar{\varepsilon}_2\delta_y}{\chi^3}(\frac{1}{\bar{k}_{22}} + \frac{1}{\bar{k}_{21}})}\}$ and $\mathbb{Z}_{v2} =$

$\{\dot{\bar{\varsigma}}_2||\dot{\bar{\varsigma}}_2| \le \sqrt{\frac{0.73\bar{\varepsilon}_2\delta_y}{\chi^3}(\frac{1}{\bar{k}_{22}} + \frac{1}{\bar{k}_{21}})}\}$, respectively.

For $\bar{\varsigma}_3$, choose Lyapunov function $V_1 = \frac{1}{2}\bar{\varsigma}_3^2$, its derivative is

$$\dot{V}_1 = -\bar{k}_{31}\bar{\varsigma}_3\tanh(\bar{\varsigma}_3) - \bar{\varsigma}_3\tanh\left(\frac{\bar{\varsigma}_3}{\bar{\varepsilon}_3}\right)\delta_z + \bar{\varsigma}_3\Delta_{fz} \tag{25}$$

From Lemma 2, $\bar{\varsigma}_3\Delta_{fz} \le |\bar{\varsigma}_3|\delta_z \le (\bar{\varsigma}_3\tanh(\frac{\bar{\varsigma}_3}{\bar{\varepsilon}_3}) + k_q\bar{\varepsilon}_3)\delta_z$, thus $\dot{V}_1 \le -\bar{k}_{31}\bar{\varsigma}_3\tanh(\bar{\varsigma}_3) + k_q\bar{\varepsilon}_3\delta_z$. Based on Lemma 1 we have $\dot{V}_1 \le -\bar{k}_{31}\chi\bar{\varsigma}_3^2 + k_q\bar{\varepsilon}_3\delta_z = -2\bar{k}_{31}\chi V_1 + k_q\bar{\varepsilon}_3\delta_z$. Integrating it yields

$$\frac{1}{2}\bar{\varsigma}_3^2 = V_1(t) \le [V_1(0) - \frac{k_q\bar{\varepsilon}_3\delta_z}{2\bar{k}_{31}\chi}]e^{-2\bar{k}_{31}\chi t} + \frac{k_q\bar{\varepsilon}_3\delta_z}{2\bar{k}_{31}\chi} \tag{26}$$

Thus $\bar{\varsigma}_3$ exponentially converges to the set $\mathbb{Z}_3 = \{\bar{\varsigma}_3||\bar{\varsigma}_3| \le \sqrt{\frac{0.2785\bar{\varepsilon}_3\delta_z}{\bar{k}_{31}\chi}}\}$.

Above sets can be made small by increasing $\bar{k}_{21}, \bar{k}_{22}, \bar{k}_{31}$ and decreasing $\bar{\varepsilon}_1, \bar{\varepsilon}_2, \bar{\varepsilon}_3$. Since $p(0) \in \bar{\mathcal{P}}_r$ and $v(0) = \mathbf{0}$, we have $\bar{\varsigma}_1(0) = \dot{\bar{\varsigma}}_1(0) = \bar{\varsigma}_2(0) = \dot{\bar{\varsigma}}_2(0) = 0$. So under the control law helicopter keeps in the neighbourhood of $\bar{\mathcal{P}}_r$.

(ii) From Remark 2 and proof of (i) it yields that $\bar{G}$ keeps invertible. When $\bar{u}_{cx} \ne 0$ or $\bar{u}_{cy} \ne 0$, $\bar{u}_c \notin \mathcal{L}$, which means no singularity occurs in deriving the desired unit-quaternion. So we only need to consider the singularity when $\bar{u}_c = [0, 0, \bar{u}_{cz}]^T$. Left-multiplying (16) by $[a_1, b_1, c_1]$ yields $\bar{u}_{cz} = m(g + \frac{\bar{v}_1}{c_1})$. From (17) $|\frac{\bar{v}_1}{c_1}| \le \frac{2\bar{k}_{11}+\delta_x}{|c_1|}$, it follows $\bar{u}_{cz} \ge m(g - \frac{2\bar{k}_{11}+\delta_x}{|c_1|})$. Since $\bar{k}_{11} < \frac{|c_1|g-\delta_x}{2}$, we get $\bar{u}_{cz} > 0$ and $\bar{u}_c \notin \mathcal{L}$, which means no singularity occurs in deriving the desired unit-quaternion with (18).

$\mathcal{P}_r$ **following** Define P to be helicopter position and $\bar{\varepsilon}$ to be a small positive constant, when $\|P - P_d\| \le \bar{\varepsilon}$, control u_c is given by (12) and v_c is designed as

$$v_c = \begin{bmatrix} -k_{11}(\tanh(\varsigma_{k1}) + \tanh(\varsigma_1)) - \tanh(\frac{\vartheta_1}{\varepsilon_1})\delta_x \\ -k_{21}(\tanh(\varsigma_{k2}) + \tanh(\varsigma_2)) - \tanh(\frac{\vartheta_2}{\varepsilon_2})\delta_y \\ -k_{31}\tanh(\varsigma_3) - \tanh(\frac{\varsigma_3}{\varepsilon_3})\delta_z \end{bmatrix} \tag{27}$$

where $\varsigma_{k1} = k_{12}\varsigma_1 + \dot{\varsigma}_1$, $\varsigma_{k2} = k_{22}\varsigma_2 + \dot{\varsigma}_2$, $\vartheta_1 = \tanh(\varsigma_{k1}) + \tanh(\dot{\varsigma}_1) + \frac{k_{12}}{k_{11}}\varsigma_1$, $\vartheta_2 = \tanh(\varsigma_{k2}) + \tanh(\dot{\varsigma}_2) + \frac{k_{22}}{k_{21}}\varsigma_2$. $k_{11}, k_{12}, k_{21}, k_{22}, k_{31}, \varepsilon_1, \varepsilon_2$ and ε_3 are positive constants satisfying

$$k_{12} \le k_{11} < \frac{|c|g - \delta_x}{2}, k_{22} \le k_{21}, \varepsilon_1 < \frac{0.382k_{12}}{k_q\delta_x}, \varepsilon_2 < \frac{0.382k_{22}}{k_q\delta_y} \qquad (28)$$

From Remark 2, ν is approximately tangent to $\mathcal{P}_r$ at switching instant. So under the control law helicopter keeps in the neighbourhood of $\mathcal{P}_r$. Based on Remark 1 it yields that G keeps invertible. The constraint for k_{11} in (28) guarantees that no singularity occurs in deriving desired unit-quaternion.

Attitude controller design Define attitude tracking error $Q_e = [\mu_e, q_e^T]^T = Q_c^{-1} \otimes Q$ [15], its derivative is

$$\dot{\mu}_e = -\frac{1}{2}q_e^T\omega_e, \quad \dot{q}_e = \frac{1}{2}(\mu_e I_3 + S(q_e))\omega_e \qquad (29)$$

where $\omega_e = \omega - R^T(Q_e)\omega_c$ is the angular velocity error and ω_c is the desired angular velocity. From (3) ω_c and its derivative $\dot{\omega}_c$ can be derived as

$$\omega_c = 2\Theta^T(Q_c)\dot{Q}_c \qquad (30)$$

$$\dot{\omega}_c = 2\Theta^T(Q_c)\ddot{Q}_c + 2\dot{\Theta}^T(Q_c)\dot{Q}_c = 2\Theta^T(Q_c)[\ddot{Q}_c - \dot{\Theta}_c\omega_c] \qquad (31)$$

where $\dot{\Theta}_c = \Theta(\dot{Q}_c)$. From Lemma 3, $\dot{Q}_c$ and $\ddot{Q}_c$ can be obtained by a command filter instead of calculating them accurately. Assign Lyapunov function

$$L = k_Q[(1 - k_\mu\mu_e)^2 + q_e^T q_e] = 2k_Q(1 - k_\mu\mu_e) \qquad (32)$$

where $k_Q > 0$, $k_\mu = 1$ when $\mu_e \ge 0$ and $k_\mu = -1$ when $\mu_e < 0$. From (29), $\dot{L} = k_Q k_\mu q_e^T \omega_e$. Define $\bar{\omega}_e = [\bar{\omega}_{ex}, \bar{\omega}_{ey}, \bar{\omega}_{ez}]^T = \omega_e + k_\omega k_\mu q_e$, where $k_\omega > 0$. Choose a Lyapunov function

$$L_1 = L + \frac{1}{2}\bar{\omega}_e^T J \bar{\omega}_e \qquad (33)$$

$$\begin{aligned}\dot{L}_1 = \bar{\omega}_e^T[-\omega \times J\omega + \tau_1 + J(S(\omega_e)R^T(Q_e)\omega_c - R^T(Q_e)\dot{\omega}_c \\ + k_\omega k_\mu \dot{q}_e) + \Delta_\tau] - k_Q k_\omega q_e^T q_e + k_Q k_\mu q_e^T \bar{\omega}_e\end{aligned} \qquad (34)$$

Design the control torque

$$\begin{aligned}\tau_1 = -k_\tau\bar{\omega}_e + \omega \times J\omega - JS(\omega_e)R^T(Q_e)\omega_c + JR^T(Q_e)\dot{\omega}_c \\ - k_\omega k_\mu J\dot{q}_e - \text{Tanh}(\frac{\bar{\omega}_e^T}{\varepsilon_4})\gamma - k_Q k_\mu q_e\end{aligned} \qquad (35)$$

where $k_\tau > 0$ and ε_4 is a small positive constant. Substituting τ_1 into (34) yields

$$\dot{L}_1 = -k_Q k_\omega q_e^T q_e - k_\tau \bar{\omega}_e^T \bar{\omega}_e - \bar{\omega}_e^T \text{Tanh}(\frac{\bar{\omega}_e^T}{\varepsilon_4})\gamma + \bar{\omega}_e^T \Delta_\tau \qquad (36)$$

Theorem 2 *Take* $k_\omega \leq \frac{4k_\tau}{\bar{\lambda}(J)}$, *the attitude controller* (35) *guarantees that the attitude tracking error* q_e *and angular velocity error* ω_e *are bounded and ultimately converge to neighbourhoods of the origin.*

Proof From Lemma 2, $\bar{\omega}_e^T \Delta_\tau \leq \sum_{i=x,y,z} |\bar{\omega}_{ei}| \gamma_i \leq \sum_{i=x,y,z} [\bar{\omega}_{ei} \tanh(\frac{\bar{\omega}_{ei}}{\varepsilon_4}) + k_q \varepsilon_4] \gamma_i = \bar{\omega}_e^T \text{Tanh}(\frac{\bar{\omega}_e}{\varepsilon_4})\gamma + d_\tau$, where $d_\tau = \sum_{i=x,y,z} k_q \varepsilon_4 \gamma_i$. Substituting it into (36) yields

$$\dot{L}_1 \leq -k_Q k_\omega q_e^T q_e - k_\tau \bar{\omega}_e^T \bar{\omega}_e + d_\tau \leq -k_Q k_\omega (1 - k_\mu \mu_e) - \frac{k_\tau}{\bar{\lambda}(J)} \bar{\omega}_e^T J \bar{\omega}_e + d_\tau \qquad (37)$$

From (33), $\dot{L}_1 \leq -\min\{\frac{k_\omega}{2}, \frac{2k_\tau}{\bar{\lambda}(J)}\} L_1 + d_\tau = -\frac{k_\omega}{2} L_1 + d_\tau$. Integrating it gives

$$L_1 \leq (L_1(0) - \frac{2d_\tau}{k_\omega})e^{-\frac{k_\omega}{2}t} + \frac{2d_\tau}{k_\omega} \leq L_1(0)e^{-\frac{k_\omega}{2}t} + \frac{2d_\tau}{k_\omega} \qquad (38)$$

Also from (33) we know that $K_Q q_e^T q_e \leq L_1$ and $\frac{\lambda(J)}{2} \bar{\omega}_e^T \bar{\omega}_e \leq L_1$. Thus q_e and $\bar{\omega}_e$ are bounded and ultimately converge to the compact sets $\mathbb{C}_q = \{q_e | \|q_e\| \leq \sqrt{\frac{2d_\tau}{k_Q k_\omega}}\}$

and $\bar{\mathbb{C}}_\omega = \{\bar{\omega}_e | \|\bar{\omega}_e\| \leq \sqrt{\frac{4d_\tau}{\lambda(J)k_\omega}}\}$. Since $\omega_e = \bar{\omega}_e - k_\omega k_\mu q_e$, we have $\|\omega_e\| \leq \|\bar{\omega}_e\| + k_\omega \|q_e\|$, so ω_e converges to the set $\mathbb{C}_\omega = \{\omega_e | \|\omega_e\| \leq \sqrt{\frac{2k_\omega d_\tau}{k_Q}} + \sqrt{\frac{4d_\tau}{\lambda(J)k_\omega}}\}$.

Remark 3 By taking $k_Q \gg k_\omega$, increasing k_Q, k_w and decreasing ε_4, the sets $\mathbb{C}_q$ and $\mathbb{C}_\omega$ can be made small.

5 Simulation

A simulation is performed to verify the proposed controller. The helicopter parameters are as follows [12]: $m = 7.4\,\text{kg}, I_x = 0.16\,\text{kgm}^2, I_y = 0.30\,\text{kgm}^2, I_z = 0.32\,\text{kgm}^2, I_{xz} = 0.05\,\text{kgm}^2, l_m = 0.01\,\text{m}, h_m = 0.14\,\text{m}, l_t = 0.95\,\text{m}, h_t = 0.05\,\text{m}, M_a = L_b = 110, A_m = 0.00452, B_m = 0.63, A_t = 0.005066, B_t = 0.008488$ and $g = 9.81\,\text{m/s}^2$. The desired reference path $\mathcal{P}_r$ is a circular curve determined by

$$f_1(x, y, z) = z - 8 = 0, \quad f_2(x, y, z) = \frac{1}{5}(x^2 + y^2) - 5 = 0$$

Define $P_0 = [11, 10, 0]^T$ and choose $P_d = [5, 0, 8]^T$, we can obtain the sphere center $\bar{O} = [11, 0, 0]^T$ and the temporary path $\bar{P}_r$ is planned by

$$\bar{f}_1(x, y, z) = 4x + 3z - 44 = 0, \quad \bar{f}_2(x, y, z) = \frac{1}{20}((x - 11)^2 + y^2 + z^2) - 5 = 0$$

where $\delta = [1, 1, 1]^T, \gamma = [0.5, 0.5, 0.5]^T$. The controller parameters are chosen as follows: $\bar{k}_{11} = \bar{k}_{12} = 1, \bar{k}_{21} = \bar{k}_{22} = 3.5, \bar{k}_{31} = 1, \bar{\varepsilon}_1 = 0.3, \bar{\varepsilon}_2 = 0.5, \bar{\varepsilon}_3 = 0.1; k_{11} = k_{12} = 2, k_{21} = k_{22} = 3.5, k_{31} = 3, \varepsilon_1 = 0.3, \varepsilon_2 = 0.8, \varepsilon_3 = 0.08; k_Q = 500, k_\omega = k_\tau = 16, \varepsilon_4 = 0.05; \xi = 0.707, \omega_n = 10, \bar{\varepsilon} = 0.01$, and they satisfy the conditions in theorems. Choose $\bar{v}_r = v_r = 2.5$ m/s.

Figure 3 shows the 3-D path following result. Figures 4, 5 and 6 illustrate the path following, attitude and angular velocity errors respectively, which are bounded and converge to neighborhoods of origin. Figure 7 shows that the speed converges to v_r. Figure 8 illustrate u_{cz} (or $\bar{u}_{cz}$ before switch) is greater than zero, implying that no singularity occurs in deriving command attitude.

Fig. 3 3-D path following

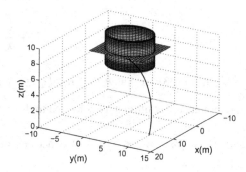

Fig. 4 Values of $\bar{\varsigma}_1, \bar{\varsigma}_2, \bar{\varsigma}_3, \varsigma_1, \varsigma_2, \varsigma_3$

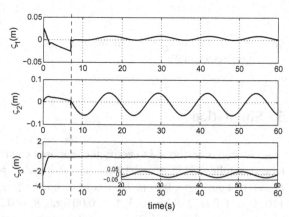

Fig. 5 Attitude tracking errors

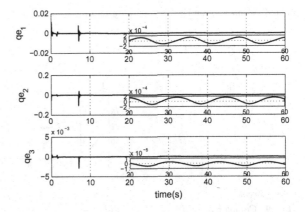

Fig. 6 Angular velocity tracking errors

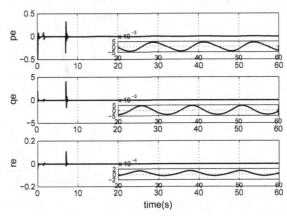

Fig. 7 Actual speed $\|v\|$

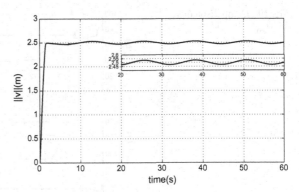

Figures 9 and 10 show comparisons of spatial distance d_s, which defines the shortest distance from actual position to the path, and thrust T_m using the proposed control law and d_{s1}, T_{m1} using the control law [1] without temporary-path generation. Obviously, T_{m1} is large in convergence process, and it would result in control saturation if further increasing parameters to reduce the error.

Fig. 8 Control $\bar{u}_{cz}$ and u_{cz}

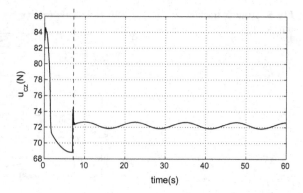

Fig. 9 Comparison of d_s and d_{s1}

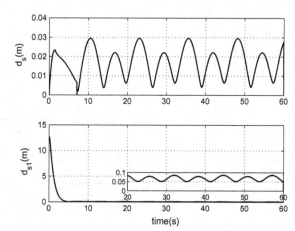

Fig. 10 Comparison of T_m and T_{m1}

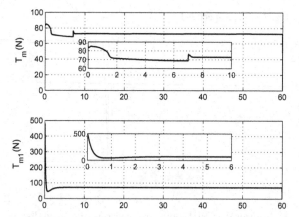

6 Conclusion

This paper presents a singularity-free path following controller for miniature unmanned helicopters. The reference path is defined by implicit expression. Numerical simulation demonstrates the effectiveness of proposed controller. In future research we will extend the controller to more general manifolds and consider the disturbances and parametric uncertainties.

Acknowledgements This work was supported by National Natural Science Foundation of China under Grant No. 61673043.

References

1. Zhu B, Huo W. 3-D path-following control for a model-scaled autonomous helicopter. IEEE Trans Control Syst Technol. 2014;22(5):1927–34.
2. Aguiar AP, et al. Performance limitations in reference tracking and path following for nonlinear systems. Automatica. 2008;44(3):598–610.
3. Huo W, Wang Q. Path following controller design for miniature unmanned helicopters. In: Control conference IEEE. 2011. p. 3537–3542.
4. Wang B, Dong X. Cascaded control of 3D path following for an unmanned helicopter. In: Cybernetics and intelligent systems IEEE. 2010. p. 70–75.
5. Wang T, et al. Combined of vector field and linear quadratic Gaussian for the path following of a small unmanned helicopter. Iet Control Theory Appl. 2012;6(6):2696–703.
6. Cabecinhas D, Cunha R, Silvestre C. A globally stabilizing path following controller for rotorcraft with wind disturbance rejection. IEEE Trans Control Syst Technol. 2015;23(2):708–14.
7. Roza A, Maggiore M. Path following controller for a quadrotor helicopter. In: American control conference IEEE. 2012. p. 4655-4660.
8. Qiang W, Huo W. Adaptive path-following control for miniature autonomous helicopter. Acta Armamentarii. 2012;33(11):1364–72.
9. Gandolfo. Path following for unmanned helicopter: an approach on energy autonomy improvement. Inf Technol Control. 2016;45(11).
10. Hua, Duc M, et al. Introduction to feedback control of underactuated VTOL vehicles: a review of basic control design ideas and principles. IEEE Control Syst. 2013;33(1).
11. Raptis IA, Valavanis KP, Moreno WA. A novel nonlinear backstepping controller design for helicopters using the rotation matrix. IEEE Trans Control Syst Technol. 2011;19(2).
12. Zou Y. Trajectory tracking ocntrol developments for miniature unmanned helicopters. PhD Dissertation. Beihang University, Beijing, China. 2016.
13. Polycarpou MM, Ioannou PA. A robust adaptive nonlinear control design. Automatica. 1996;32(3):423–7.
14. Farrell, Jay A, et al. Command filtered backstepping. IEEE Trans Autom Control. 2009;54(6):1391–1395.
15. Tayebi A. Unit quaternion-based output feedback for attitude tracking problem. IEEE Trans. Autom. Control. 2008;53(6):1516–20.

Stabilizing Quadrotor Helicopter Based on Controlled Lagrangians

Biao Zhang and Wei Huo

Abstract A stabilization controller for the quadrotor helicopter is designed based on controlled Lagrangians method in this paper. Firstly, the dynamical model of quadrotor helicopter is simplified by input transformations. With linearization at the desired equilibrium and invertible state transformations, the model can be divided to four subsystems. Then controlled Lagrangians method is applied to stabilize the two subsystems with one under-actuation degree, and the other two subsystem are stabilized by PD control laws. Finally, local asymptotic stability of the closed-loop system is proved theoretically and verified by simulation results. Different from common-used controllers, the controller presented in this paper does not depend on the outer/inner loop structure and the time-scale separate principle, which always hinder rigorous stability analysis of the controlled system and complicate control algorithm.

Keywords Quadrotor helicopter · Controlled Lagrangians · Under-actuated system · Linearization

1 Introduction

The quadrotor helicopter has draw much attention in recent years thanks to its simple mechanical structure and swift maneuverability. It has been applied to both military and civil applications, such as enemy inspection, military mapping, aerial photo and so on. The quadrotor, which has six motion degrees of freedom but only four independent control inputs, is an under-actuated and dynamically unstable system. Many researchers show great interests to studies of its control strategy and numerous control methods have been tested on quadrotor helicopter, such as sliding-mode and backstepping techniques [1], adaptive sliding mode control [2], robust control

B. Zhang · W. Huo(✉)
The Seventh Research Division, School of Automation Science
and Electrical Engineering, Beihang University, Beijing 100191, China
e-mail: weihuo@buaa.edu.cn

B. Zhang
e-mail: zhangbiao666@buaa.edu.cn

© Springer Nature Singapore Pte Ltd. 2018
Y. Jia et al. (eds.), *Proceedings of 2017 Chinese Intelligent Systems Conference*,
Lecture Notes in Electrical Engineering 460,
https://doi.org/10.1007/978-981-10-6499-9_65

approach [3] and trajectory linearization control [4], etc. Most of these control strategies depend on inner/outer loop structure which must be accord with the time-scale
separate principle. That means that convergence rate of the states in inner loop must
be much faster than that in outer loop, which is hard to be quantified. Generally, control algorithms adopting inner/outer loop structure are more complex and difficult to
rigorously analyze system stability. In this paper, a control law based on controlled
Lagrangians (CL) method is constructed to avoid these disadvantages.

The CL method is a constructive energy-based controller design technique for
mechanical systems, especially for under-actuated systems. It facilitates analyzing
stability of the closed-loop system. To apply the CL method, it is need to, firstly, construct the desire controlled kinetic energy, potential energy and generalized forces;
secondly, match the controlled system with original dynamical system and solve the
matching condition to obtain structure of the control law; finally, determine controller parameters from stabilization conditions. After introduced by [5–7], the CL
method has been applied to design control laws for many mechanical systems, such
as inverted-pendulum on a force-drive cart [8], furuta pendulum [9], planar vertical takeoff-and-landing aircraft which has 3 degrees of freedom [10], satellite with
internal actuators [11], space flexible structure [12] and so on. In this paper, the CL
method is applied to design stabilization controller for quadrotor with two under-
actuation degrees.

This paper is organized as follows: In Sect. 2, the quadrotor dynamical model
is introduced. In Sect. 3, the input transformations and linearization for quadrotor
model are presented. In Sect. 4, the detailed design steps of quadrotor helicopter
controller are given and local asymptotic stability of the closed-loop system is proved
theoretically. In Sect. 5, effectiveness of the proposed control law is verified through
a numeric simulation. In Sect. 6, some conclusions are given.

2   The Mathamatic Model of Quadrotor Helicopter

The basic structure of a quadrotor helicopter is shown in Fig. 1. Four rotors are
fixed at four endpoints of a cross. The rotations of rotors lead to thrusts and reaction torques. Rotors 1, 3 of the quadrotor rotate counterclockwise, and rotors 2, 4
rotate clockwise so they generate reaction torques with different direction. Similar
with literatures [4, 13], the quadrotor is assumed to be rigid and symmetric, and its
velocity is relatively low such that the resistance caused by movement of quadrotor and gyroscopic effect caused by rotation of rotors are neglected. As shown in
Fig. 1, $\{e: oxyz\}$ denotes an inertial frame, and $\{b: o_b x_b y_b z_b\}$ denotes a frame rigidly
attached to the quadrotor. In frame $\{e\}$, the origin is one fixed point, and the z axis
is vertical and its positive direction is opposite with the earth's core; the positive
directions of the x and the y axis point to east and north, respectively. In frame $\{b\}$,
its origin is coincide with the quadrotor's center of mass and the directions of its axis

Fig. 1 Schematic diagram
of quadrotor helicopter

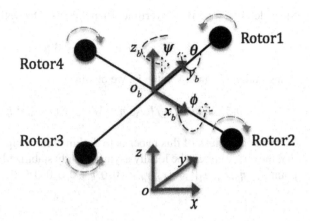

are shown as Fig. 1. According to Newton–Euler equation, the dynamical model of
quadrotor is as follows:

$$m\ddot{\xi} = -mge_3 + fRe_3, \quad J\dot{\omega} + S(\omega)J\omega = \tau \tag{1}$$

where m is the quadrotor mass, $\xi = [x, y, z]^T$ denotes position of the quadrotor helicopter's center of mass in inertial frame, g is acceleration of gravity, $e_3 = [0, 0, 1]^T$,
f denotes total thrust supplied by four rotors, R is the orientation matrix from inertial frame $\{e\}$ to body frame $\{b\}$, $J = \text{diag}\{J_1, J_2, J_3\}$ is rotary inertial matrix of
the quadrotor, $\omega = [\omega_1, \omega_2, \omega_3]^T$ is angular velocity of quadrotor expressed in body
frame. $S(\omega)$ is a skew-matrix which is defined as follows:

$$S(\omega) = \begin{bmatrix} 0 & -\omega_3 & \omega_2 \\ \omega_3 & 0 & -\omega_1 \\ -\omega_2 & \omega_1 & 0 \end{bmatrix}$$

$\tau = [\tau_1, \tau_2, \tau_3]^T$ is torque caused by four rotors expressed in body frame.

In this paper, the quadrotor attitude is described by Euler angles $\eta = [\phi, \theta, \psi]^T$,
where ϕ is the pitch angle, θ is the roll angle, ψ is the yaw angle. Then the orientation
matrix R can be presented as follows:

$$R = \begin{bmatrix} c_\psi c_\theta & s_\phi s_\theta c_\psi - c_\phi s_\psi & c_\phi s_\theta c_\psi + s_\phi s_\psi \\ s_\psi c_\theta & s_\phi s_\theta s_\psi + c_\phi c_\psi & c_\phi s_\theta s_\psi - s_\phi c_\psi \\ -s_\theta & s_\phi c_\theta & c_\phi c_\theta \end{bmatrix} \tag{2}$$

where $s_\theta = \sin(\theta)$, $c_\theta = \cos(\theta)$. The relation between ω and $\dot{\eta}$ satisfies

$$\omega = W\dot{\eta}, \quad W = \begin{bmatrix} 1 & 0 & -s_\theta \\ 0 & c_\phi & s_\phi c_\theta \\ 0 & -s_\phi & c_\phi c_\theta \end{bmatrix} \tag{3}$$

Since $\det(W) = c_\theta$, W is invertible when $\theta \neq \pm\frac{\pi}{2}$. Derivating (3) with respect to time gives

$$\dot{\omega} = W\ddot{\eta} + \dot{W}\dot{\eta} \tag{4}$$

Substituting (3) and (4) into (1), we obtain

$$m\ddot{\xi} = -mge_3 + fRe_3, \quad \ddot{\eta} = W^{-1}J^{-1}\left(\tau - J\dot{W}\dot{\eta} - S(W\dot{\eta})JW\dot{\eta}\right) \tag{5}$$

The control task of this paper is to design control inputs $u = [f, \tau_1, \tau_2, \tau_3]^T$ such that the quadrotor can be locally asymptotically stabilized to the desired equilibrium point $[\xi_d, \eta_d, \dot{\xi}_d, \dot{\eta}_d]^T = [x_d, y_d, z_d, 0, 0, \psi_d, 0, 0, 0, 0, 0, 0]^T$.

3   Simplification and Linearization of Dynamical Model

With following input transformations

$$f = \frac{m(g + \tilde{f})}{c_\phi c_\theta}, \quad \tau = J\dot{W}\dot{\eta} + S(W\dot{\eta})JW\dot{\eta} + JW\tilde{\tau} \tag{6}$$

where $\phi, \theta \neq \pm\frac{\pi}{2}$, the quadrotor dynamical model (5) can be simplified to

$$\ddot{\xi} = h\tilde{f} + kg, \quad \ddot{\eta} = \tilde{\tau} \tag{7}$$

where $h = [t_\theta c_\psi + t_\phi\frac{s_\psi}{c_\theta}, t_\theta s_\psi - t_\phi\frac{c_\psi}{c_\theta}, 1]^T, k = [t_\theta c_\psi + t_\phi\frac{s_\psi}{c_\theta}, t_\theta s_\psi - t_\phi\frac{c_\psi}{c_\theta}, 0]^T, t_\theta = \tan(\theta), \tilde{\tau} = [\tilde{\tau}_1, \tilde{\tau}_2, \tilde{\tau}_3]^T$.

Linearizing the nonlinear system (7) at the desired equilibrium point $[\xi_d, \eta_d, \dot{\xi}_d, \dot{\eta}_d]^T$ gives

$$\begin{cases} \ddot{x} = gc_{\psi_d}\theta + gs_{\psi_d}\phi, \ \ddot{y} = gs_{\psi_d}\theta - gc_{\psi_d}\phi, \ \ddot{z} = \tilde{f} \\ \ddot{\phi} = \tilde{\tau}_1, \ \ddot{\theta} = \tilde{\tau}_2, \ \ddot{\psi} = \tilde{\tau}_3 \end{cases} \tag{8}$$

Taking following invertible state transformations for linearized system (8):

$$\begin{bmatrix} \tilde{x} \\ \tilde{y} \end{bmatrix} = \begin{bmatrix} c_{\psi_d} & s_{\psi_d} \\ s_{\psi_d} & -c_{\psi_d} \end{bmatrix} \begin{bmatrix} x - x_d \\ y - y_d \end{bmatrix} \tag{9}$$

Four subsystems can be obtained:

$$\ddot{\tilde{x}} = g\theta, \quad \ddot{\theta} = \tilde{\tau}_2 \tag{10}$$

$$\ddot{\tilde{y}} = g\phi, \quad \ddot{\phi} = \tilde{\tau}_1 \tag{11}$$

$$\ddot{\psi} = \tilde{\tau}_3 \tag{12}$$

$$\ddot{z} = \tilde{f} \tag{13}$$

After input transformations, linearization and state transformation, the control target has changed to design control input $\tilde{u} = [\tilde{f}, \tilde{\tau}_1, \tilde{\tau}_2, \tilde{\tau}_3]^T$ to locally asymptotically stabilize the state $[\tilde{x}, \tilde{y}, z, \phi, \theta, \psi, \dot{\tilde{x}}, \dot{\tilde{y}}, \dot{z}, \dot{\phi}, \dot{\theta}, \dot{\psi}]^T$ of above four systems to $[0, 0, 0, 0, 0, \psi_d, 0, 0, 0, 0, 0, 0]^T$.

4 Design Stabilization Controller for the Quadrotor

Firstly, control input $\tilde{\tau}_2$ is designed to locally asymptotically stabilize the subsystem (10) at the origin. Select $q = (\tilde{x}, \theta)^T$ as generalized coordinates, the desired kinetic energy of the controlled subsystem is constructed as follows:

$$\overline{E}_k(\dot{q}) = \frac{1}{2}\dot{q}^T\overline{M}\dot{q} = \frac{1}{2}[\dot{\tilde{x}} \ \dot{\theta}]\begin{bmatrix} m_1 & m_2 \\ m_2 & m_3 \end{bmatrix}\begin{bmatrix} \dot{\tilde{x}} \\ \dot{\theta} \end{bmatrix} \tag{14}$$

where $\overline{M}$ is a constant and positive definite matrix because of the positive definiteness of the kinetic energy, i.e. $m_1 > 0, m_1 m_3 - m_2{}^2 > 0$. And the desired potential energy of the controlled subsystem is denoted by $\overline{E}_p(q)$ which gets the minimum value at the origin. So the Lagrangian of the controlled subsystem is

$$\overline{L}(q, \dot{q}) = \overline{E}_k(\dot{q}) - \overline{E}_p(q) = \frac{1}{2}\dot{q}^T\overline{M}\dot{q} - \overline{E}_p \tag{15}$$

If the desired generalized force of controlled system is represented by $\overline{u}$, the controlled Euler–Lagrangian equations are given as

$$\frac{d}{dt}\frac{\partial\overline{L}}{\partial\dot{q}} - \frac{\partial\overline{L}}{\partial q} = \begin{bmatrix} m_1 & m_2 \\ m_2 & m_3 \end{bmatrix}\begin{bmatrix} \ddot{\tilde{x}} \\ \ddot{\theta} \end{bmatrix} + \begin{bmatrix} \frac{\partial\overline{E}_p}{\partial\tilde{x}} \\ \frac{\partial\overline{E}_p}{\partial\theta} \end{bmatrix} = \overline{u} \tag{16}$$

Design $\overline{u}$ as

$$\overline{u} = -\overline{D}\dot{q} = -\begin{bmatrix} d_1 & d_2 \\ d_2 & d_3 \end{bmatrix}\begin{bmatrix} \dot{\tilde{x}} \\ \dot{\theta} \end{bmatrix} \tag{17}$$

where $\overline{D} = \overline{D}^T \geq 0$. From (16) and (17) the controlled system equation can be written as

$$\begin{bmatrix} m_1 & m_2 \\ m_2 & m_3 \end{bmatrix} \begin{bmatrix} \ddot{\tilde{x}} \\ \ddot{\theta} \end{bmatrix} + \begin{bmatrix} \dfrac{\partial \overline{E}_p}{\partial \tilde{x}} \\ \dfrac{\partial \overline{E}_p}{\partial \theta} \end{bmatrix} = -\begin{bmatrix} d_1 & d_2 \\ d_2 & d_3 \end{bmatrix} \begin{bmatrix} \dot{\tilde{x}} \\ \dot{\theta} \end{bmatrix} \tag{18}$$

Premultiplying (18) by $\overline{M}^{-1}$ gives

$$\begin{bmatrix} \ddot{\tilde{x}} \\ \ddot{\theta} \end{bmatrix} = -\frac{1}{m_1 m_3 - m_2{}^2} \begin{bmatrix} m_3 & -m_2 \\ -m_2 & m_1 \end{bmatrix} \left(\begin{bmatrix} \dfrac{\partial \overline{E}_p}{\partial \tilde{x}} \\ \dfrac{\partial \overline{E}_p}{\partial \theta} \end{bmatrix} + \begin{bmatrix} d_1 & d_2 \\ d_2 & d_3 \end{bmatrix} \begin{bmatrix} \dot{\tilde{x}} \\ \dot{\theta} \end{bmatrix} \right) \tag{19}$$

The original system (10) is equivalent to controlled system (19) if the accelerations solved from them are equal, namely:

$$\begin{bmatrix} g\theta \\ \tilde{\tau}_2 \end{bmatrix} = -\frac{1}{m_1 m_3 - m_2{}^2} \begin{bmatrix} m_3 & -m_2 \\ -m_2 & m_1 \end{bmatrix} \left(\begin{bmatrix} \dfrac{\partial \overline{E}_p}{\partial \tilde{x}} \\ \dfrac{\partial \overline{E}_p}{\partial \theta} \end{bmatrix} + \begin{bmatrix} d_1 & d_2 \\ d_2 & d_3 \end{bmatrix} \begin{bmatrix} \dot{\tilde{x}} \\ \dot{\theta} \end{bmatrix} \right) \tag{20}$$

The first row of (20) just is matching condition:

$$-m_3 \frac{\partial \overline{E}_p}{\partial \tilde{x}} + m_2 \frac{\partial \overline{E}_p}{\partial \theta} - (m_3 d_1 - m_2 d_2)\dot{\tilde{x}} - (m_3 d_2 - m_2 d_3)\dot{\theta} = (m_1 m_3 - m_2{}^2)g\theta \tag{21}$$

The second row of (20) gives the structure of control input $\tilde{\tau}_2$ as follows:

$$\tilde{\tau}_2 = \frac{m_2 \dfrac{\partial \overline{E}_p}{\partial \tilde{x}} - m_1 \dfrac{\partial \overline{E}_p}{\partial \theta} + (m_2 d_1 - m_1 d_2)\dot{\tilde{x}} + (m_2 d_2 - m_1 d_3)\dot{\theta}}{m_1 m_3 - m_2{}^2} \tag{22}$$

Considering the independence of $\dot{q}$ and q, (21) can be divided to three equations as follows:

$$-\frac{m_3}{m_2} \frac{\partial \overline{E}_p}{\partial \tilde{x}} + \frac{\partial \overline{E}_p}{\partial \theta} = \frac{m_1 m_3 - m_2{}^2}{m_2} g\theta \tag{23}$$

$$m_3 d_1 - m_2 d_2 = 0, \quad m_3 d_2 - m_2 d_3 = 0 \tag{24}$$

From [12], a particular solution of (23) is:

$$\overline{E}_p = \frac{1}{2} q^T \overline{P} q = \frac{1}{2} \begin{bmatrix} \tilde{x} & \theta \end{bmatrix} \begin{bmatrix} 1 & \dfrac{m_3}{m_2} \\ \dfrac{m_3}{m_2} & \dfrac{m_1 m_3 - m_2{}^2}{m_2} g + \left(\dfrac{m_3}{m_2}\right)^2 \end{bmatrix} \begin{bmatrix} \tilde{x} \\ \theta \end{bmatrix} \tag{25}$$

where the matrix $\overline{P}$ is positive definite, which ensure that the desired potential energy take minimum value at the origin and infer that $m_2 > 0$. The matrix $\overline{D}$ can be obtained by (24) as follows:

$$\overline{D} = \begin{bmatrix} d_1 & d_2 \\ d_2 & d_3 \end{bmatrix} = d_1 \begin{bmatrix} 1 & \frac{m_3}{m_2} \\ \frac{m_3}{m_2} & \left(\frac{m_3}{m_2}\right)^2 \end{bmatrix} \tag{26}$$

where $d_1 \geq 0$, since $\overline{D}$ is positive semidefinite. Substituting (25) and (26) into (22) gives

$$\overline{\tau}_2 = -\frac{1}{m_2}\tilde{x} - \frac{m_3 + m_1 m_2 g}{m_2^2}\theta - \frac{d_1}{m_2}\tilde{x} - \frac{d_1 m_3}{m_2^2}\dot{\theta}$$
$$= -\frac{c_{\psi_d}}{m_2}(x - x_d) - \frac{s_{\psi_d}}{m_2}(y - y_d) - \frac{m_3 + m_1 m_2 g}{m_2^2}\theta - \frac{d_1}{m_2}\tilde{x} - \frac{d_1 m_3}{m_2^2}\dot{\theta} \tag{27}$$

where $m_1 > 0, m_2 > 0, m_3 > \frac{m_2^2}{m_1}, d_1 > 0$. Select the total energy $\overline{E}(q, \dot{q})$ of controlled system as its Lyapunov function

$$\overline{E}(q, \dot{q}) = \overline{E}_k(\dot{q}) + \overline{E}_p(q) = \frac{1}{2}\dot{q}^T \overline{M}\dot{q} + \overline{E}_p = \dot{q}^T \frac{\partial \overline{L}}{\partial \dot{q}} - \overline{L} \tag{28}$$

According to (15), (16) and (28), the time derivative of $\overline{E}$ is as follows:

$$\dot{\overline{E}} = \frac{d}{dt}\left(\dot{q}^T\frac{\partial \overline{L}}{\partial \dot{q}} - \overline{L}\right) = \dot{q}^T\left(\frac{d}{dt}\frac{\partial \overline{L}}{\partial \dot{q}} - \frac{\partial \overline{L}}{\partial q}\right) = \dot{q}^T \overline{u}$$
$$= -\dot{q}^T \overline{D}\dot{q} = -\begin{bmatrix}\tilde{x} & \dot{\theta}\end{bmatrix}\begin{bmatrix}d_1 & d_2 \\ d_2 & d_3\end{bmatrix}\begin{bmatrix}\tilde{x} \\ \dot{\theta}\end{bmatrix} = -d_1\left(\tilde{x} + \frac{m_3}{m_2}\dot{\theta}\right)^2 \leq 0 \tag{29}$$

From (29) the stability of closed-loop system is proved. The asymptotic stability of the system will be analyzed as follows:

if $\dot{\overline{E}} \equiv 0$, from (29) we know

$$\tilde{x} + \frac{m_3}{m_2}\dot{\theta} \equiv 0 \tag{30}$$

Then integral and time derivative of (30) can be obtained as

$$\tilde{x} + \frac{m_3}{m_2}\theta \equiv c, \ \ddot{\tilde{x}} + \frac{m_3}{m_2}\ddot{\theta} \equiv 0 \tag{31}$$

where c is a constant. From (10) and (27) the equations of closed-loop subsystem are:

$$\ddot{\tilde{x}} = g\theta, \quad \ddot{\theta} = -\frac{1}{m_2}\tilde{x} - \frac{m_3 + m_1 m_2 g}{m_2{}^2}\theta - \frac{d_1}{m_2}\dot{\tilde{x}} - \frac{d_1 m_3}{m_2{}^2}\dot{\theta} \tag{32}$$

Substituting (31) into (32) gives

$$\theta = \frac{m_3 c}{g\left(m_2{}^2 - m_1 m_3\right)} \tag{33}$$

which shows θ is a constant, so $\dot{\theta} = 0$ and $\ddot{\theta} = 0$. Then $\tilde{x} = 0$ and $\dot{\tilde{x}} = 0$ by (30) and (31). $\theta = 0$ is obtained by the first equation of (32). Furthermore, from (33) $c = 0$ is got, then $\tilde{x} = 0$ by (30). According to LaSalle's invariance principle, the origin is the only element in the invariance set $W = \{\dot{E} \equiv 0\}$, so the asymptotic stabilization of closed-loop subsystem (32) is verified.

Similarly, the asymptotical stablization controller of subsystem (11) is designed as:

$$\begin{aligned}
\tilde{\tau}_1 &= -\frac{1}{m_5}\tilde{y} - \frac{m_6 + m_4 m_5 g}{m_5{}^2}\phi - \frac{d_4}{m_5}\dot{\tilde{y}} - \frac{d_4 m_6}{m_5{}^2}\dot{\phi} \\
&= -\frac{s_{\psi_d}}{m_5}\left(x - x_d\right) + \frac{c_{\psi_d}}{m_5}\left(y - y_d\right) - \frac{m_6 + m_4 m_5 g}{m_5{}^2}\phi - \frac{d_4}{m_5}\dot{\tilde{y}} - \frac{d_4 m_6}{m_5{}^2}\dot{\phi}
\end{aligned} \tag{34}$$

where $m_4 > 0, m_5 > 0, m_6 > \frac{m_5{}^2}{m_4}, d_4 > 0$.

The controllers of subsystem (12) is designed through PD control method:

$$\tilde{\tau}_3 = -b_1\dot{\psi} - b_2\left(\psi - \psi_d\right) \tag{35}$$

where $b_1 > 0, b_2 > 0$. Then the closed-loop system of (12) are as follows:

$$\ddot{\psi} + b_1\dot{\psi} + b_2\left(\psi - \psi_d\right) = 0 \tag{36}$$

System described by (36) is asymptotically stable at desired equilibrium points $(\psi, \dot{\psi}) = (\psi_d, 0)$ according to the theory about second-order linear systems. Similarly, the asymptotically stable controller of subsystem (13) is designed as follows:

$$\tilde{f}_3 = -c_1\dot{z} - c_2\left(z - z_d\right) \tag{37}$$

where $c_1 > 0, c_2 > 0$.

The nonlinear system (7) is locally asymptotically stabilized at the desired point by the controller $\tilde{u}$ because the closed-loop system of linear system (8), i.e. linearization equations of (7), is asymptotically stable owing to the asymptotic stability and independence of four subsystems with their own stabilization controllers. It means that the quadrotor helicopter can be locally asymptotically stabilized to the desired point by the controller u due to the invertibility between $\tilde{u}$ and u. To be specifically:

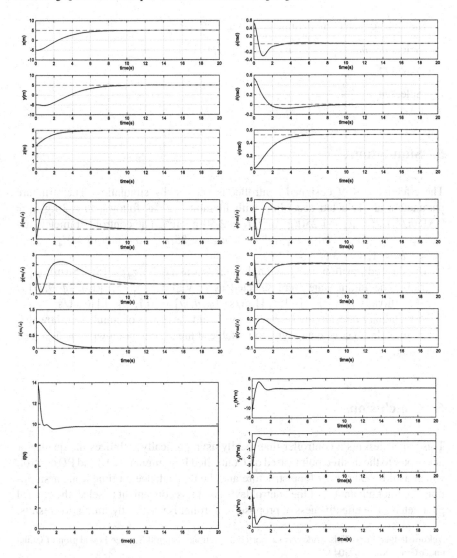

Fig. 2 Simulation results

$$f = \frac{m(g + \widetilde{f})}{c_\phi c_\theta} = \frac{m\left(g - c_1\dot{z} - c_2(z - z_d)\right)}{c_\phi c_\theta} \tag{38}$$

$$\tau = J\dot{W}\dot{\eta} + S(W\dot{\eta})JW\dot{\eta} + JW\widetilde{\tau} \tag{39}$$

where $\widetilde{\tau}$ is determined by (27), (34), (35).

The steps of selecting parameters are as follows:

(1) Select $m_1, m_2, m_4, m_5, b_1, b_2, c_1, c_2, d_1, d_4$ to be positive.
(2) Select $m_3 > \frac{m_2^2}{m_1}$.
(3) Select $m_6 > \frac{m_5^2}{m_4}$.

5 Simulation

The effectiveness of designed controller is verified by simulation. The structure parameters are selected according with literature [2] as follows: $m = 2(\text{kg}), g = 9.81(\text{m/s}^2), J_1 = J_2 = 1.25(\text{kg} \cdot \text{m}^2), J_3 = 2.5(\text{kg} \cdot \text{m}^2)$. The control parameters are chosen to be: $m_1 = m_4 = 4, m_2 = m_5 = 2.55, m_3 = m_6 = 8, d_1 = d_4 = 3.55, b_1 = 2, b_2 = 1, c_1 = 3.5, c_2 = 1.96$.

The desired position and yaw angle were set as $[x_d, y_d, z_d] = [5, 5, 5](\text{m}), \psi_d = 0.5(\text{rad})$. The initial state were selected as $[x, y, z] = [-5, -5, 3](\text{m})$, $[\phi, \theta, \psi] = [0.5, 0.5, 0](\text{rad}), [\dot{x}, \dot{y}, \dot{z}] = [-1, 1, 1](\text{m/s}), [\dot{\phi}, \dot{\theta}, \dot{\psi}] = [0.1, 0.1, 0.1](\text{rad/s})$.

The simulation results are shown as Fig. 2, and the results of simulation show that quadrotor helicopter is local asymptotic stable at the desired point with the designed controller.

6 Conclusions

This paper designs a controller that locally asymptotically stabilizes the quadrotor helicopter to the desired point based on Controlled Lagrangians (CL) and PD control, which avoids outer/inner loop structure and the dependance on time-scale assumption. Another advantage of the control law is that it is convenient to select the control parameters. The effectiveness of proposed controller is verified by simulation results.

Acknowledgements This work was supported by National Natural Science Foundation of China under Grant No. 61673043.

References

1. Bouabdallah S, Siegwart R. Backstepping and sliding-mode techniques applied to an indoor micro quadrotor. IEEE International Conference on Robotics and Automation; 2005. p. 2247–2252.
2. Sen L, Baokui L, Qingbo G. Adaptive sliding mode control for quadrotor helicopters. Chinese Control Conference; 2014. p. 71–76.
3. Bai Y, Liu H, Shi Z, Zhong Y. Robust control of quadrotor unmanned air vehicles. Chinese Control Conference; 2012. p. 4462–4467.

4. Zhu B, Huo W. Trajectory linearization control for a quadrotor helicopter. IEEE International Conference on Control and Automation; 2010. p. 34–39.
5. Bloch AM, Leonard NE, Marsden JE. Stabilization of mechanical systems using controlled Lagrangians. IEEE Conference on Decision and Control; 1997. p. 2356–2361.
6. Bloch AM, Leonard NE, Marsden JE. Controlled Lagrangians and the stabilization of mechanical systems I: the first matching theorem. IEEE Trans Autom Control. 2000;45(12):2253–2270.
7. Bloch AM, Leonard NE, Marsden JE. Controlled Lagrangians and the stabilization of mechanical systems II: potential shaping. IEEE Trans. Autom. Control. 2001;46(10):1556–1571.
8. Sandoval J, Kelly R, Santibáñez V. On the Controlled Lagrangian of an inverted pendulum on a force-driven cart. European Control Conference; 2015. p. 992–997.
9. Machleidt K, Kroneis J, Liu S. Stabilization of the furuta pendulum using a nonlinear control law based on the method of controlled Lagrangians. IEEE International Symposium on Industrial Electronics; 2007. p. 2129–2134.
10. Li Maoqing. Control design for planar vertical takeoff-and-landing aircraft based on controlled Lagrangians. Control Theory Appl. 2010;27(6):6898–994.
11. Haghi P, Ghaffari-Saadat M-H. Challenges in the stabilization of a satellite using Controlled Lagrangians I: gravitation. IEEE International Conference on Control Applications; 2009. 764–769.
12. Wan H, Huo W. Controller design of a class of space flexible structure based on Controlled Lagrangian. World Congress on Intelligent Control and Automation; 2008. p. 1359-1362.
13. M'hammed Guisser, Hicham Medromi. "A High Gain Observer and Sliding Mode Controller for an Autonomous Quadrotor Helicopter", *International Journal of Intelligent Control and Systems*, vol. 14, no.3, pp. 204-212, 2009.

Design of a New Multi-functional Humanoid Finger Sensor

Jinbao Chen, Hulin Liu, Dong Han and Meng Chen

Abstract Current robot tactile sensors have shortcoming such as low reliability and lack of uniform calibration methods. A new multi-functional humanoid finger sensor based on liquid conduction and able to measure three-dimensional force, temperature and micro vibration simultaneously and its multifunctional calibration system are designed. In order to satisfy requirements for real-time and accuracy of the calibration and use of the finger sensor, a modular software architecture is proposed, which can integrate human-computer interaction and data transmission. Finally, the finger sensor and its calibration system is proven to be stable and reliable by calibration experiments and object identification test, indicating that the finger sensor is suit to recognize the target.

Keywords Humanoid · Finger sensor · Calibration system · Object recognition

1 Introduction

Tactile sensor is an essential tool for intelligent robot to obtain tactile information. With the information provided by tactile sensor, robots are able to know object's physical properties such as three-dimensional force, temperature, pressure and micro-vibration.

Tactile sensor, improved a lot in flexibility, intelligence and reliability, can be divided into direct-perception sensor and mediate-perception sensor according to the principle of perception. Direct-perception sensor, in the name, is utilized to

J. Chen · H. Liu (✉) · D. Han
State Key Laboratory of Mechanics and Control of Mechanical Structure,
Nanjing University of Aeronautics and Astronautics, Nanjing 210016, China
e-mail: leohll_2008@163.com

J. Chen
e-mail: chenjbao@nuaa.edu.cn

M. Chen
Aerospace System Engineering Shanghai, Shanghai 201108, China

© Springer Nature Singapore Pte Ltd. 2018
Y. Jia et al. (eds.), *Proceedings of 2017 Chinese Intelligent Systems Conference*,
Lecture Notes in Electrical Engineering 460,
https://doi.org/10.1007/978-981-10-6499-9_66

apperceive three-dimensional force directly, mainly by pasting different sensitive materials with different perceptive function in human finger and manipulator. With the rapid development of technology such as resistance materials, capacitive materials, piezoelectric materials and other technologies, Direct-perception sensor tend to be more powerful and more reliable. Kim et al. [1] designed a 32×32 NiCr alloy strain array; Sohgawa et al. [2, 3] designed a 3×3 implanted the cantilever beam into the PDMS materials; SCB Mannsfeld et al. [4] utilized FET (field effect transistor) to manufacture capacitive tactile sensor; Dahiya et al. [5–8] designed a piezoelectric tactile sensor by attaching 32 micro-electrodes to the PVDF film. These tactile sensors with sensitive materials, generally, are of great flexibility and high spatial, however, their subsequent signal processing module is difficult to integrate well with the current mechanical arm system.

Mediate-perception sensor, correspondingly, apperceive three-dimensional force indirectly mainly through conductive solution. An American company called SynTouchLLC [9–12] has introduced a humanoid finger sensor based on conductive liquid, which were able to measure three-dimensional force, micro vibration and temperature simultaneously. Its sensing principle is that the internal thermal element, pressure sensitive element and electrode array indirectly perceive the external environmental characteristic through the conduction of flexible skin and conductive liquid.

Humanoid finger tactile sensor, based on conductive liquid, is of low cost and has the advantage of being easy to manufacture. However, the model of how the dielectric conduction works is not clear, moreover, there is no uniform and accurate calibration method to calibrate the sensor. Therefore, basing on the conduction of low conductivity solution, a multifunctional humanoid finger micro tactile sensor and its calibration tooling system are designed, which has the ability to perceive physical properties such as three-dimensional forces, temperature and micro vibration.

2 Mechanism of the Robot

The finger sensor Mechanism mainly includes three parts: finger phalanx, nail and flexible skin, as shown in Fig. 1. Finger phalanx is the supporter of sensitive elements and signal processing circuit; a confined space is formed between the flexible skin and phalanx, which is used to store the conductive liquid, meanwhile, the flexible skin and conductive liquid constitute the transmitting medium for finger sensor to perceive the external environment; the nail fixed the flexible skin on phalanx through screws, which, plays a role in sealing.

To show how is the internal structure of the phalange, a sectional view of the phalange is given, as shown in Fig. 2.

The liquid channel can transmit the liquid vibration to the pressure perception device equipped at the end of the tunnel, then the pressure signal and micro

Fig. 1 Whole structure

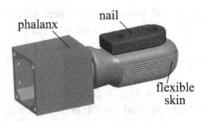

phalanx   nail

flexible
skin

Fig. 2 Sectional view of the
phalange

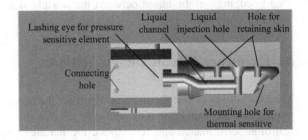

Liquid   Liquid   Hole for
Lashing eye for pressure   channel   injection hole   retaining skin
sensitive element

Connecting
hole

Mounting hole for
thermal sensitive

Fig. 3 Flexible skin texture

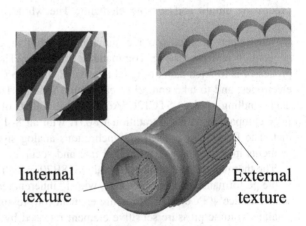

Internal
texture

External
texture

vibration signal; the connecting hole is used to connect with other equipment, so as to facilitate the calibration and the use of the finger sensor.

Flexible skin is an important medium for finger sensor to perceive the external environment. According to the high requirement in sensitivity, a silica gel able to be cured at normal temperature is selected as the skin material, of which the hardness is 50 A. The thermal conductivity of pure silica is very low, which greatly affects the temperature sensing rate. To solve this problem, a certain proportion of AlN was added into the silica gel.

In order to furthest utilize the flexible skin, the internal and external of the skin are both designed texture by imitating hand fingerprint. The external texture is composed of a plurality of semi-circular strips tightly arranged in a certain way, of which the radius is 0.4 mm, as shown in Fig. 3. There are two main function of external texture: (1) when measuring temperature, it expands contact area, increasing heat transfer rate; (2) when the finger sensor slides on the object surface, its vibration will motivate the internal liquid to produce vibration, then, micro vibration signal is obtained, which is important for object surface features gaining.

The internal texture is composed of a combination of four cines and a cuboid, as shown in Fig. 3. The size of the internal texture is 0.86 mm × 0.43 mm × 0.43 mm, and the interval of each texture is 0.43 m. The internal texture is able to improve the measuring rang of electrode impedance, because it expands the force used to edging out the liquid between the flexible skin and the electrode.

3 Hardware Circuit Design Method

The hardware system is the important part for the finger sensor to precisely percept multimodal physical quantities. In order to gain the sensing ability of accuracy, real-time performance and robustness, the hardness system, adopted hierarchical design concept, mainly includes upper monitor, microprocessor, DAQ card, signal processing circuit and sensing elements. The whole hardness frame is given in Fig. 4.

The upper monitor provide function like human-computer interaction, data acquisition and processing, command sending, etc. The main function of microprocessor is to fetch data of digital thermal element, to generate actuating signal for electrodes, and to take control of calibration tooling. The microprocessor used for data handling choices STC12C5A60S2 series SCM, of which instruction code is fully compatible with the traditional 8051, with an 8–12 times operation speed of that. The DAQ card is used to synchronous analog signal exported from sensing elements into digit signal, with high rate and accuracy.

The thermal element choices digital DS18B20, which is a digital thermometer of stable performance and simple usage, with 1° inherent temperature error. The finger sensor choice 40PC015G2A pressure element for pressure perception, which is the smallest volume pressure sensitive element released by Honeywell.

The electrode array is used to perceive the change of conductive liquid between the phalanx and the flexible skin. The electrode array can be divided into sensing

Fig. 4 Hardware system diagram

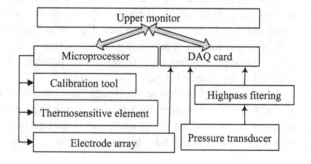

Fig. 5 Electrode array;
a electrode array of BIOTAC
sensor; electrode 1–19 are
sensing electrodes; electrode
X1–X4 are excitation
electrodes; **b** electrode array
of finger sensor in this paper;
electrode 1–15 are sensing
electrodes; electrode X1–X4
are excitation electrodes

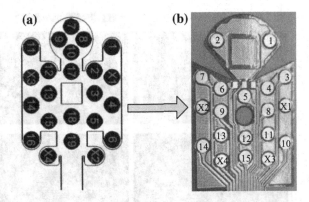

electrode and exciting electrode, in which only the sensing electrodes are to measure the conductive liquid impedance, while the excitation electrodes generate the excitation signal. The layout of electrode array is given in Fig. 5. The number of electrodes in this electrode array is four fewer than that of electrodes in BIOTAC sensor's electrode array. It's easy to see that the removed four electrodes are the two in the fingertip and the two in the root of the phalanx, which is because that in practical applications, the fingertip is mainly used for perceiving temperature and micro vibration, without excessive electrodes, moreover, in grasping operations the probability of object contacting the root of the phalanges is relatively small.

4 Software Circuit Design Method

According to the hardware system of the finger sensor, the software system can be divided into upper monitor and inferior monitor.

The upper monitor mainly contains Human Machine Interface (HMI). HMI is designed for operation staff, which unified calibration with measurement, then the problem that BIOTAC sensor cannot be calibrated quickly. On one hand, it displays the real-time measurement information, on the other hand, it controls the calibration tooling. The upper monitor's main interface is given in Fig. 6, including the following four parts: (1) Curve display window to display measurement data in real time and directly; (2) Real-time data to show the contrast of calibration results and actual values in real time; (3) Calibration tool control to control the movement of the electric telescopic rod and mobile platform; (4) Communication with inferior monitor to establish the communication interface between the upper monitor and the inferior monitor (microprocessor and DAQ card) and to initialize the system parameters.

Sub interfaces are designed for different calibration tasks. They are electrode array interface, micro vibration interface, pressure calibration interface and temperature calibration interface, as shown in Fig. 7.

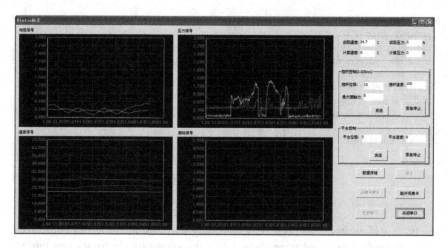

Fig. 6 HMI

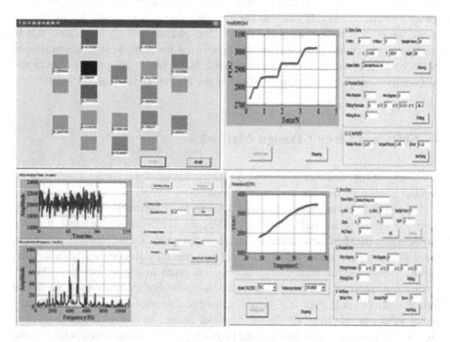

Fig. 7 Sub interfaces

The inferior monitor software means the program downloaded to the STC12C5A60S2 SCM, including program for reading data of DS18b20, AD conversion (collecting analog signal from the finger sensor), serial communication, stepper motor control and excitation signal generation, etc.

The calibration tooling is driven by two stepper motors, one of which controls the vertical movement of the electric telescopic rod, while the other one controls the horizontal movement of a mobile platform. STC 12C5A60S2 SCM control the stepper motor through connecting the CLK-port and CW-port of the stepper motor driver with its two I/O ports.

The electrodes impedance is measured by voltammetry, which requires an excitation signal input of certain frequency. Similar to the way the impulse signal produced for the stepper motor, the timer 1 generates the square wave excitation signal of fixed frequency by the interrupt function.

5 Prototype and Test Results

The installed finger sensor is shown in Fig. 8 is a contrast between the finger sensor and index finger of a normal adult male, which indicating that miniaturization of finger sensors is successful. The finger sensor will be calibrated on temperature, pressure, micro vibration and electrodes array, then experiments on object recognition are carried out from these aspects.

Figure 9 shows the contrast effect of the temperature change rate before and after the Calman filtering, which indicates that the Calman filtering can effectively cut off system noise interference and improve the stability of the software differential.

Use the finger sensor to touch homogeneous aluminum block、rock and plastic, which are in the same condition, especially at the same low temperature. Then, three temperature variation curve are easy to draw in Fig. 10. According to the Fig. 10, the conductivity of aluminum block is the best while that of plastic is poor, which agrees with the actual situation, indicating that the finger sensor is able to identify different objects with thermal conductivity at the rate of temperature change.

The calibration of pressure is shown in Fig. 11, where the electric telescopic rod is driven to move in the vertical direction, with which finger sensor will contact

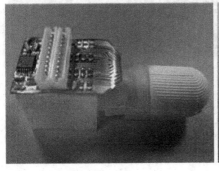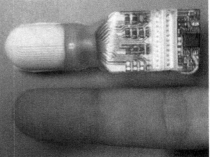

Fig. 8 The installed finger sensor

Fig. 9 Temperature test
result

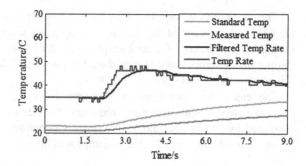

Fig. 10 Thermal
conductivity identification of
objects

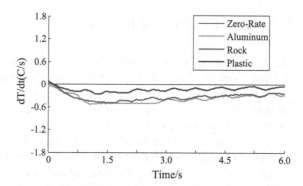

Fig. 11 Pressure calibration
of finger sensor

with standard pressure sensor meanwhile the pressure interface will record the
pressure signal of the finger sensor and the standard pressure value. Next, fit the
curves of the two sets of data with least square method, and then the relationship
between finger sensor's pressure and the actual pressure.

Fig. 12 Fitted curve

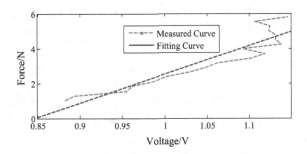

Fig. 13 Pressure calibration results

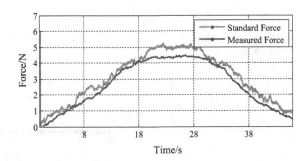

The pressure measured by the finger sensor and the fitted curve are shown in Fig. 12. When the pressure increases to about 5 N, the linearity is no more obvious, which is because the flexible skin has come into contact with the internal phalanx, resulting in a no longer linear increase of the pressure signal measured, or even a stop change. The value is related to the location of the touched point and the initial pressure inside the finger sensor.

Compare the filtered pressure with the standard pressure to verify the accuracy of calibration, as shown in Fig. 13. When the standard pressure comes to 5 N, the pressure measurement error is large. That's because linear polynomial fitting is used ignoring the nonlinear influence, which leads to the increase of measurement error.

The results of pressure calibration will be used to carry out the hardness identification experiment. Control the calibration tooling to contact objects (blankets, foam, plush toys, cartons) with different hardness and record the pressure signal.

Results of the experiment is shown in Fig. 14, which shows that the greater the hardness, the faster the pressure increase, that is the greater the pressure gradient. Therefore, objects can be identified by the pressure gradient.

Micro vibration is the key signal to obtain the surface characteristic of objects. First, fix object to be measured on the mobile platform, and then control the electric telescopic rod to move downwards until the finger sensor contacted with object. Last, take control of mobile platform to let it move horizontally and record the micro vibration signal along the process. Objects selected in this experiment are bread board, folder, mouse pad and cardboard.

Fig. 14 Hardness
identification results

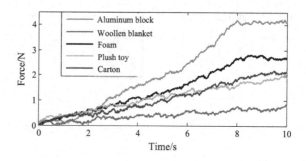

Fig. 15 Spectral
characteristics of micro
vibration signals

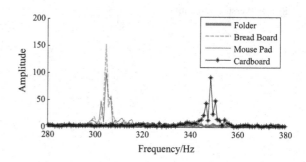

Fig. 16 Single electrode
pressure test results

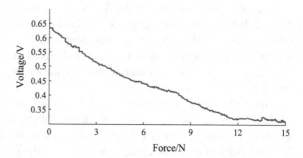

Extract the spectrum features of the micro vibration signal through fast Fourier transform algorithm, as shown in Fig. 15. Four objects have obvious differences in the spectrum. The folder's spectrum amplitude is close to 0, indicating its surface is smooth. The peak frequency of paper board is higher than that of bread board and mouse pad, which indicates the texture in the paper board surface is fine-grained. The peak value of the bread board is larger than that of the mouse pad, indicating that the roughness of the bread board is larger than that of mouse pad.

The relationship between the voltage output of each single electrode and the standard pressure is drawn in the Fig. 16. In the range of 0–12 N, they keep in good linear relationship but when the pressure is greater than 12 N, the flexible skin contacts the phalanx completely, leading to an impedance saturation between the electrodes, and the output voltage of the electrode is not change any more.

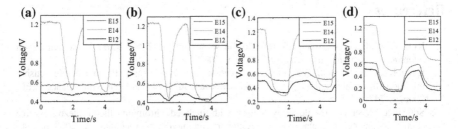

Fig. 17 Results of pressing on the finger sensor; the *red line* represents electrode 14; the *blue line* represents electrode 15; the *black line* represents electrode 12

In order to verify the recognition ability of electrode array, press the finger sensor in the same position with cylinders of different sizes, and the diameter of the cylinders are 3, 6, 10 and 18 mm. Three related electrodes' impedance is curved in Fig. 17.

In Fig. 17a–d, the output voltage of electrode 14 has the most obvious change, indicating electrode 14 is in the center of compressed area; when the cylinder's diameter increases, the change of electrode 12 and 15 becomes greater gradually. In Fig. 17a, there is no change in the voltage of electrode 12 and 15, indicating that the two electrodes were not compressed, that is to say the cylinder with the smallest diameter just compressed on one electrode. In the Fig. 17d, both the voltage of electrode 12 and 15 change obviously, indicating that the two electrodes were compressed, that is to say the cylinder with the largest diameter compressed on three electrodes. Therefore, the electrode array can identify the relative size of object.

6 Analysis of Experimental Results

A novel multi-functional humanoid finger sensor based on conductive liquid with high sensitivity, being easy to calibrate, and being able obtain tactile signal for intelligent robot is designed. Through experiments, the finger sensor can measure three-dimensional force, temperature and micro vibration at the same time, moreover, the objects with different attributes are able to be identified from the data like temperature gradient, pressure gradient and micro vibration spectrum.

Acknowledgements The authors would like to thank the anonymous reviewers for their critical and constructive review of the manuscript. This study was supported by Funding of Jiangsu Innovation Program for Graduate Education (No. KYLX16_0388), the Fundamental Research Funds for the Central Universities, National Natural Science Foundation of China (No. 51675264) and Open Foundation of Shanghai Key Laboratory of Spacecraft Mechanism.

References

1. Kim K, et al. Polymer-based flexible tactile sensor up to 32 × 32 arrays integrated with interconnection terminals. Sens Actuators, A. 2009;156(2):284–91.
2. Kang RL, Kim K, Kim YK, et al. Fabrication of polymer-based flexible tactile sensing module with metal strain gauges and inter connecter. In: IEEE conference on sensors. IEEE; 2006. p. 742–5.
3. Sohgawa M, et al. Tactile array sensor with inclined chromium/silicon piezo resistive cantilevers embedded in elastomer. In: Actuators and microsystems conference; 2009. p. 284–7.
4. Yokoyama H, Sohgawa M, Kanashima T, et al. Force intensity and direction measurement in real time using miniature tactile sensor with micro cantilevers embedded in PDMS. Sensors, p. 1–4. IEEE; 2013.
5. Mannsfeld SCB, Tee CK, Stoltenberg RM, et al. Highly sensitive flexible pressure sensors with micro structured rubber dielectric layers. Nat Mater. 2013;9(10):859–64.
6. Dahiya RS, et al. Tactile sensing arrays for humanoid robots. In: Research in microelectronics and electronics conferrence; 2007. p. 201–4.
7. Dahiya RS, Torre B, Cingolani R, et al. Voltage tunable sensitivity of piezoelectric materials based sensors and actuators. In: IEEE conference on nanotechnology, 2009. IEEE-Nano 2009. IEEE; 2009. p. 670–3.
8. Dahiya RS, Metta G, Valle M. Development of fingertip tactile sensing chips for humanoid robots. In: IEEE international conference on mechatronics. IEEE; 2009. p. 1–6.
9. Dahiya RS, Valle M, Lorenzelli L. Bio-inspired tactile sensing arrays. In: Bioengineered and bioinspired systems IV, SPIE Europe conference on micro technologies for new millennium; 2009. p. 1–9.
10. Wettels N, Santos VJ. Biomimetic tactile sensor array. Adv Robot. 2008;22(8):829–49.
11. Wettels N, Parnandi AR, Moon JH, et al. Grip control using biomimetic tactile sensing systems. IEEE/ASME Trans Mechatron. 2009;14(6):718–23.
12. Wettels N, Pletner B. Integrated dynamic and static tactile sensor: focus on static force sensing. In: Proc SPIE. 2011;8345:83454H–83454H-10. The International Society for Optical Engineering.
13. Lin CH, Erickson TW, Fishel JA, et al. Signal processing and fabrication of a biomimetic tactile sensor array with thermal, force and micro vibration modalities. In: IEEE international conference on robotics and biomimetics. IEEE; 2009. p. 129–34.

Pulp Concentration Control Based on Dynamic Matrix Control

Wen Hao, Jie Dong, Kaixiang Peng, Mengqi Jia and Qiang Wang

Abstract In the papermaking industry, the stability of the pulp concentration determines the quality of the paper. However, the change of the pulp concentration is in a state of fluctuation for a long time and the pulp concentration control system has the characteristics of large lag, non-linearity and time-varying. It is difficult to achieve the desired control effect by using the traditional PID controller in the concentration control process. Therefore, use of dynamic matrix control algorithm to design the controller can solve the problem of disturbance and modeling complexity. In this paper, dynamic matrix control algorithm and PID control simulation results show that the dynamic matrix control algorithm has better control quality. The output satisfies the constraint limit with higher probability, and the output constraints and performance index can be met.

Keywords Dynamic matrix control · PID control · Pulp concentration

1 Introduction

With the rapid development of modern economy and the expanding scale of process industry, the complexity of industrial process is also increasing. Therefore, the control requirements of the system are getting higher and higher. In the process control

This work is supported by Natural Science Foundation of China (NSFC) under Grants 61473033 and 61673032, by Beijing Natural Science Foundation (4142035), P.R. China, by special funds for basic research and operation of Central University of University of Science and Technology Beijing (FRF-BD-16-005A), by Beijing key subject construction projects (No. XK100080573).

W. Hao · J. Dong · K. Peng(✉) · M. Jia · Q. Wang
Key Laboratory for Advanced Control of Iron and Steel Process of Ministry
of Education, Beijing 100083, People's Republic of China
e-mail: kaixiang@ustb.edu.cn

W. Hao · J. Dong · K. Peng · M. Jia · Q. Wang
School of Automation and Electrical Engineering, University of Science
and Technology Beijing, Beijing 100083, People's Republic of China

© Springer Nature Singapore Pte Ltd. 2018
Y. Jia et al. (eds.), *Proceedings of 2017 Chinese Intelligent Systems Conference*,
Lecture Notes in Electrical Engineering 460,
https://doi.org/10.1007/978-981-10-6499-9_67

of pulping and papermaking, a key step is the control of pulp concentration, which can stably adjust the pulp concentration to meet the required quality standards. In the process of pulp concentration control, there exist some objective factors, such as parameter fluctuation and difficult to establish accurate mathematical model, so it is difficult to achieve the ideal control effect by using conventional PID controller. In [1], the self-tuning method of PID is proposed to adjust the control parameters online, compared with the fixed parameter PID controller, this method improves the control performance, but the large amount of calculation, it is difficult to apply to the control system of fast response. In [2], the single neuron PID module is used to adjust the pulp concentration and suppress the system interference. Although it is easy to implement, it is difficult to obtain good control effect in the complicated process control because the single layer neural network only has the ability of linear classification. In [3], a method is proposed to automatically adjust the parameters of the PID controller according to the characteristics of the object, it combines the advantages of fuzzy control and PID control to control the concentration of the pulp, but it has a longer adjustment time. Therefore, dynamic matrix control algorithm is used to control the pulp concentration, which can solve the above problems.

Predictive control algorithm as a product of industrial practice, it has been successfully applied in the oil, aviation, metallurgy and other industries, and has achieved significant economic benefits. Predictive control algorithm is a closed-loop feedback control algorithm. It optimizes the controlled object by model prediction, rolling optimization and feedback correction. Predictive control in industrial process control is not as modern control theory on the basis of accurate model and off-line calculation to get the optimal solution. It obtains a global optimal solution by online rolling optimization. In addition, it can deal with special situations such as non-minimum phase systems, and compensate for the time varying, interference and mismatch of the model by feedback correction strategy. Therefore, dynamic matrix control algorithm is used to control the pulp concentration, which can solve the above problems. The typical algorithms of model predictive control include model algorithm control (MAC), dynamic matrix control (DMC) and generalized predictive control (GPC) [4].

2   Pulp Concentration Control Modeling

In the pulp concentration control system of industrial paper, the dilution is controlled by adding dilution water to the pulp at the outlet of the storage tank. The three control points of concentration control are respectively located at the bottom of the pulping machine, the slurry tank and the front tank exit. The parts are intrinsically related to each other, so the three stage loop is adopted to control them. The concentration control loop is commonly used by the transmitter, pulp concentration adjusting valve and concentration controller is composed of three parts, and in the pulp concentration control valve between the sensor measuring the concentration will have a certain time delay. Therefore, it can be concluded that the transfer function of the pulp concentration control system can be expressed as:

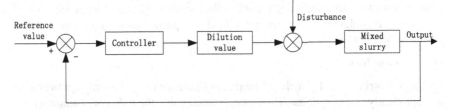

Fig. 1 Principle diagram of the pulp concentration control system

$$\frac{Y(s)}{U(S)} = \frac{k}{(T_1s + 1)(T_2s + 1)(T_3s + 1)}e^{-\tau s} \tag{1}$$

In the equation, three inertia links represent the dynamic characteristics of the valve, the dynamic characteristics of the pulp concentration transmitter and the mixing process of dilution water and pulp, and k is the ratio of the change of the pulp concentration to the change of the valve opening degree. In the above model, there are three inertia links, so it can be fitted as a inertia link [5, 6]. Thus, the transfer function of the pulp concentration control system can be simplified as:

$$\frac{Y(s)}{U(S)} = \frac{k}{(Ts + 1)}e^{-\tau s} \tag{2}$$

As the pulp concentration control system has the characteristics of a large lag and time variation, and the concentration value is in a state of fluctuation for a long time, the actual output of the system can not better track the set value. Therefore, in this paper, the PID controller and the predictive controller are used to simulate the system and compare the steps. In addition, the dynamic matrix control algorithm is used to simulate the system, and the actual output and predictive output curve of the system are obtained. The control principle is shown in Fig. 1. Using the single closed loop control, the deviation between the pulp concentration setting value and the actual output value is taken as input, and the optimum value of the concentration is calculated by the controller module.

3 Dynamic Matrix Control Algorithm

In the predictive control algorithm, the commonly used model has two kinds of parameter model and nonparametric model. Dynamic matrix control algorithm is based on step response model, and it is also a constrained multivariable control algorithm. In the pulp concentration control, the model of the controlled object contains pure hysteresis. Therefore, the dynamic matrix control algorithm is used to control the pulp concentration, which can get good control effect and weaken the influence of the delay link on the system control. The main characteristics of the dynamic matrix control algorithm are predictive model, rolling optimization and feedback correction. Using this algorithm, we can get a local suboptimal solution by repeating the online rolling optimization.

(1) Predictive Model

Dynamic matrix control is a closed loop control algorithm and is only applicable to asymptotically stable systems. Therefore, assuming that the pulp concentration control system is a asymptotically stable linear system, and the step response of the system is sampled to obtain a series of dynamic coefficients $a_1, a_2, a_3, \cdots a_N$ [7]. Where N is the length of the model, assuming that P is the predicted time domain length, M is the control time domain length, and the relation between the three is: $N \geq P \geq M$. The superposition of the linear system can be used to predict the future output of step i as follows:

$$\tilde{y}_M (k + i|k) = \tilde{y}_0 (k + i|k) + \sum_{j=1}^{\min(M,i)} a_{i-j+1} \Delta u (k + j - 1), i = 1, \cdots, N \qquad (3)$$

where a_i is the dynamic coefficient obtained from the step response, Δu is the dilution water flow control increment, and $\tilde{y}_0 (k + i|k)$ is the known pulp concentration prediction value. Therefore, at any given time k, Given the predicted initial value and the control increment, the predicted output value of the future time can be calculated.

(2) Rolling Optimization

The biggest advantage of dynamic matrix control is that the control strategy is repeatedly optimized online to avoid the deviation of the controlled quantity from the set value due to the accumulation of errors. In the pulp concentration control, select the quadratic type of optimization performance indicators as follows [8]:

$$\min J (k) = \sum_{i=1}^{P} q_i \left[\omega (k + i) - \tilde{y}_{PM} (k + i|k) \right]^2 + \sum_{j=1}^{M} r_j \Delta u^2 (k + j - 1) \qquad (4)$$

where q_i is the error weighting coefficient, indicating the degree of suppression of the tracking error, and r_j is the control weighting coefficient, indicating the degree of suppression of the control increment fluctuation.

In solving the optimization problem, the predictive model (3) is written in vector form:

$$\tilde{y}_{PM} (k) = \tilde{y}_{P0} (k) + A \Delta u (k) \qquad (5)$$

In the above equation, A is a $P \times M$ dimensional matrix whose element is the step response coefficient a_i of the system, which is called a dynamic matrix.

The quadratic performance index (4) to do the corresponding treatment, write it as vector form:

$$\min J (k) = \left\| \omega_P (k) - \tilde{y}_{PM} (k) \right\|_Q^2 + \left\| \Delta u (k) \right\|_Q^2 \qquad (6)$$

where the vector $\omega_P(k) = \left[\omega(k+1), \cdots, \omega(k+p)\right]^T$ is the set value of the pulp concentration; the values of the error weighting matrix Q and the control weighting matrix R are diagonal matrices consisting of the error weight coefficient and the control weight coefficient, Q is a P dimensional diagonal matrix, R is M dimensional diagonal matrix.

The purpose of rolling optimization is to obtain the appropriate control of the incremental sequence so that the minimum value of performance indicators. In general, the derivative of the performance index, and make the derivative $\frac{dJ(k)}{d\Delta u_M(k)} = 0$, we can get:

$$\Delta u_M(k) = \left(A^T Q A + R\right)^{-1} A^T Q \left[\omega_P(k) - \tilde{y}_{P0}(k)\right] \tag{7}$$

From the above equation can be obtained k time of the M optimal control sequence, that is, the dilution water flow control increment in the pulp concentration control system. However, when the M optimal control sequences at time k are applied to the controlled object, Because of the existence and accumulation of errors, the predicted output values deviate seriously from the set values. Thus, in the k moment, only the first control sequence is applied to the system, as shown in the following equation:

$$\Delta u(k) = c^T \Delta u_M(k) = d^T \left[\omega_P(k) - \tilde{y}_{P0}(k)\right] \tag{8}$$

$$d^T = c^T \left(A^T Q A + R\right)^{-1} A^T Q \overset{\Delta}{=} \left[d_1, \cdots, d_P\right] \tag{9}$$

where the dimension of the row vector $c^T = [1, 0, \ldots, 0]$ is M; the row vector d^T is the control vector with a dimension of P and it can be obtained by offline. After obtaining the control sequence $\Delta u(k)$, the system can be controlled by $u(k)$ and the control increment of the next time can be obtained according to the rolling optimization strategy and applied to the system.

(3) Feedback Correction

In the pulp concentration control system due to the presence of disturbances and model uncertainties and the error caused by the accumulation of the predicted values are not well follow the set value, the dynamic matrix control algorithm is a closed loop control can be eliminated to avoid the phenomenon of error feedback. The actual output of the control system and the predicted output is poor, and the error is expressed as:

$$e(k+1) = y(k+1) - \tilde{y}_1(k+1|k) \tag{10}$$

The feedback of the output of the system is corrected using the real time information $e(k+1)$ so that the predictive output is better following the set value. The error $e(k+1)$ is weighted to obtain the corrected predicted output as follows:

$$\tilde{y}_{cor}(k+1) = \tilde{y}_{N1}(k) + he(k+1) \tag{11}$$

where $\tilde{y}_{cor}(k+1)$ is the predicted output obtained after the correction at time $k+1$, and $h = [h_1, h_2 \ldots h_N]$ is the correction vector, where each element is the weighting factor for the error.

With the passage of time, in order for the prediction output of the $k+1$ moment to become the initial prediction value at the next moment, the corrected prediction output should be obtained by shifting, as shown in the following equation:

$$\tilde{y}_{N0}(k+1) = S\tilde{y}_{cor}(k+1) \tag{12}$$

where the shift matrix S is defined as:

$$S = \begin{bmatrix} 0 & 1 & & 0 \\ \vdots & \ddots & \ddots & \\ \vdots & & 0 & 1 \\ 0 & \cdots & 0 & 1 \end{bmatrix}_{N \times N} \tag{13}$$

In this way, the output prediction value $\tilde{y}_{N0}(k+1)$ at time $k+1$ is obtained, and the optimal control increment $\Delta u(k+1)$ at the time is obtained according to the above optimization method, and then the online rolling optimization is performed at the next time.

4 Parameter Tuning Principle

In the dynamic matrix control algorithm, the parameters involved in the algorithm must be adjusted in order to achieve good control effect. The choice of parameters will directly affect the performance of the prediction system, and the choice of parameters should be handled in a certain way. The following is a brief introduction to the tuning principles of related parameters.

(1) The Prediction Time Domain P and The Error Weight Matrix Q

In the dynamic matrix control algorithm, the selection of the prediction time domain P and the error weight matrix Q is based on the step response time delay characteristics and dynamic characteristics of the object's step response. It is generally necessary to determine one of the parameter values, and then determine another parameter. The individual elements of the error weight matrix Q are generally the weight coefficients of the error at each moment, The weight coefficient reflects the importance of the error at different times. In the control system, the stability and rapidity of the system are affected by the optimized time domain, and trade off is needed between them. Optimization of the time domain increases the rapidity of the system, but the stability becomes worse. Therefore, in the selection process, we should take into account both rapidity and stability.

(2) Control Time Domain M

The selection of the control time domain M indicates the dimension of the control quantity at the next time or the next time in the algorithm. Although the dynamic matrix control algorithm only takes the control of the current moment to the system, it needs to calculate M control increments at a time. The value of the control time domain M is generally defined $M \leq P$. The time domain of control will affect the maneuverability and stability of the control system. The smaller the value of the control time domain, the weaker the system's maneuverability, resulting in a greater deviation between the output predictive value and the set value, and can not achieve the desired control effect. However, the smaller the value of the predicted time domain, the control effect is often more stable and the weaker impact of the model mismatch.

(3) Control Matrix R

The control matrix R is generally a diagonal matrix consisting of individual weighting coefficients. The role of the control matrix in the quadratic performance index is used to adjust the severity of the control change, which is equivalent to a soft constraint. The stability of the system can be improved by choosing the appropriate value of the control matrix. When adjusting the stability of the system, it is common to take into account the selection of the prediction time domain P, the control time domain M and the control weight matrix value R.

5   Simulation

In second part, the model of pulp concentration control system is established. the determination of the model parameters can be derived from the experimental modeling method in [9] and to obtain the specific parameter values. As result,the mathematical model is as follows:

$$\frac{Y(s)}{U(S)} = \frac{1}{(2s+1)} e^{-3s} \tag{14}$$

By using the above model, the control performance of the system is analyzed by simulation.

(1) Comparison of PID Control and DMC Simulation

First of all, in the simulation process, the PID controller is adopted to build the closed loop simulation model of the system through Simulink. Through the implementation of the automatic optimization strategy for the PID controller, the optimal PID parameters can be obtained: $P = 0.1; I = 0.13; D = 0$. The model predictive controller is used to simulate the system. By adjusting the parameters of DMC algorithm, a better response curve can be obtained, and the specific parameter value is:

Fig. 2 Comparison of step response

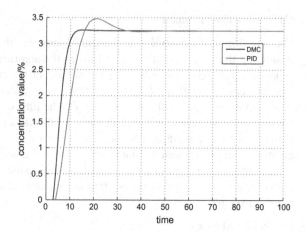

$P = 20; M = 10; q = 5; r = 10$ The setting value of the pulp concentration is 3.25%. The step response obtained using the two control strategies are shown in Fig. 2.

As can be seen from Fig. 2, when the PID controller is used, the step response curve has a large overshoot and settling time is longer. The stability of the step response obtained by the predictive controller is better. Compared with the PID control, the response of the system does not exist overshoot and settling time is obviously smaller than the adjustment time of PID control.

(2) Simulation of DMC Algorithm

DMC algorithm can compensate the disturbance, constraint and model mismatch in the system, and eliminate the influence of these factors on the system through feedback correction. Therefore, in pulp concentration control system, it is assumed that the constraint of the system output is: $0.9 \leq y \leq 1.1$, and Gaussian white noise with the mean 0 and the variance is 0.05. The simulation results are obtained by applying the rolling optimization control strategy, the change curves of the predicted output and the actual output of the system, and the change curve of the control quantity are shown in Figs. 3 and 4, respectively:

It can be seen from Figs. 3 and 4 that when the control system in the presence of disturbance and constraints using DMC algorithm can effectively avoid the influence of disturbance on the performance of the system, the prediction output curve of the system and the actual output curve is basically the same, The error is very small. The control curve changes smoothly, so the energy consumption is less. For the output value of the system, the output constraint is satisfied by a probability of 99.6%. Therefore, it can be proved that the DMC algorithm can be well applied to the control systems with disturbances and constraints.

Fig. 3 Pulp concentration
curve with disturbance

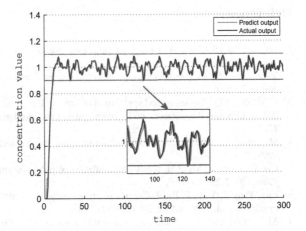

Fig. 4 Control variable
curve

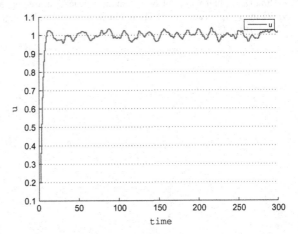

6 Conclusions

In pulp concentration control system, pulp concentration can be controlled by varying the amount of dilution water flow. In this paper, DMC algorithm can compensate
the disturbance in the system to eliminate the influence of the disturbance and avoid
the hysteresis problem in pulp concentration control. Compared with the PID control, DMC algorithm makes the response of the system no overshoot and settling
time is obviously reduced. The simulation results show that the range of concentration changes in the output range with a great probability. It can be proved that DMC
algorithm can deal with the disturbance well and improve the performance of the
system.

References

1. Ge S, Tong S, Zhou B. Design of pulp concentration control system. China Pulp Paper. 2013;21(3):50–2.
2. Ye Z, Chen M. Pulp concentration control based on dynamic matrix and single neuron PID. Autom Instrum. 2009;24(8):29–33.
3. Li M, Zhang Q. The design and application of fuzzy PID controller in pulp concentration control. 2010 International conference on logistics systems and intelligent management, vol. 2. 2010. p. 1204–7.
4. Morari M, Lee JH. Model predictive control: past, present and future. Comput & Chem Eng. 1999;23(4):667–82.
5. Cao L, Xiong Z, Muyi H. Study on simulation of pulp concentration control system. Comput Simul. 2012;29(6):176–9.
6. Li J, Yang S, Yuan P, Bai J. Control of steam pressure in coal fired boiler based on dynamic matrix control. Control Eng. 2016;23(11):1685–9.
7. Xi Y. Predictive control. Beijing: National Defence Industry Press; 1993.
8. Shu D. Predictive control system and its application. Beijing: Mechanical Industry Press; 1998.
9. Zhang J. Application of MATLAB in control system. Beijing: Publishing House of Electronics Industry; 2007.

Unknown Dynamic Estimator Based Control for Hydraulic Servo Systems

Yunpeng Li, Jing Na, Qiang Chen and Guanbin Gao

Abstract This paper investigates precise tracking control for a class of high-order hydraulic servo systems. For modeling of servo systems, the unknown nonlinearities of the hydraulic actuator and valve that may decrease the tracking performance and system stability are taken into account and then addressed via an unknown dynamic estimator with a simple structure and fewer tuning parameters. A recursive control design procedure is developed to achieve precise position tracking, where the calculations of the derivatives of the virtual control laws in traditional backstepping are remedied. The suggested controller guarantees that both the estimation error and tracking error will converge to a small set around zero. Simulation results are provided to verify the effectiveness of the proposed control strategy.

Keywords Hydraulic servo system · Tracking control · Nonlinear systems · Unknown dynamic estimator

1 Introduction

With the rapid development of modern industry, hydraulic servo systems are widely used due to their unique advantages [1]. However, some unknown nonlinear dynamics may decrease the system performance, which include the disturbances (e.g. load variation), the parametric uncertainties (e.g. hydraulic parameters) and the unmodeled dynamics (e.g. nonlinear friction, oil leakage and throttling

Y. Li · J. Na (✉) · G. Gao
Faculty of Mechanical and Electrical Engineering, Kunming University of Science & Technology, Kunming 650500, People's Republic of China
e-mail: najing25@163.com

Q. Chen
College of Information Engineering, Zhejiang University of Technology, Hangzhou 310023, People's Republic of China

© Springer Nature Singapore Pte Ltd. 2018
Y. Jia et al. (eds.), *Proceedings of 2017 Chinese Intelligent Systems Conference*, Lecture Notes in Electrical Engineering 460, https://doi.org/10.1007/978-981-10-6499-9_68

characteristics) [2–4]. Consequently, how to deal with the nonlinearities in such systems is an imperious demand. In order to guarantee safe operations and satisfactory performance, various advanced control techniques were proposed, e.g. adaptive control [5], sliding mode control [6], disturbance observer based control [7], and so on. Among those methods, adaptive control is known as an elegant strategy for systems with unknown parameters [8]. In [9], Yu et al. proposed a nonlinear controller for a class of hydraulic servo systems. Kim et al. [10] developed a nonlinear disturbance observer (DOB) to estimate the equivalent effect of modeling uncertainties from the electro-hydraulic actuators. The work [11] presented an extended state observer (ESO) to estimate not only the unmeasured states but also the lumped uncertainties. In [12], we proposed a simple and robust unknown input dynamics estimator with only one filter parameter to be selected, which is simpler than conventional DOB and ESO.

In recent years, backstepping control has been widely used in hydraulic servo systems [13–16], where the hydraulic systems can be expressed in a strict-feedback form [17]. However, aforementioned backstepping based methods have to address the derivatives of the virtual control laws [5], which exists in the backstepping procedure. As a result, one of the drawbacks of the backstepping is the so called 'explosion of complexity' problem [18], especially for the high-order systems. Nevertheless, functional approximations have been adopted in these backstepping controls for nonlinear systems, which may lead to sluggish transient responses. Thus, this paper develops a new estimator based tracking control strategy for servo systems with hydraulic actuators. By using a newly proposed unknown dynamic estimator, the derivatives of virtual control variables in the control design and the lumped unknown system dynamics can be precisely estimated. Hence, the adaptation and function approximation can be avoided. The estimation error will converge to a small set around zero and the stability of the closed-loop system can be proved. The effectiveness of the proposed control system is validated by numerical simulations.

2 System Formulation

In this paper, a hydraulic servo system shown in Fig. 1 is studied. The dynamic equation can be derived as

Fig. 1 Schematic of the hydraulic servo system

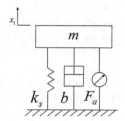

Fig. 2 Structure of the
electro-hydraulic actuator

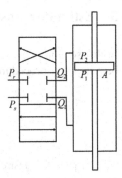

$$\begin{cases} \dot{x}_1 = x_2 \\ \dot{x}_2 = \frac{1}{m}(-k_s x_1 - b x_2 + F_a) \\ y = x_1 \end{cases} \tag{1}$$

where x_1, x_2 are the displacement and the speed of the mass m; k_s is the stiffness coefficient of the spring; b is the damping coefficient; F_a is the force produced by a hydraulic actuator shown in Fig. 2.

The dynamics of the hydraulic actuator are given as

$$\frac{V_t}{4\beta_e} \dot{P}_L = -A\dot{x}_1 - C_t P_L + Q_L \tag{2}$$

where V_t is the total volume of the cylinder; β_e is the effective bulk modulus coefficient; A is the ram area of the cylinder; $P_L = P_1 - P_2$ is the load pressure; C_t is the internal leakage factor of the cylinder and $Q_L = (Q_1 + Q_2)/2$ is the load flow.

In this paper, the magnetic directional valve is used and the load flow Q_L is controlled by the on-off operation. Thus, Q_L can be simply described as $Q_L = k_v u$, where k_v is an equivalent parameter and u is the control signal. Then we define the states as $x = [x_1, x_2, x_3]^T = [y, \dot{y}, F_a/m]$. For simplicity, we define $a = 4\beta_e/V_t$, $F_a = A P_L$, $f_1 = (-k_s x_1 - b x_2)/m$, $f_2 = -aA^2 x_2/m - aC_t x_3$, and $p = aAk_v/m$. So that we can rewrite systems (1) and (2) in a compact form as

$$\begin{cases} \dot{x}_1 = x_2 \\ \dot{x}_2 = f_1 + x_3 \\ \dot{x}_3 = f_2 + pu \end{cases} \tag{3}$$

The objective of the control design is to make the position x_1 of system (3) track a smooth reference trajectory x_{1d}, where the unknown nonlinear functions f_1, f_2 and parametric variations are handled without using conventional function approximators (e.g. neural networks (NNs) and fuzzy systems (FSs)).

3   Estimator Based Controller Design

In this section, we will develop a nonlinear estimator based controller following a backstepping procedure. We define the tracking errors as

$$
\begin{cases}
z_1 = x_1 - x_{1d} \\
z_2 = x_2 - \alpha_1 \\
z_3 = x_3 - \alpha_2
\end{cases}
\tag{4}
$$

where x_{1d} is the desired trajectory; α_1 and α_2 denote the virtual control inputs.

Step 1: The time derivative of z_1 is

$$
\dot{z}_1 = \dot{x}_1 - \dot{x}_{1d} = x_2 - \dot{x}_{1d}
\tag{5}
$$

Consider x_2 as the virtual control input to regulate z_1. The control design of x_2 is based on the Lyapunov candidate with respect to z_1. We define the Lyapunov function as

$$
V_1 = \frac{1}{2} z_1^2
\tag{6}
$$

Then we have

$$
\dot{V}_1 = z_1 \dot{z}_1 = z_1 (x_2 - \dot{x}_{1d}) = z_1 (z_2 + \alpha_1 - \dot{x}_{1d})
\tag{7}
$$

We can design a virtual control α_1 given by

$$
\alpha_1 = -k_1 z_1 + \dot{x}_{1d}
\tag{8}
$$

where $k_1 > 0$ is the control gain.

Substituting (8) into (7), the derivative of V_1 can be calculated as

$$
\dot{V}_1 = -k_1 z_1^2 + z_1 z_2
\tag{9}
$$

Step 2: It can be obtained that

$$
\dot{z}_2 = \dot{x}_2 - \dot{\alpha}_1 = f_1 + x_3 - \dot{\alpha}_1 = f_1 + z_3 + \alpha_2 - \dot{\alpha}_1 = F_1 + z_3 + \alpha_2
\tag{10}
$$

where $F_1 = f_1 - \dot{\alpha}_1$ is the lumped unknown dynamics to be handled later. We select the Lyapunov function as

$$V_2 = \frac{1}{2}z_1^2 + \frac{1}{2}z_2^2 \tag{11}$$

and design the virtual control input α_2 as

$$\alpha_2 = -k_2 z_2 - \hat{F}_1 - z_1 \tag{12}$$

where $k_2 > 0$ is the control gain, and $\hat{F}_1$ is the estimate of the lumped unknown dynamics F_1, which will be given in the following design.

Substituting (12) into (10) and the time derivative of V_2 becomes

$$\dot{V}_2 = z_1 \dot{z}_1 + z_2 \dot{z}_2 = -k_1 z_1^2 - k_2 z_2^2 + z_2 z_3 + z_2 \tilde{F}_1 \tag{13}$$

where $\tilde{F}_1 = F_1 - \hat{F}_1$ is the estimation error.

Step 3: Similarly, we take the derivative of z_3 as

$$\dot{z}_3 = \dot{x}_3 - \dot{\alpha}_2 = f_2 + pu - \dot{\alpha}_2 = F_2 + pu \tag{14}$$

where $F_2 = f_2 - \dot{\alpha}_2$ is the lumped unknown dynamics in (14) to be handled later.

The Lyapunov function for the overall system can be defined as

$$V_3 = \frac{1}{2}z_1^2 + \frac{1}{2}z_2^2 + \frac{1}{2}z_3^2 \tag{15}$$

The final control can be designed as

$$u = \left(-k_3 z_3 - z_2 - \hat{F}_2\right)/p \tag{16}$$

where $k_3 > 0$ is the feedback gain and $\hat{F}_2$ is the estimate of the lumped unknown dynamics F_2.

Then the derivative of V_3 is

$$\dot{V}_3 = z_1 \dot{z}_1 + z_2 \dot{z}_2 + z_3 \dot{z}_3 = -k_1 z_1^2 - k_2 z_2^2 - k_3 z_3^2 + z_3 \tilde{F}_2 + z_2 \tilde{F}_1 \tag{17}$$

where $\tilde{F}_2 = F_2 - \hat{F}_2$ is the estimation error.

To implement the control laws (8), (12) and (17), the estimates of unknown dynamics $\hat{F}_1, \hat{F}_2$ should be addressed. In many backstepping based control designs, the unknown dynamics F_1 and F_2 are usually estimated based on approximators such as neural networks or fuzzy logic systems, where the unknown weights or coefficients of NNs and FLSs are online updated via adaptive laws. These algorithms may impose significant computational costs and even lead to sluggish transient response. In order to address this issue, a recently proposed novel unknown dynamic estimator [12] is employed in this paper.

With loss of generality, it is assumed that $\sup_{t \geq 0} |\dot{F}_i| \leq \hbar_i, i = 1, 2$ with the bounded constants $\hbar_i > 0$. Then the estimates of F_1 and F_2 can be derived based on (10) and (14). The filtered variables of z_2, x_3, z_3, u are defined as

$$
\begin{cases}
k_s \dot{z}_{2f} + z_{2f} = z_2, & z_{2f}(0) = 0 \\
k_s \dot{x}_{3f} + x_{3f} = x_3, & x_{3f}(0) = 0 \\
k_s \dot{z}_{3f} + z_{3f} = z_3, & z_{3f}(0) = 0 \\
k_s \dot{u}_f + u_f = u, & u_f(0) = 0
\end{cases}
\tag{18}
$$

where $k_s > 0$ is the filter parameter. It should be noted that (18) can be easily implemented by imposing a low-pass filter $1/(k_s s + 1)$ on the measured variables z_2, x_3, z_3, u.

Hence, similar to those shown in [12], we can design the unknown dynamic estimators of $\hat{F}_1, \hat{F}_2$ as

$$
\hat{F}_1 = \frac{z_2 - z_{2f}}{k_s} - x_{3f}
\tag{19}
$$

$$
\hat{F}_2 = \frac{z_3 - z_{3f}}{k_s} - p u_f
\tag{20}
$$

Clearly, one can find that the proposed estimators (18)–(20) are simpler than those function approximators used in conventional backstepping based control methods.

Before theoretically proving the stability of the overall system, we first analyse the convergence property of the estimators (19) and (20). Inspired by [12], we apply a low-pass filter $1/(k_s s + 1)$ on the both sides of (10) and (14), and consider the fact $\dot{z}_{2f} = (z_2 - z_{2f})/k_s$ and $\dot{z}_{3f} = (z_3 - z_{3f})/k_s$, and then have

$$
\hat{F}_1 = F_{1f}, \hat{F}_2 = F_{2f}
\tag{21}
$$

where F_{1f}, F_{2f} is the filtered version of F_1, F_2 given by $k_s \dot{F}_{if} + F_{if} = F_i, i = 1, 2$ with zero initial condition $F_{if}(0) = 0$. Then the convergence of the estimation error can be summarized as

Lemma 1 [12] *For the proposed estimators (19) and (20), the estimation errors $\tilde{F}_i = F_i - \hat{F}_i, i = 1, 2$ will exponentially converge to a small compact set around zero, which can be given by $|\tilde{F}_i| \leq \sqrt{F_i^2(0)e^{-t/k_s} + k_s^2 \hbar_i^2}$. So that $\hat{F}_i \rightarrow F_i$ holds for $k_s \rightarrow 0$ and/or $\hbar_i \rightarrow 0$.*

Proof Based on (18)–(21), we can calculate the derivatives of the estimation errors $\tilde{F}_i = F_i - \hat{F}_i$ as

$$\dot{\tilde{F}}_i = \dot{F}_i - \dot{\hat{F}}_i = \dot{F}_i - \dot{F}_{if} = \dot{F}_i - \frac{F_i - F_{if}}{k_s} = -\frac{1}{k_s}\tilde{F}_i + \dot{F}_i \tag{22}$$

We select a Lyapunov candidate as $V_{F_i} = \tilde{F}_i^2/2$, and then it follows that

$$|\tilde{F}_i(t)| \leq \sqrt{\tilde{F}_i^2(0)e^{-t/k_s} + k_s^2\hbar_i^2} \tag{23}$$

Thus, $\tilde{F}_i(t) \to 0$ is true for $k_s \to 0$ and/or $\hbar_i \to 0$.   ◇
Then we will summarize the stability analysis of the closed-loop control.

Theorem 1 *Consider the hydraulic servo system (3) with unknown dynamics, the unknown dynamic estimators (19) and (20) and control laws (8), (12) and (16) are used. The closed-loop control system is stable, and the tracking errors z_1, z_2, z_3 and the estimation errors $\tilde{F}_1$ and $\tilde{F}_2$ will converge to a neighborhood around zero in an exponential sense.*

Proof We choose the Lyapunov candidate as

$$V = \frac{1}{2}z_1^2 + \frac{1}{2}z_2^2 + \frac{1}{2}z_3^2 + \frac{1}{2}\tilde{F}_1^2 + \frac{1}{2}\tilde{F}_2^2 \tag{24}$$

The derivation of V can be calculated along (9), (13), (17) and (22) as

$$\dot{V} = -k_1z_1^2 - k_2z_2^2 - k_3z_3^2 + z_3\tilde{F}_2 + z_2\tilde{F}_1 + \tilde{F}_1\left(-\frac{1}{k_s}\tilde{F}_1 + \dot{F}_1\right) + \tilde{F}_2\left(-\frac{1}{k_s}\tilde{F}_2 + \dot{F}_2\right)$$

$$\leq -(k_1 - \frac{1}{2k_s})z_1^2 - (k_2 - \frac{1}{2k_s})z_2^2 - (k_3 - \frac{1}{2k_s})z_3^2 - \frac{1}{2k_s}\tilde{F}_1^2 - \frac{1}{2k_s}\tilde{F}_2^2 + k_s(\hbar_1^2 + \hbar_2^2)$$

$$\leq -kV + k_s(\hbar_1^2 + \hbar_2^2) \tag{25}$$

where $k = 2\min\{(k_1 - 1/2k_s), (k_2 - 1/2k_s), (k_3 - 1/2k_s), 1/k_s\}$ is a positive constant. According to the Lyapunov Theorem, we can claim that V, $z_1, z_2, z_3, \tilde{F}_1$ and $\tilde{F}_2$ are all uniformly ultimately bounded. Specifically, the ultimate bound of V can be calculated as

$$V(t) \leq V(0)e^{-kt} + \frac{k_s}{k}(\hbar_1^2 + \hbar_2^2) \tag{26}$$

Moreover, based on Eqs. (19) and (20) and control laws (8), (12) and (16), we know that the control signals α_1, α_2, u and the estimated variables $\hat{F}_1, \hat{F}_2$ are all bounded.

Table 1 Parameters for hydraulic system

Parameter	Value
m	300 kg
k_s	15000 N/m
b	1500 N/m/s
V_t	$6.5312 \times 10^{-5} \, m^3$
A	$3.2656 \times 10^{-4} \, m^2$
C_t	4×10^{-13}
a	$4.2871 \times 10^{13} \, N/m^5$
k_v	0.00001
p	466.666

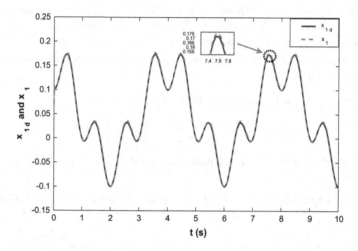

Fig. 3 Position tracking performance

4 Simulations

In this section, simulation results are presented to verify the proposed algorithm. In the simulations, the smooth reference trajectory is given as $x_{1d} = 0.1 \sin^2(\pi t) + 0.1 \cos(0.5\pi t)$, the initial states are set as $x_1 = 0.07$, $x_2 = 0$, $x_3 = 0$. The control parameters are given as $k_1 = 180, k_2 = 30, k_3 = 60, k_s = 0.001$. And other parameters of the hydraulic system (1) are listed in Table 1.

Figure 3 shows the position tracking performance. Figure 4 depicts the velocity tracking response of the control system. The trajectories of both the position and velocity are smooth. As shown in Fig. 5, the tracking errors of the stability performance are very small. So that the proposed nonlinear controller has the ability to

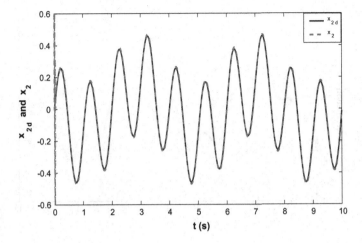

Fig. 4 Velocity tracking performance

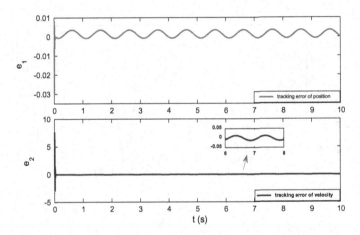

Fig. 5 Tracking errors of x_1 and x_2

compensate for the effects of unknown dynamics to a certain degree. Furthermore, the presented estimators (19) and (20) can estimate the unknown dynamics with satisfactory accuracy, which can be verified by Figs. 6 and 7, respectively. And the estimation errors of F_1 and F_2 are shown in Fig. 8. It should be noted that the peaks in the initial estimation error is due to the small k_s used in the control.

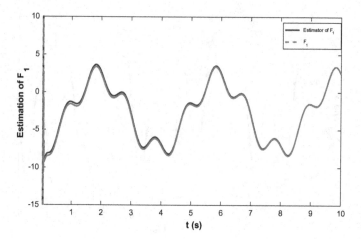

Fig. 6 Estimation of F_1

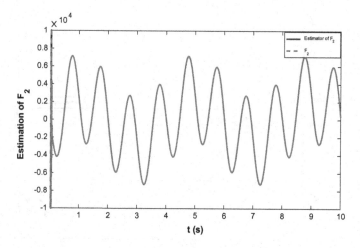

Fig. 7 Estimation of F_2

5 Conclusion

In this paper, an alternative control method is proposed for a class of hydraulic servo systems to achieve precise tracking control. A filter-based unknown dynamic estimator with only one tuning parameter is employed to address the lumped unknown nonlinear dynamics. Consequently, the 'explosion of complexity' encountered in the traditional backstepping design can be avoided. Moreover, the widely used function approximators (e.g. NNs and FLSs) and the associated adaptations are avoided. This leads to a simpler yet robust control strategies, which

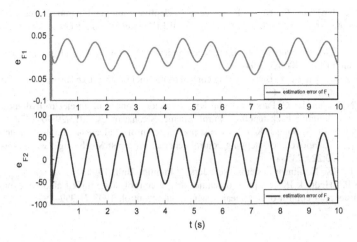

Fig. 8 Estimation errors of F_1 and F_2

can achieve improved transient control response. The convergence of both the estimation error and control error can be proved simultaneously. Simulation results are provided to verify the effectiveness of the proposed control system.

References

1. Merritt HE, Hydraulic control systems. Wiley; 1967.
2. Yao J, Jiao Z, Ma D, Yan L. High-accuracy tracking control of hydraulic rotary actuators with modeling uncertainties. IEEE/ASME Trans Mechatron. 2014;19:633–41.
3. Yao B, Bu F, Reedy J, Chiu G-C. Adaptive robust motion control of single-rod hydraulic actuators: theory and experiments. IEEE/ASME Trans Mechatron. 2000;5:79–91.
4. Milić V, Šitum Ž, Essert M. Robust H∞ position control synthesis of an electro-hydraulic servo system. ISA Trans. 2010;49:535–42.
5. Guo Q, Sun P, Yin J-M, Yu T, Jiang D. Parametric adaptive estimation and backstepping control of electro-hydraulic actuator with decayed memory filter. ISA Trans. 2016;62:202–14.
6. Huang S-J, Chen H-Y. Adaptive sliding controller with self-tuning fuzzy compensation for vehicle suspension control. Mechatronics. 2006;16:607–22.
7. Won D, Kim W, Shin D, Chung CC. High-gain disturbance observer-based backstepping control with output tracking error constraint for electro-hydraulic systems. IEEE Trans Control Syst Technol. 2015;23:787–95.
8. Na J, Mahyuddin MN, Herrmann G, Ren X, Barber P. Robust adaptive finite-time parameter estimation and control for robotic systems. Int J Robust Nonlinear Control. 2015;25:3045–71.
9. Yu H, Feng Z-J, Wang X-Y. Nonlinear control for a class of hydraulic servo system. J Zhejiang Univ Sci A. 2004;5:1413–7.
10. Kim W, Shin D, Won D, Chung CC. Disturbance-observer-based position tracking controller in the presence of biased sinusoidal disturbance for electrohydraulic actuators. IEEE Trans Control Syst Technol. 2013;21:2290–8.
11. Yao J, Jiao Z, Ma D. Extended-state-observer-based output feedback nonlinear robust control of hydraulic systems with backstepping. IEEE Trans Ind Electron. 2014;61:6285–93.

12. Na J, Herrmann G, Burke R, Brace C. Adaptive input and parameter estimation with application to engine torque estimation. In: IEEE conference on decision and control; 2015. pp. 1035–65.

13. Ursu I, Ursu F, Popescu F. Backstepping design for controlling electrohydraulic servos. J Franklin Inst. 2006;343:94–110.

14. Yagiz N, Hacioglu Y. Backstepping control of a vehicle with active suspensions. Control Eng Pract. 2008;16:1457–67.

15. Tri NM, Nam DNC, Park HG, Ahn KK. Trajectory control of an electro hydraulic actuator using an iterative backstepping control scheme. Mechatronics. 2015;29:96–102.

16. Kim H, Park S, Song J, Kim J. Robust position control of electro-hydraulic actuator systems using the adaptive back-stepping control scheme. Proc Inst Mech Eng Part I J Syst Control Eng. 2010;224:737–46.

17. Khalil HK. Nonlinear systems. New Jersey: Prentice-Hall; 1996.

18. Yip PP, Hedrick JK. Adaptive dynamic surface control: a simplified algorithm for adaptive backstepping control of nonlinear systems. Int J Control. 1998;71:959–79.

Active Disturbance Rejection Twice Optimal Control for Time Delay System

Xiaoyi Wang, Wei Li, Wei Wei and Min Zuo

Abstract Time delay makes system control be a challenge. Twice optimal algorithm is an effective approach to address time delay. However, it needs model information of a system. In order to reduce its dependence on system model, active disturbance rejection control is adopted. With the help of extended state observer, uncertainties and disturbances are able to be estimated, and those uncertainties and disturbances can be compensated by control signal in real time. By combining active disturbance rejection control and twice optimal control, active disturbance rejection twice optimal control (ADRTOC) is proposed. Comparisons between ADRTOC and PI plus Smith predictor control have been presented. Numerical results confirm that ADRTOC has better performance.

Keywords Time delay system · Active disturbance rejection control · Twice optimal control

1 Introduction

Time delay commonly exists in industrial sections. It is an infinite dimensional factor [1], and it always makes system performance be worse or even be unstable. To address time delay, researchers have paid much effort, and kinds of methods have been proposed. Smith predictor [2] is the classical approach to deal with time delay. Its basic idea is to design a predictor to eliminate the influence of time delay. However, it largely depends on model parameters, which results in performance degradation drastically when system parameters are varying. Furthermore, other approaches have been discussed, such as neural network control [3], fuzzy control [4], model free control [5] and improved algorithms based on Smith predictor [6, 7]. However, for many practical reasons, such as feasibility, the ability of disturbance

X. Wang · W. Li · W. Wei (✉) · M. Zuo
School of Computer and Information Engineering, Beijing Technology
and Business University, Beijing 100048, China
e-mail: weiweizdh@126.com

© Springer Nature Singapore Pte Ltd. 2018
Y. Jia et al. (eds.), *Proceedings of 2017 Chinese Intelligent Systems Conference*,
Lecture Notes in Electrical Engineering 460,
https://doi.org/10.1007/978-981-10-6499-9_69

rejection, and real time performance, most proposed control algorithms have not been applied in engineering.

Recently, Xiang has proposed twice optimal control (TOC) to address time delay problem [8–11]. TOC can be divided into two steps. Firstly, approximating the infinite dimensional delay factor by finite dimensional states and taking ITAE optimal control law to determine the parameters of the state observer; secondly, coming back to the delay factor. By simulations, one can find a group of optimal parameters to realize optimal control of a time-delay system. However, TOC does not completely eliminate the influence of time delay and it also needs model information of a system.

Active disturbance rejection control (ADRC) is an approach which does not rely on accurate model of a system. By combining TOC and ADRC, in this paper, we propose active disturbance rejection twice optimal control (ADRTOC) to take full advantage of the merits of TOC and ADRC. Comparisons have been made between ADRTOC and Smith predictor based PI control. Simulation results confirm ADRTOC.

2 Problem Statement

Consider first order plus time delay (FOPTD) model

$$G(s) = \frac{K}{Ts+1} e^{-\tau s} \tag{1}$$

where K is the gain of a plant, T is the time constant, τ represents time delay constant. Let $a = \frac{1}{T}, b = \frac{K}{T}$, we can rewrite system (1) as

$$\dot{y} = -ay + bu(t-\tau) \tag{2}$$

where y is system output, u is control input. From (2) we can see that control signal u needs to delay τ to take effect on system output y, i.e. u and y are out of synchronization. This is the key problem of time delay system.

3 Active Disturbance Rejection Twice Optimal Control

3.1 Twice Optimal Control

TOC is a special algorithm for time-delay systems. Its structure is given in Fig. 1 [10]. If we take a third order approximation model for time delay unity, feedback factor can be written as

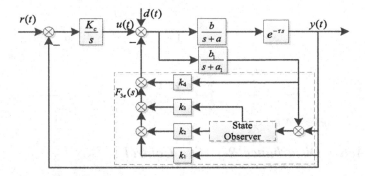

Fig. 1 Twice optimal control diagram

$$F_{3e}(s) = k_4 e^{\tau s} + \frac{(k_2 + k_3 s)\frac{6}{L^3}}{s(s + \frac{3}{L}) + \frac{6}{L^2}}(e^{\tau s} - 1) + k_1 \qquad (3)$$

where $k_i (i = 1, 2, 3, 4)$ are feedback gains of TOC.

In Fig. 1, $r(t)$ is the set value, $u(t)$ is control signal, $d(t)$ represents disturbance, a, b, τ are model parameters, a_1, b_1 are the estimated values of a, b. K_c is control gain of the closed-loop system, $y(t)$ represents system output.

Here, $k_i, i = 1, 2, 3, 4$ and K_c can be chosen as

$$\begin{cases} K_c = \dfrac{TL^3}{6K}\omega_0^5 \\[2mm] k_1 = \dfrac{\omega_0 T(\beta_1 \omega_0^3 \tau^3 - 6\beta_4)}{6K} + \dfrac{3T}{K\tau} \\[2mm] k_2 = \dfrac{\omega_0 T\tau(\beta_2 \omega_0^2 \tau^2 - 6\beta_4) + 12T}{6K} \\[2mm] k_3 = \dfrac{\omega_0 T\tau^2(\beta_3 \tau\omega_0 - 3\beta_4)}{6K} + \dfrac{T\tau}{2K} \\[2mm] k_4 = \dfrac{\beta_4 T\tau\omega_0 - \tau - 3T}{K\tau} \end{cases} \qquad (4)$$

where $\beta_i, i = 1, 2, 3, 4$ are parameters of ITAE optimal transfer function. By choosing X-Y-II ITAE optimal parameters sets, we have

$$\beta_1 = 3.381, \beta_2 = 5.33, \beta_3 = 4.78, \beta_4 = 2.795 \qquad (5)$$

By simulations, if $\tau\omega_0 = 1.1556$, we can get desired results.

TOC is able to address time delay, however, it also need model information. In order to improve the robustness of the system, ADRC is introduced.

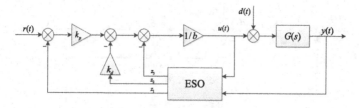

Fig. 2 Second-order LADRC system

3.2 Active Disturbance Rejection Control

Active disturbance rejection control was proposed by Han in 1990s [12, 13]. Originally, it is composed of tracking differentiator (TD), state error feedback (SEF) and extended state observer (ESO) [14, 15]. However, for simplifying the structure and applications, linear active disturbance rejection control (LADRC) has been proposed. Second order LADRC system is shown in Fig. 2.

where $r(t)$ is set value, $u(t)$ is control input, $y(t)$ represents system output, $d(t)$ is the external disturbance. z_1, z_2, z_3 are three outputs of the ESO.

ESO can be designed as [16]

$$\begin{cases} \dot{z}_1 = z_2 - \beta_{01}(z_1 - y) \\ \dot{z}_2 = z_3 - \beta_{02}(z_1 - y) + b_0 u \\ \dot{z}_3 = -\beta_{03}(z_1 - y) \end{cases} \tag{6}$$

where $\beta_{01}, \beta_{02}, \beta_{03}$ are observer parameters, b_0 is a tunable parameter. $\beta_{01} = 3\omega_o, \beta_{02} = 3\omega_o^2, \beta_{03} = \omega_o^3, \omega_o = 4\omega_c$. Here, ω_o, ω_c are the bandwidth of observer and controller, respectively. Control law can be designed as

$$\begin{cases} u = \frac{u_0 - z_3}{b} \\ u_0 = k_p(r - z_1) - k_d z_2 \end{cases} \tag{7}$$

where k_p, k_d are gains of the controller, and $k_p = \omega_c^2, k_d = 2\omega_c$.

3.3 Active Disturbance Rejection Twice Optimal Control

Combining TOC and ADRC, structure of ADRTOC is shown in Fig. 3.

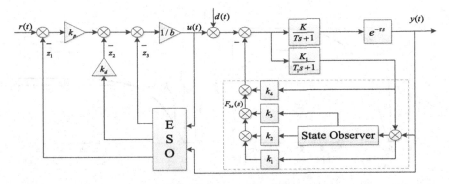

Fig. 3 Structure of the active disturbance rejection twice optimal control

4 Simulation of Active Disturbance Rejection Twice Optimal Control

Consider a FOPTD model [14]

$$G(s) = \frac{K_0}{T_0 s + 1} e^{-\tau_0 s} = \frac{0.85}{1200s + 1} e^{-1800s}$$

In this paper, three cases are considered. ADRTOC and PI plus Smith predictor control are designed, their parameters are presented in Table 1.

Case I System parameters are chosen to be nominal parameters
System response and control signals are compared and shown in Fig. 4.

Table 1 Parameters of two control approaches

	$b_0(k_p)$	$\omega_c(k_i)$	$\omega_o(k_d)$
ADRTOC	1480	0.653	2.162
PI + Smith	2.28	0.001	0

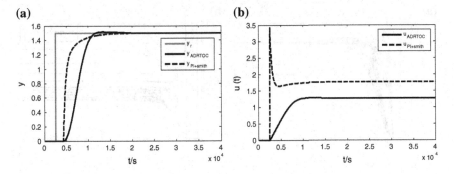

Fig. 4 Comparisons of responses and control signals for Case I

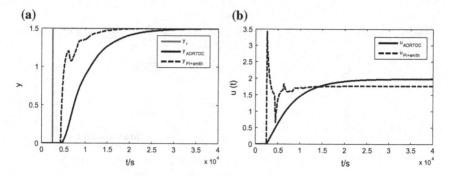

Fig. 5 Comparisons of responses and control signals for Case II

From Fig. 4, we can see that the rising time of PI plus Smith predictor control is shorter than the one of ADRTOC. However, its control signal is larger. All in all, ADRTOC has better performance.

Case II System parameters are not consistent with nominal parameters

When $K_1 = 1, T_1 = 1500, \tau_1 = 2000$. Comparisons are presented in Fig. 5.

From Fig. 5, we can see that, ADRTOC is able to achieve more stable tracking response with smoother control signal.

Case III System parameters are perturbed within a certain range

In this case, we consider the case that model parameters are perturbed within a certain range. Let $\Delta K = \pm 40\%, \Delta T = \pm 30\%, \Delta \tau = \pm 15\%$, i.e. $K = K_0(1 + \Delta K)$, $T = T_0(1 + \Delta T), \tau = \tau_0(1 + \Delta \tau)$. Here, K_0, T_0, τ_0 are nominal parameters. In this case, eight sets of parameters are randomly generated within above range. Comparisons are shown in Fig. 6.

Figure 6 shows that ADRTOC is more robust than PI plus Smith predictor control for its ability of active disturbance rejection.

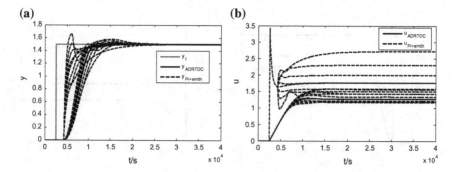

Fig. 6 Comparisons of responses and control signals for Case III

5 Conclusion

In this paper, active disturbance rejection twice optimal control (ADRTOC) is proposed. In order to make a comparison, PI plus Smith predictor control is taken as a benchmark algorithm. From numerical results, we can see that, with smaller and smoother control signal, ADRTOC is able to achieve better performance when system parameters are perturbed. Therefore, it may be a more practical way to address the control of time delay system. Further research is going on.

Acknowledgements This work is supported by National Science Technology Support Plan Projects 2015BAK36B04.

References

1. Ahuja A, Narayan S, Kumar J. Optimal two degrees of freedom decoupling smith control for MIMO systems with multiple time delays. In: 2016 IEEE 1st International Conference on Power Electronics, Intelligent Control and Energy Systems (ICPEICES);2016. p. 1–5.
2. Paor AMD. A modified Smith predictor and controller for unstable processes with time delay. Int J Control. 1985;41(4):1025–36.
3. Hongwen M, Zhou W, Ding W. Neural-network-based distributed adaptive robust control for a class of nonlinear multiagent systems with time delays and external noises. IEEE Trans Syst, Man, Cybern: Syst. 2016;46(6):750–8.
4. Shen Y, Peng S, Hongyan Y. Adaptive fuzzy control of strict-feedback nonlinear time-delay systems with unmodeled dynamics. IEEE Trans Cybern. 2016;46(8):1926–38.
5. Hyodo S, Ohnishi K. Implementation of model-free time delay compensator for bilateral control system with time delay. IEEE Int Conf Mechatron (ICM). 2015;2015:580–5.
6. Bolea Y, Puig V, Blesa J. Gain-scheduled smith proportional-integral derivative controllers for linear parameter varying first-order plus time-varying delay systems. IET Control Theor Appl. 2011;5(18):2142–55.
7. Veeramachaneni SR, Watkins JM. Robust performance design of PID controllers for time-delay systems with a Smith predictor. In: 2014 American Control Conference, Portland, OR;2014. p. 2462–7.
8. Guobo X, Yiqun Y, Qiwen Y. Twice optimal control for a class of one-capacity system with pure time-delay. Inf Control. 1995;24(4):208–14.
9. Yang Q, Yang Y, Xiang G. ITAE twice optimal control for a class of two-capacity system with pure time-delay [C]. Chinese Control Conference, 1995.
10. Guobo X. Optimal control of time delay systems. Beijing: China Electric Power Press; 2008.
11. Yan Y, Qinruo W, Guobo X, et al. Quadratic optimal control and structural shaping of second order unstable delay processes. Inf Control. 2016;02:223–8.
12. Jingqing H. A class of extended state observers for uncertain systems. Control Decis (In Chinese). 1995;10(1):85–8.
13. Jingqing H. Auto-disturbance rejection control and its applications. Control Decis (In Chinese). 1995;13(1):19–23.
14. Decui T, Zhiqiang G, Zhuhong Z. Design of predictive active disturbance rejection controller for turbidity. Control Theor Appl. 2017;34(1):101–8.
15. Qinling Z, Zhiqiang G. Predictive active disturbance rejection control for processes with delay. In: Control Conference. IEEE;2013. p. 4108–13.
16. Xing C. Actuve disturbance rejection controller tuning and its applications to thermal processes. Tsinghua University;2008.

Generating Test Data Covering Multiple Paths Using Genetic Algorithm Incorporating with Reducing Input Domain

Yan Zhang, Dunwei Gong, Xiangjuan Yao and Qiang Lu

Abstract The problem of efficiently generating test data covering multiple paths was focused on this study, and a method of generating test data covering multiple paths using a genetic algorithm incorporating with reducing the input domain of a program was presented. In this method, all target paths are first divided into several groups based on the same independent sub-path, and the input variables corresponding to the independent sub-path are determined. Then, a multi-population genetic algorithm is used to generate test data to cover these target paths, each sub-population generating test data covering target paths belonging to the same group. During the evolution, the input variables corresponding to the traversed independent sub-path are remained fixed, and the ranges of crossover and mutation operations are reduced, leading to these sub-populations' search in a reduced input domain so that the efficiency of generating test data is improved.

Keywords Software testing · Test data · Genetic algorithms · Input domain

Y. Zhang (✉) · Q. Lu
School of Computer and Information Technology, Mudanjiang Normal University,
Mudanjiang, China
e-mail: zhangyan@mdjnu.cn

Q. Lu
e-mail: 815303657@qq.com

D. Gong
School of Information and Electrical Engineering,
China University of Mining and Technology, Xuzhou, China
e-mail: dwgong@vip.163.com

X. Yao
School of Science, China University of Mining and Technology, Xuzhou, China
e-mail: yxjcumt@126.com

© Springer Nature Singapore Pte Ltd. 2018
Y. Jia et al. (eds.), *Proceedings of 2017 Chinese Intelligent Systems Conference*,
Lecture Notes in Electrical Engineering 460,
https://doi.org/10.1007/978-981-10-6499-9_70

1 Introduction

Software testing is an important means to ensure the quality of software, and a costly phase in the lifecycle of software development. Generating test data covering given paths is an important issue in software testing, because path testing can alone detect almost 65% of errors in a program [1]. Heuristic search algorithms have been used to solve the above problem in recent years, among which GA are the most widely used [2]. Genetic algorithm (GA) was widely used in solving many hard optimization problems [3]. Generating test data covering a given target path using GA needs to convert the task of generating test data to an optimization problem, and solve the optimization problem using GA, which is a robust global optimization approach that operates according to the principles of evolution, such as inheritance, selection, crossover and mutation [4]. The input domain of a program has a close relationship with the efficiency of a population's evolution, further, the larger the input domain, the slower the population searches for test data meeting a given sufficiency criterion.

Korel proposed a method of reducing the input domain of a program when generating test data covering a target path [5]. Harman et al. investigated the relationship between the input domain of a program and the performance of search-based algorithms in generating test data [6]. Recently, we proposed a method of reducing the input domain of a grogram when generating test data covering a target path using a GA [7].

The above method, however, is only suitable for the case of one target path. Many real-world programs contain hundreds or thousands of target paths. A method of generating test data covering multiple paths using a GA incorporating with reducing the input domain of a program was presented based on the work of [7].

2 Basic Concepts

Denote the program under test as Q. Some basic concepts about Q are as follow. Control flow graph (CFG), node, edge, path, the correlation between nodes and the elements of the input variable, direct correlation, indirect correlation, non-correlative, divisible path and partitioning methods of divisible path about Q are referred in [7]. For convenience in this study, a new definition is given as follows.

Independent Sub-Path: for the divisible target paths, the isolated sub-path is only associated with some components according to the above method, so the sub-path is called independent sub-path.

The correlative relationship diagram of the path P_1 and the input variable are shown as Fig. 1c. It can be seen that the path P_1 is divisible. Two independent sub-paths, say P_1_1: "n_1, n_3" and P_2_1: "$n_3, n_5, n_6, n_5, n_6, n_5, n_7$", can be isolated

```
void sumday(int year, int month, int day)
{
    int n=1, month_day_sum=0, sum=0;
    int month_day[13]={0,31,28,31,30,31,30,31,31,30,31,30,31};
n₁  if (year%100!=0&&year%4==0||year%400==0)
n₂      month_day[2]=29;
n₃  if(day>month_day[month]||day<1||month>12||month<1)
n₄      printf("error");
    else
    {
n₅      while( n<month)
n₆      {month_day_sum+=month_day[n];
         n++;
        }
n₇      sum =month_day_sum+day;
         printf("%d", sum);
    }
}
```

(a) source code of *sumday*

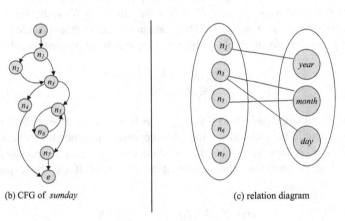

(b) CFG of *sumday*   (c) relation diagram

Fig. 1 Sumday program

from P_1, where $P_{1\_}1$: "n_1, n_3" and $P_{2\_}1$: "n_3, n_5, n_6, n_5, n_6, n_5, n_7" correspond to the element *year* and *month* as well as *day*, respectively.

Property of the independent Sub-path: if two target paths P_i and P_j have the same independent sub-path $P_{i\_}1$, suggesting that P_i and P_j are the same directions at all of branch nodes in $P_{i\_}1$, then the values of the elements corresponding to $P_{i\_}1$ of test datum which traverses P_i can be same with those of test case which traverses P_j.

3   Mathematical Model of Generating Test Data for Many Paths Coverage

A standard assumption in the classical control theory is that the data transmission required by the control or state estimation algorithm can be performed with infinite precision. However, due to the growth in communication technology, it is becoming more common to employ digital limited capacity communication networks for exchange of information between system components.

Denote the program under test as Q, its input variable is $X = (x_1, x_2, \ldots, x_n)$, in which x_i is an element of X whose domain is D_{x_i}. So the search space of the input data of Q, D, can be denoted as $D = D_{x_1} \times D_{x_2} \times \cdots \times D_{x_n}$. In addition, denote the m target paths as $P_1, P_2, \ldots, P_m$. When we use GAs to generate the test data, A fitness function is needed to distinguish different value of X, we denote the fitness function corresponding to P_i as $f_i(X)$, it includes two parts, i.e. approach level and branch distance, whose definitions were given in [8]. The bigger the value of $f_i(X)$, the closer the traversed path by X and P_i. When the value of $f_i(X)$ is the maximum, i.e. 2, the traversed path by X is just P_i. Then the mathematical model of the problem of test data generating for multiple paths coverage can be expressed as follows:

$$\begin{aligned} \max \quad & (f_1(X), f_2(X), \ldots, f_m(X)) \\ \text{s.t.} \quad & X \in D \end{aligned} \tag{1}$$

We divide m target paths into several groups based on the independent sub-path, each target path in the same group includes the same independent sub-path. Denote the number of groups as l, then the problem that we study is transformed as l sub-optimization problems, the mathematical model of it can be expressed as follows:

$$\begin{aligned} & \max (f_{11}(X), f_{12}(X), \ldots, f_{1m_1}(X)) \\ & s.t.\, X \in D \\ & \max (f_{21}(X), f_{22}(X), \ldots, f_{2m_2}(X)) \\ & s.t.\, X \in D \\ & \vdots \\ & \max (f_{l1}(X), f_{l2}(X), \ldots, f_{lm_l}(X)) \\ & s.t.\, X \in D \end{aligned} \tag{2}$$

where m_i is the number of the target paths in i-th group, and $\sum\limits_{i=1}^{l} m_i = m$.

During the process of evolution, in each sub-population, the values of elements corresponding to the traversed independent sub-path of the individual are recorded. Then the same elements of other individuals are modified by it, and the range of crossover and mutation operations is reduced, which do not include the range of fixed elements. Without loss of generality, suppose that the $x_{iq_1} \ldots x_{iq_{e_i}}$ of the i-th sub-population are fixed, leading to the sub-population search for test data in a reduced space, then the mathematical model of the problem we studied can be expressed as follows:

$$
\max \left(f_{11}(X), f_{12}(X), \ldots, f_{1m_1}(X)\right)
$$
$$
s.t.\ x_{1q_1} = c_{11}, \ldots, x_{1q_{e_1}} = c_{1e_1}
$$
$$
X \in D
$$
$$
\max \left(f_{21}(X), f_{22}(X), \ldots, f_{2m_2}(X)\right)
$$
$$
s.t.\ x_{2q_1} = c_{21}, \ldots, x_{2q_{e_2}} = c_{2e_2} \tag{3}
$$
$$
X \in D
$$
$$
\vdots
$$
$$
\max \left(f_{l1}(X), f_{l2}(X), \ldots, f_{lm_l}(X)\right)
$$
$$
s.t.\ x_{lq_1} = c_{l1}, \ldots, x_{lq_{e_l}} = c_{le_l}
$$
$$
X \in D
$$

4 Evolutionary Generation of Test Data for Multiple Paths Coverage

In this section, we first give the method for dividing the multiple paths to several groups based on the independent sub-path, and then employ a multi-population genetic algorithm to generate test data for path coverage. Each sub-population evolves to generate test data for covering the target paths in one group.

4.1 Group Target Paths Based on Independent Sub-path

For the purpose of grouping the multiple paths, we may divide every path into several independent sub-paths, and then group the multiple paths based on the same dependent sub-path, but it will be more time-consuming.

In this paper, our approach is: first, choose a target path P_i which is divisible randomly, divide P_i into several independent sub-path, denote as $P_i\_1, P_i\_2, \ldots, P_i\_g_i$, and then choose each of the independent sub-paths of P_i as the base sub-path, if a target path have the same sub-path, then put it to the same group, delete the paths which has its group from $P_1, P_2, \ldots, P_m$, employ the same method to obtain other groups from the rest target paths, until any target path has its own group.

4.2 Computing Method of Individual's Fitness in Sub-population

For the sake of clearness, we still denote the coding of a decision variable as X, fitness function includes two parts which are approach level and branch distance, the fitness of X for the i-th target path can be described as follows:

$$f_i(X) = v(P(X), P_i) + \frac{1}{d(P(X), P_i) + 1} \tag{4}$$

where $p(X)$ is the traversing path by X, $d(P(X), P_i)$ and $v(P(X), P_i)$ represent the functions related to approach level and branch distance respectively, and the corresponding mathematical expression were given in [8]. First the individual's fitness for each target path which not been traversed are calculated, and then maximized as an individual's fitness, so the fitness of X can be described as follows:

$$f(X) = \vee_{i=1}^{nm(t)} f_i(X) \tag{5}$$

where "$\vee$" denotes maximum operation, $nm(t)$ denotes the number of remaining target paths.

4.3 Algorithm Description

First, assign the values of the parameters used in this algorithm, encode the statements of the program under test, determine the target paths, and instrument the program under test; then divide the target paths using the algorithm depicted in Sect. 4.1, the steps of each sub-population are as follows.

Step 1: Initialize a population;
Step 2: Decode an individual, execute the instrumented program;
Step 3: Judge whether there exists an individual that traverses one target path. If yes, record the individual and remove the target path from the set of target paths;
Step 4: Judge whether the termination criteria are met, if yes, go to Step 8;
Step 5: Judge whether there exists an individual that traverses public independent sub-path. If yes, record the elements in the individual, modify the same elements of other individuals with it. And change the range of crossover and mutation operations, not include the range of fixed element;
Step 6: Calculate the fitness of an individual according to Eq. (5);
Step 7: perform selection, crossover, and mutation operators to generate offspring, go to Step 2;
Step 8: Stop the evolution, decode the optimal solutions, and output test data.

5 Experiments and Analysis

To validate the effectiveness of our method, we apply it to a real-world programs (sumday). We choose 3 methods for comparison: our method, the method in [7], and the method in [9]. All methods use the same calculation of an individual's fitness, the same parameter settings, and adopt the same initial population. Especially, the number of sub-population and each sub-population' size of our method is same with that of method in [9], and the sum of all sub-populations' size in our method or method in [9] is equal to the population size of the method in [7].

The same parameter settings of all experiments are as follows: roulette wheel selection, one-point crossover and one-point mutation. The probabilities of crossover and mutation are 0.9 and 0.3, respectively. Let the range of inputs to be [0, 2048], the maximal number of generations to be 2000. First, we select 10 target paths shown as Table 1.

Randomly choose the first path and divide it, obtain two independent sub-paths: one is "n_1, n_2, n_3" which correspond to the element "$year$" of the input variable; the other one is $n_3, n_5, n_6, n_5, n_6, n_5, n_6, n_5, n_6, n_5, n_6, n_5, n_6, n_5, n_6, n_5, n_6, n_5, n_6, n_5, n_7$ which correspond to the elements "$month$" and "day" of the input variable. Then we first divide the target paths which have the sub-path "n_1, n_2, n_3" into a group, obtain the first group include the path 1, 3, 6, 9 and 10. We randomly choose another path from remaining path, we choose path 2 and divide it into two independent sub-paths, "n_1, n_3" and $n_3, n_5, n_6, n_5, n_6, n_5, n_6, n_5, n_6, n_5, n_6, n_5, n_7$. After that we divide the remaining paths based on independent sub-path "n_1, n_3". Then all of 10 target paths are divided into two groups and shown as Table 2.

Since target paths are divided into 2 groups, 2 sub-populations are employed to generate test data in our method and method in [9], and the each sub-population's size is 50. Whereas the method in [7], only one population is employed, so its population size is set to be 100. Table 3 lists the mean values of time consumption (unit: s) and the number of evaluations of different methods at 15 runs to generate test data for all target paths.

Table 1 Target paths for Sumday

Path	Sequence of nodes
1	$s, n_1, n_2, n_3, n_5, n_6, n_5, n_6, n_5, n_6, n_5, n_6, n_5, n_6, n_5, n_6, n_5, n_6, n_5, n_6, n_5, n_6, n_5, n_7, e$
2	$s, n_1, n_3, n_5, n_6, n_5, n_6, n_5, n_6, n_5, n_6, n_5, n_6, n_5, n_7, e$
3	$s, n_1, n_2, n_3, n_5, n_6, n_5, n_6, n_5, n_6, n_5, n_6, n_5, n_6, n_5, n_6, n_5, n_6, n_5, n_6, n_5, n_7, e$
4	$s, n_1, n_3, n_5, n_6, n_5, n_6, n_5, n_6, n_5, n_6, n_5, n_6, n_5, n_6, n_5, n_6, n_5, n_7, e$
5	$s, n_1, n_3, n_5, n_6, n_5, n_6, n_5, n_6, n_5, n_6, n_5, n_6, n_5, n_6, n_5, n_6, n_5, n_6, n_5, n_7, e$
6	$s, n_1, n_2, n_3, n_5, n_6, n_5, n_6, n_5, n_6, n_5, n_6, n_5, n_6, n_5, n_6, n_5, n_7, e$
7	$s, n_1, n_3, n_5, n_6, n_5, n_6, n_5, n_6, n_5, n_6, n_5, n_6, n_5, n_6, n_5, n_6, n_5, n_7, e$
8	$s, n_1, n_3, n_5, n_6, n_5, n_6, n_5, n_6, n_5, n_6, n_5, n_6, n_5, n_7, e$
9	$s, n_1, n_2, n_3, n_5, n_6, n_5, n_6, n_5, n_6, n_5, n_6, n_5, n_6, n_5, n_7, e$
10	$s, n_1, n_2, n_3, n_5, n_6, n_5, n_6, n_5, n_6, n_5, n_6, n_5, n_6, n_5, n_7, e$

Table 2 Groups of the target paths by our method

Group	Path	Sequence of nodes
"n_1, n_2, n_3"	1	$s, n_1, n_2, n_3, n_5, n_6, n_5, n_6, n_5, n_6, n_5, n_6, n_5, n_6, n_5, n_6, n_5,$ $n_6, n_5, n_6, n_5, n_6, n_5, n_7, e$
	3	$s, n_1, n_2, n_3, n_5, n_6, n_5, n_6, n_5, n_6, n_5, n_6, n_5, n_6, n_5, n_6,$ $n_5, n_6, n_5, n_6, n_5, n_7, e$
	6	$s, n_1, n_2, n_3, n_5, n_6, n_5, n_6, n_5, n_6, n_5, n_6, n_5, n_6, n_5, n_6,$ n_5, n_6, n_5, n_7, e
	9	$s, n_1, n_2, n_3, n_5, n_6, n_5, n_6, n_5, n_6, n_5, n_6, n_5, n_6, n_5, n_7, e$
	10	$s, n_1, n_2, n_3, n_5, n_6, n_5, n_6, n_5, n_6, n_5, n_6, n_5, n_6, n_5, n_6,$ n_5, n_7, e
"n_1, n_3"	2	$s, n_1, n_3, n_5, n_6, n_5, n_6, n_5, n_6, n_5, n_6, n_5, n_6, n_5, n_7, e$

Table 3 Empirical results of 10 target paths

	Our method		Method in [9]		Method in [7]	
	Time consumption (s)	The number of evaluations	Time consumption (s)	The number of evaluations	Time consumption (s)	The number of evaluations
Avg.	0.038	28596.7	0.095	70466.7	0.060	59853.3

It is obvious in Table 3 that the average value of time consumption of our method is 0.038 s, the average value of the number of evaluations of our method is 28596.7, whereas these two indexes of method in [9] are 0.095 s and 70466.7 respectively, which are more than 2 times than that of our method. Indicate that compared with method in [9], our method, based on search space reduction is more advantageous in generating test data for many target paths, although both of these two methods adopt multiple sub-populations to generate test data by grouping.

6 Conclusion

We present a method of evolutionary generation of test data for multiple divisible paths. The efficiency of generating test data for covering multiple paths is enhanced. However, target paths considering only are divisible paths, and only one independent sub-path based to divide the target paths. Therefore, our future work will focus on divide the target paths based on multiple independent sub-paths.

Acknowledgements This study was jointly funded by National Natural Science Foundation of China with grant No. 61573362, Natural Science Foundation of Heilongjiang Province with grant No. F2016039 and Project of Mudanjiang Normal University No. GP201602.

References

1. Srivastaval PR, Kim T. Application of genetic algorithm in software testing. Int J Softw Eng Appl. 2009;3(4):87–95.
2. Shan JH, Wang J, Qi ZC. Survey on path-wise automatic generation of test data. Acta Electronica Sinica. 2004;32(1):109–13.
3. Holland JH. Adaptation in natural and artificial systems. Michigan: The University of Michigan; 1975.
4. Miller J, Reformat M, Zhang H. Automatic test data generation using genetic algorithm and program dependence graphs. Inf Softw Technol. 2006;48(7):586–605.
5. Korel B. Automated software test data generation. IEEE Trans Softw Eng. 1990;16(8):870–9.
6. Harman M, Hassoun Y, Lakhotia K, McMinn P, Wegener J. The impact of input domain reduction on search-based test data generation. In: Proceedings of the 6th joint meeting of the European software engineering conference and the ACM SIGSOFT symposium on the foundations of software engineering, New York, NY, USA; 2007. p. 155–64.
7. Zhang Y, Gong DW. Evolutionary generation of test data for path coverage based on automatic reduction of search space. Chin J Comput. 2012;40(5):1011–6.
8. Watkins A, Hufnagel EM. Evolutionary test data generation: a comparison of fitness functions. Softw Pract Exp. 2006;36:95–116.
9. Zhang WQ, Gong DW, Yao XJ, Zhang Y. Evolutionary generation of test data for many paths coverage. In Proceedings of 2010 Chinese control and decision conference; 2010. p. 230–35.

Reasoning for Qualitative Path Without Initial Position in VAR-Space

Xiaodong Wang , Ming Li and Shizhong Liao

Abstract Qualitative path is a basic concept in qualitative motion reasoning. In this paper, we propose an algorithm for reasoning the qualitative path in Voronoi-Adjacency-Relation Space. Different from the previous approach in which both the dynamic Voronoi edges and the initial position are required, our approach only needs the dynamic Voronoi edges. The basic idea is that the algorithm tracks all possible initial positions, gradually rules out the impossible situation and ultimately gets the correct path. Finally, experiments have been conducted and results show that the proposed algorithm is effective.

Keywords Qualitative path reasoning · Incomplete information · Voronoi diagram

1 Introduction

Qualitative approach suited to express qualitative temporal and spatial relationships between entities has gained wide acceptance as a useful way of abstracting from the real world. And the qualitative approach has the advantages of the ability to deal with uncertain or incomplete data and the high computational efficiency, so it receives the increased attention and is widely used in many fields, including Geographic Information System (GIS), high level vision, spatial propositional semantics of natural languages, and engineering design, etc. [1]. In the past few years, the attention has been extended to research on the movement of entities.

In order to apply the qualitative method to research on the movement of entities, a method is required for representing motion qualitatively, and then, the reasoning tasks can be achieved under this given motion representation. Due to the face that the movement of spatial entities may lead to changes in its spatial location, or changes in the relationship between entities, in general, the existing motion representation

X. Wang(✉) · M. Li
Mudanjiang Normal University, Mudanjiang 157012, People's Republic of China
e-mail: xdwang.mdj@gmail.com

S. Liao
Tianjin University, Tianjin 300072, People's Republic of China

© Springer Nature Singapore Pte Ltd. 2018
Y. Jia et al. (eds.), *Proceedings of 2017 Chinese Intelligent Systems Conference*,
Lecture Notes in Electrical Engineering 460,
https://doi.org/10.1007/978-981-10-6499-9_71

methods can be divided into two categories: absolute methods [2–4] and relative ones [5–7]. The former methods (the absolute representation methods) first give the qualitative spatial representation, on which, the motion can be represented qualitatively, however, the latter methods (the relative ones) first specify the qualitative relationships of (and between) spatial entities, including topology, orientation, distance, and so on, then, the motion can be represented with the changes in the spatial relationships.

So far, a variety of qualitative spatial representation methods have been proposed, which can be used to the representation of motion. Among them, the method based on the decomposition of space proposed by Forbus, which is abstracted from the quantitative details of spatial knowledge and preserves the essential properties of spatial knowledge, has been widely adopted and regarded as a seminal work in this area. Due to the advantages of decomposition of space, based on Voronoi diagram, the Voronoi-Adjacency-Relation-Space (VAR-Space) is proposed [8], in which the motion path can be described and reasoned qualitatively [9]. But the proposed method can not handle the incomplete information concerning moving objects. This paper analyzes the spatial structure and partly solves the problem.

The remainder of this paper is organized as follows. Section 2 introduces the basic concepts of Voronoi-Adjacency-Relation Space. Section 3 designs and implements an algorithm for reasoning motion path without the initial position information. Section 4 conducts the experiments and discusses the results. Finally, conclusions are reported in Sect. 5.

2 Basin Idea About VAR-Space

Due to the reasoning mechanisms depending on the representation formalism, so, the basic task in reasoning is to find an adequate formalism for representing the spatial knowledge. In qualitative spatial reasoning, generally, the continuous spatial information can be discretized by the landmarks, which separates the neighbour open intervals, resulting in discrete quantity spaces.

Since adjacency relation is an important spatial relationship, based on the adjacency in Voronoi diagram, a qualitative spatial representation is proposed [8], and its basic idea is as follows. As shown in Fig. 1a, in Voronoi diagram of a set of points in the plane, according to whether existing the common Voronoi edge in two Voronoi regions where the generators are, the adjacency relations between generators can be classified into three types: adjacent, co-circle and dis-adjacent, denoted by +, 0, and −, respectively. With generator p_1 moving upwardly, the adjacency relation between generators p_1 and p_3 changes, from dis-adjacent in Fig. 1a, to co-circle in Fig. 1b and to adjacent in Fig. 1c. In where, the co-circle is the landmark, separating the open intervals: adjacent and dis-adjacent.

In the Fig. 2a, given a set S of $n + 1$ generators in the plane. Where, generator p is movable and the others are stationary. If the generator p moves only within the hatched area (not including border), then the relation between the movable generator

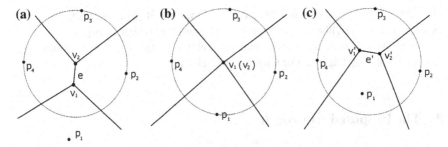

Fig. 1 Adjacency relation change in Voronoi diagram

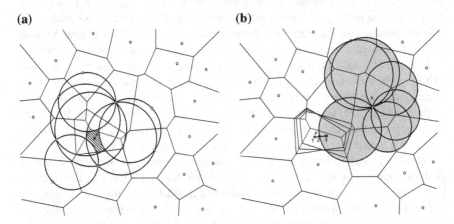

Fig. 2 Analysis of Voronoi spatial relationship. **a** p moving within the hatched area not causing the changes in Voronoi spatial relations. **b** p moving causing the changes in Voronoi spatial relations

p and the stationary generator are not change, but if p moves to the border (and outside) of the hatched area, as shown in Fig. 2b, then the relation between p and stationary generator s_i will change from dis-adjacent to co-circle (and from co-circle to adjacent). So, in a plane space with the n fixed generators, the adjacency relations (in Voronoi diagram) with these n generators can be used as an indicator, and all points in the plane, in which the generator p is, the adjacency relations between p and these n generators do not change, form a set, called a qualitative position. The collection of qualitative positions is a qualitative representation for plane.

Under the definitions in [10], the conception neighbourhood between qualitative positions can be defined as follows: two qualitative positions are (conceptual) neighbours, if they can transit into one another by continuous shift without passing another qualitative position. The qualitative positions and the transitions between them form the conceptual neighbourhoods. In the conceptual neighborhood, given a qualitative position [p], if qualitative position [q] is directly connected with [p], then we call [q] 1-order neighbor of [p], and if [q] is connected with [p] by the first-order neighbor of [p], then call [q] 2-order neighbor of $[p]_i$. Similarly, if [q] is connected with [p] by the n-1 order neighbor of [p], then call [q] n-order neighbor of $[p]_i$.

The above description is an informal account of qualitative spatial representation based on Voronoi diagram. For a formal, we refer to [8].

3 The Proposed Approach

The algorithm for qualitative path reasoning in [8] requires two conditions: (1) the initial qualitative position and (2) the dynamic Voronoi edge sequence. However, in some cases, the initial position can not be determined. In this section, we analyze the relations between the qualitative positions in transitions incident to dynamic Voronoi edges, and presents a solution for reasoning qualitative path without initial position.

In plane R^2, let S be a set of static generators, p be a moving generator, and $VD(S \cup p)$ be a Voronoi diagram. With the movement of point p, Voronoi diagram is changing, and even the edge in Voronoi diagram can degenerate into one point (shown in Fig. 1a, b) and the new edge can also generate (shown in Fig. 1b, c), i.e., the adjacency relations between generators in Voronoi diagram have changed. If this happens, it means that the qualitative position, in which point p is, changed.

Note that, many different movement of point p between the adjacent qualitative positions may cause the same change in Voronoi diagram edge, i.e., the relations between the movement of p within the adjacent qualitative positions and change in Voronoi edge, is not one-to-one, but many-to-one. As shown in Fig. 3a, only the generator 27 is movable, the others are stationary, and the circles are the circumcircles of Voronoi diagram of the stationary generators. The Voronoi vertices v_1, v_2 are the endpoints of the edge $e(3, 5)$ of Voronoi diagram of the static points, the circles c_1,

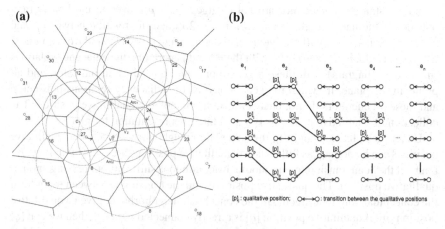

Fig. 3 Qualitative path reasoning. **a** The relationship between the edge $e(3|5)$ degenerating and the transition of neighboring qualitative positions. **b** Reasoning for qualitative path

c_2 are circumcircles incident to the vertices v_1, v_2 respectively, and arc_1, arc_2 are the arcs of the circles c_1, c_2.

It can be seen that if the generator 27 moves from the current position to the arc arc_2 of circle c_2 along the direction of the arrow, then Voronoi edge between generate 3 and generate 5 (denoted as $e(3,5)$) will generates into point, and if the generator 27 continues to move right into the circle c_2, then the new Voronoi edge $e(27,24)$ will appears. In fact, beside the above mentioned, all movement from a qualitative position adjacent to the arc arc_2 to arc arc_2 can cause the edge $e(3,5)$ to disappear. Similarly, the movement from outside to arc arc_1 is the same effect. So, as shown in Fig. 3b, each disappearance or appearance of Voronoi edge corresponds to multiple transition between adjacent qualitative positions. The Algorithm 1 can be used to identify all the transition between the two positions incident to the dynamic Voronoi Edge. Once the transitions between the neighboring qualitative positions incident to a specific Voronoi edge are identified in Voronoi diagram $VD(S \cup p)$, the initial position and the following positions, the moving p passes through, can also be reasoned by the sequence of dynamic Voronoi edges. The basic process is as follows.

Algorithm 1: Identifying All Transitions between the Qualitative Positions Incident to Voronoi Edge $e(s_i, s_j)$

Data: $VD(S)$, $e(s_i, s_j)$ of $VD(S \cup p)$, IdxTab(s_i), CNDs of $[POS]_{VD(S)}$.
Result: $T(.,.)$.
begin
 search $VD(S)$ by $e(s_1, s_3)$ for v_1, v_2, s_2, s_4
 initialize $[Pos]_{c_1}$, $[Pos]_{c_2}$, $[Pos]_{arc_1}$, $[Pos]_{arc_2}$
 forall $[pos]_i \in [Pos]_{arc_2}$ **do**
  find out $[pos]_t \in ([Pos]_{c_1} - [Pos]_{c_2})$ and next to $[pos]_i$
  $T(.,.) \leftarrow ([pos]_t, [pos]_i)$
 end
 forall $[pos]_i \in [Pos]_{arc_1}$ **do**
  find out $[pos]_t \in ([Pos]_{c_2} - [Pos]_{c_1})$ and next to $[pos]_i$
  $T(.,.) \leftarrow ([pos]_t, [pos]_i)$
 end
end

As shown in Fig. 3b, the set $\{e_1, e_2, \ldots, e_n\}$ is the dynamic Voronoi edge, which is caused by the motion of generator, and for every e_i, there may exist multiple transitions $T_{e_1} = \{t_1, t_2, \ldots, t_n\}$ between the qualitative positions that can cause its degeneration(or generation). Since the motion is continuous, in two successive transitions that moving generator pass through, the end position of the previous transition and the start position of the latter transition should be the same one, and thus the two successive transitions can be connected. Due to the start qualitative position being unknown, the successive connecting transitions between the qualitative positions with above rules, maybe produce multiple qualitative paths, but the path, which passes through the most transitions, should be the most probable path. The formal description of the algorithm for qualitative path is shown in Algorithm 2.

Algorithm 2: Reasoning Path with the Dynamic Voronoi Edges

Data: $\{e_1, e_2, \ldots, e_n\}$
Result: [Path]
begin
    identify transition $T = \{t_1, t_2, \ldots, t_m\}$ incident to e_1
    for $t_i \in T$ **do**   `/* 1 ≤ i ≤ m */`
        establish a list L_i
        add the start qualitative position and end one of transition t_i into L_i
    end
    for e_i **do**   `/* 2 ≤ i ≤ n */`
        identify transition $T = \{t_1, t_2, \ldots, t_m\}$ incident to e_i
        for $t_j \in T$ **do**   `/* 1 ≤ j ≤ m */`
            [s] ← start position of t_j, [e] ← end position of t_j
            forall L_k **do**   `/* 1 ≤ qk ≤ m */`
                if [s] *in the tail of* L_k **then**
                    add [e] into L_k
                end
            end
        end
    end
    [Path] ← select the longest list in $\{L_1, L_2, \ldots, L_m\}$
end

---

4 Experiments and Discussion

In this section, we conduct experiments to verify the effectiveness of the method proposed in this paper, and discuss the experimental results. First, we randomly input a set of stationary points $S = \{s_1, s_2, \ldots, s_{60}\}$ in plane R^2 (shown in Fig. 4a), compute Voronoi diagram $VD(S)$ of S, and can obtain the Voronoi-Adjacent-Relations Space $[\mathcal{R}]$ (shown in Fig. 4b). Then, we input a movable point s_{61}, let it move along the curve (shown in Fig. 4c), and record the changes in Voronoi edges in Table 1. Finally, we use the algorithms proposed in this paper to reason for path without the initial position, and the result of the experiment is shown in Fig. 4d.

It can be seen that although the initial position being unknown, the Algorithm 2 can reason out the path through which the movable point passes. However, we also find that when the moving distance is too short, the effect of the algorithm is not good enough and the reason is that the information about reasoning path is not enough.

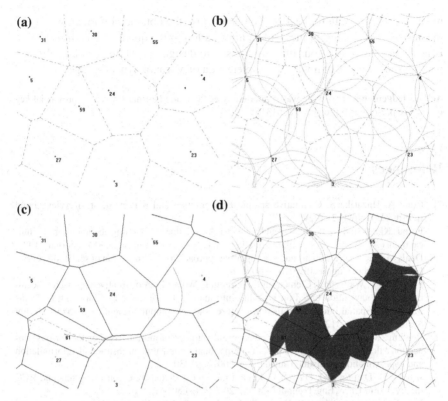

Fig. 4 Reasoning for the qualitative path of movable point. **a** stationary points $S = \{s_1, s_2, \ldots, s_{60}\}$. **b** Voronoi-Adjacent-Relations Space determined by $VD(S)$. **c** point s_{61} moving along the curve. **d** result of reasoning

Table 1 Dynamic Voronoi edges. $e^-(g_i, g_j)$: degeneration of the edge between the generators g_i and g_j, $e^+(g_i, g_j)$: generation of the edge between the generators g_i and g_j

$e^-(3, 59)$	$e^+(61, 24)$	$e^-(61, 28)$	$e^+(5, 27)$	$e^-(61, 5)$	$e^+(27, 59)$	$e^-(3, 24)$
$e^+(61, 23)$	$e^-(3, 23)$	$e^+(61, 53)$	$e^-(23, 24)$	$e^+(4, 61)$	$e^-(27, 61)$	$e^+(3, 59)$
$e^-(4, 24)$	$e^+(6, 55)$	$e^-(61, 59)$	$e^+(3, 24)$	$e^-(23, 4)$	$e^+(61, 58)$	$e^-(61, 53)$
$e^+(3, 23)$	$e^-(3, 61)$	$e^+(23, 24)$	$e^-(61, 58)$	$e^+(23, 4)$	$e^-(4, 55)$	$e^+(61, 17)$

5 Summary and Future Work

In this paper, we give an algorithm for reasoning qualitative path. Different from the previous approach, our approach just needs the dynamic Voronoi edges caused by the generator motion, and does not need the initial position of movable generator. That is to say, our approach can deals with the problem of qualitative path reasoning in the absence of initial location information. The experimental results show this approach method is effective. In the real environment, since the obtained information may be incomplete, processing of incomplete information in the qualitative path reasoning is necessary. So, one of our further works is to design and implement a qualitative path reasoning algorithm with incomplete data of Voronoi dynamic edge.

Acknowledgements This work was supported by Mudanjiang Normal University provincial key innovative pre-research project (SY201315).

References

1. Cohn A, Hazarika S. Qualitative spatial representation and reasoning: an overview. Fund Inform. 2001;46(1–2):1–29.
2. Forbus KD. A study of qualitative and geometric knowledge in reasoning about motion. Technical report AITR-615, MIT Artificial Intelligence Laboratory, Cambridge, MA, February 1981.
3. Denis M. The description of routes: a cognitive approach to the production of spatial discourse. Current Psychol Cognit. 1997;16(4):409–58.
4. Fogliaroni P, Wallgrün J, Clementini E, Tarquini F, Wolter D. A qualitative approach to localization and navigation based on visibility information. In: Hornsby K, Claramunt C, Denis M, editors. Spatial information theory. Lecture notes in computer science, vol. 5756. Berlin: Springer; 2009. p. 312–29.
5. Ibrahim ZM, Tawfik AY. An abstract theory and ontology of motion based on the regions connection calculus. Proceedings of the 7th international symposium on abstraction, reformulation and approximation, Whistler, Canada, July 2007. pp. 230–42.
6. Ji-hong O, Dan-tong O, Da-you L. Region movement model based on fuzzy sets and RCC theory. J Jilin Univ (Eng Technol Edition). 2007;37(3):591–4.
7. Van de Weghe N, Cohn AG, De Tre G, De Maeyer P. A qualitative trajectory calculus as a basis for representing moving objects in geographical information systems. Control Cybern. 2006;35(1):97–119.
8. Wang X, Liao S. Heuristic shortest path algorithm in qualitative spatial representation. J Comput Inf Syst. 2014.
9. Xiao-Dong W, Shi-Zhong L. Qualitative path reasoning based on Voronoi diagram. Pattern Recognit Artif Intell. 2013;26(5):417–24.
10. Freksa C. Temporal reasoning based on semi-intervals. Artif Intell. 1992;54(1–2):199–227.

Extreme Learning Machine Based Location-Aware Activity Recognition

Zhihao Xie, Jie Zhang, Wendong Xiao, Changguo Sun and Yifang Lu

Abstract According to the recent government reports, China has gradually entered an aging society. Pension problem is a vital problem to face. Therefore, it will be very useful to monitor the health status of elderly people who live alone at home. To evaluate the abilities of elderly people in daily life, the activities of daily living (ADL) is used. In this work, we propose a novel machine learning approach for ADL recognitions by considering the location context information of the elder. With the popularity of smart phones, motion recognition can be done by the embedded sensors such as acceleration sensors and. However different ADL models possibly have the same movement to a certain degree, which will affect the classification performance. We append the location information as an additional feature to detect ADL. Furthermore, we propose a hierarchical Extreme Learning Machine (ELM) to classify the ADL. With the experiment and test, the algorithm described in this paper can achieve obvious performance in ADL recognition.

Keywords ADL · Recognition · Smartphone · Location · ELM

Z. Xie (✉) · J. Zhang · W. Xiao · C. Sun · Y. Lu
School of Automation and Electrical Engineering,
University of Science and Technology Beijing, Beijing, China
e-mail: kula147@163.com

J. Zhang
e-mail: zhangjie2009622@163.com

W. Xiao
e-mail: wdxiao@ustb.edu.cn

C. Sun
e-mail: sunchangguo@sina.com

Y. Lu
e-mail: luyifang@ustb.edu.cn

© Springer Nature Singapore Pte Ltd. 2018
Y. Jia et al. (eds.), *Proceedings of 2017 Chinese Intelligent Systems Conference*,
Lecture Notes in Electrical Engineering 460,
https://doi.org/10.1007/978-981-10-6499-9_72

1 Introduction

In recent years, the health consciousness of people improves remarkably. China has been in an aging society since 2000. According to the sixth census results released by the government, the population aged 60 and above accounted for 13.26%, 65 yrs and older population accounts for 8.87%. The UN stipulates that countries belong to the aging countries or regions whose proportion of the elderly over the age of 65 has a total population of more than 7% or 60 with a total population of more than 10%. Thus, the problem of aging population is more and more serious which will cause increasing burden of population proportion. Burdens of an aging population on the one hand, make our country' social support rate increase; On the other hand, also bring great opportunity and challenge to the development of pension services. The concept of healthy aging is increasingly attracting the attention of the international community. If a person can still maintain a good state in the function of physical, psychological, and intellectual in the old age, the Pension problem is easy to solve.

Activities of daily living (ADL) [1] is an important index to evaluate the state of the elderly health. The monitoring of ADL is a significant issue, because it can provide clinical data for the doctor and provide the patient with health management. In the previous work, the detection of ADL is conducted by questionnaires for patients and their family members, that exists many inconveniences: the massive manpower and time, poor real-time performance, the influence of the patient's mood and so on. But with the development of modern technology, we can obtain the ADL by computer calculation, which can economize numerous costs. How to get the preciseness and low cost of ADL identification will be an area where researchers are very interested in.

A great number of researches [2–4] have been proposed in the field of pattern recognition. In [5], to quantify criteria such as ADL or AGGIR, the author proposed a monitoring model using a Health Smart Home for recognize the ADL of the subject which included infrared presence sensors (for location), door contacts (to control the use of facilities), temperature and hygrometry sensor, and microphones for sound classification. A wearable sensor which provided the motion data for pattern recognition was used to collected the data of activities. The classification of ADL is done by support vector machine, then the classification algorithm was tested by cross validation with real data.

In [6], J. Neuhaeuser and D. Proebstl proposed Eventlogger, a radio module system to record the behavior of an elderly. The instrument consists of a 2.4 GHz radio module and is controlled by a microcontroller. The system Eventlogger is different from most of the system in that, instead of evaluating in laboratories, it has already been tested in a real hospital. Then this paper discussed the results of the system. In [7], Xi Liu and Lei Liu introduce am system for human daily activity recognition using wearable and visual sensing data. The system learns a nonlinear SVM to identify 20 different activities by accelerometer and RGB-D camera data. At last the approach is proved effectively.

In the present paper, considering the user's privacy [8], we don't use camera [9] to detect ADL in the smart home [10], we develop a system based smartphone and the UWB [11] localization module to identify different ADL [12]. We don't use the wearable sensors because the monitoring objects are usually elder or disabled people, who are maybe reluctant to use equipment. Smart mobile embedded sensors as one of the necessaries of life can provide data for us. The ELM algorithm is used to recognize the low-level activities performed by object (walk, stand, et al.). Besides in order to simplify classification learning process, we divide the ADL into different groups according to the location. Then, each ADL can be labeled by two aspects: the room where the person stays and the order of the low-level activities.

2 System Description

A. The composition of the system

- A flat experimental house with several rooms: Bedroom, Living room, Kitchen, Bathroom, WC. These houses are the basic which can satisfy the needs of a person in daily life. For instance, one can cook a meal only in the kitchen, take a bath in the bathroom, excrete in the WC, and sleep in the bedroom.
- Smart Phone with sensors. We employ the three-axis accelerometer and gravity sensor embedded in the phone to recognize the patterns of activities with machine learning algorithm. The raw data packages collected by the phone can be transferred to the PC for data processing via Wi-Fi or GPRS (Fig. 1).
- Location. In order to obtain location information, we use the Ultra Wide-Band (UWB) [13] chips for indoor localization by time of arrival (TOA) [14]. TOA method measures the range through the signal propagation time. Because of the characteristics of UWB, we can achieve centimeter level accuracy in indoor localization.
- We propose a multi-layer ELM model to recognize ADL. Firstly, we take the data after feature extraction collected by the mobile phone sensor as input, and then it outputs low-level activities. Then, we consider location information as constraints, on the basis of the above, take the outputs of the last layer as the new inputs.

B. Data acquisition

In this paper, we mainly have two data sources namely motion sensors (accelerometer and gyroscope) and UWB radiofrequency integrated circuits for human activity recognition. The first one is widely embedded in smart phone, at the meantime the UWB device is easy to obtain in the market.

- Mobile sensors raw data. The smart phone is collecting data from three-axial acceleration and gyroscope sensors. Each smart phone that has an embedder accelerometer can be used for detecting the person's movements. In order to

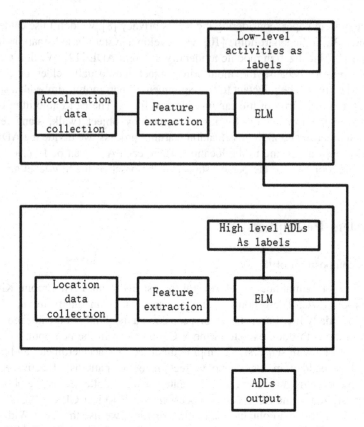

Fig. 1 Work flow diagram

achieve high accuracy, each experimenter placed the phone in the right trouser pocket.

- UWB distance information. We apply the Irish Decawave DWM1000 chips to achieve UWB ranging function. Our localization system is composed of anchors and label (both DWM1000 chips). The label corresponds with the anchors at a fixed place for distance information through TOA. Then we establish a coordinate system and map the distance information into the coordinate system so that we can get the localization information.

C. Data preprocessing

In the process of acceleration data acquisition, due to the sensor itself and communication interference, the acceleration signal will have some noise. As we can see in the figure below, the signal is noisy and unfiltered: (Fig. 2)

The acceleration value of the three-axis collected by the sensor has a certain degree of noise interference, which needs to be preprocessed to reduce the influence

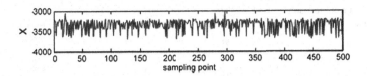

Fig. 2 Unfiltered signal plot of x-axis accelerometer data

of noise. Selecting the appropriate denoise method has certain benefits to the information processing of the whole system.

Due to the large amount of data collected by the system, the multiple motion states contained in the signal can be mistaken for one state. We use the method of the window, namely the data segmentation, to reduce the computation time. The size of window should be moderate. Therefore, the window data sampling points are selected as 150. The frequency of human motion is generally between 0.5 and 10 Hz [15], so it is necessary to filter the DC signal and the high frequency signal. In the information sequence, it can be considered that the noise is higher than that of 10 Hz, and the most widely used first order low-pass filter is selected to filter out the noise interference which is higher than 10 Hz. For the preprocessed signal, it can be seen that the pretreated acceleration signal can effectively filter out the noise interference, and the signal is more stable [16] (Fig. 3).

Besides, in the process of UWB signal propagation, there will be a lot of factors that affect the transmission of radio waves, such as multi-path effect factor, NLOS factor and so on. Because the external environment noise and noise inside the device on the signal transmission will cause interference, thus affecting the accuracy of the acquired signal, so the filtering of the signal is very necessary.

We select the widely used Kalman filter, because the Kalman filter can not only estimate the past and the current state of the signal, but also can predict the future signal state. After filtering, the error of signal can be obviously reduced (Fig. 4).

D. Feature extraction

According to previous experience, our machine learning classification algorithm cannot directly use the raw data, we need to transform the data into datasets. The features we extract are as follows:

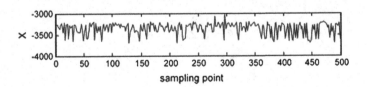

Fig. 3 X-axis accelerometer data after filtering

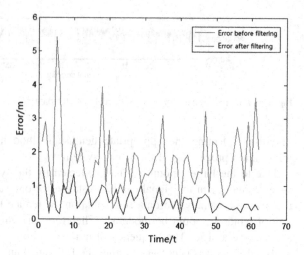

Fig. 4 Error comparison before and after filtering

- The mean of the triaxial acceleration

$$\bar{x} = \frac{x_1 + x_2 + \cdots + x_1}{n} \tag{1}$$

- Standard deviation

$$\sigma = \sqrt{\frac{1}{N} \sum_{i=1}^{N} (x_1 - \mu)^2} \tag{2}$$

- Variance

$$\sigma = \frac{1}{N} \sum_{i=1}^{N} (x_1 - \mu)^2 \tag{3}$$

- SMA

SMA (Signal Magnitude Area) indicates the extent of the acceleration signal, the greater SMA value, the more intense the movement. In Fig. 5, it shows the signal

Fig. 5 SMA values in walking and still states

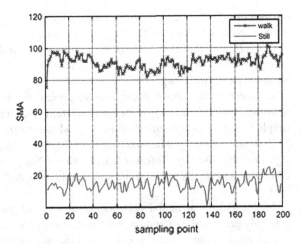

amplitude area values in still and walking. It can be seen that walking is more powerful than still.

$$SMA = \sum_{t=1}^{N} (|x(i)| + |y(i)| + |z(i)|) \qquad (4)$$

N is the number of sample points for the calculation window, $x(i)$, $y(i)$ and $z(i)$ are the acceleration value of the three-axis under the time i.

E. Classification

According to the feature extraction methods described above, we calculate characteristic values of the sensor in order to design and implement the classifier. At present, there are many kinds of classifiers used in human motion pattern recognition, such as regression classifier, Extreme learning machine (ELM) classifier, Support Vector Machine classifier and so on.

The SVM algorithm was originally designed for binary classification problems. When dealing with multi-class classification problems, there are two kinds of methods to construct SVM multi-class classifier. The first one is the direct method, which is modified directly on the objective function, and the parameters of multiple classification surfaces are combined into an optimization problem. This method appears to be simple, but its computational complexity is relatively high, therefore it is only suitable for small problems. The other one is the indirect method, mainly through the combination of a number of two classifiers to achieve the construction of multiple classifiers. The common methods are one-versus-one and one-versus-all.

One-versus-rest (OVR SVMs) [17] is the method that we need to train N binary classifiers that learned to distinguish one class from all the others. There exists an obvious drawback: the unbalanced training datasets may influence the performance of the classifier because of the existing deviation.

One-versus-one (OVO SVMs). This approach is to design a SVM between any two types of samples, so N categories of samples need to design n(n−1)/2 SVMs. So the price is quite large.

Considering the above situations, we choose the ELM classifier to detect the activities, which is known to be fast and fairly accurate. The ELM algorithm is simple and easy to implement and has good generalization performance.

Extreme Learning Machine (ELM) [18] is a simple and effective learning algorithm proposed by Professor G.B. Huang of Nanyang Technology University in 2004. It based on the study of the single hidden layer feed forward neural network [19].

In the traditional neural network, parameters of each layer network model is generally generated by iteration, but the iterative process often occupied most of the time in training process, and is easy to fall into the local optimal solution, so that the learning efficiency is low. ELM is a highly efficient single hidden layer neural network model, whose hidden layer node parameters (input weights and input bias of hidden nodes) were randomly selected and updated in the training process. Without iteration, we can calculate the optimal solution. The output weights of the hidden layer nodes are obtained by minimizing the average loss function, and then a ELM model can be obtained by combining the input weights of the hidden layer, the input bias and the output weights.

The common ELM classification steps:

- Determine the number of hidden layer nodes.
- Randomly set the input weights of each hidden layer node w and input bias b.
- Preprocessing the data to be learned.
- Divided the samples into learning samples and testing samples according to the appropriate proportion.
- Choose a suitable activation function.
- Calculate the output weight of hidden layer according to the hidden layer output and sample output.
- Classify the learning samples and test samples by using the model of learning.
- Select the appropriate model, classify the unlabeled data.

F. ADL recognition

There are two main steps in ADL recognition. The first step is to identify the basic actions (Table 1) as the low-level activities through the smart phone sensor.

After the study we found that each ADL has the regularity of location. For instance, one sleeps in the bedroom and excretes in the WC. Besides, different ADL models possibly have the same movement to some degree, which will affect the classification performance. Therefore, we divide the ADL recognition into different groups according to the room.

The second step is obtaining which the subject is staying on the basis of the UWB location information. In Table 2 we can see each ADL include a form of

Table 1 Basic actions

ID	Actions
1	Walking
2	Still
3	Walk to stand
4	Stand to sit
5	Sit to lie
6	Lie to sit
7	Sit to stand
8	Stand to walk

Table 2 ADL classification with location

Room	Activity	Action included
Bedroom	Sleep	$2^1, 4^1, 5^1, 6^1, 7^1$
Living room	Watch TV	$2^{>2}, 4^1$
Kitchen	Cook	$2^{>2}$
	Having dinner	$2^{>2}, 4^1$
Bathroom	Hygiene	$1^1, 2^1, 3^1, 8^1$
WC	Excrete	$1^1, 2^2, 3^1, 4^1, 7^1, 8^1$

Fig. 6 Single-hidden layer feed forward network

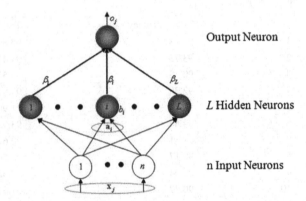

actions. The sleep and the having dinner have the same pattern. But if we add the location, it will be easy to classify. The superscript represents the time which every action occur in every 5 min. Then we take the order of the low-level activities and the location as the additional attributes of the classifier (Fig. 6).

3   Experimentation and Results

We performed experiments on 5 healthy volunteers (3 men and 2 women) for a week to build the datasets, whose average age was 23 (range 21–25). Then in order to test the performance of classifier, we implemented our ELM algorithm on the dataset which we collected by the sensors. On the selection of parameters, we chose the classification at Elm type (0 for regression, 1 for classification), the number of hidden neurons assigned to the ELM we set was 80, and the type of activation function was Sigmoidal function. To verify the stability and generalization ability of the model, we used the 10-fold cross validation to test the prediction ability of the classifier. The results of cross validation are given in Table 3.

Next, we evaluated the accuracy of ELM classification through the confusion matrix. The complete results are showed in Table 4.

What we can see from the confusion matrix is that, for example, the error for cook was mainly distributed in having dinner and excrete. The reason why the classes cook and having dinner were close and the probability to misclassify them was higher than others was that, the class cook and class having dinner were in the same room, and the low-level activities were too simple to classify.

Finally, in order to indicate the superiority of our proposed algorithm, we compared our algorithm with the common ELM for classification without Additional location information as features. Table 5 shows the performance metrics of the common ELM and the ELM algorithm we proposed classifiers. It is obvious that, the ELM algorithm we proposed performs better than the common ELM.

Table 3 Cross validation

	Train accuracy (%)	Test accuracy (%)
1	91.08	95.89
2	93.78	94.41
3	90.48	93.53
4	92.38	94.81
5	90.57	91.96
6	94.54	93.95
7	88.51	90.67
8	90.17	93.56
9	92.64	92.62
10	90.97	91.57

Table 4 Confusion matrix of the ELM algorithm

	Sleep (%)	Watch TV (%)	Cook (%)	Having dinner (%)	Hygiene (%)	Excrete (%)
Sleep	95.6	2.1	0.0	1.0	0.0	1.1
Watch TV	7.8	89.6	0.0	1.4	0.0	1.2
Cook	0.0	0.0	78.3	12.3	3.1	6.3
Having dinner	3.4	5.6	0.0	82.7	4.3	4.0
Hygiene	0.6	1.2	10.1	0.3	67.6	20.2
Excrete	0.0	0.0	0.0	1.2	6.7	92.1

Table 5 Cross validation

	Train accuracy (%)	Test accuracy (%)
Common ELM	87.63	85.63
Proposed ELM	95.75	93.33

4 Conclusion

In this study, we developed a new method for the recognition of ADL. The datasets we used to train and test came from the devices that we established through the smart phone sensors and the UWB wireless localization modules. The smart phone sensors formed the low-level activities pattern recognition, and the UWB modules provided the location information of the upper level for our proposed algorithm. The algorithm is based on ELM algorithm, which added the location as an additional feature on the basic of the fundamental motions. Finally, we applied our proposed algorithm on our collected datasets to test the performance of this work. Under 10-fold cross validation, we can consider this method as stability and strong generalization. And with comparison with ordinary ELM, the accuracy has been greatly improved. We believe that our work will help the elderly and the disabled to recognize who are not willing to carry a lot of wearable sensors.

In the future work, we plan to add time factors in the ADL recognition, that is according to the law of daily life for individual, we establish a model to help them conduct health monitoring and health management.

References

1. Roy N, Misra A, Cook D. Infrastructure-assisted smartphone-based ADL recognition in multi-inhabitant smart environments. In: 2013 IEEE international conference on pervasive computing and communications (PerCom), San Diego, CA, 2013, p. 38–46.
2. Ferretti D, Principi E, Squartini S, Mandolini L. An experimental study on new features for activity of daily living recognition. In: 2016 international joint conference on neural networks (IJCNN), Vancouver, BC, 2016, p. 3958–65.
3. Wu JK, Dong L, Xiao W. Real-time physical activity classification and tracking using wearable sensors. In: 2007 6th international conference on information, communications & signal processing, Singapore, 2007, p. 1–6.
4. Mo L, Feng Z, Qian J. Human daily activity recognition with wearable sensors based on incremental learning. In: 2016 10th international conference on sensing technology (ICST), Nanjing, China, 2016, p. 1–5.
5. Fleury A, Vacher M, Noury N. SVM-based multimodal classification of activities of daily living in health smart homes: sensors, algorithms, and first experimental results. IEEE Trans Inf Technol Biomed. 2010;14(2):274–83.
6. Neuhaeuser J, Proebstl D, D'Angelo LT, Lueth, TC. First application of behavior recognition through the recording of ADL by radio modules in a home. In: 2012 annual international conference of the IEEE engineering in medicine and biology society, San Diego, CA, 2012, p. 5841–45.

7. Liu X, Liu L, Simske SJ, Liu J. Human daily activity recognition for healthcare using wearable and visual sensing data. In: 2016 IEEE international conference on healthcare informatics (ICHI), Chicago, IL, 2016, p. 24–31.

8. Awais M, Palmerini L, Chiari L. Physical activity classification using body-worn inertial sensors in a multi-sensor setup. In: 2016 IEEE 2nd international forum on research and technologies for society and industry leveraging a better tomorrow (RTSI), Bologna, 2016, p. 1–4.

9. Avgerinakis K, Briassouli A, Kompatsiaris I. Recognition of activities of daily living for smart home environments. In: 2013 9th international conference on intelligent environments, Athens, 2013, p. 173–80.

10. Bang S, Kim M, Song Sk, Park SJ. Toward real time detection of the basic living activity in home using a wearable sensor and smart home sensors. In: 2008 30th annual international conference of the IEEE engineering in medicine and biology society, Vancouver, BC, 2008, p. 5200–3.

11. Dong F, Shen C, Zhang J, Zhou S. A TOF and Kalman filtering joint algorithm for IEEE802.15.4a UWB Locating. In: 2016 IEEE information technology, networking, electronic and automation control conference, Chongqing, 2016, p. 948–51.

12. Van Kasteren TLM, Englebienne G, Krose BJA. Human activity recognition from wireless sensor network data: benchmark and software. In: Activity recognition in pervasive intelligent environments, vol. 4 of Atlantis ambient and pervasive intelligence. Atlantis Press;2011. p. 165–86.

13. Mikhaylov K, Tikanmäki A, Petäjäjärvi J, Hämäläinen M, Kohno R. On the selection of protocol and parameters for UWB-based wireless indoors localization. In: 2016 10th international symposium on medical information and communication technology (ISMICT), Worcester, MA, 2016, p. 1–5.

14. Song L, Zhang T, Yu X, Qin C, Zhang Q. Scheduling in cooperative UWB localization networks using round trip measurements. IEEE Commun Lett. 2016;20(7):1409–12.

15. Xiao W, Lu Y. Daily human physical activity recognition based on Kernel discriminant analysis and extreme learning machine. Math Probl Eng. 2015;2015:1–8.

16. Xiao W, et al. Nonlinear optimization-based device-free localization with outlier link rejection. Sensors. 2015;15(4):8072–87.

17. Amir RB, Gul ST, Khan AQ. A comparative analysis of classical and one class SVM classifiers for machine fault detection using vibration signals. In: 2016 international conference on emerging technologies (ICET), Islamabad, Pakistan, 2016, p. 1–6.

18. Huang GB, Zhu QY, Siew CK. Extreme learning machine: theory and applications. Neurocomputing. 2006;70(1–3):489–501.

19. Huang GB, et al. Incremental extreme learning machine with fully complex hidden nodes. Neurocomputing. 2008;71(4–6):576–83.

Infrared Image Temperature Measurement Based on FCN and Residual Network

Zhengguang Xu, Jinjun Wang, Pengfei Xu and Tao Liu

Abstract The key to electrolytic aluminum infrared temperature measurement system is to establish a certain mapping relationship between the gray image of the electrolyte and the temperature on the basis of the infrared thermal imaging system. Image segmentation techniques is use to get the extraction of the electrolyte region. Based on this, this paper combined the FCN network architecture, improved the VGG19 network with relatively good performance at present. In view of the network depth deepening, this paper also incorporates the residual thought, which solves the problem of network degradation caused by the deepening of the network. The new framework model proposed in this paper can meet the requirements of industrial temperature measurement.

Keywords Temperature measurement · ResNet · FCN · Infrared physics

1 Introduction

In recent years, the DL algorithm is constantly updated. The DL algorithm achieves great achievement in natural language processing. How to apply the deep learning to the complex industrial scene is a research area to be explored. The main reason is that the amount of data needed for deep learning is relatively large, and current algorithms achieving better results are based on tagged data. In the industrial area, these data are more difficult to collect. In the field of images processing, the main works of deep learning are target detection and target recognition. Regression analyses based on the characteristics gotten by deep learning network learning are very few. This article directs at a problem in the actual industrial production, set up a network for achieving the electrolytic aluminum infrared temperature measurement system, with DL knowledge [1, 2].

Z. Xu · J. Wang (✉) · P. Xu · T. Liu
School of Automation and Electrical Engineering, University of Science
and Technology Beijing, Beijing, China
e-mail: tonyjinjun@163.com

© Springer Nature Singapore Pte Ltd. 2018
Y. Jia et al. (eds.), *Proceedings of 2017 Chinese Intelligent Systems Conference*,
Lecture Notes in Electrical Engineering 460,
https://doi.org/10.1007/978-981-10-6499-9_73

At present, major temperature measurement method in the industrial aluminum production is artificial thermocouple temperature measurement method. This kind of temperature measurement has two drawbacks. On the one hand, due to the environment of aluminum production, this temperature measurement leads to relatively large consumption of thermocouple. After several measurements, the thermocouple won't measure the temperature accurately due to wastage. Frequent replacement of thermocouples will lead to resource consumption. On the other hand, as this measurement is manual, if you do not use a standard mode of operation, it will get a large measured deviation. Even the experienced mechanic can't measure the temperature accurately. In order to solve these problems, we designed a set of electrolytic aluminum infrared temperature measurement system according to the principle of infrared temperature measurement. This paper considers how to use the constructed network to segment the area of the electrolyte during electrolytic aluminum process. And establish the corresponding relationship the electrolyte gray scale distribution and the electrolyte temperature T at the moment.

2 Related Work

Professor Zhou Huaichun [3] in Huazhong University of Science and Technology has made a lot of research on the temperature field detection based on CCD image processing technology. This paper presents a monochromatic flame temperature image detection method based on reference temperature measurement. Professor Peng Xiaoqi [4], etc. started a search on detection of temperature field based on radiation field processing, and simulated the method to test the accuracy of CCD image temperature measurement.

Simonyan and Zisserman [5] at the University of Oxford Proposed the VGG model in "Very Deep Convolutional Networks for Large-Scale Image Recognition". The model is improved based on convolution neural networks, and achieves great successes in ImageNet.

Fully CNNs (FCNNs) have recently achieved state-of-the-art performance on image segmentation tasks [6–8]. These networks build on the success of networks previously trained on image recognition tasks by converting their fully connected layers to 1_1 convolutions, producing feature maps for each output category instead of a single prediction vector.

ResNet [9, 10] gained considerable fame in 2015, The emergence of ResNet makes the network structure deeper. He Kaiming's ResNet [11] gained the best paper CVPR2016. The article solves one of the most fundamental problems: more powerful features. ResNet changed the multiplication link into the addition link in the network, so that the stability of the backward transmission is directly enhanced.

In this article we will use the knowledge of infrared images, combine FCN and ResNet to build a new network for electrolytic aluminum infrared thermal imaging system, which can establish the relationship between the gray distribution of the electrolyte region and the temperature.

3 System Description

Infrared Principle

The aluminum electrolyte infrared thermal imaging online temperature measurement system focuses on the relation between the object temperature and the radiant energy. Describe the radiation energy in a certain method at a moment to determine the object temperature at the same time. According to the Stefan-Boltzmann law and Wien displacement law [12]:

$$M(T) = \varepsilon \lambda T^4 \tag{1}$$

$$\lambda_m = \frac{b}{273 + T} \tag{2}$$

We can find that the relationship between the fourth power of temperature and radiant power is a linear relationship.

Around $900^\circ C$ of electrolyte radiation center wavelength is $\lambda_m = 2.369$. According to the curve described by The Boltzmann formula, electrolyte still has a strong radiation in $0.76\ \mu \sim 1.2\ \mu$. Because of the atmospheric window, we choose the narrow band pass filter with $\Delta\lambda = 1.1\mu \sim 1.2\mu$ band pass wavelength.

According to the above principle, we designed the thermal imaging system composed of the optical imaging system and narrowband filter system, to describe the value of the infrared radiation energy. This system forms a mapping relationship between the electrolyte temperature and the electrolyte gray scale obtained by the thermal imaging system.

Search of the Mapping Relationship

The correspondence between the temperature and the gray level of the electrolyte is the focus of our next treatment. According to our collection of images at the aluminum factory site, we found some problems.

Resource images are as shown in Fig. 1. The image has a large black border; within the black border is the area for the fire hole. We divided the fire-hole area

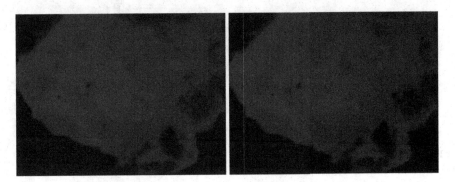

Fig. 1 Two original images

into electrolytes and covers. In general, the electrolyte gray scale is higher, the cover gray scale is low. But the appearance of the electrolyte does not have a fixed shape, and the distribution of the covering and the distribution of the electrolyte can't be differentiated well. We hope to find this rule by means of a convolution neural network, and establish the mapping relationship between electrolytic gray scale and real–time temperature.

4 Method

Network Architecture is shown as Fig. 2.

FCN for Image Segmentation

Fully convolutional networks can efficiently learn to make dense predictions for per-pixel tasks like semantic segmentation. FCN [13] classifies the images at the pixel level, to solve the semantic level of semantic segmentation problem. Because only using the last pooling feature map to do image segmentation will often lose a lot of edge details. Thus, we do up-sampling on the pooling feature map of the last three times. Then convolve the three pixel wise prediction images. The convolution produces an x_prediction, which is the final pixel-wise prediction. In this way, we set up a connection between the pixel information and the edge information in the image.

Residual Learning

Deep residual networks consist of many stacked "Residual Units". By introducing a shortcut connection between the output layer and input layer, rather than simply stacking the network, can solve the problem of gradient disappear cause by too deep network.

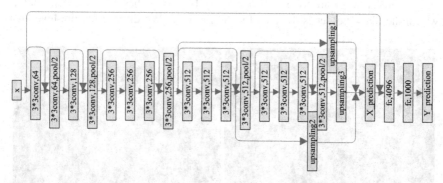

Fig. 2 Model architecture

Get the Electrolyte Information

The network we designed before does not have up-sampling structure. We directly connected to the three fc layers in up-sampling and generate the y_prediction. We found that our training was especially bad. In theory our labeling temperature and grey value correspondence should be one-to-one. But when we did actually measurement the amount of gray in the electrolyte area is not a value, but an unknown distribution. In addition, our camera images are as for standard, but the temperature measurement value has $1^\circ C \sim 3^\circ C$ deviation each time. If we the classic network, the results aren't good enough.

So we do a partition at first, and then make an intersection of the split image with the original image ($x\_prediction \cap x$). In this way, we get the image of information only in the electrolyte area. Then we can use two fully connected networks to represent the gray distribution information in the electrolyte area. Finally, we output y_prediction. According to the actual measurement requirements, the condition that the difference between the actual temperature and the prediction temperature is not more than $5^\circ C$ is to meet the requirements. We defined the following functions:

$$F(y\_prediction, y) = \begin{cases} true, & |y_{prediction} - y| \leq 5 \\ false, & |y_{prediction} - y| > 5 \end{cases} \tag{3}$$

5 Results

Dataset

Our database was built by ourselves. The data was collected in Qineng electric aluminum company in Qijiang, Chongqing, China. We took the image data from the electrolytic aluminum infrared thermal imaging system mentioned above. We made the temperature as the image label. The images we collect are distributed across the temperature range, covering the whole range. And the amount of data is pretty large, add up to 18000 images. We divided them into 12000 training images and 6000 test images.

Segmentation Results

If the network achieves better image segmentation at x_prediction, then our temperature prediction will be accurate. We use the experimental training network give the following results Fig. 3.

The above image shows that the area partition by the x_prediction image is basically in the electrolyte area. But the partition is not very accurate. On the one hand, the boundary of the electrolyte and mulch in the image is not obvious originally. On the other hand, in the process of de-convolution, we will treat the suspected coverings as electrolytes, and will also treat the suspected electrolytes as coverings.

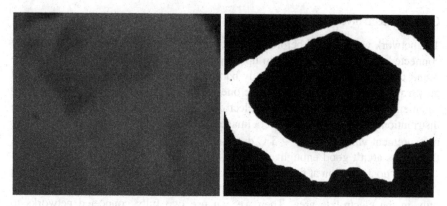

Fig. 3 Original image and its' segmentation results

Model Evaluation

To compare and verify, we trained the network we built, and test the performance of the trained model with the test set. Also, we compared the results of the VGG19 training to analyze the comparison of the two sets of results (Table 1).

In the upper table, the ΔT is the absolute value of the predicted temperature and the measured temperature difference. Both models went through 30000 iterations, and the training accuracy has not changed much. According to the comparison VGG19 model and our model, under the same ΔT condition, our temperature model has a significant performance improvement. Under the condition of meeting the actual temperature requirement $(\Delta T = 5^\circ C)$.The error of the new model is reduced by (19.7 − 12.8)/19.7 * 100% = 35.03%.

In the same model, we did a comparison experiment in different ΔT conditions. We found that with ΔT decreasing, the measurement accuracy of the model decreased dramatically. Under the condition $(\Delta T = 3^\circ C)$, our forecast has no value relative to this estimation range. We believe that the new build model basically meets our requirements.

Table 1 Comparison results

Model	ΔT (°C)	Accuracy (%)	Model	ΔT (°C)	Accuracy (%)
VGG19	5	80.30	new model	5	87.20
VGG19	4	68.60	new model	4	73.70
VGG19	3	45.70	new model	3	51.90
VGG19	2	19.30	new model	2	24.10

6 Conclusions

The model we proposed, combined with the idea of residual network, using multiple residual units to deepen the image's model, make the features of the image further enhance. It solves the problem of degradation resulting from deepening network depth. We use the last three pooling feature maps to refactor the segmentation image of the original. Compared with using only the last pooling feature map to refactor segmentation image, we deepen the details of the edge of the image. Finally, we extracted the area of the electrolyte based on the combined information of the segmentation image and the original image. We use the network to find a rule that can indicate the distribution of electrolytes.

The experiment results show that the algorithm proposed meets the requirements of industrial site application. Considering the industrial condition of electrolytic aluminum, the temperature variation of the electrolyte is relatively small for a period of continuous time. We consider using the predictive value of successive frame images to verify mutually to improve the accuracy of the algorithm. This is the direction of our next study.

References

1. Roska T, Chua LO. The CNN universal machine. J Circuits Syst Comput. 2008;12(04):377–88.
2. Kraus OZ, Lei BJ, Frey BJ. Classifying and segmenting microscopy images with deep multiple instance learning. Bioinformatics. 2016;32(12):52.
3. Huaichun Z, et al. Study on image processing of furnace flame temperature field. Chin J Electr Eng. 1995;15(5):295–300.
4. Liu Z, Peng X. Soft sensor system of high temperature melt temperature field based on CCD image sensor. Central South University. 2004.5.
5. Krizhevsky A, Sutskever I, Hinton GE. ImageNet classification with deep convolutional neural networks. International conference on neural information processing systems curran associates inc;2012. p. 1097–5.
6. Simonyan K, Zisserman A. Very deep convolutional networks for large-scale image recognition. Comput Sci. 2014.
7. Szegedy C, Liu W, Jia Y, et al. Going deeper with convolutions. Computer vision and pattern recognition. IEEE; 2015. p. 1–9.
8. Long J, Shelhamer E, Darrell T. Fully convolutional networks for semantic segmentation. Computer vision and pattern recognition. IEEE;2015. p. 3431–40.
9. Chen LC, Papandreou G, Kokkinos I, et al. Semantic image segmentation with deep convolutional nets and fully connected CRFs. Comput Sci. 2014;4:357–61.
10. He K, Zhang X, Ren S, Sun J. Deep Residual Learn for Image Recogn. 2015;1512:03385.
11. He K, Sun J. Convolutional neural networks at constrained time cost. 2014;5:353–60.
12. Ruxiang Q, Yuan T, Zongli H. Infrared imaging detection and analysis of high temperature region near coal seam. Coal Geol Explor. 2014;4:90–2.
13. Chen LC, Papandreou G, Kokkinos I, et al. Semantic image segmentation with deep convolutional nets and fully connected CRFs. Comput Sci. 2016;4:357–61.

IMA System Overall Fault Management Method Research Based on ASAAC Architecture

Kun Wang, Xiaohong Bao and Tingdi Zhao

Abstract This paper describes the architecture of the integrated modular avionics (IMA) system; summarizes the various fault management functions; and proposes an extended layered IMA system overall fault management architecture combined with fault prognosis and fault handling functions based on the European allied standard avionics architecture council (ASAAC) standards. The logic process for each level of overall fault management and the implementation method of the overall fault management of IMA system are given.

Keywords IMA · PHM · Overall fault management · GSM

1 Introduction

The development tendency of modern avionics system is integrated, generalized and modular. Fault management is the key to realize the goal of avionics system integration and ensure the safety, maintainability and economic affordability of aircraft.

Current international standards for IMA systems are the ASAAC standards and the ARINC 653 standard. ASAAC standards implement IMA system fault management through the health monitoring mechanisms and fault management mechanisms in General System Management (GSM) which located in operating system layer [1]. But the ASAAC standards do not reflect the partition mechanism. However, the development tendency of modern avionics system is to implement

K. Wang (✉) · X. Bao · T. Zhao
School of Reliability and Systems Engineering, Beihang University,
Beijing 100191, China
e-mail: wangkun1514@buaa.edu.cn

X. Bao
e-mail: baoxh@buaa.edu.cn

T. Zhao
e-mail: ztd@buaa.edu.cn

© Springer Nature Singapore Pte Ltd. 2018
Y. Jia et al. (eds.), *Proceedings of 2017 Chinese Intelligent Systems Conference*,
Lecture Notes in Electrical Engineering 460,
https://doi.org/10.1007/978-981-10-6499-9_74

partition operating systems. The fault management method in ARINC 653 standard [2] adopts process level, partition level and modular level health monitoring and fault recovery strategies. But there is no system level fault management method, so it can not cover the entire avionics system. Besides, the ASAAC standards and the ARINC 653 standard merely provide the fault management methods after failure. They lack fault prognosis function. While fault prognosis is one of an indispensable functions to support the safety, maintainability, supportability and economic affordability of aircraft.

Newest study of fault prognosis is prognosis and health management (PHM) technology. In the literature [3], the US Air Force "Joint Strike Fighter" (F-35) program PHM system and its relationship with the autonomic logistics are introduced. In the literature [4], a layered PHM architecture consisting of module level PHM, subsystem level PHM, area PHM and platform level PHM is proposed. Literature [5] gives a new generation of integrated modular avionics system PHM architecture of equipment level, subsystem level, avionics system level, but it lacks specific implementation method. In the literature [6], the main key technologies and solutions of PHM implementation are described. But this avionics PHM structure does not conform to the hierarchical division of IMA system management within ASAAC standards, namely resource level, integrated area level and aircraft level. Besides, avionics system PHM is mainly focusing on forecasting before failure. It includes very few response and recovery measures, lacking considerations to aircraft flying safety.

According to the problem of ASAAC standard fault management strategies lack fault prognosis function, this paper proposes an extended hierarchical IMA system overall fault management architecture, and introduces the specific logic process of fault management at all levels. Finally, The implementation method of the IMA system overall fault management method is given.

2   IMA System Architecture and Fault Management Methods

2.1   IMA System Architecture

Functions and tasks are accomplished through the sharing and reusing of six types of common functional modules (CFMs) in IMA system, namely signal processing modules (SPM), data processing modules (DPM), graphic processing modules (GPM), network support modules (NSM), power conversion modules (PCM) and mass memory modules (MMM). After the task is completed, if it is not necessary to continue, the occupied resources will be released. Among the six types of CFMs, SPM, DPM and GPM are processing modules, while NSM, PCM and MMM are support modules. A processing module has one or more processing elements. PCM and MMM also have a small number of processing elements. A processing element

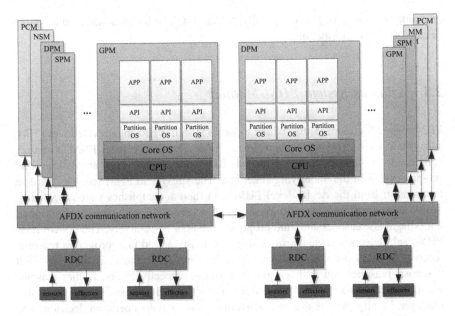

Fig. 1 IMA system architecture

can support one or more avionics applications, and partitioning mechanism of operating system will ensure that these applications run independently. The IMA system not only improves the redundancy and reconfiguration capability of avionics systems, but also facilitates the expansion and upgrading of system functions. IMA system architecture is shown in Fig. 1.

2.2 ASAAC Fault Management Strategy

IMA system fault management is achieved by health monitoring mechanisms and fault management mechanisms in GSM in ASAAC standards. The health status of applications, operating systems, module supporting components, and hardware in IMA system will be monitored by various fault detection mechanisms (FDM) respectively. Once a fault is found by fault detection mechanisms, it will be reported to the operating system health monitoring service. Then, the operating system health monitoring service responds to GSM health monitoring requests and transfers the fault information to the resource health monitor. The resource health monitor filters the fault information based on runtime blueprints (RTBPs) and then submits the confirmed fault to fault management programs. If the fault can not be handled correctly at resource level, the fault will be submitted to the upper IA level even avionics system level health monitoring programs and fault management programs according to priority. Finally, the fault will be handled based on the more detailed RTBPs information.

2.3 Avionics System PHM Method

PHM related literature [5, 6] divided avionics system PHM into device level PHM, subsystem level PHM and system level PHM. The device level PHM is mainly composed of sensors and self-tested BIT to complete the acquisition and pre-processing of the underlying state data. The subsystem PHM receives the data information from the device level PHM, and then accomplishes state monitoring, health assessment and fault prognosis by the subsystem fault detection reasoner, fault diagnosis reasoner and fault prognosis reasoner respectively. The system level PHM fault detection reasoner, fault diagnosis reasoner and fault prognosis reasoner receives the state information from the subsystem fault diagnosis reasoner, fault diagnosis reasoner and fault prognosis reasoner respectively and further reasons. Then, the reasoning result of these three reasoners will be fused by the integrated manager. Finally, reports that will assist the pilots to make operation decisions and maintenance personnel to execute condition-based maintenance will be generated.

3 IMA System Overall Fault Management Method

3.1 IMA System Overall Fault Management Functions

According to the conclusions of NASA Fault Management workshop [7] in 2008, different research groups had different meanings and understandings of fault management. Fault management related terms are fault protection, redundancy management, fault detection, isolation and recovery (FDIR) and PHM, etc. FDIR is also known as fault detection and correction (FD&C) [8]. There is no uniform terminology in the fault management field [9]. Figure 2 summarizes the relationship between the various fault management functions.

The meaning of the fault management functions in this paper is as follows:

Fault detection: Deciding that a fault exists [10].

Anomaly detection: Deciding that an anomaly exists.

Prognosis: Predicting the time at which a component will fail.

Fault isolation: Deciding the possible location of the fault and constraining the fault effects in an appropriate bounded range.

Fault identification: Deciding the causes of the fault.

Failure response: Measures to eliminate or mitigate the effects of failure.

Failure recovery: Measures to restore the system to nominal state.

Failure prevention: Operational measures to prevent a failure from occurring.

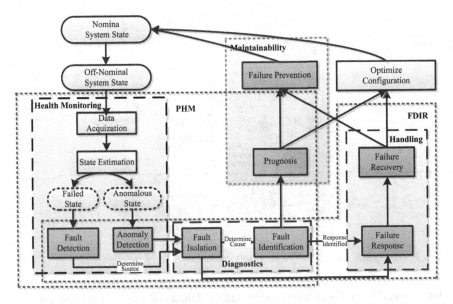

Fig. 2 IMA system overall fault management functions

PHM mainly includes three major functional modules: health monitoring, fault diagnosis and fault prognosis modules. FDIR mainly includes fault detection, fault diagnosis, failure response and failure recovery modules. Fault prognosis and failure prevention belong to maintainability. Fault management consists of the operational capabilities to predict or detect failures, and to take action to prevent or mitigate their effects [10]. So, IMA system overall fault management method of this paper is divided into four functional modules: health monitoring, fault diagnosis, fault prognosis and fault handling. These functions cover the overall fault management requirements before and after a failure occurrence.

3.2 IMA System Overall Fault Management Architecture

This section proposes a layered IMA system overall fault management architecture of resource element (RE) level, integrated area (IA) level and system level based on the ASAAC fault management mechanism, the avionics system PHM structure and the fault management functions mentioned above. IMA system overall fault management architecture is shown in Fig. 3.

Each level includes health monitoring, fault diagnosis, fault prognosis and fault handling four major fault management functional components. Data information flows sequentially within the level. Faults that can not be handled by fault handling programs in this level will be reported to the upper health monitoring manager. The functional components of the upper-level fault management receive the fault

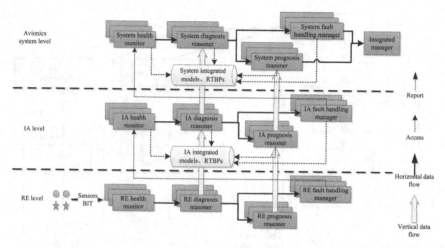

Fig. 3 IMA system overall fault management architecture

information from lower-level corresponding functional components through the vertical data flow respectively and reasons the fault information further. Finally, The system level fault prognosis and fault handling functional components information will be fused and unified reasoned by the integrated manager in avionics system level.

3.3 Overall Fault Management Logic Process

3.3.1 RE Level Overall Fault Management Logic Process

The RE level overall fault management logic process is shown in Fig. 4. Firstly, the RE level health monitoring manager extracts the data characteristics of the sensors and the FDM monitoring data, and compares the data characteristics with a given expected value or threshold value to evaluate the system health status. Then the fault filters and filtering parameters will be called according to the RTBPs to confirm the occurrence of faults. Confirmed faults will be reported to the operating system health monitoring services and an alarm and RE level fault diagnosis reasoner will be triggered. Whereas, if there is no fault, the system will continue health monitoring. Next, the RE level fault diagnosis reasoner determines the failure mode, effects, causes and possible locations of the fault according to the failure mode list, the failure mode, effect and criticality analysis (FMECA) library, and corresponding fault diagnosis models. Afterwards, a characteristic fault identification code will be generated. On the one hand, the RE level fault handling manager determines the fault response and recovery actions according to the fault identification code and the RTBPs. On the other hand, the RE level fault prognosis

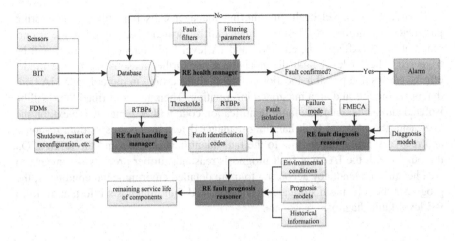

Fig. 4 The logic process of RE level overall fault management

reasoner predicts the remaining service life of the avionics components according to the environmental conditions, the prognosis models and the historical information.

3.3.2 IA Level Overall Fault Management Logic Process

Faults that can not be handled by RE level fault handling program will be reported to the IA level health monitoring manager. The IA level overall fault management logic process is shown in Fig. 5.

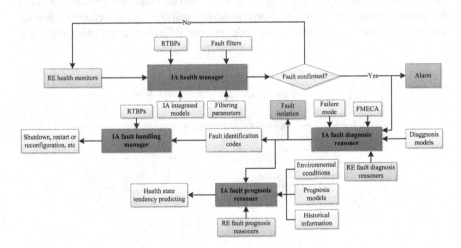

Fig. 5 The logic process of IA level overall fault management

Firstly, the IA level health monitoring manager calls fault filters and filtering parameters according to RTBPs to filter fault information and confirm the occurrence of faults. Afterwards, an alarm and IA level fault diagnosis reasoner will be triggered. The IA level fault diagnosis reasoner further reasons the failure mode, effects, causes and possible locations of the fault according to more detailed IA fault diagnosis models and area integrated information. Then, the confirmed fault will be isolated and a characteristic fault identification code will be generated. Next, on the one hand, the IA level fault handling manager determines corresponding response and recovery actions according to the fault identification code and the RTBPs. On the other hand, the IA level fault prognosis reasoner further predicts the integrated area health state tendency according to more detailed environmental conditions, the prognosis models, the historical information, and fault diagnosis information from RE level fault diagnosis reasoners.

3.3.3 System Level Fault Management Logic Process

Faults that can not be handled by IA level fault handling programs will be reported to the system level health monitoring manager. The system level overall fault management logic process is shown in Fig. 6.

Firstly, the system level health monitoring manager responds the IA level fault handling programs, report, and then filters and confirms fault according to RTBPs. Afterwards, an alarm and system level fault diagnosis reasoner will be triggered. The system level fault diagnosis reasoner further reasons the failure mode, effects, causes and possible locations according to more detailed IMA system integrated information. Then, the confirmed fault will be isolated and a characteristic fault identification code will be generated. Next, on the one hand, the system level fault handling manager determines corresponding response and recovery actions according to the fault identification code and the RTBPs. On the other hand, the

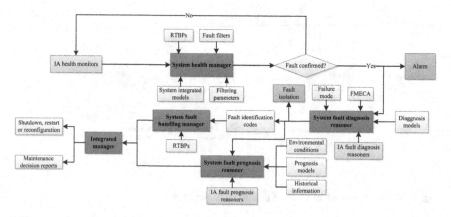

Fig. 6 The logic process of avionics system level overall fault management

system level fault prognosis reasoner further predicts the integrated area health state tendency according to more detailed area integrated information. Finally, the results of the system level fault handling manager and the system level fault prognosis reasoner will be fused and further unified reasoned by the integrated manager to determine response actions and generate health reports.

4 IMA System Overall Fault Management Implementation Method

4.1 IMA System Overall Fault Management Implementation Method

Health monitoring functional component in this method corresponds to the GSM health monitoring function, and the other three functional components correspond to the GSM extended fault management function. Figure 7 shows the implementation method of IMA system overall fault management.

Firstly, on the one hand, the distributed remote data collectors (RDC) gather and pre-process data collected by sensors and then transfer those pre-processed data to CFMs via AFDX communication network. On the other hand, a variety of FDMs within the CFMs monitor the health status of each component of the CFMs. For example, applications will be monitored by bounds checking, consistency check, performance measurement, etc.; operating system will be monitored by authorization, authentication and bounds checking; blueprints will be monitored by authorization and authentication; module supporting components will be monitored by

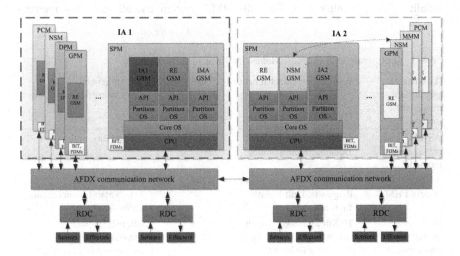

Fig. 7 The implementation method of IMA system overall fault management

BIT, parity checks, watchdog timers, etc. Operating system health monitoring services gather all information that will be delivered to the GSM health monitor when requested. Then, IMA system RE level overall fault management programs that reside on a processing element within this CFM will execute RE level health monitoring, RE level fault diagnosis, RE level fault prognosis and RE level fault handling according to IMA system RE level overall fault management logic process. Faults that can not be handled by RE level overall fault management programs will be reported to IA level health monitor. Next, IA level overall fault management programs that reside on a processing element within this IA execute IA level overall fault management according to IA level overall fault management logic process. Faults that can not be handled by IA level overall fault management programs will be reported to IMA system level health monitor. Afterwards, IMA system level overall fault management programs will be triggered. Finally, reports that will assist the pilots to make operation decisions and maintenance personnel to execute condition-based maintenance will be given.

The overall fault management programs of non-processing module such as NSMs maybe reside on other processing modules. Every processing element will host at least one overall fault management program. And some processing elements maybe host three overall fault management programs (RE level, IA level, and system level).

4.2 Comparison with Existing Methods

The IMA overall fault management method proposed in this paper has many advantages compared with existing methods, as shown in Table 1. Firstly, IMA system overall fault management method includes more fault management functionality than the other two; Secondly, IMA system overall fault management

Table 1 Comparison with existing avionics fault management methods

Avionics system fault management methods	Fault management functions	Consistent with the standard	Main general characteristics enhanced
ASAAC fault management method	Health monitoring, fault isolation, fault handling	Yes	Reliability, safety
Avionics system PHM	Health monitoring, fault diagnosis, fault isolation, fault prognosis	No	Maintainability, supportability, economic affordability
IMA system overall fault management method	Health monitoring, fault diagnosis, fault isolation, fault prognosis, fault handling	Yes	Reliability, safety, maintainability, supportability, economic affordability

method conform to ASAAC standard architecture, which means this method is feasible; Thirdly, IMA overall fault management method can simultaneously enhance the reliability, safety, maintainability, supportability and economic affordability of aircraft.

5 Conclusions and Prospects

This paper proposes an IMA system layered overall fault management architecture including fault detection, fault diagnosis, fault prognosis and fault handling functions based on the ASAAC standard fault management strategy and the avionics system PHM architecture. On the basis of analysing the logic processes of each level, the IMA system overall fault management implementation method is given. This method is of great instruction significance to implement the large passenger aircraft fault management in engineering practice. The logic process of each level will be further refined in the following research.

Acknowledgements This research has been supported by grants from the Major State Basic Research Development Program of China (973 Program) (No. 2014CB744904), and grants from a project of Ministry of Industry and Information Technology of China (No. JSZL2015601C008), and Civil Aviation Joint Funds established by National Nature Science Foundation of China and Civil Aviation Administration of China (No. U1533201).

References

1. NATO STANAG 4626 (Part VI). Final Draft of Proposed Standards for Software STANAG 4626 Part VI-Guidelines for System Issues; 2006.
2. ARINC Specification 653. Avionics Application Software Standard Interface. Aeronautical Radio Inc.; 2006.
3. Hess A, Calvello G, Dabney T. PHM a key enabler for the JSF autonomic logistics support concept. In: 2004 IEEE aerospace conference, vol. 6. IEEE; 2004. p. 3543–50.
4. Zhang L, Zhang F, Lee J, et al. Architecture of aircraft onboard prognostics and health management (PHM) system. J Air Force Eng Univ (Nat Sci Ed). 2008;9(2):6–9.
5. Zhao Ningshe, Zhai Zhengjun, Wang G. A new generation of avionics integration and prognostics and health management technology. Meas Control Technol. 2011;30(1):1–5.
6. He H, Haitao L. Application and design of PHM in integrated avionics system. Telecommun Eng. 2014;54(3):245–50.
7. Fesq LM. Current fault management trends in NASA's planetary spacecraft. In: 2009 IEEE aerospace conference. IEEE; 2009. p. 1–9.
8. Rustick J, Wisniewski M, Termohlen D. Fault management for complex space systems: assessing the use of rule-based systems. In: AIAA infotech@ aerospace conference and AIAA unmanned... unlimited conference 2009; 2009. p. 2030.
9. Fesq LM, Weitl RM. Developing a fault management guidebook for NASA's deep space robotic missions. In: AIAA infotech@ aerospace; 2015. p. 1795.
10. Johnson S, Day J. System health management theory and design strategies. In: Infotech@ aerospace 2011; 2011. p. 1493.

Printed in the United States
By Bookmasters